Inferring Phylogenies

Inferring Phylogenies

Joseph Felsenstein
University of Washington

Sinauer Associates, Inc. • Publishers
Sunderland, Massachusetts

COVER: "Far Flung Fir" 27″ × 19″ × 19″ painted plaster over rebar and recycled styrofoam. Copyright © Joan A. Rudd 1985, exhibited Oregon Biennial 1985, Portland Art Museum. Collection of the author.

INFERRING PHYLOGENIES

 For information or to order, address: Sinauer Associates, Inc., PO Box 407, 23 Plumtree Road, Sunderland, MA, 01375 U.S.A. FAX: 413-549-1118
Internet: publish@sinauer.com; http://www.sinauer.com

Printed in U.S.A.

5 4 3

Contents

Preface

Phylogenies, or evolutionary trees, are the basic structures necessary to think clearly about differences between species, and to analyze those differences statistically. They have been around for over 140 years, but statistical, computational, and algorithmic work on them is barely 40 years old. In that time there have been great advances in understanding, but much remains to be done. It's a good time to summarize this work, while it is still compact enough for a single book to cover it. Alternatively, we could put it differently: work in this field has been going on for four decades, and no book has yet summarized it; such a book is overdue.

I have tried to cover the major methods for inferring phylogenies, at a level appropriate to a graduate course for biologists interested in using numerical methods. I have also tried to cover methods of statistical testing of phylogenies, as well as some methods for using phylogenies for making other inferences. The book assumes some familiarity with statistics, some with computers, and mathematics including calculus and an elementary command of matrix algebra.

Phylogenies are inferred with various kinds of data. I have concentrated on some of the central ones: discretely coded characters, molecular sequences, gene frequencies, and quantitative traits. There is also some coverage of restriction sites, RAPDs, and microsatellites. The reader may benefit from enough familiarity with molecular biology to understand the major features of molecular sequence data, and some exposure to the theory of quantitative genetics. Other data types that are less widely used, such as DNA hybridization, are not covered.

I estimate that there are about 3,000 papers on methods for inferring phylogenies. This book refers to a small fraction of those, with less emphasis on studies that investigate behavior of methods on simulated data or real data. I hope that the reader will be able to find their way through this literature from these references, with creative use of computerized literature searches. My apologies to those hundreds of my colleagues whose best and most incisive paper was not cited.

Over the years, my understanding of phylogenies has benefitted greatly from helpful interactions with many of the people who contributed to this field. The field of inferring phylogenies has been wracked by outrageously excessive controversy, often marked by behavior that would not be condoned in other, more mature fields of science. In the midst of this there have been many biologists who strove to bring the field back to normality, who were eager for an open and friendly exchange of views. To all of them, my admiration and thanks. They know who they are (and who they aren't).

I am grateful for the frequent help of the members of my laboratory, particularly Mary Kuhner, Jon Yamato, Peter Beerli, Lindsey Dubb, Elizabeth Walkup and Nicolas Salamin. Aside from wise counsel, they frequently took extra work on themselves which allowed me time for this project. A variety of colleagues and students have made helpful comments on parts of the manuscript of this book. Thanks to all of them, particularly Peter Waddell, Jeff Thorne, Mike Steel, Doug Robinson, Barry Hall, Mike Hendy, Oliver Will, Kevin Scott, Sabin Lessard, Michael Turelli and Brian O'Meara. Anthony Edwards, Robert Sokal, Charles Michener, Peter Sneath, F. James Rohlf, John Huelsenbeck, David Swofford, Gary Olsen, Edward Dayhoff, Walter Fitch, Winona Barker, Mary Mickevich, Arnold Kluge, Vicki Funk and George Byers gave extremely helpful answers to questions that came up in writing the book. I am particularly indebted to Elizabeth Thompson and Monty Slatkin for much insight over many years. Many students in my phylogeny course have found and corrected errors and unclear passages. I am also grateful to Scott Johnston, of Vectaport, Inc., who has maintained in his Ivtools package the Idraw drawing program used for many of the figures in this book, and who was helpful in answering technical questions. Occasional new results reported in this book resulted from a number of grants to me, funded by the National Institutes of Health and the National Science Foundation.

My family made a great contribution to the writing of this book. My son Zach Rudd Felsenstein was a constant reminder that there are future generations coming who will make good, and entertaining, use of today's science. My stepson Ben Rudd Schoenberg used his skills in combinatorial geometry to arrange the points in Figure 4.3 into a symmetric pattern and point out its connection to the Petersen Graph. My wife, Joan Rudd, encouraged and inspired me, showing how to keep one's own projects alive in the midst of pressures from all sides. One of her sculptures, *Far Flung Fir*, graces the cover. She was available both as literary critic and audience. To her I dedicate this book.

Joe Felsenstein
Seattle
August, 2003

Chapter 1

Parsimony methods

Parsimony methods are the easiest ones to explain; they were also among the first methods for inferring phylogenies. The issues that they raise also involve many of the phenomena that we will need to consider. This makes them an appropriate starting point.

The general idea of parsimony methods was given in their first mention in the scientific literature: Edwards and Cavalli-Sforza's (1963) declaration that the evolutionary tree is to be preferred that involves "the minimum net amount of evolution." We seek that phylogeny on which, when we reconstruct the evolutionary events leading to our data, there are as few events as possible. This raises two issues. First, we must be able to make a reconstruction of events, involving as few events as possible, for any proposed phylogeny. Second, we must be able to search among all possible phylogenies for the one or ones that minimize the number of events.

A simple example

We will illustrate the problem with a small example. Suppose that we have five species, each of which has been scored for 6 characters. In our example, the characters will each have two possible states, which we call 0 and 1. The data are shown in Table 1.1. The events that we will allow are changes from $0 \rightarrow 1$ and from $1 \rightarrow 0$. We will also permit the initial state at the root of a tree to be either state 0 or state 1.

Evaluating a particular tree

To find the most parsimonious tree, we must have a way of calculating how many changes of state are needed on a given tree. Suppose that someone proposes the phylogeny in Figure 1.1. The data set in our example is small enough that we can find by "eyeball" the best reconstruction of evolution for each character. Figures 1.2–1.6 show the best character state reconstructions for characters 1 through

Table 1.1: A simple data set with 0/1 characters.

Species	Characters 1	2	3	4	5	6
Alpha	1	0	0	1	1	0
Beta	0	0	1	0	0	0
Gamma	1	1	0	0	0	0
Delta	1	1	0	1	1	1
Epsilon	0	0	1	1	1	0

6. Figure 1.2 shows character 1 reconstructed on this phylogeny. Note that there are two equally good reconstructions, each involving just one change of character state. They differ in which state they assume at the root of the tree, and they also differ in which branch they place the single change. The arrows show the placements of the changes, and the shading shows in which parts of the phylogeny the two states are reconstructed to exist. Figure 1.3 shows the three equally good reconstructions for character 2, which needs two changes of state. Figure 1.4 shows the two reconstructions for character 3, involving one change of state. Figure 1.5 shows the reconstructions (there are two of them) for character 4. These are the same as for character 5, as these two characters have identical patterns. They require two changes. Finally, Figure 1.6 shows the single reconstruction for character 6. This requires one change of state.

The net result of these reconstructions is that the total number of changes of character state needed on this tree is $1 + 2 + 1 + 2 + 2 + 1 = 9$. Figure 1.7 shows the reconstructions of the changes in state on the tree, making particular arbitrary

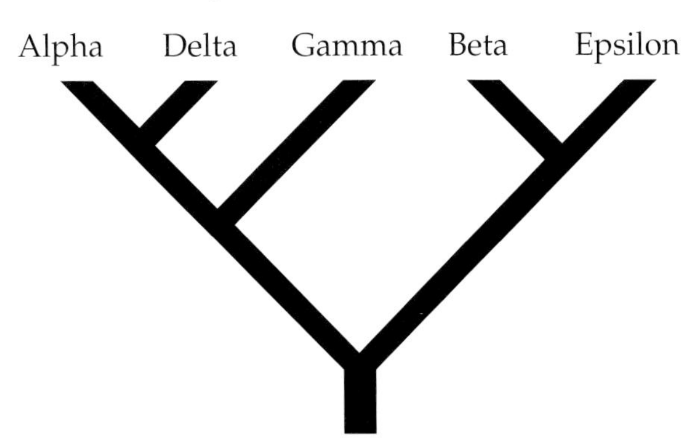

Figure 1.1: A phylogeny that we want to evaluate using parsimony.

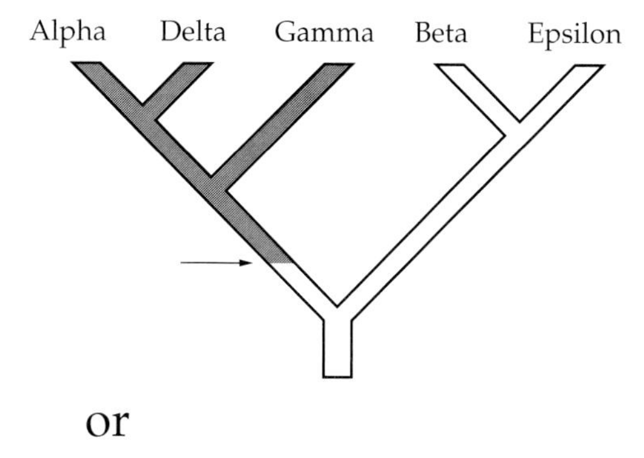

or

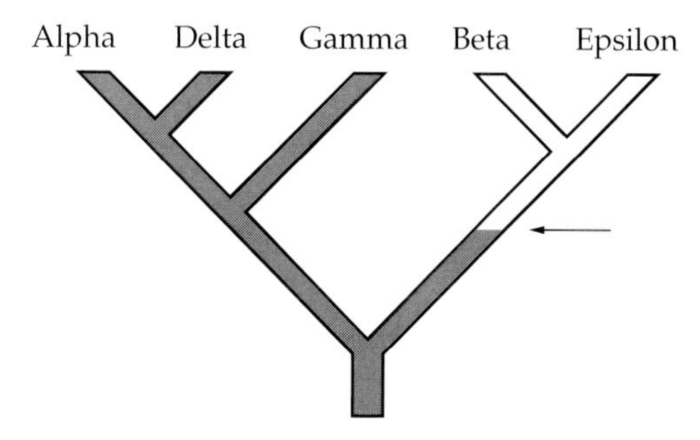

Figure 1.2: Alternative reconstructions of character 1 on the phylogeny of Figure 1.1. The white region of the tree is reconstructed as having state 0, the shaded region as having state 1. The two reconstructions each have one change of state. The changes of state are indicated by arrows.

choices where there is a tie. However, consideration of the character distributions suggests an alternative tree, shown in Figure 1.8, which has one fewer change, needing only 8 changes of state. Consideration of all possible trees shows that this is the most parsimonious phylogeny for these data. The figure shows the locations of all of the changes (making, as before, arbitrary choices among alternative reconstructions for some of the characters).

In the most parsimonious tree, there are 8 changes of state. The minimum number we might have hoped to get away with would be 6, as there are 6 characters, each of which has two states present in the data. Thus we have two "extra" changes. Having some states arise more than once on the tree is called *homoplasy*.

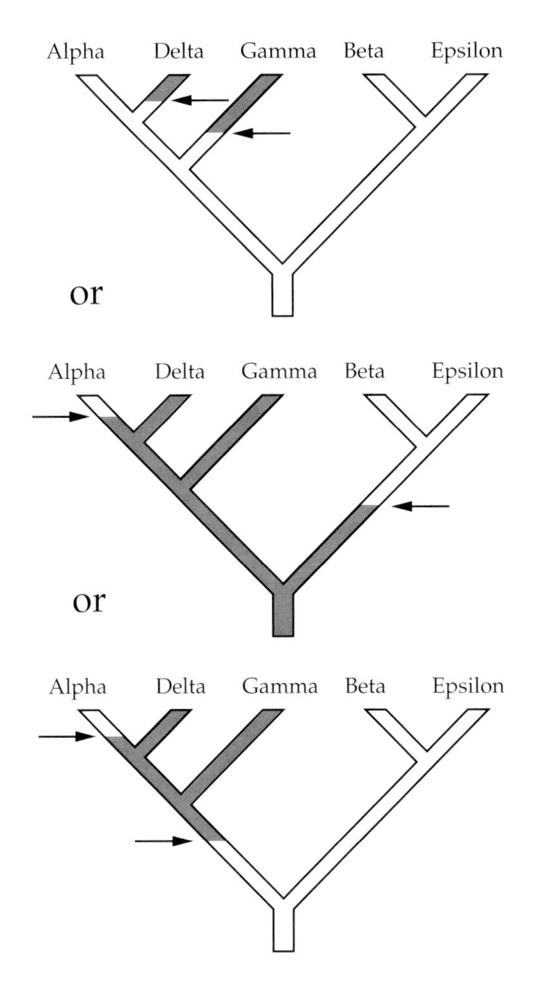

Figure 1.3: Reconstructions of character 2 on the phylogeny of Figure 1.1. The white regions have state 0, the shaded region state 1. The changes of state are indicated by arrows.

Rootedness and unrootedness

Figure 1.9 shows another tree. It also requires 8 changes, as shown in that figure. In fact, these two most parsimonious trees are the same in one important respect — they are both the same tree when the roots of the trees are removed. Figure 1.10 shows that unrooted tree. The locations of the changes are still shown (and still involve some arbitrary choices), but they are no longer shaded in to show the direction of the changes. There are many rooted trees, one for each branch of the

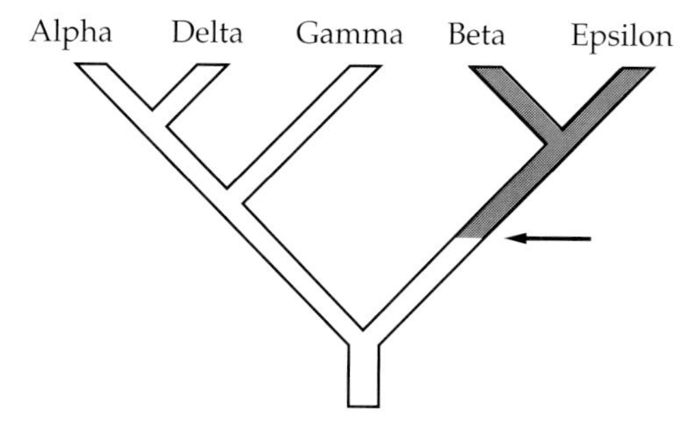

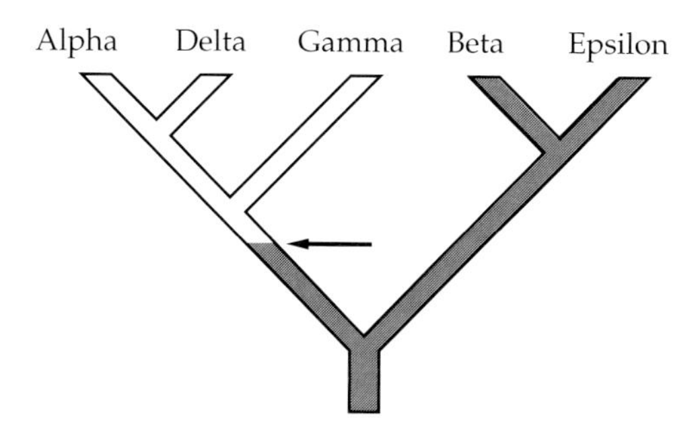

Figure 1.4: Reconstruction of character 3 on the phylogeny of Figure 1.1. The graphical conventions are the same as in the previous figures.

unrooted tree in Figure 1.10, and all have the same number of changes of state. In fact, the number of changes of state will depend only on the unrooted tree, and not at all on where the tree is then rooted. This is true for the simple model of character change that we are using ($0 \rightleftharpoons 1$). It is also true for any model of character change that has one simple property: that if we can go in one change from state a to state b, we can also go in one change from state b to state a.

When we are looking at the alternative placements of changes of state, it actually matters whether we are looking at a rooted or an unrooted tree. In Figure 1.3, there are three different reconstructions. The last two of them differ only by whether a single change is placed to the left or to the right of the root. Once the tree is unrooted, these last two possibilities become identical. So the rooted tree has three possible reconstructions of the changes of this state, but the unrooted tree has only two.

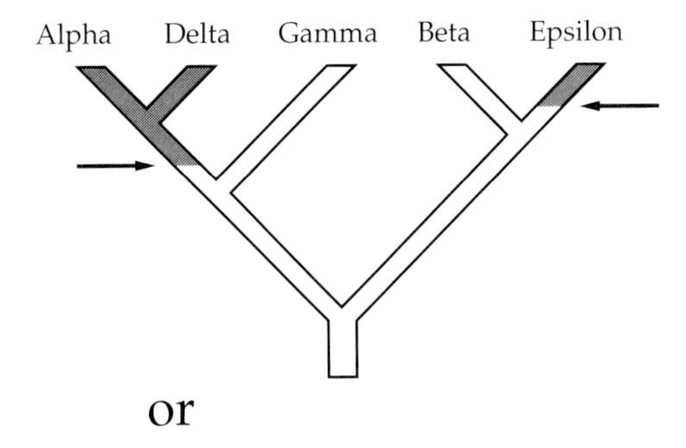

or

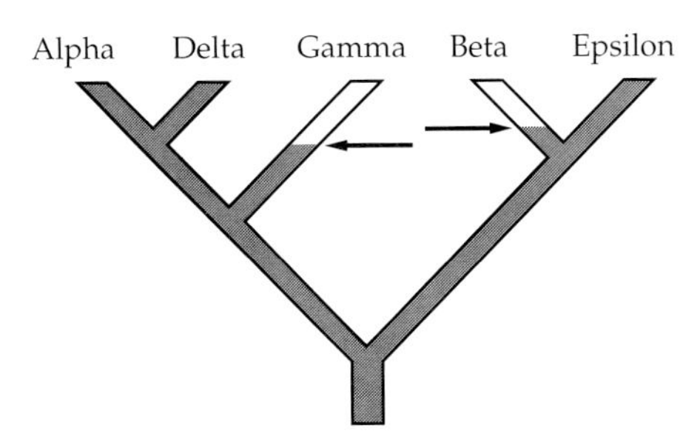

Figure 1.5: Reconstruction of character 4 on the phylogeny of Figure 1.1. This is the same as the reconstruction for character 5 as well. The graphical conventions are the same as in the previous figures.

Methods of rooting the tree

Biologists want to think of trees as rooted and thus have been interested in methods of placing the root in an otherwise unrooted tree. There are two methods: the outgroup criterion and the use of a molecular clock. The outgroup criterion amounts to knowing the answer in advance. Suppose that we have a number of great apes, plus a single old-world (cercopithecoid) monkey. Suppose that we know that the great apes are a monophyletic group. If we infer a tree of these species, we then know that the root must be on the lineage that connects the cercopithecoid monkey to the others. Any other placement would make the apes fail to be monophyletic, because there would then be a lineage leading away from the root with a subtree that included the cercopithecine and also some, but not all, of the apes. We place the root outside of the ingroup, so that it is monophyletic.

The alternative method is to make use of a presumed clocklike behavior of character change. In molecular terms, this is the "molecular clock." If an equal amount of change were observed on all lineages, there should be a point on the tree that has equal amounts of change (branch lengths) from there to all tips. With

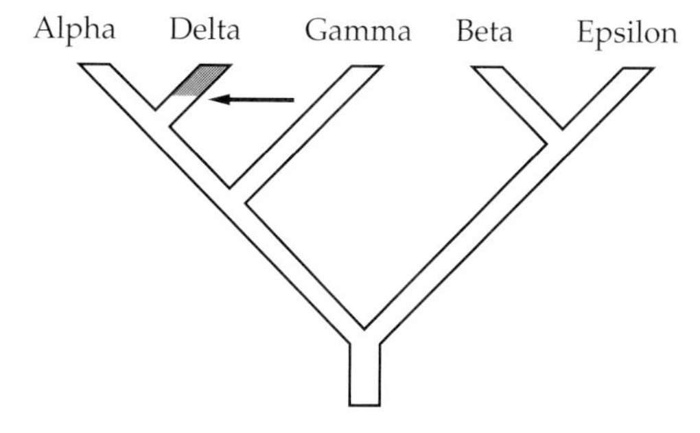

Figure 1.6: Reconstruction of character 6 on the phylogeny of Figure 1.1.

a molecular clock, it is only the expected amounts of change that are equal; the observed amounts may not be. We hope to find a root that makes the amounts of change approximately equal on all lineages. In some methods, we constrain the tree to remain clocklike by making sure that no tree is inferred that violates this constraint. If instead we infer a tree without maintaining this constraint, we can try to remedy this by finding, after the fact, a point on the tree approximately equidistant from the tips. Finding it may be difficult.

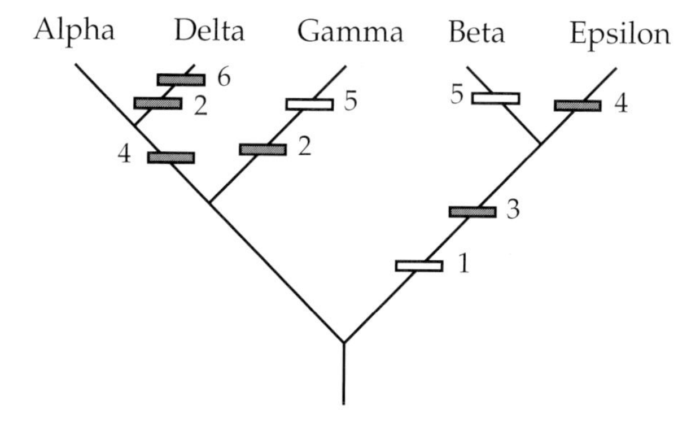

Figure 1.7: Reconstruction of all character changes on the phylogeny of Figure 1.1. The changes are shown as bars across the branches, with a number next to each indicating which character is changing. The shading of each box indicates which state is derived from that change.

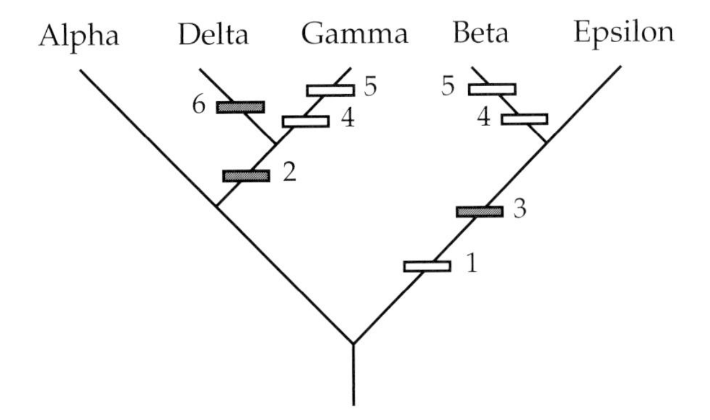

Figure 1.8: Reconstruction of all changes on the most parsimonious phylogeny for the data of Table 1.1. It requires only 8 changes of state. The changes are shown as bars across the branches, with a number next to each indicating which character is changing. The shading of each box indicates which state is derived from that change.

Branch lengths

Having found an unrooted tree, we might want to locate the changes on it and find out how many occur in each of the branches. We have already seen that there can be ambiguity as to where the changes are. That in turn means that we cannot necessarily count the number of changes in each branch. One possible alternative is to average over all possible reconstructions of each character for which there is ambiguity in the unrooted tree. This has the advantage that, although this can

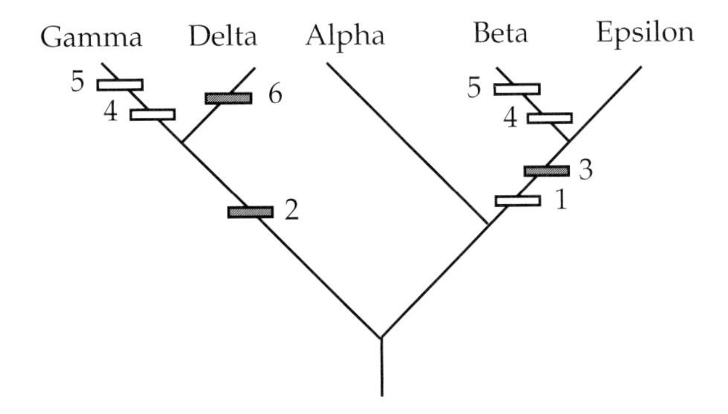

Figure 1.9: Another rooted tree with the same number of changes of state.

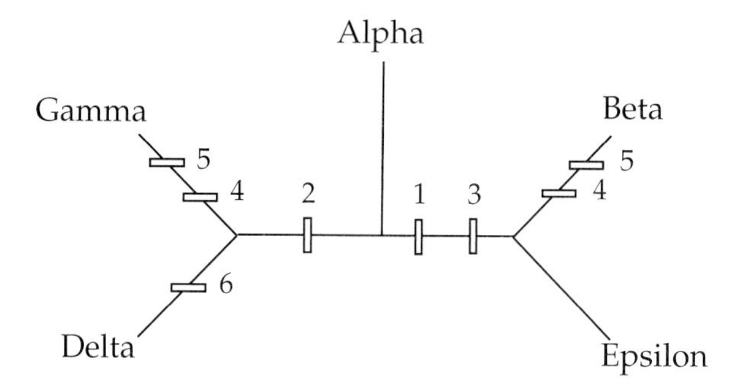

Figure 1.10: The unrooted tree corresponding to Figures 1.8 and 1.9.

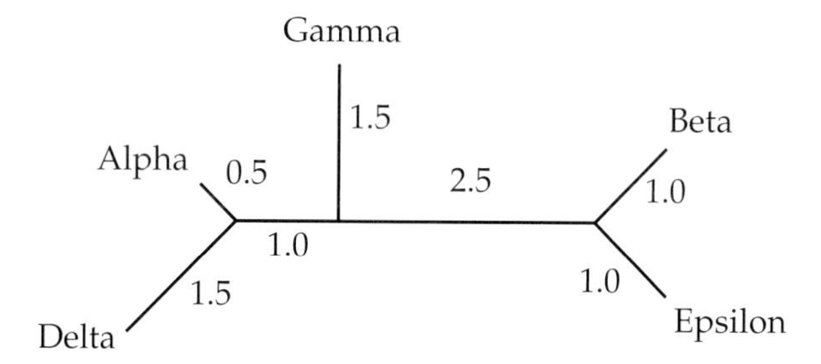

Figure 1.11: The tree of Figure 1.1 and Figure 1.7, shown as an unrooted tree with branch lengths computed by averaging all equally parsimonious reconstructions.

leave fractional numbers of changes in some branches, at least they must add up to the total number of changes in the tree. This is sometimes called the length of the tree. Figure 1.11 shows the same tree as Figure 1.7 and Figure 1.1 (not the most parsimonious tree), using these branch lengths. The lengths of the branches are shown visually and also given as numbers beside each branch.

Unresolved questions

Although we have mentioned many of the issues involved in using parsimony, we have not actually given the algorithms for any of them. In every case we simply reconstructed character states by eyeball, and, similarly, we searched the set of possible trees by informal means. Among the issues that need to be discussed are the following:

- Particularly for larger data sets, we need to know how to count the number of changes of state by use of an algorithm.
- We need to know the algorithm for reconstructing states at interior nodes of the tree.
- We need to know how to search among all possible trees for the most parsimonious ones, and how to infer branch lengths.
- All of the discussion here has been for a simple model of 0/1 characters. What do we do with DNA sequences, that have 4 states, or with protein sequences, that have 20? How do we handle more complex morphological characters?
- There is the crucial issue of justification. Is it reasonable to use the parsimony criterion? If so, what does it implicitly assume about the biology?
- Finally, what is the statistical status of finding the most parsimonious tree? Is there some way we can make statements about how well-supported a most parsimonious tree is over the others?

Much work has been done on these questions, and it is this that we cover in the next few chapters.

Chapter 2

Counting evolutionary changes

Counting the number of changes of state on a given phylogeny requires us to have some algorithm. The first such algorithms for discrete-states data were given by Camin and Sokal (1965) for a model with unidirectional changes, and by Kluge and Farris (1969) and Farris (1970) for bidirectional changes on a linear ordering of states. We will discuss here two algorithms that generalize these, one by Fitch (1971) and the other by Sankoff (1975) and Sankoff and Rousseau (1975). Both have the same general structure. We evaluate a phylogeny character by character. For each character, we consider it as a rooted tree, placing the root wherever seems appropriate. We update some information down a tree; when we reach the bottom, the number of changes of state is available. In both cases, the algorithm does *not* function by actually locating changes or by actually reconstructing interior states at the nodes of the tree. Both are examples of the class known as *dynamic programming* algorithms.

In the previous chapter we found the most parsimonious assignments of ancestral states, and did so by eyeball. In the present chapter we show how the counting of changes of state can be done more mechanically.

The Fitch algorithm

The Fitch (1971) algorithm was intended to count the number of changes in a bifurcating tree with nucleotide sequence data, in which any one of the four bases (A, C, G, T) can change to any other. It also works generally for any number of states, provided one can change from any one to any other. This multistate parsimony model was named *Wagner parsimony* by Kluge and Farris (1969). Fitch's algorithm thus works perfectly for the $0 \rightleftharpoons 1$ case as well. (In fact, Farris (1970) gave a version of this algorithm for the special case of a linear series of discrete states.) The algorithm at first seems to be mumbo-jumbo. It is only after understanding how the Sankoff algorithm works that one can see why it works, and that it is an algorithm of the same general class. We will explain the Fitch algorithm

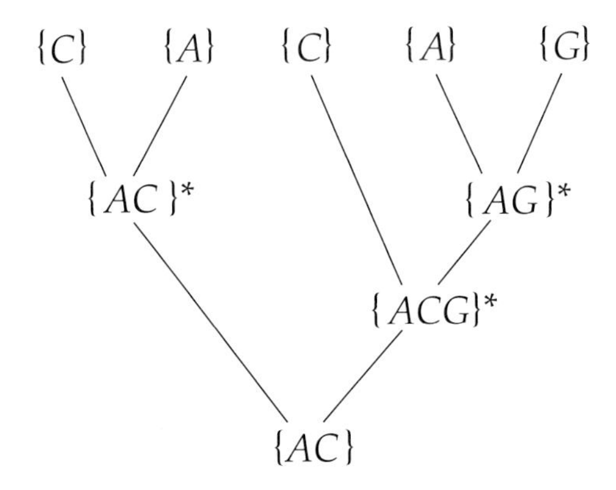

Figure 2.1: An example of the Fitch algorithm applied to a single site. The sets generated at each node are shown.

by use of an example, which is shown in Figure 2.1. The Fitch algorithm considers the sites (or characters) one at a time. At each tip in the tree, we create a set containing those nucleotides (states) that are observed or are compatible with the observation. Thus, if we see an A, we create the set $\{A\}$. If we see an ambiguity such as R (purine), we create the set $\{AG\}$. Now we move down the tree. In algorithmic terms, we do a postorder tree traversal. At each interior node we create a set that is the intersection of sets at the two descendant nodes. However, if that set is empty, we instead create the set that is the union of the two sets at the descendant nodes. Every time we create such a union, we also count one change of state.

In Figure 2.1, we are evaluating a tree with five species. At the particular site, we have observed the bases C, A, C, A, and G in the five species, where we give them in the order in which they appear in the tree, left to right. For the left two, at the node that is their immediate common ancestor, we first attempt to construct the intersection of the two sets. But as $\{C\} \cap \{A\} = \emptyset$, we instead construct the union $\{C\} \cup \{A\} = \{AC\}$ and count 1 change of state. Likewise, for the rightmost pair of species, their common ancestor will be assigned state $\{AG\}$, since $\{A\} \cap \{G\} = \emptyset$, and we count another change of state. The node below it now can be processed. $\{C\} \cap \{AG\} = \emptyset$, so we construct the union $\{C\} \cup \{AG\} = \{ACG\}$ and count a third change of state. The node at the bottom of the tree can now be processed. $\{AC\} \cap \{ACG\} = \{AC\}$, so we put $\{AC\}$ at that node. We have now counted 3 changes of state. A moment's glance at the figure will verify that 3 is the correct count of the number of changes of state. On larger trees the moment's glance will not work, but the Fitch algorithm will continue to work.

The Fitch algorithm can be carried out in a number of operations that is proportional to the number of species (tips) on the tree. One might think that we would also need to multiply this by the number of sites, since we are computing the total number of changes of state over all sites. But we can do better than that. Any site that is invariant, which has the same base in all species (such as $AAAAA$), will never need any changes of state and can be dropped from the analysis without affecting the number of changes of state. Other sites, that have a single variant base present in only a single species (such as, reading across the species, $ATAAA$), will require a single change of state on all trees, no matter what their structure. These too can be dropped, though we may want to note that they will always generate one more change of state each. In addition, if we see a site that has the same pattern (say, $CACAG$) that we have already seen, we need not recompute the number of changes of state for that site, but can simply use the previous result. Finally, the symmetry of the model of state change means that if we see a pattern, such as $TCTCA$, that can be converted into one of the preceding patterns by changing the four symbols, it too does not need to have the number of changes of state computed. Both $CACAG$ and $TCTCA$ are patterns of the form $xyxyz$, and thus both will require at least 2 changes of state. Thus the effort rises slower than linearly with the numbers of sites, in a way that is dependent on how the data set arose.

One might think that we could use the sets in Figure 2.1 to reconstruct ancestral states at the interior nodes of the tree. The sets certainly can be used in that process, but they are not themselves reconstructions of the possible nucleotides, nor do they even contain the possible nucleotides that a parsimony method would construct. For example, in the common ancestor of the rightmost pair of species, the set that we construct is $\{AG\}$. But a careful consideration will show that if we put C at all interior nodes, including that one, we attain the minimum number of changes, 3. But C is not a member of the set that we constructed. At the immediate ancestor of that node, we constructed the set $\{ACG\}$. But of those nucleotides, only A or C are possible in assignments of states to ancestors that achieve a parsimonious result.

The Sankoff algorithm

The Fitch algorithm is enormously effective, but it gives us no hint as to why it works, nor does it show us what to do if we want to count different kinds of changes differently. The Sankoff algorithm is more complex, but its structure is more apparent. It starts by assuming that we have a table of the cost of changes between each character state and each other state. Let's denote by c_{ij} the cost of change from state i to state j. As before, we compute the total cost of the most parsimonious combinations of events by computing it for each character. For a given character, we compute, for each node k in the tree, a quantity $S_k(i)$. This is interpreted as the minimal cost, given that node k is assigned state i, of all the events upwards from node k in the tree. In other words, the minimal cost of events in the subtree, which starts at node k and consists of everything above that point.

It should be immediately apparent that if we can compute these values for all nodes, we can compute them for the bottom node in the tree, in particular. If we can compute them for the bottom node (call that node 0), then we can simply choose the minimum of these values:

$$S = \min_{i} S_0(i) \tag{2.1}$$

and that will be the total cost we seek, the minimum cost of evolution for this character.

At the tips of the tree, the $S(i)$ are easy to compute. The cost is 0 if the observed state is state i, and infinite otherwise. If we have observed an ambiguous state, the cost is 0 for all states that it could be, and infinite for all the rest. Now all we need is an algorithm to calculate the $S(i)$ for the immediate common ancestor of two nodes. This is very easy to do. Suppose that the two descendant nodes are called l and r (for "left" and "right"). For their immediate common ancestor, node a, we need only compute

$$S_a(i) = \min_{j} [c_{ij} + S_l(j)] + \min_{k} [c_{ik} + S_r(k)] \tag{2.2}$$

The interpretation of this equation is immediate. The smallest possible cost given that node a is in state i is the cost c_{ij} of going from state i to state j in the left descendant lineage, plus the cost $S_l(j)$ of events further up in that subtree given that node l is in state j. We select the value of j that minimizes that sum. We do the same calculation in the right descendant lineage, which gives us the second term of equation 2.2. The sum of these two minima is the smallest possible cost for the subtree above node a, given that node a is in state i.

This equation is applied successively to each node in the tree, working downwards (doing a postorder tree traversal). Finally, it computes all the $S_0(i)$, and then (2.1) is used to find the minimum cost for the whole tree.

The process is best understood by an example, the example that we already used for the Fitch algorithm. Suppose that we wish to compute the smallest total cost for the given tree, where we weight transitions (changes between two purines or two pyrimidines) 1, and weight transversion (changes between a purine and a pyrimidine or between a pyrimidine and a purine) 2.5. Figure 2.2 shows the cost matrix and the tree, with the $S(i)$ arrays at each node. You can verify that these are correctly computed. For the leftmost pair of tips, for example, we observe states C and A, so the S arrays are respectively $(\infty, 0, \infty, \infty)$ and $(0, \infty, \infty, \infty)$. Their ancestor has array $(2.5, 2.5, 3.5, 3.5)$. The reasoning is: If the ancestor has state A, the least cost is 2.5, for a change to a C on the left lineage and no change on the right. If it has state C, the cost is also 2.5, for no change on the left lineage combined with change to an A on the right lineage. For state G, the cost is 3.5, because we can at best change to C on the left lineage (at cost 2.5) and to state A on the right lineage, for a cost of 1. We can reason similarly for T, where the costs are 1 + 2.5 = 3.5.

The result may be less obvious at another node, the common ancestor of the rightmost three species, where the result is $(3.5, 3.5, 3.5, 4.5)$. The first entry is 3.5

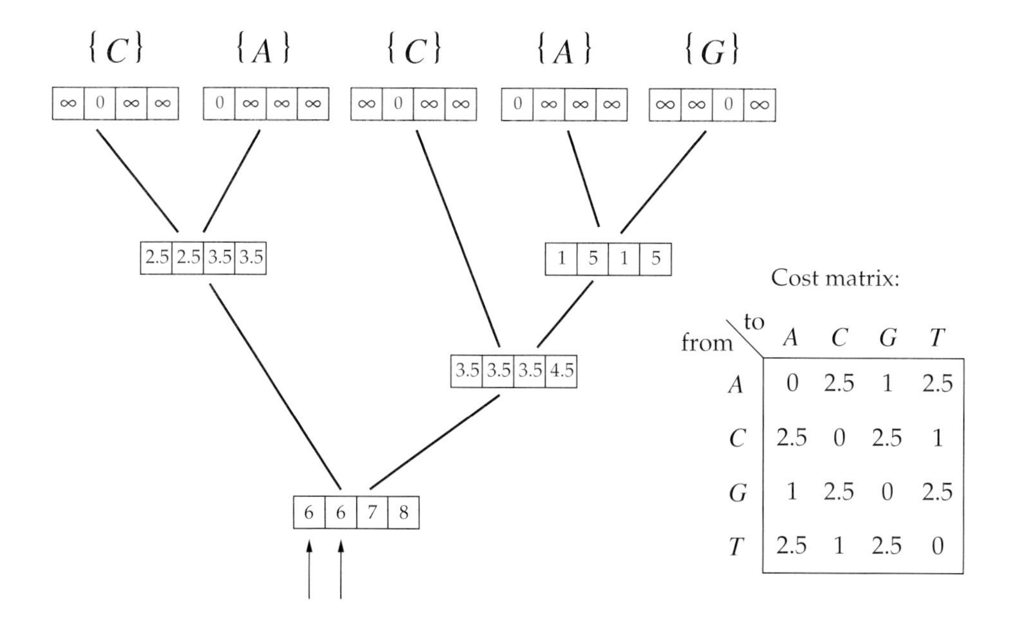

Figure 2.2: The Sankoff algorithm applied to the tree and site of the previous figure. The cost matrix used is shown, as well as the S arrays computed at each node of the tree.

because you could have changed to C on the left branch (2.5 changes plus 0 above that) and had no change on the right branch (0 changes plus 1 above that). That totals to 3.5; no other scenario achieves a smaller total. The second entry is 3.5 because you could have had no change on the left branch (0 + 0) and a change to A or to G on the right one (each 2.5 + 1). The third entry is 3.5 for much the same reason the first one was. The fourth entry is 4.5 because it could have changed on the left branch from T to C (1 + 0), and on the right branch from T to A or T to G (2.5 + 1), and these total to 4.5.

Working down the tree, we arrive at the array $(6, 6, 7, 8)$ at the bottom of the tree. The minimum of these is 6, which is the minimum total cost of the tree for this site. When the analogous operations are done at all sites and their minimal costs added up, the result is the minimal cost for evolution of the data set on the tree.

The Sankoff algorithm is a dynamic programming algorithm, because it solves the problem of finding the minimum cost by first solving some smaller problems and then constructing the solution to the larger problem out of these, in such a way that it can be proven that the solution to the larger problem is correct. An example of a dynamic programming algorithm is the well-known least-cost-path-through-a-graph algorithm. We will not describe it in detail here, but it involves gradually working out the costs of paths to other points in the graph, working outwards from the source. It makes use of the costs of paths to points to work out the costs of paths to their immediate neighbors, until we ultimately know the lengths of the lowest-cost paths from the source to all points in the graph. This does not involve working out all possible paths, and it is guaranteed to give the correct answer.

An attempt to simplify computations by Wheeler and Nixon (1994) has been shown by Swofford and Siddall (1997) to be incorrect.

Connection between the two algorithms

The Fitch algorithm is a close cousin of the Sankoff algorithm. Suppose that we made up a variant of the Sankoff algorithm in which we keep track of an array of (in the nucleotide case) four numbers, but associated them with the bottom end of a branch instead of the node at the top end of a branch. We could then develop a rule similar to equation 2.2 that would update this array down the tree. For the simple cost matrix that underlies the Fitch algorithm, it will turn out that the numbers in that array are always either x or $x + 1$. This is true because one can always get from any state to any other with penalty 1. So you can never have a penalty that is more than one greater than the minimum that is possible at that point on the tree. Fitch's sets are simply the sets of nucleotides that have the minimum value x rather than the higher value of $x + 1$. A careful consideration of the updating rule in Sankoff's algorithm in this case will show that it corresponds closely to the set operations that Fitch specified. Because it is updating the quantities at the bottom end rather than at the top end of each branch, the Fitch algorithm is not a special case of the Sankoff algorithm.

Using the algorithms when modifying trees

Views

For most of the parsimony methods that we will discuss, the score of a tree is unaltered when we reroot the tree. We can consider any place in the tree as if it were the root. Looking outward from any branch, we see two subtrees, one at each end of the branch. Taking the root to be on the branch, we can use the Fitch or Sankoff parsimony algorithms to move "down" the tree towards that point, calculating the arrays of scores for a character. There will be arrays at the two ends of our branch. This can be thought of as "views" summarizing the parsimony scores in these two subtrees, for the character. Each interior node of the tree will have three (or more) views associated with it: one for each branch that connects to that node. Thus in the tree in Figure 2.2, we see one view for the node above and to the right of the root. It shows the view up into the subtree that has the three rightmost species. But there are two other views that we could have calculated as well. One could show the view looking down at that node from the center species, and the other the view looking down at that node from the branch that leads to the

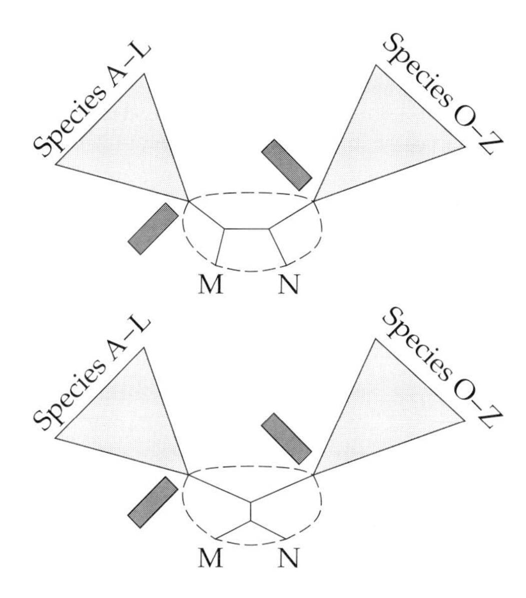

Figure 2.3: Two trees illustrating the use of the conditional scores of the Fitch and Sankoff methods in economizing on computations when rearranging a tree. The two gray rectangles stand for the views for a character in the two subtrees. When species M and N are involved in a rearrangement, the views can be used as if they summarized the data at a tip. They remain unaltered when M and N are rearranged, and the rearrangement can be evaluated by doing calculations entirely within the region outlines by the dashed curve.

two rightmost species. If the node had had four branches connecting to it, there would have been four views possible.

It is worth noting that views also exist for likelihood methods and for some algorithms for distance matrix methods.

Using views when a tree is altered

Both the Fitch and Sankoff algorithms use such views, though they only compute one view at each internal node, the one that looks up at it from below. We can calculate views anywhere in the tree, by passing inwards toward that point from tips. This can be convenient when rearranging or otherwise altering trees. Figure 2.3 shows an example. The two gray rectangles are the views for a character for the two subtrees (which are the large triangles). When we rearrange the two species M and N locally, without disrupting the structure of either subtree, we can compute the parsimony score for the whole tree by using the numbers in the rectangles and doing all of our computations within the regions enclosed by the dashed curves. This enables a fast diagnosis of local rearrangements.

This method of economizing on the effort of computing parsimony scores was first described in print by Gladstein (1997). His discussion codifies methods long in use in the faster parsimony programs but not previously described in print.

When we come to discuss likelihood methods later in the book, we will see views that play a very similar role. They allow similar economies but they are limited by the fact that as one branch length is changed, others elsewhere in the tree must also be altered for the tree to be optimal. In some least squares algorithms for distance matrix methods, there are conditional quantities that behave similarly.

Further economies

There are some additional economies, beyond Gladstein's method, that help speed up parsimony calculations. Ronquist (1998a) points out an economy that can have a large effect when we use a Fitch or Sankoff algorithm and compute views at all nodes, looking in all directions. We have been discussing the tree as if it were rooted, but in most cases it effectively is an unrooted tree.

When a tree is modified in one part, all the inward-looking views may need updating (all those that summarize subtrees that include the modified region). Ronquist points out that we do not need to go through the entirety of the tree modifying these views. As we work outward from the modified region, if we come to a view that looks back in, and that ends up not being changed when it is reconsidered, we need go no further in that direction, as all further views looking back in that way will also be unchanged. This can save a considerable amount of time. We shall see other savings when we discuss tree rearrangement in Chapter 4.

Chapter 3

How many trees are there?

The obvious method for searching for the most parsimonious tree is to consider all possible trees, one after another, and evaluate each. As we continue, we keep a list of the best trees found so far (that is, of all the trees that are tied for best). If the current tree is tied with these, it is added to the list. If one that is better is found, the list is discarded and started anew as consisting of just that tree. When the process is complete, we will have a list of all the trees that are tied for best.

The only problem with this method occurs when the list of possible trees is too large for this complete enumeration to work. In general, it is. This chapter will briefly review the work on counting phylogenies, to show that. The number of phylogenies depends on what we are calling a phylogeny and which ones we count as different. In all of the cases that we will discuss, left-right order of branching does not make any difference — we will count two trees as the same if they differ only by which subtree is on the left side of a branch and which is on the right. Figure 3.1 shows two trees that look different, but are not. They share the same "tree topology" even though they are visually different.

Among the cases that have been considered, one must distinguish between

- Rooted versus unrooted trees
- Labeled versus unlabeled trees
- Bifurcating versus multifurcating trees

Trees are described as *labeled* if their tip nodes have distinct labels. We will always consider cases in which the interior nodes do not have labels. *Bifurcating trees* are those in which every interior node is of degree 3 (it connects to three others) and every tip node is of degree 1 (it connects to only one other node). They are called bifurcating because, considered as rooted, there are two branches leading upward from each interior node. *Multifurcating trees* can have some interior nodes

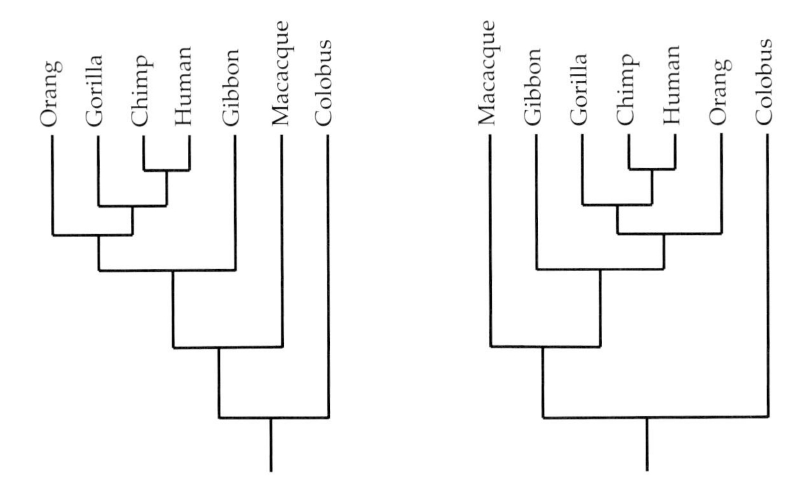

Figure 3.1: Two rooted trees that seem to be different, but are the same tree topology from the point of view of this chapter.

of higher degree. Note that multifurcating trees include all bifurcating trees — multifurcating trees are allowed to have multifurcations, but they are not forced to have them.

Rooted bifurcating trees

Figure 3.2 shows the case of bifurcating, labeled, rooted trees, for 2, 3, and 4 tips. All of the different trees are shown for these cases. But how do we know that these are all of the possibilities? In fact, there is a simple argument that allows us to compute the number of different phylogenies for this case and thus know when there are no more to look for. As elements of this argument will also appear later in other contexts, it is important to consider it in some detail.

We will consider a building up all possible trees by adding one species at a time, in a predetermined order (say, the lexicographic order of the species names). If we have a list of all possible trees of n species and add to each one of them species $n+1$, in all possible places, we will in fact generate all possible trees of $n+1$ species, each only once. Figure 3.3 shows this process of adding a new species at all possible places. Since the tree is bifurcating both before and after the addition, the new species cannot be connected to an existing interior node. It must instead be connected to a new node, which is placed in the middle of an existing branch. Thus each internal branch of a tree is the location of a possible species addition.

But how do we know that this process will lead to all possible rooted, bifurcating, labeled trees? Do we know that each such addition leads to a different such tree? In fact, both of these are true. We can see this by thinking of the process of

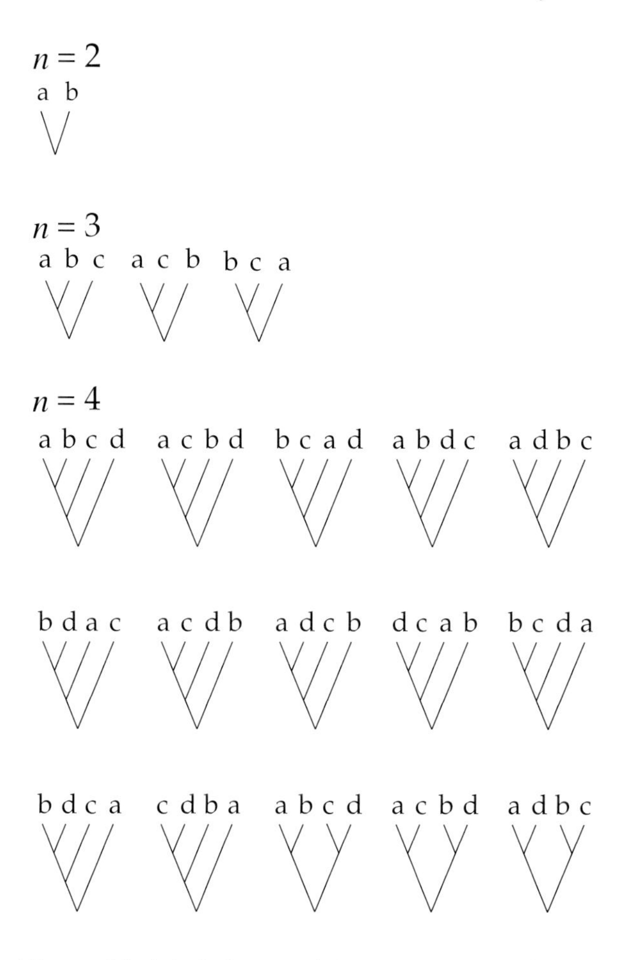

Figure 3.2: All possible labeled, rooted, bifurcating trees for 2, 3, and 4 tips.

adding species k to a tree that consists of species 1 through $k-1$. Consider also the operation of removing species k from a tree that contains species 1 through k. These two operations are inverses of each other. Suppose that we have a particular tree with n species. Remove successively species n, $n-1$, $n-2$, and so on until species $k+1$ is removed. At this point what is left must be one particular tree with species 1 through k.

Since the removal operation reverses the addition of the species, there must then be some particular sequence of places to add species $k+1$, $k+2$, ... onto that k-species tree to end up with that n-species tree. Furthermore no other k-species tree can, when those $n-k$ missing species are added, yield that particular n-species tree. If there were another k-species tree that could yield it, then that tree too would be reached by removal of those species from the n-species tree. But

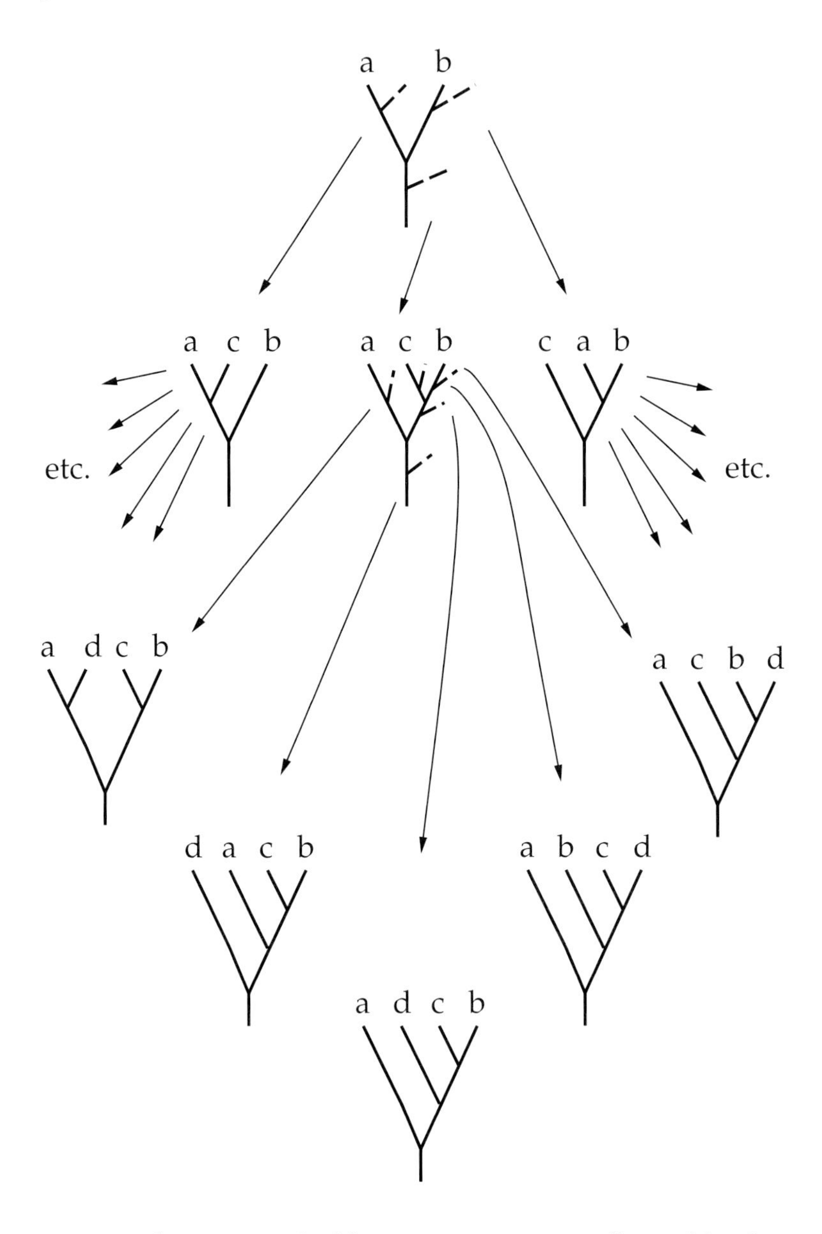

Figure 3.3: The process of adding a new species in all possible places, leading to all possible rooted bifurcating trees. For $n = 3$ the consequences of adding the fourth tip to one of the trees is shown, but for the others it is only indicated by outgoing arrows.

that is a logical impossibility, as the same sequence of removals cannot result in two different trees. Thus any n-species tree can be reached from one and only one k-species tree.

Therefore, each possible addition sequence leads to a different n-species tree, and all such trees can be generated in that way. When we add species to a tree, the number of ways in which we can do that are equal to the number of branches, including the branch at the bottom of the tree. There are 3 such branches in a two-species tree. Every time that we add a new species, it adds a new interior node, plus two new branches. Thus after choosing one of the 3 possible places to add the third species, the fourth can be added in any of 5 places, the fifth in any of 7, and so on. It will not be hard to see that the n-th can be added in any of $2n - 3$ places.

This means that there are

$$3 \times 5 \times 7 \times 9 \times 11 \times 13 \times \cdots \times (2n-3)$$

different ways to add species so as to construct an n-species tree. Each way leads to a different such tree, and together they lead to all such trees. We thus have a simple way of computing the number of rooted, bifurcating, labeled trees, without generating all of them. This is not a closed-form formula, but it is not hard to show that this is equal to

$$\frac{(2n-3)!}{2^{n-2}(n-2)!}$$

Even though that formula (sometimes called $(2n-3)!!$) looks simple, the preceding expression of it as product of successive odd integers is in practice far easier to use.

Table 3.1 shows the resulting numbers, up to 20, and approximate values for some number of species beyond that.

The immediate implication of these large numbers is that we cannot hope to examine all rooted, bifurcating, labeled trees in any algorithm for more than about 10 species. Exhaustive enumeration is probably practical up to about 10 species. This boundary of practicality will move upwards, but it will do so slowly. It will require a massively parallel approach using molecular computation methods to get up to $n = 20$, and beyond that the numbers are so much greater than Avogadro's Number that even molecular computations may not be possible. For 50 species, one is approaching Eddington's famous number, the number of electrons in the visible universe.

The counting of trees has been a mathematician's recreation since the pioneering work of Cayley (1857, 1889). Ernst Schröder (1870) was the first to compute numbers in Table 3.1. He used generating function methods, as did Cayley. The simple argument used here is due to Cavalli-Sforza and Edwards (1967). Moon (1970) has reviewed many other counting problems involving labeled trees. But partially-labeled cases like those we consider have largely been left to biologists to count.

Table 3.1: The number of rooted, bifurcating, labeled trees for n species, for various values of n. The numbers for more than 20 species are approximate.

Species	Number of trees
1	1
2	1
3	3
4	15
5	105
6	945
7	10,395
8	135,135
9	2,027,025
10	34,459,425
11	654,729,075
12	13,749,310,575
13	316,234,143,225
14	7,905,853,580,625
15	213,458,046,676,875
16	6,190,283,353,629,375
17	191,898,783,962,510,625
18	6,332,659,870,762,850,625
19	221,643,095,476,699,771,875
20	8,200,794,532,637,891,559,375
30	4.9518×10^{38}
40	1.00985×10^{57}
50	2.75292×10^{76}

Unrooted bifurcating trees

Most methods of inferring phylogenies infer unrooted trees. As each rooted tree can have its root removed, there cannot be more unrooted than rooted trees for a given number of species. In fact, there are fewer, as in an unrooted bifurcating tree with n tips there are $2n-3$ places that a root could be inserted, to give rise to rooted bifurcating trees. These are the $2n-3$ branches of the tree. If each of these were to result in different rooted tree, this would suggest that the number of unrooted trees was a factor of $2n-3$ smaller than the number of rooted trees. In fact, this supposition is true.

The easy way to see this is to try a different, and more direct, argument. An unrooted tree can always be rooted at one of its species, say, the first species. Fig-

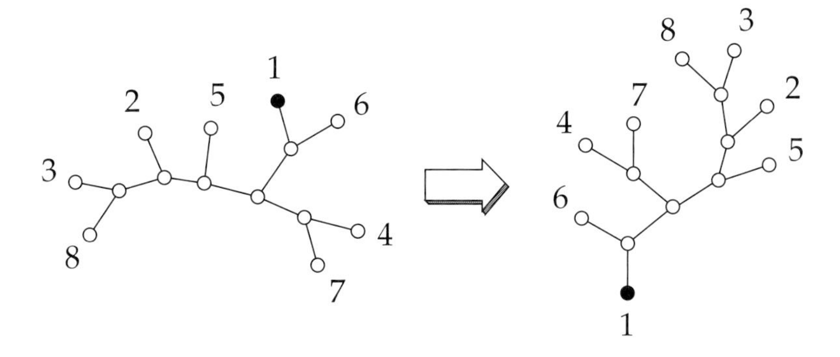

Figure 3.4: An unrooted bifurcating tree with 8 species, rooted by using species 1 as the root.

ure 3.4 shows this particular way of rooting an unrooted bifurcating tree. Suppose that we consider the rooted trees that arise by rooting at the first species in this way. With n tips on the unrooted tree, there will then be $n - 1$ tips on the resulting rooted tree (as we now can no longer consider species 1 to be a tip). We have already computed the number of rooted bifurcating trees for all possible numbers of tips. Every rooted tree with $n - 1$ labeled tips corresponds to one unrooted tree, and every unrooted tree with n tips corresponds to one rooted tree with $n - 1$ tips.

Thus there must be exactly

$$1 \times 3 \times 5 \times 7 \times \cdots \times (2n - 5)$$

unrooted bifurcating trees with n labeled tips. This is precisely the number of rooted trees with the factor $2n - 3$ removed, which is the same as the number of rooted trees with one fewer species. Thus we can consult Table 3.1 to find that with $n = 10$ the number of unrooted bifurcating trees will be 2,027,025, and with $n = 20$ it will be nearly 2.22×10^{20}. It is also possible to get the number of unrooted bifurcating trees directly from an argument that generates each tree by sequentially adding tips in all $2n - 3$ possible places, much as we did with rooted trees.

Multifurcating trees

So far, all trees have been bifurcating. Allowing for multifurcating trees introduces new complications. Ernst Schröder (1870) counted the number of rooted trees with possible multifurcations and labeled tips, using generating function methods. A simpler, if less elegant, way of getting the same numbers was given by me (Felsenstein, 1978a). It seems easier to explain than Schröder's methods.

If we were to try to use the method of adding successive species in all possible places, but allow there to be multifurcations, we run into the problem that we

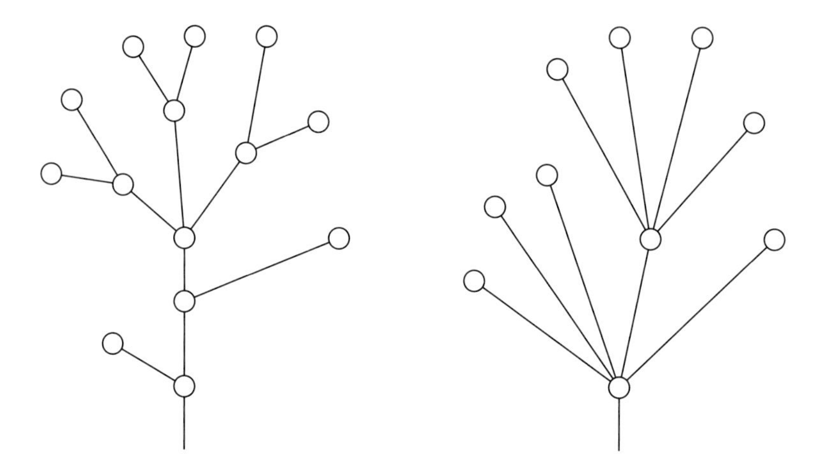

Figure 3.5: Two rooted trees with different amounts of multifurcation, showing the numbers of internal nodes and the numbers of branches each has. The tree on the left has 8 tips, 6 internal nodes, and thus 14 branches. The tree on the right has 8 tips, 2 internal nodes, and thus 10 branches.

cannot tell in how many places the next species can be added without knowing how many multifurcations there are. Figure 3.5 shows two trees with different amounts of multifurcations. If we allow a new species to be added so as to split off from any branch, there are 14 branches in the left tree and 10 in the right tree. If instead we allow the new species to increase the degree of multifurcation by splitting off from an internal node, there are, respectively, 6 and 2 of those. Thus we cannot simply use the argument that counts all placements of the next species.

The easiest way out of this dilemma is to count numbers of trees with different numbers of internal nodes. Suppose that $T_{n,m}$ is the number of rooted trees with n labeled tips and m (unlabeled) internal nodes. The number of internal nodes can be any integer from 1 to $n-1$. If we obtain all the trees with n tips and m interior nodes by adding species n to trees that had one fewer species, we must consider both the cases in which the new species was added to an internal node, creating an additional furc, and the cases in which it was added to a branch, creating a bifurcation and a new internal node. For each of the $T_{n-1,m}$ trees with $n-1$ tips and m internal nodes, there are m places to add the new species at an internal node. For each of the $T_{n-1,m-1}$ trees with $n-1$ species and $m-1$ internal nodes, there are $(n-1)+(m-1) = n+m-2$ places we can add species n. As before, each sequence of additions can be shown to generate a distinct tree, and each possible multifurcating tree can be reached by a sequence of additions.

Figure 3.6: Table of the numbers of rooted multifurcated trees with labeled tips that have different numbers of internal nodes. The flow of the calculation using the recurrence in equation 3.1 is shown for the last column The diagonal gives the number of bifurcating trees, which are included. The row at the bottom of the table is the total number of multifurcating trees for that number of species.

Number of internal nodes \ Number of species	2	3	4	5	6	7		8
1	1	1	1	1	1	1	— x 1 → ; ↘ x 8	1
2		3	10	25	56	119	— x 2 → ; ↘ x 9	246
3			15	105	490	1,918	— x 3 → ; ↘ x 10	6,825
4				105	1,260	9,450	— x 4 → ; ↘ x 11	56,980
5					945	17,325	— x 5 → ; ↘ x 12	190,575
6						10,395	— x 6 → ; ↘ x 13	270,270
7								135,135
Total	1	4	26	236	2,752	39,208		660,032

The result is the formula

$$T_{n,m} = \begin{cases} (n+m-2)\, T_{n-1,m-1} \;+\; m\, T_{n-1,m} & \text{if } m > 1 \\ T_{n-1,m} & \text{if } m = 1 \end{cases} \tag{3.1}$$

Figure 3.6 shows a table of the numbers $T_{n,m}$ with the flow of calculations shown for the rightmost column. The sum of each column is the total number of rooted trees with labeled tips, T_n. Although there is no closed-form formula for this quantity, it is easy to compute it by generating the table using equation 3.1. Table 3.2 shows these totals for moderate numbers of species. In my paper (Felsenstein, 1978a) giving this table, I also gave similar recursions and tables for

Table 3.2: Number of rooted trees with labeled tips, allowing multifurcations. The numbers are tabulated by the number of species.

Species	Number of trees
2	1
3	4
4	26
5	236
6	2,752
7	39,208
8	660,032
9	12,818,912
10	282,137,824
11	6,939,897,856
12	188,666,182,784
13	5,617,349,020,544
14	181,790,703,209,728
15	6,353,726,042,486,272
16	238,513,970,965,257,728
17	9,571,020,586,419,012,608
18	408,837,905,660,444,010,496
19	18,522,305,410,364,986,906,624
20	887,094,711,304,119,347,388,416
30	7.0717×10^{41}
40	1.9037×10^{61}
50	6.85×10^{81}
100	3.3388×10^{195}

the case in which some of the labels may be located at interior nodes of the tree. There are, of course, even more trees if we allow that.

It is possible to go further, making generating functions for these numbers (as Schröder did), formulas for the asymptotic rate at which the numbers rise, or counting the numbers of trees with some interior nodes labeled. We will not attempt to do this for any of the cases in this chapter.

Unrooted trees with multifurcations

We can extend the counting of trees that may be multifurcating from the rooted to the unrooted case by the same method as before. As we can arbitrarily root an unrooted tree at species 1, the number of unrooted trees will be the same as the number of rooted trees with one fewer species.

Tree shapes

Even without the labels being visible at the tips, trees differ in "shape." (In the terminology of Harding, 1971, these would be called *unlabeled shapes*). We may want to know how many different shapes there are for various numbers of species. We can imagine asking this about bifurcating trees and multifurcating trees, and in each of these cases, about rooted and unrooted trees.

Rooted bifurcating tree shapes

For the case in which the trees are rooted and bifurcating, the basic method of calculation was found by Wedderburn (1922) and rediscovered by Cavalli-Sforza and Edwards (1967). The key to it is that at the base of the rooted tree is a bifurcation, with m species at the tips of the left-hand subtree, and $n-m$ at the tips of the right subtree. We are not distinguishing left from right in this argument. Suppose that m happened to be 5 and $n-m$ happened to be 10. If we already know that there are S_5 different tree shapes for 5 species, and S_{10} tree shapes for 10 species, then there will be $S_5 \times S_{10}$ possible combinations of these, and each of these will be a tree of 15 species of a different shape. We can compute the total number of shapes for n species by summing over all values of m such that $m \leq n-m$. However, we must take special care when $m = n-m$, that is, when m is exactly half of n.

In that case, the number of combinations is not S_m^2, but is the number of different unordered pairs of S_m objects, which is $S_m(S_m+1)/2$. This differs from S_m^2 because that quantity would overcount by counting twice all cases where the subtrees on the two sides have different shapes, as each has the same shape as a tree with those two subtrees switched.

We can start the calculation with the obvious value $S_1 = 1$. So the algorithm is:

$$\begin{aligned} S_1 &= 1 \\ S_n &= S_1 S_{n-1} + S_2 S_{n-2} + \ldots + S_{(n-1)/2} S_{(n+1)/2} \quad \text{if } n > 1 \text{ and } n \text{ is odd} \\ S_n &= S_1 S_{n-1} + S_2 S_{n-2} + \ldots + S_{n/2}(S_{n/2}+1)/2 \quad \text{if } n > 1 \text{ and } n \text{ is even} \end{aligned} \tag{3.2}$$

It is easy to compute a table of the number of different tree shapes for this case. It is shown in Table 3.3. There are, of course, far fewer shapes than there are trees. Harding (1971) derives a generating function whose coefficients are the S_i and that can be used to study the asymptotic rate of growth of the S_i.

Figure 3.7 shows the tree shapes up to 6 species. They are arranged in order of their appearance in the terms of equation 3.2. Thus in the section for $n = 6$, we see first those having a 5 : 1 split at their base, then those having a 4 : 2 split, then those with a 3 : 3 split. Within each of these groups, the left subtrees correspond to the trees for $n = 5$, for $n = 4$, and for $n = 3$, in the order in which those appear in the figure.

Table 3.3: Number of different shapes of trees with different numbers of species, counting unlabeled rooted bifurcating trees, as computed by Cavalli-Sforza and Edwards's (1967) algorithm. Numbers for more than 20 species are shown to 5 significant figures.

Species	Number of shapes
1	1
2	1
3	1
4	2
5	3
6	6
7	11
8	23
9	46
10	98
11	207
12	451
13	983
14	2,179
15	4,850
16	10,905
17	24,631
18	56,011
19	127,912
20	293,547
30	1.4068×10^{9}
40	8.0997×10^{12}
50	5.1501×10^{16}
100	1.0196×10^{36}

I do not know of any closed-form formula for the numbers in Table 3.3, but Donald Knuth (1973, p. 388) discusses a generating function that produces these numbers, in the context of a tree enumerating problem.

Rooted multifurcating tree shapes

We can continue on to the cases in which multifurcations are allowed, and also to those where the trees are unrooted. Although the methods will be derived by extending Edwards's algorithm, these cases have not been considered anywhere in the literature, mostly from lack of interest in them. These cases will be described in less detail. When multifurcations are allowed in a rooted tree, the logic is similar to

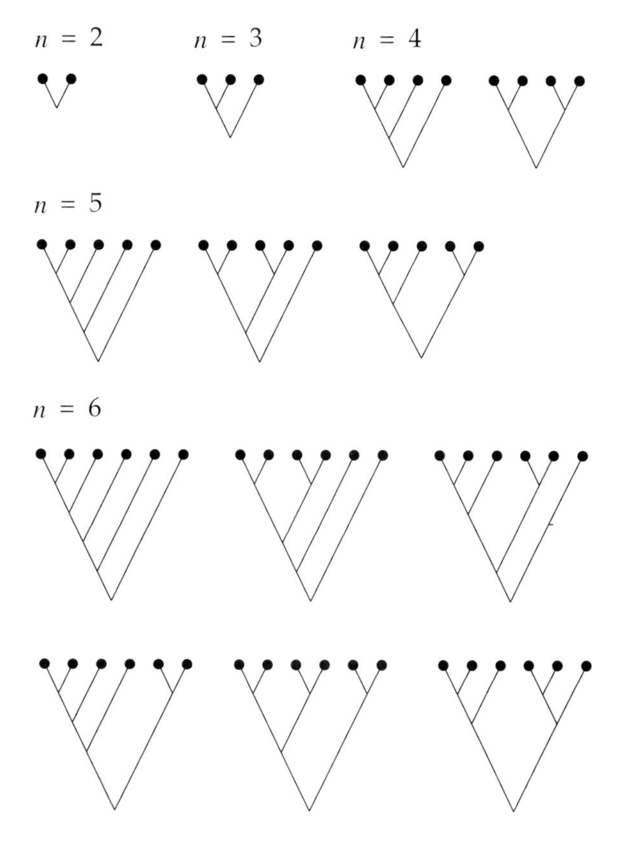

Figure 3.7: The rooted bifurcating tree shapes for 2 to 6 species.

the bifurcating case, except that at the root of the tree there can be a multifurcation (there can at interior forks as well, but that takes care of itself). For trees with n species, we then must sum over all ways that n objects can be partitioned. We are interested only in those partitions that have the larger numbers of objects on the left. If we write a partition by listing the numbers of objects in each set, for 6 objects we want to consider the partitions $(5, 1)$, $(4, 2)$, $(4, 1, 1)$, $(3, 3)$, $(3, 2, 1)$ and $(2, 2, 2)$. We would not consider the partitions $(2, 4)$ or $(3, 1, 2)$, because the order of the branches at the fork at the base of the tree is arbitrary, and to avoid overcounting cases we are keeping them in order of the number of species on the subtrees. (Strangely, in mathematics *ordered partitions* are those in which the numbers are *not* constrained to be in order of their size.)

For each such partition we have a term for the contribution it makes to the number of shapes. Suppose that we are calling the numbers of shapes of rooted multifurcating trees $T(n)$. If the sizes of the sets are all different, as is the case for

Table 3.4: Number of shapes of rooted multifurcating trees for different numbers of species. Numbers of shapes for more than 20 species are given to 4 significant figures.

Species	Shapes	Species	Shapes
2	1	16	2,253,676
3	2	17	7,305,788
4	5	18	23,816,743
5	12	19	78,023,602
6	33	20	256,738,751
7	90	30	4.524×10^{13}
8	261	40	9.573×10^{18}
9	766	50	2.237×10^{24}
10	2,312	60	5.565×10^{29}
11	7,068	70	1.445×10^{35}
12	21,965	80	3.871×10^{40}
13	68,954	90	1.062×10^{46}
14	218,751	100	2.970×10^{51}
15	699,534		

a partition like $(4, 2, 1)$, the term is the product of the numbers of shapes for each set. For that partition it would be $T(4) \times T(2) \times T(1)$. If two or more sets have the same size, we must instead use the number of different combinations of that many objects into sets of this size. So for the partition $(5, 2, 2, 2, 1)$, the term is

$$T(5)\frac{T(2)\,(T(2)+1)\,(T(2)+2)}{1 \times 2 \times 3}T(1)$$

because there are $n(n+1)(n+2)/6$ ways to write numbers in 3 boxes where each box gets a number from the range 1 through n and we are not concerned with the order of the boxes. More generally, when there are k boxes and n numbers, there are $n(n+1)(n+2)\cdots(n+k-1)/k!$ ways.

With this algorithm, the numbers of shapes of rooted multifurcating trees are as given in Table 3.4.

Unrooted Shapes

To count unrooted shapes, for either bifurcating or multifurcating trees, we need to find a fork in the tree that is uniquely defined, to temporarily root the tree there. It is not hard to show that there are at most two internal nodes in an unrooted tree whose corresponding partition has no set with more than half the species in it. Thus if we start in a tree with 10 species at a node whose partition would

Table 3.5: The numbers of bifurcating and multifurcating unrooted tree shapes. Numbers for more than 20 species are given to 4 significant figures.

Species	Multifurcating shapes	Bifurcating shapes
3	1	1
4	2	1
5	3	1
6	7	2
7	13	2
8	32	4
9	73	6
10	190	11
11	488	18
12	1,350	37
13	3,741	66
14	10,765	135
15	31,311	265
16	92,949	552
17	278,840	1,132
18	847,511	2,410
19	2,599,071	5,098
20	8,044,399	11,020
30	8.913×10^{11}	3.294×10^{7}
40	1.377×10^{17}	1.385×10^{11}

be $(8, 1, 1)$, we can always move to the node that is at the root of the 8-species subtree, which might have the partition $(4, 4, 2)$. At that point we are at a node whose largest partition does not exceed half the species. For many of the possible unrooted trees there is just one such partition, but some have two. For example if a tree has a central branch with 5 species connected to each of its ends, then the partition for the node on the left end of the branch might be $(5, 3, 2)$ and that for the node on the right end of the branch might be $(5, 2, 2, 1)$. These partitions include sets for the subtree that is at the other end of that central branch.

The algorithm for computing the numbers of shapes for these two cases (the bifurcating and multifurcating cases) consists again of listing all possible partitions of n objects into three or more sets, where the set sizes are ordered, but in this case we ignore all those whose largest set contains more than half the species. For the partitions whose largest set contains less than half the objects, we can take the same products as before, using the numbers of rooted multifurcating or bifurcating

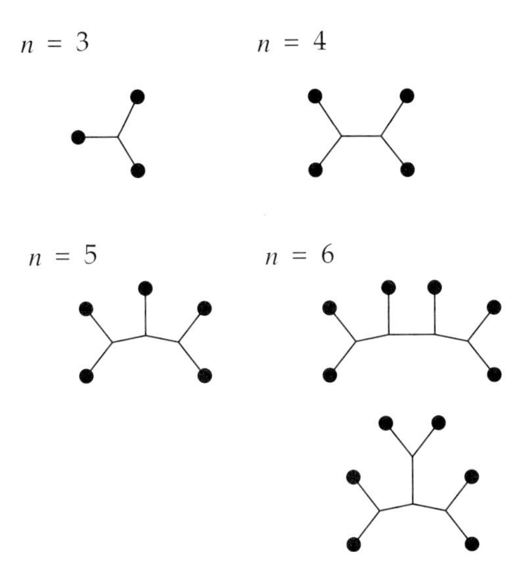

Figure 3.8: The unrooted bifurcating tree shapes for up to six species.

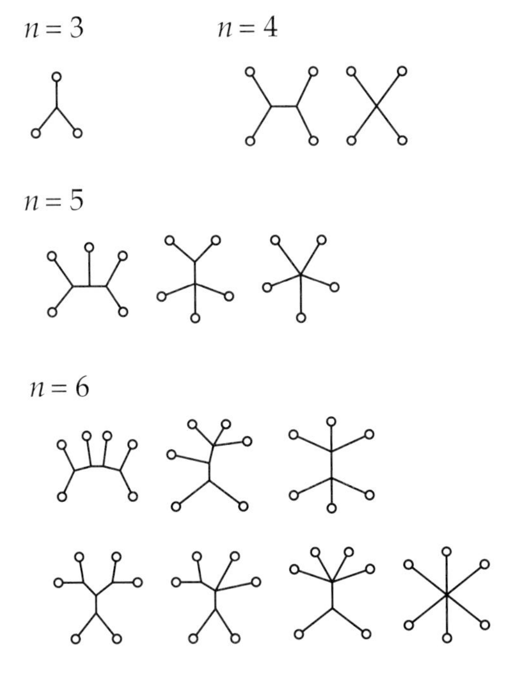

Figure 3.9: The unrooted multifurcating tree shapes for up to six species.

trees, as appropriate. For the bifurcating case, we of course consider only partitions having three sets. For the partitions whose largest set contains exactly half the species, we must realize that there is a risk of overcounting: The correct number to count for such a case is the number of ordered pairs of rooted trees that have that number of species.

Table 3.5 shows the numbers of shapes for bifurcating and for multifurcating unrooted trees.

Figure 3.8 shows the unrooted bifurcating tree shapes up to 6 species, and Figure 3.9 shows the unrooted multifurcating tree shapes up to 6 species.

Labeled histories

Usually when we consider tree topologies we do not care about the order in time of the interior nodes of the tree, except to ensure that descendants occur later than their ancestors. For some purposes connected with coalescent trees of genes within species (as in Chapter 26), priors on trees (Chapter 18), and distributions of shapes of trees (Chapter 33), we do care. Edwards (1970) defined a *labeled history* as a tree topology where we also record the order of the nodes of the tree in time. Figure 3.10 shows two trees that are the same tree topology but are different labeled histories.

Edwards also worked out formulas to count the number of bifurcating labeled histories. Working down a labeled history, each interior node brings together two of the lineages. At the top of the tree there are n lineages. There are $n(n-1)/2$ possible pairs of lineages that can be combined. Combining two of them, there are now $[n-1][(n-1)-1]/2 = (n-1)(n-2)/2$ pairs that could be combined. If we specify the pair to combine at each stage, we have specified the labeled history

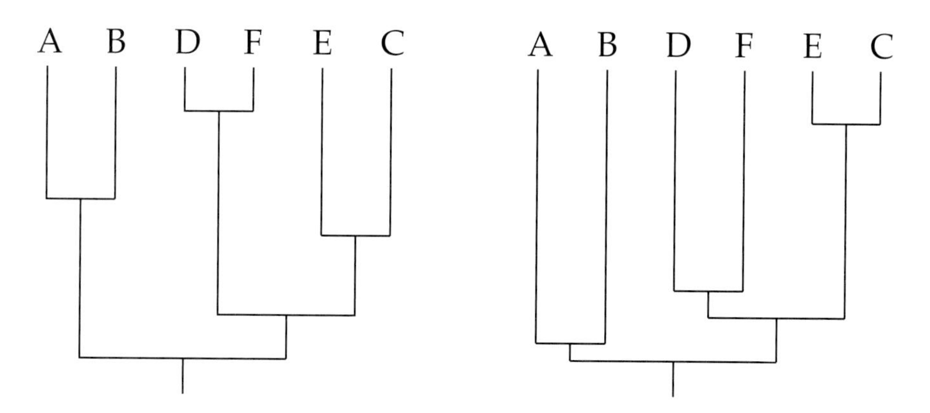

Figure 3.10: Two different labeled histories that are the same tree topology.

uniquely. The number of possible ways we could do that is the product of the number of pairs at each of the $n-1$ events, so that it is

$$\frac{n(n-1)}{2}\,\frac{(n-1)(n-2)}{2}\,\frac{(n-2)(n-3)}{2}\cdots\frac{2\times 1}{2}=\frac{n!(n-1)!}{2^{n-1}} \tag{3.3}$$

These can be very large numbers, compared to the number of tree topologies for the same size of bifurcating tree. When $n=4$, there are 15 tree topologies but 18 labeled histories, a small increase. But when $n=10$, the 34,459,425 tree topologies lead to 2,571,912,000 labeled histories, an increase by a factor of almost 75.

Perspective

Knowing exactly how many tree topologies or tree shapes there are in various cases is not particularly important, unless one is enumerating them in a computer program and wants to know whether the program has found each of them exactly once. The point is that there are very large numbers of them, and these numbers rise exponentially. This creates major difficulties for any search strategy that would work by considering all possible trees. In my 1978 paper I suggested that one use for the numbers was "to frighten taxonomists."

The number of multifurcating rooted trees for 20 species led Walter Fitch (personal communication) to exclaim that for 20 species "we have more than a gram molecular weight of evolutionary trees," as it then exceeds 6.023×10^{23}.

Some further references on counting trees and sampling random trees will be found in Gordon's (1987) review of hierarchical classification.

Chapter 4

Finding the best tree by heuristic search

If we cannot find the best trees by examining all possible trees, we could imagine searching in the space of possible trees. In this chapter we will consider *heuristic search* techniques, which attempt to find the best trees without looking at all possible trees. They are, of their very nature, a bit *ad hoc*. They also do not guarantee us to have found all, or even any, of the best trees.

The fundamental technique is to take an initial estimate of the tree and make small rearrangements of branches in it, to reach "neighboring" trees. If any of these neighbors are better, we consider them, and continue, attempting more rearrangements. Finally, we reach a tree that no small rearrangement can improve. Such a tree is at a local optimum in the tree space. However, there is no guarantee that it is a global optimum. Figure 4.1 shows the problem for the case of searching in two spatial coordinates. Trees are a rather different case, but tree space is difficult to depict in a diagram like this.

In the diagram, we are trying to maximize a quantity — trying to find the highest point on the surface. In the case of the parsimony criterion, we are actually trying to minimize the number of evolutionary changes of state. We can convert that into a maximization problem by simply placing a minus sign before the number of changes of state, so that 272 becomes -272. Or, alternatively, we could subtract it from a large number, so that 272 becomes $10,000 - 272 = 9,728$. Maximization of the resulting quantity will minimize the number of changes of state. It is easier to show the diagram as a maximization problem than as a minimization problem, as maxima are more visible than minima.

In this diagram, we imagine that we have started with a particular point on the surface and then looked at its four neighbors. One of them is higher, so we move to that point. Then we examine its neighbors. We continue this until we have climbed to the highest point on the "hill." However, as the diagram shows, this

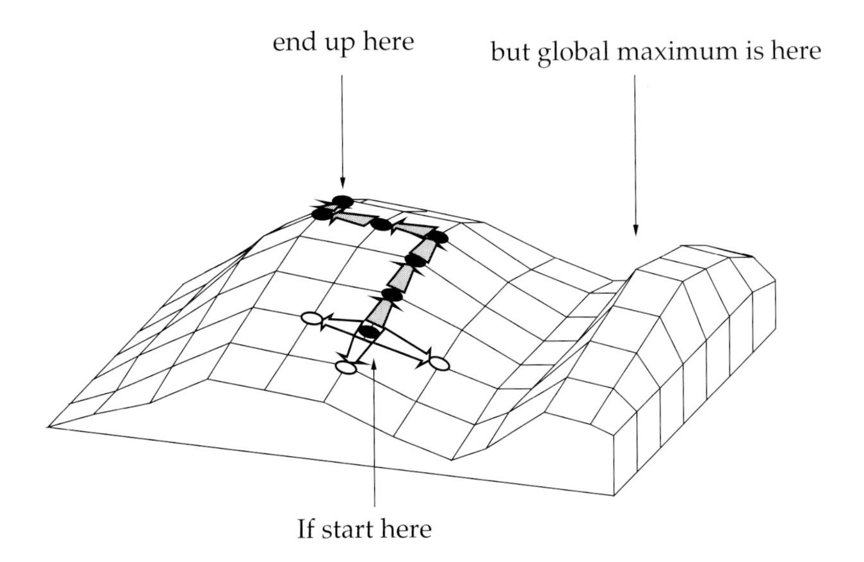

Figure 4.1: A surface rising above a two-dimensional plain (or plane). The process of climbing uphill on the surface is illustrated, as well as the failure to find a higher peak by this "greedy" method.

strategy is incapable of finding another point, one that is in fact higher, but that is not located on the hill where we started. Strategies of this sort are often called *the greedy algorithm* because they seize the first improvement that they see.

In this chapter we will examine some of the different kinds of rearrangements that have been proposed. Many others are possible. The techniques are more the result of common sense than of using any mathematical techniques. Later in the chapter we will also discuss some sequential addition strategies used for locating the starting point of the search. In the next chapter we will discuss branch and bound methods, a search technique guaranteed to find all of the most parsimonious trees.

Although the discussion here will be cast in terms of parsimony, it is important to remember that exactly the same strategies and issues arise with the other criteria for inferring phylogenies, and heuristic search techniques are employed for them in much the same way.

Nearest-neighbor interchanges

Nearest-neighbor interchanges (NNI) in effect swap two adjacent branches on the tree. A more careful description is that they erase an interior branch on the tree, and the two branches connected to it at each end (so that a total of five branches are erased). This leaves four subtrees disconnected from each other. Four subtrees

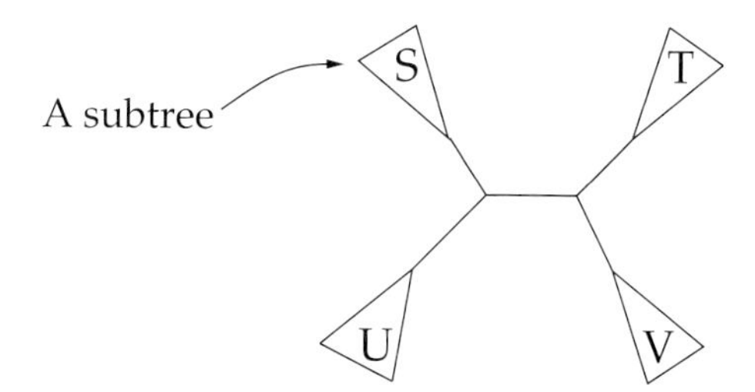

is rearranged by dissolving the connections to an interior branch

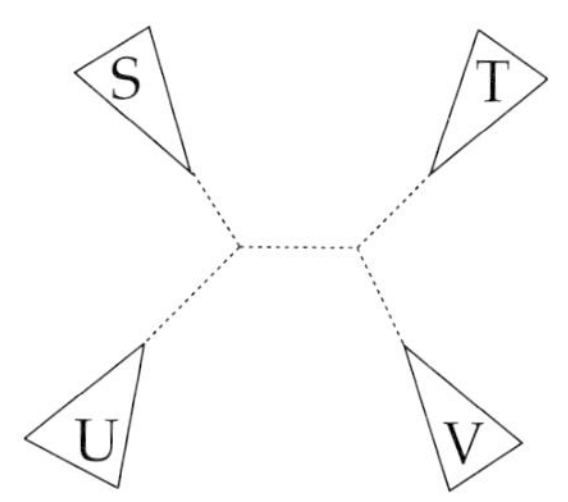

and reforming them in one of the two possible alternative ways:

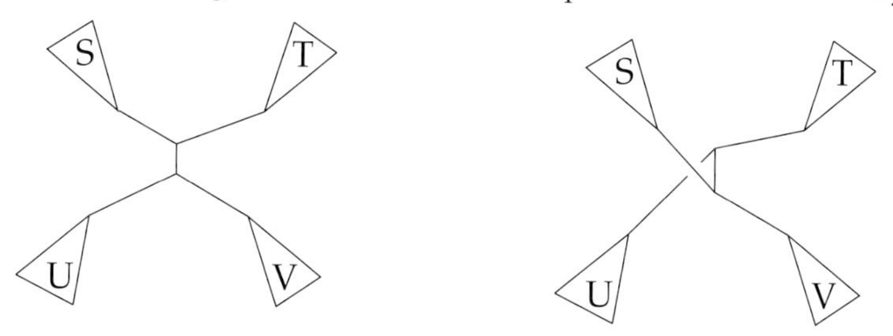

Figure 4.2: The process of nearest-neighbor interchange. An interior branch is dissolved and the four subtrees connected to it are isolated. These then can be reconnected in two other ways.

can be hooked together into a tree in three possible ways. Figure 4.2 shows the process. One of the three trees is, of course, the original one, so that each nearest-neighbor interchange examines two alternative trees. In an unrooted bifurcating tree with n tips, there will be $n - 3$ interior branches, at each of which we can examine two neighboring trees. Thus in all, $2(n - 3)$ neighbors can be examined for each tree. Thus a tree with 20 tips has 34 neighbors under nearest-neighbor interchange.

There is some ambiguity about how greedy we ought to be. If we accept the first neighboring tree that is an improvement, that will not be as good a search

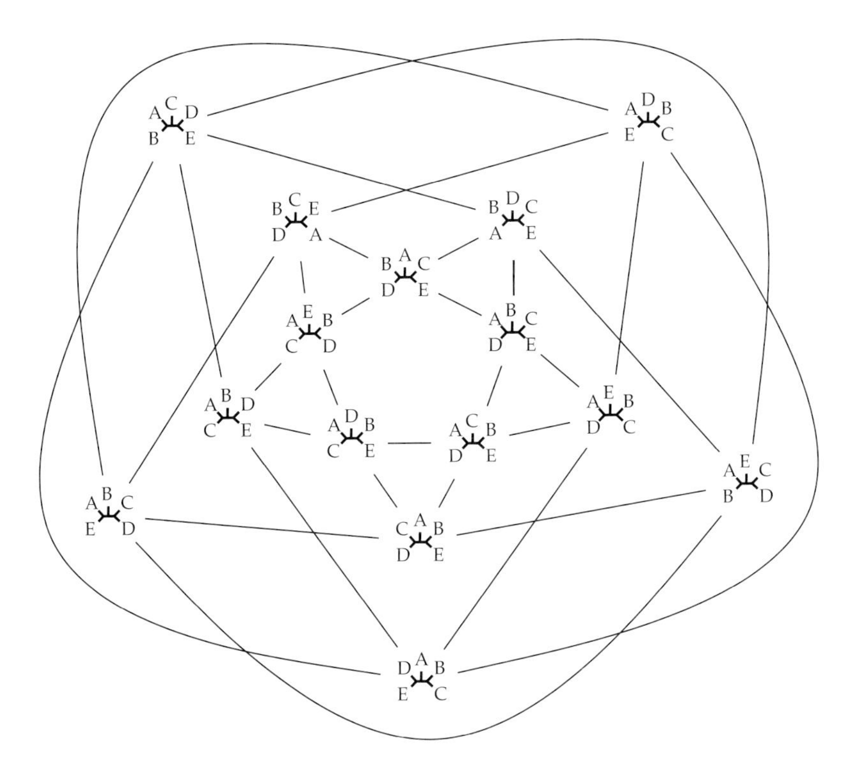

Figure 4.3: The space of all 15 possible unrooted trees with 5 tips. Neighbors are connected by lines when a nearest-neighbor interchange can convert one into the other. The labels A–E correspond to the species names Alpha through Epsilon in that data set. This symmetric arrangement of nodes was discovered by Ben Rudd Schoenberg (personal communication), and we thus denote this graph the Schoenberg graph.

method as looking at all $2(n-3)$ neighbors and picking the best one, but it will be quicker. We could also imagine trying multiple trees tied for best and evaluating rearrangements on each of them. The most sophisticated heuristic rearrangement strategies retain a list of all trees tied for best, and rearrange all of them.

Figure 4.3 shows what the space of all 15 possible unrooted trees looks like for 5 species, where trees that are adjacent by nearest-neighbor interchange are connected. Figure 4.4 shows the numbers of changes of state that are required for the data in Table 1.1 for each of these trees. Each tree has 4 neighbors. It will

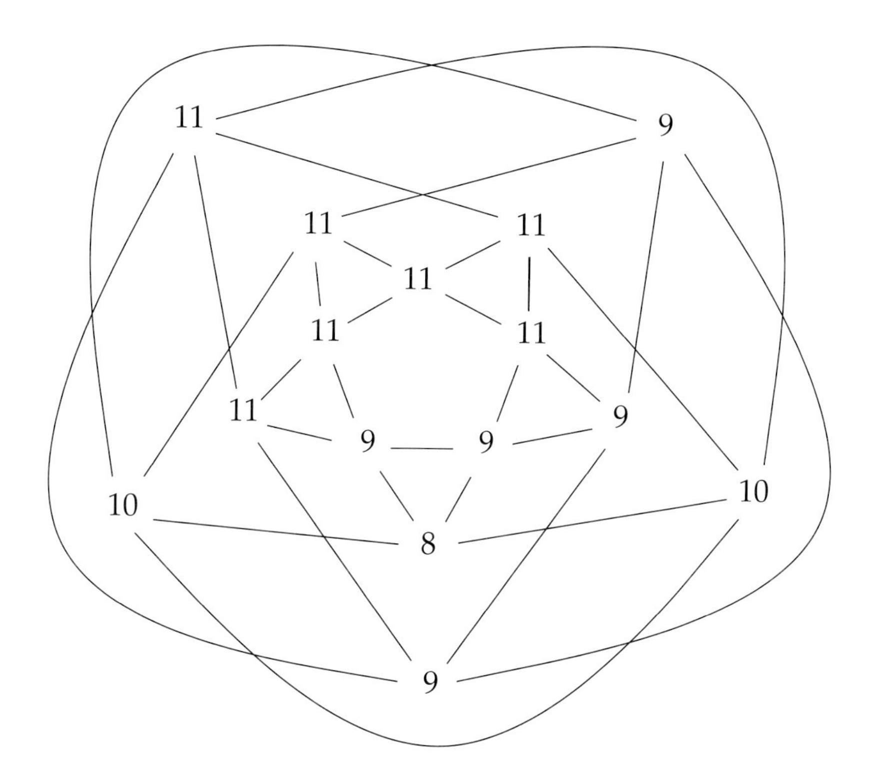

Figure 4.4: The space of all 15 possible trees, as in Figure 4.3, where the number of changes of state on the data set of Table 1.1 is shown. Nearest-neighbor interchanges search for the most parsimonious tree by moving in this graph.

be a useful exercise for the reader to pick a random starting point on this graph, and try various variations on nearest-neighbor interchange, using the lines on the graph as a guide. Does the process always find the most parsimonious tree, which requires 8 changes of state?

Subtree pruning and regrafting

A second, and more elaborate, rearrangement strategy is *subtree pruning and regrafting (SPR)*. This is shown in Figure 4.5. It consists of removing a branch from the tree (either an interior or an exterior branch) with a subtree attached to it. The subtree is then reinserted into the remaining tree in all possible places, each of which inserts a node into a branch of the remaining tree. In Figure 4.5 the 11-

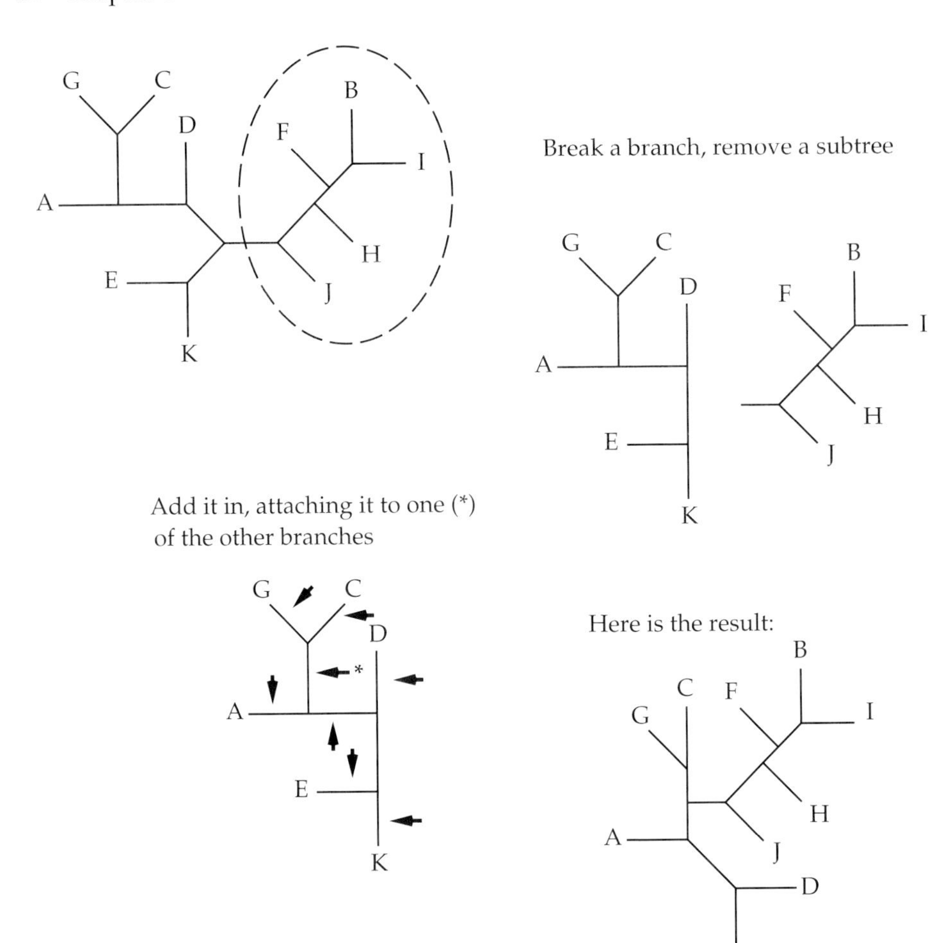

Figure 4.5: Subtree pruning and regrafting (SPR) rearrangement. The places where the subtree could be reinserted are shown by arrows. The result of one of these reinsertions (at the branch that separates G and C from the other species) is shown.

species tree has a 5-species subtree removed, and it is inserted into the remaining tree of 6 species, in one of the 9 possible places. One of these is of course the original tree. In general, if a tree of $n_1 + n_2$ species has a subtree of n_2 species removed from it, there will be $2n_1 - 3$ possible places to reinsert it. One of these is the original location. In fact, considering both subtrees (the one having n_1 species and the

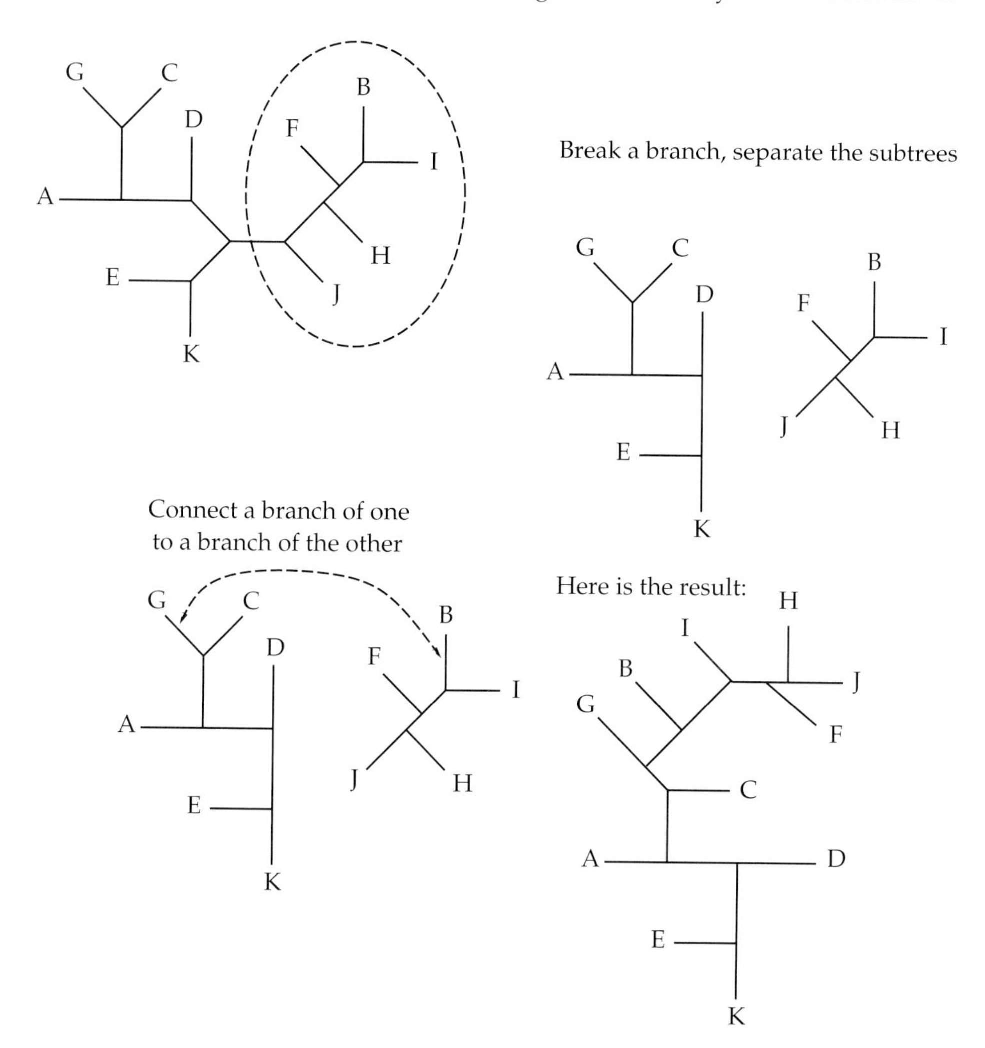

Figure 4.6: Tree bisection and reconnection (TBR). A branch is broken and the two tree fragments are reconnected by putting in branches between all possible branches in one and all possible branches in the other. One of these reconnections and its result are shown here.

one having n_2 species, there are $(2n_1 - 3 - 1) + (2n_2 - 3 - 1) = 2n - 8$ neighbors generated at each interior branch. It is also not hard to show that when an exterior branch is broken, there are $2n - 6$ neighbors that can be examined. Thus, as there are n exterior branches on an unrooted bifurcating tree and $n - 3$ interior branches, the total number of neighbors examined by SPR will be $n(2n - 6)+$

$(n-3)(2n-8) = 4(n-3)(n-2)$. However, some of these may be the same neighbor (to see that, consider $n = 4$). For the tree shown in Figure 4.5, which has $n = 11$, there are thus up to 288 neighbors under SPR. Of course, $2(n-3) = 16$ of them are the same neighbors that NNI examines. But it is clear simply from the numbers that SPR carries out a much wider search and is thus more likely to find a better peak in the space of all trees.

The issues of how greedy to be, whether to delay accepting a new tree until all SPR rearrangements have been examined, and how many tied trees to retain as the basis for further rearrangement, arise for SPR just as they do for NNI.

Tree bisection and reconnection

Tree bisection and reconnection (TBR) is more elaborate yet. An interior branch is broken, and the two resulting fragments of the tree are considered as separate trees. All possible connections are made between a branch of one and a branch of the other. One such rearrangement is shown in Figure 4.6. If there are n_1 and n_2 species in the subtrees, there will then be $(2n_1 - 3)(2n_2 - 3)$ possible ways to reconnect the two trees. One of these will, of course, be the original tree. In this case there is no general formula for the number of neighbors that will be examined. It depends on the exact shape of the tree. For the 11-species tree in Figure 4.6 (which is the same one shown in Figure 4.5), for the interior branches there can be up to 296 neighbors that will be examined. As in the other types of rearrangement, there are issues of greediness and of how many tied trees to base rearrangement on. Allen and Steel (2001) calculate how many neighbors there will be under TBR and SPR rearrangement, and calculate bounds on the maximum number of these operations needed to reach any tree from any other.

Other tree rearrangement methods

Tree-fusing

The NNI, SPR, and TBR methods hardly exhaust the possible tree rearrangement methods. The repertoire of rearrangement methods continues to expand. Goloboff (1999) has added two additional rearrangement methods. One is *tree-fusing*. This requires two trees that have been found to be optimal or nearly so, and alters them by exchanging subgroups between the two trees. This requires that both trees have a subtree on them that contains the same list of species. Thus if one tree has on it the subtree ((D,F),(G,H)) and another the subtree ((D,G),(F,H)) one could swap the subtrees. Each tree would thus propose to the other a particular resolution of that four-species group. The proposals would be expected to be better than random resolutions of that group, as they were found by heuristic search on that tree. They thus become candidates of particular interest for resolving the same group on other trees.

Genetic algorithms

Another strategy that is related to tree-fusing is use of a *genetic algorithm*. This is a simulation of evolution, with a genotype that describes the tree, and with a fitness function that reflects the optimality of the tree. Genetic algorithms (or *evolutionary computation*) have been widely used to solve complex optimization problems and are often quite effective. Their use in general optimization was inspired largely by the work of Holland (1975), though simulations of evolution by biologists and engineers date to the mid-1950s (see the historical papers reprinted by Fogel, 1998). Matsuda (1996) seems to have been first to use a genetic algorithm on phylogenies. He optimized branch lengths on each tree and used a recombination operator that swapped particularly good subtrees between trees. Lewis (1998) used an approach in which trees could mutate by changing branch lengths or doing an SPR rearrangement, and they could recombine by choosing a subtree in one tree, deleting those species from the other and inserting the subtree into it. Moilanen (1999) used a recombination operator similar to Lewis's, and also allowed heuristic searching using SPR rearrangement. Katoh, Kuma, and Miyata (2001) used mutations that were TBR rearrangements and recombinations that were swaps of subtrees containing the same set of species. It is not clear whether they did or did not also optimize branch lengths on each phylogeny. Congdon (2001) used parsimony, with a recombination operator similar to that of Lewis.

Genetic algorithms lend themselves easily to parallel computing. Brauer et al. (2002) used a separate processor for each tree with Lewis's (1998) approach, and found that efficient use of computational resources was made. Lemmon and Milinkovitch (2002) divide the trees into separate populations, which can evolve in parallel. Structures in the tree found to be shared by a number of nearby populations are protected from change, allowing search to proceed more intensively in the regions of the tree where different populations have found different structures. Charleston (2001) uses a population of trees, with a system for taking modifications that are made in one of them and propagating them to a number of others. There is no recombination operator, but rather this simultaneous adoption of successful mutations.

Genetic algorithms have been touted as a universal approach to optimization. They often can do quite well, but any evolutionary geneticist who has worked on natural selection in multilocus systems, as I have, must have doubts. How well genetic algorithms do is strongly dependent on the amount of interaction in the loci as they determine fitness. It can also depend strongly on the way in which the optimization criterion is turned into a fitness. If intelligent decisions are made, the method may do very well, but the performance is then due as much to these intelligent decisions as to the inherent strength of genetic algorithms. If, for example, we were to try to find factors of a large integer by mutating and recombining bit strings in binary numbers, genetic algorithms would be nearly worthless. That they have worked reasonably well in searches for optimal phylogenies must be put down to the wisdom of these authors in designing their genotype-phenotype mapping and their fitness scales. It is also important to realize that there is no connection between the fact that we are analyzing evolution and the use of an evolutionary algorithm to carry out the optimization. Genetic algorithms are not inherently more suited to analysis of genetics and evolution than they are to design of bridges.

Tree windows and sectorial search

We can imagine doing extensive rearrangements, not on the whole tree, but on a local region of it. Two papers (Sankoff, Abel, and Hein, 1994; Goloboff, 1999) have explored such approaches. Both take an interior node of the tree and a set of other nodes connected to it. This defines a region that they call either a *window* or a *sector*. Both approaches use parsimony to evaluate the state of the window. The branches reaching the edge of the sector or window carry summaries of the number of changes needed in each character looking outward from the window, those used in the Sankoff algorithm and discussed above in Chapter 2. This allows us to rearrange the tree locally in this window without ignoring the information outside it, and without having to retraverse the tree with each rearrangement.

Essentially these methods generalize, and broaden, the method of nearest-neighbor interchange, hoping to rearrange extensively enough to escape local optima without too great a computational burden. Sankoff, Abel, and Hein (1994) examine all possible rearrangements of the window, using the branch and bound approach discussed in Chapter 5. This extensive local rearrangement restricts their method to a window of less than 20 nodes, often less than 15. Goloboff (1999) uses the quicker but less exhaustive strategy of TBR rearrangement, and as a result his method can handle much larger windows — he suggests ones of 35 to 55 nodes. If the objective is to escape local maxima, using a larger window may be more important than exhaustively searching the window.

In both cases, when improved trees are found, the window is moved and rearrangement proceeds elsewhere in the tree.

Speeding up rearrangements

We saw in Chapter 2 that there were ways of speeding the calculation of the parsimony score of a tree when it is altered, by only recalculating the views in the part of the tree that has been changed (Gladstein, 1997). Goloboff (1993b) has suggested another saving that is particularly useful for SPR and TBR rearrangement. After a subtree is removed, we recalculate all of the views in each of the two now-separate parts of the tree. We can use the method of Ronquist (1998a) to avoid recalculating all of them. Then, when we evaluate a possible reconnection of the trees, we need only look at the views nearest the connection point to compute the parsimony score of the resulting tree. The overhead of calculating all the views after the subtree is removed results in much faster evaluation of ways that they

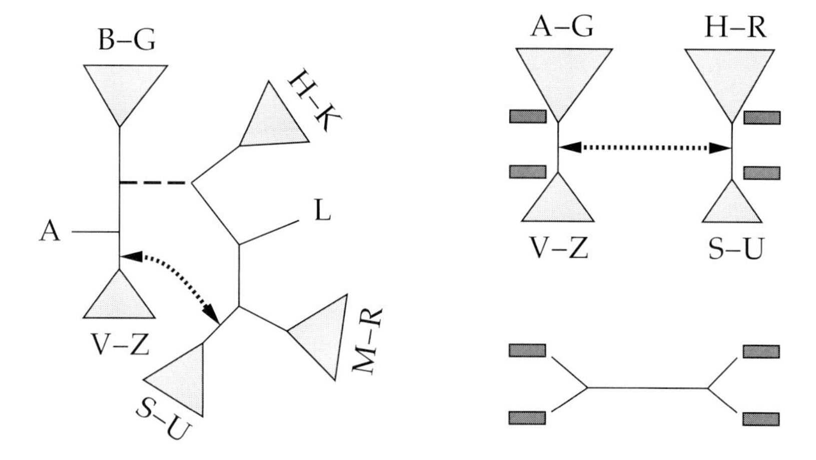

Figure 4.7: Goloboff's (1993b) economy in evaluating rearrangements. If we divide the tree into two trees by deleting the dashed line and propose to evaluate their reconnection using the dotted curved arrow, we can compute views upwards for subtrees A–G and H–R, and views inwards for subtrees V–Z and S–U (dark gray boxes). These can be used to quickly evaluate the resulting tree. It will have the same parsimony score as the 4-species tree at lower right.

may be reconnected. This results in substantial improvement. Figure 4.7 shows this method. Goloboff's (1993b) and Ronquist's (1998a) papers can be consulted for some further improvements of the speed of these algorithms.

Sequential addition

The above rearrangement strategies assume that we start with a tree. One can, of course, start with a randomly constructed tree. But most implementations have started with a tree that results from a *sequential addition* strategy. In Chapter 3, we showed that one can arrive at all possible trees by adding the species one at a time, each in all possible places. Figure 3.3 depicts the process for rooted trees. A similar process exists for unrooted bifurcating trees. An unrooted tree with 3 species has three branches. The fourth species can be added by having it branch off from the middle of any of the three branches. In the process, two more branches are added to the tree. For each of these three possibilities, there are then 5 possible ways that the next species can be added, and so on.

Suppose that we construct the single 3-species tree. Now we try to add the fourth species in all 3 possible places. We evaluate the resulting 4-species trees. But instead of following up on all of these, we simply keep the best one. Then we try to add the fifth species to all 5 possible places. Keep the one of these trees that is best. Now add the sixth species in all 7 possible places, and keep the best one. This is a greedy algorithm based on sequential addition of species. It seems as if it must always result in a most parsimonious tree. But in fact it may not, as we will see by example in the next chapter. Nevertheless, it constructs an approximation to the best tree.

Sequential addition is one of the chief methods used to obtain initial trees for rearrangement strategies. Note that the order in which the species are added is arbitrary. We can imagine adding them in the same order that they appear in the data or using a random order based on a random number generator. For the data of Table 1.1, the results of a sequential addition strategy are shown in Figure 4.8. It results in four tied trees of length 9, a step longer than the most parsimonious tree. If we try other sequences of addition, some of them do lead us to find the most parsimonious tree (examples would be B, C, D, E, A and A, B, D, E, C).

One of the most tempting orders is often the reverse of the order in which the species appear in the data. This would at first sight appear to be no different in its properties than adding them in the order in which they appear in the data. But often biologists place the most distantly related, most "primitive," and most dubiously interpretable, species first. So one is far better letting the well-interpreted species settle down into a tree and then, at the last moment, adding the dubious species in all possible places. The alternative, allowing the dubious species in early, is more likely to result in disruption of the tree structure.

The issue of how many trees tied for best to retain is present for sequential addition strategies, as it is for rearrangement strategies. In fact, the two strategies may best be combined. One can add a species, then carry out a round of rearrangements to see whether it has disrupted the existing topology. By integrating the rearrangement and sequential addition strategies, one can obtain a method that searches even more carefully for the best trees.

It is often assumed that using many different orders of species will result in a fair sampling of starting points for rearrangement. There is no actual mathematical proof of this, but it seems to behave reasonably well in practice.

Star decomposition

Rather than building up a tree by adding species one at a time, one can start with all species present, but with the tree totally unresolved. A bifurcating tree can be achieved by gradually resolving this tree by grouping two lineages at a time. Figure 4.9 shows this process and also shows that it is not unique. There may be only one way to reach each bifurcating tree by sequential addition of species in a given order, but there are multiple ways to decompose an unresolved "star" phylogeny to reach a given bifurcating tree.

These *star-decomposition* methods were first used in the clustering literature. We will see in Chapter 11 that UPGMA and neighbor-Joining methods use this approach, as does any clustering method that agglomerates groups into larger groups.

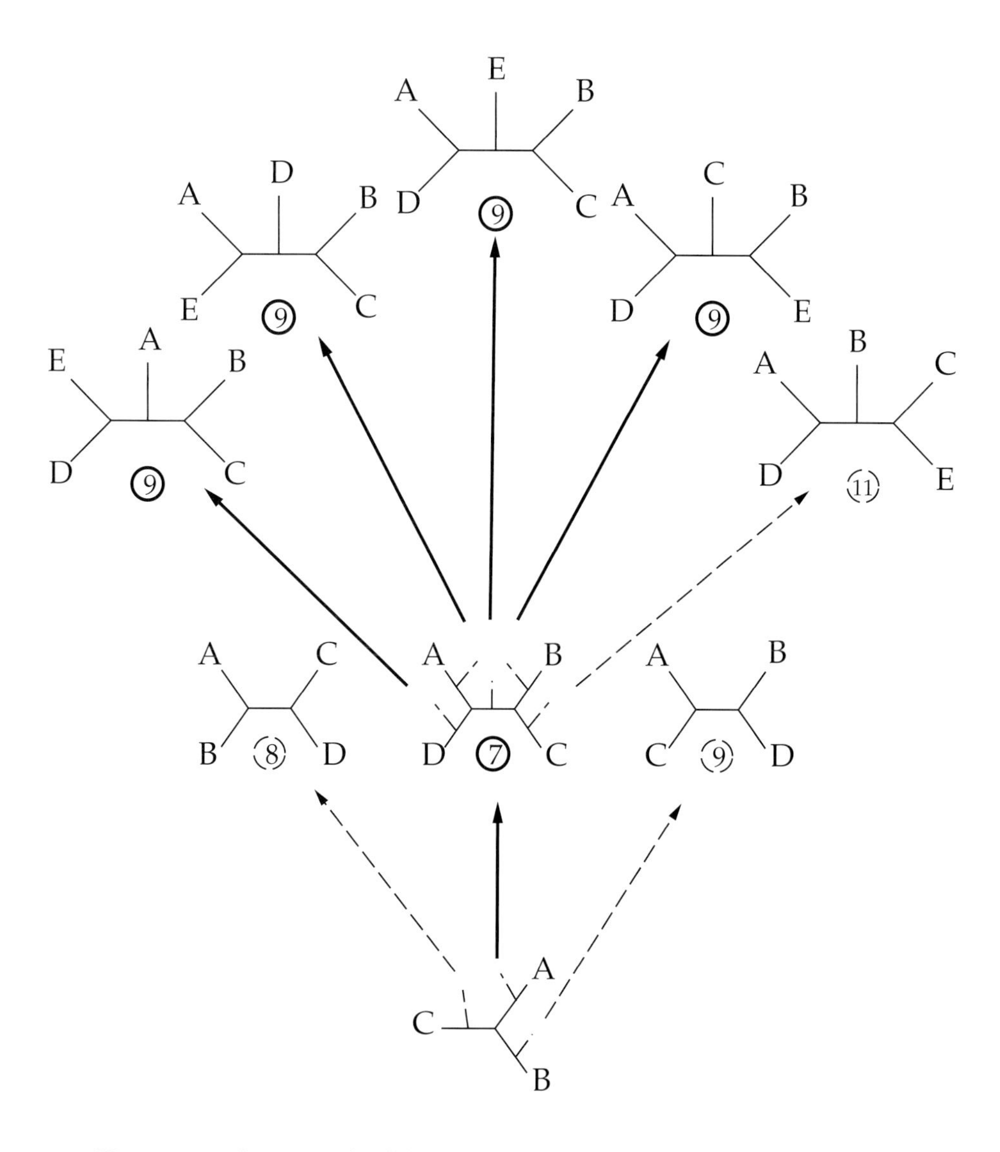

Figure 4.8: Sequential addition carried out on the data in Table 1.1. The species names A–E correspond, respectively, to the names Alpha through Epsilon in Table 1.1. Sequential addition ends up with four trees tied for best. None of these is actually the most parsimonious tree.

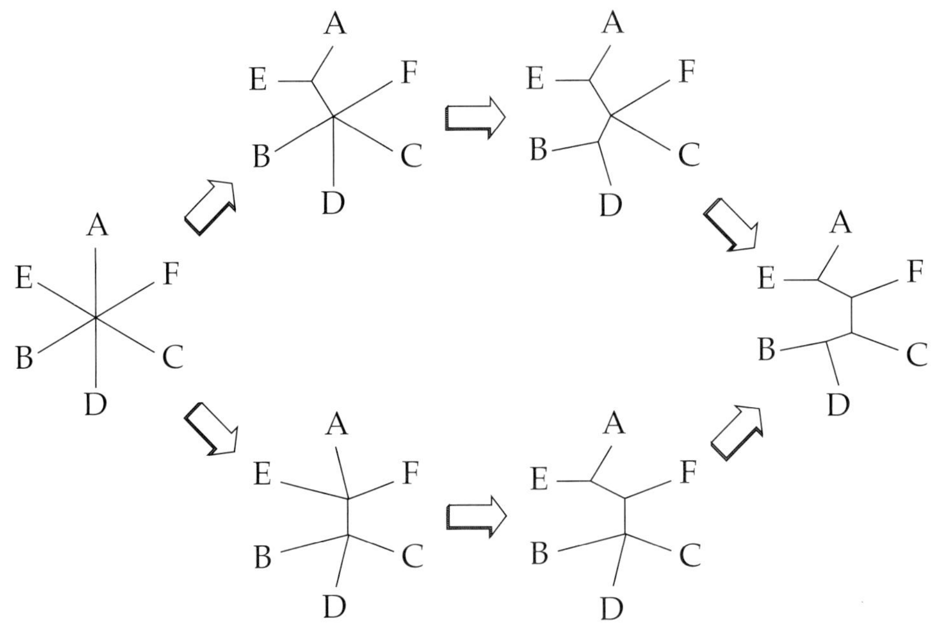

Figure 4.9: Defining a tree by star decomposition, illustrating that there is more than one way to do so.

Tree space

Rearrangement may be thought of as searching in a space of trees. We have been concerned mostly with the topology of the tree, but with other methods we will find it useful to think of the branches of the tree as having lengths, nonnegative numbers that reflect how much evolution is expected to occur on that branch. For most of the models of evolution used in this book, a branch of zero length might as well not be present. Thus if a speciation separates two lineages, and one of these immediately speciates again, this is not predicted to result in a different phenotype than a trifurcation.

Two tree topologies that have branches that can, on being shrunk to zero length, lead to the same trifurcation are adjacent in tree space. We can shrink one branch to reach the trifurcation, then insert the other branch as of zero length and then lengthen it. In that way we can move smoothly from a tree of one topology to the other. Trees that are adjacent to the same trifurcation can be reached from each other by a nearest-neighbor interchange (NNI). In this sense the NNI is not just another arbitrary rearrangement method but is fundamental to the structure of tree space. Figure 4.3 shows the pattern of sharing of subspaces for trees of 5 tips. To go from one to another in this graph, we can imagine shrinking an internal branch length until it is of zero length. At that point we have reached a subspace of trifurcating trees. We can pass through it into either of two other tree topologies, so that as we go through the looking glass we encounter two alternative universes and we have to choose which one to enter.

The graph has interesting structure. The arrangement in Figure 4.3 was found for me by my stepson, Benjamin Schoenberg. I have named it the *Schoenberg graph*. Ben pointed out another property. Each tree is part of two triangles of trees, a triangle being the three trees that can be obtained by rearrangements around one interior branch. If we consider these triangles as points in a new graph and connect those of them that share a tree, we have constructed a dual graph of the graph of trees. In the case of order 5, this dual graph has 10 points. Ben noticed that it is a famous graph, the Petersen Graph, widely used as a source of counterexamples in graph theory. It is sufficiently famous that it is depicted on the cover of the journal *Discrete Mathematics*.

We will see in the chapter on tree distances (Chapter 30) that some tree distances can be considered to be distances in this tree space. For further discussion of the geometry of tree space and the role of the Petersen Graph as a dual of 5-species tree space, see the papers by Billera, Holmes, and Vogtmann (2001) and Bastert et al. (2002) which report work independent of Ben Schoenberg's.

Search by reweighting of characters

Different characters in the data may well recommend different trees to us. To prevent the search from becoming overconcentrated on a limited set of trees, it may help to use as starting points different trees that are recommended by various subsets of characters. Nixon (1999) has suggested a method for doing this that seems to improve the effectiveness of the search. He starts from a tree, and then picks a random set of 5% to 25% of the characters for emphasis. They are emphasized by increasing the weights of the characters. We will examine character weighting in Chapter 7; in effect, all we need to know here is that Nixon's reweighting amounts to duplicating each of these characters so that each appears twice or more in the data set.

Nixon suggests starting from the initial tree and using tree rearrangement methods such as TBR with this modified data set. This will move us to a tree recommended by this reweighted data set. After we have reached it, Nixon suggests using the original data set and doing a TBR search from this tree. The effect is to carry us to a tree recommended by a subset of the data; then we search from that starting point using the full set of characters. Many such reweightings and searches are carried out — Nixon recommends 50–200 such searches. The best trees found among these searches are retained.

The method is called the *parsimony ratchet*. However it is actually not specific to parsimony methods — a similar technique can be used with any objective function based on character data, including compatibility, distance matrix, and likelihoods.

Nixon's method can be modified in many ways. We could use a variety of different reweighting methods. In Chapter 20 we will discuss the bootstrap and jackknife methods, which are reweighting methods. Although there is some concern as to exactly how these should be carried out to have their results be interpreted statistically, this need not constrain us in using them in a reweighting search. In fact, exaggerating the effect of small sets of characters appears helpful in the search. So we could use jackknife methods that select only a small set of characters, fewer than we would use when we investigate the statistical uncertainty of phylogenies.

Likewise, the search strategy can be modified. All that is necessary is to have a first stage that is some kind of search based on the reweighted characters and a second stage that starts from the result and rearranges using the original data. Nixon finds that using his strategy, more parsimonious trees can be found in the 500-species *rbcl* plant phylogeny data set.

The search strategy of Rodin and Li (2000) is related to Nixon's reweighting scheme. They use the bootstrap to choose regions of the tree where the structure is less well defined, and concentrate their tree rearrangements there. This has many of the same effects, as it entertains rearrangements to the extent that reweighting of characters occasionally suggests them. Another method, different in details, has been presented by Quicke, Taylor, and Purvis (2001). This reweights characters in a different way, one that emphasizes characters that fit the trees found so far. It remains to be seen whether this nonrandom reweighting has advantages over Nixon's random reweighting.

Simulated annealing

A well-known method of search in large combinatorial problems is *simulated annealing* (Metropolis et al., 1953). This uses the Metropolis algorithm (which we will see again when we discuss Markov chain Monte Carlo methods in Chapters 18 and 27). The Metropolis algorithm simulates statistical mechanics, in that it accepts a new state if it is better, and also sometimes when it is worse. The result is a wandering among states that is biased toward the better states. The extent of the bias depends on a "temperature" parameter. Simulated annealing uses the Metropolis algorithm with a gradually decreasing temperature. Thus the state wanders widely at first but later is more and more strongly biased to wander towards better solutions. It will wander widely but finally hill-climb towards a locally good solution. It is possible to prove that simulated annealing will find the best solution, if the temperature is lowered slowly enough. However, usually we do not know how slow is slow enough.

The first paper to describe applying simulated annealing to searches for optimal phylogenies was by Lundy (1985). She dealt with the case of minimum-length (most parsimonious) trees for continuous characters. Dress and Krüger (1987) also used simulated annealing with parsimony on molecular sequences. Their elementary operations were swaps of subtrees in the tree. Daniel Barker has produced the

LVB computer program searching for the most parsimonious tree by simulated annealing. Goloboff's (1999) method of "tree-drifting" is another implementation of simulated annealing to find most parsimonious trees. In deciding which trees to accept, he uses a Relative Fit Difference measure that can emphasize small differences in parsimony score. Salter and Pearl (2001) applied simulated annealing to maximum likelihood phylogenies.

It is early days yet in the use of randomness for searching for optimal phylogenies. It is not yet clear whether random perturbations of the tree will be more useful than random reweighting of the characters — most likely these techniques will come to be used together.

History

Local rearrangements of phylogenies were first discussed by Camin and Sokal (1965). They seem also to have been employed by Eck and Dayhoff (1966), who were first to mention a sequential addition strategy. Kluge and Farris (1969) also described a sequential addition strategy. Subtree pruning and regrafting and tree bisection and reconnection were described in print by Swofford and Olsen (1990). David Maddison (1991) has discussed the importance of multiple starts for finding isolated "islands" of most parsimonious trees. Various of these methods have been invented by others as well, used in computer programs, and circulated as oral tradition before their first description in print.

Chapter 5

Finding the best tree by branch and bound

We have already seen, in Chapter 3, that there are far too many possible trees to make it practical to search for the most parsimonious tree by simple exhaustive search, except with very few species. This is a parallel to computational problems in strategy in games such as chess, where there are far too many possible games that might be played to consider them all.

A method that was developed to allow computers to solve for the best strategy in a game can help us here, the *branch and bound* method. In effect, it discards whole classes of strategies that it has determined cannot be correct, without the need to examine all of their members one by one. The branch and bound method was discovered in the 1960s (by whom is not entirely clear). It was first applied to parsimony problems in phylogenetic inference by Hendy and Penny (1982), from whose paper modern use of branch and bound methods for inferring phylogenies has sprung.

A nonbiological example

It is not difficult to describe a branch and bound approach to solving a combinatorial optimization problem, a relative of the infamous Traveling Salesman Problem. In this problem, finding the *shortest Hamiltonian path (SHP)*, we have a map with n cities. The salesman has an airplane and he (he is always a male, for some reason) can fly directly between any two cities. The problem is to find a route (including a starting point) that will take him from one city to another until he has reached all of them, while flying the shortest total distance (he does not return to his starting point, as he does in the Traveling Salesman Problem). There are n cities that could be the starting point; for each of those, there are $n-1$ that could come next, and so on. There turn out to be $n!$ possible solutions. We could imagine tracing out all of

Table 5.1: Ten points drawn randomly from a unit square, which are the geographic coordinates of the "cities" in a shortest Hamiltonian path problem.

Point	x	y
1	0.537	0.061
2	0.274	0.222
3	0.016	0.837
4	0.871	0.400
5	0.399	0.740
6	0.815	0.531
7	0.587	0.946
8	0.992	0.733
9	0.268	0.481
10	0.895	0.068

them. In fact, there is a tree of possibilities. There are n possible choices, then for each of those $n - 1$ subsequent choices, and so on. Table 5.1 and Figure 5.1 show a typical shortest Hamiltonian path problem with 10 points. Panel (a) shows the points, which are also given (numbered arbitrarily) in Table 5.1. They are simply points drawn randomly from a two-dimensional unit square.

Panel (b) in Figure 5.1 shows a random route (one that starts at point 1, continues to point 2, to point 3, and so on in order). The total length of this route turns out to be 5.4342. We can imagine various heuristic search methods for finding better routes. One of them that is fairly obvious is to start from a point, proceed to its nearest neighbor, proceed from that to the nearest neighbor that is not yet in the path, and continue doing this until all points have been visited. This works well for a while but finally leaves you with only a single choice for the last point, and that may not be very close by. By carrying out this greedy algorithm many times, once from each point, and then choosing the best of these solutions, one can do better. In our 10-point example, the average length of the 10 greedy solutions starting from the 10 points is 3.6974, a 32% improvement over the random route. The best of the greedy solutions (actually the best two, as the solution is found twice, once from each of its ends) is of length 2.8027.

This solution is shown in panel (c) of Figure 5.1. It is close to the optimal solution, but it is not the optimal solution. The optimal solution is shown in panel (d). It is shown with lines rather than arrows, as one can traverse it in either direction. Note that it is close to the best greedy solution, with one link of the greedy solution deleted and another link added. This is further evidence that the strategy popular among algorithmists of finding solutions to the SHP by rearranging other solutions is sound. The length of the optimal solution is also close to that of the best greedy solution, as it is 2.7812, only 0.02 shorter. The greedy algorithm

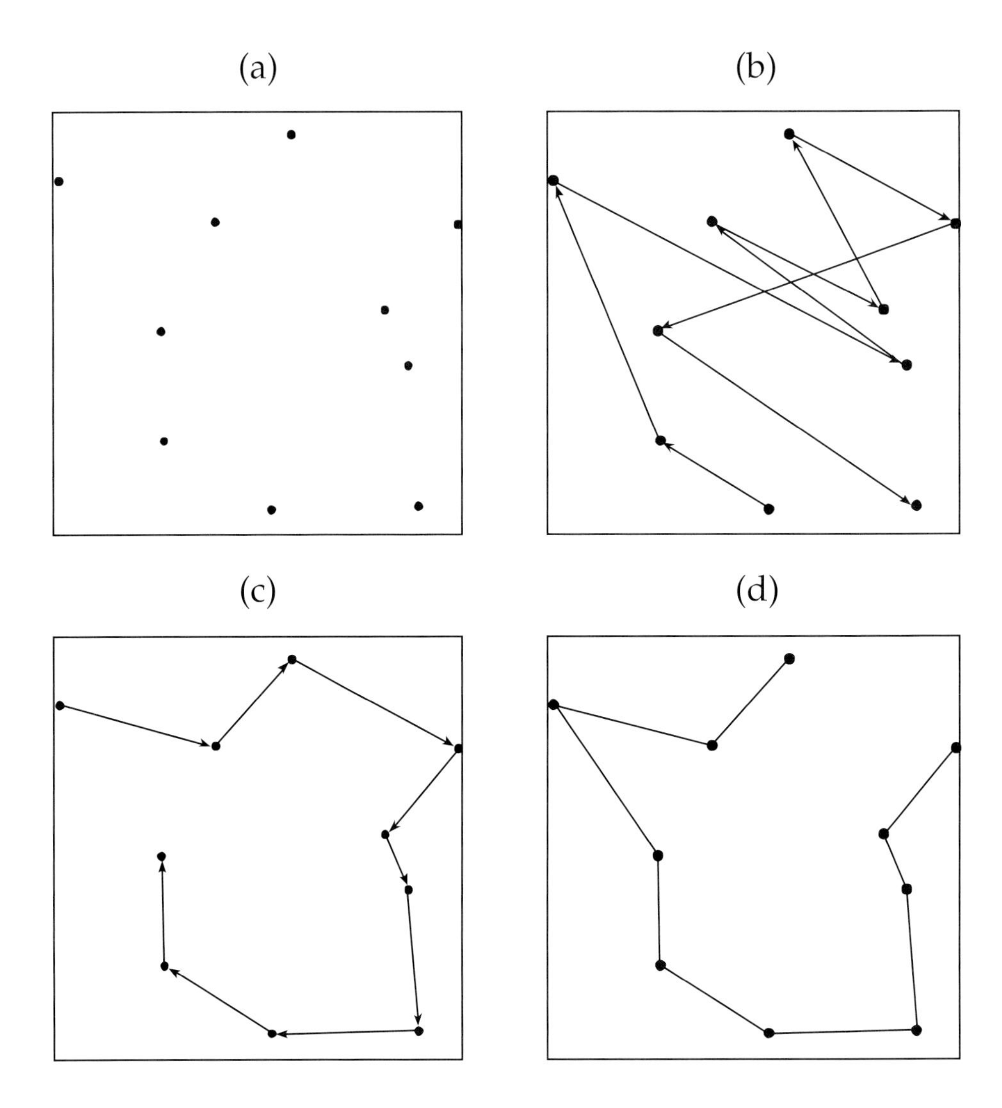

Figure 5.1: A typical shortest Hamiltonian path problem with 10 points, randomly chosen from a unit square (the ones shown in Table 5.1). Panel (a) shows the points. Panel (b) shows an arbitrary route (the points in numerical order). Panel (c) shows the shortest of the routes chosen by starting from each point and adding nearest points in greedy manner. Panel (d) shows the solution found by a branch and bound method.

with multiple starts has done very well, but it did not actually find the optimal solution.

Finding the optimal solution

We can find the optimal solution by exhaustive enumeration. In doing so we traverse a *search tree* of possibilities. The possible solutions can be indicated by the sequence of points. Thus the first solution (panel b) is (1, 2, 3, 4, 5, 6, 7, 8, 9, 10), the best greedy solution is (3, 5, 7, 8, 6, 4, 10, 1, 2, 9), and the optimal solution is (7, 5, 3, 9, 2, 1, 10, 4, 6, 8). There are in all $10! = 3,628,800$ possible solutions. We can think of arranging them in lexicographical (dictionary) order:

$$
\begin{array}{c}
(1, 2, 3, 4, 5, 6, 7, 8, 9, 10) \\
(1, 2, 3, 4, 5, 6, 7, 8, 10, 9) \\
(1, 2, 3, 4, 5, 6, 7, 9, 8, 10) \\
(1, 2, 3, 4, 5, 6, 7, 9, 10, 8) \\
(1, 2, 3, 4, 5, 6, 7, 10, 8, 9) \\
(1, 2, 3, 4, 5, 6, 7, 10, 9, 8) \\
\cdots \\
(10, 9, 8, 7, 6, 5, 4, 3, 2, 1)
\end{array}
$$

The 3.6 million rows of this table can be organized into a search tree based on sharing initial parts of the solutions. Thus the first two entries are adjacent tips on the search tree, as they share (1, 2, 3, 4, 5, 6, 7, 8). The first six entries are a cluster of six adjacent tips; they are shown in Figure 5.2. The first (leftmost) branch at the bottom of this search tree leads to all solutions that start with point 1. The next branch leads to all solutions that start with point 2, and so on. Within these subtrees there are further branchings, corresponding to what the second point in the solution is to be. This branching continues all the way up to the top of the search tree; in the figure we can see the top leftmost part of the tree with the first six solutions shown. For simplicity, we ignore the fact that the solutions come in tied pairs, as one can travel a route in either direction.

An exhaustive search of solutions can be done by traversing this tree from left to right and keeping track of the best solution found so far. When this is done for the data in our 10-point example, the solution in panel (d) of Figure 5.1 is found; it is guaranteed to be the best possible one. This took 10.85 seconds of computer time on my Digital Alphastation 400 4/233. The program needed to look at all 10! solutions to do this. Unfortunately, factorials blow up faster than exponentially as the problem size increases. For 15 points there are $15! = 1,307,674,368,000$, so that this exhaustive search would probably take about 3,909,906 seconds, or a bit over six weeks. (My computer is already long obsolete as you read this, but all you would need to do is add a few more points to the problem to overwhelm your present computers.)

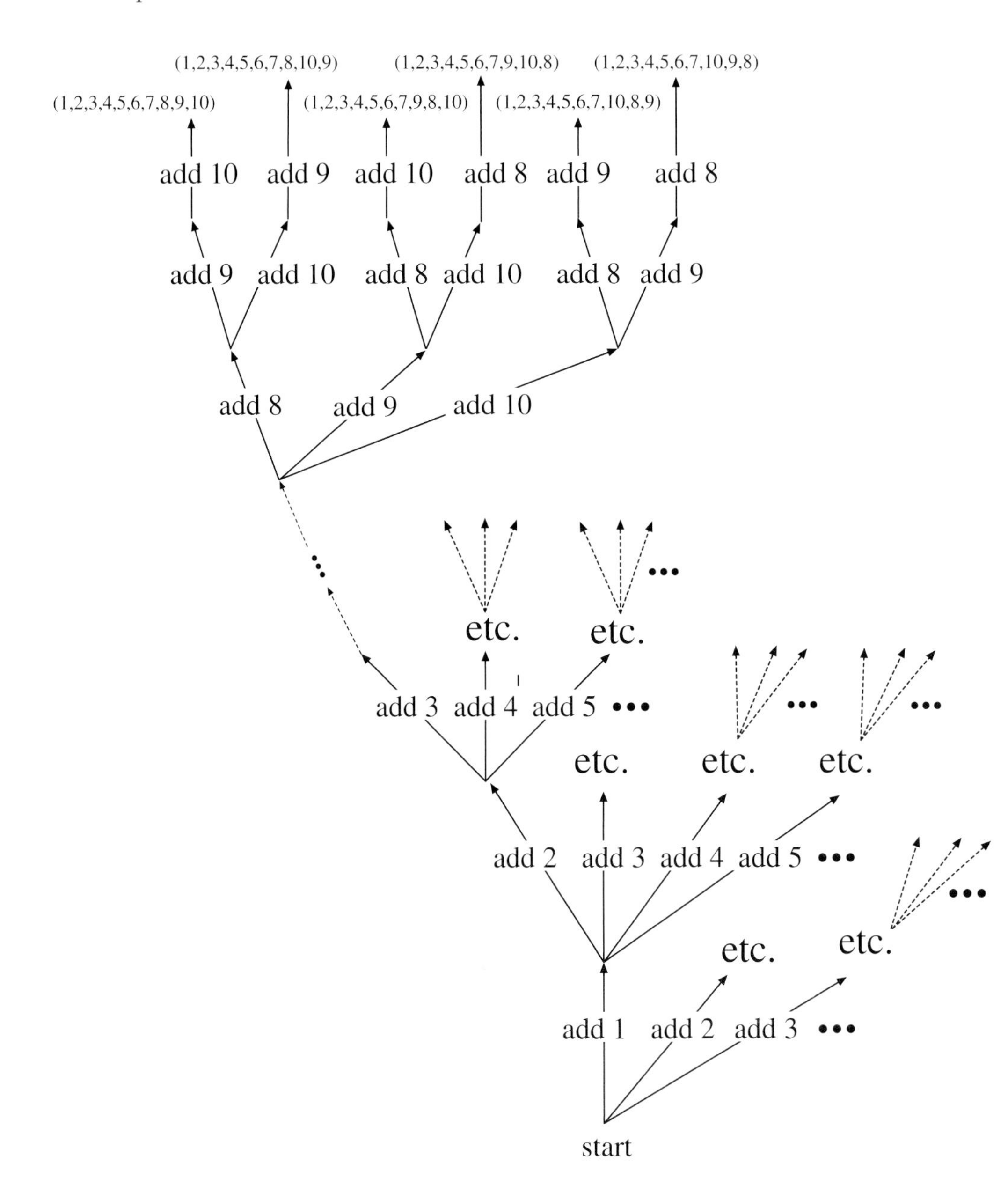

Figure 5.2: Search tree for the solutions of a 10-point shortest Hamiltonian path problem.

NP-hardness

In fact, the shortest Hamiltonian path problem is one of the best-known members of a famous class of computationally difficult problems. The number of possible solutions rises as the factorial of the number of points, n. This is faster than an exponential rise, so that when plotted on a graph whose vertical scale is logarithmic, it rises faster than linearly. That is the plot of the number of possible solutions, but we might be able to find a method of solving the SHP whose computational time does not rise that fast. For example, we would be in very good shape if we could find one whose computational time rose only linearly with n, for then when we doubled the number of points, we would only double the computational time. Even if we had an algorithm whose time rose as the square of the number of points, n^2, we would not be in such bad shape: A doubling of n would only quadruple the computational time.

We thus might want to ask, Under what circumstances does a problem have a solution whose computational time is a polynomial function of the number of points (such as $30n^3 + 2n^2 + 6$)? No matter what polynomial function it is, an exponential function like e^n will overtake it for large enough n. Thus a problem with a polynomial time is better than one with exponential time, provided one has a sufficiently large value of n.

Over the last 20 years, computer scientists have been able to establish that there is a class of problems that are all equally hard, in the sense that either they all have polynomial solutions, or none of them does. These are the *NP-complete* problems. Their theory is summarized in texts on computational complexity, such as the one by Garey and Johnson (1976). The NP-complete problems include many of the most famous problems in computer science. Another class of problems that is relevant are the *NP-hard* problems. These are problems that do not have polynomial-time solutions if the NP-complete problems do not, but which might not have them even if the NP-complete problems did.

The theory is a bit unusual, because it is not actually known whether NP-complete problems are solvable in polynomial time. They might be. If even one of them is found to be solvable in polynomial time, it can be proven that all of them can be solved in polynomial time! (Some of this proof is done by showing that there is a way of transforming an algorithm that solves one problem into an algorithm that solves another). Perhaps tomorrow morning someone will find a way of solving one of the NP-complete problems in polynomial time. That person will immediately become the most famous figure in contemporary computer science, because she will have shown that all of the NP-complete problems can be solved in polynomial time.

But so far, a lot of very crafty computer scientists have devoted a lot of time to finding a polynomial-time algorithm for some NP-complete problem, and they have all failed. To me that indicates that we probably are not going to see a polynomial-time algorithm for any of the NP-complete problems during our lifetimes. So I am going to assume that a proof that a problem is NP-complete or NP-hard is equivalent to a proof that it cannot be solved in polynomial time, and that our algorithms will run in exponential time, or worse.

The shortest Hamiltonian path problem is one of those that has been proven to be NP-hard. An associated decision problem (Yes or no — Is there a solution to this SHP that has length less than X?) is known to be NP-complete. So no algorithm that we can reasonably expect to discover can run in polynomial time. Of course that does not prevent us from discovering one that runs in exponential time but runs rather rapidly on moderate-sized cases. We might, for example, find one that runs in $0.0000001e^n$ seconds. It would run quickly on small and moderate-sized cases, but it too would ultimately be defeated by the exponential growth of execution time as the problem size grew.

Branch and bound methods

In spite of its being NP-hard, there are ways to considerably speed up the SHP. The simplest is branch and bound. We have already seen that we can search exhaustively by traversing the search tree of solutions. But we need not actually traverse all of it. As we go up the tree, building up a solution, we can keep track of the total length of that part of the solution so far. We also will be keeping track of the best solution found so far, and how long it is. Suppose that the best solution so far has length 2.932. As we go up a branch on the search tree, before we reach the end of the branch, we notice that the total length of this partial solution has reached 3.193. Any further points that we add to the solution can do nothing but increase that length. We therefore know that no solution in that subtree of the search tree can be any better. This is the "bound" in the branch and bound method. We can cut our losses by ceasing further movement into that subtree and backing out. If we have backed out when there are still a considerable number of points left to be added to the solution, we have saved a lot of work.

The result is an algorithm that branches (searching all parts of the search tree) but also uses its bound to greatly economize on the amount of work. Implementing this branch-and-bound search, we find that for the numerical example it does indeed arrive at the correct solution, and much faster than straight exhaustive search. It takes 0.46 seconds instead of 10.85, a better than 20-fold improvement.

Phylogenies: Despair and hope

Branch and bound has speeded up the solution greatly, but it has not actually escaped from the constraints of the NP-hardness proof. In fact, branch and bound algorithms too have a complexity that is exponential — it's just that they have improved the coefficient in front of the formula and maybe on the size of the exponent. (For example, they might in some case have computation time proportional to $e^{0.03n}$ instead of $e^{0.5n}$.)

The parsimony problem for nucleotide sequences is one of a number of phylogeny problems that are known to be NP-hard. (Finding the best tree or trees is

NP-hard, knowing what is the number of changes on the best tree is NP-complete.) The proof of these was given by Foulds and Graham (1982; Graham and Foulds, 1982). These phylogeny problems are examples of finding a Steiner tree in a graph. The set of all sequences is a graph, where adjacent points are connected if the sequences differ at one site. A *Steiner tree* is a tree of minimal length connecting a given set of points in a graph. Many Steiner tree problems are known to be NP-complete or NP-hard. (Generally, the problem of knowing the length of the tree is NP-complete, and the problem of finding the tree is NP-hard.) W. H. E. Day and co-workers have provided NP-completeness proofs for a variety of phylogeny criteria, most of which we introduce in later chapters. These include Wagner parsimony on a linear scale (Day, 1983), Camin-Sokal and Dollo parsimony (Day, Johnson, and Sankoff, 1986), compatibility (Day and Sankoff, 1986), least squares distance matrix methods (Day, 1986), a variant on the minimum evolution distance matrix method (Day, 1983), and polymorphism parsimony (Day and Sankoff, 1987).

There would seem to be reason for pessimism. But it is important to recall that exponential run time is not necessarily typical. The NP-hardness proof shows only that, given that no algorithm achieves polynomial time, for any problem size there are instances of it that will take exponential time. *But* these need not be biologically reasonable cases. The worst-case complexity of the problem is exponential. But what is the biological-average-case complexity?

In fact, it seems that some NP-hard problems (such as finding trees by compatibility, a method we consider later in this book) are very rapidly solved by branch and bound methods for typical biological cases. Other problems (such as parsimony) do not have such fortunate behavior.

Branch and bound for parsimony

The use of branch and bound algorithms to speed up exhaustive search for most parsimonious trees is closely analogous to the algorithm that we have just described for the shortest Hamiltonian path problem. The search tree is the tree of trees that we have already described in Chapter 4 (see also Figure 3.3). It is the tree of possibilities that results from adding the species to a tree in their numerical order, at each stage choosing one of the possible places to add that species. Thus we start with species 1 and 2 in a two-species tree, add species 3 in one of the 3 possible places, then add species 4 in one of the 5 possible places, and so on. Figure 5.3 shows this tree of trees, for a five-species case where the species are labeled A, B, C, D, and E. There are 15 possible tips, the 15 bifurcating trees, plus the interior nodes of the search trees which are 8 other incomplete trees.

We can imagine traversing this search tree. At each point on it, we have a partial (or a complete) tree. We can evaluate the number of changes that this tree requires on our data. This could be used in a branch and bound method, as was done in the SHP example. In their paper introducing the branch and bound method for phylogenies, Hendy and Penny (1982) have made some useful suggestions for

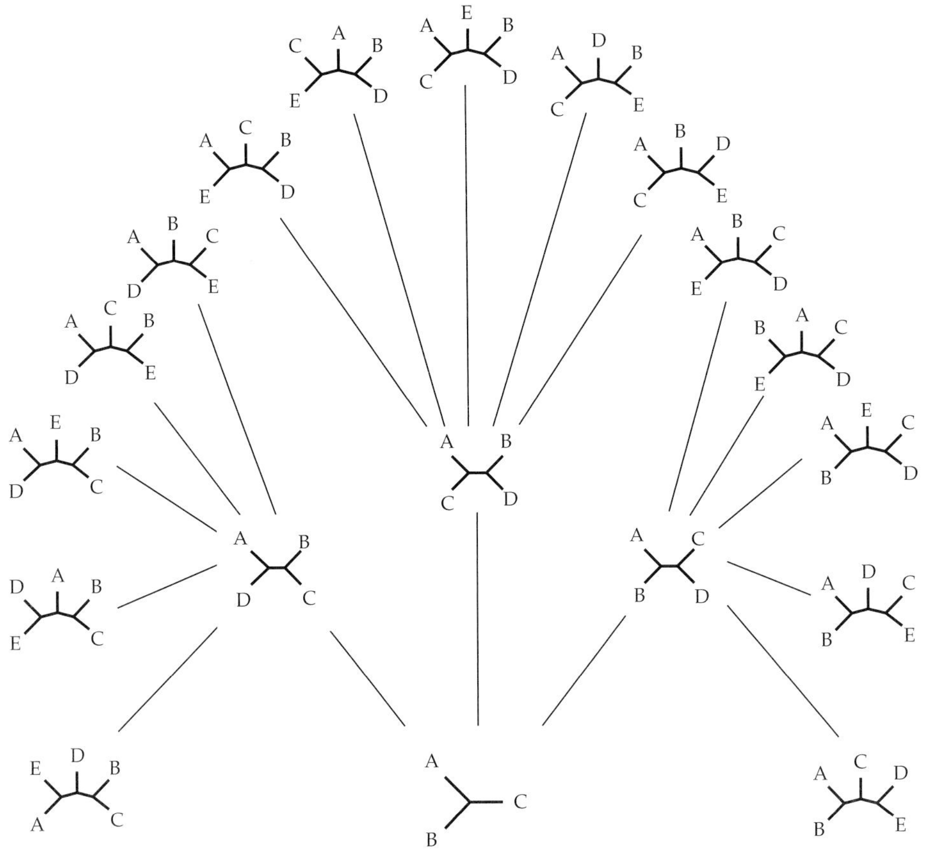

Figure 5.3: Search tree for most parsimonious tree in a five-species case.

improving the bound. Figure 5.3 shows the search tree, with all 15 unrooted bifurcating trees for 5 species. These are tied together by interior nodes that show all 3 four-species trees, and at the root is the single possible three-species tree. Figure 5.4 shows the same search tree with the trees themselves replaced by the number of changes of state that they require for the data in Table 1.1. The branch-and-bound traversal starts from the bottom of the search tree. In order to rule out as many trees as possible, as quickly as possible, it is helpful to find good trees soon. One strategy would be to search the nodes of the next level in the tree in order of the number of changes that their trees require. So we start at the bottom node (which requires 5 changes). At the next level we have nodes that require 8, 7, and 9 changes, respectively. If we make a preliminary visit to all three of them and

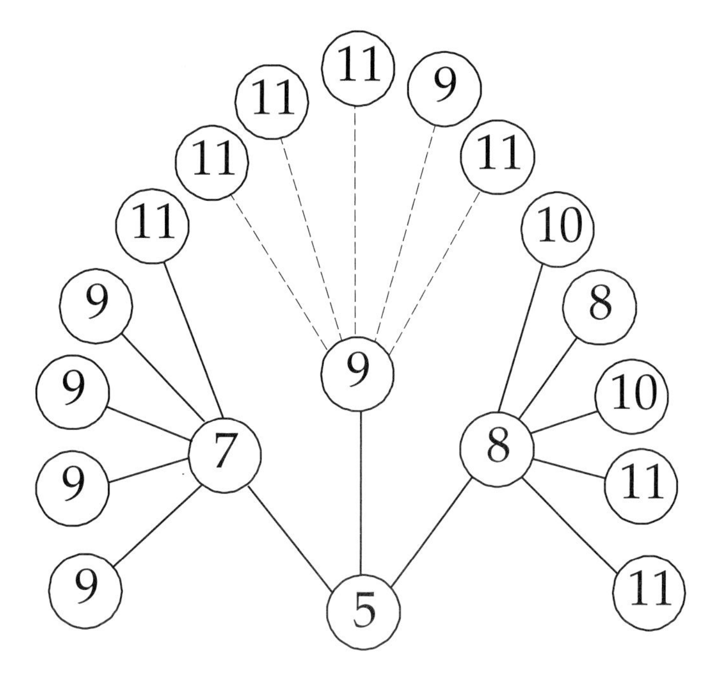

Figure 5.4: Search tree for most parsimonious tree for five species, using the data of Table 1.1. Trees are shown in Figure 5.3. Dashed lines are those not traversed by a branch and bound method. The species names in the data set correspond to labels A through E in Figure 5.3.

discover this, then we can plan to traverse the tree starting with the one that requires 7 changes. Proceeding up to it, we discover at the next level that there are 5 five-species trees, requiring 9, 9, 9, 9, and 11 changes, respectively.

Now we have candidate trees (the ones requiring 9 changes). We will be interested in any region of the tree whose bound is 9 or less. We will be uninterested in searching any region of the search tree that has all of its members requiring more than 9 changes. We proceed on to the next subtree, the one whose interior node requires 8 changes, so that its bound is 8. This has 5 five-species trees attached to it, and those require 10, 8, 10, 11, and 11 changes. Now we have a new candidate tree, requiring only 8 changes (and we discard the earlier ones that required 9). We are now interested in bounds of 8 or less. Finally, we start to examine the last of the three subtrees, whose interior node requires 9 changes. Its bound is 9. Immediately we know that none of the 5 trees attached to that interior node are of interest. All must require at least 9 changes, and we have already found a tree that requires only 8 changes. Hence we never travel along the branches of the search tree that lead beyond there (and they are therefore shown in Figure 5.4 as dashed lines). We are done, having examined only 10 of the 15 possible five-species trees.

The saving is not great in this example, but it can become enormous in larger cases. The saving is greater the less homoplasy there is in the data. In cases in which there are many conflicts between information from different characters and much parallelism and convergence, the branch-and-bound strategy does not perform particularly well.

Improving the bound

In the search tree of Figure 5.4, the bound is calculated simply by asking how many changes the partial tree at that node requires. This is a lower bound in the sense that it cannot be higher than the number of changes on any of the trees found farther out in the search tree. If we have found full trees that have as few as (say) 58 changes, then finding a partial tree that has 60 of them is sufficient reason to stop there and back out of that part of the search tree. None of the trees beyond that partial tree can have less than 60 changes, so none are candidates for being most parsimonious trees. We would like to calculate this lower bound on the number of changes so that it is as large as possible, and thus eliminate subtrees of the search tree as soon as we can, saving effort. There are further methods that help do this.

Using still-absent states

In many cases, we will be examining an interior node of the search tree corresponding to a partially constructed tree. Suppose that this tree has species A, B, D, and F on it. But species C and E have not yet been added to the tree. Suppose that the partial tree requires 48 changes. This will come from some of the characters that vary among species A, B, D, and F. But some of the characters will not vary until species C and E are added. We may be able to look at those species and see that, after they are added, there will be at least 11 more characters varying. In that case, no matter where they are added to the tree, the bound will be at least $48 + 11 = 59$. We can thus improve the bound considerably.

If we are dealing with 0/1 characters, that calculation is correct, but if the characters have multiple states, the bound can be made better by taking the multiple states into account. If a character has two states among species A, B, D, and F, but two more among C and E, then adding it will increase the number of changes by at least 2, not 1. Thus what we want to add to calculate the bound is the number of absent states, summed over all characters. This method of improving the bound is based on the paper by Foulds, Hendy, and Penny (1979). It has long been in use in branch and bound programs for inferring phylogenies, but this use was not described in print until the paper by Purdom et al. (2000).

Using compatibility

Another method of increasing the bound is to use not only the states in the individual characters but also the conflict between different characters. For two-state (0/1) characters, one can easily judge whether or not they can both have evolved on the same phylogeny with only one change each. We will cover this in more detail in Chapter 8. For now, we need only note that the two characters are *compatible* if they can evolve on the same phylogeny with only one change each, and that there is a simple test for this. If among all the species, all four of the combinations of states $(0, 0)$, $(0, 1)$, $(1, 0)$, and $(1, 1)$ are found, the characters are not compatible. If three or fewer are found, they are compatible.

That means that, when we consider the characters that are not yet varying on our partial tree, we can improve our lower bound on the number of changes of state. If we add all species to that partial tree, and in doing so now have variation in two incompatible characters that did not vary before, those characters must bring at least 3 changes of state with them. Each character individually will require one more change of state, and the pair will conflict, which means that one of them must have at least one additional change of state. If there are disjoint pairs of incompatible characters, each pair must bring with it 3 changes of state.

This method of computing the lower bound was developed by Foulds, Hendy and Penny (1979; see also Hendy, Foulds, and Penny, 1980). It was soon after applied to speeding up branch and bound methods, but the application to branch and bound search for most parsimonious phylogenies was first described by Purdom et al. (2000). If the species that remain to be added have k pairs of characters that are incompatible, and that do not now vary among the species on our partial tree, we must add $3k$ changes to the bound. Organizing the characters into pairs so that k is as large as possible can be done fairly quickly.

Increasing the bound as much as possible is important in getting a branch and bound method to run quickly. Hendy and Penny (1982) discovered that order of species was important, in particular that the most different species should be added as soon as possible. Purdom et al. (2000) describe improvements in speed by continually re-evaluating the order of addition during the search. Penny and Hendy (1987) describe a different branch and bound algorithm that adds characters one at a time rather than species.

Rules limiting the search

Another approach that has considerable promise is to rule out regions of the search tree in advance. Estabrook (1968) gave a rule which constrained the ancestral characters for the particular case of Camin-Sokal parsimony, a parsimony method that will be explained in Chapter 7. This might be used to speed branch and bound search. Estabrook's rule was rediscovered by Nastansky, Selkow, and Stewart (1973). They later (1974) presented an improved method that restricted the search further. However, these methods cannot be used with more general types of parsimony.

Andrey Zharkikh (1977; see also Ratner et al., 1995) has discovered some intriguing rules that allow us to determine that certain groups must be on all most parsimonious trees. Using them, we can reduce the size of the branch and bound search, sometimes greatly. Zharkikh has a number of interesting rules, but we will use only two of them here. They apply to parsimony problems where there are unordered states (as in the case of nucleotide sequences).

Zharkikh's algorithm follows these rules:

1. If there are two or more species in the data set that have the same characters, eliminate all but one of them. Repeat this until all species are distinct.

2. Eliminate all characters (sites) at which there are not or more two states that both occur in more than one species. These cannot affect the discovery of the most parsimonious tree and serve only to confirm that that one species is in fact a species. Now return to step 1 unless no such sites have been found.

3. Look at all states of all characters. For each one, let the state define the members of a group S. Calculate the number of states (over all characters) that are shared by all members of the group S but that do not appear anywhere else. Call this number $n_0(S)$. Compute the distances between all pairs of species i and j that are in S. The distance is in this case the number of characters that differ between the species. If the largest value of D_{ij} among all these pairs of species is less than $n_0(S)$, then the group S must appear on all most parsimonious trees. It can now be collapsed to a single fictional species, which has its state computed from a Fitch parsimony algorithm. Thus any states that are shared by all members of the group appear in the new species, and otherwise its state is an ambiguity between some of the possibilities within the group.

4. Unless all these three steps have all failed, return to step 1.

Zharkikh's paper has some additional rules that help identify pairs of species that cannot be in the same group, and he suggests that these allow us to take any character state that is shared between them and, on the assumption that this similarity is convergence, recode them as different states. The present rules can work very well when the data are relatively clean. If all characters can be reconstructed as having unique and unreversed changes on the same tree, then it will work wonders. It will, in fact, find all of the structure in the tree, without need for branch-and-bound! However, when the data set is noisy, the rules may fail to define any groups at all. It is not clear how useful these rules will be in practice, but they are viable candidates for taking advantage of structure that is present in the data to simplify the branch and bound search. It is possible that more powerful "pre-processing" rules can be developed to supplement these. Other approaches are possible: Bandelt et al. (1995) show that the most parsimonious trees are contained within their "median network", and this could form the basis of a method for finding all of them.

Note that even though Zharkikh's rules are wonderfully effective in some cases, they do not solve the NP-completeness problem. They do not work with all data sets, and thus leave us with exponential run times for many cases.

Chapter 6

Ancestral states and branch lengths

Reconstructing ancestral states

The Sankoff and Fitch algorithms get us a count of the total cost of the tree, in weighted changes of state. In and of themselves, they do not tell us what were the reconstructed states at the interior nodes of the tree. But it is possible to use the numbers in the Sankoff algorithm, or the sets in the Fitch algorithm, to make such a most parsimonious reconstruction. The states that, in the Sankoff algorithm, achieve the smallest costs at the root node of the tree are the states that parsimony would reconstruct for that node. Having assigned them, let us work out what the rule is for reconstructing the ancestral states at nodes that lie successively further up the tree.

Suppose that we have assigned a state to a node and look at one of the nodes immediately above it. Figure 6.1 shows the logic, where state 2 is the state that has been assigned. If the state reconstructed above it were the state 1, the total cost incurred would be $c_{21} + S(1)$. If it were state 2, the cost would be $c_{22} + S(2)$, (which would of course be $S(2)$, as $c_{22} = 0$). The logic is similar for the other two possibilities. The state or states that achieve the smallest total cost will be the ones that minimize $c_{2i} + S(i)$.

Thus we can assign reconstructed states in the nodes immediately above the bottommost one, on the right and on the left. If there is a multifurcation, we can assign them in all of the lineages.

Continuing in this fashion, we can backtrack up the tree, assigning nodes farther and farther up, until we reach the tips, where the states may have been observed. If a tip has an ambiguity in its state, we can use the same method to estimate its state. The only problem that will arise in this process is what to do about ties. We have acted as if they will never occur. There are basically two ways to

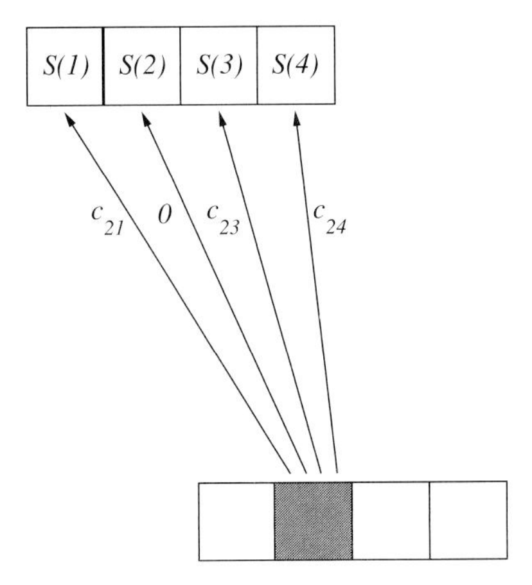

Figure 6.1: The logic of the reconstruction of ancestral states. The shaded state is the one that has been reconstructed at the lower of these two nodes. To decide what to reconstruct above it, we choose the smallest of $c_{2i} + S(i)$.

handle them. In one, whenever there is a tie, we resolve it in each of the possible ways, one after the other, each generating a traverse further up the tree. The result will be to generate all possible combinations of ancestral states. This may in some cases involve an exponential number of operations.

The other way is to keep all the possibilities, and carry out a slightly extended version of the algorithm. We assign to the higher node all of the states that can achieve minimum cost, coming from any of the allowable states at the lower node. This is not computationally burdensome but leaves us with less knowledge. We might, for example, know that a node has either state A or state C, and the one above it also either state A or state C, without knowing which goes with which. In some cases all four combinations might be possible, in some cases only two of or three of them.

Figure 6.2 shows the second type of reconstruction, for the tree in Figure 2.2 (which had a higher cost for transversions than for transitions). Note that by following the arrows one could work out all combinations of ancestral states. In this example there are three possible combinations.

We can do similar operations in the Fitch algorithm. There the assignment algorithm is simply to take the set of states at the lower node (call this set L) and the set that remains at the upper node after the Fitch algorithm was used (call this set U). The reconstructed states at the upper node will be their intersection $L \cap U$ if that set is not empty, and $L \cup U$ if it is empty. However, this will work only if L has but one element. If it has more than one, we must apply this algorithm separately

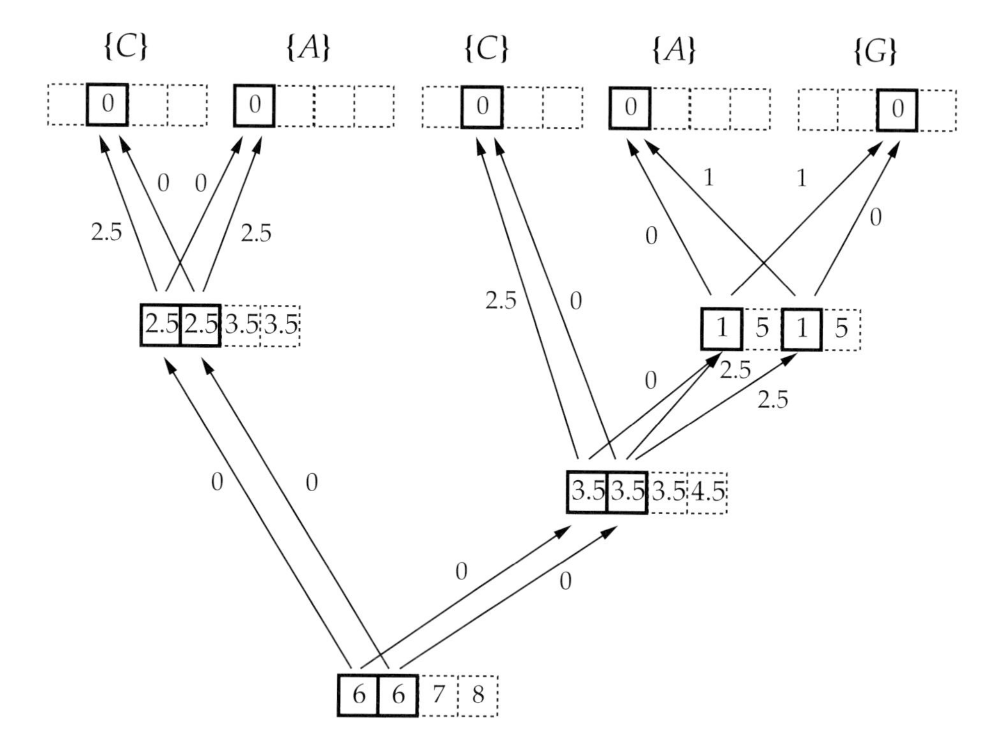

Figure 6.2: Assignment of possible states, in parsimonious state reconstructions, for the site in Figure 2.2. The parsimonious reconstructions are shown by the arrows, with the costs of the changes shown. The states that are possible at the nodes of the tree are those whose boxes in the array of numbers are solid, with the other boxes being made of dashed lines.

for each element in L. I will leave it to the reader to apply this method to the tree in Figure 2.1. The correct answer is that the interior nodes of the tree can all be either A or C, except for the upper-rightmost interior node, which can be either A or C or G.

This algorithm reconstructs states at the ancestral node more correctly than the approximate method of Maddison, Donoghue, and Maddison (1984). Even so, it can have a large statistical error. Wayne Maddison (1995) has used probability models of character evolution to compute how often parsimony reconstructions of states at interior nodes are wrong.

Accelerated and delayed transformation

In some cases it may be desirable to pick one of the tied assignments of locations of the changes. Swofford and Maddison (1987) describe *accelerated transformation* and *delayed transformation*, two methods of assignment that attempt to maximize reversals or maximize parallelisms, respectively. (The algorithm for ancestral state assignment by Farris (1970) was a form of accelerated transformation.) These have become known as ACCTRAN and DELTRAN after the names of the corresponding options in Swofford's program PAUP*. Working up from a root, ACCTRAN assigns changes as soon as possible, while DELTRAN tries to delay them until further up the tree.

These options can be implemented in our algorithm by at each stage choosing the state that is tied for best and that has changed most (or least) from the state below it. The result need not be unique — there may be more than one choice at each stage that satisfies this criterion. Accelerated transformation (ACCTRAN) forces changes to occur as far down the tree as possible, and thus maximizes reversals. Delayed transformation (DELTRAN) forces changes to occur as far up the tree as possible, and replaces reversals by parallelisms.

Branch lengths

Branch lengths are numbers that are supposed to indicate for a given branch how many changes of state have occurred in the branch. They are regarded by many biologists as simple observations. But they are never simple. In the first place, as we have already seen in Chapter 1, there may be ambiguities as to where the changes in a character actually are. When there are ambiguities, what should we do? Figure 6.3 shows an imaginary case that has many different reconstructions of changes of state. It is a bit of a worst case.

Lorentzen and Sieg (1991) have insisted that parsimony analysis should properly result in the display of all possible assignments of states to ancestral nodes (and thus all possible placements of changes of state) for each character. Their objective may find little support outside of the pulp and paper industry.

The simplest way to obtain branch lengths seems to be to average the number of reconstructed changes in each branch, averaging over all possible reconstructions of the evolution of the character. This is done for each character, and the result is summed over all characters. Figure 6.4 shows the resulting branch lengths for the tree of Figure 6.3. Such average branch lengths always sum to the number of changes of state reconstructed by parsimony.

One strange property of such average branch lengths is shown in Chapter 1, where there were three different reconstructions of the placement of changes of state in character 2 (see Figure 1.3). That tree was a rooted tree. However, when the tree is considered as unrooted, there are then only two reconstructions possible. One pair of reconstructions differed only in assigning a change to one side of the root or the other. Thus when we average the branch lengths in all possible re-

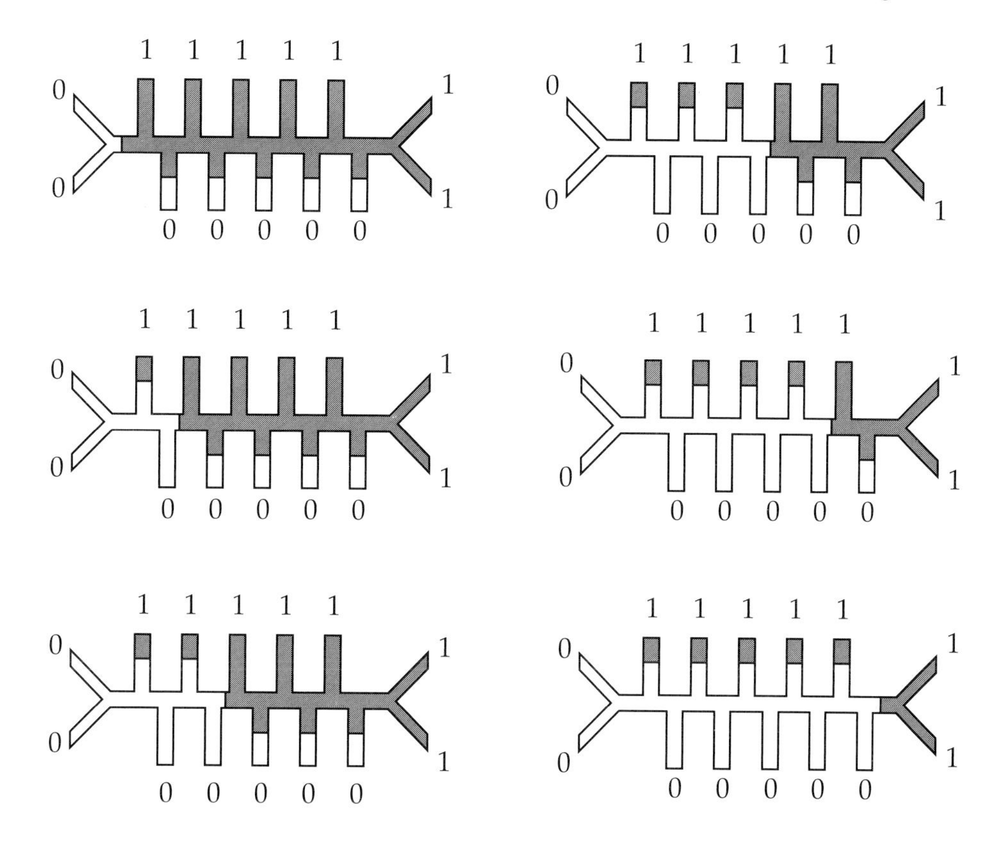

Figure 6.3: Multiple reconstructions of the placement of changes of state, all tied for most parsimonious, in a simple 0/1 character.

constructions for the rooted case, we will get different lengths than if we average them in the unrooted case!

An algorithm for computing branch lengths by averaging over all assignments of states tied for most parsimonious is given by Hochbaum and Pathria (1997). It is a dynamic programming method that makes one pass through the tree; it can re-use partial results so that one does not have to do one pass through the tree for each branch. It infers the average number of changes per branch more easily than the more complicated generating function methods of Rinsma, Hendy, and Penny (1990) and Carter et al. (1990).

Another problem with reconstructed changes is that they must frequently underestimate how many changes have occurred. If two sister species have the same derived state (with none of their other relatives having it), parsimony automatically reconstructs this as a single change of state in the lineage leading to this pair

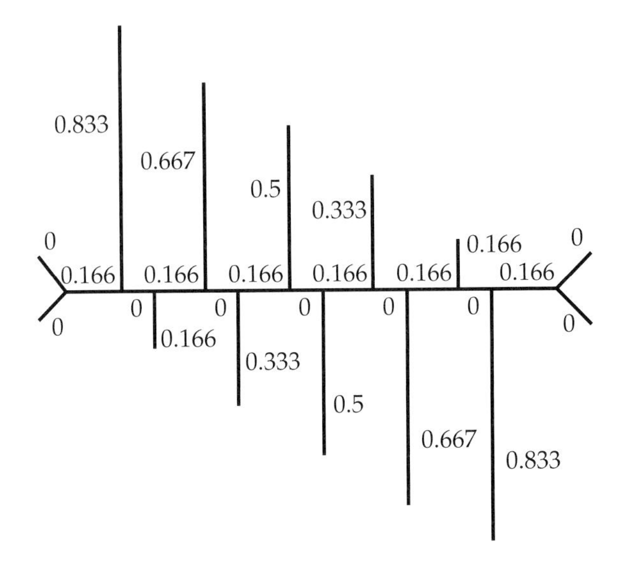

Figure 6.4: The branch lengths resulting from averaging all possible reconstructions of this character. Lengths in the diagram are proportional to the branch lengths, except for the branches of zero length.

of species. But it is also possible, if less parsimonious, for the change to have occurred in parallel in both of the lineages. It is less probable that this happened, but it is not ruled out. Instead of contributing 1 to the number of changes in the shared lineage, it might be more reasonable to contribute 0.9 to it, and 0.1 to each of the daughter lineages. Parsimony never does this. It does not correct for the unobserved changes.

In Chapter 16 we will see how probability models of character evolution can be used to infer ancestral states. Using probability models, Penny et al. (1996) and Galtier and Boursot (2000) calculate how often parsimony misses state changes in branches.

In reconstructing branch lengths, we would like to have them be the average lengths. We would like to reconstruct by averaging not over all most parsimonious reconstructions, but over all possibilities in proportion to the probability of their occurrence. We must correct the branch lengths that are reconstructed by parsimony, to allow additional events. This inflates the branch lengths beyond their parsimony values. Doing so for parsimony reconstructions is a complex task, but we shall see that in distance matrix and likelihood methods, this correction occurs automatically.

Chapter 7

Variants of parsimony

Thus far we have discussed only parsimony methods in which there is symmetrical change among two or more states. However, we have also introduced the Sankoff algorithm, which is capable of dealing with a much more diverse collection of parsimony methods. In this chapter, we will review a number of parsimony methods other than Wagner parsimony.

Camin-Sokal parsimony

Camin and Sokal (1965) introduced this, perhaps the simplest parsimony method. It assumes that we know which is the ancestral state. In its simplest form there are two states, 0 and 1, and change can only happen from state 0 to state 1; reversals are impossible. Figure 7.1 shows a 0/1 character, with its reconstruction on a given tree according to Camin-Sokal parsimony. Reconstruction of ancestral states and counting of changes of state according to Camin-Sokal parsimony are quite simple. One can use the Sankoff algorithm, with an infinite cost of $1 \rightarrow 0$ changes, but this is unnecessary. All that is necessary is to note that if a node has any 0 states in its immediate descendants, then it must have state 0. Otherwise (when all of its immediate descendants are in state 1), it must be in state 1. One need only carry out a postorder tree traversal, going down the tree. A node is assigned state 1 if some of its immediate descendants have state 1, and none of them have state 0. It is assigned an unknown state if all its immediate descendants have unknown state; otherwise, it is assigned state 0. The changes are immediately apparent: There is one for every node that has state 1 while its immediate ancestor has state 0.

One application of Camin-Sokal parsimony is in the evolution of small deletions in DNA, when we have reason to believe that they will not spontaneously revert. If we can code each deletion as present or absent in each sequence, then Camin-Sokal parsimony would be appropriate. In more complex cases, where deletions overlap and we cannot be entirely sure whether any one of them is present or absent, it would not be appropriate.

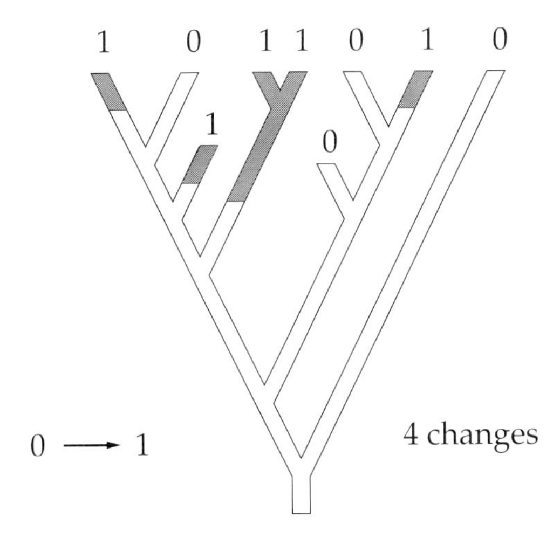

Figure 7.1: A simple 0/1 character reconstructed on a tree according to Camin-Sokal parsimony. Four $0 \rightarrow 1$ changes of state are required.

Camin-Sokal parsimony infers a rooted tree. The penalty depends on where the root is assumed to be. Thus it will favor the placement of the root in some particular part of the tree.

Estabrook (1968) gave an interesting method for delimiting the set of most parsimonious trees in Camin-Sokal parsimony (see also Nastansky, Selkow, and Stewart, 1973, 1974). This can form the basis for a branch-and-bound algorithm. As the focus of most work shifted to Wagner parsimony after this time, there was unfortunately no further development of this method.

Parsimony on an ordinal scale

Farris (1970) gave algorithms for counting changes, and for inferring ancestral states, when characters were on an ordinal scale. This is the case of discrete states arranged in a linear order, with change allowed between adjacent states and all these changes counted equally. Interestingly, Farris's algorithm also covers the case when the possible states are in a linear continuum, with changes counted by measuring the absolute values of the differences between the states at the two ends of a branch. Thus if we have states such as 0.342, 1.974, and 2.569 at the tips of the tree and we see a branch with states 0.873 at one end and 1.734 at the other, we count the total change in that branch as $|1.734 - 0.873| = 0.861$.

Farris's algorithm assigns states and counts changes in two passes through the tree, one downward and one back up. The downward pass is, in effect, an application of the Sankoff algorithm. Suppose that at a node of the tree, the function that gives the conditional cost, the function needed by the Sankoff algorithm, is $f(x)$. It turns out that $f(x)$ is always of the form:

$$f(x) = \begin{cases} c + (x_\ell - x) & x < x_\ell \\ c & x_\ell \leq x \leq x_r \\ c + (x - x_r) & x > x_r \end{cases} \tag{7.1}$$

for some values x_ℓ and x_r that are left and right ends of the interval of most parsimonious state assignments at that node. In other words, the conditional cost is constant within an interval $[x_\ell, x_r]$ and rises linearly from both ends of the interval, with slopes of -1 and 1.

In that case, we can use the Sankoff algorithm to show that when two adjacent nodes in a bifurcating tree have intervals $[x_1, x_2]$ and $[x_3, x_4]$, with $x_3 \geq x_1$, if the two sets overlap ($x_3 \leq x_2$), then the immediate ancestor can be assigned interval $[\max(x_1, x_3), \min(x_2, x_4)]$. If they do not overlap ($x_3 > x_2$), then the interval is the one that separates the two disjoint intervals ($[x_2, x_3]$). (In the former case, one counts zero change. In the latter case, one counts an amount of change $x_3 - x_2$.) With multifurcating trees, the rules are more complicated but still result in a single interval being constructed on the downward pass at each internal node of the tree.

Swofford and Maddison (1987) gave a detailed exposition of Farris's algorithm and provided a proof of its correctness. They discussed the issue of reconstructing ancestral states when there are multiple possible reconstructions and described the accelerated transformation and delayed transformation (ACCTRAN and DELTRAN) reconstructions. Farris's reconstruction, which was an ACCTRAN method, has become known as *Farris optimization*.

Dollo parsimony

Walter Le Quesne (1974) suggested a parsimony method that is based on "Dollo's Law" (Dollo, 1893). This law, which Dollo called the "Law of Phylogenetic Irreversibility," states (in one form) that a complex character, once attained, cannot be attained in that form again. Thus once a complex character is lost, it cannot re-evolve, except in noticeably different form. There are many exceptions to this law, and also many different statements of it in the literature. Le Quesne suggested algorithms that do not quite implement what he intended to; this was pointed out and the necessary algorithms given by Farris (1977a).

In the simplest form, Dollo parsimony assumes that there are two states, 0 and 1, with 1 playing the role of the complex derived state. 0 is the ancestral state. 1 is allowed to evolve only once, but it is allowed to revert to state 0 multiple times. The number of these reversions is the quantity being minimized. This can be approximately implemented in Sankoff's algorithm by assigning a large cost to $0 \rightarrow 1$ changes and a considerably smaller cost to $1 \rightarrow 0$ changes. It is possible to carry it out more directly as well, using an algorithm similar to the Fitch algorithm, but with two passes, one down the tree and one back up.

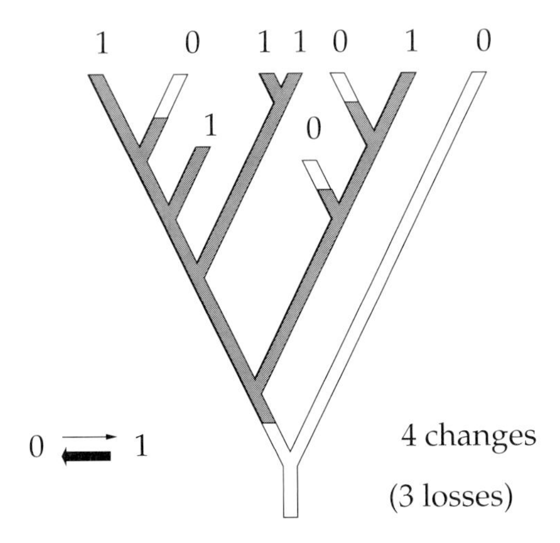

Figure 7.2: A simple 0/1 character reconstructed on a tree according to Dollo parsimony. Three $1 \rightarrow 0$ losses are required.

Figure 7.2 shows the same tree and data as in the previous figure, reconstructed according to Dollo parsimony. In the branch leading to the upper-left from the bottommost fork, it is necessary to assume that state 1 arises. All occurrences of state 0 above that are assumed to arise by losses $1 \rightarrow 0$. There are 3 losses required to explain these data. Dollo parsimony has been most widely applied as a crude model of the gain and loss of restriction sites in DNA. We will discuss that application and its validity in a later chapter.

Like Camin-Sokal parsimony, Dollo parsimony is inherently a rooted method. It assigns different penalties to differently rooted trees, so that it allows us to infer not only the unrooted tree topology, but the placement of the root as well.

Polymorphism parsimony

A third variant of parsimony assumes that apparent parallel changes of state are not really independent. They occur because the alleles that are needed are already segregating in both populations. The parsimony method based on this assumes that a state of polymorphism for two alleles is attained in a population; beyond that point, all occurrences of either state 1 or state 0 are to be explained by losses of one allele or the other. Polymorphism is assumed to arise only once in each character. But this parsimony method does not minimize the number of losses. Instead, it assumes that retention of the polymorphism along branches of the tree is to be minimized. One counts the number of branches that are reconstructed as being polymorphic for both states and minimizes the sum of that number over all characters.

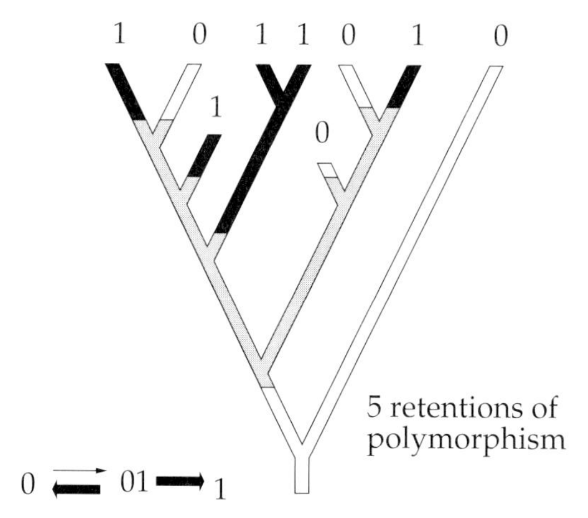

Figure 7.3: A simple 0/1 character reconstructed on a tree according to polymorphism parsimony. Five occurrences of retention of polymorphism are required. The region of the tree that has the polymorphic state is shaded gray.

The polymorphism parsimony method was first used in a numerical example computed by me for the use of Inger (1967); it was called by the less euphonious name "heteromorphism parsimony." It was first described completely in print by Farris (1978) and by me (1979). Figure 7.3 shows the reconstruction of the character that polymorphism parsimony would make.

Polymorphism parsimony can be used for morphological characters, but it finds its most natural application in explaining changes in karyotype due to chromosome inversions. It is very unlikely that the same inversion will arise twice, but inversions can coexist in the same population, and one would explain apparent parallel gains of the same inversion by assuming that the inversion arose only once but that polymorphism might have persisted long enough to explain different events.

The algorithms necessary for implementing the polymorphism parsimony method are discussed in the references given above. Interestingly, it is not simple to implement polymorphism parsimony using the Sankoff algorithm. One can have a state corresponding to polymorphism, but one must be able to assign a penalty to the retention of polymorphism. Thus one must use a version of the Sankoff algorithm that will allow us to assign a penalty for persistence in a state without change!

Unknown ancestral states

All three of the variants described in this chapter require us to know which state is the ancestral one in each character. What if we do not know this? We can imagine inferring the ancestral state by parsimony. We could count, for character i, $N_0^{(i)}$, the number of changes if 0 is the ancestral state, and $N_1^{(i)}$, the number of changes if 1 is the ancestral state. This requires two evaluations per character. Then we simply choose the smaller of these two numbers. Thus the total number of changes in the tree is

$$N = \sum_i \min\left(N_0^{(i)}, N_1^{(i)}\right) \tag{7.2}$$

The estimate of the ancestral state for each character is, of course, whichever state s gives the smaller value of $N_s^{(i)}$. It should be immediately apparent that the amount of effort involved in inferring ancestral states is twice as great if there are two possible states.

Multiple states and binary coding

Camin and Sokal's (1965) original method assumed that each character could have multiple states, arranged in a linear order, such as

$$-1 \leftarrow 0 \rightarrow 1 \rightarrow 2$$

A corresponding assumption for undirected change would be

$$-1 \leftrightarrow 0 \leftrightarrow 1 \leftrightarrow 2$$

Both assumptions can be addressed by the proper choice of costs in Sankoff's algorithm. For the first, we would use the cost matrix:

From : \ To :	-1	0	1	2
-1	$-$	∞	∞	∞
0	1	$-$	1	∞
1	∞	∞	$-$	1
2	∞	∞	∞	$-$

while for the latter, we would instead assume:

From : \ To :	-1	0	1	2
-1	$-$	1	∞	∞
0	1	$-$	1	∞
1	∞	1	$-$	1
2	∞	∞	1	$-$

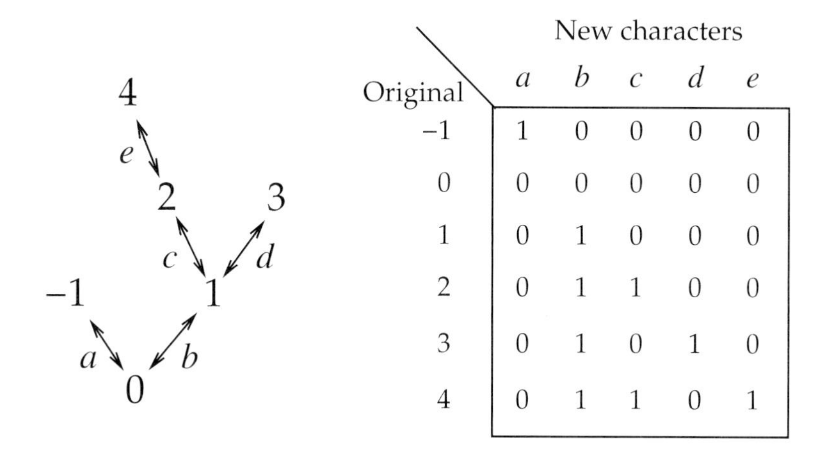

Original \ New characters	*a*	*b*	*c*	*d*	*e*
−1	1	0	0	0	0
0	0	0	0	0	0
1	0	1	0	0	0
2	0	1	1	0	0
3	0	1	0	1	0
4	0	1	1	0	1

Figure 7.4: A character with 6 states, connected in a character state tree that shows which states can be reached from which other states. The set of binary (0/1) characters that is equivalent to this is shown. These binary characters will have the same number of character state changes as the original multistate character. The new characters are labeled with letters, which are also shown next to the corresponding branches on the original character state tree.

Similar cost matrices can be set up for other cases with multiple states.

However, there is an alternative method of treating such cases without using the Sankoff algorithm. Sokal and Sneath (1963) invented *additive binary coding* for states that are arranged in a linear order. This was generalized by Kluge and Farris (1969) to cope with branching sequences of states, which have been called *character state trees*. The general idea is that a set of 0/1 characters can be produced that always have the same number of changes of state as do the original characters, when evaluated on the same tree. Figure 7.4 shows a simple character state tree and the set of binary characters, called *binary factors*, that are equivalent to it. The general method of producing this binary recoding is to make a new 0/1 character for each branch in the character state tree. Each state of the original multistate character is then assigned a 1 in this 0/1 character if it is connected more closely to one end of the branch, a 0 if it is connected more closely to the other. Thus the new state c has states 2 and 4 connected to one end of branch c on the character state tree, and states −1, 0, 1, and 3 connected to the other end of that branch. States 2 and 4 are then assigned state 1 in the new character c, and states −1, 0, 1, and 3 are assigned state 0. Similarly, the new state b has the original states 1, 2, 3, and 4 connected to one end of branch b, and −1 and 0 connected to the other end of the branch b. It should be immediately apparent that no two of the original character

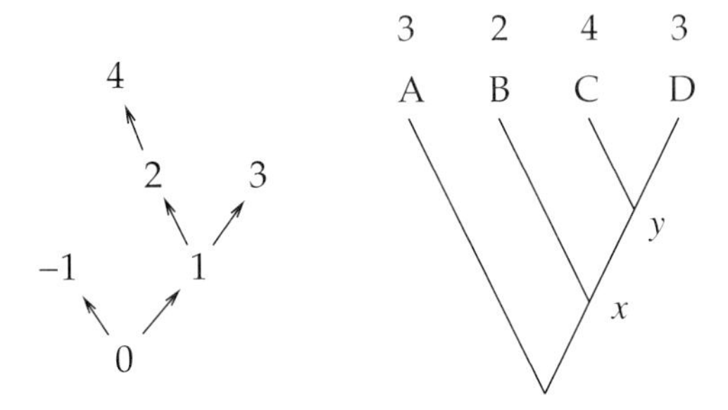

Figure 7.5: A character with six states, connected in a rooted character state tree. Also a phylogeny with the states of this character given at the tips. There is no way to assign states x and y and satisfy the requirements of Dollo parsimony.

states will have the same set of binary factors. When one moves along a branch in the character state tree, exactly one of the new characters will change, namely the one corresponding to that branch.

In the above example, all the arrows on the character state tree were bidirectional, and the binary factors were evaluated by Wagner parsimony. If change is unidirectional along any arrow on the character state tree, one should evaluate the corresponding binary factor by Camin-Sokal parsimony instead.

Unfortunately, binary recoding is not possible for unordered multistate characters such as bases in DNA, where every state can change to every other, producing loops in the character state tree.

Dollo parsimony and multiple states

An interesting paradox arises when we attempt to extend Dollo parsimony to characters with multiple states. Recall that Dollo parsimony is inherently a rooted method. Figure 7.5 shows a character state tree (essentially a rooted version of the one in the preceding figure). It also shows a phylogeny that we wish to evaluate, along with the states at the four tips of the tree. How can we reconstruct states such as x and y so as to satisfy the requirement that each derived state originate no more than once? In fact, we can't. If states x and y are taken to be 1, for example, then states 2 and 3 must both have arisen twice. If x and y are both taken to be 2, then state 3 must have arisen twice (and 2 must have reverted to 1 an extra time as well). A similar problem arises if we take both x and y to be state 3. The reader can try other possibilities. All have similar problems, or worse.

The only way out is to relax some of the requirements of Dollo parsimony and to assume that a state may have arisen more than once. However, then we must have some way of deciding how much to penalize a tree for an extra origin, compared to extra losses. Interestingly, if we try to escape from the problem by using binary recoding, it will reconstruct some states in the ancestor that are nonexistent. These problems were discussed by me (Felsenstein, 1979) in a paper on the relationship between methods.

Polymorphism parsimony and multiple states

With polymorphism parsimony, there are fewer complications when characters have multiple states. The main problem comes with counting the number of states segregating. Consider the situation in Figure 7.4. Suppose that we have a branch that has states 3 and 4 segregating in it. Do we count this as one polymorphism or more? The character state tree seems to imply that if the genetic material for states 3 and 4 are present, states 1 and 2 will also be present. Presumably, in that case we will want to count the polymorphism threefold, for the 3 extra states that are segregating.

If we use binary factors, we can do this automatically. Using the binary coding in Figure 7.4, we would have states 01010 and 01101 present at these five binary factors. This means that there would be 3 polymorphisms instead of one.

Transformation series analysis

While multiple states are often analyzed by Wagner parsimony, we may frequently suspect that the states in morphological or behavioral characters reflect some underlying determination that implies a character state tree. Sometimes the details of the phenotypes suggest features of the tree; sometimes they do not. In the latter case we might want to find that character state tree that would allow evolution of the character on the given phylogeny with the fewest changes of state. We may even want to go further, and adjust the phylogeny and the character state tree to fit each other.

This has been proposed by Mickevich (1982), who called this *transformation series analysis*. She had a specific algorithm that involved assessing which states were nearest neighbors on the tree, and connecting those to make an estimate of the character state tree. Starting with initial guesses of the character state trees, one would infer an initial phylogeny, infer new character state trees from that, and proceed iteratively until the estimates of both trees ceased to change.

For the phylogeny in Figure 7.5, the three states 2, 3, and 4 are observed. The character state tree $2 \longleftarrow 3 \longrightarrow 4$ would fit that phylogeny, with no homoplasy. One might alternatively not want to force change to be unidirectional, so that an undirected character state tree $2 - 3 - 4$ would be inferred instead.

The papers by Mickevich and Weller (1990) and Mickevich and Lipscomb (1991) discuss further issues, including use of prior biological information about the characters to constrain the inference of the character state tree. An analogous approach has been suggested to biogeography (Mickevich, 1981) and to electrophoretic alleles (Mickevich and Mitter, 1981). When the character values exist on a numerical scale that constrains their order, transformation series analysis will confront the character coding problems which are discussed in Chapter 24.

As trees become larger and morphological and behavioral characters are analyzed in greater detail, there are bound to be more algorithms proposed to find the combination of character state trees and phylogeny that, taken together, result in the fewest changes of state. Statistical modeling approaches should also be possible. The algorithmic challenges may be formidable, given the large number of character state trees that are possible with even a modest number of states. It seems best to reserve the name transformation series analysis for the general problem rather than a particular algorithm.

Weighting characters

In all parsimony methods discussed so far, we have counted changes equally no matter in what character they occurred. Systematists have frequently discussed "weighting" characters. In parsimony methods weighting assumes a concrete form. We can have a set of weights, $w_1, w_2, w_3, \ldots, w_n$ for the n characters. A change in character k is counted by adding an amount w_k to the total penalty.

This approach has two uses. If we want to drop a set of characters from the analysis (for example, to examine the effect that they are having on the result), we could set $w_i = 0$ for the indices i of all of these characters and leave $w_j = 1$ for all other characters j. Or we may want to allow our method to be differently sensitive to different characters. We shall see in a later chapter that these different weights correspond to different rates of evolution. For example, we might be analyzing DNA coding sequences, and we might wish to take most seriously change in position 2 of each codon, take change in position 1 a bit less seriously, and take change in the often-synonymous third codon position a lot less seriously. We might then assign weights 3, 4, and 1 to the three codon positions. The sequence of weights of sites would then be: 3, 4, 1, 3, 4, 1, 3, 4, 1, 3, 4, 1, ...

This would cause the method to avoid changes in the second codon position as much as possible, avoid changes in the first codon almost as strongly, but place much less emphasis on avoiding change in the third codon position. Another heterogeneity that is often used to weight changes differentially is the difference in rate between transitions and transversions. It has long been recognized that transversions occur at a lower rate than transitions. *Transversion parsimony* is based on the concern that, across deeper branchings, transitions may have reached saturation, leaving the transversions bearing most of the information. It is the parsimony algorithm in which all transitions have weight 0, and all transversions have weight 1. Weighting methods have also been empirically derived, based on the distribution of numbers of changes of state across a random sample from all possible phylogenies (the *profile parsimony* of Faith and Trueman, 2001), and across a distribution of phylogenies obtained by bootstrap sampling of all characters (the weighting method of Kjer, Blahnik, and Holzenthal, 2001).

Algorithms for weighting characters were first clearly discussed by Farris (1969b) in the first paper on numerical weighting algorithms.

Successive weighting and nonlinear weighting

It is also possible to weight changes of state differentially according to how many of them there are. This possibility is raised, and an algorithm given, by Farris (1969b). If the weight on character i were w_i, and there were n_i changes of state of that character on a particular tree, we would normally expect to penalize the tree $n_i w_i$. Farris's approach is to use a function $f(n)$ and penalize the character $f(n_i)$. This raises the possibility of diminishing returns when penalizing additional changes of state on a character, or even of increasing returns that penalize each additional state more. Farris recommends the use of weights that drop rather rapidly with increasing probability of change of the character.

Another way of thinking of this scheme is that when there are n_i changes, we penalize each of them by an amount $g(n_i) = f(n_i)/n_i$.

Successive weighting

Farris suggested a *successive weighting* algorithm for searching for most parsimonious trees under these nonlinear weighting schemes. One starts with equal weights for all changes and searches for the tree T_1 that minimizes the total number (and hence the weighted number of changes as well). Looking at T_1, one calculates a new weight per change for each character, based on the total number of times it changes on T_1. Thus if character 37 changes 4 times, one sets the weight per change in that character to $w_{37} = g(4)$. Armed with this new set of weights per change (one for each character) we set out to find the tree T_2 that minimizes the weighted sum of changes. Having found it, we now count how many changes each character shows on tree T_2 and use it to set new weights $w_i = g(n_i)$. These are then used to search for a new tree, T_3.

The process continues until it does not change the tree. This generally happens fairly quickly. The final tree has the property that the weights based on the number of changes of each character on that tree cause us to find the same tree again. Table 7.1 shows the process for the simple example of the data in Table 1.1. There are 15 possible trees for these 5 species, as the trees are unrooted. Considering the number of changes in each character, there are 5 patterns of changes. Thus 3 of the 15 trees have 2, 2, 2, 1, 1, and 1 changes on characters 1 through 6. No matter which of these 3 trees we are looking at, the weights that will be assigned to the characters for the next round of search will be the same: $g(2)$, $g(2)$, $g(2)$, $g(1)$, $g(1)$, and $g(1)$. We call these trees type IV and represent them by one row in Table 7.1. In this example we use the function $g(n) = 1/(n+2)$ (though, as there are only 2s and 1s, it actually does not matter what weighting function we use).

Starting with any tree (say one of the 3 trees of type IV), we use the weights implied by that tree, which would be (0.25, 0.25, 0.25, 0.333, 0.333, 0.333) and com-

Table 7.1: A simple example of successive weighting using the data of Table 1.1. There are 15 possible unrooted bifurcating trees, which fall into 5 types according to how many changes they have in each character. The table shows the total weighted number of changes when each tree type is evaluated using the weights implied by the 5 different tree types.

Number of trees	Have pattern of changes	Type of tree	Tree type used for weights I	II	III	IV	V
1	(1,1,1,2,2,1)	I	2.333	2.250	2.167	2.417	2.083
2	(1,2,1,2,2,1)	II	2.667	2.500	2.500	2.667	2.333
2	(2,1,2,2,2,1)	III	3.000	2.917	2.667	2.917	2.583
3	(2,2,2,1,1,1)	IV	2.833	2.667	2.500	2.500	2.333
7	(2,2,2,2,2,1)	V	3.333	3.167	3.000	3.167	2.833

pute the total sum of weighted changes for each of the 5 types of tree. For example, type II will require

$$0.25 + 2(0.25) + 0.25 + 2(0.333) + 2(0.333) + 0.333 = 2.667$$

weighted changes. We can consider all the tree types using these character weights by looking down column IV of this table. The tree that has the smallest sum of weights is the first one, tree type I. In fact, in all 5 columns, tree type I is the best. Thus if we start successive weighting with tree type IV, we arrive at tree I. As tree type I recommends itself, the process terminates there. Farris (1969b) discussed using for starting weights a formula based on work by Le Quesne (1969). The formula considered how many other characters a given character was compatible with. In the present case, this would yield the same result as starting with unweighted characters, and would start with tree type I.

With such a small example, we cannot see many aspects of the behavior of the method. For example, successive weighting can make it difficult to detect ties. Suppose that we had a case where two characters were in conflict. On tree I character 1 had one change, and character 2 had 2 changes. On tree II character 1 had 2 changes, and character 1 had one change. Therefore, if we initially look at tree I, we are told character 1 is of high weight, and this causes us to continue to prefer tree I to tree II. The case of Table 7.2 shows this problem. For four species there are three possible unrooted trees. If we have two 0/1 characters that favor ((A,B),(C,D)) and two that favor ((A,D),(B,C)), we get the situation in this table. Starting with tree III we have an even choice between tree I and tree II. If we choose tree I, it recommends itself over tree II. Similarly, tree II recommends itself over tree I. The final outcome depends on the initial tree, and once one has reached one of trees I or II, the other seems less desirable.

Table 7.2: An example of successive weighting that would show the difficulty it has in detecting ties. The table shows the total weighted number of changes when each tree type is evaluated using the weights implied by the different tree types.

Number of trees	Pattern of changes:	Type of tree	Tree type used for weights I	II	III
1	(1,1,2,2)	I	1.667	1.833	1.5
1	(2,2,1,1)	II	1.833	1.667	1.5
1	(2,2,2,2)	III	2.333	2.333	2

Nonsuccessive algorithms

An alternative to successive weighting is to allow the weights to be functions of the number of changes in the character, but to do the search nonsuccessively. Goloboff (1993a, 1997; see also David Maddison, 1990) has proposed using functions much like those Farris used, but choosing the weights used for a tree based on the number of changes on that tree. Thus if tree i has n_{ij} changes in character j, the total penalty for tree i will be

$$\sum_j n_{ij}\, g\left(n_{ij}\right)$$

This corresponds to evaluating each tree using its entry in the diagonal of Table 7.1. Going down that diagonal from upper-left to lower-right, we find that the penalties of the trees are 2.333, 2.500, 2.667, 2.500, and 2.833. The best of these is 2.333, giving the same final result as before. In general, the results need not be the same, though if we start at the best tree by this criterion (as Goloboff, 1993a, asserts), then it can be shown that the successive method will not choose any other tree, provided $g(n)$ is a decreasing function of n.

Farris (1969b) and Goloboff (1993a, 1997) argue that we should prefer functions $g(n)$ that decrease with n. One example of a specific weighting function that is decreasing and that was applied by a nonsuccessive weighting algorithm is threshold parsimony (Felsenstein, 1981a), which uses the function

$$g(n) = \min[n, T]/n$$

where T is a threshold value. This is the same as counting all characters that have more than T changes as if they had exactly T changes. We shall see in Chapter 9 that there is a rationale for weighting functions like this, when one uses a maximum likelihood framework for inferring phylogenies. The *minimum phylogenetic number* criterion of Goldberg et al. (1996) is a nonsuccessive weighting procedure, unusual for putting its weight on the character with the highest number of changes of state. Most biologists would feel happier with an emphasis on the fit of the characters that change most slowly. The Strongest Evidence weighting procedure of Salisbury (1999) is also, in effect, a nonsuccessive weighting procedure, as it evaluates the weight of each character with reference to the distribution of numbers of changes of state obtained if we permute taxa on the present tree.

Chapter 8

Compatibility

As we have seen, some systems of character weighting have as their objective to discount information from the noisier characters and to emphasize the information from those characters that have the least homoplasy. The simplest such system would be to use only information from characters that have no homoplasy — that fit the tree perfectly. Of course, we cannot predict in advance which characters those will be.

One method of finding out is to take each tree, compute the number of changes of state that each character requires on it, and score each character as either compatible with the tree or not. We define a character as being compatible with a tree if it can evolve on that tree with no homoplasy. We then find that tree that maximizes the number of compatible characters. This criterion was first proposed (though somewhat implicitly) by Le Quesne (1969). If a character has k states in our data, then it is easy to assess whether it is compatible with a tree: we simply see whether it requires $k - 1$ changes of state on that tree. It must require at least $k - 1$ because one of the states will be ancestral, and the other $k - 1$ must arise at least once each. If it requires more than $k - 1$, then one of the states will arise more than once on the tree.

Viewed this way, compatibility is a close relative of parsimony methods. In certain cases, it is also derivable from weighted parsimony methods. We have seen, in the previous chapter, that threshold parsimony involves counting the number of changes up to a threshold value, T, and beyond that, counting the character as having penalty T. If the characters all have two states and we set the threshold to $T = 2$, the threshold method will actually be the same as a compatibility method. This is easy to show: On any given tree there might be n_0 characters that require no changes, n_1 that require 1, and $n - n_0 - n_1$ that require more than one change. The total penalty will then be $n_1 + 2(n - n_0 - n_1)$. That turns out to be $2n - 2n_0 - n_1$. As the numbers n and n_0 are always the same on all trees that we might look

Table 8.1: The data set of Table 1.1 with an added species all of whose characters are 0.

	Characters					
Species	1	2	3	4	5	6
Alpha	1	0	0	1	1	0
Beta	0	0	1	0	0	0
Gamma	1	1	0	0	0	0
Delta	1	1	0	1	1	1
Epsilon	0	0	1	1	1	0
Omega	0	0	0	0	0	0

at (characters that require 0 changes having only one state, and thus requiring 0 changes no matter what the tree), it follows that in minimizing the thresholded number of changes of state, we are necessarily maximizing the number of characters n_1.

Thus an appropriately weighted parsimony method, one in which the weights of changes drop away strongly with the number of changes in that character, is the same as a compatibility method.

The problems of searching among all possible trees, and of evaluating the thresholded number of changes of state, are the usual ones, and we will not go into them further here. However, in some cases we can use a table of pairwise compatibility among characters to go more directly to the correct tree. We need to look first at the way compatibility among characters is tested.

Testing compatibility

In a number of cases, a different approach can be used. In the case of two states (such as 0 and 1) with no ambiguous or missing data, we can test directly whether two characters could be compatible with the same tree. This then leads to a very different algorithm for finding the best tree. The test was introduced by E. O. Wilson (1965). The test is extremely simple. For any pair of characters that we wish

Table 8.2: The compatibility test for characters 1 and 2 of the data of Table 8.1

	0	1
0	X	
1	X	X

Table 8.3: The compatibility test for characters 1 and 4 of the data of Table 8.1

	0	1
0	X	X
1	X	X

to test, we consider each species in turn. Four possible combinations of states are possible, as the first of these characters can be either 0 or 1, and the second can also be either 0 or 1. We note which of the four combinations have occurred. Table 8.2 shows the four combinations, for characters 1 and 2 of the data in Table 8.1. This table adds one species to the simple example with 0/1 characters that we used in Chapter 1. There are three of the state combinations that occur. Wilson's test says that if all four boxes are marked, then the two characters are not compatible, in the sense that they cannot co-occur on the same phylogeny without at least one of the characters changing twice. If three or fewer boxes are marked, then the characters are compatible: There exists a phylogeny with which both of them are compatible. Table 8.3 shows the compatibility test for characters 1 and 4 in the data of Table 8.1. These characters are not compatible. Proceeding in this way, we can test every pair of characters for compatibility. Figure 8.1 shows the resulting *compatibility matrix*, a table showing for all pairs of characters which ones are compatible with each other.

The above discussion has assumed a Wagner parsimony model, with the ancestral state unknown. If we know which state is ancestral in both characters, the test changes slightly. If state 0 were the ancestral state in both characters, we would mark the box for the state combination (0,0) whether or not it was present in the data. After that the test is the same.

The correctness of Wilson's test is easy to establish. If three of the boxes in the table are marked, then we can connect them together by two links. This corresponds to a tree with the species grouped into three clusters, connected by those two links. Each link corresponds to the origin of one state.

Thus a link from (1,0) to (0,0) corresponds to state 0 arising in the first character from state 1 (or the other way around). The reader will quickly see that if one box is checked, no states need arise; if two or three of them are checked, no more than one origin of each state need occur, but if all four boxes are checked, then of necessity one of the states in one of the characters must arise twice.

The Pairwise Compatibility Theorem

The chief reason for making a compatibility matrix is to use it to choose as large as possible a set of characters, all of which can be compatible with the same tree. The compatibility matrix does not at first sight tell us that; it shows us which pairs of characters are compatible, although without ensuring us that the different pairs that are compatible with each other are all compatible with the same tree.

In fact, there is a remarkable theorem that guarantees us that. It does not have a standardized name, but I like to call it the

> **Pairwise Compatibility Theorem.** A set S of characters has all pairs of characters compatible with each other if and only if all of the characters in the set are jointly compatible (in that there exists a tree with which all of them are compatible).

The theorem is true for 0/1 characters or any characters having at most two states per character. It is not true, as we shall see, for data with missing states or for nucleotide sequence data having four states. The Pairwise Compatibility Theorem has been proven for a number of different cases in three papers by George Estabrook, F. R. McMorris, and their colleagues (Estabrook, Johnson, and McMorris, 1976a, 1976b; Estabrook and McMorris, 1980). The last of these papers provides a particularly simple proof. For characters with two states it was earlier proven by Buneman (1971).

If we consider the set of species in which one of the states (say, 1) occurs, the test of compatibility between two characters amounts to saying that the two sets of species (one for each character) S and T are either disjoint, so that $S \cap T = \emptyset$, or $S \subseteq T$, or $S \supseteq T$. Let us say that in this case the two sets are part of an hierarchical structure. If the two sets overlap but neither is contained within the other, then all regions of the set of species exist, so that all four state combinations exist. The Pairwise Compatibility Theorem then amounts to the assertion that if all pairs of subsets are able to fit into an hierarchical structure, then when taken together they are all part of an hierarchical structure. This is nearly obvious, although a formal proof is not a bad idea (see Buneman, 1971).

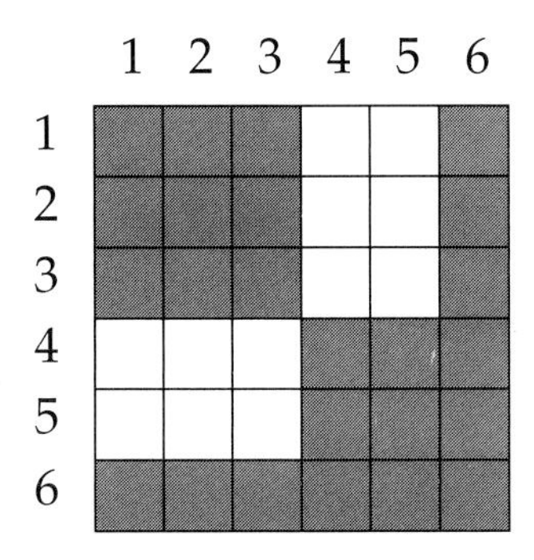

Figure 8.1: A compatibility matrix for the data set of Table 1.1. Shaded boxes are those for which the pair of characters are compatible.

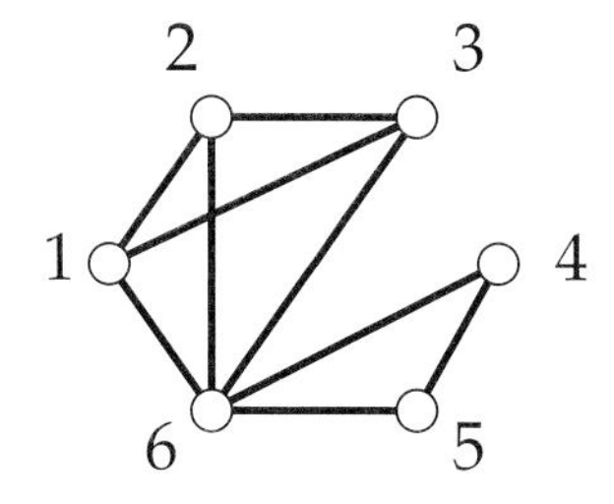

Figure 8.2: The graph corresponding to the compatibility matrix in Figure 8.1. Points are characters, and lines between them indicate that that pair is compatible.

Cliques of compatible characters

Figure 8.2 shows a graph corresponding to the compatibility matrix in Figure 8.1. Each character is represented by a point, and lines are drawn between each point and all others with which that character is compatible. A set of characters that are all pairwise compatible corresponds in this graph to a set of points that form a *clique*, points that are all mutually connected. A *maximal clique* is a clique to which no point can be added and have the result be a clique. In the graph in Figure 8.2, the set of characters $\{1, 2, 3\}$ is a clique but not a maximal clique (since 6 can be added). $\{1, 2, 3, 6\}$ is a maximal clique. The other maximal clique in the graph is $\{4, 5, 6\}$. Le Quesne's criterion directs us to find the largest maximal clique, which in this case is $\{1, 2, 3, 6\}$. By the pairwise compatibility theorem, there must be a tree for each of these cliques, and the characters that are in the clique all are compatible with that tree.

We need to find the largest clique. In the graph of Figure 8.2 we can do this by inspection. More generally we need to use an algorithm. One such is Bron and Kerbosch's (1973) clique-finding algorithm, which is a branch-and-bound procedure. It makes use of an incidence matrix of the graph: In effect, this is just the compatibility matrix. It may find cliques that are tied for size, and will find all such.

The finding of trees by compatibility using cliques has led to the method sometimes being called a *clique method*. This is often done by opponents of the technique, who prefer that name because it implies that the proponents of it are a mere clique.

In principle, the finding of the largest clique is not easy: Day and Sankoff (1986) have shown that the task is NP-hard. Practice is, however, different. A bad case for parsimony methods is (for two-state data) a data set that is a box filled with random 0s and 1s. But for compatibility that case is an easy one. Few if any pairs of characters will then be compatible. The cliques will be small and will easily be found by algorithms such as Bron and Kerbosch's. When the data sets are very clean (the opposite case), then almost all characters will be in a single large clique, which the algorithm again finds easily. In fact, biological data rarely, if ever, generate compatibility matrices that cause the algorithm any difficulty. Thus in practice, though not in theory, compatibility methods run much more quickly than do parsimony methods.

It may be possible to go even faster. Gusfield (1991) has presented an algorithm for testing a set of 0/1 characters for joint compatibility and creating the tree from them if they are compatible. It requires an amount of computation only proportional to the product of the number of species and the number of characters. This is less effort than checking the compatibilities of all pairs of characters. However, it requires that we know which set of characters we want to check; it does not solve the problem of finding the largest clique of characters.

Finding the tree from the clique

Once we have the largest clique or cliques, we still need to estimate the phylogeny. This is done simply by using each character to successively subdivide the species. This algorithm was described by Meacham (1981) and is known as *tree popping*. Each character defines a partition of the set of species into those that have state 0 and those that have state 1 for the character. In the data set of Table 8.1, with the clique {1, 2, 3, 6}, the first character divides the species into two sets according to the state of that character. It divides the species into {Alpha, Gamma, Delta} and {Beta, Epsilon, Omega}. Implicitly, there is to be a branch on the tree between these two sets, with character 1 changing once along that branch. The second character sets apart the set {Gamma, Delta}; in effect it subdivides the first set of that pair. Because character 2 separates {Gamma, Delta} from the other species, it implicitly leaves {Alpha} connected with {Beta, Epsilon, Omega}. We then continue this process of subdivision of one set or the other. Each such subdivision can introduce at most one branch. Some characters may not divide the species sets further, but serve to reinforce them and place one more change along a branch. When a new branch is inserted, the character shows which species are separated by that new branch.

Figure 8.3 shows the process, resulting in a tree. This is not the same as the most parsimonious tree. It is not hard to show that with five or fewer species and 0/1 data with unknown ancestral states, the parsimony and compatibility trees will always be the same. This case has 6 species and is the simplest one I know in which the parsimony and compatibility methods yield different trees.

In general, compatibility trees and most parsimonious trees will not be the same when the weights of changes do not depend on how many there are in the character.

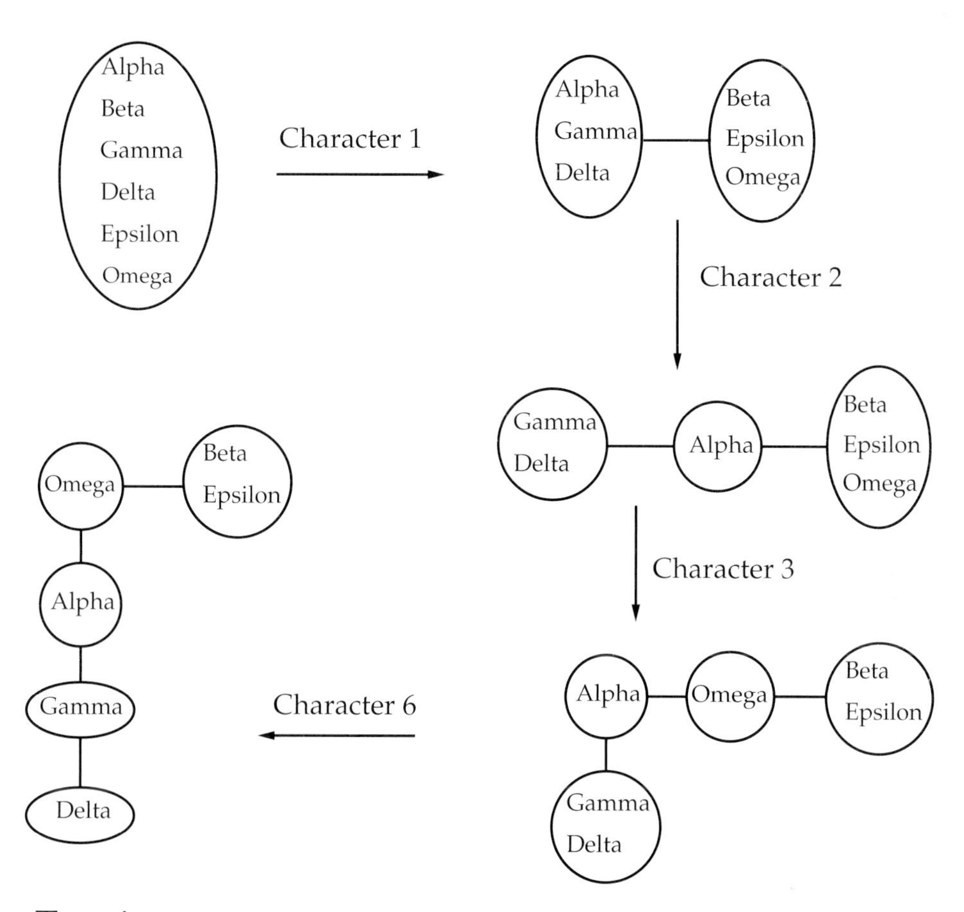

Tree is:

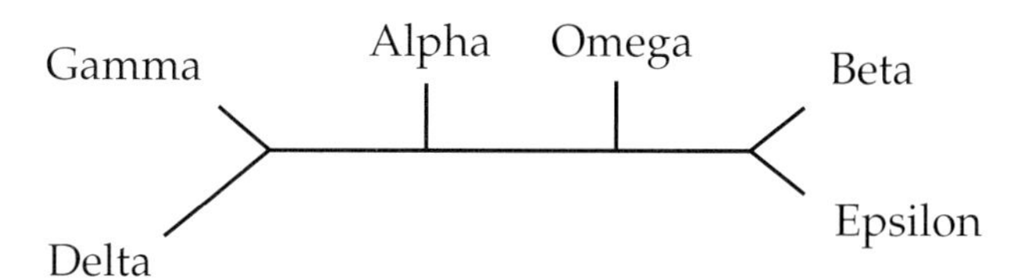

Figure 8.3: The "tree popping" process of sequentially subdividing the set of species, according to the characters in a clique. The result in this case is a tree that is compatible with all the characters in the clique.

Table 8.4: A data set that has all pairs of characters compatible, but that cannot have all characters compatible with the same tree. This violates the Pairwise Compatibility Theorem, owing to the unknown ("?") states.

Alpha	0	0	0
Beta	?	0	1
Gamma	1	?	0
Delta	0	1	?
Epsilon	1	1	1

Other cases where cliques can be used

Estabrook, Johnson, and McMorris (1976b) have shown that when we have multistate characters, where the states are related by a character state tree, the Pairwise Compatibility Theorem works. We have already seen (in Chapter 7) that in such cases we can recode the characters into a set of 0/1 characters, called the binary factors. Persuade yourself that two characters are compatible if and only if their binary factors are all pairwise compatible. Then if a set of characters are all pairwise compatible, so are all pairs of binary factors of those characters. It follows that the Pairwise Compatibility Theorem works for characters that have binary factors.

This makes compatibility methods usable on many kinds of morphological characters, including the important case of characters that are in a linear series of states.

Where cliques cannot be used

The clique framework cannot be used if there are missing data or if there are more than two states that are not related in a character state tree, and hence cannot be recoded into binary factors. That includes the important case of nucleotide sequences. Table 8.4 shows a data set for which all pairs of characters are compatible. It has three missing states (coded "?"). In computing the compatibility of each pair of characters, we ignored each species that had a "?" in either member of the pair of characters. There is in fact no tree with which this set of characters are all compatible. I leave it to readers to persuade themselves of that. For that matter, perhaps they can find a case with fewer "?" states that would serve as an example here. I can't.

Benham et al. (1996) have investigated computational methods for cases with ambiguous states. For the case of a series of states in a directed linear scale, they describe a polynomial-time algorithm for testing the compatibility of a set of such

Table 8.5: Fitch's set of nucleotide sequences that have each pair of sites compatible, but that are not all compatible with the same tree.

Alpha	A	A	A
Beta	A	C	C
Gamma	C	G	C
Delta	C	C	G
Epsilon	G	A	G

characters. For the more general case, they show that the problem is NP-complete. Bonet et al. (1999) give algorithms for testing joint compatibility that allow for the possibility of polymorphic states in some of the species.

For the case of nucleotide sequences Estabrook and Landrum (1976) have demonstrated a test for compatibility of two nucleotide sites. But Fitch (1975, pp. 200–205) had already given the crucial counterexample, showing a set of nucleotide sequences for which all pairs of sites are compatible, but for which there is no tree with which all of the sites can be compatible. The data set is shown in Table 8.5.

In such a case we cannot make use of the Pairwise Compatibility Theorem, and thus cannot use cliques of graphs to find largest cliques of jointly compatible characters. We might use the cliques to suggest candidates for largest cliques of jointly compatible characters, as jointly compatible characters must always be pairwise compatible as well. But as the relationship does not run the other way, the largest clique of jointly compatible characters might be a subset of a clique of pairwise compatible characters, even of one that is not the largest clique in the graph.

In such cases we could always revert to treating compatibility as a kind of parsimony. For any proposed tree we can count the number of characters that can evolve with the minimal number of changes on that tree. This number of compatible characters is then used as the criterion that we try to maximize by searching tree space in the usual way.

Perfect phylogeny

The unavailability of the Pairwise Compatibility Theorem for the case of unordered multistate characters has led to much work on the *perfect phylogeny* problem. This involves testing whether a set of such characters are jointly compatible, and if so, constructing the tree they imply. Bodlaender, Fellows, and Warnow (1992) and Steel (1992) proved that finding out whether characters are jointly compatible is NP-complete (see also earlier work by Buneman, 1974, and Meacham, 1983).

Algorithms have been found to test compatibility of a set of unordered multistate characters and construct the tree, in cases where the number of possible states is bounded. In those cases the problem is solvable in polynomial time. Agarwala and Fernández-Baca (1994) found an algorithm for the case with no more than r states that scales as $2^{3r}(nk^3 + k^4)$. Kannan and Warnow (1994) found an algorithm for the four-state case that scales as n^2k when there are n species and k characters. They later (Kannan and Warnow, 1997) found one for r states that scales as $2^{2r}k^2n$. As they point out, improvement of the exponent of k is particularly important, as sequence lengths in real data sets can be long. In all of these cases, the algorithms evaluate the joint compatibility of a given set of characters. They do not search for the largest set of jointly compatible characters. Finding the largest set of jointly compatible characters is still NP-hard. Dress, Moulton, and Steel (1997) have defined a generalized notion of compatibility, which they call strong compatibility. When it holds between all pairs of characters, the Pairwise Compatibility Theorem is applicable, and they can be proven to be jointly compatible. In that case, a clique algorithm would be applicable.

Once the set of jointly compatible characters is found, finding the tree can be done in time linear in the number of compatible characters. Bonet et al. (1998) produce algorithms to refine a given partially-unresolved tree using a set of compatible characters.

Using compatibility on molecules anyway

Although the failure of the Pairwise Compatibility Theorem for nucleotide sequence data seems to force us to use the more difficult perfect phylogeny methods, one can also use compatibility in a simpler way. One can simply score each site as compatible with the given tree or not, and then try to find the tree that maximized the number of compatible sites. This is slower than perfect phylogeny algorithms, but is simpler. But there is another way to escape all these problems and use the Pairwise Compatibility Theorem on nucleotide sequences.

Recall that the rationale for discarding incompatible sites is that they are inferred to be ones that change at a high rate and thus have little information. Even one extra change of state is enough to convict a site in this case. If so, then why do we allow sites that have three different nucleotides present to remain? A site that has some species A, some C, and some T will require two changes of state, even when it is compatible with the tree. If we are interested in using a compatibility framework, I do not see why we would not want to discard that site as well. If we do so, then the only sites on which we base our analysis are going to be those that have two (or fewer) different nucleotides present. And for those, the Pairwise Compatibility Theorem will work! Thus we simply reduce our data to the sites that have two states, use the original Wilson compatibility test, make a compatibility matrix, and find largest cliques. The result will be a procedure that runs quickly (if there are no ambiguities in the data such as gaps in the sequence). Whether it is biologically reasonable will, of course, depend on whether we want to convict a site on that small an amount of evidence.

Chapter 9

Statistical properties of parsimony

To understand why parsimony methods should or should not be used, we need to consider them, not as arbitrary algorithms, but as statistical methods, with investigable statistical properties. Although many justifications for parsimony methods base themselves in philosophical frameworks that are nonstatistical, in this chapter we shall consider the statistical framework only. There are two general ways we could proceed. One would be to ask whether the parsimony method is a known statistical estimator and therefore has the desirable statistical properties of that method. The other is to consider the statistical properties of parsimony directly. We will start with the former approach.

Likelihood and parsimony

All attempts to show that parsimony methods are derivable from a known statistical estimation method have used the same one: maximum likelihood. The first such framework was introduced by Farris (1973b). I have argued (1973b) that it does so by introducing too many quantities that need to be estimated, and that therefore it does not guarantee us that the method will be statistically well-behaved. A second argument by Farris (1977a) introduces fewer quantities, but still too many for comfort.

I have introduced a different argument (Felsenstein, 1981a; see also Felsenstein, 1979) that avoids this problem and yields some insight into character weighting and when parsimony is expected to work. Suppose that we have a set of characters, each of which has two states, 0, and 1. We have a tree that has branches whose lengths in time are given, and we also have branch lengths, which result from multiplying the length of branch i in time by a multiplier, to allow different branches to have different evolutionary rates. In effect, branch length is a pseudo-time scale that reflects the rate of change of the characters. Let the probability of change of character i in branch j of the tree be r_i per unit branch length, and let the branch length be t_j. If we have a symmetric model of change between the two states, it is not hard to show that the net probability, at the end of branch j, that this branch shows a change from 0 to 1 (or from 1 to 0, whichever is appropriate) is

$$\text{Prob}\,(1\,|\,0, t_j) \;=\; \frac{1}{2}\left(1 - e^{-2r_i t_j}\right) \tag{9.1}$$

When $r_i t_j$ is small, this is to a good approximation simply $r_i t_j$, and the probability that the state does not change is approximately $1 - r_i t_j$. When $r_i t_j$ is large, the probability of change to the other state approaches 1/2. Thus we have a symmetrical random process of change between the two states. Once one reaches a state, there is a constant chance of change, and the probability of change, as well as the probabilities of different kinds of change, do not depend on how you reached that state or how long ago you reached it. A random process that has this property is known as a *Markov process*. In more general versions of two-state Markov processes, one can have unequal probabilities of being in the two states, and the rate of change can differ among characters and among branches of the tree.

The likelihood of a tree is the probability of the data given that tree. We will discuss it more extensively in Chapter 16. Let us calculate the likelihood of a tree on a given data set. We shall see that in certain limiting cases, this becomes closely related to the total number of weighted changes of state. That gives a justification of the use of parsimony.

We start by computing the likelihood, L, the probability of the data (given the tree and the model):

$$\begin{aligned} L &= \text{Prob}\,(\text{Data}\,|\,\text{Tree}) \\ &= \prod_{i=1}^{\text{chars}} \sum_{\substack{\text{recon-}\\ \text{structions}}} \left(\frac{1}{2} \prod_{j=1}^{B} \begin{cases} r_i t_j & \text{if this character changes} \\ 1 - r_i t_j & \text{if it does not change} \end{cases} \right) \end{aligned} \tag{9.2}$$

The first product is over characters. We assume that the evolutionary processes that effect character change in different characters are independent, so that the likelihood is simply the product of a series of terms. The terms have different values of i, the index for the characters. Each of these terms in the product is a sum over all possible ways that states can be assigned to the interior nodes of the tree (the hypothetical ancestors). This is indicated by the word *reconstructions* under the summation sign, to indicate this complicated summation over many possibilities. The summation is used because these alternative possibilities are mutually exclusive events.

Within the summation we calculate the probability of the events that occur in that particular reconstruction. The reconstruction starts with a 1/2 because that is the probability of the particular state (whether state 0 or state 1) that occurs at the bottom of the tree. Then there is a term for each branch in the tree. The branches are indexed by j, which runs from 1 to B. We are also assuming that the events in different branches are independent. Given the starting state at the base of each branch, the occurrence of a change is independent of whether there is a change in some other branch. Thus the probability is a product of terms, each $r_i t_j$ or $1 - r_i t_j$, depending on whether there is or is not a change in that branch.

To show the correspondence between this expression and the number of changes of state in parsimony reconstructions, we have to make some more assumptions. We will assume that all the $r_i t_j$ are small. That, in turn, implies that we can approximate the innermost product of (9.2) by

$$\prod_{j=1}^{B} (r_i t_j)^{n_{ij}} \tag{9.3}$$

where n_{ij} is the number of changes of state (there will be 0 or 1 of them) in branch j of character i. This can be done because the terms $1 - r_i t_j$ are then close to 1. To the degree of approximation we need, they can be replaced by 1.

Another approximation involves the terms in the sum over reconstructions. If all the $r_i t_j$ are small, then all the terms in the sum are small, and they will differ a great deal in magnitude. If there are no two of them tied in size, then most of the value of the summation will be contributed by one or a few terms. Terms involving a product of many small quantities $r_i t_j$ will tend to make a tiny contribution compared to terms with one or two of these quantities. Thus we will replace the sum over reconstructions with a single term, involving the one reconstruction that achieves the highest probability for that character. If there are two or more tied terms, we need to multiply by a quantity T_i, the number of tied terms in character i. Though we will not dwell on it, it can be shown that these quantities will not affect the argument that we will make about parsimony.

With a single reconstruction chosen for each character, the likelihood in equation 9.2 now looks much simpler. Dropping the factors of 1/2 that are the same for all trees,

$$L \approx \prod_{i=1}^{\text{chars}} \prod_{j=1}^{\text{branches}} (r_i t_j)^{n_{ij}} \tag{9.4}$$

We take the logarithm, which is useful because it changes products into sums. Finding the tree that maximizes the log of the likelihood is equivalent to finding the tree that maximizes the likelihood, as the larger a quantity is, the larger its logarithm is. Taking the negative of the logarithm gives us

$$-\ln L \approx \sum_{i=1}^{\text{chars}} \sum_{j=1}^{\text{branches}} n_{ij} \left[-\ln(r_i t_j)\right] \tag{9.5}$$

Suddenly, we have reached parsimony. Note that this is simply a weighted parsimony sum. In maximizing the likelihood, we minimize the negative of the log-likelihood. And that quantity is the sum, over all the changes on the tree (whose locations are indicated by the n_{ij}) of penalties of the form $-\ln(r_i t_j)$. Thus we have made enough approximations to prove that the weighted parsimony tree with these weights is always the same tree as the maximum likelihood tree.

The weights

In this case we have a formula for the weights. They are no longer arbitrary but now can be related to probabilities of events. The weight of a given change is now the negative logarithm of the probability that that particular character changes in that particular branch. Note that if the r_i is small for a character, it thereby has a higher weight (as its logarithm is smaller, and therefore the negative of its logarithm is larger). This dependence on the logarithm of rate of change contradicts an "obvious" method of assignment of weights that is often used. It seems natural, if a character has half as large a rate of change as another, to assign it twice the weight. But that would be a weighting function of $1/r_i$, not $-\ln(r_i)$. In fact, it is when a character has a probability of change as low, in a given branch, as the square of the probability of change of another, that we assign it twice the weight. We do that because $\ln(r^2) = 2\ln(r)$. This is actually intuitively sensible: we assign a change in one character twice the weight of another when it has the same probability for one change as two changes in the other character do.

But notice another, rather horrifying, fact about the weights. They depend not just on the rates of change, r_i, but also on the branch lengths, t_j. Thus we are supposed to accord higher weights to changes that occur in shorter branches of the tree. This too makes sense. We find a tree less implausible if the changes occur in long branches rather than in short ones, and this weighting expresses that understanding. However, when we evaluate a tree we typically do not have branch lengths for it. That leaves us somewhat uncertain how to proceed in practice.

Farris (1969b), in the first modern paper on character weighting, discussed which functions of probability of change would be desirable for use as weights. He did not derive weights from a relationship between likelihood and parsimony as done above.

Unweighted parsimony

If the rates of change become very small, another simplification emerges. As they become smaller, the ratio of the weights to each other becomes more equal. In the limit, the method is simply unweighted parsimony. Suppose, for example, that we have three characters, with different rates of change. One of them has two changes, the other two have one change. The $r_i t_j$ and the resulting weights (their negative logarithms) are given in Table 9.1. We can see that two changes in character 2 incur less total penalty than one change in character 3. In this case, a weighted parsimony method might prefer a tree with more changes, provided that they were located in character 2 rather than in character 3. When we consider the same sort of case, identical except with only 10% as great a probability of change

Table 9.1: Probabilities of change and resulting weights for an imaginary case

Character	$r_i t_j$	Changes	Total weight
1	0.01	1	4.605
2	0.01	2	9.210
3	0.00001	1	11.519

in each character (in Table 9.2), characters 2 and 3 now have equal total penalties. With another reduction of the probability of change by a factor of 10 (in Table 9.3), character 3 incurs less total penalty. In fact, it is easy to show that with enough reduction of the rates, all by the same factor, one must in the limit have the ratios of the weights for a single change approach 1. In this case, the ratios of the weights of single changes in characters 1 and 3 started out at 2.5, with reduction by a factor of 10 became 2, and with a further reduction became only 1.75. With enough reduction of rates of change, these ratios approach 1.

Thus unweighted parsimony receives its justification from the assumption of very low rates of change. Note that we did not assume that rates of change in different characters were equal. Unweighted parsimony will make a maximum likelihood estimate of the tree if rates of change (or branch lengths) are small enough, even though the different characters may have very different rates of change. This is a very good property to have, because we are often unable to say what the relative rates of change of different characters will be in advance.

Limitations of this justification of parsimony

The main problem with this justification for parsimony is that it assumes a low rate of change in all characters. That corresponds to our intuition — that if we are trying to find an explanation that minimizes the number of something, we must find that event an implausible one. But the difficulty is that in many data sets, after the parsimony method is used, the number of changes found is too large to be consistent with this view.

Table 9.2: The same case as in Table 9.1 with one-tenth the rate of change in each character

Character	$r_i t_j$	Changes	Total weight
1	0.001	1	6.908
2	0.001	2	13.816
3	0.000001	1	13.816

Table 9.3: The same case as in Table 9.2 with one-tenth less again

Character	$r_i t_j$	Changes	Total weight
1	0.0001	1	9.210
2	0.0001	2	18.421
3	0.0000001	1	16.118

For morphological characters one might imagine that the high rate of observed change arises from ignoring characters that do not vary in our data set or in a larger data set from which it is drawn. But even this supposition will not wash. For if that were the only reason for the observation of so much change, we would expect the change to occur only once per character. In other words, we would expect to find data sets that had all characters perfectly compatible with each other (or most of them perfectly compatible, at any rate). This is not found. If it were, there would be little need to explore different algorithms for inferring phylogenies.

Farris's proofs

The pioneering attempt to connect parsimony with likelihood was by J. S. Farris (1973a). I have discussed (Felsenstein, 1973b) my reasons for believing that his argument does not entirely succeed. Farris's proofs are like the one sketched here, except that they do not sum over all possible reconstructions of the states at the interior nodes of the tree. They also do not need the assumptions that I have made about low rates of change in each character. They arrive at formulas for the character weights that resemble those presented here. What Farris is estimating is not the tree, but the tree together with the states at the interior nodes and at a great many points along the branches (the "evolutionary hypothesis"). This introduces a great many additional quantities that are being estimated.

The strength of these proofs is that, if they are successful, parsimony inherits the known good behavior of likelihood methods. The limitation of Farris's proof is that likelihood often is found to misbehave as the number of quantities being estimated rises, especially if it is proportional to the amount of data. We shall see in the remainder of this chapter a particular misbehavior that characterizes parsimony and that occurs even in some cases to which Farris's proof applies. This suggests that his proofs are not enough to serve as a general justification for parsimony.

I (Felsenstein, 1973b) and Farris (1977a) put forward alternative arguments, ones that assume that the probability of change in each branch of the tree is small for all characters. They are thus similar to the argument above. Like it, they assume rarity of change, and thus they do not prove a general correspondence between maximum likelihood and parsimony estimates of the phylogeny. Goldman

(1990) has discussed Farris's (1977a) argument in some detail, giving a similar proof.

Although I have described Farris's (1973b, 1977a) arguments as based on maximum likelihood, they are actually Bayesian. However Farris (1973b) assumes a flat prior distribution, and then argues that the maximum posterior probability in a Bayesian argument yields the same estimate as maximum likelihood.

No common mechanism

Penny et al. (1994) and Tuffley and Steel (1997) have made an illuminating connection between parsimony and likelihood. They loosened the assumptions about the rate of change to the maximum extent possible. They allowed each character to have a different probability of change in each branch of the tree. This is called the case of *no common mechanism*. There is no rate of change for a branch that applies across all characters, and no rate of change for a character that applies to all branches. Instead, the rate of change is arbitrarily different in each combination of branch and character. Thus when there are n species and p sites, there will be $(2n-3)p$ parameters in all, one for each character in each branch.

In general the proof of the no-common-mechanism result is not obvious, but in one particular case it is easy to see the connection between parsimony and likelihood. This is the case with symmetric change between two states (states 0 and 1) and where there is only one possible assignment of states to interior nodes in the most parsimonious tree. When the likelihood is computed under such a model, it will have a form quite close to equation 9.2:

$$L = \prod_{i=1}^{\text{chars}} \left(\frac{1}{2} \prod_{j=1}^{B} \left\{ \begin{array}{ll} p_{ij} & \text{if this character changes} \\ 1 - p_{ij} & \text{if it does not change} \end{array} \right. \right) \tag{9.6}$$

B is the number of branches in the tree. In the previous case, there was a common mechanism at work in all branches and all characters, with rates of change specific to characters but not to combinations of branches and characters. In that case, the probability of change was given by equation 9.1. In the case of no common mechanism, instead of $r_i t_j$ we have a quantity subscripted by the combination ij. It predicts a net probability of change along the branch, which is some number in the interval $[0, 1/2]$.

As equation 9.1 shows the probability of change to be a monotonic function of the appropriate branch length, maximizing the likelihood with respect to that branch length will yield the same result as maximizing it with respect to the net probability of change p_{ij}, provided that we keep that quantity in the interval $[0, 1/2]$. This invariance is well-known as the "functional property of maximum likelihood." Examining equation 9.6 as a function of one of the p_{ij} shows immediately that it is a multiple of either p_{ij} or $1 - p_{ij}$. It is linear in p_{ij}, so that the maximum of the likelihood with respect to p_{ij} will be either at 0 or at $1/2$.

Thus the p_{ij} for all branches that show a change should be 1/2, and those for the branches that do not change should be 0. The upshot is that (with the extra factor of 1/2, which is for the initial state) the maximum likelihood is

$$L = \left(\frac{1}{2}\right)^{m+1} \tag{9.7}$$

where m is the number of changes of state in the particular reconstruction of states at the interior nodes of the tree. To maximize the likelihood, we find the tree that minimizes the parsimony score. Thus in this case, likelihood and parsimony will always recommend the same tree.

When there is more than one possible reconstruction of the character, things are more complicated. As the likelihood terms that contain one of the p_{ij} are now summed over different reconstructions of that character, the likelihood is still linear in p_{ij}, so that it is still the case that the maximum likelihood values of p_{ij} are either 0 or 1/2 (or else are all values including those). By choosing one reconstruction that is one of the most parsimonious ones and letting those branches that have changes have $p_{ij} = 1/2$ and all others have $p_{ij} = 0$, we can again achieve the likelihood shown above in equation 9.7. In that case, all other reconstructions contribute nothing to the likelihood. What is less obvious is whether, by having more branches have nonzero p_{ij}, and thus having contributions from more possible state reconstructions, we can make the likelihood higher.

The proof by Penny et al. (1994) and Tuffley and Steel (1997) is elegant and not simple, and I cannot explain it here (or anywhere, for that matter). It does rule out higher values of the likelihood in the two-state and multistate cases. In the r-state case, the maximum likelihood is

$$L = \left(\frac{1}{r}\right)^{m+1} \tag{9.8}$$

where m is the number of changes reconstructed by parsimony. As in the two-state case, this maximum value can be achieved by choosing one of the most parsimonious state reconstructions and setting $p_{ij} = 0$ in all branches that do not have a change of state and setting $p_{ij} = (r-1)/r$ in those that do have a change.

The no-common-mechanism result is a remarkable connection between likelihood and parsimony; it is important to know what it does and does not mean. It does show that there is a statistical model in which likelihood and parsimony always infer the same trees. Previously, we knew only that they would choose identical trees when expected rates of change were small. In the no-common-mechanism model, rates of change can be anything but small. As we will see below, in some situations parsimony methods can have undesirable statistical properties such as inconsistency (convergence to the wrong tree as the number of characters increases). Likelihood with the no-common-mechanism model will share these undesirable properties. The number of parameters inferred in that model is high: the product of the number of branches and the number of characters. Thus, as the number of characters rises, so does the number of parameters. This is the "infinitely many parameters" problem that causes misbehavior of likelihood methods in other case as well. Thus the identification of parsimony with likelihood is not enough to ensure that it behaves well.

The no-common-mechanism model has been developed only for cases in which the r states are symmetric. If we have asymmetric change between states whose equilibrium frequencies are unequal, this cannot be accommodated in the present no-common-mechanism model. Thus that model is not totally general. We so far lack a clear understanding of how much further this model can be generalized and still preserve this connection between likelihood and parsimony. For a cautious assessment of the connections between likelihood and parsimony, see the paper by Steel and Penny (2000). An earlier attempt by Sober (1985) to prove, for a 3-species rooted tree, that maximum likelihood will always choose the same tree as parsimony, has been criticized as invalid by Forster (1986). He pointed out that the argument implicitly assumed that the trees compared always had the same interior branch length.

Likelihood and compatibility

We have shown that there is a likelihood justification of parsimony methods when rates of change are small. There is a similar one for compatibility (Felsenstein, 1979, 1981a). Recall that in compatibility methods any character that requires more than one step on a given tree is counted as not fitting that tree, in effect lending no valid phylogenetic support to that tree. The way to have such a character in likelihood inference is to have a possibility that a character is pure noise. This could be so by having it be massively misinterpreted, but a simpler possibility is that it has a very high rate of evolution. In this section, we will argue backwards from the desired result to obtain the model that gives it.

Consider a model with two states, 0 and 1, and probabilities of change given by equation 9.1. We previously explored the effects of having small rates of change. Suppose that for some character the rate of change r_i is very large. The equation then shows that the probability of changing from 0 to 1 in a branch of length t_j is $1/2$, regardless of the size of t_j. It follows that the probability of changing from 0 to 0 is also $1/2$ (as the two probabilities must add up to 1). From this we can quickly show that no matter what the pattern of 1s and 0s in a character, the probability of that pattern is $1/2^n$, where n is the number of species. This is true whatever the tree (as long as at least $n - 1$ of the terminal branches of the tree are nonzero in length).

Suppose that each character has one of two possible rates, r_i and ∞, where r_i is small, and that we do not know in advance which rate it has. If the probability that the character has the infinitely high rate of change is p, then the probability of the observed pattern of data $D^{(i)}$ at that character is

$$\text{Prob}\,(D^{(i)}|T) \;=\; p\left(\frac{1}{2}\right)^n + (1-p)\;\text{Prob}\,(D^{(i)}|T,r_i) \tag{9.9}$$

where the rightmost term has already been given by equation 9.2. Now suppose that of the two terms on the righthand side of equation 9.9, for any given tree, in each character one or the other contributes almost all of the likelihood, and that the term for the high rate is the one that contributes most of the probability whenever there is more than one step in the character. Recall that when its rate of change r_i is small, the contribution of the character to the likelihood is

$$\text{Prob}\,(D^{(i)}|T,r_i) \;\approx\; \frac{1}{2}\prod_{j=1}^{\text{branches}} (r_i t_j)^{n_{ij}} \tag{9.10}$$

If it is always true that for any tree that has two changes in a character, that character is better explained by having a high rate of change, then for any two branches j and k (the ones that have the changes)

$$(1-p)\;\frac{1}{2}\,(r_i t_j)\,(r_i t_k) \;\ll\; p\,\left(\frac{1}{2}\right)^n \tag{9.11}$$

Thus if the term $\frac{1}{2^n}$ is smaller than any one of the rt terms but larger than the product of any two of them, it will turn out that the likelihood for the full set of characters, for any tree, is

$$L \;=\; \prod_{i=1}^{\text{chars}} \begin{cases} \frac{1}{2}(1-p) & \text{if this character does not change on tree T} \\ \frac{1}{2}(1-p)\,r_i t_j & \text{if this character changes only once, in branch } j \\ p\left(\frac{1}{2}\right)^n & \text{if it changes more than once} \end{cases} \tag{9.12}$$

An argument similar to the previous one then shows that the maximum likelihood tree is the one that selects the largest set of compatible characters. It is equivalent to a threshold parsimony method with threshold set to 2. As we have already seen, this finds the largest set of characters that can evolve with only one change. Note that this does not allow a character that has more than two states to be counted as compatible.

This argument justifying compatibility works only when the probability p that a character has infinite rate of change is small.

Parsimony versus compatibility

In general, compatibility methods assume that most characters will change only once, having a low rate of evolution. A small fraction will change quite frequently and thus have almost no phylogenetic information. When this is true, the largest clique of characters will be quite large, and almost no characters will show 2 changes.

Thus compatibility assumes that the quality of characters varies greatly, most of them being quite good. By contrast, parsimony assumes that although rates of change are low, some characters that still have phylogenetic information will show homoplasy.

Strictly speaking, these justifications work only for data sets in which the fraction of characters showing homoplasy is small. If there are many homplasious characters (ones showing parallelisms or reversals), we are not in the situation assumed by these proofs of equivalence between either parsimony or compatibility and likelihood.

Consistency and parsimony

Character patterns and parsimony

Although no property of a statistical estimator is accepted by all statisticians as essential, one of the more important ones is consistency. An estimator is *consistent* if, as the amount of data gets larger and larger (approaching infinity), the estimator converges to the true value of the parameter with probability 1. If it converges to something else, we must suspect the method of trying to push us toward some untrue conclusion. In 1978 I presented (Felsenstein, 1978b) an argument that parsimony is, under some circumstances, an inconsistent estimator of the tree topology. At about the same time James Cavender (1978) found the same worst case for parsimony, though he did not say so very loudly.

In investigating this, we do not try to establish whether parsimony is or is not a maximum likelihood method. We simply accept that it is a statistical estimator of some sort, and try to establish its properties. The easiest way to do this is to consider its behavior in a simple model case. We start with a four-species tree, on which a series of characters are all evolving, independently, according to exactly the same model of evolution. If the characters are sites in a nucleotide sequence, each species can exhibit one of the states A, C, G, or T. There are then $4^4 = 256$ possible outcomes of a site. These range from AAAA to TTTT. They have come to be called *patterns*. If we had characters with two states (0 and 1) instead of four states, we would instead have $2^4 = 16$ possible patterns, ranging from $0000, 0001, 0010, 0011, 0100, \ldots, 1111$.

With four species, there are only three unrooted tree topologies. For each of them we can imagine working out how many changes of state are necessary for that pattern to evolve on that tree. Figure 9.1 shows part of the table for four

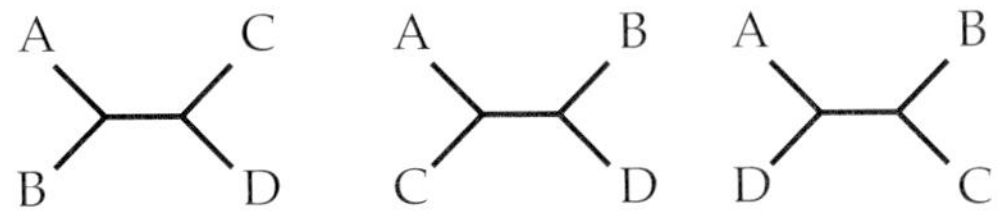

	(A,B),(C,D)	(A,C),(B,D)	(A,D),(B,C)
AAAA	0	0	0
AAAC	1	1	1
AAAG	1	1	1
AAAT	1	1	1
AACA	1	1	1
AACC	**1**	**2**	**2**
AACG	2	2	2
AACT	2	2	2
AAGA	1	1	1
AAGC	2	2	2
AAGG	**1**	**2**	**2**
AAGT	2	2	2
AATA	1	1	1
AATC	2	2	2
AATG	2	2	2
AATT	**1**	**2**	**2**
ACAA	1	1	1
ACAC	**2**	**1**	**2**
ACAG	2	2	2
ACAT	2	2	2
ACCA	**2**	**2**	**1**
ACCC	1	1	1
⋮	⋮		
TTTT	0	0	0

Figure 9.1: The table of nucleotide patterns (some of the 256 possible patterns are shown) for four species. For each, the number of changes of state needed on each of the three possible unrooted tree topologies are shown. Those patterns that have different numbers of changes of state on different tree topologies are highlighted.

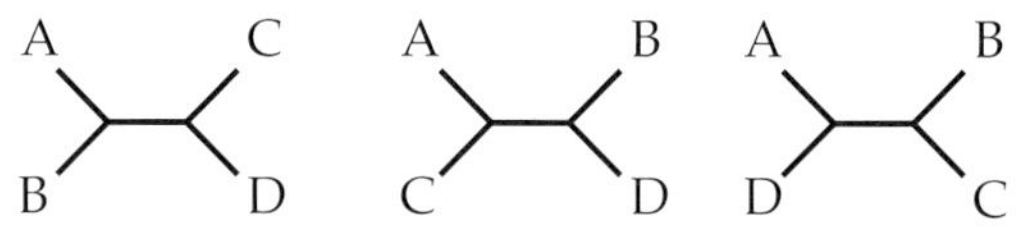

	A C / B D	A B / C D	A B / D C
0000	0	0	0
0001	1	1	1
0010	1	1	1
0011	**1**	**2**	**2**
0100	1	1	1
0101	**2**	**1**	**2**
0110	**2**	**2**	**1**
0111	1	1	1
1000	1	1	1
1001	**2**	**2**	**1**
1010	**2**	**1**	**2**
1011	1	1	1
1100	**1**	**2**	**2**
1101	1	1	1
1110	1	1	1
1111	0	0	0

Figure 9.2: The table of character patterns for a two-state character for four species. For each, the number of changes of state needed on each of the three possible unrooted tree topologies are shown. Those patterns that have different numbers of changes of state on different tree topologies are highlighted.

nucleotide states and the three tree topologies. Note that for most of the nucleotide patterns, there is no difference between the number of changes of state on different tree topologies. In the figure, the rows that do have different numbers of changes of state for different tree topologies are emphasized by larger type size and bolder type face. There are 36 such rows in the full table. Figure 9.2 shows the same table, showing all of it this time, for two states. Here too, the rows that show differences between tree topologies are emphasized. All 6 of them are visible.

Note that the patterns that are of the form $xxyy$, $xyxy$, or $xyyx$ (where x and y are any two symbols) are the only ones that can affect the count of the numbers of changes of state in a parsimony method. These are commonly called *phylogenetically informative* characters, but that terminology is somewhat misleading. The other characters are informative when methods such as distance matrix methods or likelihood methods are used. With nucleotide sequences, patterns like $xxyz$ turn out not to affect a parsimony method. That one, for example, can evolve on any tree with 2 changes of state.

Observed numbers of the patterns

Having made the tables in either of these figures, we can imagine collecting data. Each site (or character) in our data table will show one of the possible patterns. The effect of the site (character) on the outcome of a parsimony method will depend only on which pattern it has, and not at all on where it is in the data. It follows that all that we need to know about a data set, in order to determine which phylogeny it leads us to estimate, is how many times each pattern occurs in it. We can summarize a data set by a list of 256 (or in the 0/1 discrete character case, 16) numbers. The numbers are integers that add up to the size of the data set. Having those numbers, we can figure out from them the number of changes of state that each tree requires. Suppose we know, for example, that pattern AACC requires 1 change on the first tree and 2 changes on the second tree. If we know that there are 12 instances of this pattern observed in our data set, we know that they contribute $12 \times 1 = 12$ changes to the first tree and $12 \times 2 = 24$ changes to the second tree. In this way, we can sum the total changes for each tree, given the list of numbers of times that each pattern is seen.

Note that the three classes of patterns $xxyy$, $xyxy$, and $xyyx$ each can be combined. Since AACC, AAGG, CCTT, and all other patterns of the form $xxyy$ require 1, 2, and 2 changes of state on the three trees, we need not pay attention to anything but the total number of times that $xxyy$ occurs, the total number of times $xyxy$ occurs, and the total number of times $xyyx$ occurs. Let us call these n_{xxyy}, n_{xyxy}, and n_{xyyx}. To discover which tree is favored by parsimony, we need only count the changes of state for the sites with these classes of patterns. Then the changes of state for the first tree are

$$n_{xxyy} + 2n_{xyxy} + 2n_{xyyx} = 2(n_{xxyy} + n_{xyxy} + n_{xyyx}) - n_{xxyy} \tag{9.13}$$

Similarly, the total contribution of changes of state to the other two trees turns out to be

$$2n_{xxyy} + n_{xyxy} + 2n_{xyyx} = 2(n_{xxyy} + n_{xyxy} + n_{xyyx}) - n_{xyxy} \tag{9.14}$$

and

$$2n_{xxyy} + 2n_{xyxy} + n_{xyyx} = 2(n_{xxyy} + n_{xyxy} + n_{xyyx}) - n_{xyyx} \tag{9.15}$$

Since the first term (on the right side) of each of these expressions is the same, it follows that the one that is smallest will depend on which of the three numbers n_{xxyy}, n_{xyxy}, and n_{xyyx} is largest. That determines which tree will be preferred by parsimony. If there is a tie, then two (or three) trees will be tied.

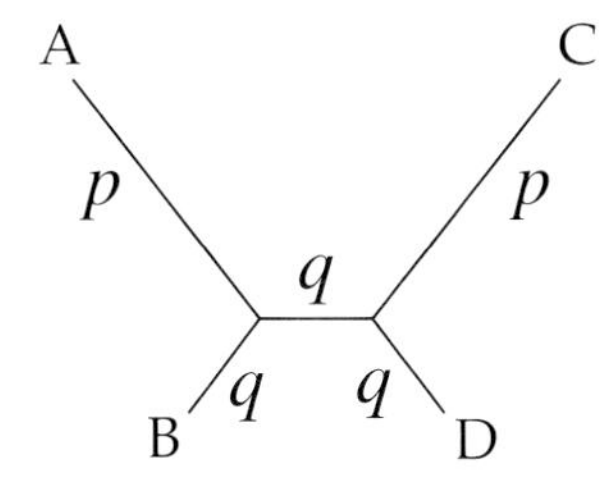

Figure 9.3: A tree with probabilities of change next to each branch. This tree is the example that will used to investigate conditions for parsimony to be consistent.

Observed fractions of the patterns

Note that we could have done the same calculations using frequencies rather than numbers. If we calculated the fraction of all sites (characters) that showed patterns of class $xxyy$, and similarly for the other two classes, we would concentrate on the three observed fractions of characters: $f_{xxyy} = n_{xxyy}/N$, $f_{xyxy} = n_{xyxy}/N$, and $f_{xyyx} = n_{xyyx}/N$, where N is the total number of sites (or characters) in the data. The tree that the parsimony method selects is simply determined by which of the three fractions f_{xxyy}, f_{xyxy} and f_{xyyx} is largest.

Expected fractions of the patterns

Imagine that we knew the true tree, and that it was not merely a tree topology, but a tree with branch lengths. Suppose further that we have a probabilistic model of evolutionary change of the characters. We consider what happens when this same model operates independently in each character. From such a model one can calculate the expected frequencies of each of the 16 (or 256) character patterns.

For the simple case with two states, 0 and 1, and a symmetric model of change between them, we can write the formula for the probability that the character changes in a branch of the tree, given the length of the branch. We have already seen this formula (equation 9.1), but there is a way of avoiding it that is convenient. Suppose that we know for each branch, not its length, but the net probability that the character will change in that branch. Figure 9.3 shows an unrooted tree which will be the critical example for this argument. Next to each branch is the net probability of change along it, in this case either p or q. Note that p and q cannot be greater than 0.5, as even an infinitely long branch has a chance of only 1/2 that the state at the end of the branch is different than the state at the beginning.

We have not specified the position of the root of the tree. It turns out that the probability of any character pattern on this tree is the same no matter where the root is put. With this tree and this simple probabilistic model of character change, we can calculate the fraction of times that each character pattern is expected to be seen. For pattern 0011, we can start at the leftmost of the two interior nodes of the

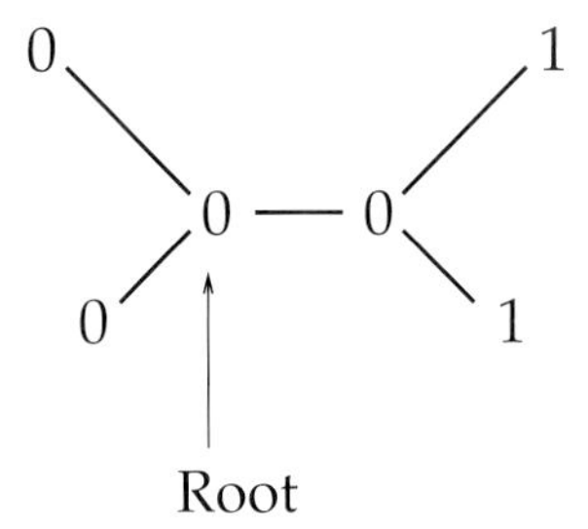

Figure 9.4: One way in which the character pattern 0011 can be achieved. Starting at the root (arrow), which has state 0, there is no change in any of the left three branches of the tree. There is change in both of the right two branches. There are three other scenarios for achieving 0011, corresponding to other assignments of states to the unobserved interior nodes of the tree.

tree. Consider the four possible ways that we could assign states to the left and right interior nodes. These are, of course, 00, 01, 10, and 11. The probability of the pattern 0011 will be the sum of the probabilities of all four of these ways of achieving this pattern. Figure 9.4 shows this scenario. Starting at the root, which is the left interior node of the tree, the probability that we find 0 there is 1/2 (as the model is symmetric, and has presumably been operating in all previous evolution). The probability of no change in the upper-left branch is $(1 - p)$, of no change in the lower-left branch is $(1 - q)$, and of no change in the interior branch is $(1 - q)$. Given the state (0) of the right interior node, there has been change in both of the right branches. These have probabilities p and q.

We are assuming independence of the evolutionary processes in different lineages and in different segments of a single lineage, given the states at the start of each branch. So we can combine these probabilities by multiplication to get

$$\frac{1}{2}(1-p)(1-q)(1-q)pq$$

There are three other combinations of states that could have existed at the unobserved interior nodes. For each we can derive a probability in an analogous way. The result is

$$P_{0011} = \frac{1}{2}\left[(1-p)(1-q)^2pq + (1-p)^2(1-q)^2q + p^2q^3 + pq(1-p)(1-q)^2\right] \tag{9.16}$$

This is the probability of 0011, but there is also the pattern 1100. Together they make up the class of patterns $xxyy$. They have (by the symmetry of our model) equal probabilities, so the total probability of pattern $xxyy$ is given by doubling the quantity in equation 9.16, which simply removes the 1/2.

We can also do the same for the other two patterns classes $xyxy$ and $xyyx$. The result is:

$$\begin{aligned}
P_{xxyy} &= (1-p)(1-q)[q(1-q)(1-p)+q(1-q)p]+pq[(1-q)^2(1-p)+q^2p] \\
P_{xyxy} &= (1-p)q[q(1-q)p+q(1-q)(1-p)]+p(1-q)[p(1-q)^2+(1-p)q^2] \\
P_{xyyx} &= (1-p)q[(1-p)q^2+p(1-q)^2]+p(1-q)[q(1-q)p+q(1-q)(1-p)]
\end{aligned} \tag{9.17}$$

It is not hard to find, after a bit of tedious algebra, which of these is largest. After we compute $P_{xyxy} - P_{xyyx}$, it turns out to simplify into $(1-2q)\left[q^2(1-p)^2 + (1-q)^2p^2\right]$. This is never negative and is positive as long as $q < 1/2$, and either $q > 0$ or $p > 0$, all of which apply except in trivial cases. So pattern $xyyx$ can never have the highest expected frequency. Taking the difference $P_{xxyy} - P_{xyxy}$, the condition that this is positive simplifies (after a struggle) into

$$(1-2q)\left[q(1-q)-p^2\right] > 0 \tag{9.18}$$

The condition $1-2q > 0$ being trivial, this basically simplifies to $q(1-q) > p^2$.

Inconsistency

We now have a condition under which, for our simple tree with an idealized model of evolution, the expected proportion of $xxyy$ patterns is greater than that of $xyxy$ or $xyyx$ patterns. This becomes very relevant when the number of characters (or sites) becomes very large. For it is in that case that the observed frequency of each pattern is expected to converge to its expected frequency. To be more precise, the Law of Large Numbers guarantees us that, as the number of characters grows infinitely large, that the probability becomes 1 that $xxyy$ is the most frequent of these patterns.

Thus in this case, when $q(1-q) > p^2$, we can guarantee that with enough characters, we will arrive at an estimate of the tree that has the correct tree topology. But note what happens when this condition does not hold. In that case, $xyxy$ patterns have the highest expected frequency, and we can guarantee that with enough characters, the tree estimate is certain to be wrong!

Figure 9.5 shows the regions of values of p and q that guarantee consistency or inconsistency.

Note that we are not simply saying that in some cases parsimony methods can give wrong answers. Any method is subject to statistical error. But an inconsistent method becomes more and more certain to give a particular kind of wrong answer as more characters are collected. It is pulling us toward the wrong answer.

The intuitive explanation of what is happening here is fairly simple. With long branches leading to species A and C, the probability of parallel changes that arrive at the same state (which is roughly p^2) becomes greater than the probability of

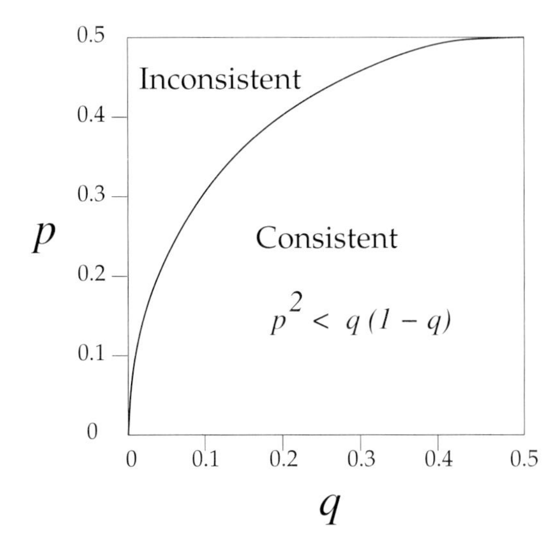

Figure 9.5: Values of p and q that guarantee consistency or inconsistency of the estimate of the tree topology by parsimony. Values of 0.5 of either parameter correspond to infinitely long branches.

an informative single change in the interior branch of the tree (which is roughly q). Thus we have, in these cases, two changes in long branches that are, taken together, actually more probable than one change in a short branch. The situation may be described as one in which "long branches attract" each other. The region of the parameter space in which this occurs is sometimes called the *Felsenstein zone* (Huelsenbeck and Hillis, 1993; this is something like having a black hole named after you).

Penny, Hendy, and Steel (1991) have given the general conditions for inconsistency of parsimony in the 4-species, 2-state model, with all branch lengths allowed to be different.

When inconsistency is not a problem

Note that as you approach the lower-left corner of Figure 9.5, along any diagonal at any angle other than vertically, you ultimately find yourself inside a region of consistency. This happens because if we multiply p by α and q also by α, p^2 is multiplied by α^2 while $q(1-q)$ is multiplied by α (this is more nearly true as q gets small). Thus ultimately, with a small enough value of α, the consistency condition will hold.

This argument is slightly oversimplified. Reducing the rate of change of a character by multiplying it by a factor α is not quite the same as multiplying p by α. For the symmetrical two-state case, equation 9.1 shows how p depends on the rate of change r_i. To be more precise, we should have two different branch lengths, t_1

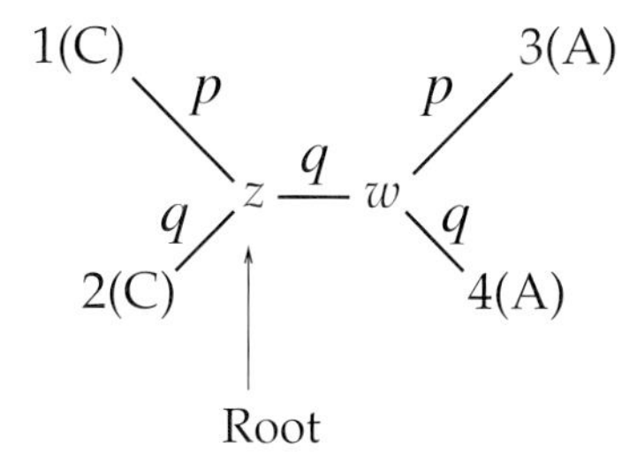

Figure 9.6: A tree showing the species and the nucleotides that are present at its tips and the hypothetical nucleotides that existed at that site at its interior nodes. The net probabilities of change are given next to each branch.

and t_2. If we reduce the overall rate of change r_i, which we here assume is the same for all characters, we can use equation 9.1 to trace out a curve of pairs of (p, q) values, using t_1 to compute p and t_2 to compute q. However the net result is the same, since for small rates of change equation 9.1 shows that the probability of change is approximately $r_i t_j$. Thus simply multiplying p and q by a factor α is approximately correct when all these quantities are small.

In other words, if the tree is short enough, even large ratios of the length of the long to the short branches do not cause inconsistency. This is in accord with what our maximum likelihood derivations showed: When branches are short, parsimony is a maximum likelihood method, and it shares the property of consistency that likelihood methods will have.

The nucleotide sequence case

A similar proof can be made in the case of nucleic acid sequences, which have 4 states. The answer is qualitatively the same, but the region of inconsistency is smaller. This happens because parallel changes along the two long branches are less likely to result in the same base. To investigate the case, we need a probability model of base change. The simplest one, as we will see in Chapters 11 and 13, is the Jukes-Cantor model. This simply assumes that when a base changes, it is equally likely to change to each of the three alternatives. In Chapter 11 we will develop a formula that is the counterpart of equation 9.1. For the moment we don't need it. Instead of branch lengths, we can assign to each branch of the tree its probability p or q of a net change. Equation 11.17 will show how those depend on branch length. All we need to do for the moment is keep in mind that when two changes occur, starting from the same state, there is probability 1/3 that the result is the same.

Figure 9.6 shows an unrooted tree with tips whose pattern is of the class $xxyy$. To calculate the probability of getting this pattern of nucleotides at the tips, we could sum over all 16 possibilities for z and w. For example, the probability of starting with $z = C$ and evolving the pattern given that $w = A$ is $(1/4)(1-p)(1-q)(q/3)(1-p)(1-q)$. The symmetry of the model means that some choices of z and w have the same probability. For example, if z and w are, respectively, C and G, this yields the same probability as if they are respectively C and T, and also the same probability as if they are T and A. Taking all of these into account the probability of the pattern turns out to be

$$\begin{aligned}\text{Prob}\,[CCAA] &= \frac{1}{18}(1-p)(1-q)^2pq + \frac{1}{27}pq^2(1-p)(1-q) + \frac{1}{162}p^2q^2(1-q) \\ &\quad + \frac{7}{972}p^2q^3 + \frac{1}{12}(1-p)^2(1-q)^2q \qquad (9.19)\end{aligned}$$

There are 11 other patterns that make up the pattern type $xxyy$, and each of those has the same pattern probability, so that

$$\begin{aligned}\text{Prob}\,[xxyy] &= (1-p)^2q(1-q)^2 + \frac{2}{3}p(1-p)q(1-q)^2 + \frac{4}{9}p(1-p)q^2(1-q) \\ &\quad + \frac{2}{27}p^2q^2(1-q) + \frac{7}{81}p^2q^3 \qquad (9.20)\end{aligned}$$

Similarly, we can work out the probabilities of the pattern classes $xyxy$ and $xyyx$. These turn out to be

$$\begin{aligned}\text{Prob}\,[xyxy] &= \frac{1}{3}(1-p)^2q^2(1-q) + \frac{2}{9}p(1-p)q^2(1-q) + \frac{4}{27}p(1-p)q^3 \\ &\quad + \frac{1}{3}p^2(1-q)^3 + \frac{2}{9}p^2q^2(1-q) + \frac{2}{81}p^2q^3 \qquad (9.21)\\ \text{Prob}\,[xyyx] &= \frac{1}{81}(1-p)^2q^3 + \frac{2}{3}p(1-p)q(1-q)^2 + \frac{4}{9}p(1-p)q^3 \\ &\quad + \frac{1}{9}p^2q(1-q)^2 + \frac{6}{27}p^2q^2(1-q) + \frac{2}{81}p^2q^3 \qquad (9.22)\end{aligned}$$

Equations 9.20, 9.21, and 9.22 can be shown to be equivalent to expressions that I have published (Felsenstein, 1983a).

The condition for consistency of the estimate of the phylogeny for this case can be obtained, but is not very illuminating:

$$p < \frac{-18\,q + 24\,q^2 + \sqrt{243\,q - 567\,q^2 + 648\,q^3 - 288\,q^4}}{9 - 24\,q + 32\,q^2} \qquad (9.23)$$

Figure 9.7 shows the regions of consistency and inconsistency for this case. Note that near the lower-left corner of the square the region of inconsistency is noticeably narrower than in Figure 9.5. This comes about because, while the condition for consistency in the two-state case is approximately $p^2 < q$, the condition for the Jukes-Cantor DNA model is approximately $p^2 < 3q$, which is easier to sat-

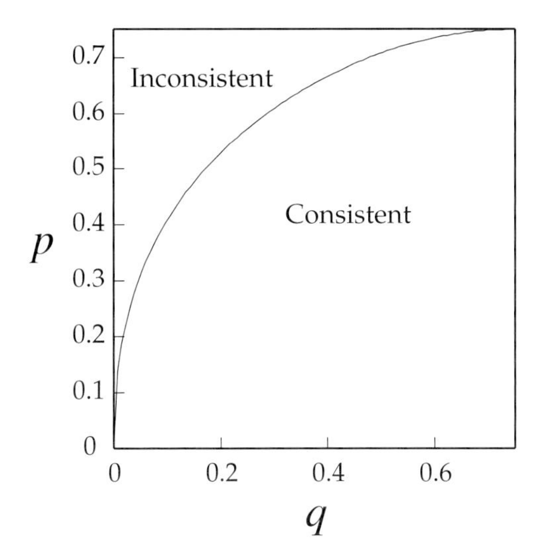

Figure 9.7: The regions of consistency and inconsistency for the Jukes-Cantor DNA model. Branch lengths are expressed in terms of the probability that the state is different at one end of the branch from the other. This probability has a maximum value of 0.75.

isfy. The explanation for this difference is simply that the probability of arriving at the same state by independent evolution in the two-state case is approximately p^2 when p and q are small, while in the Jukes-Cantor case it is $p^2/3$, as parallel changes arrive at the same state only one-third of the time. It should be evident that with more states, the conditions for inconsistency are harder to satisfy.

Steel and Penny (2000) have given a fairly general proof that as the number of possible states in each character rises, ultimately parsimony will be consistent. Their proof encompasses any number of species and fairly asymmetrical models of change.

Other situations where consistency is guaranteed

The above conditions for inconsistency suggest that long branches and unequal branch lengths predispose towards inconsistency. This has been examined more precisely in some special cases.

Kim (1996) examined the consistency of inference of topology in a region of a tree surrounded by four subtrees. He examined a number of special cases, finding that there were exceptions to many proposed generalizations about when parsimony would be consistent. His examples suggested that inconsistency somewhere in the tree could occur more easily, the more species there were in the analysis. Steel (2001) has given a more generalized proof, for the model of symmetric change among r states, that with sufficiently short branches the parsimony and

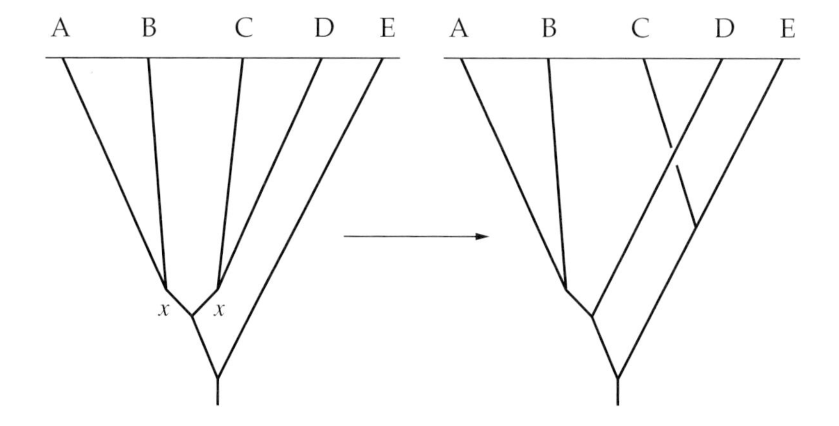

Figure 9.8: A clocklike tree that generates data on which parsimony is not consistent. The true tree is shown and one of the trees that will tend to be found as the number of characters becomes large. There are four of these trees, each with a different lineage attracted to species E.

compatibility methods will be consistent. His conditions for consistency are sufficient conditions—they do not rule out consistency in other cases, with longer branches. They become more difficult to satisfy as the number of species is made larger. I had earlier shown (Felsenstein, 1979) that in the limit as all branch lengths become small at the same rate, parsimony and likelihood will pick the same tree topology.

Does a molecular clock guarantee consistency?

From the examples given above, it would be a tempting generalization to conclude that parsimony can be inconsistent only if there is no molecular clock. The molecular clock, which is discussed in more detail later in this book, is the assumption that lineages have evolved at equal rates. Under a molecular clock, the true tree has branch lengths that cause the tips to all lie equally distant from the root. I have suggested (Felsenstein, 1983b) that imposing a molecular clock is sufficient to assure us that parsimony is consistent.

Hendy and Penny (1989) have shown that this is not so. There are clocklike trees on which parsimony is inconsistent. Their example is of the sort shown in Figure 9.8. One or more of the long branches leading to the species A, B, C, or D becomes attracted to the long branch that leads to species E. Using their Hadamard transform method of calculation (which we will discuss in Chapter 17), they could compute the probabilities of all 32 possible patterns of 0/1 data. They could show that if branch length x is short enough, that many other trees, including ones in which these long branches attract, will be more parsimonious than the true tree.

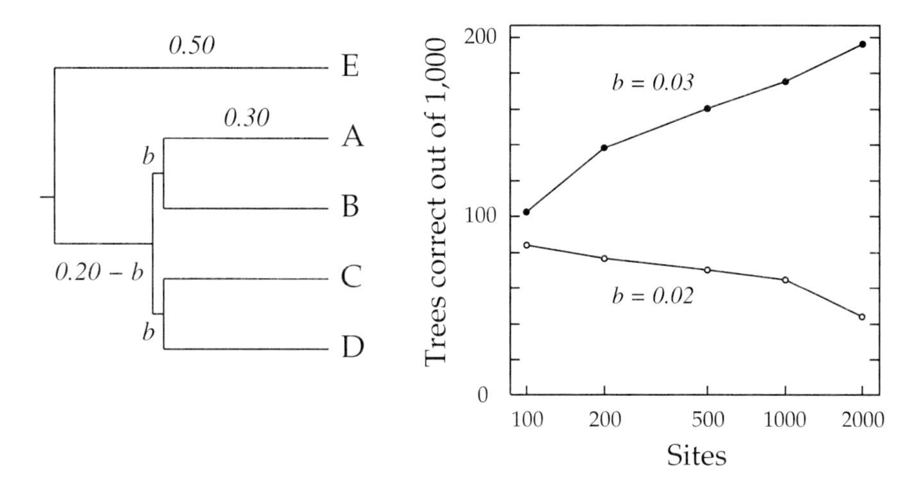

Figure 9.9: A clocklike five-species tree with a pair of short internal branches, making a case similar to that of Figure 9.8. For branch lengths $b = 0.02$ and $b = 0.03$, evolution of various lengths of DNA were simulated using a Jukes-Cantor model. The graph on the right shows the number of times that parsimony found the correct unrooted tree when analyzing 1,000 replicate data sets. For $b = 0.02$, the inference appears to be inconsistent.

Figure 9.9 shows a computer simulation to verify this phenomenon. It involves a clocklike five-species tree with the same topology as the first tree in the previous figure. One pair of internal branches has length $b = 0.02$ or $b = 0.03$, with other branches adjusting their lengths accordingly. I have simulated the evolution of different lengths of DNA molecule, using these branch lengths and a Jukes-Cantor model of evolution. All sites evolved at the same rate.

The graph in Figure 9.9 shows different lengths of DNA for these two values of the branch length, plotting the number of times that the correct tree was obtained when using parsimony to analyze the data. There were 1,000 replicates for each combination of branch lengths and number of sites of DNA. The figure shows how many of these resulted in the correct unrooted tree topology. For the two different branch length values, 0.02 and 0.03, the results are noticeably different. When the true branch length is 0.03, the fraction of the time that parsimony produces the correct tree increases gradually as we consider cases with more and more sites. If it continued to rise to 100%, parsimony would be consistent. When the true branch length is 0.02, the fraction of times that parsimony obtains the true tree falls. This implies that parsimony would be inconsistent in this case. In both bases the curves change slowly—it will take a great amount of sequence data to see the limiting behavior. Presumably this is because correctly and incorrectly interpreted sites are nearly equal in their effects, and it is only with large amounts of data that

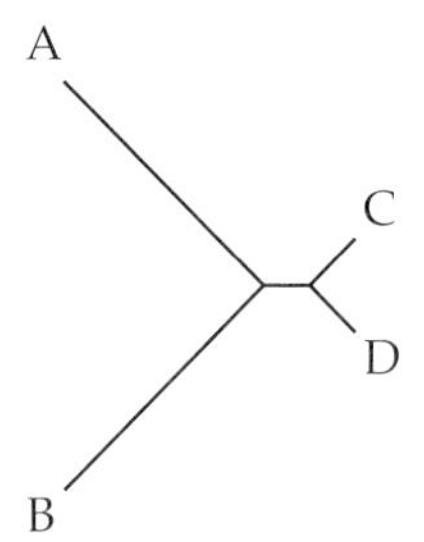

Figure 9.10: A tree in the Farris zone. The two long branches are on the same side of the interior branch. Long branch attraction will cause the topology to be inferred correctly by parsimony more often than by likelihood; in this case, the bias happens to be in favor of the correct topology.

the influence of one overcomes the other. With larger values of the branch length b, the fraction of correct trees rises much more rapidly towards 100%.

This simulation confirms the pattern found by Hendy and Penny (1989). Zharkikh and Li (1993) have given a different proof of Hendy and Penny's result, and have made more extensive numerical study of which five-species trees are inferred inconsistently by parsimony. It will take much more experience before we have a good understanding of the cases of a molecular clock in which parsimony is misleading, but for the moment we can suggest that they are ones in which long branches are separated by short ones, as in the nonclocklike cases.

The Farris zone

Waddell (1995, pp. 391 ff.), Yang (1996), Siddall (1998b), and Steel and Penny (2000) have pointed out a case in which parsimony methods outperform likelihood methods. Siddall named the region of tree space where this effect occurs the *Farris zone*. (This case has also been called the "anti-Felsenstein-zone" by Waddell, 1995, and the "inverse-Felsenstein zone" by Swofford et al., 2001). Figure 9.10 shows a tree that displays this behavior. In this zone, the tree has long branches that are connected to the same node. As Yang (1996) has noted, long branch attraction in a parsimony method helps guarantee that this relationship is correctly inferred. In effect, the inherent bias of parsimony happens to be pointing in the right direction, toward the correct tree topology. Siddall (1998b), Farris (1999), and Pol and Siddall (2001) consider this a case favoring parsimony methods over likelihood. Swofford et al. (2001) argue persuasively that in this case the evidence favoring the correct topology is being given too much weight by parsimony but is evaluated correctly by likelihood.

They note that when the interior branch in Figure 9.10 has infinitesimal length (such as 0.000001), no characters will actually change in that branch. In their sim-

Table 9.4: Properties of the Felsenstein and Farris zones, showing the difference between them

	Felsenstein zone	Farris zone
Parsimony consistent?	No	Yes
Likelihood consistent?	Yes	Yes
Correct tree if very short internal branch		
... using parsimony?	No	Yes
... using likelihood?	Random	Random

ulations, with 10,000 sites of DNA sequence, parsimony infers the correct tree almost 100% of the time, in spite of a total lack of evidence as to the tree topology. By contrast, likelihood infers the correct tree topology 1/3 of the time, as would happen if the tree topology were chosen randomly from among the three possibilities. Pol and Siddall (2001) present simulation results for 10 species trees that have 4 long and 6 short branches. In their trees the long branches are in adjacent pairs. This results in parsimony doing better than likelihood. As they lengthen the long branches, both do worse, as expected. At a certain length, parsimony falls prey to a long branch attraction effect, causing likelihood to outperform it beyond that length. They argue that their results show evidence of "long branch repulsion," but I find their evidence unconvincing. They do not find evidence of inconsistency of maximum likelihood.

One is tempted to think of the two zones as counterparts, one favoring likelihood and distance methods, the other favoring parsimony methods. Siddall (1998b) has viewed them this way. But they are very different phenomena. Table 9.4 shows some of their properties: If the two zones were counterparts, the table would show the same pattern when we switched zones while at the same time switching methods. They are not counterparts. One is a zone where parsimony has the disadvantage of inconsistency, the other a zone where that method has the advantage of bias toward a tree that happens to be correct. Neither of these is true for likelihood in the opposite zone. We will see in Chapter 16 that likelihood methods will be consistent in both zones; likelihood also does not push us to the correct answer even when there is little chance of having any relevant evidence.

Some perspective

The inconsistency of parsimony has been the strongest challenge to its use. It becomes difficult to argue that parsimony methods have logical and philosophical priority, if one accepts that consistency is a highly desirable property. Some schools of statistical thought (notably Bayesians) reject the relevance of consistency and might not be troubled by this argument, though they do insist on use of probabilistic models. If we accept the relevance of consistency, the resulting picture has a pleasing coherence. The arguments of the earlier part of this chapter, as to what assumptions lead a likelihood method to become the same as a parsimony method, suggest that it is low probability of change in the branches of the tree that are needed. These are the same assumptions that work against inconsistency. Likelihood recommends parsimony, and parsimony is consistent if the rate of evolutionary change per branch of the tree is sufficiently small.

If it escapes the clutches of long branch attraction, parsimony is a fairly well-behaved method. It is close to being a likelihood method, but is simpler and faster. It is robust against violations of the assumption that rates of change at different sites are equal. (It shares this with its likelihood *doppelganger*.) Thus parsimony will work particularly well for recently diverged species whose branch lengths are not long.

But when the inconsistency caused by long branch attraction is a problem, then if one wants to continue using parsimony, one will need an alternative logical framework.

Chapter 10

A digression on history and philosophy

It will be useful to pause at this point and ask about the history and philosophical underpinnings of the process of inferring phylogenies. Phylogenies have been inferred by systematists ever since they were discussed by Darwin and by Haeckel, but we concentrate here only on algorithmic methods, those that are well-defined enough to be carried out by a computer. It will not be surprising that these developed only once computing machinery was available.

How phylogeny algorithms developed

Sokal and Sneath

Sustained numerical work on phylogenies started in about 1963. Computers had been available to biologists for about six years by then, in the form of centralized "mainframes" that took input from punched cards and printed the results on paper printouts. During that period a number of lines of work had started that were influential in the development of numerical phylogenies. Chief among these was the development of numerical taxonomy by Peter Sneath and Robert Sokal (Figure 10.1). Starting in the late 1950s, as mechanical methods of computation (initially, punched card machines) first became available to academics, they independently developed numerical clustering methods for biological classification (Michener and Sokal, 1957; Sokal and Michener, 1958; Sneath, 1957a, 1957b). In the original paper of Michener and Sokal (1957), the purpose of the clustering was not simply to classify, but to infer the phylogeny. There is thus a good case to be made that this was the first paper on numerical inference of phylogenies (Figure 10.2 is from that paper). The interpretation as a phylogeny was made by Michener; Sokal saw it as a classification that did not necessarily have any validity as a

Figure 10.1: Peter Sneath (left) in Madison, Wisconsin in 1959, and Robert Sokal (right) at the International Entomological Congress in 1964. Sneath was a medical microbiologist at the University of Leicester and Sokal was an evolutionary entomologist at the University of Kansas when they co-founded numerical taxonomy and introduced and popularized many of its techniques and concepts. (Photos courtesy of Peter H. A. Sneath and Robert R. Sokal.)

phylogeny. In their subsequent work Sokal and Sneath did not interpret their trees as phylogenies.

They soon combined forces to explain and promote the use of numerical classification, especially in their widely noticed book *Numerical Taxonomy* (Sokal and Sneath, 1963). Explaining many methods and providing examples, they argued that classification should be based on phenetic principles, with measures of overall similarity of organisms used to make the classification, without any consideration of phylogenetic relationships. Their book was an important early exposition of clustering methods; it is regarded as a founding work by mathematicians, statisticians, and psychometricians interested in mathematical clustering.

In systematics their views were regarded by many as outrageous and oversimplified; there was an intense debate between them and proponents of the more traditional "evolutionary systematics" school of classification, notably Ernst Mayr and George Gaylord Simpson. Few of the latter were converted and phenetic classification remained a minority view. But a variety of people interested in numerical approaches to evolution were influenced by their methods, which had an important effect in preparing people to think algorithmically.

Numerical approaches to morphological evolution were increasingly being attempted (e.g. Olson and Miller, 1958). The late 1950s also saw great progress in molecular biology. The first wave of protein sequences were determined almost

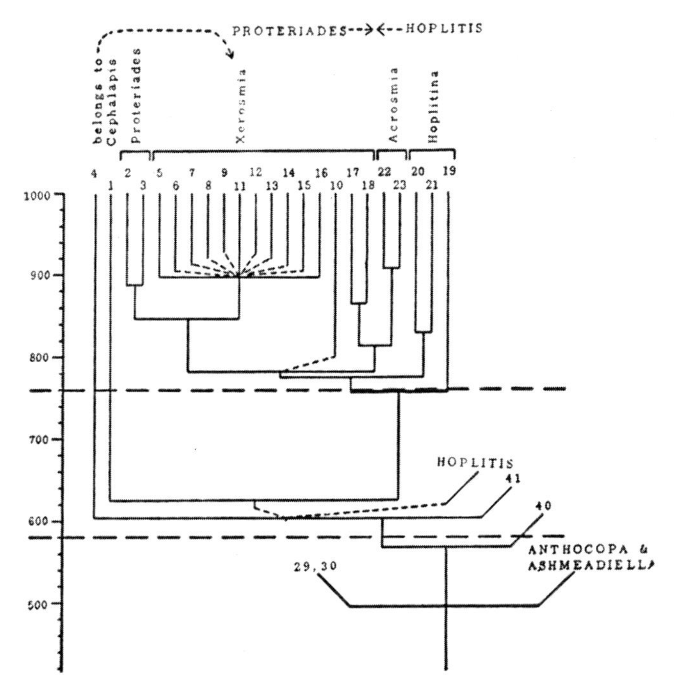

FIG. 5. Diagram of relationships for the genus *Proteriades* obtained by the weighted variable group method.

Figure 10.2: The first phylogeny inferred by numerical methods (Michener and Sokal, 1957). The tree, of morphological characters of bees, was inferred by a clustering method. (Reprinted by permission of the authors and of the Society for the Study of Evolution).

simultaneously with the birth of numerical classification. It was soon recognized that these sequences could be a source of information on the phylogenies of organisms and of genes (Zuckerkandl and Pauling, 1962).

(I am indebted to Robert Sokal and Charles Michener for discussing their joint work. This section is based in part on those recollections.)

Edwards and Cavalli-Sforza

One of the foundations of numerical work on phylogenies was the remarkably creative work of Anthony Edwards and Luca Cavalli-Sforza (Figure 10.3) (Edwards and Cavalli-Sforza, 1963, 1964). Both had been students of the famous statistician and population geneticist R. A. Fisher. They were trying to make trees of human populations from gene frequencies of blood group alleles. It was natural in population genetics to think of gene frequencies as establishing a system of coordinates

Figure 10.3: Luca Cavalli-Sforza (standing) with Anthony Edwards in Italy in 1963, and Anthony Edwards (right, in Cambridge, England, in 1970). Edwards and Cavalli-Sforza were both at the University of Pavia when they collaborated in founding numerical phylogenetics, seeing it as a problem in statistical inference and introducing the parsimony, likelihood, and distance matrix methods for inferring phylogenies. (Left photo by Motoo Kimura, courtesy of Mrs. Hiroko Kimura; right photo by the author.)

in a space, and for evolutionary forces to create a distribution in this space. Assuming a branching, treelike genealogy of human populations, the two co-workers arrived at different methods for inferring the tree. Edwards thought of the space of gene frequencies; he realized that the points that represented the populations could be connected by a tree, and that the branches of the tree would correspond to paths in that space, connecting both the known points (the tips of the tree) and the unknown ones. He wondered whether the best tree would be the one that tied these points together with the minimum amount of string.

Cavalli-Sforza took a different approach. He had been working on divergence of gene frequencies in local populations in the Po valley of northern Italy, where random genetic drift seemed to be the main force bringing about different gene

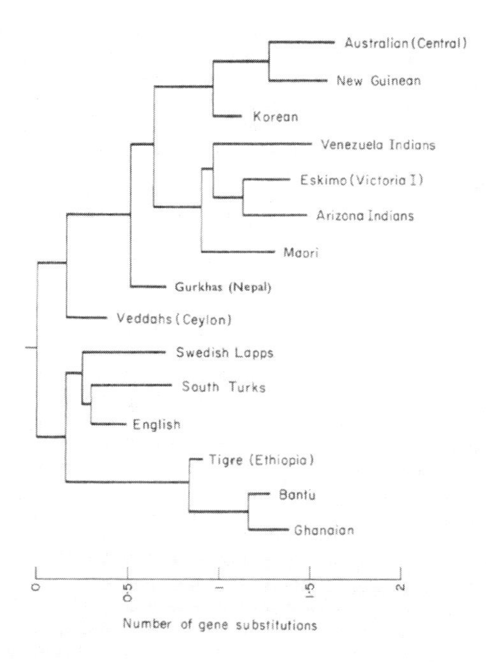

FIG. 5. Tree computed from blood group gene frequencies.

Figure 10.4: The first numerical phylogeny produced by parsimony. The tree of human populations, inferred by Cavalli-Sforza and Edwards (1965) from human blood group polymorphism gene frequencies by parsimony. This tree was presented at the 1963 International Congress of Genetics and was printed in its Proceedings volume two years later, and is reprinted by permission of the authors.

frequencies in different locales. He was interested in pairwise distances between the gene frequencies of local populations, calculated so as to take into account that genetic drift would more easily cause large differences in gene frequency for alleles that had intermediate gene frequencies. As small local effective population sizes would cause a lot of genetic drift, he wanted to allow a treelike genealogy with branch lengths that could vary greatly from branch to branch and would predict the amount of genetic drift on that branch. The result was a least squares method, in which the tree predicted a table of distances between populations, and these were compared to the actual distances by a least squares measure, which was to be minimized.

Having arrived at two different methods, Cavalli-Sforza and Edwards were puzzled and tried to find a way to reconcile them. As both were students of R. A. Fisher, they immediately wondered whether Fisher's great method of maximum likelihood could be employed. If so, it would surely validate one method or the

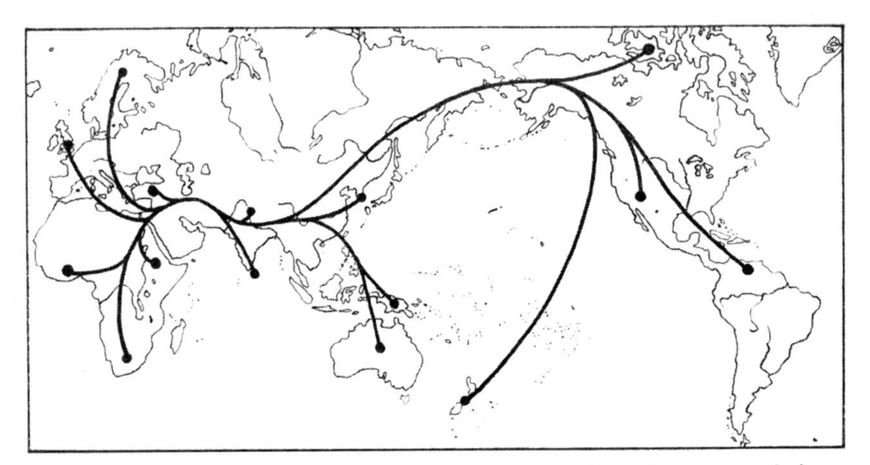

FIG. 1. Topology of the minimum-evolution tree uniting fifteen human populations; constructed on the basis of the frequency of blood-group alleles.

Figure 10.5: The 1963 gene frequency tree inferred by parsimony, as it appears when the tree topology is plotted onto a map of the world (Edwards and Cavalli-Sforza, 1964). Reprinted by permission of the Systematics Association.

other. They worked out a likelihood approach to the problem, and were startled to discover that it was not equivalent to either of their two methods!

In their 1963 abstract, they stated the parsimony method for the first time. In the 1964 paper, they presented their parsimony and likelihood methods, discussing mostly the likelihood method. They deferred the least squares method to a later paper (Cavalli-Sforza and Edwards, 1967). Figure 10.4 shows the tree of human populations that they inferred by parsimony, presented at the International Congress of Genetics in 1963, and which was published in 1965. Figure 10.5 is the same tree, plotted onto a map of the world (and thus losing much of its branch length information). This was the first publication of a parsimony tree.

Edwards and Cavalli-Sforza's paper of 1964 is remarkable in that it introduces the parsimony method, the likelihood method, and the statistical inference approach to inferring phylogenies, all in one paper. It could have introduced the distance matrix method as well, but did not. Although Michener and Sokal (1957) had published earlier, this paper has at least an equal claim to be the founding paper for the numerical inference of phylogenies. However, it did not discuss the algorithmics of parsimony, and the likelihood method turned out to be unworkable in the form they presented.

[Anthony Edwards was kind enough to discuss the history of his work with Luca Cavalli-Sforza with me, and this section is partly based on his recollections. He has also published his account of this work (Edwards, 1996).]

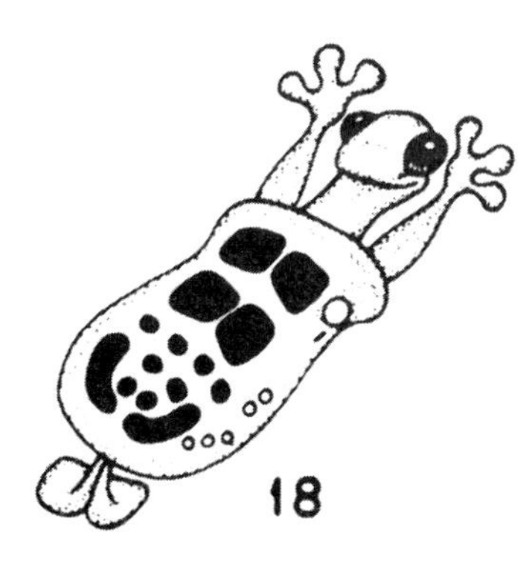

Figure 10.6: Left, Joseph H. Camin in the mid-1970s. Photo courtesy of the Snow Entomological Division, Natural History Museum, University of Kansas, thanks to George W. Byers. Right, a member of the Caminalcules, from the paper by Sokal (1983). Reprinted by permission of the Society of Systematic Biology.

Camin and Sokal and parsimony

Although Edwards and Cavalli-Sforza introduced parsimony, modern work on it springs from the paper of Camin and Sokal (1965). In Sokal and Sneath's (1963) book, they had maintained that phylogenies could not be inferred reliably enough to be the bases for classifications. As part of his studies of classification methods, Sokal wanted to have a set of organisms whose true phylogeny was known. The best way to have them seemed to be to have a practicing systematist evolve some artificial organisms, whose complete history would then be known.

A University of Kansas entomological systematist, Joseph Camin, agreed to do this, and evolved the Caminalcules, cartoon organisms with affinities to schmoos (see Figure 10.6). As they evolved, they were traced from one sheet of paper to another, the ancestors being carefully labeled and filed. Camin and Sokal prepared a data matrix encoding the forms of the Caminalcules as a series of characters with discrete states, as had been advocated in Sokal and Sneath's book. These would be given to students and to systematists and their taxonomic decisions studied. Camin noticed that the students who seemed to be doing best in reconstructing the known phylogeny were those who minimized the number of changes of state of the characters. Sokal and he then published (Camin and Sokal, 1965) a description of the algorithms necessary to evaluate the number of changes of a given tree, and to construct and rearrange the tree to search among topologies. This was the first reasonably complete account of a parsimony method; it was widely noticed and stimulated most further work on parsimony. A FORTRAN computer program written by Roland Bartcher was available (on decks of punched cards) and listings were published (Bartcher, 1966) and some copies were sent to other researchers — though this was the era in which different computers often had incompatible FORTRAN compilers.

Camin and Sokal's parsimony method assumed that one might have a character with multiple states, that these were arranged in a linear order, and that one knew which state was the ancestral one. It also assumed that change was irreversible. Thus the sequence of states would look something like this:

$$-3 \leftarrow -2 \leftarrow -1 \leftarrow 0 \rightarrow 1 \rightarrow 2 \rightarrow 3 \rightarrow 4 \rightarrow$$

Their paper is the first to apply the word "parsimony" to the method, and it states that "the correctness of our approach depends on the assumption that nature is, indeed, parsimonious," an assertion that has been rejected by most subsequent workers.

Camin and Sokal's parsimony method seems to have been derived independently of Edwards and Cavalli-Sforza's. Although Edwards discussed his parsimony method with Sokal at the International Congress of Genetics in 1963, the connection to the Camin-Sokal method may not have been obvious. Sokal points out that Camin, who was not aware of Edwards's work, suggested the criterion to him when they were working on this project.

It is interesting that Sokal, who was skeptical that numerical phylogenies could be of any value, nevertheless played such a central role in the development of numerical phylogenetic methods. One is reminded of the role the statistician Karl Pearson played in the development of quantitative genetics. Pearson did not believe that Mendelian genetics could explain variation in quantitative characters; to bolster this view he and his students worked out consequences of variation at Mendelian loci, in order to show that the resulting patterns did not fit the data. They ended up contributing to the successful explanation of variation in quantitative characters by the effects of Mendelian genetics.

(I am indebted to Robert Sokal for discussions of the history of the Camin and Sokal paper, on which this section is partly based.)

Eck and Dayhoff and molecular parsimony

In the 1960s the molecular sequence data that were available were mostly protein sequences. As these sequences accumulated, Margaret Dayhoff (shown in Figure 10.7) at the National Biomedical Research Foundation began to accumulate them in a database that was distributed in printed form, the *Atlas of Protein Sequence and Structure*. Her work was the ancestor of the modern sequence databases (together with the DNA sequence database project of Walter Goad). From the very outset she took a determinedly evolutionary view of this information. She was interested in developing methods for inferring phylogenies. In the second edition of the *Atlas*, in 1966, she and R. V. Eck (Eck and Dayhoff, 1966) published a description of algorithms for the parsimony analysis of protein sequences. These were based

Figure 10.7: Left, Margaret O. Dayhoff in about 1966. A pioneer of molecular sequence databases and an early researcher on gene families, she and R. V. Eck published in 1966 the first molecular phylogeny produced by numerical methods. Right, Walter Fitch in 1975. He published the first major paper on distance matrix methods, invented the algorithm for counting changes in DNA parsimony, and has made many other contributions to the study of molecular evolution. (Photos courtesy of Edward Dayhoff and Walter Fitch.)

on a model in which each of the 20 amino acids was allowed to change to any of the 19 others in a single step.

They described a sequential addition strategy that used an approximate evaluation of the merit of connecting the sequence to each pre-existing branch, and they also describe what seems to be a phase of local rearrangements. Although the algorithm for counting changes was described only sketchily, this is not only the first molecular sequence parsimony method, it is the first introduction of a parsimony method with unordered states, in which each state is allowed to change to each other in one step.

Fitch and Margoliash popularize distance matrix methods

Just as the parsimony method was most effectively publicized, not by its first description, but by the subsequent publication by Camin and Sokal, so too distance matrix methods were popularized most effectively, not by Cavalli-Sforza and Edwards, but by the work of Fitch and Margoliash (1967). Emanuel Mar-

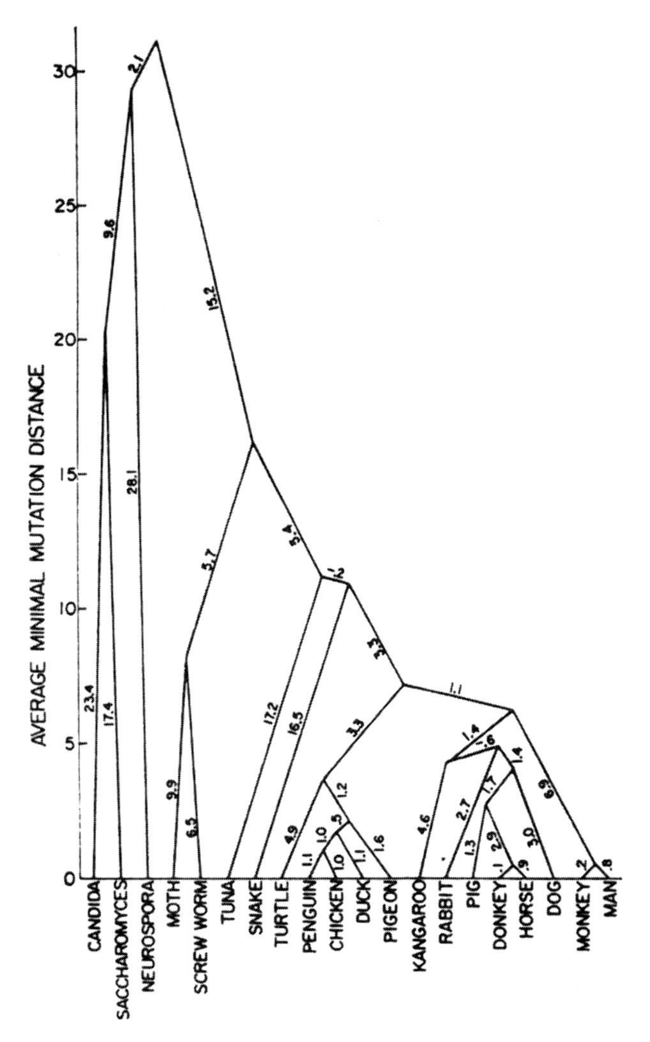

Fig. 3 (right above). A gene phylogeny as reconstructed from observable mutations in several heme-containing globins. See Fig. 2 for details. The percent "standard deviation" (7) for this tree is 1.33.

Figure 10.8: The phylogeny inferred by Fitch and Margoliash (1967) using their distance matrix method on cytochrome sequences. This is one of the first distance matrix phylogenies published. (Reprinted with permission from *Science*, vol. 155, issue 3760, page 282. Copyright 1967 American Association for the Advancement of Science.)

goliash was interested in analyzing evolutionary relationships of cytochrome c sequences, which were becoming numerous. Walter Fitch (Figure 10.7), a biochemist at the University of Wisconsin, developed a distance matrix method based on least squares. The distances were fractions of amino acids different between the particular pair of sequences. The least squares was weighted, with greater observed distance given less weight. Although not discussed in explicitly statistical terms, this was implicitly based on the realization that large distances would be more prone to random error owing to the stochasticity of evolution.

Fitch and Margoliash described not only the least squares criterion that they used, but some of the details of the algorithm. This involved a method of clustering populations based on their distances, which was justified on grounds of the molecular clock. In subsequent references to this work, "the Fitch-Margoliash method" has sometimes been considered to be any method making use of their weighted least squares criterion to choose among trees, and sometimes it has been considered to be their detailed algorithm. Thus, sometimes the "Fitch-Margoliash method" has been criticized as sensitive to departure from a molecular clock, and sometimes it has been praised for its insensitivity to the assumption of a molecular clock. Both views are correct, depending on which "Fitch-Margoliash method" is being discussed.

I prefer to refer to the Fitch-Margoliash *criterion*, the weighted formula. Their detailed algorithm is not widely used, but the criterion has prospered. Figure 10.8 shows the phylogeny published in Fitch and Margoliash's original paper: It is remarkably accurate, with some problems involving rattlesnakes and primates.

Some further confusion has resulted from some comments in the Fitch and Margoliash paper identifying parsimony as the basis of their method; it is not at all a parsimony method.

It is interesting that Fitch, after introducing a major distance matrix method, went on to make important developments in application of parsimony to molecular sequences, while Dayhoff, who had done the first molecular parsimony paper, went on to use distance matrix methods for most of her subsequent work.

Wilson and Le Quesne introduce compatibility

The noted behavioral ecologist E. O. Wilson, an active student of systematics of ants, contributed to the development of numerical phylogenetics in a brief note (Wilson, 1965), which showed how one could test whether two 0/1 characters were compatible, in the sense that both could evolve without reversals or parallelism on the same phylogeny. This was the test that looks across species to see whether all four combinations of the two characters (00, 01, 10, and 11) occur. If they do not, the two characters are compatible. Camin and Sokal (1965) had a more complex way of assessing compatibility and used it in some of the steps of their parsimony method.

A chemical engineer who was an amateur entomologist, Walter Le Quesne, built on this to suggest that one should find the tree that had as many characters as possible compatible with it (Le Quesne, 1969). He did not use the word "compatibility" or actually state the criterion explicitly, though he made it clear that the number of "uniquely derived characters" was the measure of the extent to which a proposed phylogeny fit the data. Like Camin and Sokal, he used a table of compatibilities between all pairs of characters. The algorithm he used was approximate, involving dropping those characters that were incompatible with the most other characters. This is not guaranteed to find the largest possible set of mutually compatible characters. The detailed algorithmics of doing so were worked out later, by George Estabrook and his colleagues (see Chapter 8).

Jukes and Cantor and molecular distances

Although Fitch and Margoliash (1967) had made the first molecular application of distance matrix methods, they used a distance that was a simple fraction of amino acids that differed between the species. This was uncorrected for the effects of multiple replacements. The first distance correcting for this was the DNA sequence distance of Jukes and Cantor (1969). They did so in the midst of an extensive discussion of the evolution of proteins. In that era, it was hard to get methodological papers published, but much easier to publish extensive discussions of data, so no separate paper was written discussing the formulas.

The authors were at the University of California, Berkeley. Thomas Jukes was a pioneer of molecular evolution, one of the great experts on the evolution of the genetic code. Charles Cantor, who later became well-known for his work in developing pulse-field gel electrophoresis, was at that time Jukes's graduate student.

Farris and Kluge and unordered parsimony

Although Camin and Sokal had given impetus to work on parsimony, their method assumed irreversible change. Most users of morphological and molecular parsimony would prefer not to assume either irreversibility, or that it was known which character state was the ancestral one. However, the barrier to having a parsimony method that allowed reversible change and did not assume that the ancestral state was known was mostly algorithmic. Eck and Dayhoff (1966) had apparently had an algorithm but did not explain it precisely. With Camin-Sokal parsimony it is easy to assign states to hypothetical ancestors. If a node has two descendants, whose states are s_1 and s_2, then one assigns to the node the state that is the most recent common ancestor of those two states. For example, if the states are ordered $0 \rightarrow 1 \rightarrow 2 \rightarrow 3$ and 0 is the ancestral state, and the two descendants have states 2 and 3, then their common ancestor must have state 2 in a parsimony reconstruction.

With unordered parsimony the matter is much less clear, and this stymied development of these methods. This problem was overcome by Kluge and Farris

Figure 10.9: James S. Farris in 1983. His work in 1969 and 1970 gave algorithms for counting the number of changes on a tree with characters from a linear scale that did not assume that the ancestral state was known. He gave the first clear description of numerical character weighting and has made many other contributions to inference of phylogenies. He has been the main figure in founding the Willi Hennig Society and persuading phylogenetic systematists that the parsimony criterion plays a central role in the philosophical justification of their field. (Photo courtesy of Vicki Funk.)

(1969) and Farris (1970) in two papers that presented algorithms for reconstructing changes on a given tree, as well as algorithms for searching among trees for the most parsimonious tree. They named the criterion *Wagner parsimony* in honor of Herb Wagner, whose *groundplan divergence method* (Wagner, 1961) helped stimulate work on phylogeny algorithms.[1] Arnold Kluge was a faculty member of the Museum of Zoology and the Department of Zoology of the University of Michigan, in Ann Arbor. J. S. Farris (shown in Figure 10.9) was a student in that department, although by the time the papers were published he had become a faculty member in the Department of Biological Sciences of the State University of New York at Stony Brook.

Kluge and Farris described algorithms, which were primarily due to Farris, for evaluating the smallest number of changes of state required by a data set on a

[1]Wagner may also be the only contemporary systematist who has been mentioned in a Hollywood film: *A New Leaf* (1971), starring Walter Matthau and Elaine May.

given tree. Farris's algorithm is similar to the one later introduced by Fitch (1971) except that it is limited to states that are arranged in a linear order. Kluge and Farris also described a sequential addition strategy for adding species to a tree.

Farris (1969b) also gave the first clear description of numerical weighting procedures, introducing a "successive weighting" method that reconsidered the weights as the tree construction proceeded.

Fitch and molecular parsimony

It remained for Walter Fitch (1971) to provide an algorithm for evaluating the number of changes of state on a tree when there were four unordered states, namely the states A, C, G, and T in a DNA sequence. We have already described Fitch's algorithm. His paper completed the description of methods for construction of phylogenies from nucleotide sequences by parsimony.

Further work

It is interesting to note that, from the first papers by Sneath and by Sokal to Fitch's work of 1971, only 14 years had elapsed (from Edwards and Cavalli-Sforza's 1963 abstract, only 8 years). Although many of these authors were unaware of each others' work, the nearly-simultaneous rise of computers and of molecular biology had created the conditions for numerical phylogenetic techniques to be introduced and rapidly applied.

For the subsequent history of the field, the reader is referred to the references in this book. For some sense of the bitter controversies that arose, see the book by Hull (1988), the reaction to it (Farris and Platnick, 1989), and my own brief account (Felsenstein, 2001a).

What about Willi Hennig and Walter Zimmerman?

The story I have told above is not a very well-known one, though it reflects the influences as I remember them. Many systematists are likely instead to attribute the development of parsimony methods to Willi Hennig (1950, 1966) rather than to Edwards and Cavalli-Sforza or Camin and Sokal. They feel that parsimony is introduced implicitly, or even explicitly, in Hennig's book.

Hennig is justly famous for his strong and clear advocacy of phylogenetic classification and for clearly stating a method for reconstructing phylogenies based on morphological characters. His methods spring from the earlier paper of the botanist Walter Zimmerman (1931). An account of Zimmerman's life and his phylogenetic work is given by Donoghue and Kadereit (1992).

Hennig was the major advocate of monophyletic classification, and his work had an important effect in clarifying thought about classification and inferring phylogenies. However, neither he nor Zimmerman specified what to do if there was conflict between evidence from different characters. Neither introduced the parsimony method, or any other algorithmic approach. Hennig (1966, p. 121) did say that "it becomes necessary to recheck the interpretation of [the] characters" to determine whether parallelism has occurred, or whether the characters in different species are not homologous. But he does not give any specific algorithm. He does say (1966, p. 121) that "the more certainly characters interpreted as apomorphous (not characters in general) are present in a number of different species, the better founded is the assumption that these species form a monophyletic group." (Farris, Kluge, and Eckardt (1970) argue, giving the original German, that Hennig ought to have been translated as saying that "the more characters certainly interpretable as apomorphous ..."). In any case, Hennig gave no numerical method for assessing this.

Some systematists have asserted that the parsimony method is implied by Hennig's (1966, p. 121) "auxiliary principle." For instance, Farris (1983, p. 8) says that

> I shall use the term in the sense I have already mentioned: most parsimonious genealogical hypotheses are those that minimize requirements for ad hoc hypotheses of homoplasy. If minimizing ad hoc hypotheses is not the only connotation of "parsimony" in general useage, it is scarcely novel. Both Hennig (1966) and Wiley (1975) have advanced ideas closely related to my useage. Hennig defends phylogenetic analysis on the grounds of his auxiliary principle, which states that homology should be presumed in the absence of evidence to the contrary. This amounts to the precept that homoplasy ought not be postulated beyond necessity, that is to say parsimony.

Hennig's discussion of his auxiliary principle is concerned with the case in which "only one character can certainly or with reasonable probability be interpreted as apomorphous." He was concerned with whether one ought to infer a relationship based only on a single character, and says (1966, p. 121) that

> In such cases it is impossible to decide whether the common character is indeed synapomorphous or is to be interpreted as parallelism, homoiology, or even as convergence. I have therefore called it an "auxiliary principle" that the presence of apomorphous characters in different species "is always reason for suspecting kinship [i.e. that the species belong to a monophyletic group], and that their origin by convergence should not be assumed a priori" (Hennig 1953). This was based on the conviction that "phylogenetic systematics would lose all the ground on which it stands" if the presence of apomorphous characters in different species were considered first of all as convergences (or parallelisms), with proof to the contrary required in each case.

One can have considerable sympathy for Hennig's position here, without interpreting it as a rule for reconciling conflicts among characters. Indeed, in this case it is directed at cases where there is only one character providing the evidence, and hence no possible conflict. Hennig is concerned with whether one ought to accept the evidence of one character, where there is no other character providing support, and concludes that to be self-consistent, one ought to accept its evidence at face value. It is not obvious how to get from this "auxiliary principle" to the parsimony criterion.

Farris, Kluge, and Eckardt (1970) were the first authors to attempt a formal connection between Hennig's methods and numerical parsimony methods. After careful discussion of other steps in the logic, they confronted the issue of how to deal with character conflict. They cited as their axiom AIV Hennig's statement, cited above, that the more apomorphic (derived) characters a group shares, the better founded is the assumption that it is monophyletic. But then they note (Farris, Kluge, and Eckardt, 1970, p. 176) that

> Unfortunately, AIV is not sufficiently detailed to allow us to select a unique criterion for choosing a most preferable tree. We know that trees on which the monophyletic groups share many steps are preferable to trees on which this is not so. But AIV deals only with single monophyletic groups and does not tell us how to evaluate a tree consisting of several monophyletic groups. One widely used criterion—parsimony—could be used to select trees. This would be in accord with AIV, since on a most parsimonious tree OTUs [tips] that share many states (this is *not* the same as the OTUs' being *described* by many of the same *states*) are generally placed together. We might argue that the parsimony criterion selects a tree most in accord with AIV by "averaging" in some sense the preferability of all the monophyletic groups of the tree. Other criteria, however, may also agree with AIV.

This honest assessment may serve as a caution to those who wish to derive parsimony directly from Hennig's work.

Hennig and Zimmerman did not invent parsimony. But they did put forward clear principles for inferring phylogenies when there was no conflict between different characters. And they were the primary figures in placing monophyletic classification at the forefront of taxonomic thinking. As such, they played an essential role in preparing systematists for algorithmic methods.

Different philosophical frameworks

This book has been written from a statistical viewpoint. Methods have been evaluated according to their properties as statistical estimators, with due consideration of criteria such as consistency. There are many scientists (particularly systematists) who reject this as the proper framework for evaluating methods of inferring phylogenies. It is worth briefly examining their reasoning, as otherwise the reader might mistake these frameworks for a statistical one. These nonstatistical views have tended to be held by some systematists of the "phylogenetic systematics" school.

Hypothetico-deductive

Many of the early expositions of phylogenetic systematics in English adopted a hypothetico-deductive view. According to it, characters falsify potential phylogenies if they cannot evolve in a unique and unreversed fashion on them. We can also say that two characters falsify each other if there is no tree on which they can both evolve in unique and unreversed fashion.

In an influential article, Wiley (1975) identified this approach with the scientific methods advocated by Karl Popper (1968a, b), and with Hennig's phylogenetic systematics. Although he mentioned parsimony only in passing, he declared (p. 243) that "the phylogenetic hypothesis which has been rejected the least number of times is preferred over its alternates."

A more detailed discussion of this view was given by Gaffney (1979), who derived parsimony from the hypothetico-deductive method, which he describes as exemplified in the work of Popper (1968a, b). He says that "the use of derived character distributions as articulated by Hennig (1966) appears to fit the hypothetico-deductive model best." When he deals with character conflict, Gaffney (1979, pp. 98–99) finds parsimony to be directly derivable from his hypothetico-deductive approach:

> In any case, in a hypothetico-deductive system, parsimony is not merely a methodological convention, it is a direct corollary of the falsification criterion for hypotheses (Popper, 1968a, pp. 144–145). When we accept the hypothetico-deductive system as a basis for phylogeny reconstruction, we try to test a series of phylogenetic hypotheses in the manner indicated above. If all three of the three possible three-taxon statements are falsified at least once, the least-rejected hypothesis remains as the preferred one, not because of an arbitrary methodological rule, but because it best meets our criterion of testability. In order to accept an hypothesis that has been successfully falsified one or more times, we must adopt an *ad hoc* hypothesis for each falsification .... Therefore, in a system that seeks to maximize vulnerability to criticism, the addition of *ad hoc* hypotheses must be kept to a minimum to meet this criterion.

To Gaffney (1979, p. 98) this ought not be a controversial matter: "It seems to me that parsimony, or Ockham's razor, is equivalent to 'logic' or 'reason' because any method that does not follow the above principle would be incompatible with any kind of predictive or causal system."

Eldredge and Cracraft (1980, p. 69) are careful to point out that

> "Falsified" implies that the hypotheses are proven false, but this is not the meaning we (or other phylogenetic systematists) wish to convey. It may be that the preferred hypothesis will itself be "rejected" by some synapomorphies.

The hypothetico-deductive approach to parsimony is also the basis for the discussion in the book by Wiley (1981, p. 111):

> In other words, we have no external criterion to say that a particular conflicting character is actually an invalid test. Therefore, saying that it is an invalid test simply because it is unparsimonious is a statement that is, itself, an ad hoc statement. With no external criterion, we are forced to use parsimony to minimize the total number of ad hoc hypotheses (Popper, 1968a: 145). The result is that the most parsimonious of the various alternates is the most highly corroborated and therefore preferred over the less parsimonious alternates.

It is also invoked by Farris (1983, p. 8):

> Wiley [(1975)] discusses parsimony in a Popperian context, characterizing most parsimonious genealogies as those that are least falsified on available evidence. In his treatment, contradictory character distributions provide putative falsifiers of genealogies. As I shall discuss below, any such falsifier engenders a requirement for an ad hoc hypothesis of homoplasy to defend the genealogy. Wiley's concept is then equivalent to mine.

One might note that these discussions do not distinguish clearly between parsimony and compatibility methods. With small numbers of species, there is no difference between these methods. (For example, with 0/1 characters where ancestral character states are not specified, parsimony and compatibility methods will be identical unless there are at least 6 species.) When Wiley (1981, p. 111) speaks of accepting the hypothesis that "requires the fewest ad hoc hypotheses about invalid character tests," we are faced with the issue of how to count invalid character tests. If we count an entire column of the character state table (a character which can take alternative character states) as valid or invalid, then in maximizing the number of valid tests we are carrying out a compatibility method. However, to most phylogenetic systematists a "character" is a unique derivation of a character state. When a character state arises three times on a phylogeny, the issue is whether we are to count that as one invalid character test or two, and whether the decision is implicit in the works of Popper, William of Ockham, or Hennig. This question is not directly dealt with in any of the philosophical writings of phylogenetic systematists.

Phylogenetic systematists have tended to back parsimony and denounce compatibility. This seems to come, not from any philosophical principle, but from the feeling that compatibility discounts a character's value too rapidly, that there is still good information to be had from characters that have been observed to change more than once on a tree. It has also been a result of the greater readiness of advocates of parsimony to ally themselves with phylogenetic systematists in the taxonomic wars.

Logical parsimony

Beatty and Fink (1979) took a different approach to the logical foundations of parsimony methods. They discussed the application of Popper's framework and were skeptical that it was the proper justification for parsimony. They concluded (p. 650) that

> We can account for the necessity of parsimony (or some such consideration) because evidence considerations alone are not sufficient. But we have no philosophical or logical argument with which to justify the use of parsimony considerations — a not surprising result, since this issue has remained a philosophical dilemma for hundreds of years.

In effect they propose relying on parsimony as its own justification, though they do suggest that the ultimate criterion is predictiveness of classifications, and that this will be settled by empirical experience.

Kluge and Wolf (1993, p. 196) seem to come to the same conclusion. Listing a series of methods that they have been criticizing, they comment:

> Finally, we might imagine that some of the popularity of the aforementioned methodological strategies and resampling techniques, and assumption of independence in the context of taxonomic congruence and the cardinal rule of Brooks and McLennan (1991), derives from the belief that phylogenetic inference is hypothetico-deductive (e.g. Nelson and Platnick, 1984: 143–144), or at least that it should be. Even the uses to which some might put cladograms, such as "testing" adaptation (Coddington, 1988), are presented as hypothetico-deductive. But this ignores an alternative, that cladistics, and its uses, may be an abductive enterprise (Sober, 1988). We suggest that the limits of phylogenetic systematics will be clarified considerably when cladists understand how their knowledge claims are made (Rieppel, 1988; Panchen, 1992).

Kluge and Wolf have thus cut loose from the hypothetico-deductive framework, but they continue to consider parsimony as the foundation of their inferences. Their position can be described as a "logical-parsimony" view, as they take parsimony itself as the basic principle, rather than deriving it from other (Popperian, falsificationist, or hypothetico-deductive) arguments.

Sober (1988), whom Kluge and Wolf cite with approval, does not take parsimony as its own justification, but justifies parsimony in terms of statistical inference, presenting a derivation that he believes shows that parsimony is generally a maximum likelihood method. He is quite pointedly critical of Popperian falsificationism. His basic criticism (p. 126) of Popper is that

> Popper's philosophy of science is very little help here, because he has little to say about *weak* falsification. Popper, after all is a hypothetico-*deductivist*. For him, observational claims are deductive consequences

> of the hypothesis under test .... Deductivism excludes the possibility of probabilistic testing. A theory that assigns probabilities to various possible observational outcomes cannot be strongly falsified by the occurrence of any of them. This, I suggest, is the situation we confront in testing phylogenetic hypotheses. (AB)C is logically consistent with all possible character distributions (polarized or not), and the same is true of A(BC). [Emphasis in the original]

Thus, cut loose from a Popperian foundation, parsimony must either rely on a statistical justification, or stand on its own. Sober chooses the former; Kluge and Wolf, the latter.

Logical probability?

More recently Kluge (1997a) has preferred to describe his position as using "logical probability." In spite of the name, it is distinct from any parametric statistical framework. Siddall and Kluge (1997) have argued against "probabilism," the statistical approach to inferring phylogenies, identifying it as "verificationist," whereas they prefer to be "refutationist." Kluge (1997a, b, 1998) prefers to base inferences on the "degree of corroboration," a measure due to Karl Popper. Popper's formula includes terms such as $\mathrm{Prob}\,(D\,|\,T)$ and $\mathrm{Prob}\,(D)$, where D is the data, and T the hypothesis about the tree. (I have omitted the symbol b for the "background knowledge" because it appears in every term.) The first term is the likelihood. The second cannot be computed unless we sum over all possible trees, weighting each by its prior probability. Thus Popper's formula assumes a Bayesian inference framework, as only in that case are prior probabilities of trees assumed to be available. As Popper was an opponent of Bayesianism (Eliott Sober, personal communication) his corroboration formula seems fundamentally at odds with his other views.

De Queiroz and Poe (2001) and Faith and Trueman (2001) have argued against Kluge's use of Popper's measure of degree of corroboration. De Queiroz and Poe conclude that likelihood is compatible with Popper's approach, but that parsimony can only be justified by it if further assumptions allow us to compute the relevant probabilities. They did not discuss whether Popper's measure requires a Bayesian framework, but they do note a statement by Popper (1959) that likelihood is an adequate measure of the degree of corroboration when the term $\mathrm{Prob}\,(D)$ is small enough to be ignored. They argue that $\mathrm{Prob}\,(D)$ can be ignored as it does not affect which hypothesis is preferred. Their view has been opposed by Kluge (2001), who quotes Popper (1959) saying that he intends his calculation of corroboration to be applied only to "the severest tests we have been able to design." Kluge cites Tuffley and Steel's no-common-mechanism result as establishing a direct connection between parsimony and likelihood. He does not give any direct argument that evaluating parsimony constitutes the severest test available.

Faith and Trueman (2001) make a broader argument that Popper's corroboration formula is compatible with the use of many measures of goodness-of-fit, including likelihood, parsimony and others. They reject the notion that there is a direct connection between Popperian corroboration and parsimony. They pay considerable attention to the term for $\mathrm{Prob}\,(D)$. They do not see it as requiring a Bayesian approach, but propose that the PTP randomization procedure of Archie (1989) and Faith and Cranston (1991) be used to evaluate it, as the probability that the goodness-of-fit for a tree would occur when the data are randomized so that phylogenetic structure is eliminated. This randomization procedure is discussed further in Chapter 20. It is not obvious to me that Popper's term for $\mathrm{Prob}\,(D)$ is intended to allow this type of randomization, rather than a Bayesian calculation.

In Kluge's framework, shared derived states (synapomorphies) are regarded as improbable when not predicted by a tree, and thus the tree that requires the fewest of them has the highest value of Popper's corroboration measure. De Queiroz and Poe point out that these probabilities cannot be calculated unless more is known about probabilities of change in the characters in various branches of the tree. Faith and Trueman (2001), in a similar discussion, point to $\mathrm{Prob}\,(D)$ as ill-defined in the logical parsimony framework. I would add that it is not clear that it can be calculated in any framework other than a Bayesian one.

Consideration of the cases in which parsimony is inconsistent will make it apparent that the probability of a synapomorphy given the wrong tree can sometimes be higher than its probability given the right tree. Therefore, a count of the number of synapomorphies cannot, by itself, allow us to calculate $\mathrm{Prob}\,(D\,|\,T)$ or Popper's measure. We also would need to know whether we are in one of these inconvenient cases, and we would need to consider other aspects of the data D in addition to the number of synapomorphies. This "logical probability" lacks the details necessary to make it actually be a probability. In their absence, it would be better for the approach to be called a logical-parsimony approach.

Farris (1999, 2000a) invokes Popper's corroboration measure, arguing that it is maximized when likelihood is maximized. He then points to Tuffley and Steel's (1997) no-common-mechanism result and argues that when a sufficiently realistic model of variation of evolutionary rates among sites is adopted, parsimony obtains the same tree as likelihood and hence the tree favored by Popper's measure. I have already noted that in such cases the inference can be inconsistent. In such a case Popper's formula is corroborating the wrong tree! If more is known about the distribution of evolutionary rates, one might be able to use a more specific model that achieved consistency. In that case the likelihood method would not be identical to parsimony.

Criticisms of statistical inference

Advocates of the hypothetico-deductive and logical-parsimony frameworks are united in one important respect: They reject statistical inference as a correct model for inferring phylogenies. The basic objection most often heard is that statistical approaches require us to know too much about the details of the evolutionary process. For example, Farris (1983, p.17) declares that:

> The statistical approach to phylogenetic inference was wrong from the start, for it rests on the idea that to study phylogeny at all, one must first know in great detail how evolution has proceeded.

Siddall and Kluge (1997) make a similar argument. Siddall (2001, p. 396), opposing the views of De Queiroz and Poe (2001), makes a strong distinction between "*frequency* probability (i.e. 'implicit probabilistic assumptions') and *logical* probability [i.e., that the hypothesis of fewer *ad hoc*isms is the one in which we should have a higher *degree of rational belief*]."

One can have similar doubts about any statistical inference: If we toss coins, are the different tosses really independent and really identical processes? We must always temper our detailed statistical conclusions with a skepticism of the model from which they arise. In the case of tossing coins, the model may be so close to true that we accept it as given. Systematics lies close to the other end of the scale: The models are only rough approximations of reality, and it is worth remembering that and worrying about it. Of the statistical methods we use, some (such as maximum likelihood) make use of all details of the model. Others, such as bootstrapping, use empirical information about the level of conflict of the characters, and thus they may rely on the characters being independent and chosen from some pool of characters, but they rely less on the details of a probability model of evolution.

However, there is always some reliance on the model. Critics of the statistical approach from the logical-parsimony school usually believe that they have a method (parsimony) that makes only noncontroversial assumptions. When parsimony is examined as a statistical method, this does not prove to be the case — there are implicit assumptions about rates of change in different lineages. From within the logical-parsimony framework it seems difficult to examine the assumptions of parsimony.

A second criticism of statistical inference rejects the use of at least some kinds of statistical methods, based on the fact that evolutionary events are historical, and therefore not repeatable:

> "As an aside, the fact that the study of phylogeny is concerned with the discovery of historical singularities means that calculus probability and standard (Neyman-Pearson) statistics *cannot* apply to that historical science ...." (Kluge, 1997a).

In a later paper, Kluge (2002) expands on this argument, declaring that "the probabilities of the situation peculiar to the time and place of the origin of species are *unique*."

One wonders whether this position is tenable. Suppose that we toss a coin 100 times and get 58 heads. We can regard the experiment as repeatable and infer the probability of heads. But suppose that, after we finish tossing, the coin rolls to the floor and then down a drain and disappears forever. Are not the 100 tosses now historical singularities? Yet clearly nothing important has changed that prevents us from inferring the probability of heads!

Although I have made clear where my own loyalties lie, in the end questions like these must be settled by the readers of this book.

The irrelevance of classification

So far in this book I have said little or nothing about classification. Almost all systematists have considered taxonomy, the naming of organisms and their placement in an hierarchical classification, to be the basic task of systematics. The construction and maintenance of a system of classification has been loudly declaimed as the most important objective of systematics, its unifying theme. This emphasis has been made by systematists of all three major schools, evolutionary-systematic, phylogenetic, and phenetic. Textbooks emphasize the point, and after-dinner speakers concentrate on it. And yet ... attending the annual meeting of a contemporary systematic society, such as the Society of Systematic Biology, will reveal that few of the speakers are concerned with classification. They spend their time making estimates of the phylogeny and using them to draw conclusions about the evolution of interesting characters. They use phylogenies a great deal. But, having an estimate of the phylogeny in hand, they do not make use of the classification.

This is a major shift in interest, and textbooks, after-dinner speeches, historians of science, and philosophers of science have not yet caught up. There has been a major shift away from interest in classification. The after-dinner speakers themselves do not practice what they preach. The delimitation of higher taxa is no longer a major task of systematics, as the availability of estimates of the phylogeny removes the need to use these classifications. Thus the outcome of the wars over classification matters less and less. A phylogenetic systematist and an evolutionary systematist may make very different classifications, while inferring much the same phylogeny. If it is the phylogeny that gets used by other biologists, their differences about how to classify may not be important.

I have consequently announced that I have founded the fourth great school of classification, the It-Doesn't-Matter-Very-Much school. Actually, systematists "voted with their feet" to establish this school, long before I announced its existence.

The terminology is also affected by the lingering emphasis on classification. Many systematists believe that it is important to label certain methods (primarily parsimony methods) as "cladistic" and others (distance matrix methods, for example) as "phenetic." These are terms that have rather straightforward meanings when applied to methods of classification. But are they appropriate for methods of inferring phylogenies? I don't think that they are. Making this distinction implies that something fundamental is missing from the "phenetic" methods, that they are ignoring information that the "cladistic" methods do not. In fact, both methods can be considered to be statistical methods, making their estimates in slightly different ways.

Similarly, we might infer the mean of a normal distribution from the sample mean or from the sample median. These differ in their statistical properties, but both are legitimate statistical estimates. Surprisingly many systematists use terminology for phylogeny methods which denies a similar legitimacy to distance matrix methods. Unfortunately, the passions that animate debates over classification have carried over into the debates over methods of inferring phylogenies. In this book we will give the terms "cladistic" and "phenetic" a rest and consider all approaches as methods of statistical inference of the phylogeny.

Chapter 11

Distance matrix methods

A major family of phylogenetic methods has been the *distance matrix methods*, introduced by Cavalli-Sforza and Edwards (1967) and by Fitch and Margoliash (1967; see also Horne, 1967). They were influenced by the clustering algorithms of Sokal and Sneath (1963). The general idea seems as if it would not work very well: calculate a measure of the distance between each pair of species, and then find a tree that predicts the observed set of distances as closely as possible. This leaves out all information from higher-order combinations of character states, reducing the data matrix to a simple table of pairwise distances. One would think that this must leave out so many of the subtleties of the data that it could not possibly do a reasonable job of making an estimate of the phylogeny.

Computer simulation studies show that the amount of information about the phylogeny that is lost in doing this is remarkably small. The estimates of the phylogeny are quite accurate. Apparently, it is not common for evolutionary processes (at least not the simple models that we use for them) to leave a trace in high-order combinations of character states without also leaving almost the same information in the pairwise distances between the species.

The best way of thinking about distance matrix methods is to consider distances as estimates of the branch length separating that pair of species. Each distance infers the best unrooted tree for that pair of species. In effect, we then have a large number of (estimated) two-species trees, and we are trying to find the n-species tree that is implied by these. The difficulty in doing this is that the individual distances are not exactly the path lengths in the full n-species tree between those two species. They depart from it, and we need to find the full tree that does the best job of approximating these individual two-species trees.

Branch lengths and times

In distance matrix methods, branch lengths are not simply a function of time. They reflect expected amounts of evolution in different branches of the tree. Two branches may reflect the same elapsed time (as when they are sister lineages in a rooted phylogeny), but they can have different expected amounts of evolution. In effect, each branch has a length that is a multiple r_i of the elapsed time t_i. The product $r_i t_i$ is the branch length. This allows different branches to have different rates of evolution.

The least squares methods

We start by describing the *least squares methods*, which are some of the best-justified ones statistically. The distances themselves also need some discussion, as they must have particular mathematical and statistical properties to work with these methods. We also describe one variant, the minimum evolution methods, and two quicker but more approximate distance matrix methods: UPGMA clustering and the neighbor-joining method.

The fundamental idea of distance matrix methods is that we have an observed table (matrix) of distances (D_{ij}), and that any particular tree that has branch lengths leads to a predicted set of distances (which we will denote the d_{ij}). It does so by making the prediction of the distance between two species by adding up the branch lengths between the two species. Figure 11.1 shows a tree and the distance matrix that it predicts. We also have a measure of the discrepancy between the observed and the expected distances. The measure that is used in the least squares methods is

$$Q = \sum_{i=1}^{n} \sum_{j=1}^{n} w_{ij}(D_{ij} - d_{ij})^2 \tag{11.1}$$

where the w_{ij} are weights that differ between different least squares methods. Cavalli-Sforza and Edwards (1967) defined the unweighted least squares method in which $w_{ij} = 1$. Fitch and Margoliash (1967) used $w_{ij} = 1/D_{ij}^2$, and Beyer et al. (1974) suggested $w_{ij} = 1/D_{ij}$. We are searching for the tree topology and the branch lengths that minimize Q. For any given tree topology it is possible to solve for the branch lengths that minimize Q by standard least squares methods.

The summation in equation 11.1 is over all combinations of i and j. Note that when $i = j$, both the observed and the predicted distances are zero, so that no contribution is made to Q. One can alternatively sum over only those j for which $j \neq i$.

Least squares branch lengths

To find the branch lengths on a tree of given topology using least squares we must minimize Q. The expression for Q in equation 11.1 is a quadratic in the branch lengths. One way that it can be minimized is to solve a set of linear equations. These are obtained by taking derivatives of Q with respect to the branch lengths, and equating those to zero. The solution of the resulting equations will minimize

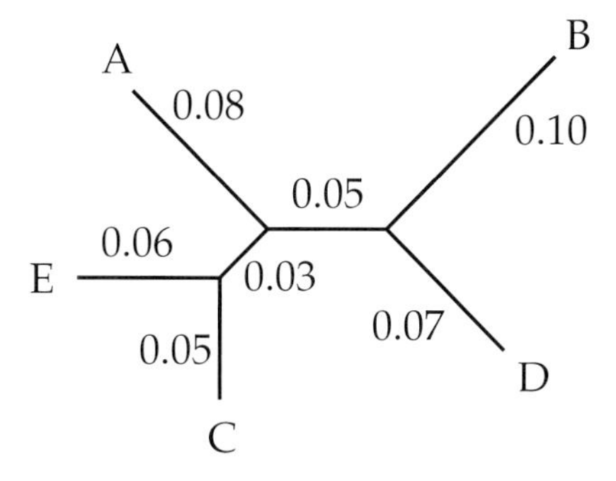

	A	B	C	D	E
A	0	0.23	0.16	0.20	0.17
B	0.23	0	0.23	0.17	0.24
C	0.16	0.23	0	0.15	0.11
D	0.20	0.17	0.15	0	0.21
E	0.17	0.24	0.11	0.21	0

Figure 11.1: A tree and the distances it predicts, which are generated by adding up the lengths of branches between each pair of species.

Q. In equation 11.1 the d_{ij} are sums of branch lengths. Figure 11.2 shows the same tree with variables for the branch lengths. If the species are numbered in alphabetic order, d_{14} will be the expected distance between species A and D, so that it is $v_1 + v_7 + v_4$. The expected distance between species B and E is $v_{2,5} = v_5 + v_6 + v_7 + v_2$.

Suppose that we number all the branches of the tree and introduce an indicator variable $x_{ij,k}$, which is 1 if branch k lies in the path from species i to species j and 0 otherwise. The expected distance between i and j will then be

$$d_{ij} = \sum_k x_{ij,k}\, v_k \tag{11.2}$$

Equation 11.1 then becomes

$$Q = \sum_{i=1}^{n} \sum_{j:j \neq i} w_{ij} \left(D_{ij} - \sum_k x_{ij,k} v_k \right)^2 \tag{11.3}$$

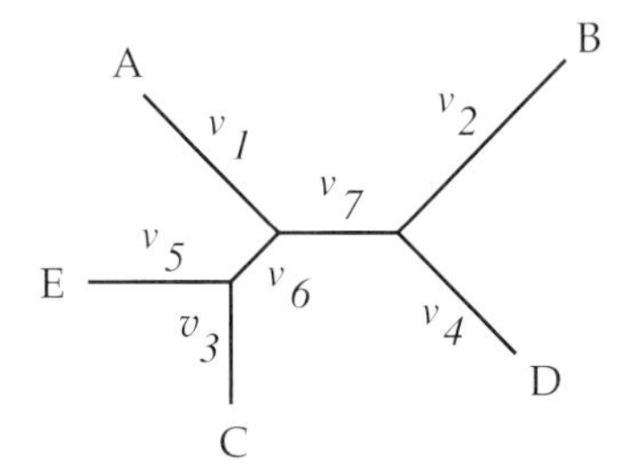

Figure 11.2: The same tree as the previous figure, with the branch lengths as variables.

If we differentiate Q with respect to one of the v's such as v_k, and equate the derivative to zero, we get the equation

$$\frac{dQ}{dv_k} = -2 \sum_{i=1}^{n} \sum_{j:j \neq i} w_{ij} \, x_{ij,k} \left(D_{ij} - \sum_k x_{ij,k} v_k \right) = 0 \tag{11.4}$$

The -2 may be discarded.

One way to make a least squares estimate of branch lengths is to solve this set of linear equations. There are both exact and iterative methods for doing this. In the case of Cavalli-Sforza and Edwards's original unweighted least squares methods, where the weights w_{ij} are all 1, the equations are particularly simple. This will lead us to a nice matrix form, and the more general case can then be put in that form. (The reader who is prone to panic attacks at the sight of matrices should skip the rest of this subsection and the one on generalized least squares as well.) For the unweighted case, for the tree in Figures 11.1 and 11.2, the equations are:

$$\begin{aligned}
D_{AB} + D_{AC} + D_{AD} + D_{AE} &= 4v_1 + v_2 + v_3 + v_4 + v_5 + 2v_6 + 2v_7 \\
D_{AB} + D_{BC} + D_{BD} + D_{BE} &= v_1 + 4v_2 + v_3 + v_4 + v_5 + 2v_6 + 3v_7 \\
D_{AC} + D_{BC} + D_{CD} + D_{CE} &= v_1 + v_2 + 4v_3 + v_4 + v_5 + 3v_6 + 2v_7 \\
D_{AD} + D_{BD} + D_{CD} + D_{DE} &= v_1 + v_2 + v_3 + 4v_4 + v_5 + 2v_6 + 3v_7 \\
D_{AE} + D_{BE} + D_{CE} + D_{DE} &= v_1 + v_2 + v_3 + v_4 + 4v_5 + 3v_6 + 2v_7 \\
D_{AC} + D_{AE} + D_{BC} & \\
+ D_{BE} + D_{CD} + D_{DE} &= 2v_1 + 2v_2 + 3v_3 + 2v_4 + 3v_5 + 6v_6 + 4v_7 \\
D_{AB} + D_{AD} + D_{BC} & \\
+ D_{CD} + D_{BE} + D_{DE} &= 2v_1 + 3v_2 + 2v_3 + 3v_4 + 2v_5 + 4v_6 + 6v_7
\end{aligned} \tag{11.5}$$

Now suppose that we stack up the D_{ij}, in alphabetical order, into a vector,

$$\mathbf{d} = \begin{bmatrix} D_{AB} \\ D_{AC} \\ D_{AD} \\ D_{AE} \\ D_{BC} \\ D_{BD} \\ D_{BE} \\ D_{CD} \\ D_{CE} \\ D_{DE} \end{bmatrix} \tag{11.6}$$

The coefficients $x_{ij,k}$ can then be arranged in a matrix, each row corresponding to the D_{ij} in that row of $\mathbf{d}$ and containing a 1 if branch k occurs on the path between species i and j. For the tree of Figures 11.1 and 11.2,

$$\mathbf{X} = \begin{bmatrix} 1 & 1 & 0 & 0 & 0 & 0 & 1 \\ 1 & 0 & 1 & 0 & 0 & 1 & 0 \\ 1 & 0 & 0 & 1 & 0 & 0 & 1 \\ 1 & 0 & 0 & 0 & 1 & 1 & 0 \\ 0 & 1 & 1 & 0 & 0 & 1 & 1 \\ 0 & 1 & 0 & 1 & 0 & 0 & 0 \\ 0 & 1 & 0 & 0 & 1 & 1 & 1 \\ 0 & 0 & 1 & 1 & 0 & 1 & 1 \\ 0 & 0 & 1 & 0 & 1 & 0 & 0 \\ 0 & 0 & 0 & 1 & 1 & 1 & 1 \end{bmatrix} \tag{11.7}$$

Note that the size of this matrix is 10 (the number of distances) by 7 (the number of branches). If we stack up the v_i into a vector, in order of i, equations 11.5 can be expressed compactly in matrix notation as:

$$\mathbf{X}^T\mathbf{d} = \left(\mathbf{X}^T\mathbf{X}\right)\mathbf{v} \tag{11.8}$$

Multiplying on the left by the inverse of $\mathbf{X}^T\mathbf{X}$, we can solve for the least squares branch lengths:

$$\mathbf{v} = \left(\mathbf{X}^T\mathbf{X}\right)^{-1}\mathbf{X}^T\mathbf{d} \tag{11.9}$$

This a standard method of expressing least squares problems in matrix notation and solving them. When we have weighted least squares, with a diagonal matrix of weights in the same order as the D_{ij}:

$$\mathbf{W} = \begin{bmatrix} w_{AB} & 0 & 0 & 0 & 0 & 0 & 0 & 0 & 0 & 0 \\ 0 & w_{AC} & 0 & 0 & 0 & 0 & 0 & 0 & 0 & 0 \\ 0 & 0 & w_{AD} & 0 & 0 & 0 & 0 & 0 & 0 & 0 \\ 0 & 0 & 0 & w_{AE} & 0 & 0 & 0 & 0 & 0 & 0 \\ 0 & 0 & 0 & 0 & w_{BC} & 0 & 0 & 0 & 0 & 0 \\ 0 & 0 & 0 & 0 & 0 & w_{BD} & 0 & 0 & 0 & 0 \\ 0 & 0 & 0 & 0 & 0 & 0 & w_{BE} & 0 & 0 & 0 \\ 0 & 0 & 0 & 0 & 0 & 0 & 0 & w_{CD} & 0 & 0 \\ 0 & 0 & 0 & 0 & 0 & 0 & 0 & 0 & w_{CE} & 0 \\ 0 & 0 & 0 & 0 & 0 & 0 & 0 & 0 & 0 & w_{DE} \end{bmatrix} \tag{11.10}$$

then the least squares equations can be written

$$\mathbf{X}^T\mathbf{W}\mathbf{d} = \left(\mathbf{X}^T\mathbf{W}\mathbf{X}\right)\mathbf{v} \tag{11.11}$$

and their solution

$$\mathbf{v} = \left(\mathbf{X}^T\mathbf{W}\mathbf{X}\right)^{-1}\mathbf{X}^T\mathbf{W}\mathbf{d} \tag{11.12}$$

Again, this is a standard result in least squares theory, first used in least squares estimation of phylogenies by Cavalli-Sforza and Edwards (1967).

One can imagine a least squares distance matrix method that, for each tree topology, formed the matrix $\mathbf{X}^T\mathbf{X}$ (or $\mathbf{X}^T\mathbf{W}\mathbf{X}$), inverted it, and obtained the estimates in 11.9 (or 11.12). This can be done, but it is computationally burdensome, even if not all possible topologies are examined. The inversion of the matrix $\mathbf{X}^T\mathbf{W}\mathbf{X}$ takes on the order of n^3 operations for a tree of n tips. In principle, this would need to be done for every tree topology considered. Gascuel (1997) and Bryant and Waddell (1998) have presented faster methods of computation that compute the exact solutions of the least squares branch length equations, taking advantage of the structure of the tree. They cite earlier work by Vach (1989), Vach, and Degens (1991), and Rzhetsky and Nei (1993). For a tree with n tips these fast methods save at least a factor of n (and for the unweighted cases, n^2) operations.

I have presented (Felsenstein, 1997) an iterative method for improving branch lengths. It uses a "pruning" algorithm similar to the one which we will see in the next chapter for likelihood. It computes distances between interior nodes in the tree and tips, and between interior nodes. These distances depend on the current estimates of the branch lengths. Using these new distances, improved estimates of branch lengths can then be obtained. The method is of the "alternating least squares" type, in which least squares estimates of some variables are obtained, given the values of the others, and this is done successively for different variables (branch lengths, in the present case). They converge fairly rapidly on the correct values. Although they are iterative, they do enable us to constrain the branch lengths to be nonnegative, which may be helpful as negative branch lengths have no biological interpretation.

This algorithm uses, at each node in the tree, arrays of distances from there to each other node. These can play the role that the conditional score arrays play in the Fitch and Sankoff algorithms for computing the parsimony score of a tree. Like those, these arrays can be used to economize on computations when rearranging the tree. This is of less use in the least squares distance matrix methods than it is in the parsimony methods, because the branch lengths in a subtree typically do not remain completely unaltered when other regions of the tree are changed. We will see similar quantities when we discuss likelihood methods.

Finding the least squares tree topology

Being able to assign branch lengths to each tree topology, we need to search among tree topologies. This can be done by the same methods of heuristic search that were discussed in Chapter 4. We will not repeat that discussion here. No one has yet presented a branch-and-bound method for finding the least squares tree exactly. Day (1986) has shown that finding the least squares tree is an NP-complete problem, so that polynomial-time algorithms for it are unlikely to exist.

Note that the search is not only among tree topologies, but also among branch lengths. When we make a small change of tree topology, the branch lengths of the resulting tree should change mostly in the regions that are altered, and rather little elsewhere. This means that the branch lengths from the previous tree provide us with good starting values for the branch lengths on the altered tree. My own iterative algorithm for estimating branch lengths (Felsenstein, 1997) retains partial information at interior nodes of the tree. Thus we not only retain the previous branch lengths, but we do not need to recompute the partial information at the interior nodes, at least not the first time they are used. Another iterative algorithm for estimating branch lengths is described by Makarenkov and Leclerc (1999).

We defer coverage of the highly original least squares method of De Soete (1983) until the next chapter, as it uses quartets of species.

The statistical rationale

The impetus behind using least squares methods is statistical. If the predicted distances are also expected distances, in that each distance has a statistical expectation equal to its prediction on the true tree (equal to the sum of the intervening branch lengths), then we can imagine a statistical model in which the distances vary independently around their expectations and are normally distributed around them. If this were true, the proper least squares estimate would minimize the sum of squares of the standardized normal deviates corresponding to the different distances. The deviation of an individual distance from its expectation would be

$D_{ij} - \mathbb{E}(D_{ij})$, and the variance of this quantity would be $\text{Var}(D_{ij})$. We can make a squared standardized normal variate by dividing the square of the deviation by the variance. The sum of squares would then be

$$Q = \sum_{i=1}^{n} \sum_{j:j \neq i} \frac{[D_{ij} - \mathbb{E}(D_{ij})]^2}{\text{Var}(D_{ij})} \tag{11.13}$$

The expectation $\mathbb{E}(D_{ij})$ is computed from the predicted distance, the result of summing branch lengths between the species. The variance in the denominator depends on the details of the process that produced these distances. In effect, Cavalli-Sforza and Edwards's least squares methods are assuming equal variances for all the distances, and Fitch and Margoliash are assuming that the error (and hence the standard deviation) is proportional to the distance. Fitch and Margoliash approximate the variance (the square of that standard deviation) by using the square of the observed distance.

The problem with this framework is the assumption that the observed distances vary independently around their expectations. If the distances are derived from molecular sequences, they will not vary independently, as random evolutionary events on a given internal branch of the tree can simultaneously inflate or deflate many distances at the same time. The same is true for distances for restriction sites and gene frequencies. DNA hybridization techniques would seem to be likely to satisfy the assumption, however. Their errors have much more to do with experimental error than with random evolutionary events. But alas, DNA hybridization values are computed by standardizing them against hybridizations of a species against its own DNA, and those standards are shared by multiple hybridization values. The result is a lack of independence even in this case.

Fortunately, it can be shown that least squares methods that do not have corrections for the correlations among data items will nevertheless at least make consistent estimates, that they will converge to the true tree as the size of data sets becomes large, even if the covariances are wrongly assumed to be zero and the variances are wrongly estimated. All that is necessary is that the expectations be correct.

I have discussed this approach to justifying distance matrix methods (Felsenstein, 1984), pointing out that it does not require that there be any paths through the data space to the observed data that exactly achieve the estimated branch lengths. For a contrary view see Farris's arguments (Farris, 1981, 1985, 1986) and my reply (1986).

Generalized least squares

The least squares methods as formulated above ignore the correlations between different distances. It is possible to modify the methods, in straightforward fashion, so that they take the correlations into account. This should be statistically preferable. However, one pays a large computational cost for taking the correlations into account. Chakraborty (1977) presented a least squares method for trees under a molecular clock. He assumed that the covariances of distances were proportional to the shared path length on the paths connecting the two pairs of species. This would be true if evolution were a Poisson process, in which there were random events occurring along the paths, with variances and covariances determined by the number of events. This is approximately true for small amounts of divergence. However, he estimated the divergence times by ordinary unweighted least squares, using the covariances only in the computation of the standard errors of the divergence times.

Hasegawa, Kishino, and Yano (1985) used an explicit model of DNA evolution to derive expressions for the variances and covariances of the distances, and they based a generalized least squares method on this. Bulmer (1991) used the Poisson process approximation, basing a generalized least squares analysis on it.

These methods require more computation than ordinary least squares. The equations are similar to 11.12 and 11.9, but the diagonal array of weights, $\mathbf{W}$, must be replaced by the inverse of the covariance matrix of the distances:

$$\mathbf{X}^T\mathbf{V}^{-1}\mathbf{d} = \left(\mathbf{X}^T\mathbf{V}^{-1}\mathbf{X}\right)\ \mathbf{v} \tag{11.14}$$

and their solution

$$\mathbf{v} = \left(\mathbf{X}^T\mathbf{V}^{-1}\mathbf{X}\right)^{-1}\mathbf{X}^T\mathbf{V}^{-1}\mathbf{d} \tag{11.15}$$

The inverse of the covariance matrix $\mathbf{V}$ is inversion of an $n(n+1)/2 \times n(n+1)/2$ matrix. For 20 species, for example, this would be a 190×190 matrix. This must be done for each tree topology examined. Matrix inversion requires an effort proportional to the cube of the number of rows (or columns) of the matrix. Thus the naive cost of finding branch lengths for a least squares tree of given topology would be proportional to n^6. However, Bryant and Waddell (1998) have described a more efficient algorithm that reduces the cost to n^4.

Distances

In order for distances that are used in these analyses to have the proper expectations, it is essential that they are expected to be proportional to the total branch length between the species. Thus, if in one branch a distance X is expected to accumulate and on a subsequent branch a distance Y, then when the two branches are placed end-to-end the total distance that accumulates must be expected to be $X+Y$. It need not be $X+Y$ in every individual case, but it must be in expectation. It is not proper to use any old distance measure, for this property may be lacking. If the distances do not have this linearity property, then wrenching conflicts between fitting the long distances and fitting the short distances arise, and the tree is the worse for them.

We will give an example of how distances may be computed to make them comply with this requirement, using DNA sequences as our example.

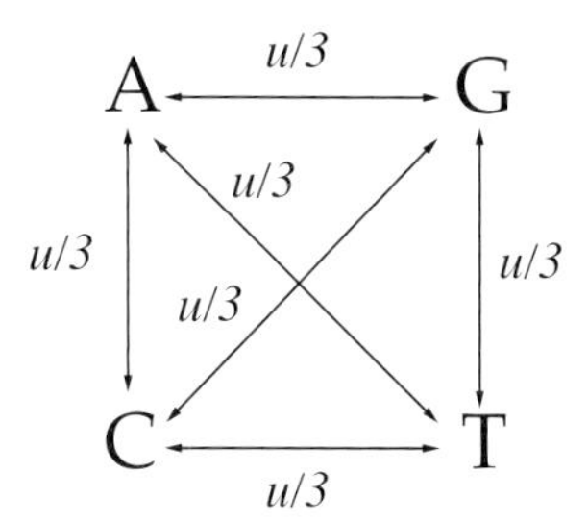

Figure 11.3: The Jukes-Cantor model of DNA change. The rate of change between all pairs of nucleotides is $u/3$ per unit time.

The Jukes-Cantor model—an example

The simplest possible model of DNA sequence evolution is the model of Jukes and Cantor (1969). In this model, each base in the sequence has an equal chance of changing. When one changes, it changes to one of the three other bases with equal probability. Figure 11.3 shows a diagram of this model. The result is, of course, that we expect an equal frequency of the four bases in the resulting DNA. The quantity u that is the rate of change shown on all the arrows is the rate of substitution between all pairs of bases. Note that although this is often miscalled a rate of "mutation," it is actually the rate of an event that substitutes one nucleotide for another throughout a population, or at any rate in enough of the population that it shows up in our sampled sequence. In certain cases of neutral mutation, the rates of substitution and of mutation will be the same.

To calculate distances we need to compute the transition probabilities in this model. Note that this does *not* mean the probabilities of transition rather than transversion; it is much older mathematical terminology, meaning the probability of a transition from one state (say C) to another (say A). The easiest way to compute this is to slightly fictionalize the model. Instead of having a rate u of change to one of the three other bases, let us imagine that we instead have a rate $\frac{4}{3}u$ of change to a base randomly drawn from all four possibilities. This will be exactly the same process, as there then works out to be a probability $u/3$ of change to each of the other three bases. We have also added a rate $u/3$ of change from a base to itself, which does not matter.

If we have a branch along which elapsed time is t, the probability in this fictionalized model that there are no events at all at a site, when the number expected to occur is $\frac{4}{3}ut$, is the zero term of a Poisson distribution. We can use that distribution because we take time to be continuous, and the branch of time t consists then of a vast number of tiny segments of time dt each, each having the small probability $\frac{4}{3}u\ dt$ of an event. The probability of no event is then

$$e^{-\frac{4}{3}ut}$$

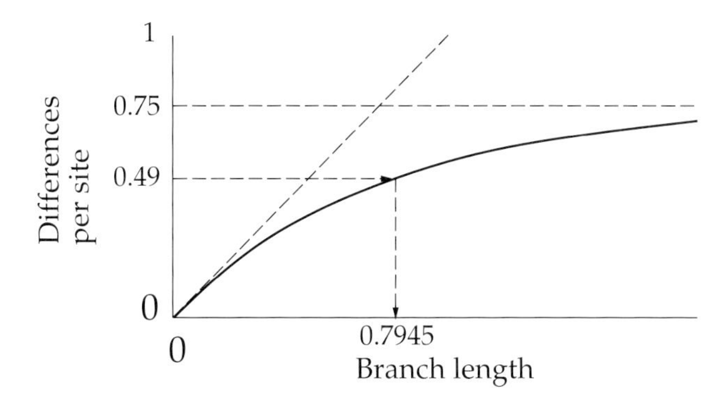

Figure 11.4: The expected difference per site between two sequences in the Jukes-Cantor model, as a function of branch length (the product of rate of change and time). The process of inferring the branch length from the fraction of sites that differ between two sequences is also shown.

The probability of at least one event is the complement of this,

$$1 - e^{-\frac{4}{3}ut}$$

If there is an event, no matter how many there are, the probability that the last one resulted in a particular nucleotide is then 1/4. So, for example, the probability of C at the end of a branch that started with A is

$$\text{Prob}\,(C|A, u, t) \;=\; \frac{1}{4}\left(1 - e^{-\frac{4}{3}ut}\right) \tag{11.16}$$

As there are three other nucleotides to which the A could have changed, the probability that this site is different at the two ends of the branch is the sum of three such quantities, being

$$D_S \;=\; \frac{3}{4}\left(1 - e^{-\frac{4}{3}ut}\right) \tag{11.17}$$

Figure 11.4 shows this curve of difference against ut. Note that it plateaus at 3/4. This is what we expect; when a sequence changes by so much that it is unrelated to its initial sequence, there are still 1/4 of the sites at which it happens to end up in the same state as when it started.

Note that if we try to use the difference per site, which is the vertical axis of Figure 11.4, it will certainly rise linearly with branch length. As it flattens out at 3/4, it will accumulate less and less difference with successive branches traversed. If we have two branches that each would, individually, lead us to expect 0.20 difference from one end of the branch to the other, when combined they will in fact only lead to an expected distance of 0.34666. The differences will not be additive at all.

The easiest way to find an additive distance is simply to use the difference per site to estimate ut itself. The value ut is the product of the rate of change and the time. It is the expected number of changes along the branch, counting both those that end up being visible to us, and those that do not. We call the value of ut for a branch the *branch length*. The values of ut on each branch will, by definition, add perfectly. Figure 11.4 shows this estimation. Starting with a value 0.49 of the difference, the dashed arrows show the process of estimating ut. The resulting estimate, 0.7945, is in effect the difference corrected for all the events that are likely to have occurred, but could not be seen. They include changes that overlay others, or even reversed their effects.

The formula for this estimation is easily derived from equation 11.17. It is:

$$D \;=\; \widehat{ut} \;=\; -\frac{3}{4}\ln\left(1-\frac{4}{3}D_S\right) \tag{11.18}$$

This is actually a maximum likelihood estimate of the distance, it turns out. Its one tiresome aspect is that it becomes infinite when the difference between sequences becomes greater than 3/4. That cannot occur in the data if infinitely long sequences follow the Jukes-Cantor model, but it can certainly happen for finite-length sequences, simply as a result of random sampling.

Although the result of these calculations is called a distance, it does not necessarily satisfy all the requirements that mathematicians make of a distance. One of the most important of these is the Triangle Inequality. This states that for any three points A, B, and C,

$$D_{AC} \;\leq\; D_{AB} + D_{BC} \tag{11.19}$$

A simple example of violation of the Triangle Inequality is three DNA sequences 100 bases in length, with 10 differences between the A and B, and 10 differences between B and C, those being at different sites. Thus A and C differ at 20 sites. Using equation 11.18, $D_{AB} = D_{BC} = 0.107326$ and $D_{AC} = 0.232616$, which violates the inequality. Thus we can call the number a distance in a biological sense, but not in a mathematical sense. Fortunately, most distance matrix methods do not absolutely require the Triangle Inequality to hold.

Why correct for multiple changes?

The Jukes-Cantor distance does not simply compute the fraction of sites that differ between two sequences. Like all the distances we will encounter, it attempts a correction for unobserved substitutions that are overlain or reversed by others. Why is this necessary? The first impulse of many biologists is to use the uncorrected differences as distances. This is dangerous.

An example is given in Figure 11.5. The original tree is shown on the left. Under the Jukes-Cantor model, the uncorrected fractions of sequence difference predicted from this tree are shown in the table in the center. If these are used with the unweighted least squares method, the tree on the right is obtained. It has the

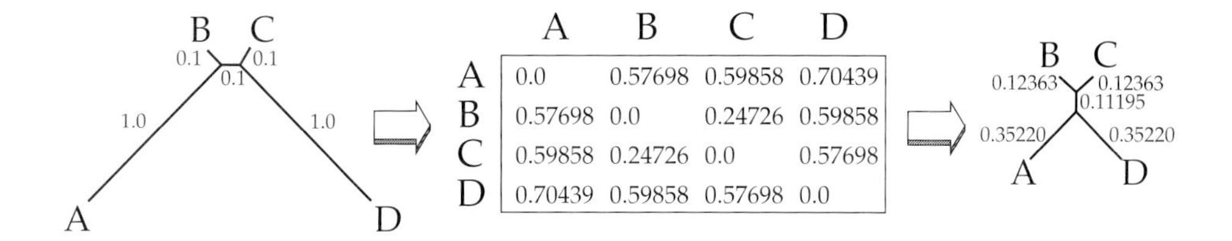

	A	B	C	D
A	0.0	0.57698	0.59858	0.70439
B	0.57698	0.0	0.24726	0.59858
C	0.59858	0.24726	0.0	0.57698
D	0.70439	0.59858	0.57698	0.0

Figure 11.5: An example of distortion of tree topology when uncorrected distances are used. The true tree is on the left, the expected uncorrected sequence differences under a Jukes-Cantor model are shown in the table in the center. The least squares tree from those differences is shown on the right. It incorrectly separates B and C from A and D.

wrong topology, most likely because the tips A and D are trying to get close to each other harder than either is to get close to B or C. There is a battle between the long and short distances, with the lack of correction making the long distances try harder to shorten the corresponding branches.

The example is what we would see if we used infinitely long sequences, but without correction of the distances for multiple changes. Despite the infinitely long sequences, we get an incorrect topology. Of course, if the corrected Jukes-Cantor distance were used, there would be a perfect recovery of the true tree, as the distances would be the sums of branch lengths along that tree. By contrast, if we use the Jukes-Cantor correction, we approach the true branch lengths as more and more DNA is sequenced, and the correct left-hand tree is found.

One case in which correction is unnecessary is when the trees are clocklike. The deeper is the common ancestor of two species, the greater will be the expected difference between their sequences. Correction for multiple changes will not alter the ranking of the distances, and distance matrix methods that assume a clock will tend to find the same topology whether or not there is correction of the distances. Rzhetsky and Sitnikova (1996) show this in simulations, where failure to correct distances has serious consequences in nonclocklike cases, but does not cause serious problems when there is a molecular clock.

Minimum evolution

Having seen the computational methods and biological justification of the least squares methods, we now look at distance matrix methods that do not use the least squares criterion. Some use other criteria; others are defined by an algorithm for constructing the tree and do not use an explicit criterion.

The *minimum evolution method* (ME) uses a criterion, the total branch length of the reconstructed tree. It is not to be confused with the "minimum evolution" method of Edwards and Cavalli-Sforza (1964) which was the first parsimony method. One might think that the minimum evolution tree should simply be a tree all of whose branches are of length 0. That would be the case if the tree were unconstrained by the data. In the minimum evolution method the tree is fit to the data, and the branch lengths are determined, using the unweighted least squares method. The least squares trees are determined for different topologies, and the choice is made among them by choosing the one of shortest total length. Thus this method makes partial use of the least squares criterion. In effect it uses two criteria at the same time, one for choosing branch lengths, another for choosing the tree topology.

This minimum evolution method was first used by Kidd and Sgaramella-Zonta (1971), who used the sum of absolute values of branch lengths. Its present-day use comes from its independent invention by Rzhetsky and Nei (1992, 1993, 1994). They used the sum of branch lengths. Trees with negative branches thus tend to attract the search, and heuristic tree rearrangement may spend considerable time among them. Kidd and Sgaramella-Zonta suggested that if there is any tree topology that has all positive estimated branch lengths, then the best solution by their method would also have no negative branch lengths.

Rzhetsky and Nei showed that if the distances were unbiased estimates of the true distance (many distances are not unbiased) then the expected total length of the true tree was shorter than the expected total length of any other. However, this is not the same as showing that the total length is always shortest for the true tree, as the lengths vary around their expectation. It would be impossible for it to be true that the total length is always shorter for the true tree, as that would establish that this particular criterion always triumphs over statistical noise! Their result is meaningful if one reduces all the information in the data to one quantity, the estimated length of the tree. Even then, having its expectation be least for the true tree is not the same as showing that the use of the minimum evolution criterion makes a maximum likelihood estimate given the tree length. For that we would need to know that this quantity was normally distributed, and had equal variances for all tree topologies. It is not clear whether minimum evolution methods always have acceptable statistical behavior. Gascuel, Bryant, and Denis (2001) have found cases where minimum evolution is inconsistent when branch lengths are inferred by weighted least squares or by generalized least squares.

Minimum evolution requires an amount of computation similar to least squares, since it uses least squares to evaluate branch lengths for each tree topology. The methods of Bryant and Waddell (1998) for speeding up least squares calculations will thus speed up minimum evolution methods as well. Kumar (1996) has described search methods that improve on Rzhetsky and Nei's. Rzhetsky and Nei (1994) describe the use of bootstrap support of branches (which I describe in Chapter 20) to guide the search for branches where the tree topology should be reconsidered. Desper and Gascuel (2002) have found that using a "greedy" search of tree topologies and a somewhat approximate version of minimum evolution led to great increases in speed with good accuracy of the resulting trees.

An early variant on Minimum Evolution that did not use least squares to infer the branch lengths was given by Beyer et al. (1974; Waterman et al., 1977). They instead required that the path lengths between all pairs of species remain longer than, or equal to, the observed distances. This makes the inference of branch lengths a linear programming problem. Their inequality is justified in the case of closely related molecular sequences, where the total branch length will approximate a parsimony criterion. Like a parsimony criterion, their method may fit branch lengths that are substantially shorter than is plausible when the sequences are quite different.

Clustering algorithms

The methods mentioned so far optimize a criterion such as the sum of squares, searching among all trees for the tree with the best value. Another class of distance matrix methods does not have an explicit criterion, but instead applies a particular algorithm to a distance matrix to come up with a tree more directly. This can be quite a lot faster, but it has the disadvantage that we are not sure that the distance information is being used fully, and we are not sure what are the statistical properties of the method.

These methods are derived from clustering algorithms popularized by Sokal and Sneath (1963). Chief among them is the UPGMA method, whose name is an acronym for its name in their classification of clustering methods. UPGMA can be used to infer phylogenies if one can assume that evolutionary rates are the same in all lineages.

UPGMA and least squares

One can constrain the branch lengths so that they satisfy a "molecular clock." Trees that are clocklike are rooted and have the total branch length from the root up to any tip equal. They are often referred to as being *ultrametric*. When a tree is ultrametric, it turns out to be extremely simple to find the least squares branch lengths. The total branch length from a tip down to any node is then half the average of the distances between all the pairs of species whose most recent common ancestor is that node. Thus if a node leads to two branches, one of which leads on upwards to all mammals and the other on upwards to all birds, the estimate of the total branch length down to the node is half the average of the distances between all (bird, mammal) pairs. The weights w_{ij} are used to weight this average.

The branch lengths are then the differences between these total branch lengths. If they give a negative branch length, it may be necessary to set that branch length to zero, which combines two nodes, and recompute the associated sums of branch lengths. Farris (1969a) was the first to note this relationship between averages and least squares branch lengths.

A clustering algorithm

Done this way, finding ultrametric trees has the same search problems as other phylogeny methods. However, there is a simple algorithm that can be used to quickly construct a clocklike phylogeny—the *UPGMA* or average linkage method. It is not guaranteed to find the least squares ultrametric phylogeny, but it often does quite well. This algorithm was introduced by Sokal and Michener (1958) — it belongs to the class of phenetic clustering methods that were predecessors of most modern phylogeny methods. It has been rather extensively criticized in the phylogeny literature, but if a clock is thought to be a reasonable assumption (and it often is if the species are closely related), then UPGMA is a well-behaved method.

The algorithm works on a distance matrix and also keeps track, for each species or group, of the number, n_i, of species in the group. These are initially all 1. The steps in the algorithm are:

1. Find the i and j that have the smallest distance, D_{ij}.
2. Create a new group, (ij), which has $n_{(ij)} = n_i + n_j$ members.
3. Connect i and j on the tree to a new node [which corresponds to the new group (ij)]. Give the two branches connecting i to (ij) and j to (ij) each length $D_{ij}/2$.
4. Compute the distance between the new group and all the other groups (except for i and j) by using:

$$D_{(ij),k} = \left(\frac{n_i}{n_i + n_j}\right) D_{ik} + \left(\frac{n_j}{n_i + n_j}\right) D_{jk}$$

5. Delete the columns and rows of the data matrix that correspond to groups i and j, and add a column and row for group (ij).
6. If there is only one item in the data matrix, stop. Otherwise, return to step 1.

This method is easy to program and takes about n^3 operations to infer a phylogeny with n species. Each time we look for the smallest element in the distance matrix, we need a number of operations proportional to n^2, and we do this $n - 1$ times. However, we can speed things up by a large factor by simply retaining a list of the size and location of the smallest elements in each row (or column). Finding the smallest element in the matrix then requires a number of operations proportional to n rather than n^2. With each clustering, this list of minima can be updated in a number of operations proportional to n, so that the whole algorithm can be carried out in a number of operations proportional to n^2. It can be shown never to give a negative branch length.

An example

Using immunological distances from the work of Sarich (1969) we can show the steps involved in inferring a tree by the UPGMA method. The amount of work needed is so small that we can carry out the clustering by hand. Here is the original distance matrix, which has the distances corrected logarithmically to allow for a presumed exponential decay of immunological similarity with branch length.

	dog	bear	raccoon	weasel	seal	sea lion	cat	monkey
dog	0	32	48	51	50	48	98	148
bear	32	0	26	34	29	33	84	136
raccoon	48	26	0	42	44	44	92	152
weasel	51	34	42	0	44	38	86	142
seal	50	29	44	44	0	24	89	142
sea lion	48	33	44	38	24	0	90	142
cat	98	84	92	86	89	90	0	148
monkey	148	136	152	142	142	142	148	0

We start by looking for the smallest distance. In this table it is marked by a box, and the elements of those rows and columns are indicated in boldface and by asterisks at the borders of the table.

		dog	bear	raccoon	weasel	* **seal**	* **sea lion**	cat	monkey
	dog	0	32	48	51	**50**	**48**	98	148
	bear	32	0	26	34	**29**	**33**	84	136
	raccoon	48	26	0	42	**44**	**44**	92	152
	weasel	51	34	42	0	**44**	**38**	86	142
*	**seal**	**50**	**29**	**44**	**44**	**0**	**[24]**	**89**	**142**
*	**sea lion**	**48**	**33**	**44**	**38**	**[24]**	**0**	**90**	**142**
	cat	98	84	92	86	**89**	**90**	0	148
	monkey	148	136	152	142	**142**	**142**	148	0

Combining the rows for seal and sea lion, we average their distances to all other species. After we do this, we infer the immediate ancestor of seal and sea lion to be 12 units of branch length from each, so that the distance between them is 24. The new combined row and column (marked SS) replaces the seal and sea lion rows and columns. This reduced table has its smallest element marked by a box. It involves bear and raccoon, and those rows and columns are boldfaced and indicated by asterisks:

		dog	* **bear**	* **raccoon**	weasel	SS	cat	monkey
	dog	0	**32**	**48**	51	49	98	148
*	**bear**	**32**	**0**	**26**	**34**	**31**	**84**	**136**
*	**raccoon**	**48**	**26**	**0**	**42**	**44**	**92**	**152**
	weasel	51	**34**	**42**	0	41	86	142
	SS	49	**31**	**44**	41	0	89.5	142
	cat	98	**84**	**92**	86	89.5	0	148
	monkey	148	**136**	**152**	142	142	148	0

Again, we average the distances from bear and from raccoon to all other species, and we infer their common ancestor to have been 13 units of branch length below each of them. We replace their rows and columns by a new one, BR:

		dog	* **BR**	weasel	* **SS**	cat	monkey
	dog	0	**40**	51	**49**	98	148
*	**BR**	**40**	**0**	**38**	**37.5**	**88**	**144**
	weasel	51	**38**	0	**41**	86	142
*	**SS**	**49**	**37.5**	**41**	**0**	**89.5**	**142**
	cat	98	**88**	86	**89.5**	0	148
	monkey	148	**144**	142	**142**	148	0

The smallest element in this table was 37.5, between BR and SS. The ancestor of these two groups is inferred to be 18.75 units of branch length below these four species. It is thus 5.75 below the ancestor of bear and raccoon, and 6.75 below the ancestor of seal and sea lion. You should refer to Figure 11.6 to see the branches and branch lengths that are added to the tree by each step. Each of the groups BR and SS is a group with two species, so the proper average is again a simple average of their distances to other species:

		dog	* **BRSS**	* **weasel**	cat	monkey
	dog	0	**44.5**	**51**	98	148
*	**BRSS**	**44.5**	**0**	**39.5**	**88.75**	**143**
*	**weasel**	**51**	**39.5**	**0**	**86**	**142**
	cat	98	**88.75**	**86**	0	148
	monkey	148	**143**	**142**	148	0

Now the smallest distance, 39.5, is between BRSS and weasel. One is a group of four species, the other a single species. In averaging their distances to all other species, we do not do a simple average, but weight the distance to BRSS four times as much as the distance to weasel. For example, the distance of the new group to dog is $(4 \times 44.5 + 51)/5 = 45.8$. The new row and column are called BRSSW and replace BRSS and weasel.

		* dog	* BRSSW	cat	monkey
*	**dog**	**0**	**45.8**	**98**	**148**
*	**BRSSW**	**45.8**	**0**	**88.2**	**142.8**
	cat	**98**	**88.2**	0	148
	monkey	**148**	**142.8**	148	0

Now dog joins BRSSW, and the average of those rows and columns is again a weighted average, weighting BRSSW five times more heavily than dog.

		* DBRWSS	* cat	monkey
*	**DBRWSS**	**0**	**89.833**	**143.66**
*	**cat**	**89.833**	**0**	**148**
	monkey	**143.66**	**148**	0

With only three groups left, cat joins up next. Finally, we have only two groups which must, of course, join one another.

	DBRWSSC	monkey
DBRWSSC	0	144.2857
monkey	144.2857	0

The final tree is shown in Figure 11.6. It is fairly close to biological plausibility.

UPGMA on nonclocklike trees

The main disadvantage of UPGMA is that it can give seriously misleading results if the distances actually reflect a substantially nonclocklike tree. Figure 11.7 shows a small set of imaginary distances, which are derived from a nonclocklike tree by adding up branch lengths. Also shown is the resulting UPGMA tree. It first clusters species B and C, which creates a branch (the one separating them from A and D) that is not on the true tree. For this problem to arise, evolutionary rates

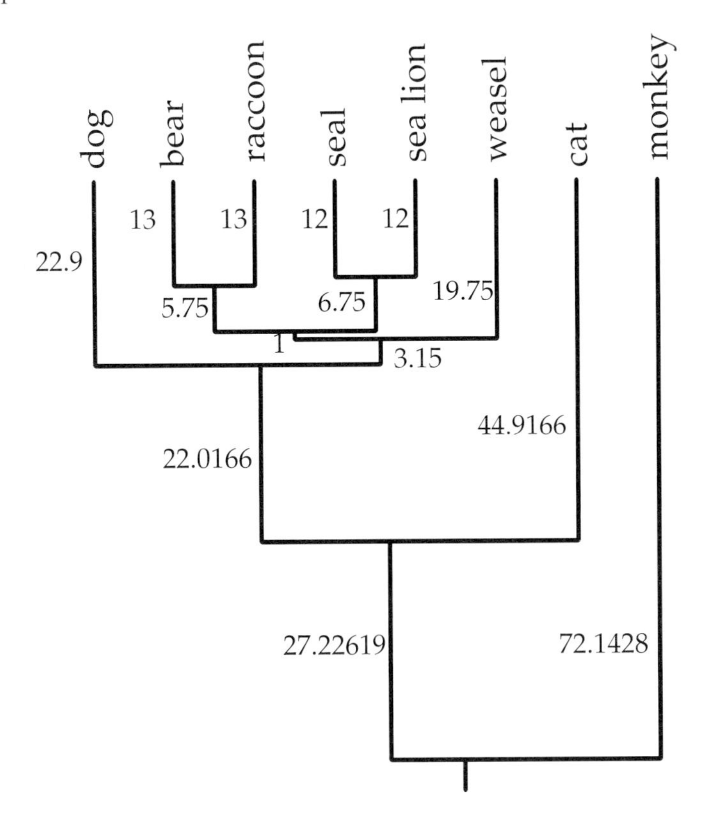

Figure 11.6: The tree inferred by UPGMA clustering of the Sarich (1969) immunological distance data set.

on different branch lengths must differ by at least a factor of two. Note that this is not long branch attraction. In fact, it is short branch attraction: B and C are put together because they are similar in not having changed.

Neighbor-joining

The *neighbor-joining* (NJ) algorithm of Saitou and Nei (1987) is another algorithm that works by clustering. It does not assume a clock and instead approximates the minimum evolution method. (It may also be thought of as a rough approximation to least squares.) The approximation is in fact quite good, and the speed advantage of neighbor-joining is thus not purchased at much cost. It is practical well into the hundreds of species.

Neighbor-joining, like the least squares methods, is guaranteed to recover the true tree if the distance matrix happens to be an exact reflection of a tree. Thus for

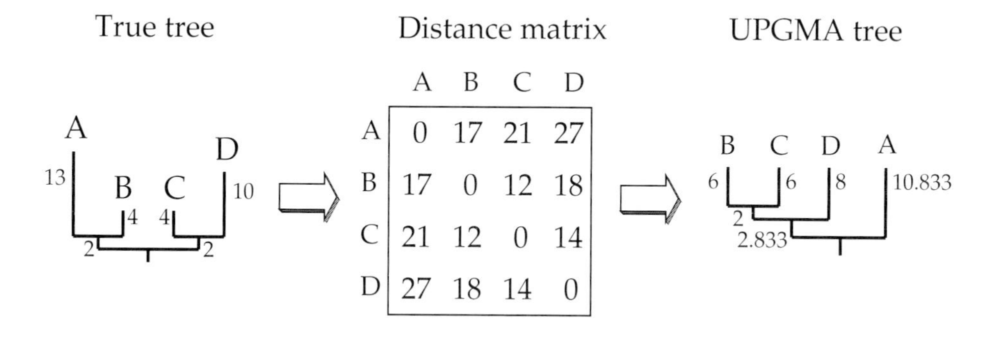

Figure 11.7: A four-species, nonclocklike tree and the expected data matrix it yields, when distances are the sums of branch lengths. The tree estimated by applying the UPGMA method to this distance matrix is shown—it does not have the correct tree topology. In both trees the branch lengths are proportional to the vertical length of the branches.

the data matrix of Figure 11.7 it simply recovers the true tree. The algorithm is (as modified by Studier and Keppler, 1988):

1. For each tip, compute $u_i = \sum_{j:j \neq i}^{n} D_{ij}/(n-2)$. Note that the denominator is (deliberately) not the number of items summed.

2. Choose the i and j for which $D_{ij} - u_i - u_j$ is smallest.

3. Join items i and j. Compute the branch length from i to the new node (v_i) and from j to the new node (v_j) as

$$\begin{aligned} v_i &= \tfrac{1}{2}D_{ij} + \tfrac{1}{2}(u_i - u_j) \\ v_j &= \tfrac{1}{2}D_{ij} + \tfrac{1}{2}(u_j - u_i) \end{aligned}$$

4. Compute the distance between the new node (ij) and each of the remaining tips as

$$D_{(ij),k} = \left(D_{ik} + D_{jk} - D_{ij}\right) \Big/ 2$$

5. Delete tips i and j from the tables and replace them by the new node, (ij), which is now treated as a tip.

6. If more than two nodes remain, go back to step 1. Otherwise, connect the two remaining nodes (say, ℓ and m) by a branch of length $D_{\ell m}$.

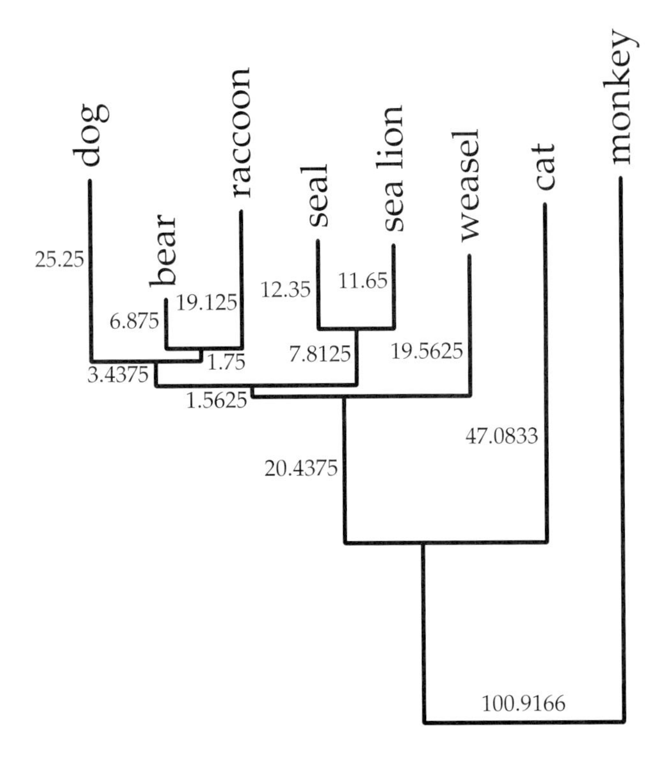

Figure 11.8: The neighbor-joining tree for the data set of Sarich (1969) rooted at the midpoint of the longest path between species. It may be compared with Figure 11.6.

We will not show the steps in detail for the Sarich data set but only the result of applying neighbor-joining to that data set, in Figure 11.8. The midpoint rooting method used is due to Farris (1972). Unlike the UPGMA algorithm, the neighbor-joining algorithm is not carried out in a number of operations proportional to n^2. Current algorithms use a number of operations proportional to n^3, owing to the necessity of updating the u_i and subtracting them from the distances.

Performance

Computer simulation studies have shown that neighbor-joining performs quite well. Although it has sometimes been claimed to perform better than the Fitch-Margoliash method, this seems not to be the case, although the difference is not great (Kuhner and Felsenstein, 1994). When the decision is made as to which pair of tips to cluster, ties are possible. Backeljau et al. (1996) have raised this issue and examined how well various implementations of neighbor-joining cope with ties. Farris et al. (1996) noted that when neighbor-joining is used together with bootstrap resampling, an arbitrary resolution of ties can produce the appearance of strong support for a grouping when there is none. Takezaki (1998) finds that this problem is serious only for short sequences or closely related species and can be avoided if one randomly chooses among tied alternatives. This can be done by ensuring that each bootstrap replicate is analyzed with a different input order of species.

Using neighbor-joining with other methods

Neighbor-joining is also useful to rapidly search for a good tree that can then be improved by other criteria. Pearson, Robins, and Zhang (1999) use it while retaining nearly-tied trees, choosing among them by minimum evolution or least squares criteria. Ota and Li (2000, 2001) use neighbor-joining and bootstrapping to find an initial tree and identify which regions of it are candidates for rearrangement (as Rzhetsky and Nei, 1994, did for minimum evolution). They then use maximum likelihood (which is described in Chapter 16) for further search. This results in a substantial improvement in speed over pure likelihood methods.

Relation of neighbor-joining to least squares

There is a relationship between neighbor-joining and other methods, though this is not immediately obvious from the algorithm. It belongs to the class of clustering methods that are defined by the precise algorithm rather than by a criterion. If we made a change in the neighbor-joining algorithm that resulted in somewhat different trees, we could not argue that we were doing better at accomplishing the objectives of neighbor-joining. By definition, anything that results in a different tree is not neighbor-joining. In comparison, methods that minimize a criterion always allow us the possibility that we could find an algorithm that does a better job of searching for the tree that achieves the minimum value.

The relation of neighbor-joining to other methods may not be clear from its presentation in the papers of Saitou and Nei (1987) and Studier and Keppler (1988). In their equations, observed distances are used as if they were equal to the sums of branch lengths, when in fact only their expectations are. There is acknowledgment in these papers that observed distances do differ from these expectations, but little more than that. Nevertheless, Saitou and Nei (1987) do establish one connection. In an extensive appendix they show that, for a given pair of species that is to be clustered, the branch lengths inferred are those that would be assigned by unweighted least squares. At first sight this appears to establish neighbor-joining as a more approximate version of least squares. In Chapter 4 I mentioned star-decomposition search, in which an unresolved starlike tree is gradually resolved by clustering pairs of species. We might think that neighbor-joining is simply a star-decomposition search using the unweighted least squares criterion.

It would be, if it used least squares to choose which pair of tips to join. But although it computes their branch lengths by least squares, it does not use the sum of squares as the criterion for choosing which pair of species to join. Instead it uses the total length of the resulting tree, choosing that pair that minimizes this length. This is a form of the minimum evolution criterion. But we also cannot identify neighbor-joining as a star-decomposition search for the minimum evolution tree. Neighbor-joining allows negative branch lengths, while minimum evolution bans them. Thus neighbor-joining has some relation to unweighted least squares and some to minimum evolution, without being definable as an approximate algorithm for either. It would be of interest for someone to see whether a star-decomposition algorithm for least squares, or one for minimum evolution, could be developed that was comparable in speed to neighbor-joining.

Gascuel (1994) has pointed out a relationship between neighbor-joining and the quartets method of Sattath and Tversky (1977).

Weighted versions of neighbor-joining

Two modifications of neighbor-joining have been developed to allow for differential weighting in the algorithm to take into account differences in statistical noise. Gascuel (1997) has modified the neighbor-joining algorithm to allow for the variances and covariances of the distances, in a simple model of sequence evolution. This should correct for some of the statistical error. Gascuel's method, called *BIONJ*, thus comes closer to what generalized least squares would give, though it is, of course, still an approximation. The weights are applied at step 3 of the neighbor-joining algorithm given above.

Subsequently, Bruno, Socci, and Halpern (2000) developed weighted neighbor-joining (the *weighbor* method) which uses weights in the formulas at somewhat different steps, and in a different way. They are used in steps 2 and 3. The weighbor method is justified by appeal to a likelihood argument, but it is not the full likelihood of the data but a likelihood calculated separately for each of a series of overlapping quartets of species, and under the assumption that distances are drawn from a Gaussian (normal) distribution.

I will not attempt a detailed justification of the terms in either BIONJ or weighbor, both for lack of space and because I believe that both are approximate. It would be better to start with a likelihood or generalized least squares method, and then show that a weighted version of neighbor-joining is an approximation to it. This has not been done in either case.

Nevertheless, both methods seem to improve on unweighted neighbor-joining. In Gascuel's BIONJ method, the variances and covariances of the distances are taken to be proportional to the branch lengths. The variance of D_{ij} is taken as proportional to branch length between species i and j. The covariance of D_{ij} and $D_{k\ell}$ is taken to be proportional (with the same constant) to the total shared branch length on the paths i–j and k–ℓ. This is usually a good approximation provided the branch lengths are not too long. Gascuel (2000) presents evidence that BIONJ can outperform minimum evolution.

Bruno, Socci, and Halpern's weighbor method uses the exact formula for the variance of a Jukes-Cantor distance instead. This is approximate for other models of DNA change, but more correctly copes with the very high variances of distances when tips are far apart on the tree. The cost paid for this greater accuracy is that some additional approximations are needed to keep calculation to order n^3. These authors argue that weighbor can find trees more accurately than BIONJ because it is less affected by noise from very large distances. BIONJ should do well when no distances are large, and both should do better than neighbor-joining.

There seems more left to do in developing weighted versions of neighbor-joining that properly reflect the kinds of noise that occur in biological sequence data.

Other approximate distance methods

Before the neighbor-joining method, a number of other approximate distance methods were proposed. Like it, they were defined by their detailed algorithms, not by an explicit criterion. The earlier ones have since largely been superseded by neighbor-joining.

Distance Wagner method

The earliest is Farris's (1972) *distance Wagner method.* This is closely related to his earlier WISS (weighted invariant shared steps) method (Farris, Kluge, and Eckardt, 1970) and his "Wagner method" algorithm (Kluge and Farris, 1969; Farris, 1970) for approximate construction of a most parsimonious tree. The distance Wagner method is intended as an approximation to construction of a most parsimonious tree. Species are added to a tree, each in the best possible place. This is judged by computation of the increase in the length of the tree caused by each possible placement of that species.

The intention is thus similar to that of minimum evolution, but the details are different. Instead of using a least squares reconstruction of the branch lengths, the lengths are computed from distances between pairs of nodes. The distances between the tip species are given, but those between a tip and an interior node, or between two interior nodes, are also computed approximately. These approximate distances between interior nodes, and between interior nodes and tips, determine the branch lengths. The approximation used assumes the Triangle Inequality. Unlike many other distance matrix methods, this restricts the use of the distance Wagner method to distances that satisfy the Triangle Inequality, which many biological distance measures do not.

An exposition of the details of the distance Wagner method will be found in the book by Nei (1987, pp. 305-309). Modifications of the distance Wagner method have also been proposed (Tateno, Nei, and Tajima, 1982; Faith, 1985).

A related family

Another family of approximate distance matrix methods (Farris, 1977b; Klotz et al., 1979; Li, 1981) uses a reference species to correct distances between species for unequal rates of evolution. If C is the reference species and A and B are two species, then the branch length of the branches leading from C to the node x where the paths leading to A and B separate is

$$D_{Cx} = (D_{AC} + D_{BC} - D_{AB})/2 \tag{11.20}$$

If D_{Cx} can be calculated for all pairs of species (other than the reference species), the largest such distance can be used to choose a candidate pair to be clustered. This corrects for unequal rates of change in the lineages. The transformation of the distance was also used in the WISS method of Farris, Kluge, and Eckardt (1970). It is related to similar steps in the neighbor-joining methods.

The more recent "Harmonic Greedy Triplets" method of Csűrös and Kao (2001) and Csűrös (2002) builds a tree rapidly by using some of the possible triplets of species. It can build a tree in a time proportional to n^2. This could be quite useful in studies with large numbers of species.

Minimizing the maximum discrepancy

Farach, Kannan, and Warnow (1995) were able to find a fast algorithm to find a tree that minimizes the L_∞ norm in the case of a molecular clock. This norm is simply the absolute value of the largest discrepancy between the predicted and the observed distances.

Atteson (1999) has shown that if the shortest branch in a tree has length ε, then if the distances are all within $\varepsilon/2$ of the true, treelike distances, a number of methods including neighbor-joining will recover the true tree topology. This is also true of the BIONJ modification of neighbor-joining. He shows that no method can achieve a better multiplier than $1/2$, and that there is a similar multiplier α, the edge L_∞ radius, such that any branch whose length is greater than that multiple of ε is correctly reconstructed. For neighbor-joining $\alpha = 1/4$. Some other methods such as a modification of the algorithm of Sattath and Tversky (1977) achieve values of α as great as $1/2$. He shows that no method can have α greater than that.

Erdős et al. (1997a, 1997b, 1999) have produced another algorithm. They argue that it is superior in quality of result to neighbor-joining. I will defer discussion of this algorithm to the next chapter, as it uses quartets of species. Similarly the "neighborliness" method of Sattath and Tversky (1977) and Fitch (1981), the four point metric method of Buneman (1971), and the innovative tree search method of De Soete (1983) are best discussed in that chapter.

Two approaches to error in trees

One of the interesting aspects of these computer scientists' approaches to investigating the effect of statistical error in trees is that they approach the problem very differently from statisticians. This difference is on display in the work of Farach, Kannan, and Warnow (1995), Erdős et al. (1997a), and Atteson (1999). A statistician takes a problem like the inference of phylogenies, considers the joint distribution of the data given an (unknown) true tree, and asks how the statistical variation in the data creates statistical variation in the estimates of the phylogeny. The computer scientists take a very different tack. They assume that one somehow knows the extent to which the data depart from a perfect reflection of the true phylogeny. For example, they assume that they know by how much the observed distances depart from the true expected distances. They then investigate what bounds can be placed on the departure of the predicted distances for the estimated tree from the true distances.

The counterpart would be a linear regression fitted through a set of points. The statistician asks how the variation and covariation of the observations create variation and covariation of the estimates of the slope and intercept of the line. The computer scientists' approach would correspond to assuming that we knew the value of the largest departure of an observation from the true line, and would then place a bound on the departure that would result between the estimated line and the true line. They would prefer that method that made this value as small as possible.

In effect, the computer scientists have found a different way of addressing the issue of statistical noise. This is true, not only of the work on phylogenies, but of the body of work on computational learning theory from which it is derived. In this book we will mostly concern ourselves with the more familiar statistician's approach, but it is worth noting, both that the other approach exists, and that although stated in very different terms, it is not necessarily incompatible with statistics.

A puzzling formula

For some of the distance methods there have been recent bounds on how many characters are needed to ensure a given level of accuracy of the resulting tree. These are inspired by theory on "Probably Approximately Correct" (PAC) methods in "computational learning theory." Erdős et al. (1997a) give two versions of these bounds. One (cited as a personal communication from Sampath Kannan) is that for a symmetric 0/1 model of change, with k characters and n species, when we apply the L_∞ norm method of Farach, Kannan, and Warnow (1995) we will have high probability of finding the correct tree if

$$k > \frac{c \log n}{f^2 \, (1-2g)^{\, 2 \, diam(T)}} \tag{11.21}$$

where c is a constant, f and g are lower and upper bounds on the net probability of change on the branches of the true tree, and $diam(T)$ measures the largest number of branches in any path across the tree. Steel and Székely (1999, 2002) have used a quite general statistical estimation framework to show that the dependence of the number of characters on $1/f^2$ is necessary as well as sufficient.

The result is surprising because it seems to imply that we need only have a number of characters proportional to the logarithm of the number of species, in order to be quite likely get an accurate estimate of the tree topology. I am not going to discuss the derivations or even the exact conditions for these inequalities. Similar inequalities, for a number of distance methods, will be found in papers by Atteson (1997), Ambainis et. al. (1997), Erdős et al. (1999), Farach and Kannan (1999), Huson, Nettles, and Warnow (1999), Cryan, Goldberg, and Goldberg (2001), and Csűrös (2002). To get this logarithmic dependence on the number of species, we would have to be considering a series of cases with larger and larger numbers of species, while holding quantities like f, g, and diam(T) constant.

A number of people (so far not in print) have pointed out that the logarithmic dependence is misleading. Can quantities such as f and g remain constant as the number of species increases? In fact, they cannot. For example, suppose that we obtained our true trees from the outcome of a birth process, where lineages had a constant probability of branching. As we get trees with n species, the probability of change on the shortest branch, f, will be shrinking as $1/n$. Thus the bound on k will grow, not proportionally to $\log n$, but at least at the rate $n^2 \log n$. The issue is complicated, because there are different inequalities for different methods, and more than one way to define the process that produces the true trees.

Consistency and distance methods

Properly formulated distance methods do have the property of consistency. This is remarkably easy to establish. Suppose that our distance measure uses the correct model of evolution. As we collect more data (for molecular sequences, as we collect longer sequences), the estimate of the branch length between a pair of species converges to the true branch length. For example, with the Jukes-Cantor model, the fraction of sites different between the two species converges to the expected fraction of sites different. If the expected fraction is 0.49, as in Figure 11.4, then we get closer and closer to having 0.49 of the sites different as we collect longer sequences. That in turn means that the estimated distance becomes closer and closer to 0.7945. Similar proofs can be made for other distances, though the proofs become more complex for other distance measures.

Thus all the distances converge to their true values. If it is true that our distance matrix method reconstructs the correct tree when fed the true total branch lengths between each pair of species, then we can be assured of consistency. For the least-squares methods, the proof is fairly trivial. When the observed distances equal the expected distances, the sum of squares is necessarily 0. Even if some other tree were to also have the sum of squares be 0, the true tree would be tied with it and would therefore be among those that must be included in the set of best trees. For neighbor-joining the corresponding proof has been given by Saitou and Nei (1987). The corresponding proof for UPGMA is trivial and is left as an exercise to the reader. For minimum evolution, Rzhetsky and Nei (1993) gave the proof for four-species trees. However, Gascuel, Desper, and Denis (2001) show some cases with inconsistency of minimum evolution when weighted least squares or generalized least squares is used to infer branch lengths.

Thus we can be sure of consistency if the convergence of the distances to their true values occurs and if a close enough set of distances implies a close enough tree. For cases in which the true tree is bifurcating, this seems easy to prove for all the algorithms mentioned. I will leave this issue of continuity to the mathematicians.

For cases in which the distances have converged and become perfectly additive, it is possible to recover the true tree and do so quickly. Neighbor-joining and minimum evolution do so, but it can be done even more quickly. Waterman et al. (1977) gave an algorithm whose effort is proportional to n^2; Culberson and Rudnicki (1989) have improved this to $n \log(n)$.

Of course, if not, then not. If the distances are not computed according to the correct model, they will converge to wrong values, and the tree inferred by distance matrix methods will then be wrong, perhaps in topology but certainly in branch lengths. (See, for example, DeBry, 1992.) If the model is nearly correct, then the distances will be nearly correct. Violation of consistency for distance matrix methods is thus rather easy to investigate. One need only compute the limits to which the distances tend as more data is collected and see which tree is selected when those limits are fed into the distance matrix method. It will be true that all reasonable distance matrix methods will be equally consistent with a given method of calculating the distances. Unreasonable distance matrix methods, such as UPGMA applied to nonclocklike data, will not be consistent, but again this is rather easy to investigate without need for consideration of the rates at which the distances approach their limiting values.

A limitation of distance methods

There has been much work on distance methods. They are probably the easiest phylogeny methods to program, and they certainly can be very fast. Although this guarantees continuing popularity, they have an inherent limitation that is worrisome. When evolutionary rates vary from site to site in molecular sequences, distances can be corrected for this variation, as we shall see in Chapter 13. A similar correction is possible in likelihood methods, as will be explained in Chapter 16. When variation of rates is large, these corrections become important. In likelihood methods, the correction can use information from changes in one part of the tree to inform the correction in others. Once a particular part of the molecule is seen to change rapidly in the primates, this will affect the interpretation of that part of the molecule among the rodents as well. But a distance matrix method is inherently incapable of propagating the information in this way. Once one is looking at changes within rodents, it will forget where changes were seen among primates. Thus distance matrix methods must use information about rate variation substantially less efficiently than likelihood methods. This casts a cloud over their use, one which may prove hard to dispel.

Chapter 12

Quartets of species

We saw that distance matrix methods can be considered to be constructing the tree as the best possible compromise between estimates of all possible two-species trees. This involves some loss of information. In DNA sequences, for example, there are 4^n possible nucleotide patterns when there are n species. But distance matrix methods first reduce the 4^n numbers down to $n(n-1)/2$ numbers. Could we do better if we discarded less information?

It would be possible to estimate three-species trees and then find the full tree as the best possible fit to them. There would be $n(n-1)(n-2)/6$ possible triples. If we had 10 species, the original 1,048,576 possible DNA patterns would be reduced to $10 \times 9/2 = 45$ distances. There are $10 \times 9 \times 8/6 = 120$ different triples. All of these triples have the same unrooted tree topology because there is only one possible unrooted tree topology when there are three species. It would be interesting to know whether methods that construct an overall tree from trees of all triples would give noticeably better results than distance methods. As far as I know, only Grishin (1999) has tried to develop a triples method of combining unrooted three-species trees into a full n-species tree. He develops definitions of two-, three- and four-species distances, and fits them by least squares. Another possible candidate, the *harmonic greedy triplets* method of Csűrös and Kao (2001) and Csűrös (2002) is really a distance method in which a few of the triplets of species are examined.

However, if we go on to *quartets* of species, we find an active literature with many methods. The greater interest in quartets occurs because they have different tree topologies — some of the literature attempts to combine the tree topologies and infer from them an overall tree topology. When conflicts arise between the topologies of the quartets, some of the methods make a network that is not a tree, but tries to summarize as much as possible of the conflicting information. The quartets approaches are not exactly parallel to the distance matrix methods. They do not spend much effort trying to measure the goodness of fit between the quartet trees and the four-species subtrees that come from the full tree. In addition to methods that use all possible quartets (or a sample from them), there are

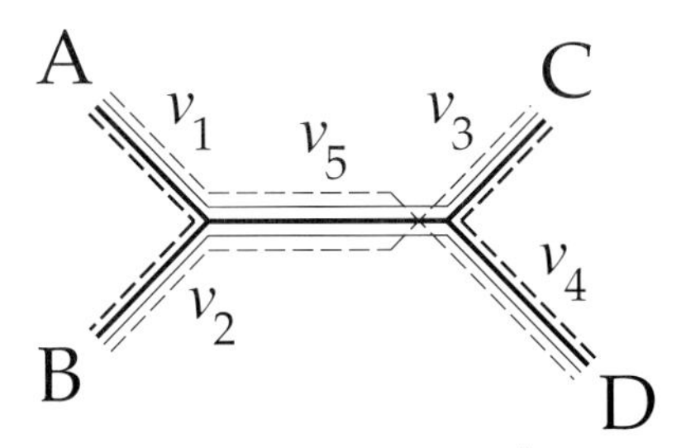

Figure 12.1: A four-species tree with branch lengths. If distances are determined by adding up the branch lengths between the species, the four point metric condition will hold. The thin solid lines show the paths for the distances d_{AC} and d_{BD}, the thin dashed lines show the paths for d_{AD} and d_{BD}, and the thick dashed lines show the paths for d_{AB} and d_{CD}.

also methods that use one outgroup species and all possible combinations of three other species. One is the *three-taxon statement* (3TS) method which is more closely related to parsimony methods than to distance matrix methods.

The four point metric

All of the quartets methods are based on the *four-point metric* (FPM) condition. This was discovered by Zaretskii (1965), and independently by Buneman (1971). (For references to other rediscoveries see the paper by Gascuel and Levy, 1996.) Consider the four-species tree ((A,B),(C,D)), as shown in Figure 12.1. If the distances were exactly those implied by the tree, we could get the distances by adding up the branch lengths on the paths from one to another. You can easily see from the figure that

$$\begin{aligned}
d_{AC} &= v_1 + v_5 + v_3 \\
d_{BD} &= v_2 + v_5 + v_4 \\
d_{AD} &= v_1 + v_5 + v_4 \\
d_{BC} &= v_2 + v_5 + v_3 \\
d_{AB} &= v_1 + v_2 \\
d_{CD} &= v_3 + v_4
\end{aligned} \tag{12.1}$$

The first two equations are for distances that have thin solid lines in their paths in Figure 12.1. Adding them,

$$d_{AC} + d_{BD} = v_1 + v_2 + 2\,v_5 + v_3 + v_4 \tag{12.2}$$

The third and fourth equations correspond to the thin dashed lines and add to the same sum:

$$d_{AD} + d_{BC} = v_1 + v_2 + 2\,v_5 + v_3 + v_4 \tag{12.3}$$

The thick dashed lines correspond to the last two equations, and add to

$$d_{AB} + d_{CD} = v_1 + v_2 + v_3 + v_4 \tag{12.4}$$

These three quantities all contain $v_1 + v_2 + v_3 + v_4$, and the first two also contain $2v_5$. Then

$$d_{AB} + d_{CD} \le d_{AC} + d_{BD} = d_{AD} + d_{BC} \tag{12.5}$$

These three terms are the three possible ways of grouping all four species into two pairs.

The unrooted trees for four species include three bifurcating trees: ((A,B),(C,D)), ((A,C),(B,D)), and ((A,D),(B,C)). We can get the second and third of these by simply switching species B and C, and then switching species C and D. So there will be conditions exactly analogous to equation 12.5 for them, with the appropriate species switched. The case where the tree is not bifurcating is the one where $v_5 = 0$ in any one of these trees.

Buneman (1971) defined a set of distances as being a four point metric if for any four species i, j, k, and ℓ the three quantities $D_{ij}+D_{k\ell}$, $D_{ik}+D_{j\ell}$ and $D_{i\ell}+D_{ij}$ have the two largest ones equal to each other. This is simply a condition that guarantees that there is a tree whose path lengths add up to the distances.

The split decomposition

Buneman's four point metric condition tells whether four species can have their distances exactly represented by a tree. What do we do if there are more then four species? We could check whether all quartets of species satisfy the FPM condition, but this does not guarantee that there is one tree that implies all of those individual FPM conditions. The most extensively studied method for combining quartet trees is the *split decomposition method* of Bandelt and Dress (1992b).

They start by noting that the length of the interior branch of the tree implied by a quartet (such as ABCD) can be computed as

$$\begin{aligned} \tfrac{1}{2}\ [&\max\left(D_{AB} + D_{CD},\ D_{AC} + D_{BD},\ D_{AD} + D_{BC}\right) \\ &- \min\left(D_{AB} + D_{CD},\ D_{AC} + D_{BD},\ D_{AD} + D_{BC}\right)] \end{aligned} \tag{12.6}$$

A look at Figure 12.1 or equations 12.2–12.4 will verify this. Bandelt and Dress then consider all possible splits of the set of species. A *split* is an unordered partition of the species into two sets. (Usually we are interested only in those that have at least two species in each of the two sets). If there are seven species, A through G, one possible split is $\{ACDF \mid BEG\}$. For each split, they consider all quartets

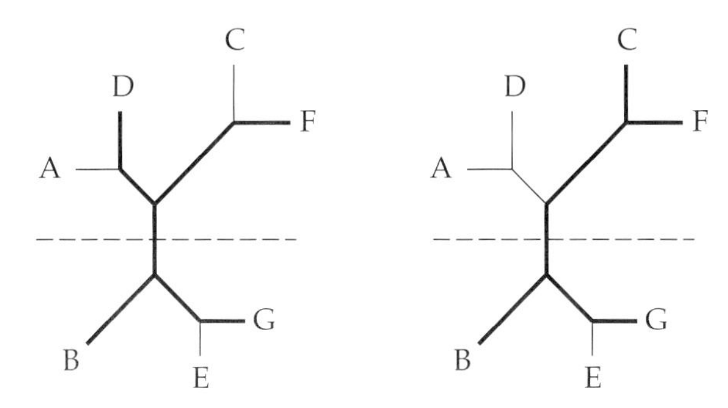

Figure 12.2: A seven-species tree with a split separating sets ACDF and BEG. The tree is shown twice, with two of the quartets that have two of the species from each side of the split. The branches connecting the members of the quartet are emphasized with darker lines.

of species that have two species from one set and two from the other. For this split, there are 18 such quartets, ranging from ACBE, through ACBG, to DFEG. But ABEG is not one of these, since three of its species come from the same set of the partition.

For each of these 18 quartets we can see whether it is compatible with the partition, as judged by whether the smallest of the three distance sums is the one implied by the partition. This is a weakening of the original Four-Point Metric condition, as it does not require that the two larger sums be exactly equal. It was suggested by Sattath and Tversky (1977) and Fitch (1981). In our seven-species example, when we look at quartet ADBE we should have

$$\max\left(D_{AB} + D_{DE},\ D_{AE} + D_{DB}\right) \;>\; D_{AD} + D_{BE} \tag{12.7}$$

for that quartet to be compatible with the split $\{ACDF \mid BEG\}$. The distance sum whose two pairs of species are each within sets of the partition should not be the largest of the three distance sums for the quartet to support that split.

For each possible split, Bandelt and Dress compute a weight that is an estimate of the length of the interior branch that separates the sets of the partition. Figure 12.2 shows a split on a seven-species tree (our example) and two of the quartets (connected by bold lines) that will be compatible with it. Note that the four-species tree for one of these quartets has its internal branch as the one that separates the two sets in the partition for the split. The other has a longer internal branch, including that branch and one other. Bandelt and Dress compute the weight of the split by having it be zero if any of the quartets does not support the split. If they all support it, the interior branch length for each quartet is taken, and the weight is the smallest of these interior branch lengths.

If the distances between species were all computed from the tree with no error, this would simply compute the length of the branch that separates the sets of the

Table 12.1: Distances between a set of seven species for 232 nucleotides near the D-loop region of mitochondrial DNA, in a data set assembled by Masami Hasegawa

	Bovine	Mouse	Gibbon	Orang	Gorilla	Chimp	Human
Bovine	0.0000	1.6866	1.7198	1.6606	1.5243	1.6043	1.5905
Mouse	1.6866	0.0000	1.5232	1.4841	1.4465	1.4389	1.4629
Gibbon	1.7198	1.5232	0.0000	0.7115	0.5958	0.6179	0.5583
Orang	1.6606	1.4841	0.7115	0.0000	0.4631	0.5061	0.4710
Gorilla	1.5243	1.4465	0.5958	0.4631	0.0000	0.3484	0.3083
Chimp	1.6043	1.4389	0.6179	0.5061	0.3484	0.0000	0.2692
Human	1.5905	1.4629	0.5583	0.4710	0.3083	0.2692	0.0000

partition for the split. In practice, the lengths will be computed with some error, so this minimum branch length will often be smaller than the actual branch length.

There are a great many possible splits. With n species there are $2^n - 2n - 2$ nontrivial splits, so that for 10 species there are 1,002 possible splits. But Bandelt and Dress (1992a) have shown that at most $\binom{n}{2} = n(n-1)/2$ of them can have a positive weight (these they call the d-splits). This is a much smaller number — for $n = 10$ it is 55. Bandelt and Dress (1992b) give an algorithm for searching for the d-splits by starting with a few species and adding the remaining species. Having the d-splits for fewer species, one can find the d-splits when one more species is added by trying adding the new species to each of the partitions of the existing splits. The computational effort for finding all the d-splits for n species rises at a rate of at most n^6. Although this is a polynomial increase, it could still be fairly fast. However, on moderately clean data sets the actual computational burden rises more slowly than n^6.

A single unrooted tree with n species has $2n - 3$ internal branches. If the splits found by Bandelt and Dress's method are all compatible with the same tree, there will be no more than $2n - 3$ of them. Frequently there are more, so that the splits specify a set of trees. Bandelt and Dress (1992b) note that it is frequently possible to draw a graph that is not a tree that summarizes all the splits and makes it rather easy to see which trees need to be considered.

As an example of the splits method, we will take a distance matrix of seven species, five of them great apes, for 232 sites in or near the D-loop of mitochondrial DNA. If we compute the distances among the sequences using the F84 distance (for which see Chapter 13) with the empirical base sequences and transition/transversion ratio 2 we get the distances shown in Table 12.1. Using Huson and Dress's program `Splitstree` we find that the following splits are supported (aside from the trivial splits that have one species separated from the rest):

Split	Weight
{Bovine Mouse | Gibbon Orang Gorilla Chimp Human }	0.4029
{Bovine Mouse Gibbon | Orang Gorilla Chimp Human }	0.03675
{Chimp Human | Bovine Mouse Gibbon Orang Gorilla}	0.0306
{Gorilla Chimp Human | Bovine Mouse Gibbon Orang }	0.021
{Bovine Gorilla | Mouse Gibbon Orang Chimp Human }	0.0103
{Mouse Gibbon | Bovine Orang Gorilla Chimp Human }	0.01015
{Gibbon Human | Bovine Mouse Orang Chimp Gorilla }	0.00975

These splits taken together do not make a single tree, as some of them conflict. In fact, we can simply use the concept of compatibility. Two splits are compatible if they could be splits of the same tree. This will be possible if one split divides one of the two sets of the other, but does not divide the other set. In fact, if we make up an imaginary character for each split, with states 0 and 1 for each species depending on which set of the split contains that species, then the characters will be compatible (see Chapter 8) if the splits are compatible. We can find, from this set of splits, the maximal cliques of splits. Each specifies a tree.

Figure 12.3 shows the trees implied by this set of splits. The branches leading to the tips are given lengths corresponding to the seven trivial splits, which can be inferred by the split decomposition method but which I have not shown in the table of splits here. They include the one that parsimony, distance matrix, and likelihood methods infer from these data (upper left), as well as others with less sensible groups uniting Bovine and Gorilla or Gibbon and Human. As can be seen in the set of splits, these have small interior branch lengths.

The branch lengths are inferred from the weights for each split, including branch lengths for the trivial splits that separate each species from all the others. (I have omitted those from the above table of splits.) Note that there are a number of different ways to infer the interior branch lengths for a single quartet, and a number of possible ways to combine those to infer the weight of a split. Thus to some extent the branch lengths inferred by the split decomposition are arbitrary. Bandelt and Dress (1992a) have shown that if we predict each distance using the sum of all those splits that separate that pair of species, we get on average an underestimate. They prove that the average of the predictions of all the distances in the distance matrix is no greater than the average of the distances themselves. This involves using all splits, even incompatible ones that could not be present together on the same tree. When only the ones compatible with a particular tree are used, it follows that there must be even more underestimation. The branch lengths in the split decomposition are not inferred by optimizing some measure of goodness of fit between the observed and the expected distances.

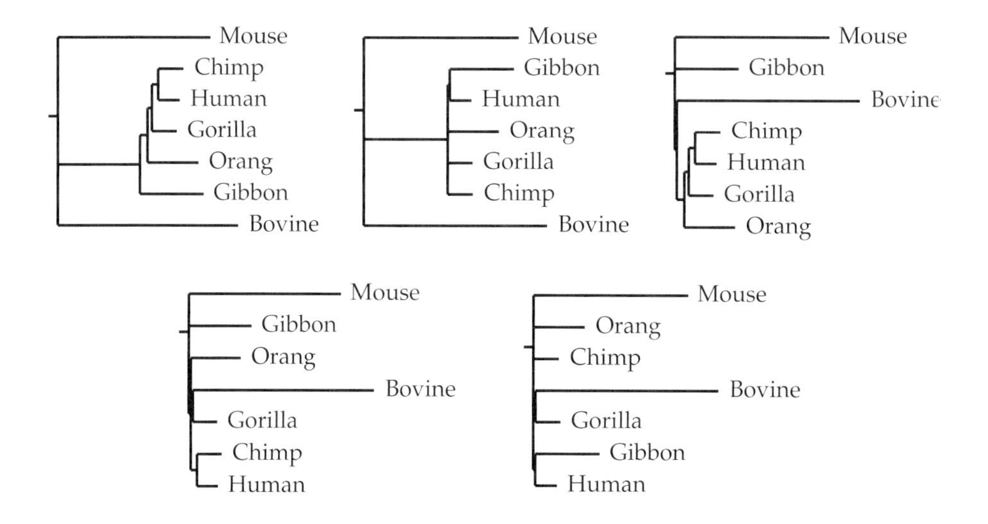

Figure 12.3: The trees implied by the maximal cliques of splits from the list of splits for the example. The split weights are used as the branch lengths of the corresponding branch.

Related methods

A useful way to summarize the splits is the splits-graph introduced by Eigen, Winkler-Oswatitsch, and Dress (1988). It is reviewed by Dress, Huson, and Moulton (1996) and by Nieselt-Struwe (1997).

The split decomposition method was preceded by a method of combining quartets due to Buneman (1971), which tended to lose more resolution. Bryant and Moulton (1999) described the refined Buneman tree, a variant of it that loses less resolution. Berry and Gascuel (2000) have suggested finding the largest set of quartets that all are compatible with the same tree, and finding that tree. They have presented an algorithm to do this that requires effort proportional to n^4. Bryant and Steel (2001) present methods to infer a tree from a set of quartets constrained to reflect those splits implied by the sites in a set of molecular sequences.

Short quartets methods

A limitation of the split decomposition method is that it will succumb rather readily to noise in the data. A split has a weight (an estimate of the branch length separating the two sets in the partition of species) that is the minimum of the internal branch lengths for all quartets that have two members in each of the sets in the partition. If even one of these quartets infers the wrong tree topology for those four species, the weight of the split becomes zero. Thus if the split is $\{ABCD \mid EFGHI\}$ and quartet ACFG happens to give the tree $((A, F), (C, G))$, then the whole split is effectively eliminated.

The noise can easily arise if some of the species are rather distant from each other. This is also a serious problem with distance matrix methods such as neighbor-joining and those using the Fitch-Margoliash criterion. The weighting of large distances in the data set by these methods is not correct—large distances are paid too much attention. The large distances are quite noisy and can wreak havoc with the tree.

To correct for this, Erdős et al. (1997a, 1997b, 1999) have put forward the *short quartets method*. This reconstructs a tree from quartets that do not involve any of the larger distances. This method uses a threshold value of the distance and accepts only those quartets that do not have any of the distances between their members greater than this threshold. Thus if a quartet *abcd* is being considered, it will be used only if all of the distances D_{ab}, D_{ac}, D_{ad}, D_{bc}, D_{bd}, and D_{cd} are smaller than the threshold. Inferring trees for these "short" quartets, they then combine them to make an estimate of the overall tree. The method of combination used is complete compatibility of the quartets. If some of the short quartets are incompatible with each other, the method returns only the result Inconsistent. If the short quartets are too few to specify the tree, the method returns the result Insufficient. In the former case, the threshold value is decreased, in the latter case, it is increased. A tree may result, or it may be discovered that all values of the threshold result in either Insufficient or Inconsistent.

Figure 12.4 shows a tree with the short quartets circled. (Here the actual branch lengths are used rather than the observed distances.) If all these quartets had the proper four-species tree topology, this would be sufficient to specify this tree topology.

The method of tree reconstruction used in the short quartets method is called the *dyadic closure method* (DCM). Its efficient implementation in an algorithm that can construct a tree in time proportional to n^5 is discussed in their paper. It involves maintaining a list of all possible quartet splits and updating it for each of the quartets in turn.

The disk-covering method

The short quartets method will fail to find a tree if there is any inconsistency in the topologies from different quartets. Huson et al. (1998) and Huson, Nettles, and Warnow (1999) have extended it to become the *disk-covering method*, or DCM (somehow that acronym seems familiar). This differs from short quartets in two major ways: It allows us to assemble trees from pieces larger than quartets, and it does not insist that all the estimates that are combined be compatible. It is thus able to respond more easily to well-defined regions of the tree and also to cope more easily with noise in the data.

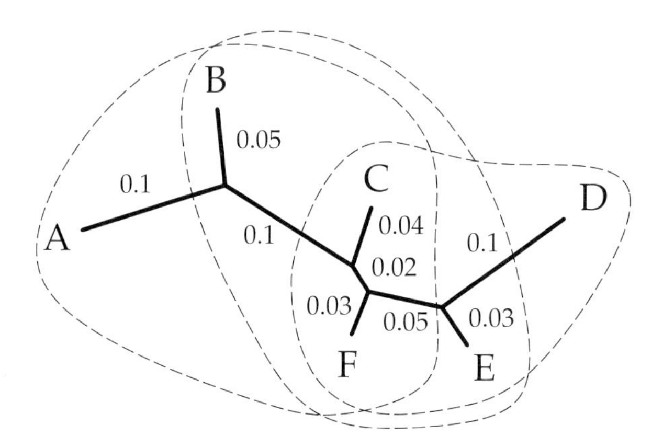

Figure 12.4: A tree with the short quartets circled. The shortness is in this example judged by the true path lengths between species (rather than, as in actual practice, by estimates of the distances between species). There are 15 possible quartets for these 6 species; here only those whose distances do not exceed 0.25 are circled. This is the smallest threshold that leads us to infer a connected tree.

The criterion used to divide the list of species into overlapping sets of species is again a threshold value of the distance between species. We make a graph in which the species are points, and lines are drawn between all pairs of species whose distances are below the threshold. The graph is then *triangulated*, by drawing in extra lines so that every loop in the graph has all of its points directly connected. This is hard to do in the worst case, but easy for most actual biological data.

The graph now has maximal cliques. Each of these is taken as a set of species whose trees will be inferred, then combined to form the full tree, using a supertree method like those covered later in this chapter. All $n(n-1)/2$ possible distance values are used as threshold values, and a consensus of the the resulting trees used as the final result.

Unlike the short quartets method, the disk-covering method does not specify what method is to be used to infer the trees for the sets of species. Any method can be used, not necessarily distance matrix methods; parsimony and maximum likelihood are also possibilities. Once the trees have been inferred, they are combined by making strict consensus supertrees (see later in this chapter). The DCM also leaves the user some flexibility in choosing the threshold value of the distance. Huson et al. (1998) report some experience with different methods of inferring the trees of the sets of species and with different choices of threshold values.

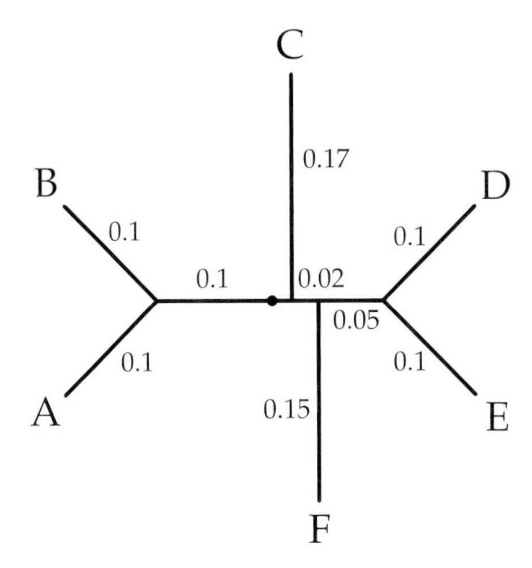

Figure 12.5: A tree of the same unrooted topology as in Figure 12.4 but with a clock. The root is at the black circle. Of the 15 possible quartets, all but one cross the root, and all 14 of these have the same maximum distance, 0.39. In clocklike trees there is no set of short quartets that crosses the root and allows construction of the tree. As in Figure 12.4, the quartets are here evaluated by their expected distances rather than their observed distances.

Challenges for the short quartets and DCM methods

The disk-covering method is valuable in that it tries to make the divide-and-conquer approach to reconstructing large trees a well-defined algorithm. By dividing the set of species into smaller sets, it allows use of methods of inference that would be unacceptably slow on the full data set. The hope is that the loss of accuracy in doing so will not be unacceptable. By using a threshold value of distance, an effort is made to find sets of species whose trees can be determined with some accuracy. The hope is that large trees can be stitched together out of smaller ones that have a reasonably small "width."

This is not always possible. If the tree is close to clocklike, the deepest divergences in the tree will not be easy to span with short quartets, or even with "disks" with more species. Figures 12.4 and 12.5 show trees, one nonclocklike and one clocklike. If we use the expected distances produced by each tree and apply the disk-covering method, we will find that the nonclocklike tree can be spanned by four- and five-species sets. But any clocklike tree will not have any disks that can cover it, except for the trivial one in which the disk contains all the species. This will be true when we use the expected distances, as we must for such an example. In real cases where the observed distances vary around the expected distances, the situation will not be as bad as this, but it should be nearly as bad. The more closely it is true that all distances between species on opposite sides of the root are equal to each other, the more likely that we will be unable to find a small tree that crosses the root and can be used in the short quartets method or in the DCM.

The ill-behavior of clocklike trees in the disk-covering method raises the question of how often it will be able to effectively cope with large trees, either actual or simulated.

Three-taxon statement methods

In the methods mentioned so far, quartets are analyzed by distance methods or analyzed by likelihood methods, and these subtrees combined to infer a full tree. Parsimony methods can be used as well, and this has been proposed by Nelson and Platnick (1991). It will not immediately be obvious that their *three-taxon statement* (3TS) method could be a quartets method, but it is. The three-taxon statements are rooted trees, so that they make a statement about a quartet of species, one of which is always the root of the full tree (or, in practice, the outgroup).

The 3TS method is to take all triples of species, add the root (if we know the ancestral states) or the outgroup (if we do not), and for each of these infer the tree topology by parsimony. In such a case the parsimony analysis is trivial. Nelson and Platnick count each character as making a three-taxon statement about the topology of the full tree. Thus if species B, C, F, and the outgroup have states 0, 1, 1, and 0, respectively, the statement favors the tree topology (B,(C,F)). That statement votes for any tree that has that topology when all species other than B, C, F, and the outgroup are removed. The full tree is given a score according to how many of the three-taxon statements support it.

The computational burden of the 3TS method is considerable. One can, for each possible three-taxon statement, make up a data set with all other species having unknown ("?") character states and only those three species having their actual character states. Thus for the triple B, C, F, all species other than those three would have state "?". There must be one of these characters created for each original character for each possible triple. Thus if we had p characters and n species we would need to create a data set with $p \times n(n-1)(n-2)/6$ characters. Evaluating these data with a parsimony method would select the tree or trees favored by the 3TS method. For example, with 50 characters and 10 species (one of them the outgroup), one would need to create a data set with 4,200 characters. This is not impossible, and ways could be found to considerably simplify the task. (With integer character weights, one could get the number of characters down to 252 for a 10-species data set.) So far there has been no work on speeding up the algorithm, nor any attempts to approximate it by considering fewer than the full set of triples of species.

Most of the literature has instead been concerned with disputing whether or not the method is valid or consistent with the principle of parsimony (Kluge, 1994; Platnick et al., 1996; Farris and Kluge, 1998; De Laet and Smets, 1998; Scotland and

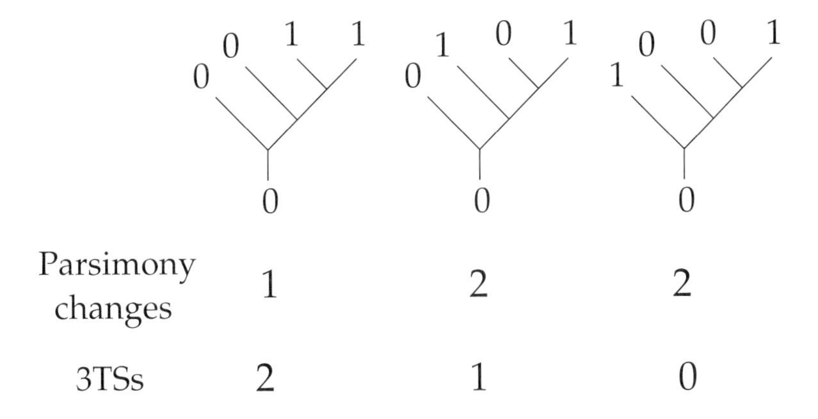

Figure 12.6: Three trees with 0/1 data, showing that the 3TS method makes a distinction that Wagner parsimony does not.

Carine, 2000). Some of the reaction has been extraordinarily strong, including the circulation of an anti-3TS paper as a kind of petition for signature by 30 authors (Farris et al., 1995a; see historical comments and response by Platnick in Platnick et al., 1996, p. 248). The issue of whether 3TS is or is not precisely consistent with parsimony need not detain us here; it is not exactly the same, although it can be regarded as a quartets approximation to it.

One argument against a connection between parsimony and the 3TS method has been that the latter cannot use reversal of a derived state as evidence (Farris, 2000b). For example, in a full tree we may see tetrapod limbs arise among vertebrates, only to be lost in a group such as snakes. Could 3TS ever use the secondary absence of limbs as a derived character, as a full tree could? It could not, as originally formulated. However, De Laet and Smets (2000) note that a 4TS method could do so.

In Figure 12.6 are three trees with 0/1 data. The left tree requires 1 change under Wagner parsimony, and the right two 2 changes. The three-taxon statements that could support a tree on this character would have to have both species that are 1s. In the left tree either of the species having 0 would then make a three-taxon statement that supported the tree. In the center tree one of the species, the lower one that has a 0, would lead to a three-taxon statement supporting the tree. In the rightmost tree, no three-taxon statement would support the tree. In this case the 3TS method is rewarding the tree for placing parallel origins of state 1 near each other on the tree. This is reminiscent of the behavior of Dollo parsimony, although these two methods are not identical. Although most of the literature supporting the use of the 3TS method insists that it is about classification rather than phylogeny, some (e.g., Platnick et al., 1996) seem to find closer location of parallel changes a feature recommending the use of the 3TS criterion.

Table 12.2: Table of the 10 pairs of values of RAS and E for a data set of five species with the first two having all states 1, and the last three having all states 0

Species pair		RAS	E
1	2	15	5
1	3	0	0
1	4	0	0
1	5	0	0
2	3	0	0
2	4	0	0
2	5	0	0
3	4	10	10
3	5	10	10
4	5	10	10

Other uses of quartets with parsimony

Quartets arguments with parsimony can also be used for testing for significant historical structure in a data set. This is the objective of the RASA (*relative apparent synapomorphy analysis*) method of Lyons-Weiler, Hoelzer, and Tausch (1996). It has also been put by them to other uses, including identifying best outgroups (Lyons-Weiler, Hoelzer, and Tausch, 1998) and detecting situations likely to lead to long branch attraction (Lyons-Weiler and Hoelzer, 1997).

RASA measures the signal in a data set by plotting the relationship between a measure of similarity between two species and a measure of apparent synapomorphy. If we have a data set with discrete characters that have two or more states, we can consider only the variable characters, and measure for a pair of species (i, j) the number of characters E_{ij} in which the two species have the same state. We can also measure RAS_{ij}, the number of times, across all characters, that these two species have the same state while another species has a different state. Thus if we had five characters, all of which have 1s in species A and B and 0s in species D, E, and F, $E_{12} = 5$. For each character there are 15 instances (species D, E, and F in each of the five characters) in which A and B have the same state while the other species has another state.

RASA plots RAS_{ij} against E_{ij} and seeks to test whether the regression is significant. Table 12.2 (which I suppose can be called a tabula rasa) shows the values of RAS and E for all pairs of species. Lyons-Weiler, Hoelzer, and Tausch (1996) compare the slope of the regression of RAS_{ij} on E_{ij} by a t-test with predicted values of either $\sum_{ij} RAS_{ij} / \sum_{ij} E_{ij}$ or with values in the same matrix with the columns scrambled separately in each character. In this example, the predicted slope is 2.25 from the first method, so that there is judged to be no signal in the data set.

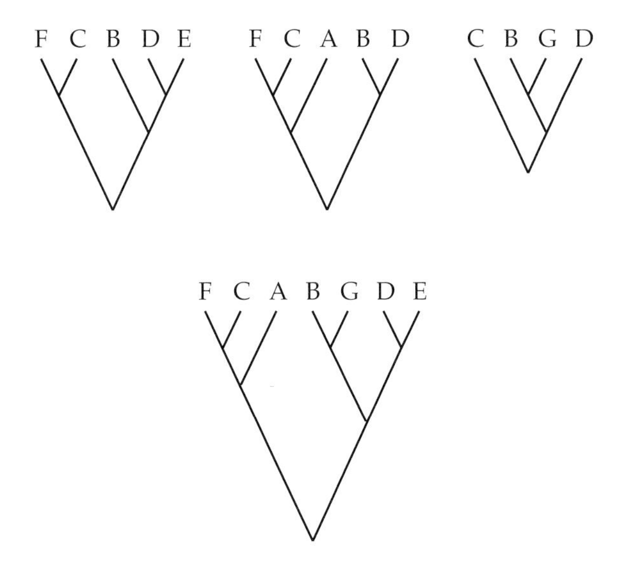

Figure 12.7: A consensus supertree (bottom), computed from three trees with different sets of tips.

It is difficult to say whether RASA is really a quartets method. It looks only at triples, but the RASA statistic looks at more than one triple at the same time, where these share one pair in common. In effect it is looking at the split separating that pair from all other species. Because it does not make any attempt to use the structure among those other species, it can count data from closely-related species as if it were independent information. For this reason many objections have been raised to RASA (Simmons et al., 2002; Faivovich, 2002; Farris, 2002).

Consensus supertrees

Methods like disk covering assemble a large tree from smaller trees. This step is a use of *consensus supertrees*. When not all of the input trees have the same species on them, we may want to find which tree that has all of the species is most consistent with them. This active area is well reviewed by Bininda-Emonds, Gittleman, and Steel (2002). I will only touch on some of the more widely-used methods here. Gordon (1986) has suggested finding the tree which has all of the input trees within it, in the sense that one can get them by dropping different sets of species. Figure 12.7 shows three trees, with different and overlapping sets of species, and their consensus supertree. Supertrees have become of increasing interest as data sets with large numbers of species have been analyzed. When computationally intensive methods of inferring trees are used, it may be helpful to analyze a series of subsets of the species. One then will want to combine the inferred phylogenies for the subsets and see whether a consensus supertree can be

Table 12.3: The set of fictional characters implied by the three input trees in Figure 12.7. Each character corresponds to one internal branch of one tree. Species missing from a tree have "?" states for its characters. In this example, the trees are considered to be rooted, so an outgroup O with all 0s is also included.

	Tree 1			Tree 2			Tree 3	
O	0	0	0	0	0	0	0	0
A	?	?	?	0	1	0	?	?
B	0	1	0	0	0	1	1	1
C	1	0	0	1	1	0	0	0
D	0	1	1	0	0	1	1	0
E	0	1	1	?	?	?	?	?
F	1	0	0	1	1	0	?	?
G	?	?	?	?	?	?	1	1

made. Algorithms for constructing supertrees are difficult; Gordon presented only approximate methods. The nonadditive binary recoding technique introduced by Kluge and Farris (1969) and used to encode phylogenies by Farris (1973a) can be used in one method of constructing supertrees. We have seen in Chapter 7 that one can construct characters corresponding to branches in a phylogeny (there it was done for a character state tree). Each branch's character has a 1 for species connected to one end of the branch and 0 for species connected to the other end. If the tree is rooted the set connected to the upper end of the branch gets state 1. We can omit characters for the branches connected directly to tips, as they will not affect the calculation of the supertree.

For the calculation of consensus supertrees, we make these characters for each tree. If a species is missing from the tree, its state for all of that tree's characters is taken to be "?". The characters for all of the input trees are appended into one data set. If we are using rooted trees to build the supertree, an outgroup that has all states "0" should also be included.

Table 12.3 shows the resulting data set for the three trees in Figure 12.7. The groups of fictional characters for each of the three trees is indicated. We can imagine evaluating the changes of these characters according to Camin-Sokal parsimony (if the trees are rooted) or Wagner parsimony (if they are unrooted). If a tree is proposed to be the consensus supertree, these characters should each change only once. Thus the characters should all be compatible. Unfortunately, we cannot use the Pairwise Compatibility Theorem to find the tree, as the data have "?" states in them, and as we have seen in Chapter 8, this makes the theorem inapplicable. However, we can do an approximate search for the consensus supertree by inferring the tree by Camin-Sokal parsimony (or in the unrooted case, Wagner parsimony).

Even when all of the trees are compatible with each other, there may be many different supertrees, each of which contains all of the trees. The strict consensus of all of these is called the *strict consensus supertree*. Steel (1992) showed a polynomial time algorithm for finding it directly in the rooted case, more quickly than with Gordon's algorithm. The BUILD algorithm of Aho et al. (1981) tests compatibility of the set of trees and, if they are compatible, finds one of the supertrees.

The recoding method can be extended to deal with cases in which not all the trees are consistent with each other. Finding the most parsimonious tree then creates a compromise between conflicting information from different trees. This method, which is called MRP (*matrix representation with parsimony*) was suggested by Baum (1992) and by Ragan (1992b). There have been a number of suggestions of ways to weight the trees so as to more effectively incorporate information about the support for various parts of the trees (Ragan, 1992a; Purvis, 1995; Ronquist, 1996b; Rodrigo, 1996).

Direct methods for computing supertrees in polynomial time when not all of the trees are compatible include the MinCutSupertree method of Semple and Steel (2000) and the modification of it by Page (2003).

A consensus supertree method, the *average consensus* method, has been proposed for trees with branch lengths by Lapointe and Cucumel (1997). They suggest computing the path lengths between each pair of species and averaging across all trees that contain that path. The consensus supertree is then computed by a least squares fit of the tree to these average path lengths.

Note that the quartets methods described above are not the same as any of these supertree methods. Most of these methods will produce the same tree when the individual trees are completely consistent with each other, and all make different compromises when they are not. All of them may be compared with the result of assembling a combined data set, with "?" states for characters not scored in particular species. Analyzing the combined data set is a "total evidence" approach and may yield different results. I will discuss the controversy between total evidence and consensus tree methods later, in Chapter 30.

Note that the depressing results of Steel, Dress, and Böcker (2000) (the discussion will be found in Chapter 30) include some for supertrees.

Neighborliness

Sattath and Tversky (1977), Fitch (1981), and Bandelt and Dress (1986) had earlier developed methods based on quartets, using measures of *neighborliness* to choose tips to join on the tree. Sattath and Tversky looked at every pair of species and every quartet that includes that pair. For each quartet they asked whether that pair was supported as being neighbors. If they were considering species 3 and 7, they asked, for each other pair of species i and j, whether the tree inferred for that quartet placed 3 and 7 on the same side of the interior branch. The trees were inferred by considering whether $D_{37} + D_{ij}$ was the smallest of the three distance sums (the others are $D_{3i} + D_{7j}$ and $D_{3j} + D_{7i}$). They counted the number of times this has occurred, to get a neighborliness score for each pair of species. That pair was then connected and replaced by a single species, whose distances to the others were computed by averaging. The process was continued until the tree was constructed. The use of a simple average of the distances implies that the tree is clocklike.

Fitch (1981) used instead a neighborliness that was an average of the branch lengths inferred. Gascuel (1994) discussed these strategies in the context of the neighbor-joining algorithm, and showed that Fitch's method gives essentially the same result as neighbor-joining. In effect it is a distance method rather than a quartets method. This underscores the close relationship between distance methods and quartets methods.

De Soete's search method

The most original quartets method of all, and one of the most unusual approaches to searching for trees is that of De Soete (1983). He starts with an observed matrix of distances (D_{ij}) and attempts to construct a tree that will most nearly approximate them. He does this by trying to achieve a matrix of expected distances (d_{ij}) that are those predicted by a tree. If all of the expected distances satisfy the four-point metric (FPM) condition, it can be shown that there is a tree that predicts these distances. De Soete gradually modifies the matrix of expected distances and ensures that it comes closer and closer to satisfying the FPM condition. The examination of whether the individual FPM conditions are satisfied makes this a quartets method.

De Soete computes an overall measure of the form

$$L(D) = \sum_{ij} (D_{ij} - d_{ij})^2 + \rho \, P(d) \tag{12.8}$$

which is the usual sum of squares, plus a function $\rho P(d)$ that penalizes departures of the fitted distances from adhering to the Triangle Inequality and the four-point metric. For each quartet there is a term in $P(d)$ of the form

$$(d_{ik} + d_{j\ell} - d_{i\ell} - d_{jk})^2 \tag{12.9}$$

where $d_{ij} + d_{k\ell}$ is the smallest of the three distance sums for that quartet. It is the square of the difference between the two larger distance sums.

This penalty function has the property that it is positive unless the FPM is precisely satisfied, at which point it becomes zero. De Soete envisages fitting distances d_{ij} that minimize $L(D)$ for a given value of ρ. The bigger the constant ρ, the more weight is placed on the penalty function, and thus the more effort the minimization will put into reducing $P(d)$. Taking a series of increasing values of ρ, the method will force $P(d)$ closer and closer to zero. Thus the expected distances d_{ij} will become more and more treelike as ρ increases.

Note that the initial expected distance matrix does not satisfy the FPM condition, and neither do any of the later ones, although they come closer and closer to satisfying it. Thus we do not have estimates of trees that we alter to wander in tree space. What we have are entities that are not quite trees. Imagine that all of tree space is a flat surface, divided into regions for the different tree topologies. De Soete starts, not in that surface, but in the air above it. As the penalty for departing from the FPM condition becomes greater, the method in effect falls out of the sky toward the space of trees, being attracted to particular regions of the space. De Soete's method does not crawl through tree space like all the others but instead parachutes onto it from above. Some further developments of similar techniques have been made by Roux (1988) and by Gascuel and Levy (1996). This is a unique approach that deserves more attention.

Quartet puzzling and searching tree space

The above methods have used the distance information, never returning to the original data. Quartets can also be used to generate trees as candidates for search using other criteria, including criteria that measure the goodness of fit to the full n-species data. Strimmer and von Haeseler (1996) have introduced the *quartet puzzling method*, which uses quartets to infer a number of estimates of the full tree, and then chooses among them using likelihood. We will introduce likelihood inference of phylogenies in Chapter 16. For the moment we need only know that it is a criterion for choosing among trees. In fact, although Strimmer and von Haeseler specify likelihood as the criterion for quartet puzzling, other criteria such as parsimony, or even least squares distance criteria, could as easily be used.

In their method, one starts by taking all quartets, and using likelihood to infer the four-species tree for each. This is different from the split decomposition and neighborliness methods, in which the quartets are always inferred using their distances. The likelihood is computed using the full molecular sequences for those four species. The species are taken in a random order. The quartet for the first four is used as a starting point. For each species in turn:

1. Take all triples of species that are already on the tree.
2. Add the next species to the triple and infer the tree for that quartet.
3. If the quartet is (say) ACDG, and D is the new species, then if the tree comes out ((A,D),(C,G)), this is considered a vote against having D located on any branch between C and G.
4. Place the votes-against on those branches.
5. Accumulate the votes for all these triples.
6. Place D on the tree by connecting it to the branch with the fewest votes-against (choosing one of the placements at random if there is a tie).

Figure 12.8 shows some of the steps in this process. Although the overall criterion, likelihood, is used to infer the trees, there are some arbitrary steps in the

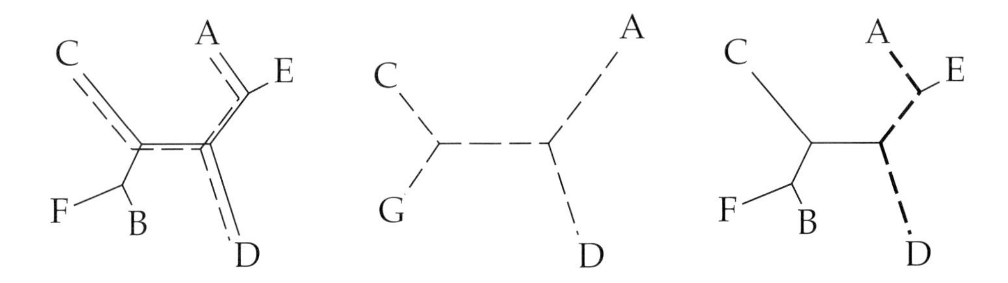

Figure 12.8: Steps in quartet puzzling, showing the evaluation of a triple, ACD, which becomes a quartet ACDG (dashed lines) when we add to it a new species G whose placement on the six-species tree (left tree) is being evaluated. The inferred tree for the quartet is shown in the center tree. The branches that this quartet "votes against" as places to add G are shown as dashed lines on the righthand tree.

resulting algorithm. In particular, the votes are all-or-none; they do not reflect the amount of difference in likelihood between different resolutions of the quartet. In the example in the figure, the placement of species G in the quartet is far enough to the left that it would also seem to vote against the center branch of the original tree, but this is not counted. Strimmer, Goldman, and von Haeseler (1997) have described a modification of the quartet puzzling algorithm, using Bayesian posterior probabilities of quartets as weight. The modification addresses this arbitrariness and seems to improve the effectiveness of the search for the maximum likelihood tree. Ranwez and Gascuel (2002) have invented another weighted method of combining quartets that also seems to improve on quartet puzzling.

A particularly interesting use of quartets for searching is given by Willson (1998). He uses parsimony, and for each tree is able to measure the extent to which quartets within the tree disagree with the tree topology. The measure is not a count of quartets, but is weighted to measure the strength of the disagreement. With the parsimony criterion, he is able to prove that no other tree achieves a lower value of the total strength of this disagreement. This proves that the tree is one of those optimizing this criterion, without the need for an exhaustive search.

Lake's (1995) "bootstrapper's gambit" method is a resampling method that builds trees from bootstrap sampled data sets, insisting that no quartets conflict. It will be described further in Chapter 20 when we consider bootstrap sampling.

Perspective

Quartets methods lead to a lot of nice mathematics, but we should keep in mind a basic limitation — that they treat the quartets as a sufficient representation of the original data. This can be dangerous. Distance matrix methods have been developed that weight departures of the tree from the individual distances in ways that reflect the statistical noise arising from the stochastic processes of evolution. Quartets methods are at an earlier stage of development; they do not make any attempt to do this, with the exception of some of the modifications of the MRP method and the Bayesian weighting for quartet puzzling. It is going to be necessary to take the data into account. Consider a simple example. We have various quartets inferred from data on bird species, and we combine them to make a tree. In the process, the information from one particular species of sparrow is weighted by consideration of the quartets in which it appears.

Now suppose that a second, nearly identical species of sparrow is included in the data set. If it is closely related to the first species of sparrow, parsimony or likelihood methods will take this into account, and the presence of the second species of sparrow will have little effect. It will always be found on the tree next to the first species and will scarcely change the assessment of which events have occurred in evolution. But in a quartets method or in some distance matrix methods, the effect of including the second species of sparrow can be more dramatic. In effect, each quartet that includes the first species is joined by another that reaches the same conclusion and includes the second species. When the quartets are combined to make the overall tree, the sparrow quartets will loom larger and have a greater influence on the tree, more influence than is really appropriate. By ignoring the data once the quartets are available, the quartets methods lose the ability to properly assess the noise in the inference. This may be changed by future developments of these methods. It is not clear that this will be easy to do.

Chapter 13

Models of DNA evolution

We have just introduced distance matrix methods, which use models in the formulas for the distances. In later chapters we will cover likelihood and invariants methods, which are also dependent on having a probabilistic model of evolution. This is a good point to review some of the models of evolution that are used in many methods of inferring phylogenies. We will cover in this chapter the models of change of DNA (and RNA) sequences. In subsequent chapters we will cover other molecular models. We defer to a later chapter the use of the Brownian motion approximation to model change of gene frequencies and quantitative characters, and the issue of genetic distances based on gene frequencies. As we cover various models of DNA change, we will also comment on methods for obtaining distances from them.

We have already introduced the simplest model of DNA evolution, the Jukes-Cantor model (in Chapter 11). This model assumes, in effect, that the frequencies of all four bases at equilibrium are equal and that there is no difference of rates of substitution between transitions and transversions. A (large) number of alternative models have since been introduced to remedy this. We will review some of them. An important objective in each case will be to compute, for each model, the probability of change from a given state to a given other state in a branch of length t. These are called *transition probabilities*. The name has nothing to do with the transition/transversion distinction that molecular biologists make — it is older terminology from probability theory. The reader will have to be alert to whether the word *transition* is being used in the molecular biological sense, or to denote any change of state (whether it is a transition or a transversion).

Kimura's two-parameter model

Kimura (1980) introduced a model that allows a transition/transversion inequality of rates, while still being very symmetrical. It is shown in Figure 13.1. The two parameters, α and β, of the *Kimura two-parameter (K2P) model* allow us to vary not

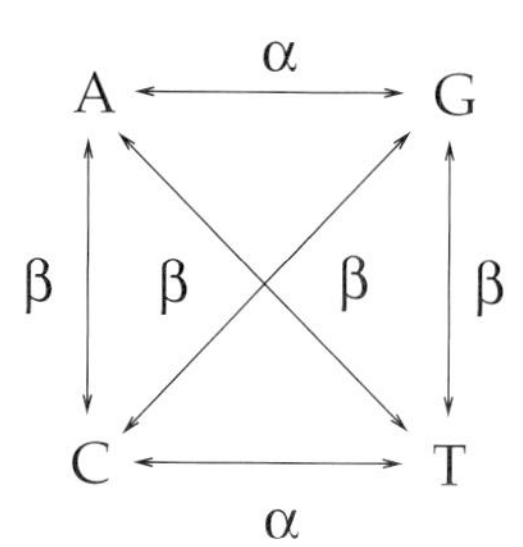

Figure 13.1: Kimura's two-parameter model. Rates differ between transitions and transversions, but are otherwise equal.

only the overall rate of substitution per unit time, but also the fraction of them that are transitions as opposed to transversions. It should be apparent that, from any nucleotide, there is one change, at rate α, that causes a transition, and two, at rate β, that cause transversions. The ratio of transitions to transversions, which we will call R, will be $\alpha/(2\beta)$. The total rate of change will be $\alpha + 2\beta$.

The model is symmetrical, and one can immediately see that, after enough time has elapsed, it will be equally likely for the base to be a purine or a pyrimidine. Within these two categories there is also complete symmetry between the two possible nucleotides, so the equilibrium frequencies of all four bases under Kimura's model are, as they were in the Jukes-Cantor model, $\pi_A = \pi_G = \pi_C = \pi_T = 1/4$. Note that the Jukes-Cantor model is simply the particular case of Kimura's two-parameter model in which $\alpha = \beta$, so that $R = 1/2$.

In the next subsection we will see formulas for the transition probabilities of a more general model that has Kimura's model as a special case. To express these in a meaningful way, we would like to consider a site that has rate of change 1 per unit time (which means that we are scaling branch length in expected nucleotide substitutions per site). This would require that $\alpha + 2\beta = 1$. Given the transition/transversion ratio R, that in turn means that

$$\begin{aligned} \alpha &= \frac{R}{R+1} \\ \beta &= \left(\frac{1}{2}\right) \frac{1}{R+1} \end{aligned} \tag{13.1}$$

as only those values guarantee that the expected transition/transversion ratio is R and the total rate of change is 1. It is possible to derive the transition probabilities for Kimura's two-parameter model. For now we just note that, for a branch of length t, the probabilities that the net result is a transition (change from, say, A to G) and that the net result is any transversion (change from A to C or to T) turn out to be

$$\begin{aligned} \text{Prob}\,(\text{transition}\,|\,t) &= \tfrac{1}{4} - \tfrac{1}{2}\exp\left(-\tfrac{2R+1}{R+1}\,t\right) + \tfrac{1}{4}\exp\left(-\tfrac{2}{R+1}\,t\right) \\ \text{Prob}\,(\text{transversion}\,|\,t) &= \tfrac{1}{2} - \tfrac{1}{2}\exp\left(-\tfrac{2}{R+1}\,t\right) \end{aligned} \tag{13.2}$$

Note that the probability of a particular transversion (say, A to T) is half of the value given in equation 13.2. These probabilities are not the probabilities of single changes but the net result of whatever number of changes occur. A change from A at one end of a branch to C at the other may involve multiple changes of base in between.

Figure 13.2 shows these quantities plotted against time (t) for two different values of R, 10 and 2. In both cases the number of transition differences rises to a maximum and then begins a slow decline toward an ultimate value of 1/4. In both cases the expected number of transversion differences rises slowly toward an ultimate maximum of 1/2. Note that beyond the point where the overall fraction of differences between the sequences is expected to be about 50%, there will be little information about the distance conveyed by the transitions. This might be thought to argue for ignoring transitions and simply basing our inferences on the transversions. But note that in these same figures, the transitions do contribute meaningful information below 50% divergence (in the case of $R = 10$, most of the information). The figures also suggest that no one ratio of weights will suffice to use this information in an optimal way. Above 50% divergence, transversions should be heavily weighted. Below it, they should not be.

Calculation of the distance

With the Jukes-Cantor distance, we could compute the distance by expressing the fraction of difference between sequences in terms of the distance, and then solving that equation for the distance. With Kimura's two-parameter distance, a similar procedure can be followed. In effect, we have two observations, the fraction P of transition differences between the two sequences, and the fraction of transversion differences (Q). Solution of the two equations 13.2 yields:

$$\begin{aligned} \hat{t} &= -\tfrac{1}{4}\ln\left[(1-2Q)(1-2P-Q)^2\right] \\ \hat{R} &= \frac{-\ln(1-2P-Q)}{-\ln(1-2Q)} - \frac{1}{2} \end{aligned} \qquad (13.3)$$

the first of these being given by Kimura (1980). For each pair of sequences, we are estimating both t and R. One difficulty with doing things this way is that we may obtain wildly different transition-transversion ratios R for different pairs of sequences, when in fact all of them have evolved on the same phylogeny and thus share important parts of their evolution. Another is that any pair in which either $Q > 1/2$ or $2P + Q > 1$ will have complex values of R and t, which does not make sense.

An alternative method of computing the distance is to treat the matter as a statistical estimation. It can be shown that the quantities P and Q are in fact "sufficient statistics," so that any estimation of t need only make use of those quantities plus the number of sites n. All other observations, such as how many of the differences are $A \leftrightarrow G$ transitions as opposed to $C \leftrightarrow T$ transitions, do not increase the

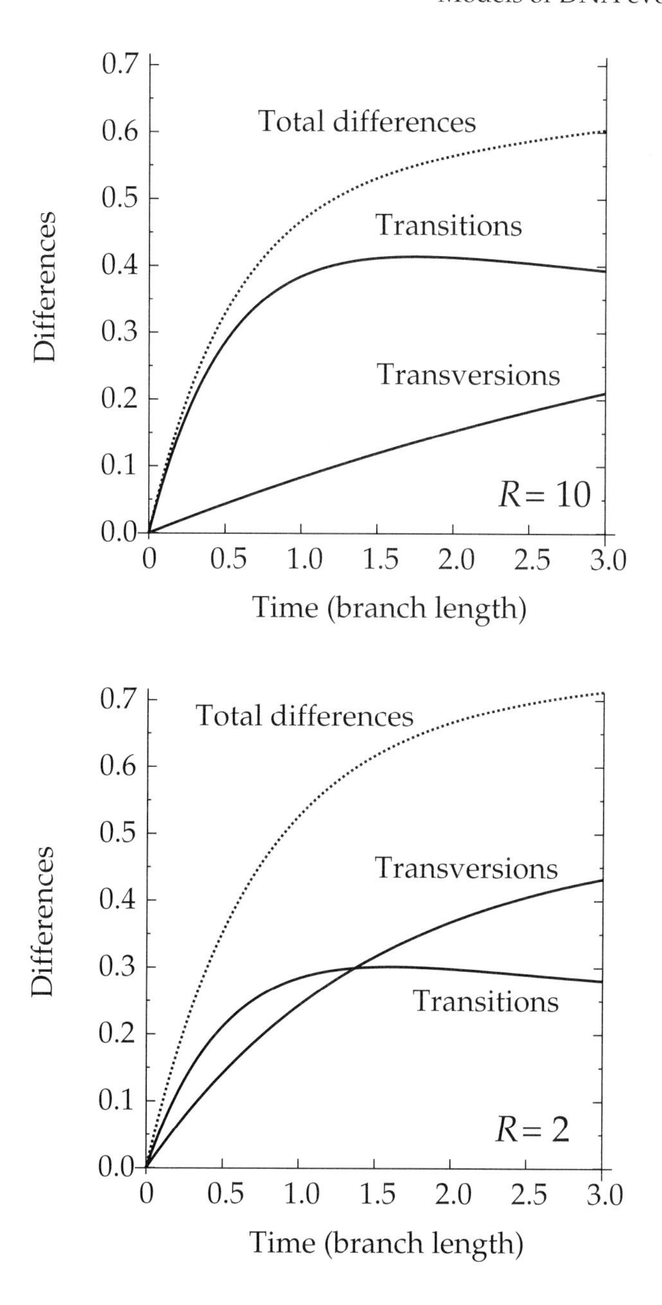

Figure 13.2: Net probabilities of transition and transversion change in Kimura's two-parameter model, with two different values of the transition/transversion ratio, $R = 10$ and $R = 2$. The first is typical of some mitochondrial DNAs, the second of mammalian nuclear DNA.

power of the estimation, and taking them into account may actually hurt. It is not hard to show that equations 13.3 in fact make maximum likelihood estimates of t and R, when they have real-valued solutions. We can also imagine making maximum likelihood estimates of t when R has a fixed value. Suppose that we think we know R. The likelihood, the probability of the data given the model and t, is the product of the probabilities of the differences at each site. At each site the prior probability of the observed base in the first sequence is $1/4$, and the conditional probability of the base observed in the other sequence is either $1-P-Q$, P, or $Q/2$ (there are two of the latter). If there are n sites in all, n_1 of which have transition differences, and n_2 transversion differences, the product of terms will be:

$$L \;=\; \text{Prob}\,(\text{data}\,|\,t, R) \;=\; \left(\frac{1}{4}\right)^n (1-P-Q)^{n-n_1-n_2}\, P^{n_1} \left(\frac{1}{2}Q\right)^{n_2} \qquad (13.4)$$

The quantities P and Q are given by the two equations 13.2. They are expressions in R and t. Our objective here is to maximize L with respect to t, holding R at the value that we have fixed. One might imagine solving this by differentiating equation 13.4 with respect to t, equating the derivative to 0, and solving for t. The reader who attempts this will find that no closed-form solution is possible. However, this need not stop us from solving it numerically in a computer, by, for example, Newton-Raphson iteration. The result is a rapid and well-behaved estimation method that has the advantage of preserving the consistency of R across all pairs of sequences. For each pair we must maximize its L with respect to t. An additional advantage of this framework is that it shows us that when $Q > 1/2$ or $2P + Q > 1$, the maximum likelihood estimate of t is infinite. One can also relate it to Kimura's formula, by differentiating equation 13.4 with respect to both t and R, equating both derivatives to 0, and showing that the resulting equations are precisely equations 13.2.

The Tamura-Nei model, F84, and HKY

The Kimura two-parameter model and the Jukes-Cantor model place great restrictions on the DNA sequences, in the interest of ease of computation. One assumption of the models that it would be nice to relax is that all four bases have equal expected frequencies. Two of the most widely used models that relax this assumption, and allow arbitrary base frequencies, are the *F84* and *HKY* models. The former has been in use in my phylogeny package PHYLIP since 1984, but was not published until Kishino and Hasegawa (1989) described it (with my assent). Felsenstein and Churchill (1996) have explained it further. The HKY model was introduced by Hasegawa, Kishino, and Yano (1985). These models are generically similar. Both extend the Kimura two-parameter model to asymmetric base frequencies, in slightly different ways. Both have five parameters. Rather than present them separately, we shall examine a slightly more generalized model, with six parameters, which includes both of these as a special case. It was introduced by Tamura and Nei (1993). Our notation will differ from theirs.

Table 13.1: Instantaneous rates of change in the Tamura-Nei model

From : / To :	A	G	C	T
A	$-$	$\alpha_R \pi_G/\pi_R + \beta\pi_G$	$\beta\pi_C$	$\beta\pi_T$
G	$\alpha_R \pi_A/\pi_R + \beta\pi_A$	$-$	$\beta\pi_C$	$\beta\pi_T$
C	$\beta\pi_A$	$\beta\pi_G$	$-$	$\alpha_Y \pi_T/\pi_Y + \beta\pi_T$
T	$\beta\pi_A$	$\beta\pi_G$	$\alpha_Y \pi_C/\pi_Y + \beta\pi_C$	$-$

In effect, the *Tamura-Nei model* imagines that we have at any site a constant probability of two kinds of event. If a site has a purine (A or G), it has a constant probability α_R per unit time of an event of type I, which replaces the base with a random base drawn from a pool of purines. It also has a constant probability of replacing it by a random base drawn from a pool of all four bases. If the site has a pyrimidine, it has a constant probability α_Y of replacing the base by a random base drawn from a pool of pyrimidines, and also a constant probability of replacing it by a base drawn from the pool of all four bases.

The pool of all four bases is imagined to contain them at some frequencies π_A, π_C, π_G, and π_T. It will turn out that these are precisely the equilibrium frequencies that we expect under this model. The purine pool and the pyrimidine pool are each set up with a mix of these bases in precisely the same ratios as in the overall pool. Thus in the purine pool, we have A and G in the ratio $\pi_A : \pi_G$. If we call the total frequency of purines in the overall pool $\pi_R = \pi_A + \pi_G$, the frequencies of A and G in the purine pool are then π_A/π_R and π_G/π_R. Similarly, the pyrimidine pool has these two bases in a ratio of $\pi_C : \pi_T$ and a total frequency of π_Y and thus frequencies of π_C/π_Y and π_T/π_Y.

The two kinds of events, in a tiny slice of time that has length dt, occur with probabilities $\alpha_R\, dt$ and $\beta\, dt$, when the original base is a purine. They occur with probabilities $\alpha_Y\, dt$ and $\beta\, dt$ when it is a pyrimidine. Table 13.1 shows the instantaneous probabilities per unit time of the 12 possible changes of base. Note that the order of bases is not alphabetical. These are rates, that is, when multiplied by dt they become the probabilities of change in a very small interval of length dt.

If this model has $\alpha_R = \alpha_Y$, it is the F84 model. If $\alpha_R/\alpha_Y = \pi_R/\pi_Y$, it is the HKY model. Both of these are special cases of the Tamura-Nei model.

The parameterization in terms of α_R, α_Y, and β is not the most useable one. We would like to express these in terms of the expected transition/transversion ratio R, and we would also like to scale the α_R, α_Y, and β so that the average rate of change per site is 1 per unit time. Noting that the four rows of Table 13.1

will apply π_A, π_G, π_C, and π_T of the time, respectively, we can add up the total expected rates of transitions and transversions. The total rate of transitions is:

$$T_s = 2\alpha_R \pi_A \pi_G / \pi_R + 2\alpha_Y \pi_C \pi_T / \pi_Y + \beta\,(2\pi_A \pi_G + 2\pi_C \pi_T) \tag{13.5}$$

and the total rate of transversions is:

$$T_v = 2\beta \pi_R \pi_Y \tag{13.6}$$

We would like to have $T_s/T_v = R$ and $T_s + T_v = 1$. Substituting 13.5 and 13.6 into those, and denoting the ratio of α_R to α_Y as ρ, we end up with equations that can be solved to yield the α's and β. We get:

$$\beta = \frac{1}{2\,\pi_R\,\pi_Y\,(1+R)} \tag{13.7}$$

$$\alpha_Y = \frac{\pi_R\,\pi_Y\,R - \pi_A\,\pi_G - \pi_C\,\pi_T}{2(1+R)\,(\pi_Y\,\pi_A\,\pi_G\,\rho + \pi_R\,\pi_C\,\pi_T)} \tag{13.8}$$

and

$$\alpha_R = \rho\,\alpha_Y \tag{13.9}$$

We could substitute these into Table 13.1, but this results in unexciting expressions, so we resist the temptation.

With all this done, it remains to find the transition probabilities (in the probabilistic rather than the molecular sense of the phrase) for a branch of length t. These are needed, as in the other cases, as raw material for distance or likelihood calculations.

The nice property that this family of models has is that they let us calculate the transition probabilities easily. The property that allows that is that each kind of event erases the effect of some events earlier. For example, suppose that we have a series of events of type I and type II that follow each other, the last event being of type II. Then, as that replaces the base by one randomly drawn from the nucleotide pool, it does not matter at all what other events preceded it: We know that the net probability of obtaining an A is π_A, of obtaining a C is π_C, and so on. If an event of type II is followed by an event of type I, the result is the same — it is only the type II event that matters. So a type II event anywhere along a branch results in the nucleotide at the end of the branch effectively being drawn at random from the nucleotide pool.

Likewise, if a branch has a series of events of type I, but none of type II, it does not matter how many events of type I it had: The probability of getting an A is π_A/π_R (if the lineage started with a purine). So type I events erase the effect of previous and subsequent type I events, and type II events erase the effect of any previous and subsequent events. Thus all we need to know of a branch is whether it has had any type I events, and whether it has had any type II events. If an event

Table 13.2: Probabilities of different numbers of Type I and Type II events in the model

If the branch starts with a purine:	
No events	$\exp(-(\alpha_R + \beta)t)$
Some type I, no type II	$\exp(-\beta t)\ (1 - \exp(-\alpha_R t))$
Some type II	$1 - \exp(-\beta t)$
If the branch starts with a pyrimidine:	
No events	$\exp(-(\alpha_Y + \beta)t)$
Some type I, no type II	$\exp(-\beta t)\ (1 - \exp(-\alpha_Y t))$
Some type II	$1 - \exp(-\beta t)$

had rate λ, the probability that there are none in a branch of length t is $\exp(-\lambda t)$. The net probabilities for a branch of length t are shown in Table 13.2.

From this table, from the rules about events "erasing" each other's effect, and from the composition of the nucleotide pools, one can easily work out the transition probabilities. For example, if we start with an A, a G could arise (1) if the last event is one of type I that results in a G in a lineage that has no events of type II, or (2) if there is an event of type II and the last event happens to yield a G. In the first case, the probability of a G is π_G/π_R, and in the event of the second, it is π_G. Putting these together,

$$\text{Prob}\,(G \mid A, t) \;=\; \exp(-\beta t)\,[1 - \exp(-\alpha_R t)]\,\frac{\pi_G}{\pi_R} \;+\; [1 - \exp(-\beta t)]\,\pi_G \tag{13.10}$$

We could write out 16 expressions like this, but we can write them more generally. Suppose that we use two notational devices when dealing with subscripts i and j that range over the four bases A, G, C, and T. They are the standard Kronecker delta function δ_{ij}, which is 1 if $i = j$ and 0 otherwise, and the "Watson-Kronecker" equivalent, ε_{ij}, which is 1 if i and j are both purines or if they are both pyrimidines, but is 0 otherwise. We also have α_i, which is either α_R or α_Y depending on whether i indexes a purine or a pyrimidine. With these notational conveniences, the transition probability can be written as:

$$\begin{aligned}\text{Prob}\,(j \mid i, t) \;=\;& \exp(-(\alpha_i + \beta)t)\ \delta_{ij} \\ &+ \exp(-\beta t)\ (1 - \exp(-\alpha_i t))\left(\frac{\pi_j \varepsilon_{ij}}{\sum_k \varepsilon_{jk}\pi_k}\right) \\ &+ (1 - \exp(-\beta t))\,\pi_j\end{aligned} \tag{13.11}$$

Note that the sum involving ε_{jk} in the denominator of the fraction of the term on the right side is simply π_R or π_Y, whichever is appropriate.

It must be kept in mind that these "events" are purely fictional, and are of use as a means of easily calculating transition probabilities. They are not to be confused with the actual processes of substitution of bases, since some of them result in no change. The substitution probabilities occur as if these events were happening.

There are no simple formulas for estimating distances from the Tamura-Nei model. However a numerical maximum likelihood approach is easy to construct using the formulas for the transition probabilities. Using the equilibrium frequencies π_i and the transition probabilities $\mathrm{Prob}\,(j|i,t)$ we note that the fraction of sites at which we have in our two species bases i and j, respectively, will be $\pi_i\,\mathrm{Prob}\,(j|i,t)$. The likelihood for DNA sequences n bases long that have base m_i at site i in the first sequence and base n_i at that site in the second sequence will then be

$$L = \prod_{i=1}^{n} \pi_{m_i}\ \mathrm{Prob}\,(n_i \mid m_i, t) \tag{13.12}$$

We can compute L for various values of t using the transition probability formulas, and use standard methods of searching for the maximum of a curve. This likelihood makes use of the fact that the joint probability of bases m_i and n_i does not depend on where their common ancestor is along the lineage connecting the two species. We can arbitrarily choose one to be ancestral to the other, without affecting the result. This comes from the property of reversibility.

Schadt, Sinsheimer, and Lange (1998) have presented a model with two more parameters than the Tamura-Nei model. They show how to calculate transition probabilities explicitly for this model. When constraints are imposed on it to ensure reversibility, it becomes the Tamura-Nei model. Schadt, Sinsheimer, and Lange's model is the most complex one so far for which transition probabilities can be explicitly calculated.

The general time-reversible model

All of the models so far have been *reversible*. If the equilibrium frequencies of the bases are π_A, π_C, π_G, and π_T, then a model is reversible if

$$\pi_i\ \mathrm{Prob}\,(j \mid i, t) \ = \ \pi_j\ \mathrm{Prob}\,(i \mid j, t) \tag{13.13}$$

This means that the probability of starting with i at one end of the branch, and ending with j at the other, is the same as the probability of starting with j and evolving to i. That is, if we see base i at one end of a branch, and base j at the other end, there is no way to decide which end was the ancestor and which the descendant. An example of a nonreversible model would be one in which A was most likely to change to G, G to C, C to T, and T to A. Thus if we saw many Cs at one end of a branch, and many Ts at the other, this would be a hint that the Cs were in the ancestor and the Ts in the descendant.

Table 13.3: The general time-reversible model of DNA evolution

From : \ To :	A	G	C	T
A	$-$	$\pi_G\,\alpha$	$\pi_C\,\beta$	$\pi_T\,\gamma$
G	$\pi_A\,\alpha$	$-$	$\pi_C\,\delta$	$\pi_T\,\varepsilon$
C	$\pi_A\,\beta$	$\pi_G\,\delta$	$-$	$\pi_T\,\eta$
T	$\pi_A\,\gamma$	$\pi_G\,\varepsilon$	$\pi_C\,\eta$	$-$

Reversibility is mathematically convenient; it is the basic reason why we usually are not able to place the root of a tree, and are consequently inferring unrooted trees. But there is no biological reason why models of DNA change should be reversible — it is a convenience rather than a reality. However, some of the reversible models come close enough to fitting real data that a realistic model may be nearly reversible, so that we have rather little statistical information enabling us to place the root of the tree.

The instantaneous rates of change for the most *general time-reversible (GTR)* model for DNA are shown in Table 13.3. This form of the rates is due to Lanave et al. (1984).

The rates as given above need to be adjusted so that one unit of time is the time in which we expect to see one change per base. As the π_i are the equilibrium frequencies of the bases, the total rate of changes will be the sum of the off-diagonal elements of Table 13.3, each multiplied by the probability that one would start with that base. Thus the rates should be standardized so that

$$2\pi_A\pi_G\alpha + 2\pi_A\pi_C\beta + 2\pi_A\pi_T\gamma + 2\pi_G\pi_C\delta + 2\pi_G\pi_T\varepsilon + 2\pi_C\pi_T\eta \;=\; 1 \qquad (13.14)$$

As with all models of DNA change, if we have the rate matrix $\mathbf{A}$ we can calculate the transition probability matrix $\mathbf{P}$ from it for any branch length t. In this case, the rate matrix is

$$\mathbf{A} = \begin{bmatrix} -(\alpha\pi_G + \beta\pi_C + \gamma\pi_T) & \alpha\pi_A & \beta\pi_A & \gamma\pi_A \\ \alpha\pi_G & -(\alpha\pi_A + \delta\pi_C + \varepsilon\pi_T) & \delta\pi_G & \varepsilon\pi_G \\ \beta\pi_C & \delta\pi_C & -(\beta\pi_A + \delta\pi_G + \eta\pi_T) & \eta\pi_C \\ \gamma\pi_T & \varepsilon\pi_T & \eta\pi_T & -(\gamma\pi_A + \varepsilon\pi_G + \eta\pi_C) \end{bmatrix} \qquad (13.15)$$

Note that this is the transpose of Table 13.3, so that a_{ij} is the rate at which base j changes into base i.

In principle, the transition probability matrix $\mathbf{P}$ can be computed from $\mathbf{A}$ by matrix exponentiation, so that

$$\mathbf{P}(t) \;=\; e^{\mathbf{A}t} \qquad (13.16)$$

In practice, this must be done numerically, as there is no convenient formula for the elements of $\mathbf{P}(t)$. For the other models we have been describing, formulas do exist, for example equations 11.16, 13.2, and 13.11 for the Jukes-Cantor, Kimura two-parameter, and Tamura-Nei models, respectively. In the case of the general time-reversible model, one must instead take the rate matrix $\mathbf{A}$ and find its eigenvalues λ_i and eigenvectors, decomposing it into the product

$$\mathbf{A} = \mathbf{T}\,\mathbf{\Lambda}\,\mathbf{T}^{-1} \tag{13.17}$$

where $\mathbf{\Lambda}$ is a diagonal matrix of the λ_i, and $\mathbf{T}$ is a matrix whose columns are the right eigenvectors of $\mathbf{A}$. The matrix exponential is then obtained by replacing the elements of $\mathbf{\Lambda}$ by $\exp(\lambda_i t)$. In place of $\mathbf{A}$ we will then have $\mathbf{P}(t)$. The eigenvalues and eigenvectors of $\mathbf{A}$ could be taken in advance and used repeatedly to evaluate $\mathbf{P}(t)$ for different values of t.

$\mathbf{A}$ is not a symmetric matrix, so that one might expect some difficulty getting the eigenvalues and eigenvectors. But it is the product of two symmetric matrices:

$$\mathbf{A} = \begin{bmatrix} \pi_A & 0 & 0 & 0 \\ 0 & \pi_G & 0 & 0 \\ 0 & 0 & \pi_C & 0 \\ 0 & 0 & 0 & \pi_T \end{bmatrix} \begin{bmatrix} -W & \alpha & \beta & \gamma \\ \alpha & -X & \delta & \varepsilon \\ \beta & \delta & -Y & \eta \\ \gamma & \varepsilon & \eta & -Z \end{bmatrix} \tag{13.18}$$

where W, X, Y, and Z are slightly messy but can easily be worked out from the diagonal entries in equation 13.15. We can call the two matrices on the right side of equation 13.18 $\mathbf{D}$ and $\mathbf{B}$. It is easily proven that if we can find the eigenvectors of the symmetric matrix $\mathbf{D}^{1/2}\mathbf{B}\mathbf{D}^{1/2}$, and the matrix whose columns are its right eigenvectors is called $\mathbf{U}$, that the eigenvectors and eigenvalues of $\mathbf{A}$ can be obtained from

$$\mathbf{A} = \left(\mathbf{D}^{1/2}\mathbf{U}\right)\mathbf{\Lambda}\left(\mathbf{U}^T\mathbf{D}^{-1/2}\right) = \left(\mathbf{D}^{1/2}\mathbf{U}\right)\mathbf{\Lambda}\left(\mathbf{D}^{1/2}\mathbf{U}\right)^{-1} \tag{13.19}$$

so that the eigenvectors of $\mathbf{A}$ can be obtained from those of $\mathbf{D}^{1/2}\mathbf{B}\mathbf{D}^{1/2}$ by premultiplication by the diagonal matrix $\mathbf{D}^{1/2}$, and the eigenvalues of $\mathbf{D}^{1/2}\mathbf{B}\mathbf{D}^{1/2}$ are the same as those of $\mathbf{A}$. Thus a program for finding the eigenvalues and eigenvectors of a symmetric matrix can be used.

Since there are only three nonzero eigenvalues of these matrices, we could also have used the closed-form solution of the cubic equation to write expressions for them. These are so tedious that there seems no point in doing so.

Distances from the GTR model

We can of course use the matrix machinery to compute distances from the general time-reversible model numerically. We could try different values of t, using some numerical optimization scheme, and compute the likelihood for t for the particular

Table 13.4: Pairs of nucleotides for 500 sites of sequence that has diverged according to a Kimura two-parameter model with $R = 2$ for a branch length of 0.2. The columns are the bases in the first sequence.

	A	G	C	T	total
A	93	13	3	3	112
G	10	105	3	4	122
C	6	4	113	18	141
T	7	4	21	93	125
total	116	126	140	118	500

pair of sequences. If the base at site i is m_i in one sequence, and n_i in the other, then the likelihood is:

$$L = \prod_i \pi_{n_i} P_{m_i n_i}(t) \tag{13.20}$$

We then have to find the value of t that maximizes this likelihood. If there are ambiguous nucleotides, the expression is a bit more complicated: Instead of a single pair of nucleotides at that site, we have a sum over all pairs that are compatible with the ambiguity.

This is numerically tedious. There is hope for a more direct approach, though it gets into difficulties too. Looking at equation 13.20, using the expressions in Table 13.3, we will find that there are only 10 different quantities. (This is because we can't distinguish between $A \rightarrow G$ and $G \rightarrow A$, for example.) The likelihood will be maximized if we can set the parameters of the model, and t, so that the expected frequencies of the 10 events are exactly equal to their observed frequencies. Can we do this? The number of parameters is the number of parameters in Table 13.3, plus one for t. This at first appears to be 11 (for the six Greek letters plus four equilibrium frequencies of bases, plus one for the branch length t). However, there is a constraint that all the π_i must add to 1, and another constraint, which is that equation 13.14 must hold. They reduce the number of parameters to exactly nine.

It then seems that we need only find the values of these parameters that fit the observed fractions of the 10 events exactly. Often this can be done, but sometimes not. As an example of a case in which it can be done, here are results from a computer simulation of two sequences that have diverged for a total branch length of 0.2 according to a Kimura two-parameter model with $R = 2.0$. Although we have simulated sequences evolving according to that simpler model, we are using the more general GTR model to analyze the resulting sequences. We can then see whether the individual rates of change between bases are inferred to be close to those in the K2P model.

The pairs of nucleotides can be tabulated across the 500 sites to give Table 13.4. Under the general time-reversible model, we can see that the number of sites at which the first sequence has (say) a G and the second a C has the same

Table 13.5: The same table of nucleotides as in Table 13.4, with symmetrical off-diagonal elements averaged.

	A	G	C	T	total
A	93	11.5	4.5	5	114
G	11.5	105	3.5	4	124
C	4.5	3.5	113	19.5	140.5
T	5	4	19.5	93	121.5
total	114	124	140.5	121.5	500

expectation as the number at which the first has a C and the second a G. We can therefore combine these cases, which we do by averaging the table with its transpose, getting Table 13.5. A relevant estimate of the equilibrium frequencies of the bases is obtained by dividing the column sums by 500 to get $(\pi_A, \pi_G, \pi_C, \pi_T) = (0.228, 0.248, 0.281, 0.243)$. Dividing each column by its sum, we get an estimate of the net transition matrix along the branch:

$$\widehat{\mathbf{P}} = \begin{bmatrix} 0.815789 & 0.0927419 & 0.0320285 & 0.0411523 \\ 0.100877 & 0.846774 & 0.024911 & 0.0329218 \\ 0.0394737 & 0.0282258 & 0.80427 & 0.160494 \\ 0.0438596 & 0.0322581 & 0.13879 & 0.765432 \end{bmatrix} \tag{13.21}$$

Noting from equation 13.16 that this is expected to be the matrix exponential of $\mathbf{A}t$, we can estimate $\mathbf{A}t$ by taking its matrix logarithm:

$$\widehat{\mathbf{A}t} = \log\left(\widehat{\mathbf{P}}\right) = \begin{bmatrix} -0.212413 & 0.110794 & 0.034160 & 0.046726 \\ 0.120512 & -0.174005 & 0.025043 & 0.035554 \\ 0.0421002 & 0.028375 & -0.236980 & 0.205579 \\ 0.0498001 & 0.034837 & 0.177778 & -0.287859 \end{bmatrix} \tag{13.22}$$

The matrix logarithm is the inverse of the matrix exponential. One takes the eigenvalues and eigenvectors of the matrix, takes the logarithms of the eigenvalues, and then reconstitutes the matrix with those. This may be impossible if any of the eigenvalues are not positive.

In this case we have an estimate of $\mathbf{A}t$. Can we separate the t from the $\mathbf{A}$? We can if we have some way of standardizing the rate matrix $\mathbf{A}$. The matrix $\mathbf{A}t$ is promising. The rates off the diagonal are all positive, and it can be verified that they do indeed have the property of reversibility. The standardization we seek is the requirement that the total rate of changes be 1 per unit time. This fixes $\mathbf{A}$. The changes from state j to state i are $\pi_j a_{ij}$ per unit time. It is possible to show rather easily that as the columns of $\mathbf{A}$ sum to zero, that if we let $\mathbf{D}$ be a diagonal matrix whose elements are the π_i, we require that the trace of the product $\mathbf{AD}$ satisfy

$$-\text{trace}(\widehat{\mathbf{A}}\,\widehat{\mathbf{D}}) = 1 \tag{13.23}$$

It follows that

$$\widehat{t} \;=\; -\mathrm{trace}(\widehat{\mathbf{A}t}\;\widehat{\mathbf{D}}) \;=\; -\mathrm{trace}\Big[\log(\widehat{\mathbf{P}})\;\mathbf{D}\Big] \tag{13.24}$$

This is the distance for the general time-reversible model. It is made by estimating the base frequencies π_i and the rates a_{ij}, and finding ones that exactly predict the observed net transition matrix $\mathbf{P}$. The estimate is only allowable if all the a_{ij} are nonnegative, and can only be computed if all the eigenvalues of $\mathbf{P}$ are positive.

Note that we also get, with the estimate of t, an estimate of $\mathbf{A}$. In symbolic terms this works out to:

$$\widehat{\mathbf{A}} \;=\; -\log(\widehat{\mathbf{P}})\;/\;\mathrm{trace}\Big[\log(\widehat{\mathbf{P}})\;\mathbf{D}\Big] \tag{13.25}$$

In the present numerical example the distance is estimated to be $\widehat{t} = 0.228125$. This is close to the true value of 0.2. The estimate of $\mathbf{A}$ works out to be:

$$\widehat{\mathbf{A}} = \begin{bmatrix} -0.931124 & 0.485671 & 0.149741 & 0.204826 \\ 0.528274 & -0.762764 & 0.109776 & 0.155852 \\ 0.184549 & 0.124383 & -1.038820 & 0.901168 \\ 0.218302 & 0.152710 & 0.779302 & -1.261850 \end{bmatrix} \tag{13.26}$$

This is moderately close to the rate matrix that is implied by the Kimura two-parameter model, which for this case would be

$$\mathbf{A} \;=\; \begin{bmatrix} -1 & 2/3 & 1/6 & 1/6 \\ 2/3 & -1 & 1/6 & 1/6 \\ 1/6 & 1/6 & -1 & 2/3 \\ 1/6 & 1/6 & 2/3 & -1 \end{bmatrix} \tag{13.27}$$

(For example, the average of the diagonal elements is –0.9986385, the average of the four parameters for transitions is 0.67360175, and the average of the eight parameters for transversions is 0.162517375.) However, an exactly fitting estimate of the rate matrix and distance is not always possible. As an example, consider an imaginary data set of 400 sites. Table 13.6 shows the nucleotide pairs at the two ends of the branch. There are 100 of each of the four bases in each of the sequences. 208 of the 400 sites show the same base in both sequences, so that 50.2% of the bases have not changed. Some strange features are seen (for example, an `A` is more likely to have changed than not) but even those might well be expected occasionally as the result of a fluctuation due to random sampling.

On more detailed examination it will be found that it is impossible for any set of rates in equation 13.16 to lead to precisely these empirical joint frequencies. It can do so only if we allow negative rates, which is impossible. The constraint against negative rates prevents us from fitting each and every empirical result exactly. The symptom of this problem is that when we attempt to take the matrix logarithm when evaluating equation 13.24, we encounter a negative eigenvalue

Table 13.6: An imaginary data set of two sequences 400 bases in length. They are tabulated according to the base that appears in the first sequence (vertical) and in the second sequence (horizontal).

	A	G	C	T	total
A	32	40	8	20	100
G	40	40	8	12	100
C	8	8	76	8	100
T	20	12	8	60	100
total	100	100	100	100	400

whose logarithm would be an imaginary number. The correct value of the distance in this case would be infinite (Waddell and Steel, 1997).

The general time-reversible model is more apt to give infinite distances than are the more restrictive models that we have discussed previously. This can create problems for their use in situations such as bootstrap sampling of sites, in which the tables of pairs of nucleotides can vary in many ways.

The general time-reversible model distance in equation 13.24 was introduced by Lanave et al. (1984). The equations in this section were derived by Rodríguez et al. (1990).

The general 12-parameter model

If we do not assume that the stochastic process is reversible, it is still possible to use most of the preceding section. We would then have a general Markov process, though we would assume that it was the same throughout the tree, and that it had an equilibrium distribution. There would be 12 parameters, as from each of the four states there would be four probabilities of change, which had to add up to 1. So there would be three parameters for each column of the transition matrix. We would have equation 13.20 for the likelihood, and we would reduce the data to a table such as Table 13.4. Now, however, we would not symmetrize that table, but would use it as it stands. Dividing the numbers in its columns by the sum of that column, we would estimate $\widehat{\mathbf{P}}$.

However, it is not quite this simple. In estimating the equilibrium frequencies π_A, π_G, π_C, and π_T, as well as the 12 rates of change, we cannot simply use the empirical frequencies and the matrix logarithm of the conditional probability matrix, as we did in the previous section. These might, in this case, give us estimates that were impossible under the model, such as base frequencies that would not result from the estimated rates of change. We have to maximize the likelihood in equation 13.20 subject to constraints. These are that the frequencies of the bases add to 1, that the columns of the transition matrix $\mathbf{P}$ add to 1, and that the base frequencies be the equilibrium frequencies for this transition matrix.

There are no formulas for what the estimates will be, or what the resulting distance will be. Rodríguez et al. (1990) pointed out that when we have a molecular clock, with two branches of equal length leading from an ancestor to these two descendants, the 12 parameters are confounded and we cannot estimate them all. This is not quite the same as saying that we cannot estimate the distance, but there the matter lies. This model needs more work to determine whether there is any way of using it.

A serious issue is that we cannot rely on reversibility. Thus if we have a pair of species i and j, the probability of the 16 possible combinations of bases depends, not simply on the total branch length between them, but on where along that branch their latest common ancestor was. We would be considering this simultaneously for all pairs of species. Even if we were willing to estimate this location, would we not want to keep it consistent between all pairs? Thus if the sequences for human and oak tree are best explained by having their common ancestor be close to the oak, shouldn't the branch between chimpanzee and oak tree also have it there?

LogDet distances

We have argued that there is work to be done to see whether the general 12-parameter model is useable. However, there is one body of work that seems to deliver even more than that. This is on the *LogDet* or *paralinear* distance measure. A version of it was introduced by Barry and Hartigan (1987) and it reached full development in the work of Lake (1994), Steel (1994a), and Lockhart et al. (1994). The LogDet distance can be defined as (Steel, 1994a)

$$\hat{t} = -\ln \det(\mathbf{F}) \tag{13.28}$$

where $\det$ is the determinant and $\mathbf{F}$ is the "divergence matrix" whose (i, j) element is the fraction of sites at which the first sequence has base i and the second has base j. However, this form of the LogDet distance has the disadvantage of having a nonzero value when the two sequences are identical. To cure this problem both Lockhart et al. (1994) and Lake (1994) rescale $\mathbf{F}$ according to the base composition. Lockhart et al. obtain:

$$\hat{t} = -\frac{1}{4}\left[\ln \det(\mathbf{F}) - \frac{1}{2}\ln \det(\mathbf{D}_x \mathbf{D}_y)\right] \tag{13.29}$$

where $\mathbf{D}_x$ is the diagonal matrix of the observed base frequencies in the first sequence, and $\mathbf{D}_y$ is the diagonal matrix of the observed base frequencies in the second. Lake's (1994) paralinear distance is the same except for the factor of 1/4.

The argument is that this distance applies to models in which the process of base change differs at different points along the tree. One wonders whether it would be equivalent to estimating rates in different regions of the tree, and then

Table 13.7: The table of joint frequencies nucleotides for two sequences that evolve from a common ancestor according to HKY models with different equilibrium base frequencies

	A	G	C	T	total
A	0.101572	0.0328061	0.0239532	0.0151002	0.173431
G	0.058329	0.106783	0.0354118	0.0239532	0.224477
C	0.074999	0.0609347	0.106783	0.0328061	0.275523
T	0.091669	0.074999	0.058329	0.101572	0.326569
total	0.326569	0.275523	0.224477	0.173431	1.0000

even though these are confounded with each other, somehow extracting the correct estimate of the distance anyway. The LogDet distance does not quite deliver as much as this. For example, consider an ancestor in which the base composition is $(1/4, 1/4, 1/4, 1/4)$. One descendant lineage is assumed to be 0.5 long, and along this we imagine the sequence evolving according to an HKY model with $R = 2/3$ and equilibrium base frequencies for A, G, C, and T of $(0.4, 0.3, 0.2, 0.1)$. The other descendant lineage is also 0.5 in branch length, and also has an HKY model operating on it with $R = 2/3$, but for this model the equilibrium base frequencies are taken as $(0.1, 0.2, 0.3, 0.4)$. If we observe enough sites, we will find the joint base frequencies in the two sequences to be as given in Table 13.7. Using these joint frequencies, the marginal frequencies in the two sequences, and equation 13.29, we obtain the distance 1.04418. The true distance is a bit smaller, 1.0. This is no mere sampling fluctuation—we have assumed an infinite number of sites to sample from. Waddell (1995) noted this overestimation when the frequencies of the bases are unequal.

Suppose that we have two successive branches that separate a pair of species. The base composition starts out with its diagonal matrix being $\mathbf{D}^{(0)}$. Then it evolves with transition probability matrix $\mathbf{P}^{(1)}$. After that its base composition becomes $\mathbf{D}^{(1)}$. Then a different transition matrix comes into play for the second branch, $\mathbf{P}^{(2)}$. It is straightforward to show that the divergence matrix will be

$$\mathbf{F} = \mathbf{P}^{(2)} \mathbf{P}^{(1)} \mathbf{D}^{(0)} \tag{13.30}$$

The divergence matrices for the two successive branches, if we could observe them, would be

$$\mathbf{F}^{(1)} = \mathbf{P}^{(1)} \mathbf{D}^{(0)} \tag{13.31}$$

and

$$\mathbf{F}^{(2)} = \mathbf{P}^{(2)} \mathbf{D}^{(1)} \tag{13.32}$$

It can be shown that the LogDet distances for the two individual branches do not sum up to give the LogDet distance for the two branches one after the other. The exception is when the frequencies at the end of branch 1 are $(1/4, 1/4, 1/4, 1/4)$. Then the LogDet distances sum properly.

The upshot of all this is that LogDet distances do not have the magical properties that they seem to, unless we somehow constrain all the base frequencies to be equal. However, they do give us distances that are "tree-additive" (Steel, 1994a; Lockhart et al., 1994). In the limit as the sequences become infinitely long, they can be fit perfectly by an additive tree. However, the problems with the base frequencies can lead to the branch lengths being wrong. The problem arises because we cannot assume that the expected base frequencies at the start of a branch are the equilibrium frequencies under the Markov process that operates along that branch. Since earlier branches have other processes, there is no necessity that they generate those base frequencies.

Thus LogDet distance also does not solve the problem of how to get a distance that applies to the general 12-parameter model. Nevertheless, LogDet distances are frequently applied to correct for inequalities of base composition across a tree. Swofford et al. (1996), in their excellent short summary of this issue, argue on the basis of experience that LogDets are no worse than other distances in handling this case, as the other distances also have assumptions that are violated. Our numerical example can serve to check this: If we were to fit an HKY model to the table of joint frequencies (Table 13.7) it turns out that the overall frequency of bases has frequency 0.25 for each base, and we are actually fitting a K2P model. Using the equations 13.3, we find that the maximum likelihood estimate of the distance is 1.1302. Thus LogDet is doing better, with less than one-third as much bias.

Galtier and Gouy (1995) have described another method for coping with base composition variation between lineages. It is less general, as it allows only GC versus AT variations, and it is rather approximate, as it assumes that the observed base compositions in two lineages are the equilibrium base compositions for the processes of change since their common ancestor. As far as I know there has been no comparison of this method with the LogDet method.

Other distances

There are many more distances, but there is no space here to discuss most of them. Zharkikh (1994) has a good review covering many distances and explaining relationships between them. Yang, Goldman, and Friday (1994) carry out a likelihood-based evaluation of the fit of several models to real DNA data.

In reading the distance literature, it is good to keep in mind the distinction between statistics and their expectations, a distinction ignored by many authors. Much of the distance literature treats the distance as a quantity that has some intrinsic meaning, rather than as an estimate of the branch length in a two-species tree. As a result, distances are often not presented as estimates, and their formulas are frequently not derived from any statistical principle. This has led to a literature which is quite difficult to read. Nevertheless, most of the resulting formulas

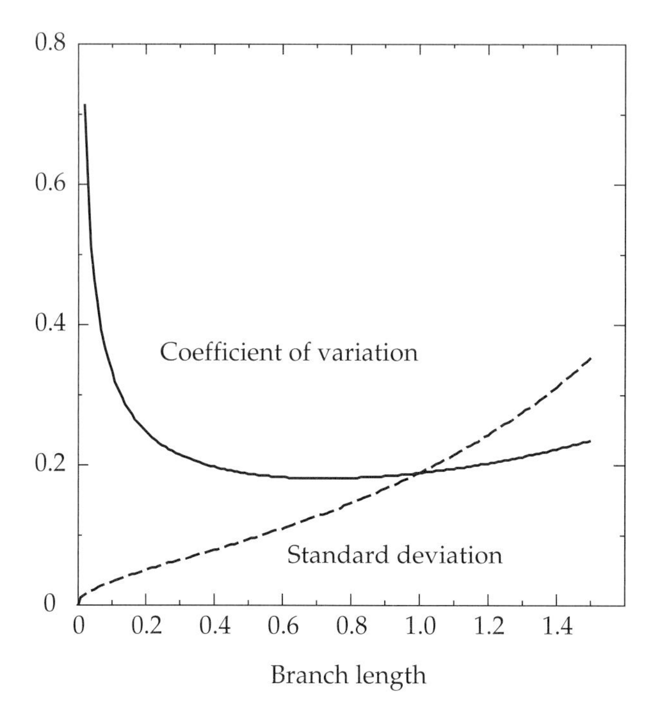

Figure 13.3: The large-sample standard deviation of the Jukes-Cantor distance (dashed curve) and its coefficient of variation (solid curve), for different values of the true branch length, when the sequences are 100 bases long

can be shown to be well-behaved, being at least consistent when considered as statistical estimators.

Variance of distance

Since distances are estimates of the branch length in two-species phylogenies, we can use the likelihood curve for each distance to examine its statistical error (as will be explained further for phylogenies in Chapter 19). For some of the simpler distances we can get closed-form expressions for the large-sample variance of the distance. For the Jukes-Cantor distance we can note that it is computed from the variance of the proportion p of nucleotides that differ between the sequences. This has the usual binomial variance $p(1-p)/n$ when there are n sites. The large-sample "delta method" approximation to the variance of the distance is obtained by multiplying this by the square of the derivative of the function relating distance to p. This function is given in equation 11.18 (where p is called D_S). The result is that the variance is

$$\mathrm{Var}(D) = \frac{9\,p\,(1-p)}{(3-4p)^2\;n} \tag{13.33}$$

When equation 11.17 is used with $u = 1$ to express this in terms of a branch length t, this becomes

$$\mathrm{Var}(\,\hat{t}\,) = \frac{3}{16n}\left(e^{4t/3}+3\right)\left(e^{4t/3}-1\right) \tag{13.34}$$

The large-sample variance of the Jukes-Cantor distance was first given by Kimura and Ohta (1972) for the more general case of an arbitrary number of states.

Figure 13.3 shows the standard deviation and the coefficient of variation (ratio of standard deviation to mean) for sequences of length 100. The standard deviation rises at first as the square root of branch length, quickly passes through an inflection point, and then starts to rise faster than linearly. The result is that the coefficient of variation is remarkably level for branch lengths between 0.2 and 1.4. This gives some support to using a power of 2.0 in the least squares weights $w_{ij} = 1/D_{ij}^2$. But above branch lengths of 1.4 the variance rises more rapidly, introducing more noise than this weighting would imply. The variance rises faster than any power of branch length, implying that large distances can introduce substantial noise into the phylogeny unless appropriate weights are used.

The coefficient of variation calculations confirm our suspicion that we have the most accurate picture of the amount of divergence when it is intermediate. The branch length that is best estimated is 0.719, which corresponds to sequences 46.2% different. Much shorter sequences lead to too few differences to estimate accurately, and much longer sequences have too many sites with multiple substitutions obscuring each other.

It is a startling fact that the actual variance is infinite. There is a small (often astronomically small) probability that the sequences will be more than 75% different, in which case the estimated branch length is infinite. The large sample approximation ignores this dismaying possibility.

Rate variation between sites or loci

Different rates at different sites

In all the distance calculations so far in this chapter, we have assumed that the rates of change at all sites are equal. We have always taken the rate of change to be 1 per unit branch length. If we have sites with different rates of change, we will want to standardize branch lengths so that the average rate of change across sites is 1 per unit branch length. Thus if we have six sites with rates of change that stand in the ratios: 2.0 : 3.0 : 1.5 : 1.0 : 4.4 : 0.8, we should divide these by their average (which is 2.11666) to get the rates as 0.94489 : 1.41732 : 0.70866 : 0.47244 : 2.078740 : 0.377953. These now have a mean of 1.

In any of these models, a site that has rate of change r, and that changes along a branch of length t, simply behaves as if it were changing with rate 1 along a branch of length rt. Thus, given that we know the rate of change of a site, and given that we can compute the transition probability matrix for any branch length, $\mathbf{P}(t)$, we need only compute the transition probability $\mathbf{P}(rt)$ instead. This can be done site by site, using the rates for each site. For example, if we thought that rates at first, second and third positions in a protein coding sequence were in the ratio of 2: 1.2 : 5.0, the rates we would use for those positions would be 0.7317, 0.4390, and 1.8293, which average to 1. The probability of change in a given site in a branch of length t would then be either $\mathbf{P}(0.7317t)$, $\mathbf{P}(0.4390t)$, or $\mathbf{P}(1.8293t)$, depending on which codon position the site was in.

Distances with known rates

We can use the maximum likelihood approach to compute distances, for any of the reversible models of change given in this chapter, if we know the relative rates of change at all the sites in the sequence. We first standardize the rates, as we discussed in the previous paragraphs. Then we can use a modification of equation 13.20. If r_i is the relative rate of change at site i, we need only modify equation 13.20 a bit to write the likelihood as

$$L = \prod_i \pi_{n_i} P_{m_i n_i}(r_i t) \tag{13.35}$$

and then find the value of t that maximizes it. This can be done for the Jukes-Cantor, Kimura two-parameter, Tamura-Nei, and general time-reversible models. The resulting distance will be scaled so that one unit is one expected change per site, and it will correctly take into account the fact that some sites are expected to change more quickly than others.

Distribution of rates

Of course, often we know that there are unequal rates of change at different sites, but we do not know in advance which sites have high and which low rates of change. One possible approach would be to estimate the rate at each site. This requires a separate parameter, r_i, for each site. Such an approach has been used by Gary Olsen (Swofford et al., 1996). As the number of sites increases, the number of parameters being estimated rises correspondingly. This is worrisome: in such "infinitely many parameters" cases maximum likelihood often misbehaves and fails to converge to the correct tree as the number of sites increases. A likelihood ratio test allowing different rates among segments of DNA has been proposed by Gaut and Weir (1994). They found by simulation that it failed to follow the proper distribution when segment sizes were small.

A better-behaved alternative is to assume a distribution of rates and that each site has a rate drawn from that distribution at random, and independently of other sites. If we are computing the distance, for example, we must for each site integrate over all possible rates, multiplying the contribution of the site to likelihood for this pair of species, evaluated at that rate, by the prior probability density $f(r)$ of the rate. Then

$$L(t) = \prod_{i=1}^{\text{sites}} \left[\int_0^\infty f(r)\, \pi_{n_i} P_{m_i n_i}(rt)\, dr \right] \tag{13.36}$$

This can be done numerically for any density of rates $f(r)$. For some special density functions there are useful analytical results as well.

Gamma- and lognormally distributed rates

Rate variation was first explicitly modeled in molecular evolution by Uzzell and Corbin (1971) (see also Nei, Chakraborty, and Fuerst, 1976). They used a gamma distribution, as it was analytically tractable, varied from 0 to ∞, and had parameters that would allow the mean and variance to be controlled.

Olsen (1987) modeled rate variation by a lognormal distribution (a distribution in which the logarithm of the rates is normally distributed). He was unable to get analytical formulas for the distance and so used numerical integration.

The gamma model was further developed by Jin and Nei (1990), who were able to use the analytical tractability of the gamma distribution to get formulas for the Jukes-Cantor distance with gamma distributed rates. Although it is possible to carry out the integration with respect to the rate density $f(r)$ for the transition probability $P_{ij}(rt)$, this is of utility mostly for computing distances. We shall see that the transition probabilities are used in maximum likelihood estimation of phylogenies. However, to integrate the probability for each branch against the density function of $f(r)$ is to assume that the rate varies, not only from site to site, but from branch to branch. By integrating separately for each branch, we in effect assume that the rates vary independently from branch to branch. This is too wild a rate variation. A more realistic model would have rates varying from site to site in an autocorrelated fashion, and from branch to branch also in an autocorrelated fashion. Such a pattern was envisaged by Fitch and Markowitz (1970) in their model of "covarions."

Invariant sites. In addition to gamma-distributed rate variation, it is also often biologically realistic to assume that there is a probability f_0 that a site is invariant, that it has zero rate of change (Hasegawa, Kishino, and Yano, 1987). It is quite easy to do calculations for such a model, and frequently such a class of invariant sites is included with the rest of the sites having rates drawn from a gamma distribution.

Distances from gamma-distributed rates

The one tractable case of interest is the computation of distances using the gamma distribution of rates. The gamma distribution has density function

$$f(r) = \frac{1}{\Gamma(\alpha)\,\beta^\alpha}\, r^{\alpha-1}\, e^{-\frac{r}{\beta}} \tag{13.37}$$

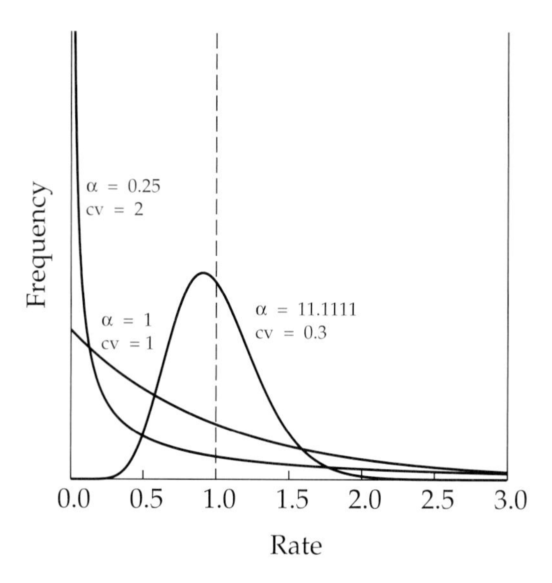

Figure 13.4: The density function of a gamma distribution of rates when the mean of the rates is 1 and there are various amounts of variance in the rates. For each of the three densities, the value of α and the coefficient of variation (cv) of the rates is shown.

The mean of the distribution is $\alpha\beta$ and its variance is $\alpha\beta^2$. If we want the gamma distribution of rates r to have a mean of 1, we take the density in this equation and we set $\beta = 1/\alpha$ and have the density function

$$f(r) = \frac{\alpha^\alpha}{\Gamma(\alpha)} r^{\alpha-1} e^{-\alpha r} \tag{13.38}$$

Figure 13.4 shows three gamma distributions, for different values of the parameter α. The "shape parameter" α is the inverse of the squared coefficient of variation of the rates. Thus if the rates have coefficient of variation 0.3, so that the standard deviation of rates is 30% of their mean, the value of α is $1/0.09 = 11.1111$. Figure 13.4 shows the density function of the rates for three different values of α, and corresponding coefficients of variation of the rates. I find the coefficient of variation a more meaningful description of the degree of rate variation, but α is the parameter commonly used in the literature.

When the coefficient of variation is small, the distribution of rates varies around 1 in what is nearly a normal distribution. As the coefficient of variation goes to zero, so that α goes to infinity, the distribution becomes a single spike centered on 1 (where the dashed line is). As the coefficient of variation rises, the dis-

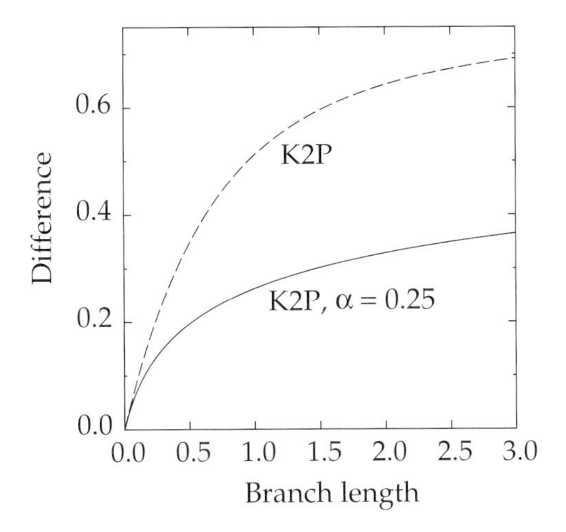

Figure 13.5: Comparison of the rise of sequence difference with time in Kimura 2-parameter models (R = 3) with no rate variation (dashed curve) and with gamma-distributed rate variation with $\alpha = 0.25$ (solid curve). In the latter case the high-rate sites saturate quickly, and the sequence difference then rises much more slowly as the low-rate sites gradually accumulate differences.

tribution of rates spreads out, to have many of the sites with rates that are nearly 0, and the rest with rates that are quite large.

There is nothing about the gamma distribution that makes it more biologically realistic than any other distribution, such as the lognormal. It is used because of its mathematical tractability.

Jin and Nei's distance. Jin and Nei (1990) have used this approach to derive a formula for the Jukes-Cantor distance with gamma-distributed rates. In the case of the Jukes-Cantor distance, the sufficient statistic is the fraction of sites that differ between two sequences. The distance will be the one that predicts this fraction exactly (unless the fraction exceeds 3/4, in which case the distance is infinity. Taking equation 11.17, altering it slightly to accommodate the rate r as a multiplier of ut, and integrating it over a gamma distribution of rates, gives

$$D_S = \int_0^\infty f(r)\frac{3}{4}\left(1 - e^{-\frac{4}{3}r\,ut}\right)\,dr \qquad (13.39)$$

where $f(r)$ is the gamma distribution. This can be done exactly. The result is related to the moment-generating function of the gamma distribution, which is the expectation of $\exp(\theta r)$. When the integral is evaluated and the resulting formula solved for ut, the result is

$$D = -\frac{3}{4}\alpha\left[1-\left(1-\frac{4}{3}D_S\right)^{-1/\alpha}\right] \tag{13.40}$$

ML distances for other models. For most more complex models, there is no analytically tractable formula for the distance once gamma-distributed rates are allowed, but it is not hard to obtain formulas for the likelihood, and this can be maximized numerically. In cases such as the Kimura two-parameter model and the Tamura-Nei model (including its subcases the HKY and F84 models), the quantities $P_{ij}(t)$ can be computed. In each of these models there are formulas (equations 11.16, 13.2, and 13.11) of the form

$$P_{ij}(t) = A_{ij} + B_{ij}e^{-bt} + C_{ij}e^{-ct} \tag{13.41}$$

(for appropriate choices of As, Bs, and Cs). Looking at equation 13.36, and substituting in equation 13.41 evaluated at rt, we need only to be able to carry out the integration. A bit of consideration of moment-generating functions will show that all that needs to be done is to replace the exponentials such as $\exp(-bt)$ by their expectation under the gamma distribution, which is

$$\mathbb{E}_r\left[e^{-brt}\right] = \left(1+\frac{1}{\alpha}bt\right)^{-\alpha} \tag{13.42}$$

Thus we can get transition probabilities for all of these models, as needed. For the general time-reversible model one can do an analogous operation (Gu and Li, 1996; Waddell and Steel, 1997). After the transition probabilities are available, the likelihood is straightforward to evaluate and maximize. Figure 13.5 shows the large effect that variation of rates among sites can have on distance calculations. In that figure, a Kimura two-parameter model with transition/transversion ratio 3 has gamma-distributed variation of rate of substitution with $\alpha = 0.25$, and that model is compared to one with no rate variation. This is one of the curves shown in Figure 13.5. The figure plots the difference between sequences (vertical axis) against the true distance (horizontal axis). At first the sequence differences rise at the same rate, but once the high-rate sites have saturated, difference afterwards accumulates much more slowly owing to the slow process of substitution elsewhere. Both curves will ultimately plateau at 0.75.

A general formulation. Gu and Li (1996) and Waddell and Steel (1997) discuss more general methods that can allow other distributions of rates. They give general formulas for the GTR model, which, instead of the matrix logarithm, apply the inverse of the moment generating function of the distribution of rates across sites. This is not always easily accessible, but when it is, their formulas will allow computation of the transition probabilities, and thus the estimation of the maximum likelihood distances. As all the other models we have discussed are special cases of the GTR model, they too can be treated in this way.

Models with nonindependence of sites

Rates can be correlated between nearby sites; we will see in Chapter 16 this can be handled by hidden Markov model methods. In some cases the correlations will be stronger than that—the occurrence of substitutions in one site will affect the occurrence in another. The most dramatic cases of this correlation are the *compensating substitutions* that occur in ribosomal RNA molecules. Where two sites are paired in the stem of a secondary structure, natural selection will favor them continuing to be paired. If an A is replaced by a G, natural selection will favor replacing the U with a C in its pairing partner.

Assuming that we know the structure, and that it has not changed during this process, we could imagine a model in which a pair, AU is replaced by another pair, GC. GU or UG pairs can also occur, though they are less favored. Such a model has been developed by Tillier (1994; see also Tillier and Collins, 1995, 1998, and related models by von Haeseler and Schöniger, 1995, and Muse, 1995). Her model allows changes between AU and GU, and also between GC and GU, with a lower rate of change between AU and GC. At a different rate, any of these pairs can change into any of the reverse pairs UA, UG, and CG. Use of such models has confirmed the existence of compensating changes (Rzhetsky, 1995; Tillier, and Collins, 1998).

Models with compensating substitutions have unusual properties, as has been pointed out by Carl Woese (1987). Two independently-evolving sites would both change in the same branch with a probability that depends on $(ut)^2$. But if there are compensating substitutions, after one has occurred, its bearers are at a fitness disadvantage s. Unless "rescued" by the compensating change, they will be eliminated from the population. After the first mutation, we expect only about $1/s$ copies of that mutation ever to exist before it is eliminated by natural selection. Thus the occurrence of the first substitution has probability that depends on ut, but the occurrence of the compensating mutation has probability dependent on u/s. Thus the probability that both changes occur is a function of u^2t/s rather than u^2t^2. Woese points out that this provides us with the opportunity to separate u from t, whereas they are normally confounded. If we can do so, we can use the t's for the different branches of the tree to locate its root.

Context-dependent models of DNA change have also been made, usually without a specific model of the mechanism of the dependency. Schöniger and von Haeseler (1994) and Jensen and Pedersen (2000) have modeled the dependence of rates of change on sequence at neighboring sites and have discussed methods of estimation of the pattern of rates. The number of possible parameters being large, approaches that use models of the expected dependency will have an advantage.

Chapter 14

Models of protein evolution

The first molecular sequences available were protein sequences, so it should not be surprising that some of the earliest phylogeny methods were designed to work with protein sequences. Eck and Dayhoff (1966) made the first numerical molecular phylogeny with protein sequences. Fitch and Margoliash (1967) designed their distance matrix methods with protein sequences in mind, and Neyman (1971) made the first molecular use of likelihood methods on protein sequences.

Amino acid models

Eck and Dayhoff (1966) used a symmetric Wagner parsimony method whose states were the 20 amino acids. Any amino acid could change to any other with equal ease. Although Eck and Dayhoff's model was not formulated probabilistically, Neyman (1971) explicitly used an m-state model with symmetric change between states, with the case of $m = 20$ in mind. However, this is not a very realistic model, as it allows changes between amino acids that may be very remote from each other in the genetic code and that also might not be observed to change to each other very often in actual sequences.

The Dayhoff model

Dayhoff and Eck (1968, chap. 4) had already formulated a probabilistic model of change among amino acids, based on tabulating changes among closely related sequences (it is usually known as the *Dayhoff model*). Although there was no explicit reference to the genetic code, it was hoped that the empirical frequencies of change would reflect the constraints caused by the code. In its more complete development, Dayhoff, Shwartz, and Orcutt (1979) tabulated 1,572 changes of amino acid inferred by parsimony in 71 sets of closely related proteins. From this they compiled a "mutation probability matrix" that showed the probabilities of changes from one amino acid to another, for different amounts of evolutionary divergence.

The most important of these tables is the mutation data matrix, the PAM 001 table. PAM stands for *probability of accepted mutation*. PAM 001 (Table 14.1) is the probability of changing from one amino acid to another along a branch short enough that 1% of the amino acids are expected to have changed. Each column shows the probabilities of change to different amino acids. It is noticeable that the entries of the PAM 001 table have been rounded off, so that the columns do not add exactly to 10,000. Even if they had not been rounded, the probabilities would also not have been reversible. The trees from which the mutation probabilities were tabulated were apparently rooted, so that it could be seen whether a given change was $A \rightarrow R$ or $R \rightarrow A$, and these are tabulated separately. If the trees had been considered to be unrooted, it would have been necessary to make the matrix be reversible.

The PAM 001 model can be turned into a transition probability model for larger evolutionary distances. A widely used one is the PAM 250 matrix, which is, in effect, the result of raising the PAM 001 matrix to the 250th power, corresponding to an expected difference of amino acid sequences of about 80%.

PAM matrices are derived by taking a large number of pairs of sequences from closely related species. For each, it is assumed that the sequences are close enough that no multiple substitutions have occurred. If in a pair of species, a protein differs at 1 out of 100 positions, and we see an Alanine in one species and a Threonine in the other, we count that as one difference and tabulate it for that pair of amino acids. But if the proteins differ at 2 out of 100 positions, and one has an Alanine opposite a Threonine, this is counted as only 0.5 of a difference in tabulating the number of changes from one amino acid to the other. It is assumed that the branch length is twice as long in this case, so that each observed difference indicates less change.

We will not go further into explaining how Dayhoff, Schwartz, and Orcutt (1979) derived their PAM 001 matrix, as this is not essential. The point is that it was the first major attempt to derive an empirically-based probabilistic model of amino acid substitution.

Other empirically-based models

As further data have accumulated, others have redone the PAM tabulation, with a much larger base of data. Jones, Taylor, and Thornton (1992) have described their procedure for tabulating mutation data matrices from protein databases, and have derived a PAM 250 matrix from this larger set of data. In addition, it is noticeable that these mutation data matrices are averages over many different contexts. Jones, Taylor, and Thornton (1994a, b) have tabulated a separate mutation data matrix for transmembrane proteins. Koshi and Goldstein (1995, 1997, 1998) have described the tabulation of further context-dependent mutation data matrices. Dimmic, Mindell, and Goldstein (2000) have made one of the most interesting tabulations, using estimated "fitness" values for different amino acids, that affect

Table 14.1: The PAM 001 table of transition probabilities from one amino acid to another along a branch short enough that 1% of the amino acids are expected to have changed. Values shown are multiplied by 10,000. The original amino acids are shown across the top, their replacements down the side. The columns thus add to (nearly) 10,000. The amino acids are arranged alphabetically by their three-letter abbreviations; the one-letter codes are also shown.

		A	R	N	D	C	Q	E	G	H	I	L	K	M	F	P	S	T	W	Y	V
		ala	arg	asn	asp	cys	gln	glu	gly	his	ile	leu	lys	met	phe	pro	ser	thr	trp	tyr	val
A	ala	9867	2	9	10	3	8	17	21	2	6	4	2	6	2	22	35	32	0	2	18
R	arg	1	9913	1	0	1	10	0	0	10	3	1	19	4	1	4	6	1	8	0	1
N	asn	4	1	9822	36	0	4	6	6	21	3	1	13	0	1	2	20	9	1	4	1
D	asp	6	0	42	9859	0	6	53	6	4	1	0	3	0	0	1	5	3	0	0	1
C	cys	1	1	0	0	9973	0	0	0	1	1	0	0	0	0	1	5	1	0	3	2
Q	gln	3	9	4	5	0	9876	27	1	23	1	3	6	4	0	6	2	2	0	0	1
E	glu	10	0	7	56	0	35	9865	4	2	3	1	4	1	0	3	4	2	0	1	2
G	gly	21	1	12	11	1	3	7	9935	1	0	1	2	1	1	3	21	3	0	0	5
H	his	1	8	18	3	1	20	1	0	9912	0	1	1	0	2	3	1	1	1	4	1
I	ile	2	2	3	1	2	1	2	0	0	9872	9	2	12	7	0	1	7	0	1	33
L	leu	3	1	3	0	0	6	1	1	4	22	9947	2	45	13	3	1	3	4	2	15
K	lys	2	37	25	6	0	12	7	2	2	4	1	9926	20	0	3	8	11	0	1	1
M	met	1	1	0	0	0	2	0	0	0	5	8	4	9874	1	0	1	2	0	0	4
F	phe	1	1	1	0	0	0	0	1	2	8	6	0	4	9946	0	2	1	3	28	0
P	pro	13	5	2	1	1	8	3	2	5	1	2	2	1	1	9926	12	4	0	0	2
S	ser	28	11	34	7	11	4	6	16	2	2	1	7	4	3	17	9840	38	5	2	2
T	thr	22	2	13	4	1	3	2	2	1	11	2	8	6	1	5	32	9871	0	2	9
W	trp	0	2	0	0	0	0	0	0	0	0	0	0	0	1	0	1	0	9976	1	0
Y	tyr	1	0	3	0	3	0	1	0	4	1	1	0	0	21	0	1	1	2	9945	1
V	val	13	2	1	1	3	2	2	3	3	57	11	1	17	1	3	2	10	0	2	9901

their probabilities of substitution. They also allow for there being a mixture of categories of sites.

Many other tabulations have also occurred (e.g., Gonnet, Cohen, and Benner, 1992; Henikoff and Henikoff, 1992). Most of these compute log-odds for use in aligning sequences or searching for local matches (see the paper by Altschul, 1991). Our objective here is to concentrate on the use of probabilistic models of amino acid substitution to obtain distances and maximum likelihood methods. The log-odds matrices are of little use for this purpose.

There has also been recent progress in refining methods for inferring mutation data matrices. Adachi and Hasegawa (1996) used maximum likelihood inference to create their mtREV model from the complete mitochondrial sequences of 20 vertebrate species. They used a phylogeny for the species, and chose the matrix that maximized the likelihood for the proteins of these mitochondria. Adachi et al. (2000) made a similar matrix for plastids. Whelan and Goldman (2001) have used likelihood methods to improve on the Jones-Taylor-Thornton mutation matrix. Müller, Spang, and Vingron (2002) have examined the adequacy of three methods, Dayhoff's, a maximum likelihood method, and a "resolvent" method of their own (Müller and Vingron, 2000) that approximates maximum likelihood. They find the latter two to be superior, with maximum likelihood slightly better but considerably slower. Cao et al. (1994) and Goldman and Whelan (2002) have advocated adjusting the matrices for the equilibrium distribution of amino acids found in the particular data under study.

Models depending on secondary structure

As the substitution model can depend on the secondary structure, one can also use such models to infer secondary structure. This has been done by Thorne, Goldman, and Jones (1996; see also Goldman, Thorne, and Jones, 1996). They assume that each amino acid position has a secondary structure state (they use alpha-helix, beta-sheet, and "loop" states). These states are hidden from our view. They use the *hidden Markov model* (HMM) approach, which will be described in more detail in Chapter 16. The essence of the method is that the overall likelihood is the sum over all possible assignments of states to positions, weighted by the probability of that assignment given a simple stochastic model that tends to assign similar states to neighboring positions. The likelihood of a phylogeny can be computed efficiently, even though we are summing over all assignments of states to positions. Afterwards, the fraction of the likelihood contributed by the assignments can be used to infer the secondary structure assignment at each position. Thorne, Goldman, and Jones use amino acid models inferred from sequences whose secondary structure assignments are known.

The secondary structure HMM approach has been applied to transmembrane proteins (Liò and Goldman, 1999). It seems to increase accuracy of prediction of secondary structure.

Codon-based models

An important limitation of the purely empirical approach is that it does not force the resulting model to have any relationship to the genetic code. It is possible to predict events that require two or even three base changes in a codon. This is not necessarily a bad thing; such events do occur. If most events are one-base changes, that will be reflected in the empirical tables of amino acid changes. Benner et al. (1994) argue that for short evolutionary distances the structure of the genetic code dominates the changes that occur in evolution. Yang, Nielsen, and Hasegawa (1998) formulate an empirical model of amino acid change, REV0. It is a general reversible model that disallows changes that would require more than one base change in the genetic code.

However, simply tabulating the amino acid changes and making an empirical table of probabilities cannot fully reflect the effects of the genetic code. The most dramatic case is Serine, which has two "islands" of codons in the code: (TCA, TCG, TCC, TCT) and (AGC, AGT). One cannot pass between the two by a single base change. Over a short evolutionary time we might find evidence of change from Tryptophane (TGG) to Serine, by a change of the second codon position (TGG $\rightarrow$ TCG). We might also find amino acid positions that had changed from Serine to Asparagine (AGT $\rightarrow$ AAT). But that does not mean we are very likely to find changes from Tyrosine to Asparagine, even when enough time has elapsed that two base changes are expected. The solitary codon for Tryptophane (TCG) differs by three bases from the codons for Asparagine (AAC and AAT), so at least three base substitutions, not two, would be needed.

This suggests that we must take the code into account if we believe that it is relevant. I have described (Felsenstein, 1996) a parsimony method that has states for the amino acids, including two for Serine, and that allows changes only between amino acids can be reached from each other by one base change. An observation of Serine is treated as an ambiguity between the two Serine states.

Goldman and Yang (1994) and Muse and Gaut (1994) have made the important step in formulating a codon-based model of protein change. They assume that changes are proposed by a standard base substitution model, but that the probability of acceptance of changes from one amino acid to another is given by a formula that has the probability of rejection increase as the chemical properties of the two amino acids become different. The coefficients of the formula for the probability of acceptance can be estimated by empirical study of amino acid changes or base changes. The result is a 64×64 matrix of probabilities of change among codons (actually 61×61 because the three nonsense codons are omitted). The transition probabilities that result from such a model will much more accurately reflect the probabilities of longer-term change than will amino-acid-based models such as Dayhoff's PAM matrix. The difficulty is largely computational—the spectral decomposition of a 61×61 matrix and computation of its powers is considerably slower than for a nucleotide model or an amino acid model. Since computational effort in computing the eigenvalues and eigenvectors of a matrix rises as the cube of the number of rows or columns, the effort should be $61^3/4^3 \approx 3,547$ times greater for a codon than for a nucleotide model, and $61^3/20^3 \approx 28$ times greater for a codon model than for an amino acid model.

Nevertheless codon models are attractive, both in taking into account the constraints due to the genetic code and in fitting the transition probabilities for amino acids with fewer parameters. With a PAM matrix model that is reversible, there are 208 quantities that are estimated in compiling the matrix (Yang, Nielsen, and Hasegawa, 1998). By contrast, Goldman and Yang (1994) used an HKY model of nucleotide sequence change, with one additional parameter that converts a measure of amino acid difference into the probability of acceptance of the amino acid change. They allow for different base frequencies at the three codon positions, so that they end up with only 11 parameters. Their extensive empirical study suggests that the resulting model fits the data better than empirical models. Schadt, Sinsheimer, and Lange (2002) have extended the codon model by grouping amino acids into four classes, and allowing three different probabilities of acceptance between different classes. This has the potential to improve the fit further.

Inequality of synonymous and nonsynonymous substitutions

Yang and Nielsen (2000) use a codon model similar to that of Muse and Gaut (1994), generalized to allow for an HKY model of nucleotide substitution. It has one parameter (ω) for the ratio of the rates of nonsynonymous and synonymous substitution. If natural selection is discriminating against changes of amino acid, ω will be less than 1. If positive natural selection favors change of amino acids, ω will be greater than 1. Neutral change will have $\omega = 1$. Likelihood ratio tests (which will be described in Chapter 19) can be used to test for departure from neutrality. These approaches build on the work by Miyata and Yasunaga (1980), Perler et al. (1980) , Li, Wu, and Luo (1985), and Nei and Gojobori (1986), which use pairs of sequences rather than full phylogenies, and use statistical methods other than likelihood.

The problem with the likelihood ratio test approach with phylogenies is that it assumes that positive (or negative) selection applies to all amino acid positions, everywhere in the tree. It is most likely that natural selection favors change at certain positions, in certain lineages, with change at most other positions being resisted. A change of enzyme substrate in one lineage may favor modifications in the active site of the protein, while at most other positions natural selection maintains the secondary and tertiary structure of the protein. Thus there is a danger that the codon model will miss positive selection, as the evidence for it could be outweighed by negative selection at most amino acid positions in most lineages.

Lineage specificity. Muse and Gaut (1994) allowed rates of nonsynonymous (or synonymous) substitution to vary between lineages. Yang (1998) and Yang and Nielsen (1998) presented a likelihood ratio testing framework for the inequality of the ratio of nonsynonymous to synonymous substitution between prespecified sets of lineages. Methods that allow the ratio of nonsynonymous to synonymous substitution to be the same for most lineages, but to differ on lineages that are not prespecified, might be developed using hidden Markov models (see Chapter 16); these are presumably not far off.

Site specificity. Nielsen and Yang (1998) and Yang et al. (2000) allowed the ratio of nonsynonymous to synonymous substitution to differ among sites, without assuming that it was known which sites differed from which. They allowed some sites to be neutral, some to be under negative selection, and some under positive selection. Which sites were which could be identified by computing posterior probabilities in a Bayesian framework. This is analogous to the method using maximum posterior probabilities, which will be described in Chapter 16 when we discuss hidden Markov models of variation of evolutionary rates among sites.

Both varying. Yang and Nielsen (2002) have combined site variation and lineage variation in the ratio of nonsynonymous to synonymous substitution rate, and Bustamante, Nielsen, and Hartl (2002) have made specific application of such techniques to inferences involving pseudogene evolution.

Protein structure and correlated change

There has been considerable interest in using covariation of amino acids in evolution as a source of information about the three-dimensional structure of proteins. Simply using the correlation of amino acid states across sequences is dangerous, since some of the correlation seen may simply reflect the phylogeny, which makes the individual sequences nonindependent samples. (For a more detailed discussion of this issue, see Chapter 25.) Pollock and Taylor (1997) show in a simulation study that failing to take phylogenies into account can lead to false indications that protein sites are correlated in their evolution. Wollenberg and Atchley (2000) used a statistic that did not correct for the phylogeny, but used parametric bootstrapping (for which see Chapter 20 and Chapter 19) to correct the distribution of their statistic for the effects of the phylogeny.

A method of detecting correlated substitutions that attempted to take the phylogeny into account was developed earlier by Shindyalov, Kolchanov, and Sander (1994). It uses a statistic comparing the number of times a pair of sites has simultaneous substitutions on the same branch of the tree. Tufféry and Darlu (2000) examined its behavior by simulation, and argued that error in the inference of the phylogeny can be a source of false positive signals. One might hope to make a simple likelihood ratio test of the nonindependence of substitution at two protein sites. (For an introduction to such tests, see Chapter 19.) There are not only many pairs of possible sites, there also would need to be nearly 20×20 parameters for the association of all possible pairs of amino acids at each pair of sites. We would drown in possible parameters. To avoid this trap, Pollock, Taylor, and Goldman (1999) classified amino acids by size or charge into two classes, greatly reducing the number of parameters needed. They were able to detect covariation between sites in simulated and real data, using a likelihood ratio test developed for discrete characters by Pagel (1994).

The search for signal in correlated substitutions at different sites is at an early stage. Much more work will be needed before we have a clear picture of how much information about protein sequence can be had from sequence comparisons across species.

Chapter 15

Restriction sites, RAPDs, AFLPs, and microsatellites

Much of the early work on variation in DNA sequences used variation in restriction sites rather than attempting full sequencing. It is necessary to have a way to model the variation in restriction sites and restriction fragment length polymorphisms in order to interpret these data. More recently, RAPDs (randomly amplified polymorphic DNA) and AFLPs (amplified fragment-length polymorphisms) have been developed, which use the PCR reaction to detect variation at certain sites, without the need to sequence. For within-species work, microsatellite loci are widely used, both because they can be assessed without sequencing and because the mutation rates between size classes are much higher than point mutation rates. They thus have increased genetic variation and this makes them quite useful. In this chapter, we will consider models for these data.

Restriction sites

The literature on statistical treatment of restriction sites started by considering how to compute a distance between restriction fragment patterns (Upholt, 1977; Nei and Li, 1979; Kaplan and Langley, 1979; Gotoh et al., 1979; Kaplan and Risko, 1981). Probabilistic models were used in these and other papers (Nei and Tajima, 1981, 1983; Tajima and Nei, 1982). A major simplification in the case of a Jukes-Cantor symmetrical model of DNA change was first noted by Nei and Tajima (1985). We will consider it and its implications, returning afterwards to the treatment of restriction fragment length differences.

Nei and Tajima's model

Suppose that a restriction enzyme has a recognition site r bases long, with a single sequence that is recognized. Out of 4^r possible sequences of this length, one is the restriction site. One could use any model of DNA change and use it to predict the change of the restriction site, but to do so we would have to keep track of all possible sequences of length r. The observation that a site is present specifies that the recognition sequence (say ACTAGT) is present. The observation that it is not present is, in effect, ambiguous among all the other $4^r - 1$ sequences. To make a transition probability matrix for the sequence, we would need to make one with 4^r states. Thus for a 6-base recognition sequence we would need to compute transition probabilities among $4^6 = 4,096$ states! One might think that as we cannot tell the difference between all the states that are not the recognition sequence, we could simply lump them all into one state. However, that cannot be done with most DNA mutation models. A state that differs by one site from the recognition sequence could be, say, ATTAGT or CCTAGT. If these have somewhat different probabilities of changing to the recognition sequence, the two states cannot be lumped.

Nei and Tajima (1985) realized that in the case of the Jukes-Cantor model, symmetries reduce the state space enormously. In the Jukes-Cantor model any two states that are the same number of mutations away from the recognition site have the same probability of changing to it. Thus ATTAGT and CCTAGT both have the same probability of mutating to a recognition sequence along a given branch of a tree. In the Jukes-Cantor case, lumping of states is possible. For an r-base recognition site, there are states that are 0, 1, 2, up to r bases away from the recognition sequence. In each of these categories, all the states can be lumped together. We can compute the probability of changing from one of the sequences k bases away from the recognition sequence to one ℓ bases away, without having to ask what the exact sequences involved are. The number of states that are k bases away from the recognition sequence will be

$$N_k = \binom{r}{k} 3^k \tag{15.1}$$

For a 6-base cutter, the numbers of states that are k bases away are given in Table 15.1

Nei and Tajima (1985) gave formulas for the transition probabilities from k to ℓ bases away from the recognition sequence. Their result is straightforward to derive (we use a notation slightly different from theirs). If a sequence changes from k to ℓ bases away, it can do so in a number of ways. Initially there are k bases in which the sequence differs from the recognition sequence, and $r - k$ bases in which it does not. To change to being ℓ bases away, the sequence could have $\ell + m$ of the sites that originally did not differ change their sequence, while m of the sites that did differ changed to become similar to the recognition sequence. The result is a sequence that differs by ℓ.

If the time (or branch length) that elapses is enough that we have a probability p that a site will change to another base, the probability that a site that is different from the recognition sequence changes to become similar is $p/3$. Thus the probability of changing to have ℓ differences from the recognition sequence is

Table 15.1: Number of nucleotide sequences of length 6 which differ by k bases from a 6-base restriction site

k	Number
0	1
1	18
2	135
3	540
4	1215
5	1458
6	729

the probability that $\ell - k + m$ of the $r - k$ sites change, times the probability that m of the k sites change. These are binomial probabilities with probabilities p and $p/3$, respectively. It follows that we can sum the product of the probabilities of the binomial probabilities:

$$P_{k\ell} = \sum_{m=a}^{b} \binom{r-k}{\ell-k+m} p^{\ell-k+m} (1-p)^{r-k-(\ell-k+m)} \binom{k}{m} \left(\frac{p}{3}\right)^m \left(1-\frac{p}{3}\right)^{k-m} \tag{15.2}$$

The limits a and b need careful attention. The smallest possible value of m is either 0 (if $\ell > k$) or $k - \ell$ (otherwise). The largest possible value of m is, for similar reasons, the smaller of k and $r - \ell$. The resulting expression is

$$P_{k\ell} = \sum_{m=\max[0,k-\ell]}^{\min[k,r-\ell]} \binom{r-k}{\ell-k+m} p^{\ell-k+m} (1-p)^{r-\ell+m} \binom{k}{m} \left(\frac{p}{3}\right)^m \left(1-\frac{p}{3}\right)^{k-m} \tag{15.3}$$

To express the transition probabilities in terms of time and the rate of base change per unit time, we need only substitute the expressions for p for the Jukes-Cantor model (equations 11.16 and 11.17). There is no more compact analytical expression — the transition probability must be computed by doing the summation. But this is not hard to do. We shall see in the next chapter that maximum likelihood methods of inferring the phylogeny can be based on transition probabilities, if these can be computed for arbitrary pairs of states and arbitrary branch lengths.

For more general models of base change the computation becomes too difficult. One would wish to allow for differences in rate between transitions and transversions. The Kimura two-parameter model does this, while otherwise treating the bases symmetrically. Li (1986) computed transition probabilities analogous to the present ones using that model. But he did so only for patterns involving a few species. In general, instead of having $r + 1$ states for an r-base cutter, it requires us to specify for each r-base sequence by how many transitions and how many transversions it differs from the recognition sequence. This means that there are $(r+1)(r+2)/2$ possible states. For a 6-base cutter that would be $7 \times 8/2 = 28$ states, so that the transition matrices would be 28×28. This is an increase of a factor of 16 in matrix size (and hence an increase of a factor of 64 in computation) over the case of the Jukes-Cantor model. For models with asymmetric base composition the matter is much more serious. I cannot see any way to do the computations without making a separate state for each of the 4^r possible r-base sequences. For a 6-base cutter, that is 4,096 states, an increase in computational effort of more than 3×10^5 over the Jukes-Cantor model.

Distances based on restriction sites

Most work using probability models of restriction sites has been focussed on obtaining distances for use in distance matrix methods. With two sequences there are four possible observations at each restriction site location: the four combinations of presence and absence in each of the two sequences. We can denote these $++$, $+-$, $-+$, and $--$. We can use the models to calculate the probabilities of each of these. Suppose that the branch length is t between two species (A and B), that the enzyme is an r-base cutter, and that the model of DNA change is taken to be the Jukes-Cantor model. Under that model, the probability of a restriction site at any location is, of course:

$$\pi_0 = \left(\frac{1}{4}\right)^r \tag{15.4}$$

This will be the sum of the probabilities of $++$ and $+-$, and it will also be the sum of the probabilities of $++$ and $-+$. If we can find a way to compute the probability $++$, then by subtraction we will be able to compute the probabilities of $-+$ and $+-$. Since all four probabilities should add up to 1, we can then compute the probability of $--$ as well. The critical quantity is then seen to be P_{++}, the probability of $++$.

This was first computed by Nei and Li (1979). Our notation will differ from theirs. The probability that a sequence of r bases is a restriction site in sequence A is π_0, and the probability that all r bases continue to code for a restriction site at the other end of a branch of length t is $(1-p)^r$, where p is the probability that an individual base is different at the other end of the branch. Thus

$$P_{++} = \pi_0 \,(1-p)^r \tag{15.5}$$

The value of p can be computed from the Jukes-Cantor formula 11.17. Taking the rate of substitution as 1 per unit time, this will give

$$P_{++} = \left(\frac{1}{4}\right)^r \left(\frac{1+3\,e^{-\frac{4}{3}t}}{4}\right)^r \tag{15.6}$$

and by the requirements for the probabilities to sum up, we then immediately see that

$$P_{+-} = P_{-+} = \left(\frac{1}{4}\right)^r \left[1 - \left(\frac{1+3\,e^{-\frac{4}{3}t}}{4}\right)^r\right] \tag{15.7}$$

and

$$P_{--} = 1 - \left(\frac{1}{4}\right)^r \left[2 - \left(\frac{1+3\,e^{-\frac{4}{3}t}}{4}\right)^r\right] \tag{15.8}$$

Issues of ascertainment

It would seem straightforward to use these probabilities to compute the distance by maximizing a likelihood with respect to t. However, we have to consider whether we see all the cases that are $--$. We typically have not chosen a restriction site independently of its having the restriction recognition sequence. In fact, we are likely to have examined only those sites that showed variation in some sample of individuals. Thus we have to correct for a bias toward finding $+$ sites. This would be called an *ascertainment bias* in fields such as human genetics.

It is not easy to characterize all of the different ways that we could have chosen the sites for study. We might have a phylogeny, and look only at those sites that varied somewhere in the phylogeny. Thus we might see $--$ in these two species, but somewhere else on the tree there would have to be a $+$ at this position. Requiring that reduces the probability of a $--$. Alternatively, we might have chosen the sites based on the phylogeny of some other, relatively unrelated organisms. In a paper on maximum likelihood estimation of phylogenies from restriction sites, I have (Felsenstein, 1992b) shown how to correct for some kinds of these ascertainment effects, as we will see in the next chapter.

If we are at a loss to know what to do with the ascertainment effects, we can simply drop the $--$ sites and make our estimate of distance from the relative numbers of $++$, $+-$ and $-+$ sites. The probabilities of these three outcomes, given that a site is not $--$, are $P_{++}/(P_{++}+P_{+-}+P_{-+})$, $P_{-+}/(P_{++}+P_{+-}+P_{-+})$, and $P_{+-}/(P_{++}+P_{+-}+P_{-+})$. The likelihood for the numbers of sites n_{++}, n_{+-} and n_{-+} will be

$$\begin{aligned} L &= \binom{n_{++}+n_{+-}+n_{-+}}{n_{++}\;\; n_{+-}\;\; n_{-+}} \left(\frac{P_{++}}{P_{++}+P_{+-}+P_{-+}}\right)^{n_{++}} \\ &\quad \times \left(\frac{P_{+-}}{P_{++}+P_{+-}+P_{-+}}\right)^{n_{+-}} \left(\frac{P_{-+}}{P_{++}+P_{+-}+P_{-+}}\right)^{n_{-+}} \end{aligned} \tag{15.9}$$

the P's being functions of t. The combinatorial term in front can be dropped, as it is a constant independent of t for any data set. We have a reversible model of DNA evolution, so that $P_{+-} = P_{-+}$. Then we can combine terms and get

$$L = \left(\frac{P_{++}}{P_{++}+P_{+-}+P_{-+}}\right)^{n_{++}} \left(\frac{P_{+-}}{P_{++}+P_{+-}+P_{-+}}\right)^{n_{+-}+n_{-+}} \tag{15.10}$$

This is of the form

$$L = x^{n_{++}} \left[\frac{1}{2}(1-x) \right]^{n_{+-}+n_{-+}} \tag{15.11}$$

where

$$x = \frac{P_{++}}{P_{++} + P_{+-} + P_{-+}} \tag{15.12}$$

Taking the logarithm of L, differentiating with respect to x, equating the derivative to zero, and solving, we find that the maximum is at

$$x = \frac{n_{++}}{n_{++} + n_{+-} + n_{-+}} \tag{15.13}$$

This is reasonable: The empirical frequencies of the site patterns match their expectations. We can use equations 15.6 and 15.7 to express x in terms of t. The result is

$$\left(\frac{1 + 3e^{-\frac{4}{3}t}}{4} \right)^r = \frac{n_{++}}{n_{++} + \frac{1}{2}\left(n_{+-} + n_{-+}\right)} \tag{15.14}$$

which then can easily be solved to give the distance

$$D = \hat{t} = -\frac{3}{4} \ln \left[\frac{4}{3} \left(\frac{n_{++}}{n_{++} + \frac{1}{2}\left(n_{+-} + n_{-+}\right)} \right)^{1/r} - \frac{1}{3} \right] \tag{15.15}$$

This formula is quite similar to that derived by Nei and Li (1979). (It differs by a factor of two in the branch length, as it asked a slightly different question, and by their assumption that all sites that are shared between two species were present at a common ancestor halfway between them.) Nei (1987, pp. 97ff.) derived a similar distance by allowing the frequencies of nucleotides in the sequence to be unequal, using them in the formula, but then also using the Jukes-Cantor model which assumes their equality. In practice it is unlikely that the resulting formula will differ much from the one given here, especially for closely related species.

Li (1986) used the Kimura two-parameter model for restriction site inferences. His equations could be adapted above instead of using equations 15.6 and 15.7, but they would not lead to a closed-form formula. However, the resulting equations can easily be solved numerically. Nei and Li (1979) also show how the assumption of a gamma distribution of evolutionary rates among sites affects the formula.

Parsimony for restriction sites

Although a full treatment of restriction sites requires a stochastic model, we can consider whether there are parsimony methods that approximate the behavior of the full model. DeBry and Slade (1985) discussed a model with three states, 0, 0′ and 1. Of these, 1 is the presence of the site, 0′ is any sequence that is one base away from it, and 0 groups all other sequences. They derived expressions for transitions between the three states using a Jukes-Cantor model. They approximated the probabilities of change and used them to argue that the most probable scenarios for homoplasy would be situations in which the site was gained once but then lost multiple times.

We have seen that Dollo parsimony is based, implicitly, on a model in which gain is quite improbable but loss is not as rare. DeBry and Slade argue that Dollo parsimony will be an acceptable approximation to use of the full likelihood model.

Albert, Mishler, and Chase (1992) have argued against the use of Dollo parsimony for this purpose. They find that Wagner parsimony comes closer to approximating the probabilities of change of restriction sites; that the total avoidance of parallel origins of the restriction recognition sequence in Dollo parsimony is too strong a constraint. In using Wagner parsimony, they specify that the ancestral state is unknown.

Which method, Dollo or Wagner parsimony, does a better job at approximating restriction sites change may depend on the degree of divergence between species. If the species are closely related, parallel gains may be strongly outweighed by parallel losses. If the species are not so closely related, explanation by parallel gains can become important. We shall see below that RAPDs and AFLPs can in many respects be treated as large restriction sites. These will have strong asymmetries of change that may make Dollo parsimony more defensible that it proved to be in the cases that Albert et al. considered.

We can illuminate these asymmetries by using the formulas given above. Suppose that we have an ancestor, whose state is unknown, giving rise to two species; each are diverged t units of branch length from that ancestor, and both possess the restriction site. Given this, what is the probability that the ancestor also had the restriction site? The transition probability from + to + we know to be $(1-p)^r$. If the probability that a base differs after t units of branch length is p, we can use an argument like that used for equation 15.5 to show that the joint probability of the two tips and the ancestor all being + is $\pi_0(1-p)^r(1-p)^r$. This is

$$P_{+++} = \left(\frac{1}{4}\right)^r \left(\frac{1+3\,e^{-\frac{4}{3}t}}{4}\right)^{2r} \tag{15.16}$$

The probability that both tips are + is given by equation 15.6, with t replaced by $2t$. The conditional probability that the ancestor is + is computed from the ratio of P_{+++} to that quantity. It will be apparent from Figure 15.1 that when two species are distant, the appearance of the same restriction site in both is not strong evidence for their ancestor having had that site, though it may have had a DNA sequence that matches many of the bases in the recognition sequence. These calculations are similar to ones made by Alan Templeton (1983).

Modeling restriction fragments

So far we have discussed only restriction sites. Many studies using restriction digests analyze the presence or absence of restriction fragments instead. There is no

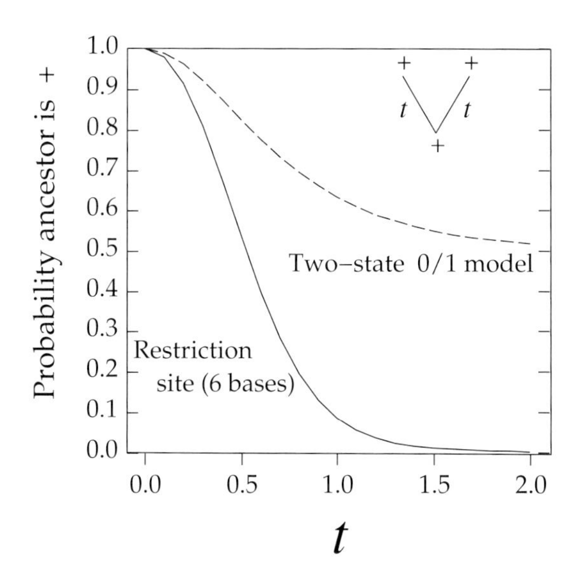

Figure 15.1: Fraction of the time the ancestor of two species who both have a restriction site is expected to have that site also. The two species are each diverged from the ancestor by a branch of length t. The curves for a 6-base cutter (solid curve) and a symmetric two-state 0/1 character are shown.

full probabilistic treatment of restriction fragment variation, but Nei and Li (1979) have constructed a distance for restriction fragment data. If a given fragment of length L nucleotides is found in one species, the probability that it will also be found at another, which is a branch length t away, can be calculated approximately. One can work out approximate formulas for this using the mathematics of restriction sites by noting that a fragment will be retained if the restriction sites at its two ends are preserved, and also if no new restriction site arises between them (Upholt, 1977; Nei and Li, 1979).

Nei and Li (1979) have given a formula that does not depend on L. This they do by computing the average probability that a restriction fragment will be retained after a branch of length t. This averages over all lengths of sequences and thus does not depend on L. They then compute the distance by, in effect, finding the value of t that fits the observed fraction of restriction fragments that are retained. If the restriction recognition sequence is r bases long, the fraction of fragments that are m bases long will be (approximately)

$$(1-a)^{m-r}\, a \tag{15.17}$$

where a is $1/4^r$, the probability of a recognition sequence at a random point on the DNA. Suppose that b is the probability that a sequence that is not a recognition sequence changes to one that is by the end of a branch of length t. Nei and Li point out that the probability that a restriction recognition sequence arises at a given site among the $m - r$ that are in the interior of the fragment is b. Although the $m - r$ possible such events are not completely independent, if we treat them as such, we should get a good approximation. If Q is the probability that a restriction site is present after branch length t, the probability that the two restriction sites at the ends of the fragment are both still present after branch length t is Q^2. Thus the probability that a fragment of length $m + r$ is preserved is, to good approximation,

$$Q^2(1-b)^{m-r} \tag{15.18}$$

Averaging this over all values of m, weighting by the probabilities of fragments of all lengths, we get

$$F = \sum_{m=r}^{\infty} (1-a)^{m-r} a\, Q^2 (1-b)^{m-r} \tag{15.19}$$

which can be summed as a geometric series to get

$$F = \frac{a\, Q^2}{1-(1-a)(1-b)} \tag{15.20}$$

We have the value of a (above), and it turns out that

$$b = a\, \frac{(1-Q)}{(1-a)} \tag{15.21}$$

Substituting these into equation (15.20), we get the average fraction of restriction fragments retained as

$$F = \frac{Q^2}{2-Q} \tag{15.22}$$

This is in slight disagreement with Nei and Li who define Q and t differently and reach a slightly different result for b. If we use the Jukes-Cantor model to compute

$$Q = \left[\frac{1 + 3\exp\left(-\frac{4}{3}t\right)}{4} \right]^r \tag{15.23}$$

we can note that equation (15.22) is a quadratic equation in Q and solve it for Q, and then we can solve equation (15.23) for $\hat{t}$.

One might wonder whether it is not possible to do better by explicitly taking into account the length of each individual fragment and by making a maximum likelihood estimate of the branch length from the full set of fragments. Aside from the daunting probability and statistical problems involved, the data may not fit the model well enough for this approach to succeed. The total length of all fragments ought to be to same in both species, and frequently this is not the case. We would have to accommodate in the model the inaccuracy of our measurements of fragment sizes, as well as the disappearance from our data of fragments that were too small or too large.

Parsimony with restriction fragments

One can also use parsimony methods, coding each fragment as a character which is 1 when present, 0 when absent. Although a given change from 0 to 1 is likely to be a strongly asymmetric event, we cannot know whether the gain of a fragment represents the gain or loss of a site. A fragment may appear as a result of gain of a restriction site, which cuts a larger fragment in two. At the same time, the gain of the site results in the loss of the larger fragment. The loss of that site would cause the gain of that fragment. Thus there is no simple rule as to whether the appearance or the disappearance of a restriction fragment corresponds to the origin or loss of a restriction site. Each event is asymmetric, but it is hard to know in which direction the asymmetry is. We could use Wagner parsimony, ignoring the asymmetry of the changes. If we want to use Dollo parsimony, we must ensure that for each character the ancestral state is taken to be unknown. If that is done, the algorithm will choose the direction of the asymmetry, based on the fit to the data on each tree.

Note that as a single base change can cause the disappearance of one fragment and the simultaneous appearance of two smaller ones (or vice versa), the 0/1 characters do not evolve independently. It thus becomes doubtful whether techniques such as bootstrapping, jackknifing, or paired-sites tests can be used to evaluate the strength of the resulting inferences. As we will see in the chapters that cover these methods, all of them assume that the characters evolve independently.

By contrast, with restriction sites (rather than fragments) the changes in the individual sites occur in different nucleotides, and can be regarded as independent unless the sites overlap. This makes it easier to apply statistical methods that require independence.

RAPDs and AFLPs

In many organisms, RAPDs or AFLPs have become widely used in a manner similar to restriction fragments. How can RAPD or AFLP data be analyzed in phylogenetic inference? In both cases a given band is present when a PCR primer sequence is present at both ends of a piece of DNA. When either of these mutates enough to cease to be a primer, or when another site arises within the fragment, the band that is seen on the gel moves enough that it seems to disappear.

For example, if we were doing RAPD analysis with primers of length 10 nucleotides, and under conditions in which all 10 bases must match the primer for a PCR reaction to recognize it, then a change in any of the 20 bases could cause the fragment to disappear from its place on the gel. If the two PCR primers are close enough to each other to successfully result in a PCR reaction, they have few enough sequences of length 10 between them that it is very unlikely that any sequences exist there that are only one base different from being a primer. Thus the main mechanism of change of RAPD bands will be mutation in the two PCR primers. AFLPs are detected differently, but have essentially the same properties, but with longer primers (about 20 bases long at each end) and three additional nucleotides at each primer site.

In effect, we then have associated with each fragment a total of 20 (or 46) bases that must be in a particular sequence for the fragment to be seen. The 20 bases are not entirely contiguous, but this does not matter. They act like a 20-base restriction site. We can thus make a statistical model of the presence and absence of RAPD or AFLP bands by simply using the model given above for restriction sites. In spite of their detection as fragments, the statistical behavior of RAPD and AFLP bands is more like that of restriction sites.

Restriction sites distances can be used, as can parsimony methods. The recognition sequence is so long that we should expect strong asymmetries between presence and absence of the sites. A quantitative examination has not yet been made, but I would expect it to favor the use of Dollo parsimony rather than Wagner parsimony in this case. Backeljau et al. (1995) have discussed reasons for believing that no parsimony method conforms adequately to the properties of the evolution of RAPDs. It would also be possible to use the restriction sites model for maximum likelihood inference of RAPD or AFLP phylogenies.

The issue of dominance

One problem with use of restriction fragments data, RAPDs, and AFLPs is that these markers are dominant. Thus in a diploid individual that has a fragment present in one of its two haplotypes for the relevant region of the genome, the fragment is scored as present, and its absence from the other haplotype is not noted. For genomes that have diverged substantially this may not be a problem—fragments that differ between species may tend to have their presence or absence fixed within each species. For within-species inferences the problem is more serious, as we are likely to be concentrating on those fragments that show differences within the species, and these will not necessarily show fixed differences between local populations. Clark and Lanigan (1993) have described a correction for distances from RAPDs to allow for the effect of dominance.

Unresolved problems

The models for AFLP and RAPD data presented here are oversimple. In such data, one can have more complex phenomena that are not reflected in these models. For example, two PCR primers could lie near each other on the DNA, so that one PCR primer is present at the left end of a fragment, but two lie near the right end, resulting in two bands that share their left primer but have different right primers. Mutation of the left PCR primer sequence could then eliminate both bands. This and other phenomena create additional sources of noise in RAPD and AFLP data. The users of these methods have been slow to recruit theoreticians to address these issues, preferring instead to forage among existing methods, which are developed for other kinds of data. If the communities of users of RAPD and AFLP techniques are to be well-served by methods of data analysis, they will need to take the initiative to ensure that these problems are addressed.

Microsatellite models

Microsatellite (VNTR) loci are another class of partial sequence information. A microsatellite locus has a simple sequence, such as a di- or trinucleotide [such as $(ACT)_n$] repeated tandemly a number of times. We typically know only the number of copies of the repeat. RAPDs, AFLPs, and restriction sites are assumed to have a normal mutational process for noncoding DNA, but microsatellites are distinguished by having mutational processes that are related to their status as repeats, such as slippage. One event can cause addition or removal of some of the repeats. Although restriction sites and PCR primers are only known to be such by our laboratory procedures, tandem repeats have an effect on the biology of replication, and this leads to having their own distinct mutational processes.

These mutational processes have high rates of change, compared to sitewise mutation rates in noncoding DNA. Typical rates of mutation range up to 0.0001 per locus per generation, many orders of magnitude larger than rates of point mutation in DNA. The resulting high rates of change has led to the growing use of microsatellite loci as markers within species. However, their use for inferring phylogenies between species has lagged, because it is frequently found that a marker that is highly variable within a species cannot be found at all in a neighboring species. As we shall be concerned with coalescent trees of genes as well as with phylogenies, we need to consider models for the mutational change of microsatellite loci.

The one-step model

The simplest model is one in which, in each generation, there is a small probability that the number of repeats increases by 1, and an equal probability that it decreases by 1. Things cannot be this simple if the number of copies reaches 1, for it cannot decrease below that value. If one copy is taken to be an absorbing state (a state in which the system becomes stuck once it is reached), then it is possible to show that there is a higher and higher chance that the locus is in this state, the remaining copies of the locus having higher and higher numbers of copies.

The *one-step model* has been used by Valdes, Slatkin, and Freimer (1993) and Goldstein et al. (1995a). They took the sensible step of ignoring the issue of what happens when there is one copy, on the grounds that we are primarily interested in probabilities of change among moderate numbers of copies, and the behavior of the locus when there is only one copy will have a scarcely noticeable effect on that. We can closely approximate their model by assuming that time is continuous (rather than discrete generations) and that there is a constant risk of taking a step, with equal probability of stepping up or down. This is a model well-known as the *randomized random walk* (Feller, 1971, p. 59). Feller shows that the transition probabilities can be written in terms of the modified Bessel function; we derive a clumsier expression for simplicity.

In a branch whose length is t, with mutation rate μ, the number of mutations occurring will be drawn from a Poisson distribution with expectation μt. If we end up changing the number of copies by i (where i can be any integer, positive or negative), this can occur in a number of ways. We can have $i+k$ steps upwards, and k steps downwards, for a total of $i+2k$ steps and a net displacement of i steps. The probability that there are $i+2k$ steps is the Poisson probability

$$e^{-\mu t}\left(\mu t\right)^{i+2k}/(i+2k)! \tag{15.24}$$

and the probability that $i+k$ of these are steps upwards is the binomial probability of getting $i+k$ heads in $i+2k$ tosses:

$$\binom{i+2k}{i+k}\left(\frac{1}{2}\right)^{i+k}\left(\frac{1}{2}\right)^{k}=\binom{i+2k}{i+k}\left(\frac{1}{2}\right)^{i+2k} \tag{15.25}$$

Taking the product of these we get the net transition probability by summing it over all possible values of k:

$$\begin{aligned} p_i(\mu t) &= \sum_{k=0}^{\infty} e^{-\mu t}\frac{(\mu t)^{i+2k}}{(i+2k)!}\binom{i+2k}{i+k}\left(\frac{1}{2}\right)^{i+2k} \\ &= e^{-\mu t}\sum_{k=0}^{\infty}\left(\frac{\mu t}{2}\right)^{i+2k}\frac{1}{(i+k)!\,k!} \end{aligned} \tag{15.26}$$

Although Feller's Bessel-function version of this formula is more compact, simply computing the present form by summing equation 15.26 over relevant values of k is probably the most effective way.

Figure 15.2 shows the transition probabilities to different numbers of copies, starting with 10 copies, and with $\mu t = 2$ and $\mu t = 10$. Note that with the larger numbers of expected changes, the transition probability becomes nearly Gaussian in shape. For branches longer than this, some chromosomes will approach the limit of one copy, and we would expect this model to become inaccurate.

Microsatellite distances

From models such as one-step models one can make distances. This has been done by Slatkin (1995) and by Goldstein et al. (1995a, b). Their distances are intended to measure the divergence of populations within species. We shall see in the chapter on coalescents that two genes in a random-mating population of constant finite effective population size N_e are expected to be separated by $4N_e$ generations, the mean time back until they have a common ancestor. Under a one-step model the change in copy number has a mean of 0 and a variance which is equal to the branch length times μ. The difference in copy number between two genes from

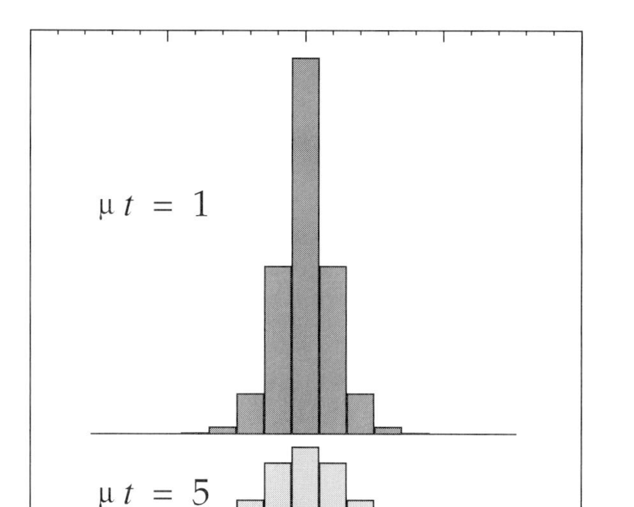

Figure 15.2: Transition probabilities for numbers of copies of a microsatellite locus under the one-step model of change. The histograms show the relative probabilities of different numbers of changes in repeat number, for a branch long enough that 1 change is expected (upper) and long enough that 5 changes are expected (lower). The areas of the histograms are equal.

the same population will then have mean 0 and variance $4N_e\mu$. If two populations separated t generations ago, the number of generations separating a gene from one population from a gene from the other is $2t + 4N_e$, since they trace back to separate copies in the ancestral population, and those are separated by about $4N_e$ generations. The situation is depicted in Figure 15.3.

The difference in copy number between two genes, one from one population and one from the other, has mean 0 and variance approximately $(2t + 4N_e)\mu$. (This is only approximate since two genes from a population vary around $4N_e$ generations in their time of separation.) That also implies that the mean squared difference in copy number is approximately $(2t + 4N_e)\mu$. One is immediately tempted to make the mean squared difference in copy number between the populations the measure of distance between them. However, we would like to subtract the $4N_e\mu$ so that we get a distance that is proportional to t. Goldstein et al. (1995b) have

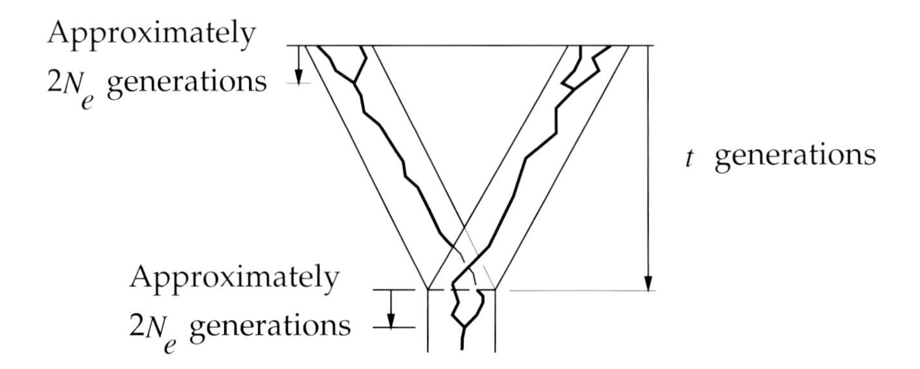

Figure 15.3: Genealogy of copies where a pair of gene copies is drawn from each of two species, which have a common ancestor.

accomplished this correction by noting that the mean squared difference between copies within a population is $4N_e\mu$ (set $t = 0$ to see this). We might then want the distance to be the mean squared difference between copies in the two populations, less the mean squared difference between copies within populations.

An algebraic simplification is available, as it can be shown (Goldstein et al., 1995a) that for large values of N_e this simply equals

$$(\delta\mu)^2 \;=\; (\mu_A - \mu_B)^2 \tag{15.27}$$

The derivation makes use of the assumption that the effective population sizes of the two populations are equal and also equal to the effective population size of their common ancestor. Without that, the result is not exact.

Note that this estimate comes from equating variances to their expectations. This is not a maximum likelihood estimate. If the difference in number of copies between two populations were normally distributed, it would be a maximum likelihood estimate. Thus, to some extent it will be not be a fully powerful statistic, but it does have some robustness. The derivation assumes that the variance of copy number that accumulates is μ per unit time, but it does not actually rely on the assumption that the changes of copy number are one-step changes. Thus the $(\delta\mu)^2$ distance should be applicable to cases in which there could be multiple-step mutations.

A Brownian motion approximation

In fact, a Brownian motion approximation can be made (Felsenstein and Beerli, in prep.; see also Zhivotovsky and Feldman, 1995). It is not yet clear how good an approximation this will be in practice. We simply assume that the copy number changes along a branch of the tree according to a Brownian motion, rather than by the more realistic processes of one-step or multistep models of copy number

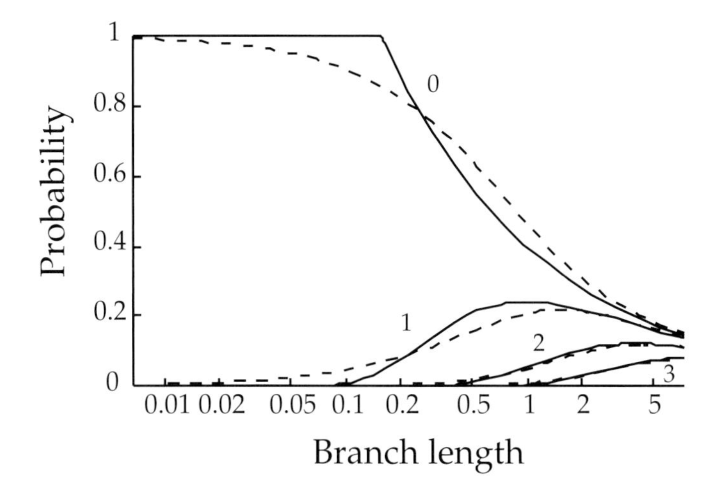

Figure 15.4: Probabilities of different numbers of changes of copy number in a microsatellite locus, for different values of the branch length μt. The dashed curves are the probabilities from the one-step model, and the solid curves are the probabilities from the truncated Brownian motion approximation.

change. The transition probability for changing from copy number m to $m+k$ in a branch of length t can be computed, given the constant μ. This can be computed as the probability of changing by an amount between $k-1/2$ and $k+1/2$ when the expected change is 0 and variance of change is μt, where the amount of net change comes from the normal distribution with this mean and variance.

An integral like this must be computed numerically, but we can approximate it by the probability density of at k, times the width of the interval around k (which is 1). However this can give a probability greater than 1. To avoid this we can truncate it:

$$\text{Prob}\,(m+k\,|\,m) \;\approx\; \min\left[1,\; \frac{1}{\sqrt{\mu t}\,\sqrt{2\pi}}\exp\left(-\frac{1}{2}\frac{k^2}{\mu t}\right)\right] \tag{15.28}$$

Figure 15.4 shows probabilities of different numbers of copy number changes, for the one-step model and with this approximation. The scaled branch length μt is the horizontal axis, and the curves give the probabilities of a net change of 0, 1, 2, ... (the probabilities of negative numbers of changes –1, –2, ... are the same).

It will be apparent from the figure that the approximation is less than perfect. When μt is near 0.1, it expects less change than actually would occur in the one-step model. When μt is near 0.7, it expects somewhat more change. At longer branch lengths, the approximation is quite good.

Models with constraints on array size

The one-step model cannot be exactly true, for it predicts some chance of having negative numbers of copies! Change cannot pass zero copies. In fact, empirical studies have suggested stronger constraints than this on the number of repeats. Models of constrained variation in microsatellites have been made (Garza, Slatkin, and Freimer, 1995; Feldman et al., 1997; Pollock et al., 1998). If there were advance knowledge of the constraints, this could be used to advantage, but if we do not know for each locus at what number of copies they are constrained, it is hard to see how to estimate the limits to change in a way that can be used practically (as pointed out by Nielsen and Palsboll, 1999). We would have to estimate the strength of the constraints separately at each locus, so that as the number of loci in the study increased, the ratio of the amount of data to the number of parameters would not increase indefinitely.

Multi-step and heterogeneous models

There has also been some examination of whether the single-step model is adequate (DiRienzo et al., 1994). In general it is often possible to show that it is not. It is difficult to calculate transition probabilities for a multi-step model. However, with the Brownian motion approximation we need only know the variance of the change of allele number per generation, and this is insensitive to whether the changes are by one repeat or many.

There does seem to be evidence that changes in loci with large numbers of repeats are more frequent, or larger, than in loci with few repeats. If we were to assume that the standard deviation of change in repeat number was proportional to the repeat number, we could approximate this by taking the logarithm of the repeat number. It would have constant variance of the change on its scale, and could be approximated by a Brownian motion model.

For distance matrix methods, where exact transition probabilities are not needed, one can go further. Zhivotovsky, Feldman, and Grishechkin (1997) derive formulas for the expected distance as a function of time when there are biases in mutation and heterogeneity of rates of change from locus to locus. Zhivotovsky, Goldstein, and Feldman (2001) extend these results even further.

Snakes and Ladders

The models mentioned so far all assume that changes in the number of repeats are not very large. Two groups have investigated models in which the number of repeats can undergo large decreases. Kruglyak et al. (1998, 2000) modeled the occurrence of point mutations. A point mutation in one of the repeat units of a microsatellite locus divides the tandem array into two tandem arrays. One may even be so small that it ceases to have slippage events. Thus point mutations can cause large decreases in array size. Calabrese, Durrett, and Aquadro (2001) have investigated in more detail the consequences of the model for the distribution of number of repeats.

Falush and Iwasa (1999) have made another model that allows large decreases in the number of repeats. They do not specify the mechanism of the decreases, but simply specify the distribution of the number of repeats after mutation. They use an asymmetric triangular distribution that has an upward bias in change, but a variance that is proportional to the square of the number of repeats. In such a model, one might imagine that the upward bias of change would cause the number of repeats to increase without limit. The fascinating thing about their model is that this does not happen. Number of repeats tends to rise for a while, but sooner or later the number of repeats decreases to a low number. Computer simulations and a diffusion-equation approximation both verify that this collapse of the number of repeats occurs often enough to keep the number of repeats from blowing up.

Both models predict large decreases in the number of repeats. Chambers and McAvoy (2000) analogize such processes to the game of Snakes and Ladders, where the player is constantly at risk of plunging backwards toward the start of their journey.

The difficulty with both of these interesting models is that they do not easily lend themselves to calculation of long-term transition probabilities of the number of repeats. It may be necessary to adopt Markov chain Monte Carlo methods to use them in statistical inference within or between species (cf. Nielsen, 1997; Wilson and Balding, 1998).

Complications

For all that has been done to model microsatellite loci, the biology is sufficiently complicated that much more work is needed. Enough is known to motivate more work. Sibly, Whittaker, and Talbot (2001) have shown that the change of repeat number rises with repeat number, and ceases when there are only a few repeats. Even that is not an adequate description of microsatellite mutational processes. Ellegren (2000) reviews the depressing results of pedigree studies that observe new mutations. Noor, Kliman, and Machado (2001) also find a complex pattern of changes in long-term evolutionary studies. The issue that must be faced is not whether one-step models are close to being true—they are not—but under what circumstances they can be used as an approximate model for inference.

Chapter 16

Likelihood methods

Having models of evolution for a character, we can use standard statistical methods to make estimates of the phylogeny. Perhaps the most standard framework of all is maximum likelihood, which we have encountered earlier. It was invented by R. A. Fisher (1912, 1921, 1922). Many of the usual statistical estimates that we know and love are really maximum likelihood estimates, including the average as the estimate of the mean of a normal distribution, the observed fraction of "heads" as an estimate of the parameter of a binomial distribution, and the least squares fit of a straight line to a series of (x, y) points when y is normally distributed around the true regression line.

Likelihood methods for phylogenies were introduced by Edwards and Cavalli-Sforza (1964), for gene frequency data. The first application of likelihood methods to molecular sequences was by the famous statistician Jerzy Neyman (1971). Ironically, he was a well-known skeptic of the use of likelihood. Kashyap and Subas (1974) extended Neyman's work. I showed (1981b) how to make likelihood computations for nucleotide sequences practical for moderate numbers of sequences.

Maximum likelihood

Although likelihood is of central importance in statistics, it has usually been omitted from the "cookbook" statistics courses that biologists take, and therefore is unfamiliar to most biologists. For this reason it is important to describe likelihood here. Using the laws of conditional probability, we can easily show for any two hypotheses H_1 and H_2 about a set of data D that since $\mathrm{Prob}\,(H|D) = \mathrm{Prob}\,(H \text{ and } D)/\,\mathrm{Prob}\,(D) = \mathrm{Prob}\,(D|H)\,\mathrm{Prob}\,(H)/\,\mathrm{Prob}\,(D)$, then

$$\frac{\mathrm{Prob}\,(H_1|D)}{\mathrm{Prob}\,(H_2|D)} = \frac{\mathrm{Prob}\,(D|H_1)}{\mathrm{Prob}\,(D|H_2)}\,\frac{\mathrm{Prob}\,(H_1)}{\mathrm{Prob}\,(H_2)} \tag{16.1}$$

This expresses the "odds ratio" in favor of hypothesis 1 over hypothesis 2 as a product of two terms. The first is the ratio of the probabilities of the data given the two hypotheses. The second is the ratio of the prior probabilities of the two hypotheses (the odds ratio favoring H_1 over H_2) before we look at the data.

If we have the odds favoring H_1 over H_2, equation 16.1 shows how to take into account the evidence provided by the data, and come up with a valid posterior odds ratio. This formula is the odds ratio form of Bayes' theorem. The quantity $\mathrm{Prob}\,(D|H)$ is called the likelihood of the hypothesis H. Note that, in spite of common English usage, it is not the probability of the hypothesis. That would be $\mathrm{Prob}\,(H|D)$. It is instead the probability of our data, given the hypothesis.

If we have independent observations, then

$$\mathrm{Prob}\,(D|H_i) \;=\; \mathrm{Prob}\,(D^{(1)}|H_i)\;\mathrm{Prob}\,(D^{(2)}|H_i)\;\cdots\;\mathrm{Prob}\,(D^{(n)}|H_i). \tag{16.2}$$

It follows that

$$\frac{\mathrm{Prob}\,(H_1|D)}{\mathrm{Prob}\,(H_2|D)} = \left(\prod_{i=1}^{n}\frac{\mathrm{Prob}\,(D^{(i)}|H_1)}{\mathrm{Prob}\,(D^{(i)}|H_2)}\right)\frac{\mathrm{Prob}\,(H_1)}{\mathrm{Prob}\,(H_2)} \tag{16.3}$$

In equation 16.1 and 16.3 we can see that if there is a large amount of data, the right side of the equation will be dominated by its first term, the likelihood ratio of the two hypotheses.

Bayesian statisticians try to come up with valid prior probabilities and use formulas such as equation 16.1 to infer valid posterior probabilities of the various hypotheses. Non-Bayesians are skeptical of our ability to come up with valid prior probabilities. They may prefer that hypothesis that maximizes the likelihood $\mathrm{Prob}\,(D|H)$. This may not end up being the hypothesis that has the largest posterior probability, but if the amount of data is large, the chance that it is is good. As the amount of data increases, this maximum likelihood estimate will become more and more likely to be the best estimate as well, as equation 16.3 becomes dominated by the quantity in the large parentheses. Fisher showed (1922) that maximum likelihood estimates have a variety of good properties, including consistency (converging to the correct value of the parameter) and efficiency (having the smallest possible variance around the true parameter value) as the amount of data grows large.

Other statistical frameworks are possible as well, such as least squares. These estimates are sometimes easier to compute than maximum likelihood, but as likelihood is extremely efficient in extracting information, and as faster computers have removed the barriers to doing it, most of the development of statistical methods for inferring phylogenies has concentrated on likelihood.

An example

To make likelihood methods concrete, consider the estimation of the heads probability of a coin that is tossed n times. If we assume that the tosses are all independent, and all have the same unknown heads probability p, then on observing the

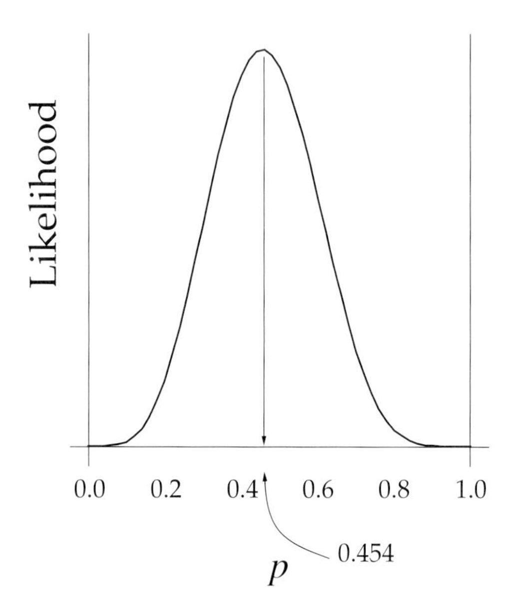

Figure 16.1: The likelihood for the heads probability p for a series of 13 independent coin tosses that results in 5 heads and 6 tails. Note that the maximum of this curve is at the point $\widehat{p} = 5/11$.

sequence of tosses HHTTHTHHTTT, we can easily calculate the probability of these data. It is

$$L = \text{Prob}\,(D|p) = pp(1-p)(1-p)p(1-p)pp(1-p)(1-p)(1-p) = p^5(1-p)^6 \quad (16.4)$$

where we can take the product because the tosses are assumed independent. Figure 16.1 shows the likelihood curve, plotting this L against p. Note that although this looks rather like a distribution, it is not. It plots the probabilities of the same data D for different values of p. Thus it does not show the probabilities of different mutually-exclusive outcomes, and the area under this curve need not be 1.

The maximum likelihood is at $p = 0.454545$, which is $5/11$. This can be verified by taking the derivative of L with respect to p:

$$\frac{dL}{dp} = 5p^4(1-p)^6 - 6p^5(1-p)^5 \quad (16.5)$$

equating it to zero, and solving:

$$\frac{dL}{dp} = p^4(1-p)^5\,[5(1-p) - 6p] = 0 \quad (16.6)$$

which yields as the position of the maximum (the point in the interior where the slope is zero) $\widehat{p} = 5/11$.

Likelihoods are often maximized by maximizing their logarithm. This maximization works more easily:

$$\ln L \;=\; 5\,\ln p + 6\,\ln(1-p) \tag{16.7}$$

whose derivative is

$$\frac{d(\ln L)}{dp} \;=\; \frac{5}{p} - \frac{6}{(1-p)} \;=\; 0, \tag{16.8}$$

which again yields $\widehat{p} = 5/11$.

Computing the likelihood of a tree

We now examine how to compute the likelihood of a given tree. This will be done with DNA sequences as the example, but the procedure used is actually general to all discrete-characters models, and it can be related closely to the methods for continuous characters as well.

Suppose that we have a set of aligned DNA sequences, with m sites. We are given a phylogeny with branch lengths, and an evolutionary model that allows us to compute probabilities of changes of states along this tree. In particular, the model allows us to compute transition probabilities $P_{ij}(t)$, the probability that state j will exist at the end of a branch of length t, if the state at the start of the branch is i. Note that t measures branch length, not time. We will make two assumptions that are central to computing the likelihoods:

1. Evolution in different sites (on the given tree) is independent.

2. Evolution in different lineages is independent.

The first of these allows us to take the likelihood and decompose it into a product, one term for each site:

$$L \;=\; \text{Prob}\,(D|T) \;=\; \prod_{i=1}^{m} \text{Prob}\left(D^{(i)}|T\right) \tag{16.9}$$

where $D^{(i)}$ is the data at the ith site. This means that we need only know how to compute the likelihood at a single site. Suppose that we have a tree, and the data at a site. An example is shown in Figure 16.2 The likelihood of the tree for this site is the sum, over all possible nucleotides that may have existed at the interior nodes of the tree, of the probabilities of each scenario of events:

$$\text{Prob}\left(D^{(i)}|T\right) \;=\; \sum_x \sum_y \sum_z \sum_w \text{Prob}\,(A, C, C, C, G, x, y, z, w|T) \tag{16.10}$$

each summation running over all four nucleotides.

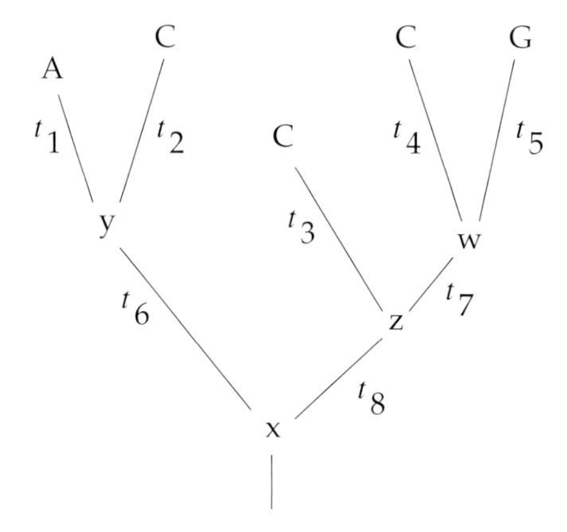

Figure 16.2: A tree, with branch lengths and the data at a single site. This example is used in the text to describe calculation of the likelihood.

The assumption that we have made about evolution in different lineages being independent allows us to decompose the probability on the right side of equation 16.10 into a product of terms:

$$\begin{aligned}\text{Prob}\,(A, C, C, C, G, x, y, z, w|T) \;=\; & \\ \text{Prob}\,(x) \quad \text{Prob}\,(y|x, t_6) \quad & \text{Prob}\,(A|y, t_1) \quad \text{Prob}\,(C|y, t_2) \\ \text{Prob}\,(z|x, t_8) \quad & \text{Prob}\,(C|z, t_3) \\ & \text{Prob}\,(w|z, t_7) \quad \text{Prob}\,(C|w, t_4) \quad \text{Prob}\,(G|w, t_5)\end{aligned} \tag{16.11}$$

The probability of x may be taken to be the probability that, at a random point on an evolving lineage, we would see base x (where $x = A, C, G,$ or T). If we are allowed to assume that evolution has been proceeding for a very long time according to the particular model of base substitution that we are using, it is reasonable to take $\text{Prob}\,(x)$ to be the equilibrium probability of base x under that model. The other probabilities are derived from the model of base substitution. The change in each lineage is independent of that in all other lineages, once the bases at the start of each lineage have been specified.

The expression in equation 16.11 still looks difficult to compute. The individual probabilities are not hard to compute (for example, we can use one of equations 11.16, 13.2, or 13.11, depending on which of these models of DNA change we happen to be using). The problem is that there are a great many terms in equation 16.11. For each site, it insists that we sum $4^4 = 256$ terms. That does not sound too hard, but with larger numbers of species the problem increases. The number of terms rises exponentially with the number of species. On a tree with n species, there are $n-1$ interior nodes, and each can have one of 4 states. So we need 4^{n-1} terms. For $n = 10$ this is 262,144; for $n = 20$ it is 274,877,906,944. That is definitely too many.

Economizing on the computation

Fortunately, there is an economy that makes the whole computation practicable. I introduced this (Felsenstein, 1973b, 1981b), calling it "pruning." (It was also independently invented by Gonnet and Benner, 1996.) It is simply a version of the "peeling" algorithm invented by Hilden (1970), Elston and Stewart (1971), and Heuch and Li (1972) for rapidly computing likelihoods on pedigrees in human genetics. That in turn may also be regarded as a version of Horner's rule, which makes us able to compute the values of polynomials rapidly. This method was first published by Horner (1819), though it seems to have been used much earlier by Isaac Newton. It is a particular case of a method well-known in computer science as "dynamic programming."

The method may be derived simply by trying to move summation signs in equation 16.10 as far right as possible and enclose them in parentheses where possible. Equation 16.10 can be written:

$$\begin{aligned}\text{Prob}\left(D^{(i)}|T\right) \;=\; \sum_x\sum_y\sum_z\sum_w \;& \text{Prob}\,(x)\;\text{Prob}\,(y|x,t_6)\;\text{Prob}\,(A|y,t_1)\\ & \text{Prob}\,(C|y,t_2)\\ & \text{Prob}\,(z|x,t_8)\;\text{Prob}\,(C|z,t_3)\\ & \text{Prob}\,(w|z,t_7)\;\text{Prob}\,(C|w,t_4)\;\text{Prob}\,(G|w,t_5)\end{aligned} \tag{16.12}$$

and when we move the summation signs as far right as possible

$$\begin{aligned}\text{Prob}\left(D^{(i)}|T\right) \;=\;& \\ & \sum_x \text{Prob}\,(x)\left(\sum_y \text{Prob}\,(y|x,t_6)\;\text{Prob}\,(A|y,t_1)\;\text{Prob}\,(C|y,t_2)\right)\\ & \quad\times\left(\sum_z \text{Prob}\,(z|x,t_8)\;\text{Prob}\,(C|z,t_3)\right.\\ & \qquad\left.\times\left(\sum_w \text{Prob}\,(w|z,t_7)\;\text{Prob}\,(C|w,t_4)\;\text{Prob}\,(G|w,t_5)\right)\right)\end{aligned} \tag{16.13}$$

We may note that the pattern of parentheses and terms for tips in this expression is

$$(A, C)\,(C, (C, G))$$

which has an exact correspondence to the structure of the tree. The flow of computations in equation 16.13 is from the inside of the innermost parentheses outwards.

This suggests a flow of information down the tree, and indeed, an algorithm to compute equation 16.13 is easily found that works in this way.

It makes use of a quantity that we may call the *conditional likelihood* of a subtree. We will call this $L_k^{(i)}(s)$. It is the probability of everything that is observed from node k on the tree on up, at site i, conditional on node k having state s. In equation 16.13 the term

$$\text{Prob}\,(C|w, t_4)\;\text{Prob}\,(G|w, t_5)$$

is one of these quantities, being the probability of everything seen at or above that node (the node that lies below the rightmost two tips), given that the node has base w. There will be four such quantities, corresponding to different values of w. The key to the pruning algorithm is that, once these four numbers are computed, they need not continually be recomputed.

The algorithm is most easily expressed as a recursion that computes the $L^{(i)}(s)$ at each node on the tree from the same quantities in the immediate descendant nodes. Suppose that node k has immediate descendants ℓ and m, which are at the top ends of branches of length t_ℓ and t_m. Then we can compute

$$L_k^{(i)}(s) \;=\; \left(\sum_x \text{Prob}\,(x|s, t_\ell)\,L_\ell^{(i)}(x)\right)\left(\sum_y \text{Prob}\,(y|s, t_m)\,L_m^{(i)}(y)\right) \qquad (16.14)$$

The logic of this equation is straightforward: The probability of everything at or above node k, given that node k has state s, is the product of the events taking place on both descendant lineages. In the left lineage, it sums over all of the states to which s could have changed, and for each of those computes the probability of changing to that state, times the probability of everything at or above that node (node ℓ), given that the state has changed to state x. Nothing more complicated than simple probability bookkeeping is involved. The conditional likelihoods at nodes ℓ and m fit easily into the calculation. The extension to multifurcating trees is immediate, simply involving more factors on the right side of equation 16.14.

To start the process, we need values of the $L^{(i)}$ at the tips of the tree. If state A is found at a tip, the values of the $L^{(i)}$ at that tip will be

$$\left(L^{(i)}(A), L^{(i)}(C), L^{(i)}(G), L^{(i)}(T)\right) \;=\; (1, 0, 0, 0) \qquad (16.15)$$

Whichever base is seen at the tip has the corresponding value of $L^{(i)}$ set to 1, and all others are 0.

This algorithm is applied starting at the node that has all of its immediate descendants being tips (there will always be at least one such node). Then it is applied successively to nodes further down the tree, not applying it to any node until all of its descendants have been processed. The result is the $L_0^{(i)}$ for the bottommost node in the tree. We then complete the evaluation of the likelihood for this site by making a weighted average of these over all four bases, weighted by their prior probabilities under the probabilistic model:

$$L^{(i)} = \sum_x \pi_x L_0^{(i)}(x) \tag{16.16}$$

Working your way in equation 16.13 from the innermost terms to the outer ones, you can verify that this recursion computes the likelihood correctly. The effort needed is not excessive: For each site it is carried out $n - 1$ times, and each time it is four calculations, each the product of two terms, each having a sum of four products. Thus with n tips on the tree, and sequences p bases long with b different possible bases, the effort required is proportional to $p(n - 1)b^2$. Actually, it can be reduced below this if some sites have the same pattern of nucleotides, as we need only compute the probability for one of those sites and can reuse the same quantity for the other site.

Once the likelihood for each site is computed, the overall likelihood of the tree is the product of these, as noted in equation 16.9.

Handling ambiguity and error

We have assumed that at each tip, we have a precise observation of which nucleotide is present. The present framework actually has no trouble at all handling ambiguities. For example, we might have observed an R rather than an A (which means we observed only that this nucleotide was a purine, and hence might either be an A or a G). In that case, given that we were capable only of seeing R or Y, the four values of $L^{(i)}$ at that tip should be $(1, 0, 1, 0)$, because our observation (R) would have probability 1 given that the truth was A or G, and would have probability 0 otherwise.

Similarly, when we fail to observe the nucleotide, so that the base must be encoded as N in the standard IUB one-letter code, the $L^{(i)}$ would be $(1, 1, 1, 1)$, because the observation of total ambiguity would have probability 1, no matter what the true nucleotide, given that we did not observe at all.

With these values at the tips, likelihoods can be computed straightforwardly using the algorithm for data sets that have ambiguous nucleotides. Note that the $L^{(i)}$ at a tip do not necessarily add up to 1, as they are not probabilities of different outcomes but probabilities of the same observation conditional on different events. You may be sorely tempted to make the quantities $(\frac{1}{4}, \frac{1}{4}, \frac{1}{4}, \frac{1}{4})$, but you should resist this temptation.

Another situation that is easily accommodated is sequencing error. Suppose that our sequencing has probability $1 - \varepsilon$ of finding the correct nucleotide, and $\varepsilon/3$ of inferring each of the three other possibilities. Consideration of the meaning of the $L^{(i)}$ then shows that when an A is observed, the four values should be $(1 - \varepsilon,\ \varepsilon/3,\ \varepsilon/3,\ \varepsilon/3)$. If a C is observed, they should be $(\varepsilon/3,\ 1 - \varepsilon,\ \varepsilon/3,\ \varepsilon/3)$, and so on. Although this uses a particularly simpleminded model of sequencing error, any other model could be incorporated in this manner. The effect of sequencing error is similar to lengthening the branches of the tree: If there is a 1% rate of sequencing error, this can look rather like a lengthening of all exterior branches by 0.01. But it is better to treat it by altering the $L^{(i)}$. If we tried instead to remove 0.01 from the length of each exterior branch, that might make some of them of negative length. Also, it is not obvious that parameters like transition/transversion rate are the same for sequencing error as for evolution.

Unrootedness

The trees inferred by maximum likelihood appear from this description to be rooted trees. If the model of base substitution is reversible, as most of them are, the tree is actually unrooted. Consider the region near the root of the tree in Figure 16.2. Using the conditional likelihoods, we can write the likelihood as

$$L^{(i)} = \sum_y \sum_z \sum_x \text{Prob}(x)\, \text{Prob}(y|x, t_6)\, \text{Prob}(z|x, t_8)\, L_6^{(i)}(y)\, L_8^{(i)}(z) \quad (16.17)$$

But reversibility of the substitution process also guarantees us that

$$\text{Prob}(x)\, \text{Prob}(y|x, t_6) = \text{Prob}(y)\, \text{Prob}(x|y, t_6) \quad (16.18)$$

Substituting that into equation 16.17, we get

$$L^{(i)} = \sum_y \sum_z \sum_x \text{Prob}(y)\, \text{Prob}(x|y, t_6)\, \text{Prob}(z|x, t_8)\, L_6^{(i)}(y)\, L_8^{(i)}(z) \quad (16.19)$$

so that the root can be shifted to the node to its upper left, without any change in the likelihoods of the tree at the individual sites. In fact, it is easy to show, using the Chapman-Kolmogorov formula for transition probabilities, that the root can be shifted anywhere in between as well: If we add an amount u to t_6 and subtract the same amount u from t_8, we do not alter the likelihoods at all.

Once the root has reached the upper-left or upper-right descendants, it can in fact move on beyond them, using the same argument. In fact, it can be placed anywhere in the tree. The tree is therefore actually an unrooted tree, without any information present as to where the root is. Of course outgroup information can help root it, and so can considerations of a molecular clock.

Finding the maximum likelihood tree

The pruning algorithm for updating likelihoods along a tree also greatly simplifies the task of finding the maximum likelihood tree, though perhaps not enough to make it entirely easy. In general, the task is very much the same as with distance matrix methods. We are searching in a space of trees with branch lengths. We need to find the optimum branch lengths for each given tree topology, and we also need to search the space of tree topologies for the one that has a set of branch lengths that gives it the highest likelihood.

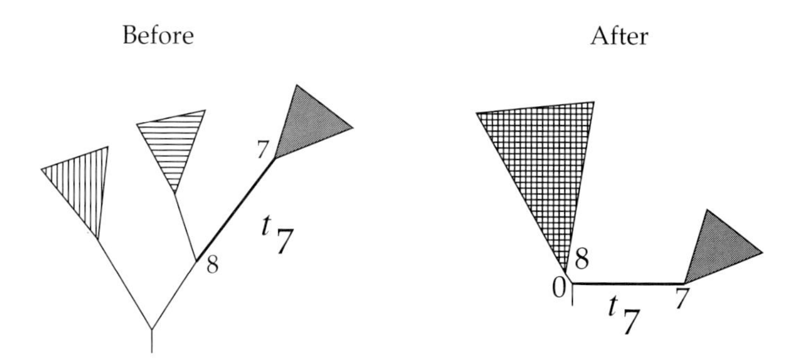

Figure 16.3: The region around branch 7 in the tree of Figure 16.2, when a new root (node 0) is placed in that branch. The change in the notation and structure of the tree is shown. The subtrees are shown as shaded triangles.

It is the former task that is simplified by the pruning algorithm. There is no analytical solution to the problem of finding the optimal branch lengths for a given tree topology. Unlike the case of least squares distance matrix methods with negative branch lengths allowed, there is no matrix equation that can solve for the best branch lengths. However, we can try to find our way to a maximum of the likelihood curve by proceeding uphill and by hoping that this is also the highest such maximum for that tree topology. In general, we have no assurance that there are not multiple maxima, but they are rarely seen in practice. So this strategy works fairly well. (See Steel, 1994b, for a case in which it can be shown that there are multiple maxima for a given tree topology, and Chor et al., 2000, who use Hadamard transform techniques to find data sets that have multiple optima and even ridges of tied trees. Simulation studies by Rogers and Swofford, 1999, found less cause for concern.)

The pruning algorithm speeds the iteration of branch lengths. To see this, imagine that we are iterating the branch length of branch 7 in the tree in Figure 16.2. Provided that we have a reversible model of evolution, we can, without affecting any of the likelihoods, take the root of the tree to be at the beginning of this branch. Figure 16.3 shows the reconfiguration. We have in effect added a new root, node 0, just to the right of node 8, with a zero-length branch connecting it to node 8. If we do that, the likelihood for site i will turn out to be

$$L_0^{(i)}(z) = L_8^{(i)}(z)\,\text{Prob}\,(z)\,\text{Prob}\,(w|z,t_7)\,L_7^{(i)}(w) \tag{16.20}$$

We can summarize all the likelihoods for the part of the tree connected to the left end of branch 7 by $L_7^{(i)}$, which can be computed from that subtree. The rerooting of the tree has changed the meaning of that quantity — it now summarizes that part of the tree. We already had $L_8^{(i)}$, and it has not changed in meaning. Note that as the root is at the left-hand end of branch 7, $\text{Prob}(z)$ is simply multiplied by $L_8^{(i)}(z)$ as there is no time for the state to change between nodes 0 and 8 (and so $\text{Prob}(y|z,0) = 0$ unless $y = z$).

In effect, once the information from the subtrees at the end of a branch is "pruned" down to that node, we are left with a tree with only two tips. The branch length for that tree can then be calculated. Its optimal value has not been altered by any of the rerooting and pruning.

The likelihood then needs to be maximized with respect to t_7 by finding the best length for this branch in the two-species tree. This can be done by many methods; the Newton-Raphson method seems to be a good one for this, but EM algorithms also work, as does direct search with quadratic interpolation. However, once one branch has changed length, there is no guarantee that the others are still at their optimal lengths. It is necessary to return to all the other branches, improving their lengths. This process continues indefinitely, but it must converge. At each step the likelihood of the tree increases. The process will not fall into an endless cycle, because in such a cycle the likelihood would have to decrease as well as increase.

Schadt, Sinsheimer, and Lange (1998) show how to prune derivatives of likelihoods with respect to branch lengths along a tree, enabling use of the Newton-Raphson method to update all branch lengths simultaneously. In practice, improving the length of one branch at a time works quite well, as there are not very strong or complex interactions between branch lengths. After a few sweeps through the tree improving the likelihood, it does not improve much more, and the tree can then be evaluated.

For the search among tree topologies, we do not gain much because we use likelihoods. The usual issues of local and global rearrangements are present. Branch-and-bound search is possible, but for the moment the bounds available are far too loose for it to work effectively. There is room for improvement there. There has also been no proof that the problem is NP-hard (as there has been for many other methods). It seems likely that it is some sort of "NP-nasty" problem, but there is actually no formal examination yet of its computational difficulty.

However, there are hints that we may be able to do better than conventional heuristic search. Friedman et al. (2002) suggest a "structural EM algorithm" for searching tree space in a way suggested by the data. This unusual algorithm starts by computing conditional likelihoods for all nodes in a tree, then produces a matrix of log-likelihoods for imagined connections of all pairs of nodes in the tree (including interior nodes). A new tree is produced by reconnecting the nodes using a minimum spanning tree and then resolving it into a bifurcating tree.

It is amazing that such a strategy works at all. Interior node i may end up connected to j, and k to ℓ, when previously k was part of the section of the tree that contributed to node i's conditional likelihoods. However Friedman et al. have theorems showing that the likelihood is guaranteed not to decrease by doing this. In effect, this is a data-driven tree rearrangement that has some minimal guarantees. Friedman et al. do find it necessary to add a simulated annealing step to their search (for which see Chapter 4) to avoid getting stuck on local optima. It is too soon to know how effective such approaches will be, but the availability of data-driven search methods that can guarantee never to give a worse tree is encouraging.

Inferring ancestral sequences

The ability to reroot the tree anywhere also gives us a simple computational method for finding ancestral states for sites with a given tree. If we have some node in the tree for which we want to know the ancestral states, we need only root the tree at that node. As we prune likelihood down that tree, we end up with quantities $\pi_s L_0^{(i)}(s)$ for all states s at each site i. Each computes the contribution that each state s makes to the overall likelihood of the tree at that site. The state that makes the largest contribution to this likelihood is the one to estimate (Yang, Kumar, and Nei, 1995). Although we could do this successively for each interior node, one can speed the process by reusing the views of likelihood as we go. This is much the same process for likelihood as it is for parsimony. The result is that we can compute the estimates of ancestral sequences by two passes through the tree, in time linear in the number of species and no worse than linear in the number of sites. Koshi and Goldstein (1996) suggested inference of ancestral states by a method equivalent to this, and so did Schluter et al. (1997) for 0/1 characters that evolved according to a simple stochastic model.

One potential difficulty is that if we choose the best ancestral sequence at each internal node of the tree, this may turn out to be different from the combination of states that together make the highest contribution to the likelihood. In equation 16.10 that would correspond to finding the combination $xyzw$ that maximized the term inside the summations. This would yield a joint estimate of the states at all interior nodes of the tree. The previous method is a marginal estimate at each node. Usually they will yield the same estimated states, but occasionally they will not. Figure 16.4 shows a simple example of discrepancy between the joint and the marginal reconstructions. It shows the same tree as in our example of likelihood calculations, with branch lengths specified and a Jukes-Cantor model used. Considering the ancestral nodes left-to-right, the states that have highest marginal probabilities are A, C, and G. The overall likelihood is 0.0000124065. The largest contribution to that likelihood from one joint ancestral reconstruction is 0.0000041594, more than one-third of the total. It is for the reconstruction A, G, G, which conflicts with the marginal information. There is no paradox — the other joint reconstructions that contribute to the likelihood include many that have a C at the bottommost node.

Yang, Kumar, and Nei (1995) first suggested doing the joint calculations. Pupko et al. (2000) gave a dynamic programming algorithm that avoids the necessity of searching over all combinations of interior node states. One can modify

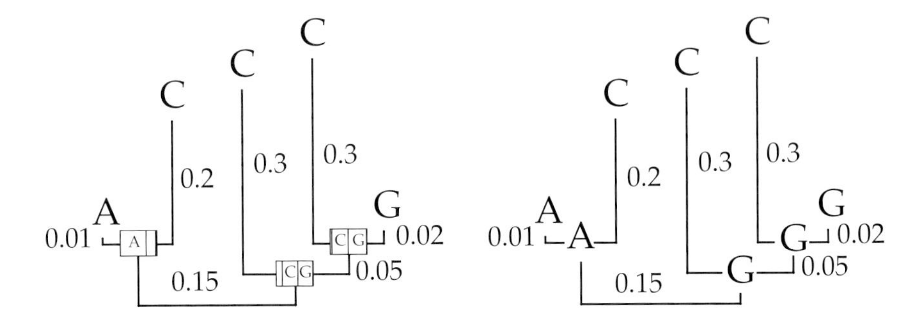

Figure 16.4: A five-species unrooted tree with one site (the same as in Figure 16.2 with branch lengths added). Under a Jukes-Cantor model of base change, the marginal (left) and joint (right) reconstructions of states at the ancestral nodes are shown. They differ in the bottom-most node, which has state G in the joint reconstruction, but which assigns slightly more probability to C in the marginal reconstruction. The marginal reconstructions are shown as boxes divided vertically into regions proportional to the contribution each base makes to the overall likelihood.

equation 16.14 by replacing the summation signs by $\max$. The resulting recursion computes at each node the probability contributed by the best combination of states from that point up:

$$L_k^{(i)}(s) = \left(\max_x \text{ Prob } (x \mid s, t_\ell)\, L_\ell^{(i)}(x) \right) \left(\max_y \text{ Prob } (y \mid s, t_m)\, L_m^{(i)}(y) \right) \quad (16.21)$$

At the root of the tree one can use π_s to weight the different states and then choose the one with the highest value. From there one can backtrack up the tree in a straightforward way, choosing the interior node states. The logic is similar to the algorithm given in Chapter 6 above. This is equivalent to Pupko et al.'s algorithm. Bayesian approaches to joint reconstruction are also possible (Schultz and Churchill, 1996). See Chapter 18 for Markov chain Monte Carlo methods for Bayesian inference of phylogenies. Pagel (1999b) suggests finding the combination of states, together with a set of branch lengths, that makes them together have the highest likelihood. This would compare different state combinations using for each a tree with different branch lengths.

Rates varying among sites

We have already seen in Chapter 13 that rates of evolution can vary among sites in a nucleotide (or protein) sequence. In calculating distances we discussed modeling the rate variation by using a gamma distribution of rates across sites (the gamma distribution is described in Chapter 13). In principle we could do the same for likelihoods on trees. If we have m sites, and the density of the gamma distribution of rates is $f(r; \alpha)$, the overall likelihood would be

$$L = \prod_{i=1}^{m} \left[\int_0^\infty f(r_i; \alpha)\, L^{(i)}(r_i)\, dr_i \right] \tag{16.22}$$

where $L^{(i)}(r_i)$ is the likelihood of this tree for site i given that the rate of evolution at site i is r_i. This approach has been advocated by Yang (1993). The difficulty is that we need to compute the $L^{(i)}(r_i)$ for all possible r_i. Or at least, so it seems. We would then have to evaluate the likelihood at site i for a densely spaced set of evolutionary rates, greatly increasing the amount of computation.

There might seem to be hope of getting around this problem. The L's are themselves sums of products of expressions, each of which is an exponential in the length of the branch. With the gamma distribution, the integral of an exponential function of the rate, averaged over the gamma distribution, is analytically tractable. If we can express the likelihood for each site as a linear combination of exponentials, then we can integrate this expression termwise. This raises the possibility of exactly calculating the likelihood. However, when we try to do this, we find that for a tree with n tip species, there turn out to be 3^{2n-3} terms that need to be evaluated for each site. For example, for 10 species, we need $3^{17} = 129,140,163$ terms, and for 20 tips there are 4.5×10^{17}. This is too many.

An alternative that works for trees whose branches are not long is given by Steel and Waddell (1999). Taking the pattern probabilities for a tree as functions of the rate of evolution at the character, they approximate that function by a power series. They obtain quite simple formulas for the pattern probabilities in terms of the mean and variance of the evolutionary rates. Their approximations work well when amounts of divergence are small.

The best alternative, which has been adopted by Yang (1994a), is to approximately evaluate the integral at each site by numerical integration. Thus if we evaluate the likelihood for each site at a series of k different rates $r_1, r_2, \ldots, r_k$, we can approximate the likelihood at the ith site by the weighted sum

$$L^{(i)} = \int_0^\infty f(r_i; \alpha)\, L^{(i)}(r_i)\, dr_i \approx \sum_{j=1}^{k} w_k\, L^{(i)}(r_k) \tag{16.23}$$

The weights w_k and the points r_k must be chosen so that the proper approximation is made. Yang uses for r_k the average over the gamma distribution of the values in the corresponding quantile. (For example, if $k = 10$, for r_3 he averages the ordinates between the 2/10 and the 3/10 points of the gamma distribution.) A possibly more accurate method would be to consider the process as Gauss-Laguerre quadrature, with the k points being chosen as the zeros of the generalized Laguerre polynomials (Felsenstein, 2001b).

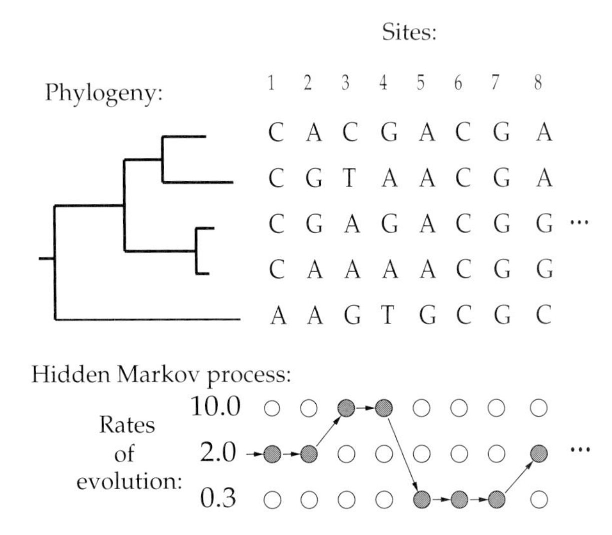

Figure 16.5: Diagram of a hidden Markov model (HMM). Nature assigns rates to sites according to a Markov process that is hidden from our view. We can observe the resulting patterns of nucleotides at the sites. To compute the likelihood based on these, we must sum over all combinations of rates at sites.

The value of α is needed; if not known in advance, it can be inferred by maximizing the likelihood with respect to this parameter by numerical evaluation for different values. Gu, Fu, and Li (1995) have discussed ways of doing this, also in the presence of invariable sites.

Hidden Markov models

Computing the likelihood of a tree for each site at a number of different rates of substitution opens the door to more complex and realistic models. We can assume that the rate of evolution is laid down by a Markov process operating along the sequence. Each site's rate of change is determined by a random process that depends on which rate of change was chosen at the previous site. The process of choosing rates of change is done by nature; it is hidden from us. The outcome of evolution at each site is then dependent on the rate chosen by this hidden Markov process.

Such a model is called a *hidden Markov model* (HMM). These have been first introduced in ecology (Baum and Petrie, 1966; Baum et al., 1970) but have become most widely used in communications engineering and speech recognition. More recently (Churchill, 1989) they have invaded molecular biology. Ziheng Yang, Gary Churchill and I have applied them to inferring phylogenies, modeling the assignment of rates to sites (Yang, 1995; Felsenstein and Churchill, 1996). Suppose that we have specified some stochastic process that assigns rates to sites, and we can compute the probability of any given combination of rates. The overall likelihood of a tree T is the probability of the data, making a weighted average over all possible combinations of rates:

$$\text{Prob}\,(D \mid T) \;=\; \sum_{i_1}\sum_{i_2}\ldots\sum_{i_m} \text{Prob}\,(r_{i_1}, r_{i_2}, \ldots r_{i_m})\; \text{Prob}\,(D \mid T, r_{i_1}, r_{i_2}, \ldots r_{i_m}) \tag{16.24}$$

The number of rate combinations is vast. Even with (say) four rates possible at each site, there would be 4^m terms, one factor of 4 for each site.

However, a simple algorithm, rather like the one we use to compute likelihood along a tree reduces the computation—it is well-known as the *backward algorithm* to users of HMMs. Note that the only nonindependence of sites is due to the possible correlation of rates among neighboring sites. Given the rates, then conditional on the rates, we can factor

$$\text{Prob}\,(D \mid T, r_1, r_2, \ldots, r_m) \;=\; \prod_{i=1}^{m} \text{Prob}\,(D^{(i)} \mid T, r_i) \tag{16.25}$$

Using this factorization and moving the summation over i_m as far right as possible in the expression on the right side of equation 16.24, we get

$$\begin{aligned}\text{Prob}\,(D \mid T) \;=\; & \sum_{i_1}\sum_{i_2}\cdots\sum_{i_{m-1}} \text{Prob}\,(r_{i_1}, r_{i_2}, \ldots r_{i_{m-1}}) \prod_{i=1}^{m-1} \text{Prob}\,(D^{(i)} \mid T, r_i) \\ & \sum_{i_m} \text{Prob}\,(r_{i_m} \mid r_{i_{m-1}}) \quad \text{Prob}\,(D^{(m)} \mid T, r_{i_m})\end{aligned} \tag{16.26}$$

The factorization of the probabilities of rates into

$$\text{Prob}\,(r_{i_1}, r_{i_2}, \ldots r_{i_m}) \;=\; \text{Prob}\,(r_{i_1}, r_{i_2}, \ldots r_{i_{m-1}})\; \text{Prob}\,(r_{i_m} \mid r_{i_{m-1}}) \tag{16.27}$$

is possible because of the Markov model of assignment of rates to sites.

It turns out that the Markovian nature of the hidden stochastic process allows a computational economy very similar to the recursive evaluation of likelihood down a tree in the "pruning" algorithm. Recall that in that algorithm, the essential quantities were the probabilities of all data seen at or above a point in the tree, given the (unknown) state of the site at that point of the tree. In this case, the quantity needed is the probability of all data seen at sites $j, j+1, \ldots, m$, given the i_jth rate of evolution r_{i_j} at site j. This we define as

$$\text{Prob}\,(D^{[j]} \mid T, r_{i_j}) \;=\; \text{Prob}\,(D^{(j)}, D^{(j+1)}, \ldots, D^{(m)} \mid T, r_{i_j}) \tag{16.28}$$

and, as in the pruning case, this quantity is easily shown to satisfy the recursive algorithm

$$\text{Prob}\,(D^{[j]}\,|\,T, r_{i_j}) \;=\; \text{Prob}\,(D^{(j)}\,|\,T, r_{i_j}) \sum_{i_{j+1}} \text{Prob}\,(r_{i_{j+1}}|r_{i_j})\;\text{Prob}\,(D^{[j+1]}\,|\,T, r_{i_{j+1}}) \tag{16.29}$$

The algorithm starts at site m where it is easy to see that

$$\text{Prob}\,(D^{[m]}\mid T, r_{i_m}) \;=\; \text{Prob}\,(D^{(m)}\mid T, r_{i_m}), \tag{16.30}$$

a quantity we already know. Equation 16.29 can then be used recursively to compute the quantities $\text{Prob}\,(D^{[j]}|T, r_{i_j})$ for sites $m-1, m-2, \ldots,$ until we have computed the quantities $\text{Prob}\,(D^{[1]}|T, r_{i_1})$. These are the probabilities of the entire data set given the unknown rates at site 1, and they can be combined by weighting them by the prior probabilities (under the HMM) of those rates:

$$\text{Prob}\,(D\mid T) \;=\; \sum_{i_1} \pi_{i_1}\;\text{Prob}\,(D^{[1]}\mid T, r_{i_1}) \tag{16.31}$$

The recursive algorithm economizes because we are summing probabilities of all possible paths through the array of sites by rates. If two of these paths share all but their first rate, then we do not need to calculate the contributions from both paths separately, but we save the terms that are shared and reuse them.

Thus there is a recursive algorithm for computing the likelihood with the HMM model, and it merely requires for a given tree that we compute the probability for each site given each possible rate at that site. Then we combine them in a recursion that needs computational effort equal to the product of the number of sites and the number of rates. In general, since the computational effort is dominated by computing the likelihoods for each rate at each site, if there are (say) 5 rates, we compute the likelihood of the tree, not with 5^m times as much effort, but only with 5 times as much effort, as in the case of a single rate of evolution. The algorithm updates partial information from the last site forward to the first. It is equally possible to make it go the other way.

Autocorrelation of rates

The hidden Markov model approach allows us to model not only the variation of rates among sites, but the autocorrelation of rates. Yang (1995) used an autocorrelated gamma distribution and approximated it by having discrete categories. He used as the rates for the categories the means of rates between quantiles of the gamma distribution. The transition matrix from one rate category to another is here represented as $\text{Prob}\,(r_{i_j}|r_{i_{j-1}})$; he computed its elements from a discrete approximation to the bivariate gamma distribution.

My own paper with Gary Churchill (Felsenstein and Churchill, 1996) used a simpler transition matrix. The rates r_i and their probabilities π_i are allowed to be arbitrary, which gives great flexibility, but sometimes more flexibility than the user wants. We assumed that with probability λ a new rate is drawn from this distribution; otherwise the rate at the next site remains the same as at the current site. Note that drawing a new rate could result in getting the same rate again. The transition probabilities are then

$$\text{Prob}\,(r_i \mid r_j) \;=\; (1-\lambda)\;\delta_{ij} + \lambda\;\pi_i, \tag{16.32}$$

where δ_{ij} is the usual Kronecker delta, a bookkeeping device that is 1 when $i = j$ and 0 otherwise.

Any other Markov chain for change of rates could be used instead. There is nothing particularly magical about the gamma distribution; for example, the lognormal might do about as well.

HMMs for other aspects of models

Although we have presented the HMM framework with the hidden states being different evolutionary rates, there is nothing to prevent them from being any aspect of the model of change (Churchill, 1989; Felsenstein and Churchill, 1996). For example, the states could be purine-richness of the region in which the site was located, or AT-richness, or the secondary structure of the region in which an amino acid position is located. Goldman, Thorne, and Jones (1996) have made just such a model. They find that when the estimates of the states are examined, that this model, which uses phylogenies, increases the accuracy of assessment of secondary structure by about 10%.

Estimating the states

We have concentrated on the estimation of the phylogenies, summing over all possible combinations of the hidden states at the sites. One can also make an estimate of the hidden states. Two possibilities are evident (Felsenstein and Churchill, 1996). One is to make an estimate of the single combination of states across sites that makes the largest contribution to the likelihood. The other is to estimate, for each site, which state has the largest total probability added up over all combinations of states at other sites. The first of these finds that, for example, 1, 2, 1, 1, 3, 3, 2, 2, 1 is the combination of states across all 9 sites that makes the single largest contribution. Alternatively, we might find that more of the total probability of the data is contributed by state 1 at the first site than is contributed by states 2 and 3. At the second site, we might find that, overall, combinations of states having state 2 there make the greatest contribution. In general we would hope that the combination of states that resulted would be the same either way, but there is no guarantee that this will happen.

Models with clocks

The algorithms we have discussed in this chapter have allowed each branch to assume any nonnegative length. That implies that the tree can be far from clock-like. It is of great interest to be able to find the maximum likelihood tree under a molecular clock. In doing so, we have to maintain the branch lengths under a set of constraints that causes all of the tips to be equidistant from the root. The easiest way to do this seems to be to iterate, not the branch lengths themselves, but the times of the interior nodes of the tree. The tips are all assigned times of 0, and the interior nodes then will have negative times.

In this scheme one can also use pruning to economize. If we consider the branches (usually three, but two for the root) that connect to an interior node, we can re-evaluate the time of the node. We use pruning to obtain conditional likelihoods for the three nodes to which this node connects. Then we iterate the time of this interior node, with only as much computational effort as is needed on a three-species tree. We must be careful never to allow the iteration to carry a node's time beyond that of its descendants or of its immediate ancestor. As with the nonclock-like case, one must iteratively improve different parts of the tree, returning after one time has changed to see if the others need changing.

Relaxing molecular clocks

The notion of a molecular clock has been controversial almost since its introduction by Zuckerkandl and Pauling (1965). Morphological systematists, who see no counterpart to the molecular clock, have been at the forefront of a wave of skepticism about it. One frequently hears the assertion that there simply are no molecular clocks. Certainly, when we have organisms of greatly different biology, there is no reason to expect that their molecules will evolve at the same rate. Their generation times, replication and repair mechanisms, and constraints of natural selection will differ greatly, and thus so will their overall rates of molecular evolution. At the opposite pole, members of the same species should have very similar biology, and we would expect the rates of molecular evolution to be quite similar. Essentially all analyses within species use the molecular clock assumption. We shall see such analyses in Chapters 26 and 27.

The issue is then not whether there is any molecular clock, but how far beyond the species level one can go before the clock breaks down seriously. Recognizing this, a number of researchers have developed methods using relaxed molecular clocks, ones whose rates change slowly enough to still enable them to be used. Some of these methods start by inferring the numbers of changes in each branch for a molecule, and then fit a tree to these in such a way to have as little departure from a clock as possible. Sanderson (1997, 2002) used branch lengths inferred by parsimony, and then fitted divergence times to them in such a way as to minimize a measure of difference between the rates observed in successive branches as one moves up the tree. He found that the dates of origin of groups obtained this way fitted the geological evidence better than when a single rate was used. Cutler (2000a) also used numbers of changes inferred by parsimony or distance methods. He allowed rates to vary across branches as if drawn from a normal distribution, with negative rates taken to be zero. He too found that the trees fitted with the relaxed clock were superior.

Thorne, Kishino, and Painter (1998) used a Bayesian treatment with a prior distribution of rates of evolution among branches. (Bayesian methods will be covered in Chapter 18.) They assumed that each branch had a rate that was drawn from a lognormal distribution. The logarithms of the rates were assumed to be autocorrelated, with a correlation that decayed exponentially with the difference between the midpoint times of the branches. To integrate the likelihood over all possible combinations of rates they used a Markov chain Monte Carlo (MCMC) integration. MCMC methods will be described below in Chapters 18 and 27. Confining attention to a single tree topology, they were able to infer the distributions of ancestral node times. Kishino, Thorne, and Bruno (2001) improved this method by having the rate in each branch drawn from a lognormal distribution with autocorrelation of the logarithms of rates, with the correlations depending not on the difference of midpoints of branches, but on the length of the earlier branch.

In all of these treatments, rates of evolution were assumed to be constant within each branch. Huelsenbeck, Larget, and Swofford (2000) present a model for rate change along trees that has two parameters. One is the rate of a Poisson process that places events of rate change along the tree. The other is the variance of a gamma distribution, which has expectation 1. At each rate-change event, the previous rate is multiplied by a quantity drawn from this gamma distribution. They too use a Bayesian MCMC approach, and they too find that allowing relaxation of the molecular clock improves their results.

Models for relaxed clocks

John Gillespie has done the pioneering work on relaxed molecular clocks. In 1984 he proposed a model of protein evolution with rates of substitution varying continuously through time according to a stationary stochastic process. (For a wider perspective, see Gillespie, 1991.) One of the models that fits his framework is a two-state Markovian model with a rate that is either 0 or a nonzero value, with changes at random times. This is similar to the model of Huelsenbeck, Larget, and Swofford (2001) in that the rate can change at any time. Cutler (2000b) uses neutral population genetic models to show that fluctuations in their parameters can only cause noticeable variation in rates of substitution when the fluctuations occur on a very long time scale. Zheng (2001) has shown that in some DNA models a violation of clockness is expected, but this usually leads to only minor departures from a clock. Bickel and West (1998) find that a fractal renewal process determining intervals between substitutions in proteins fits data, though they do not attempt a full likelihood analysis.

Table 16.1: The covarion model of Penny et al. (2001), a particular case of the covarion model of Galtier (2001). There are two rates, one of which is 0 and the other is 1. Note that transitions between the rates can occur at any time, but transitions between the 4 nucleotide states can occur only when the rate is nonzero, in this case according to a Kimura three-state model. Diagonal elements are omitted.

		To:							
		Rate = 1				Rate = 0			
		A1	G1	C1	T1	A0	G0	C0	T0
From:	A1	—	α	β	γ	δ	0	0	0
	G1	α	—	γ	β	0	δ	0	0
	C1	β	γ	—	α	0	0	δ	0
	T1	γ	β	α	—	0	0	0	δ
	A0	$\kappa\delta$	0	0	0	—	0	0	0
	G0	0	$\kappa\delta$	0	0	0	—	0	0
	C0	0	0	$\kappa\delta$	0	0	0	—	0
	T0	0	0	0	$\kappa\delta$	0	0	0	—

Covarions

As mentioned above, Fitch and Markowitz (1970) put forward the notion that evolutionary rates change not only along the molecule but also along the phylogeny. They called these changing regions "concomitably variable codons" or *covarions*. Making a stochastic model of such a process is doubly difficult because one wants to maintain the autocorrelation of rates along the molecule. Thus the rates at each site cannot change independently along the tree, without becoming uncorrelated along the molecule. The clusters of sites that have high rates of change will move back and forth along the molecule as evolution proceeds. Fitch and Markowitz used a rough method, looking for a momentary excess of multiple changes at a site as an indication that it was one of the few that could change substantially.

If one does not require that rates be autocorrelated along the molecule, it becomes possible to make a likelihood analysis of a covarion model. Galtier (2001) and Penny et al. (2001) have developed covarion models of this sort. The two models are quite similar, Galtier's being more general. They assume that at each site there are two stochastic processes, one making changes among a number of rates of evolution, the other making changes according to the current rate of evolution among a number of states (such as bases). We can only see the states, not the rates, so our observations are treated as ambiguous observations of the (state, rate) pairs. Table 16.1 shows the model of Penny et al., which has two rates, one of them 0. There are then 2 rates and 4 states, for a total of 8 (state, rate) pairs. In the table the rates of change between all 8 pairs are shown. The top four of these (A1, G1, C1 and T1) are states with nonzero rate, the bottom four (A0, G0, C0, T0) states with rate of change zero.

The states will be spend κ times as much time with a nonzero rate of change as as it will with a zero rate of change, and in each unit of time a fraction δ of all sites will change between the two rates. The transitions between a site being invariant and site being variable occur at rates δ and $\kappa\delta$. Given that the rates are in the nonzero state, the transitions of the bases occur according to one of the standard models, in this case the Kimura three-parameter model. An observation of a C is treated as an ambiguity between two states, C1 and C0, since we can only see the base, not the rate. Using the usual methods for inferring ancestral sequences, one can infer also the sites that are variable at each node in the tree.

A covarion model of this sort is approximately four times slower in computation time than a conventional model, because its transition matrix has four times as many entries. If we wanted to take into account the correlation of rates at neighboring sites, it would probably be necessary to use an MCMC method, with rate-change events visible in the tree.

Empirical approaches to change of rates

Gu (1999, 2001) has taken a more empirical approach to change of rates among sites in different parts of the phylogeny. His interest is in gene duplications, in discovering whether rates of evolution at a site differ in the two duplicate loci. The model assumes that the rates do not change within each locus, but have a probability θ_λ of changing between these two subtrees. In the 1999 paper, Gu uses corrected parsimony reconstructions of numbers of changes in each branch of the tree. He then makes a maximum likelihood estimate of θ_λ given these numbers. In the 2001 paper, he uses a full maximum likelihood approach, taking the likelihood to be the product of the likelihoods of the two subtrees. As he notes, this does not use information about the part of the tree that precedes both subtrees.

Using a similar model, Susko et al. (2002) test whether the rates across sites are different in two subtrees. Under the alternative hypothesis the rates are assigned separately to both subtrees, under the null hypothesis they are assigned in common. They obtain posterior probabilities of rates for sites in both subtrees, and use a parametric bootstrap or a regression technique to test whether the subtrees have different rates.

Are ML estimates consistent?

The issue of consistency must arise for maximum likelihood as it does for parsimony. There are some general proofs in the literature of statistics that maximum likelihood estimates are consistent. In my paper (1973b) on maximum likelihood phylogenies from discrete characters, I said that the proof by Wald (1949) could be used to prove consistency. A number of evolutionary biologists have argued that phylogenies may not satisfy the conditions for likelihoods to be compared between topologies, or the conditions for the consistency proofs to apply.

Comparability of likelihoods

Nei (1987) argued that "the likelihood computed in this method is conditional for each topology, so that it is not clear whether or not the topology showing the highest likelihood has the highest probability of being the true topology when a relatively small number of nucleotides are examined." Saitou (1988) agreed that "the ML values for different topologies are conditional and cannot be compared in the usual statistical sense." Li and Gouy (1991) said that "the ML values for different topologies would be equivalent to the ML values computed under different hypotheses and thus cannot be compared in the traditional sense." Because of a similar concern that the topology could not be considered a parameter, Yang (1996) preferred to call the method the "maximum maximum likelihood method".

In fact, the likelihoods of trees of different topologies *can* be compared. Each is the probability of the same event (the data), computed conditional on different phylogenies. As such they are on the same scale, and one number being larger than another, the probability of the data is higher given that tree than given the other. If we had prior probabilities for two trees, and these were equal, then equal likelihoods for the trees would also imply equal posterior probabilities. Thus the likelihoods are precisely comparable. What cannot be done is to use the conventional likelihood ratio test to compare these phylogenies, as we will see in Chapter 19. But this does not mean that the likelihood scales for different trees are incommensurable.

A nonexistent proof?

Yang (1996) argued that Wald's proof required "the continuity and differentiability of the likelihood function with respect to the topology parameter. Such concepts are not defined." Siddall (1998b) argued (citing a lecture by Farris) that the applicability of Wald's proof

> cannot be true. Among Wald's (1949) criteria for consistency were requirements for independence and identical distributions, which sequenced nucleotides cannot have, and that the likelihood function is everywhere continuous and continuously differentiable with respect to the parameter of interest. Cladograms being discrete, it has yet to be explained how that condition can be satisfied or indeed what it would mean in this case ....

Farris (1999) concurred with Siddall, and concluded that "Felsenstein's claim was incorrect," and that "it had never occurred to me that Felsenstein would base his position on a nonexistent proof."

The proof is *not* nonexistent. Although I did not formalize it in 1973, I examined Wald's conditions and convinced myself that they would hold. Since then such proofs have been constructed more formally. Chang (1996b) gave a Wald-like proof for binary trees with evolution according to the same Markov process at all characters. Rogers (1997) independently reached the same conclusions for a somewhat more general class of tree topologies. Swofford et al. (2001) have argued that Wald's conditions *are* met in the case of phylogenies, that they do not include the differentiability and continuity conditions that Yang, Siddall, and Farris cited. Rogers (2001) gives a detailed argument that the Wald proof applies for maximum likelihood estimation of phylogenies with a DNA model with a class of invariant sites and a gamma distribution of rates across sites.

A simple proof

For the purposes of this book, it will suffice to sketch a more simplified proof of the consistency of maximum likelihood inference of phylogenies. A proof much like this one was given by Yang (1994b). Let us assume that characters evolve independently according to the same Markov process. For a given tree T, the likelihood will be

$$L = \text{Prob}\,(D \mid T) = \prod_{i=1}^{m} \text{Prob}\,(x_i \mid T)^{n_i} \tag{16.33}$$

where m is the number of character patterns, x_i the ith of these, and n_i the number of times the ith character pattern is observed. Taking logarithms,

$$\ln L = \sum_{i=1}^{m} n_i \ln \text{Prob}\,(x_i \mid T) \tag{16.34}$$

If we divide this by the number of characters, n, we get

$$\frac{1}{n} \ln L = \sum_{i=1}^{m} f_i \ln(q_i) \tag{16.35}$$

where f_i is the observed fraction of characters that have pattern i, and q_i is the fraction expected to have pattern i given tree T. This is the log-likelihood per character. Its expectation is obtained by calling the expectation of f_i under the true tree p_i:

$$\mathbb{E}\left[\frac{1}{n} \ln L\right] = \sum_{i=1}^{m} p_i \ln(q_i) \tag{16.36}$$

Note that p_i is the fraction of pattern i expected under the true tree, whereas q_i is the fraction expected under tree T, which may or may not be the true tree. Suppose that we also consider the expectation of the log-likelihood per character under the true tree. This will be

$$\mathbb{E}\left[\frac{1}{n} \ln L\right] = \sum_{i=1}^{m} p_i \ln(p_i) \tag{16.37}$$

It is an old and well-known inequality in probability theory (sometimes called Kullback's inequality) that for two sets of probabilities p_i and q_i,

$$\sum_{i=1}^{m} p_i \ \ln(p_i) \ > \ \sum_{i=1}^{m} p_i \ \ln(q_i) \tag{16.38}$$

unless all the q_i are equal to all the p_i, in which case, of course, the two quantities are identical. So the expected log-likelihood per character is greater for the true tree than for any other, with equality only if there happens to be another tree that gives exactly the same predicted frequencies of patterns.

As we examine greater and greater numbers of characters, all evolving with the same Markov process, the observed frequencies f_i will converge (with probability 1) to the true frequencies p_i. It is then easy to show that, with probability 1, the log-likelihood per character converges to the value expected on the true tree. The log-likelihood per character for any other tree will then, in the limit, be less. The sole exception is when two trees predict exactly the same pattern frequencies. We might think that this completes the proof. What this argument has established is that the tree is becoming one whose expected pattern frequencies are closer and closer to those expected on the true tree. That is not quite the same thing as establishing that the tree itself is becoming closer and closer to the true tree. But some rather weak continuity conditions will ensure that, and we are going to leave them undiscussed. (In fact, that is essentially what Wald's proof does.) The interested reader can consult the detailed arguments in the papers cited above.

In the case where two trees, T_1 and T_2, predict the same pattern frequencies we might think that things will go badly wrong. But they actually do not, because if we evolve on tree T_1 but then arrive at tree T_2 as the estimate, T_1 is also the maximum likelihood estimate. We define the maximum likelihood estimate as the set of all trees that maximize the likelihood. In fact, there won't be two such trees in any interesting case. Only trees that differ in branches of zero length will predict the same pattern frequencies.

There have been several examinations of this "identifiability" issue. Chang (1996b) showed under what conditions on the transition matrices of the Markov chain the expected pattern frequencies allow not only convergence to the correct tree topology but reconstruction of the transition matrices on the branches as well.

Misbehavior with the wrong model

Likelihood is usually consistent if we use the correct stochastic model in our analysis. When we use the wrong model, there are few guarantees. Suppose that the model we use in the analysis predicts pattern frequencies Q_i on the true tree and pattern frequencies q_i on another tree. We could be guaranteed of consistency only if

$$\sum_{i=1}^{m} p_i \ \ln(Q_i) \ > \ \sum_{i=1}^{m} p_i \ \ln(q_i) \tag{16.39}$$

Since we have said nothing about how our model differs from the true one, we cannot guarantee that this inequality is true. If the model differs only slightly from the true model, the Q_i should be close to the true expected pattern frequencies p_i. In that case the true tree will probably be preferred to the untrue tree. But if under our model the untrue tree itself also has pattern frequencies that are close to the true ones, it may be preferred.

In some cases the pattern frequencies may not be changed when the model is changed. One example would be any model that had the same pattern frequencies but allowed some autocorrelation between characters. In a simple (and artificial) example in which each character is perfectly correlated with one other, as if we had accidentally photocopied the data so as to double its length, the pattern frequencies are unchanged even though there are now correlations among characters. In such a case a maximum likelihood inference that failed to account for the correlations would nonetheless be consistent.

Note that the wrong tree that is preferred may have the same tree topology as the true tree, differing only in branch lengths. When our main objective is to infer the tree topology, we can hope that getting the model slightly wrong will not make much difference. Gaut and Lewis (1995), Waddell (1995, pp. 377–385), Chang (1996a), Sullivan and Swofford (1997), and Huelsenbeck (1998) have all discussed the inconsistency of likelihood under various departures from the model. Kuhner and Felsenstein's (1994) simulations showed signs of inconsistency of maximum likelihood when there was unacknowledged rate variation from site to site.

Chang (1996a) showed a particularly interesting case in which the pattern of branch lengths differed in two sets of characters. Using a symmetrical two-state model of evolution, he then showed that the pattern frequencies for four species were identical to those predicted for a different tree topology from a model that had only one set of branch lengths! This establishes neatly that when one analyzes assuming one rate of character change, one can be substantially misled if the pattern of branch lengths differs greatly between characters. But note that it does *not* prove that analysis with the correct model would be inconsistent, either with the correct assignment of branch-length patterns to characters, or with a model allowing each character to evolve with one of two sets of branch lengths, chosen at random. Steel, Székely, and Hendy (1994) earlier presented a proof that with a general form of variation among characters in the rate of evolution, one may get the same expected pattern frequencies for two different trees. They provide sufficient (but not necessary) conditions for the consistency of maximum likelihood estimation of the phylogeny. Farris (1999) has argued that their result "sharply restricts the variety of circumstances under which maximum likelihood can guarantee consistency." In fact, their result does not show that this problem will occur for all distributions of evolutionary rates that do not satisfy their sufficient conditions. Even when two trees predict the same pattern frequencies, this means only that in the limit these two trees will both be found to be maximum likelihood estimates. If one is the true tree, it will be one of the estimates found.

Better behavior with the wrong model

Sometimes having the wrong model can actually improve our chances of recovering the true tree. We saw (above in Chapter 9 that long branch attraction can cause a parsimony method to perform better, if the long branches happen to be adjacent on the tree. This happens because there is no correction for parallel changes on the two branches, and those changes are reconstructed as occurring on the branch ancestral to the two long branches. With a wrong model of base change, a similar phenomenon can occur with likelihood (Bruno and Halpern, 1999). Yang (1997b) has shown that analyzing DNA sequence data without allowing for variation in the rate of evolution from site to site can have a higher probability of inferring the true tree, even when there is rate variation. The method Yang tested undercorrects for multiple substitution, compared to the true method. When long branches are adjacent on the tree, it will increase the chance that they are correctly placed. Yang's simulations show this pattern, but also that the wrong model makes inference less efficient when long branches are separates by short branches. This result is as expected. Pol and Siddall (2001) also find long branch attraction when using a Jukes-Cantor model to analyze data generated by an HKY model. As they place the long branches in adjacent pairs, this attraction results in improved trees. Since we do not know at the outset which situation we are in, Yang's and Pol and Siddall's results should not be taken as a general recommendation to use the wrong model. Bruno and Halpern (1999) have presented further simulations and arguments to this effect.

Chapter 17

Hadamard methods

The New Zealand school of Michael Hendy, David Penny, Michael Steel, and their co-authors (Hendy and Penny, 1989, 1993; Hendy, 1989, 1991; Hendy and Charleston, 1993; Hendy, Penny, and Steel, 1994; Steel et al., 1992; Steel, Hendy, and Penny, 1993) have brought a set of powerful mathematical techniques to bear on the reconstruction of phylogenies by methods similar to likelihood, for models of change that are symmetrical and that have modest numbers of species. This is one of the nicest applications of mathematics to phylogenies so far. A more recent review of the mathematics involved will be found in the paper by Steel, Hendy, and Penny (1998).

Their method makes use of the observed and expected frequencies of the different patterns of data. (Character patterns were introduced in Chapter 9 in the discussion of the consistency of parsimony methods.) If we have nucleotide sequence data for four species (say *a*, *b*, *c*, and *d*), for example, a site might turn out to have any of 256 different patterns, $AAAA, AAAC, AAAG, AAAT, AACA, \ldots, TTTT$. If instead we had a character with two states, 0 and 1, as we might have if DNA data had been recoded by classifying nucleotides into purines (R) and pyrimidines (Y), then there are 16 possible patterns, $0000, 0001, 0010, \ldots, 1111$. The data can be summarized by reporting the fraction of sites that had each of the patterns. If the model of evolution is symmetrical change between the two states, the pattern 1111 and the pattern 0000 convey the same information. The 16 patterns can be summed in complementary pairs to form 8 *partition frequencies*, $P_{xxxx}, P_{xxxy}, P_{xxyx}, P_{xxyy}, P_{xyxx}, P_{xyxy}, P_{xyyx}$, and P_{xyyy}. Note that these partitions are in the same orders as the binary numbers $0000, 0001, \ldots, 0111$, where 0 is replaced by x and 1 is replaced by y. The vector of these partition frequencies Hendy and co-workers call the *sequence spectrum*. In summarizing the data in this way, we know how many times each partition pattern occurs, but we lose sight of where the patterns are in the molecule. Hence, without further development, this analysis cannot deal with hidden Markov models of rate variation across sites, for any case in which rates at neighboring sites are correlated.

The methods make use of a transform of the partition frequencies, the Hadamard transform. It is a special case of the Fourier transform. The Hadamard transform for any size can be generated by taking the 2×2 matrix

$$\mathbf{H}_1 = \begin{bmatrix} 1 & 1 \\ 1 & -1 \end{bmatrix} \tag{17.1}$$

and forming Kronecker products of it. The Kronecker product will be further introduced in Chapter 23. For our purposes we need only note that taking any matrix $\mathbf{A}$ and taking its Kronecker product with $\mathbf{H}_1$ yields

$$\mathbf{H}_1 \otimes \mathbf{A} = \begin{bmatrix} \mathbf{A} & \mathbf{A} \\ \mathbf{A} & -\mathbf{A} \end{bmatrix} \tag{17.2}$$

In other words, make the product by producing a matrix with twice as many rows and twice as many columns as $\mathbf{A}$, with the matrix $\mathbf{A}$ in each of its four corners, but with the lower-right one having the signs of its elements reversed. If we take Kronecker products of $\mathbf{H}_1$ with itself we get the Hadamard transform matrices:

$$\mathbf{H}_2 = \mathbf{H}_1 \otimes \mathbf{H}_1 = \begin{bmatrix} \mathbf{H}_1 & \mathbf{H}_1 \\ \mathbf{H}_1 & -\mathbf{H}_1 \end{bmatrix} = \left[\begin{array}{cc|cc} 1 & 1 & 1 & 1 \\ 1 & -1 & 1 & -1 \\ \hline 1 & 1 & -1 & -1 \\ 1 & -1 & -1 & 1 \end{array}\right] \tag{17.3}$$

Proceeding in this fashion one can construct the 8×8 matrix $\mathbf{H}_3 = \mathbf{H}_1 \otimes \mathbf{H}_2$, which is

$$\mathbf{H}_3 = \begin{bmatrix} 1 & 1 & 1 & 1 & 1 & 1 & 1 & 1 \\ 1 & -1 & 1 & -1 & 1 & -1 & 1 & -1 \\ 1 & 1 & -1 & -1 & 1 & 1 & -1 & -1 \\ 1 & -1 & -1 & 1 & 1 & -1 & -1 & 1 \\ 1 & 1 & 1 & 1 & -1 & -1 & -1 & -1 \\ 1 & -1 & 1 & -1 & -1 & 1 & -1 & 1 \\ 1 & 1 & -1 & -1 & -1 & -1 & 1 & 1 \\ 1 & -1 & -1 & 1 & -1 & 1 & 1 & -1 \end{bmatrix} \tag{17.4}$$

When this matrix is used to make the Hadamard transform, in effect it computes differences between one set of patterns and another. The lone exception to this is the first row, which computes the sum of all the partition frequencies, a number that must always be 1. The second row computes the sum of all the odd-numbered patterns, minus the sum of all the even-numbered patterns. In the 0/1 case this is

$$(P_{xxxx} + P_{xxyx} + P_{xyxx} + P_{xyyx}) - (P_{xxxy} + P_{xxyy} + P_{xyxy} + P_{xyyy}) \tag{17.5}$$

A moment's consideration will show that it is simply the total of all pattern frequencies that have the same state in species d as in species a, less the total of all those that have different states in those two species. If we denote the probability that $a \neq d$ by d_{ad}, this is $(1 - d_{ad}) - d_{ad}$ or $1 - 2d_{ad}$.

The Hadamard transform method uses the transform to make a direct measurement of how much support there is in the data for various partitions of species. It is a remarkable fact that this can be done. I will not give a proof of this here but will explain some of the context.

Suppose that we consider the three possible tree topologies for four species, under this symmetric model. In each case the four exterior branches of the tree are numbered 1 through 4, and the interior branch is numbered 5. We can characterize the branch length by using the net probabilities of change p_1, p_2, p_3, p_4, and p_5 in each branch. Calculating the 16 expected probabilities of each pattern in each topology using these, forming the 8 expected partition frequencies from these, and taking the Hadamard transform of these 8 probabilities for each topology, we get the results shown in Table 17.1. Their order is the same as the order of rows in equation 17.4. The sets in the fourth column of the table are in binary order for the first three species names c, b, bc, etc., but with species d added to those that have an odd number of members. All of these quantities will be expected to be positive if none of the probabilities of change $p_1, \ldots, p_5$ are as large as 1/2.

We are assuming a symmetric model of change among the two states. Note that the entries in the Hadamard transform of the sequence spectrum are products of terms each of which is one of the $(1 - 2p_i)$. With the symmetric two-state model of character change we can recall from equation 9.1 that each of the net probabilities of change p_i is, as a function of the branch length,

$$p_i = \frac{1}{2}\left(1 - e^{-2v_i}\right) \tag{17.6}$$

which means that

$$1 - 2p_i = e^{-2v_i} \tag{17.7}$$

Thus a term like $(1 - 2p_1)(1 - 2p_5)(1 - 2p_2)$ can be written as an exponential of a sum of branch lengths:

$$(1 - 2p_1)(1 - 2p_5)(1 - 2p_2) = \exp\left[-2\left(v_1 + v_5 + v_2\right)\right] \tag{17.8}$$

It becomes natural to think of taking the logarithms of these quantities.

The result, for the three trees of our example, is shown in Table 17.2.

Each term is a multiple of a sum of branch lengths. Hendy (1989, 1991) has shown that if we divide each of these sets of species into pairs of species that have the paths between them on the tree disjoint, in that no two of these paths have a branch in common, their branch lengths add to these sums. For example, in the tree $((a, c), (b, d))$ the set a, b, c, d consists of pairs (a, c) and (b, d). Although there are two other ways of dividing this set of four species into two pairs, this one has

Table 17.1: The Hadamard transform of the expected sequence spectrum $P_{xxxx}, \ldots, P_{xyyy}$. This is shown for all three unrooted bifurcating trees with four species. The next-to-rightmost column shows the set S of species for which each term calculates the difference between the probabilities that there are an even and an odd number of symbols in set S. Additional interpretations of these quantities are given in the rightmost column.

Topology			Set	Interpretation
((a, b), (c, d))	((a, c), (b, d))	((a, d), (b, c))		
1	1	1	$\emptyset$	$P_{....}$
$(1-2p_3)(1-2p_4)$	$(1-2p_3)(1-2p_5)(1-2p_4)$	$(1-2p_3)(1-2p_5)(1-2p_4)$	$\{cd\}$	$1-2\,d_{cd}$
$(1-2p_2)(1-2p_5)(1-2p_4)$	$(1-2p_2)(1-2p_4)$	$(1-2p_2)(1-2p_5)(1-2p_4)$	$\{bd\}$	$1-2\,d_{bd}$
$(1-2p_2)(1-2p_5)(1-2p_3)$	$(1-2p_2)(1-2p_5)(1-2p_3)$	$(1-2p_2)(1-2p_3)$	$\{bc\}$	$1-2\,d_{bc}$
$(1-2p_1)(1-2p_5)(1-2p_4)$	$(1-2p_1)(1-2p_5)(1-2p_4)$	$(1-2p_1)(1-2p_4)$	$\{ad\}$	$1-2\,d_{ad}$
$(1-2p_1)(1-2p_5)(1-2p_3)$	$(1-2p_1)(1-2p_3)$	$(1-2p_1)(1-2p_5)(1-2p_4)$	$\{ac\}$	$1-2\,d_{ac}$
$(1-2p_1)(1-2p_2)$	$(1-2p_1)(1-2p_5)(1-2p_2)$	$(1-2p_1)(1-2p_5)(1-2p_2)$	$\{ab\}$	$1-2\,d_{ab}$
$(1-2p_1)(1-2p_2)$ $\times(1-2p_3)(1-2p_4)$	$(1-2p_1)(1-2p_2)$ $\times(1-2p_3)(1-2p_4)$	$(1-2p_1)(1-2p_2)$ $\times(1-2p_3)(1-2p_4)$	$\{abcd\}$	Prob[$abcd$ even]− Prob[$abcd$ odd]

Table 17.2: The logarithms of the Hadamard transform of the expected sequence spectrum $P_{xxxx}, \ldots, P_{xyyy}$. This is shown for all three unrooted bifurcating trees with four species. The rightmost column shows the set S of species for which the Hadamard transform calculated the difference between the probabilities that there are an even and an odd number of symbols in set S.

Topology			Set
((a, b), (c, d))	((a, c), (b, d))	((a, d), (b, c))	
0	0	0	$\emptyset$
$-2(r_3+r_4)$	$-2(r_3+r_5+r_4)$	$-2(r_3+r_5+r_4)$	$\{cd\}$
$-2(r_2+r_5+r_4)$	$-2(r_2+r_4)$	$-2(r_2+r_5+r_4)$	$\{bd\}$
$-2(r_2+r_5+r_3)$	$-2(r_2+r_5+r_3)$	$-2(r_2+r_3)$	$\{bc\}$
$-2(r_1+r_5+r_4)$	$-2(r_1+r_5+r_4)$	$-2(r_1+r_4)$	$\{ad\}$
$-2(r_1+r_5+r_3)$	$-2(r_1+r_3)$	$-2(r_1+r_5+r_4)$	$\{ac\}$
$-2(r_1+r_2)$	$-2(r_1+r_5+r_2)$	$-2(r_1+r_5+r_2)$	$\{ab\}$
$-2(r_1+r_2+r_3+r_4)$	$-2(r_1+r_2+r_3+r_4)$	$-2(r_1+r_2+r_3+r_4)$	$\{abcd\}$

the path between a and c consist of branches 1 and 3, the path between b and d consist of branches 2 and 4, and there is no branch shared between them. The set consisting of all of these branches defines the terms in Table 17.2: The entry for this tree for set a, b, c, d consists of the sum of the branch lengths r_i for the four branches in the set of nonoverlapping paths between pairs, in this case $r_1+r_2+r_3+r_4$. Note that it does *not* consist of the sum of the r_i for the set of branches connecting the set of species a, b, c, d; it instead uses the set of branches connecting those pairs of species whose connecting paths are mutually exclusive.

Thus for each entry in this vector, we could compute rather easily what its value will be, without going through making a Hadamard transform of the expected pattern frequencies. However, we will not have to do this, as there is one more surprise ahead.

The edge length spectrum and conjugate spectrum

Hendy and Penny (1989) write these sums of branch lengths as a vector $\boldsymbol{\rho}$, They also denote the Hadamard transform of the expected sequence spectrum as the vector $\mathbf{q}$, and the expected sequence spectrum as the vector s so that

$$\mathbf{q} = \mathbf{Hs} \tag{17.9}$$

Taking logarithms of all the entries in $\mathbf{q}$,

$$\boldsymbol{\rho} = \ln(\mathbf{q}) = \ln(\mathbf{Hs}) \tag{17.10}$$

This equation can of course be solved for the expected sequence spectrum by exponentiating and then using the inverse of the transformation matrix $\mathbf{H}$:

$$\mathrm{s} = \mathbf{H}^{-1}\mathbf{q} = \mathbf{H}^{-1}\exp(\boldsymbol{\rho}) \tag{17.11}$$

If we take the inverse Hadamard transform of the the vector $\boldsymbol{\rho}$, we get the surprising outcome that the result is a vector that has one positive entry for each branch in the tree, with the rest being zero except for the initial entry, which is minus the sum of the other entries. Hendy and his co-workers call this the *edge length spectrum*. (They have sometimes also called it the *tree spectrum*.) From equations 17.10 and 17.11 we find by taking the Hadamard transform that the formula for the edge length spectrum is

$$\mathbf{g} = \mathbf{H}^{-1}\ln(\mathbf{H}\mathbf{s}) \tag{17.12}$$

This is called a *Hadamard conjugation*. It involves taking a Hadamard transform, applying a function to the result, and then undoing the Hadamard transform. It is the *Hadamard log conjugation*. It may be inverted by doing the *Hadamard exponential conjugation*:

$$\mathrm{s} = \mathbf{H}^{-1}\exp(\mathbf{H}\mathbf{g}) \tag{17.13}$$

We can also make a Hadamard conjugation of the observed sequence spectrum, the vector of observed partition frequencies. We get the *conjugate spectrum*

$$\gamma = \mathbf{H}^{-1}\ln(\mathbf{H}\widehat{\mathrm{s}}) \tag{17.14}$$

We can obtain the edge length spectrum in our test case of a four-species tree. Taking the Hadamard transform of the vectors in Table 17.2, we get the edge length spectra, in Table 17.3. Note that where the tree topologies differ, the pattern of zero and nonzero entries in the table differ. If the observed partition frequencies are close to the expected partition frequencies, we can hope that the edge length spectra computed from them would show a similar pattern.

The zeros in this table correspond to algebraic expressions in the partition frequencies that remain 0 for all trees of a given topology. In Chapter 22 we will see that these are called the invariants of the tree, and that they are of great interest for future work on inferring phylogenies. The Hadamard machinery gives us a means to compute all of these invariants.

A numerical example of the Hadamard conjugation may be helpful. If we have the four-species case, with the true tree being ((a,b),(c,d)), and $r_1 = r_2 = r_3 = r_4 = 0.1$, and $r_5 = 0.05$, then if we simulate a 200-character set of data along this tree, we find in one simulation that the sequence spectrum (the vector of partitions) is

$$(158/200, 7/200, 13/200, 4/200, 9/200, 2/200, 0, 7/200)$$

and its Hadamard transform is

$$(1, 0.8, 0.76, 0.78, 0.82, 0.8, 0.72, 0.64)$$

Taking logarithms of that vector, we get

$$(0, -0.223144, -0.274437, -0.248461, -0.198451, -0.223144, -0.328504, -0.446287).$$

Table 17.3: The edge length spectra (tree spectra) for the three trees, computed from the expected partition frequencies for those trees. The entry is the length of the branch that separates the set (shown in the right column) from the other species.

((a, b), (c, d))	((a, c), (b, d))	((a, d), (b, c))	Set
$-(r_1 + r_2 + r_3 + r_4 + r_5)$	$-(r_1 + r_2 + r_3 + r_4 + r_5)$	$-(r_1 + r_2 + r_3 + r_4 + r_5)$	$\emptyset$
r_4	r_4	r_4	$\{d\}$
r_3	r_3	r_3	$\{c\}$
r_5	0	0	$\{cd\}$
r_2	r_2	r_2	$\{b\}$
0	r_5	0	$\{bd\}$
0	0	r_5	$\{bc\}$
r_1	r_1	r_1	$\{bcd\}$

Applying the inverse Hadamard transform one gets
(–0.48561, 0.084911, 0.16324, 0.039007, 0.11259, 0.013673, –0.013360, 0.085552).
Examining Table 17.3 we can see that the elements whose expectations are zero under different tree topologies are the 4th, 6th, and 7th elements, which are respectively 0.039, 0.014, and –0.013. The correct values are 0.05, 0, and 0. Thus simple inspection may reveal which partitions correspond to the true tree. Figure 17.1 shows an example, using a six-species tree on which 1,000 sites were simulated with a K2P model with transition/transversion ratio of 2. The sites were then scored using the states purine or pyridine. The figure shows the histogram of partition frequencies, its Hadamard transform, and the resulting conjugate spectrum. It can be seen that the Hadamard conjugation cleans up the signal, bringing most classes close to zero. The numbers along the bottom axis represent the partitions when they are seen in binary notation. Thus class 19 is 010011, and when the species A–F are indicated by 0s and 1s, it is partition $ACD|BEF$. The dark arrows point to the signals for the interior branches, the light arrows the exterior branches. Note that in this example, the signal for this particular partition, which has the shortest branch in the tree, is nearly lost.

The closest tree criterion

The ideal use of the Hadamard transform would then be to create an edge length spectrum from which the branches could simply be read off. This is not necessarily easy. We have shown an example in which there is only one possible internal branch, and the other two entries in the vector have expectation 0. For larger numbers of species the pattern may be harder to discern. For 10 species, for example,

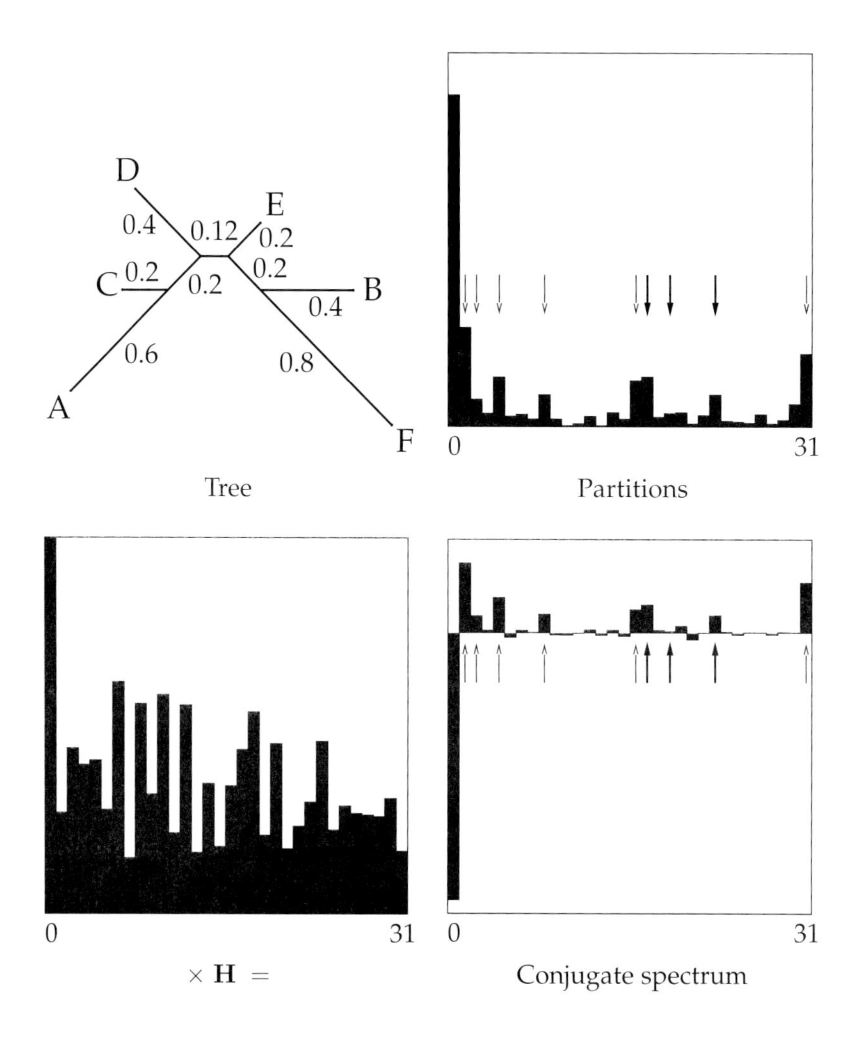

Figure 17.1: An example of the use the Hadamard conjugation. 1,000 sites simulated on the tree at above left and coded for purine versus pyrimidine yield the partition frequencies in the upper-right histogram, their Hadamard transform (at the lower left), and the conjugate spectrum at the lower right. Classes that correspond to external branches in the true tree are shown with lighter arrows, internal branches with dark arrows.

Table 17.4: Conjugate spectrum for the numerical example (leftmost column) together with the edge length spectra for the three bifurcating tree topologies in this case. The least squares best fit error $\Gamma^2(T,\gamma)$ is shown for each.

γ	$((a,b),(c,d))$	$((a,c),(b,d))$	$((a,d),(b,c))$	Set
-0.4856071	-0.4856071	-0.4856071	-0.4777432	$\emptyset$
0.0849109	0.08497342	0.09004024	0.09277486	$\{d\}$
0.163238	0.16330052	0.16836734	0.17110196	$\{c\}$
0.0390072	0.03906972	0	0	$\{cd\}$
0.112586	0.11264852	0.11771534	0.12044996	$\{b\}$
0.0136731	0	0.01880244	0	$\{bd\}$
-0.0133605	0	0	0.00000000	$\{bc\}$
0.0855524	0.08561492	0.09068174	0.09341636	$\{bcd\}$
	0.000365457	0.00183162	0.00312386	

there are $2^9 = 512$ entries in the sequence spectrum, of which 10 correspond to terminal branches of the tree, one corresponds to the empty set, and the other 501 to different possible internal branches. We need to select a set of branches that are compatible with each other (in the sense of Chapter 8, that the partitions they define are compatible). We cannot depend on finding them by taking the largest entries in the vector, as some of those might be incompatible. There can be short branches in the true tree that would show small entries in the edge length spectrum, and those might be ignored while entries that were due to sampling error were not. One wants to look at all sets of branches that correspond to trees.

Hendy (1989) has proposed inferring phylogenies by finding the closest match between the observed and expected edge length spectra. This is not the same as maximum likelihood, but neither is it the same as distance matrix criteria, as it uses some higher-order information. If there are n species, there will be $n(n-1)/2$ distances, but there will be 2^{n-1} entries in the tree spectrum. Thus it is a much more sensitive criterion than any that uses distances. For example, with 10 species there will be 45 distances but 512 nonzero entries in the edge length spectrum.

The criterion Hendy used is least squares, using only those entries of the edge length spectrum that are expected to be nonzero. If the entries in the observed tree spectrum are denoted by γ_i, and the entries of the expected tree spectrum are denoted by g_i, the criterion is

$$\Gamma^2(T,\gamma) = \sum_i (\gamma_i - g_i)^2 \tag{17.15}$$

A complication in using this criterion is that for each tree topology we must find a set of branch lengths that minimize $\Gamma^2(T,\gamma)$. This can be done by least squares fitting, and rather easily. Hendy (1989, 1991; Hendy and Penny, 1993) has shown that the least squares branch lengths can be found by taking the sum γ_T of all of the γ_i entries (those that would have nonzero g_i for that tree topology, including γ_0), dividing by $B+1$, where B is the number of branches in the tree, and then subtracting this from each of those γ_i. This, if positive, gives an estimate of the branch length v_j that one gets by minimizing Γ^2 in equation 17.15. (If negative, the branch length is 0.) Note that for each possible tree topology we need only start from the conjugate spectrum — it need only be computed once. So the process of fitting branch lengths can be quite fast, permitting many topologies to be checked.

For our numerical example, Table 17.4 shows the estimates of branch lengths obtained for the three tree topologies, together with their values of $\Gamma^2(T, \boldsymbol{\gamma})$. The tree $((a, b), (c, d))$ is the closest tree. Note that tree $((a, d), (b, c))$ has its internal branch set to length zero, rather than allow it to be negative.

Although the closest tree criterion is roughly similar to a likelihood criterion, departures from expectation for different entries in the conjugate spectrum can vary, and can covary. The closest tree criterion does not weight their departures from expectation differently and does not correct for covariation. Waddell et al. (1994) have shown how to calculate the covariance matrix of the entries in the conjugate spectrum. This is a fair amount of work — $(n-1) \times 2^n$ operations — but can lead to a more realistic assessment of the goodness of fit of a tree.

DNA models

I have discussed Hadamard methods as being done on two-state data, but actually some four-state DNA models are available. Steel et al. (1992) have extended the Hadamard methods to four-state DNA sequences, assuming a Kimura 3ST model (Kimura, 1981). This is an extension of the Kimura two-parameter model that adds an additional rate. The pattern is shown in Table 17.5. Instead of bipartitions of the set of species, Steel et al. (1992) count *quadripartitions*, which in this case are pairs of bipartitions. The four bases (A, G, C, T) are assigned the binary digits (00, 01, 10, 11) and the partitions correspond to the first and second digits of these pairs. (Thus the first distinguishes between purines and pyrimidines.) For a data set with n species, there will be 2^{2n-2} different quadripartitions possible. Many

Table 17.5: The Kimura 3ST model of DNA base change. The instantaneous rates of change per unit branch length are shown.

	To :			
From :	A	G	C	T
A	—	α	β	γ
G	α	—	γ	β
C	β	γ	—	α
T	γ	β	α	—

of the same results go through, and one can compute closest trees. However, the results cannot be generalized to models less symmetrical than the Kimura 3ST model, which is a rather special symmetric model of DNA change.

Computational effort

The amount of computation needed to infer trees by the closest tree criterion differs substantially from that needed by maximum likelihood. With n species, there will be 2^{n-1} bipartitions possible. The Hadamard matrix is then a $2^{n-1} \times 2^{n-1}$ matrix. It might seem that doing the Hadamard transform would thus require 2^{2n-2} operations. But the Hadamard transform for 2^{n-1} points is a special case of the Fourier transform; as in the Fourier case, there is a fast Hadamard transform algorithm that speeds things up a great deal. To multiply by the matrix in equation 17.4 one need only do the following:

- Take adjacent pairs of entries in the vector of bipartitions.
- Replace the first by their sum, the second by their difference.
- Then take all pairs of entries 2 apart and do the same. As you do so, skip along the vector as needed so as not to involve any element in more than one of these pairs.
- Then take all pairs of entries 4 apart and do the same, and continue. After each pass double the size of the interval between entries until it becomes 2^{n-2} for the last pass.

This accomplishes the matrix multiplication of the vector by the appropriate Hadamard matrix. The inverse of this transform is a simple multiple $(1/2^{n-1})$ of the transform, so it too can be done by essentially the same algorithm.

To carry out the Hadamard conjugation should require $n \times 2^n$ operations. Thus for 10 species, only $10 \times 1024 = 10,240$ operations are needed. For 20 species, one needs $20 \times 1,048,576 = 20,971,520$ operations. Once the Hadamard conjugation is done, different tree topologies can be evaluated with only about 2^{n-1} operations each, as one does not have to do the Hadamard conjugation each time.

For four-state data using the Kimura 3ST model, the computational requirements are much heavier. The number of operations for the Hadamard conjugation becomes $n \times 2^{2n-2}$. With 10 species this is 2,621,440 operations; with 20 species it is about 5.5×10^{12}. Thus at present one starts running into computational limits between 10 and 20 species.

We may contrast this with ordinary maximum likelihood. A limitation of Hadamard methods is that they must compute terms for all bipartitions or quadripartitions, whether or not they occur in the sequences. Ordinary ML requires that we compute terms for only those patterns that show up in the data. This can be substantially smaller than the number of sites, if the number of species is small or if the species are closely related. Thus a 20-species DNA data set with 1,000 sites and differences of about 25% in sequence between species may turn out (as one did in a simulation I ran) to have 526 distinct patterns in the DNA. Unlike the closest tree algorithm, maximum likelihood does require iteration of branch lengths for each tree topology. This must be balanced against the large number of operations required for one Hadamard conjugation. It seems clear that Hadamard methods with DNA sequences will run into a limit near 20 species, while maximum likelihood can deal with larger cases than this, though not of course with nearly as many species as distance matrix methods can. If Hadamard algorithms could be found that do not spend time on patterns that do not occur in the sequences, this would make these methods much more practical on data sets of moderate size. If any way could be found to extend them to nonsymmetric models of base change, that too would be a great advance.

Extensions of Hadamard methods

Although Hadamard methods inherently cannot cope with autocorrelations of rates or of sequence patterns among sites, they can be adapted to correct for unequal rates of evolution at different sites, as long as the rates are assigned independently to the sites. Steel et al. (1993) have shown that if the distribution of rates r among sites is $f(r)$, then we must replace the logarithm in equations 17.10 and 17.12 by the inverse of the moment-generating function of $f(r)$. This is

$$M^{-1}(r) = \int_0^\infty \ln(\lambda r)\, f(\lambda)\, d\lambda \qquad (17.16)$$

If this function can be evaluated (which it can for distributions such as gamma distributions) then its use instead of the logarithm allows us to have a Hadamard conjugation that works for the model with varying rates among sites. Waddell, Penny, and Moore (1997) have discussed this method further, giving inverse moment-generating functions for a number of relevant distributions and discussing which distributions of rates among sites might give similar fit to data.

Hadamard methods can also be used to carry out a distance matrix method (Hendy and Penny, 1993), and to correct parsimony and compatibility methods so that they are not inconsistent (Steel, Hendy, and Penny, 1993). The latter authors show that if in the conjugate spectrum we take the "parsimony partitions," namely those that correspond to internal branches of trees, the entries for these have the branch lengths as their expectations. We can interpret their sum as an estimate of the expected numbers of changes for a tree, and we expect that the smallest value will correspond to the correct tree. Of course this expectation is met only asymptotically, but it does establish that this method, in effect a parsimony method corrected for multiple changes, is consistent. Thus Steel et al. argue that the source of the inconsistency of ordinary parsimony is not the parsimony criterion per se, but its lack of a correction for multiple changes.

In the numerical example that we used above, the three different internal branches had corrected lengths 0.0390072, 0.0136731, and –0.0133605. Steel,

Hendy, and Penny (1993) suggest using the absolute values of these. For the tree ((a,b),(c,d)) the first partition is compatible and requires one change. The other two partitions are not compatible and require two changes. All other partitions require one change. Thus if we used these corrected lengths, we could sum over all the nonempty partitions and obtain the total corrected parsimony score for the tree ((a,b),(c,d)) (divided by the number of sites) of

$$\begin{aligned} S_1 &= 0.0849109 \times 1 + 0.163238 \times 1 + 0.0390072 \times 1 + 0.112586 \times 1 \\ &\quad + 0.0136731 \times 2 + (-0.0133605) \times 2 + 0.0855524 \times 1 \\ &= 0.4859197 \end{aligned} \tag{17.17}$$

The other two trees, evaluated in an analogous way, give total parsimony scores of 0.5112538 for ((a,c),(b,d)) and 0.5382874 for ((a,d),(b,c)). The correct tree does have the smallest corrected parsimony score in this example. This is not one of the examples where parsimony is misleading — uncorrected parsimony also leads to the same conclusion.

Chapter 18

Bayesian inference of phylogenies

Bayesian methods are closely related to likelihood methods, differing only in the use of a prior distribution of the quantity being inferred, which would typically be the tree. Use of a prior enables us to interpret the result as the distribution of the quantity given the data. Bayesian methods date to 1790, and controversy among statisticians over their appropriateness is almost that old. Recently, the use of Markov chain Monte Carlo methods has given a new impetus to Bayesian inference. In this chapter I hope to introduce Bayes' theorem, as well as the Markov chain Monte Carlo (MCMC) methods used to implement it. We will see MCMC methods again in Chapter 27.

After a description of the literature on Bayesian inference of phylogenies, I will spend much of the remainder of the chapter considering whether Bayesian inference is appropriate. It may be enough to say that the arguments were old long before anyone thought of using Bayesian approaches to inferring phylogenies. Nothing that biologists say is going to settle the matter. A recent review of work on Bayesian methods for phylogenies is given by Huelsenbeck et al. (2001).

Bayes' theorem

We have already seen in Chapter 16 that Bayes' theorem can be put into an odds-ratio form (equation 16.1). It is appropriate to give it in its usual form here. Given an hypothesis H (such as a possible tree) and some data D, the probability of the hypothesis given the data is

$$\mathrm{Prob}\,(H \mid D) \;=\; \frac{\mathrm{Prob}\,(H \;\&\; D)}{\mathrm{Prob}\,(D)} \tag{18.1}$$

The joint probability of H and D, $\text{Prob}(H \,\&\, D)$, can itself be written as a product of the probability of H and the conditional probability of D given H:

$$\text{Prob}(H \,\&\, D) = \text{Prob}(H)\ \text{Prob}(D \mid H) \tag{18.2}$$

Substituting this expression into equation 18.1 we get

$$\text{Prob}(H \mid D) = \frac{\text{Prob}(H)\ \text{Prob}(D \mid H)}{\text{Prob}(D)} \tag{18.3}$$

This is Bayes' theorem in its simplest form. The denominator $\text{Prob}(D)$ is the sum of the numerators $\text{Prob}(H \,\&\, D)$ over all possible hypotheses H and is the quantity that is needed to normalize them so that they add up to 1. This leads to the more usual form of the theorem:

$$\text{Prob}(H \mid D) = \frac{\text{Prob}(H)\ \text{Prob}(D \mid H)}{\sum_H \text{Prob}(H)\ \text{Prob}(D \mid H)} \tag{18.4}$$

The theorem is a simple exercise in conditional probabilities. As such it is uncontroversial. Figure 18.1 shows the Bayesian inference of the heads probability with 11 (left column) and with 44 (right column) tosses of a coin, both resulting in a 5:6 ratio of heads to tails. The prior that is assumed, a truncated exponential distribution with mean 0.34348, is at the top of the figure. The posteriors, at the bottom, show that the influence of the prior is substantial with 11 tosses but has less influence with 44 tosses. In the left-hand case, the mode of the posterior is displaced from 0.4545 to below 0.40, but in the right case the displacement is much less, as the likelihood curve is providing more of the information and the prior less.

The odds-ratio form of Bayes' theorem (which we saw in equation 16.1) is easiest to use. The odds favoring one hypothesis over another are the odds the person gave them initially (the prior odds), multiplied by the ratio of the likelihoods under the data. Suppose that there are two possible hypotheses, and in advance we favor H_1 over H_2, giving them odds of 3:2. Now some data are examined. The likelihood ratio $\text{Prob}(D \mid H_1) \,/\, \text{Prob}(D \mid H_2)$ turns out to be 1/2, so that the data are half as probable given hypothesis 1 as it is given hypothesis 2. Bayes' theorem tells us to compute the posterior odds ratio by multiplying these two to get $(3/2) \times (1/2) = 3/4$. After looking at the data we now give odds in favor of H_1 of only 3:4.

This calculation too is uncontroversial; there is no statistician who would not use it, given the correctness of the prior odds. What is controversial is whether usable prior odds exist.

When not in odds-ratio form, Bayes' theorem allows us to turn a prior distribution into a posterior distribution. It is enormously attractive because it computes what we most need, the probabilities of different hypotheses in the light of the data.

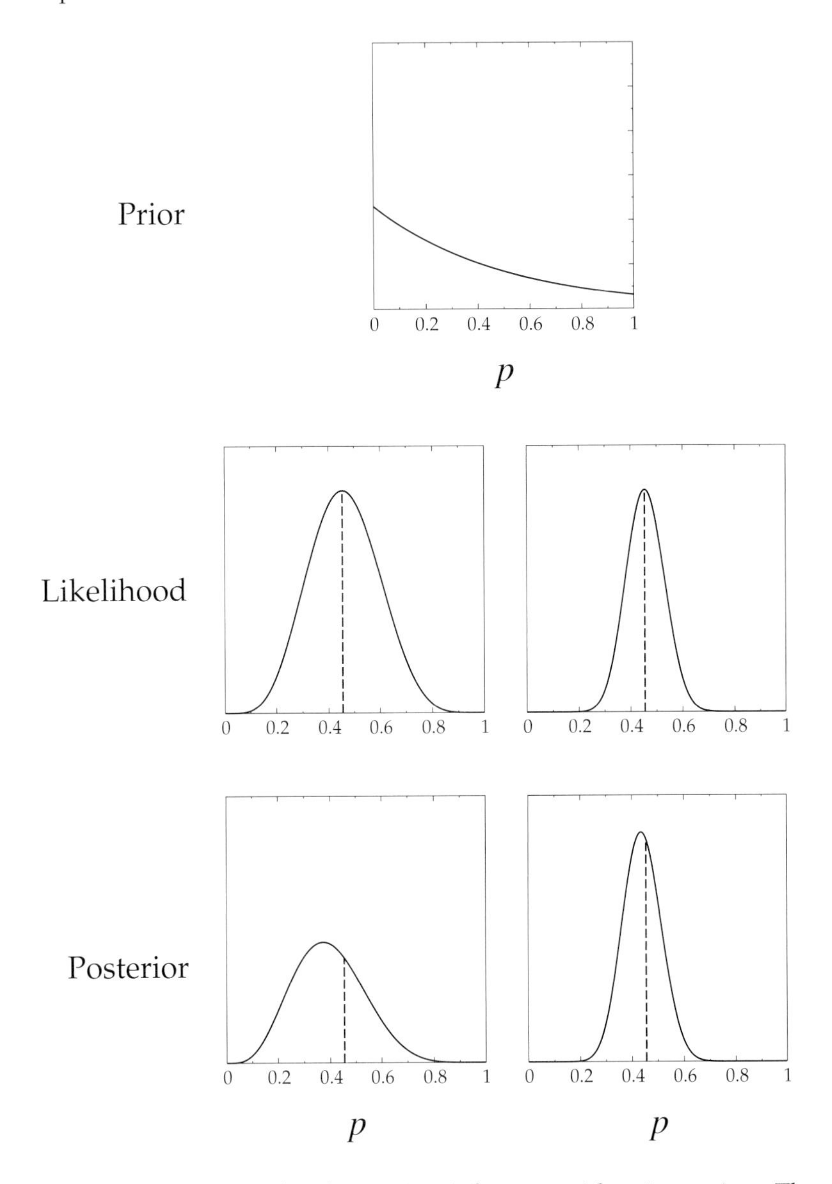

Figure 18.1: An example of Bayesian inference with coin tossing. The probability of heads is assumed to have a prior (top) that is a truncated exponential with mean 0.34348 on the interval (0,1). The left column shows the likelihood curve and the posterior on the probability of heads when there are 11 tosses with 5 heads. The right column shows the case of 44 tosses with 20 heads. The vertical dashed lines are located at the observed fraction of heads.

Bayesian methods for phylogenies

Early in the history of numerical work on phylogenies, Bayesian arguments were in use, but usually without a fully Bayesian inference of phylogenies. Edwards (1970) discussed the possibility of using random models of branching and extinction to place priors on trees but concluded that this was not computationally practical. The influential unpublished manuscript by Gomberg (1968) adopted a Bayesian approach to inferring phylogenies from characters that changed according to a Brownian motion process. Farris (1973a, 1977a) based his arguments in favor of parsimony on a Bayesian foundation. Harper (1979) used Bayesian arguments to compute probabilities that a group of taxa that shared a derived state were truly a clade. Ward Wheeler (1991) assumed that each step in a parsimony method reduced the likelihood by a factor of e; he argued from this that the proper assessment of phylogeny from different data sets is the tree chosen by overall parsimony. Smouse and Li (1987) used a more fully Bayesian approach. For a case with a rooted tree with three species, they placed equal prior probabilities on each of the three tree topologies, then computed posterior probabilities using the likelihood function for each of these. However, they did not place priors on the times of the interior nodes of the tree. Instead, they maximized the likelihoods over these and took the results as the likelihoods for the tree topologies. This was almost, but not quite, a fully Bayesian analysis of phylogenies. Another early effort by Sinsheimer, Lake, and Little (1996) used the patterns of nucleotides across species, evaluating whether the phylogenetic invariants had their expected values. These invariants will be discussed in Chapter 22. Lake's invariants are computed from sums of pattern frequencies. Sinsheimer, Lake, and Little placed equal prior probabilities on the three possible unrooted tree topologies with four species, but they did not place priors on branch lengths or node times. They instead placed priors on the sums of pattern frequencies.

Zander (2001) has argued that one can compute posterior probabilities of particular alternative rearrangements of the tree topology around an interior branch. He uses the number of steps reconstructed by parsimony on a branch, and on the corresponding interior branches in the two trees obtained by doing a nearest-neighbor interchange at that branch. I will argue in Chapter 19 that the statistical test of significance he uses for the branch has some merit, but the posterior probabilities he derives are more doubtful. They do not seem to be obtained using any particular prior distribution, so that it is unclear how they could be correct posteriors.

Computational difficulties prevented most of these authors from doing a more fully Bayesian inference of phylogeny. Rannala and Yang (1996) attempted a fully Bayesian inference of phylogenies. They used a birth-and-death process prior on the trees in an analysis of DNA sequences using a molecular clock. For a fixed interval of time since the start of the process, they inferred the birth, death, and substitution rates, as well as the transition-transversion ratio for the HKY model of base substitution, by finding values that maximized the posterior probabilities summed over all trees. Fixing these parameters at their estimated values, they used the probability contributed to the posterior by each tree topology as its posterior probability. They showed calculations indicating that these posterior probabilities did not depend much on the nuisance parameters.

Rannala and Yang's approach used numerical integration of the posterior probabilities over all interior node times for each given tree topology. Owing to the large number of topologies and the need to integrate over many dimensions, this approach was feasible only for a modest number of species.

Markov chain Monte Carlo methods

In Bayesian inference, the expression for the posterior distribution has a denominator that can be very difficult to compute, as it involves summing over all possible hypotheses. Fortunately, samples from the posterior distribution can be drawn using a Markov chain that does not need to know the denominator. In the late 1990s the increasing feasibility of Bayesian inference by these Markov chain Monte Carlo methods led to their more widespread use, and they came to be used in phylogenetic inference. We will see this method again in Chapter 27, where it will be introduced again for likelihood inference rather than for Bayesian inference. I will describe such a method here and show that it achieves the desired distribution.

Markov chain Monte Carlo methods in Bayesian inference draw a random sample from the posterior distribution of hypotheses (in this case, trees). It thus becomes possible to make probability statements about the true tree. If 96% of the samples from the posterior distribution of trees have {Human, Chimp} as a monophyletic group, then we can say that the probability that these are a monophyletic group is 96%. Of course, there is some uncertainty if the number of samples is not large; this is reflected in use of the name of the famous gambling casino at Monte Carlo. In estimating phylogenies we have the same hope as the house does in the casino: that enough samples will lead to the expected results.

The Metropolis algorithm

One of the most widely used Markov chain Monte Carlo (MCMC) methods is the Metropolis algorithm (Metropolis et al., 1953). It is usually used in a modified form called the Metropolis-Hastings method, the modification being due to Hastings (1970). The idea of MCMC methods is to wander randomly in a space of trees in such a way as to settle down into an equilibrium distribution of trees that has the desired distribution (in this case, the Bayesian posterior). Imagine all possible trees, connected to each other so as to form a connected graph. Suppose we know the distribution of trees $f(T)$ that we want. We will later see that we don't need to know the denominator of this distribution.

The Metropolis algorithm involves the following steps:

1. Start at some tree. Call this T_i.

2. Pick a tree that is a neighbor of this tree in the graph of trees. Call this the proposal T_j.
3. Compute the ratio of the probabilities (or probability density functions) of the proposed new tree and the old tree:

$$R = \frac{f(T_j)}{f(T_i)}$$

4. If $R \geq 1$, accept the new tree as the current tree.
5. If $R < 1$, draw a uniform random number (a random fraction between 0 and 1). If it is less than R, accept the new tree as the current tree.
6. Otherwise, reject the new tree and continue with tree T_i as the current tree.
7. Return to step 2.

This algorithm never terminates. It is a Markov chain because it is a random process in which the next change depends only on the current state and not on where the process was previously. Assume that the probability of proposing T_j when we are currently at T_i is the same as the probability of proposing T_i when we are at T_j. (If this is not true, we need to make Hastings' correction, which we omit for now—it will be described in Chapter 27 when we use the Metropolis algorithm in another context.)

Its equilibrium distribution

The Metropolis algorithm is a Markov chain that has an equilibrium distribution π_{T_i}. To verify that it is the desired one, we need to compute the ratio of the probabilities of forward and backward change $\mathrm{Prob}\,(T_j \mid T_i)$ and $\mathrm{Prob}\,(T_i \mid T_j)$. Suppose that $f(T_i) > f(T_j)$. Using the rules above, one can see that when the tree is T_j and T_i is proposed, it will be accepted with probability 1. When the tree is T_i and T_j is proposed, the ratio R will be $f(T_j)/f(T_i)$. The probability that the random number that is drawn will be less than R will be R. So the probability that tree T_j will be accepted will be R. A similar proof draws the analogous conclusion when $f(T_i) \leq f(T_j)$.

The upshot is that

$$\frac{\mathrm{Prob}\,(T_j \mid T_i)}{\mathrm{Prob}\,(T_i \mid T_j)} = \frac{f(T_j)}{f(T_i)} \tag{18.5}$$

so that

$$f(T_i)\,\mathrm{Prob}\,(T_j \mid T_i) = f(T_j)\,\mathrm{Prob}\,(T_i \mid T_j) \tag{18.6}$$

Figure 18.2 illustrates that this equation implies "detailed balance" in which the numbers of times there is a transition from state i to state j equals the number of times there is a transition from state j to state i. If we sum both sides of this equation over all T_i, on the right we are summing the probabilities of transition to all possible T_i. The transition probabilities on the right sum to 1, and we get

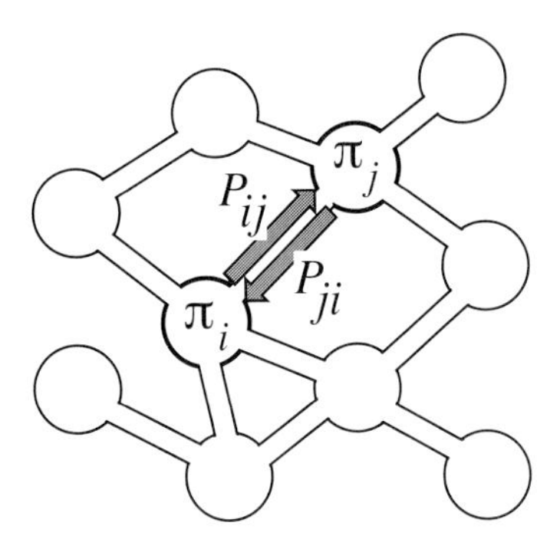

Figure 18.2: If we have a distribution π_i of states, and if the transition probabilities $P_{ij} = \text{Prob}(T_j \mid T_i)$ from tree T_i to tree T_j and the probabilities of transition in the reverse direction P_{ji} satisfy equation 18.6, then the amount of probability moving along the two shaded arrows in the next step will be equal. If this is true for all paths between adjacent states, then the equilibrium distribution $\pi_i = f(T_i)$ will be maintained unchanged.

the equation that establishes that the $f(T_i)$, if scaled so they add up to 1, are the equilibrium probabilities of the process:

$$\sum_{T_i} f(T_i) \, \text{Prob}(T_j \mid T_i) = f(T_j) \tag{18.7}$$

Starting in the distribution $f(T)$, we see from this equation that if we sum up all the ways that one could arrive at tree T_j, counting for each the probability contributed by that way, the resulting probabilities remain the $f(T)$, which is thus the equilibrium distribution.

Thus for any distribution $f(T)$ the Metropolis algorithm is a stochastic process (in fact, a reversible one) whose equilibrium is that function. This would not be much help for the cases we want to treat, because we don't know the denominator of the distribution. Fortunately, in computing the acceptance ratio R we don't need to know the functions $f(T_i)$ and $f(T_j)$. All that matters is their ratio — if they have the same denominators, these cancel out. Thus, knowing only the numerators, we can carry out the algorithm.

Bayesian MCMC

Adapting the Metropolis algorithm to Bayesian inference involves sampling from the posterior distribution (equation 18.4). If we can compute the prior probabilities of trees $\text{Prob}(T)$ and the likelihoods $\text{Prob}(D \mid T)$, then we can carry out the

Metropolis algorithm. Although we don't know the denominator in equation 18.4, it cancels out in the expression for the *acceptance ratio*:

$$R = \frac{\text{Prob}\,(T_j)}{\text{Prob}\,(T_i)} \frac{\text{Prob}\,(D \mid T_j)}{\text{Prob}\,(D \mid T_i)} \tag{18.8}$$

The acceptance ratio is the ratio of the prior probabilities of the proposed tree and the current tree, multiplied by the likelihood ratio of these trees. We already know how to compute the likelihoods of trees. We need only choose a prior distribution for which the probabilities can be computed.

Once these are available, we can start with an estimate of the tree and propose changes in it, using the Metropolis algorithm to accept or reject those changes. We start with a "burn-in" period to allow the process to settle into its equilibrium distribution. After that, we record the tree every S steps, where S is chosen so that we have enough trees but not too many. For example, we might use 5,000 steps as a burn-in period, and then record every 100th tree for 100,000 steps. This sample of 1,000 trees can be assumed to be from the posterior distribution. In deciding when to sample, it is important to count both steps in which we have accepted the proposed tree, and steps in which we have rejected it and retained the old tree. This might even mean repeatedly sampling the same tree.

Bayesian MCMC for phylogenies

Three groups of authors introduced Bayesian MCMC methods into phylogenetic inference, essentially independently. They were Yang and Rannala (1997), Mau and Newton (1997) (extended by Mau, Newton, and Larget, 1999, and by Larget and Simon, 1999), and Li, Pearl, and Doss (2000). All of these authors used Metropolis algorithms to implement Markov chain Monte Carlo methods. We can focus on several issues that differ among their methods:

1. The prior distribution each assumed
2. The proposal distribution each used for changes in the phylogeny
3. The kind of summary of the posterior distribution each used

Priors

The methods differ as to whether the tree was assumed to arise from a stochastic process of random speciation, extinction and sampling, or whether it was simply to have been drawn from a uniform distribution of all possible rooted trees. In all these cases the assumption was that the trees are clocklike.

Yang and Rannala (1997) used the birth-and-death process from their 1996 Bayesian approach to place a prior on the phylogeny. They assumed that in an interval of time of known length t_1, a birth and death process operates and produces S species. The observed number of s species is then sampled at random from these. They do not make it clear what happens if $S < s$, so that there are too few species to sample from. They present a distribution of times of speciation conditional on having sampled s species. They place a uniform prior on all the clocklike tree topologies (without node times specified).

Mau and Newton assumed that all possible labeled histories are equiprobable. They seem also to have assumed that for each labeled history, the placement of its interior nodes came from a uniform distribution. Li, Pearl, and Doss assumed instead that all possible rooted tree topologies were equiprobable, which is a somewhat different distribution, with less of its probability concentrated on symmetric topologies. Such distributions of trees are technically inadmissible, since if there is no limit on the antiquity of the earliest node, there cannot be a uniform distribution of node times. Mau and Newton do not make it clear how they limit the prior to prevent inadmissibility. The subsequent papers of Mau, Newton, and Larget and of Larget and Simon do not clarify the matter. Larget and Simon also mention a method that does not assume a clocklike tree, but they do not say what prior it assumes.

Li, Pearl, and Doss allow the interior nodes of their trees to be uniformly distributed, subject to the constraints on them by the topology and the times of the contemporary nodes and the root. They do fix the time of the root at a time estimated in a rough starting tree, and never change it. To scale all the branch lengths, they estimate the rate of substitution, which is equivalent to changing the time of the root node. Their procedure for estimating it is not derived from an MCMC method.

It is not clear whether the resulting uncertainties about the exact prior used in each case are serious problems, or whether they are simply unimportant consequences of the difficulty of describing exactly what was done when the procedure is complex. As these are founding papers of Bayesian inference of phylogenies, the lack of a clear explanation of what prior was used is an agonizing omission.

Proposal distributions

The three methods also differ in the proposal distributions that they use for wandering by MCMC through the space of trees. In principle the methods could use any proposal distribution, as long as it could reach all trees from any starting tree. It would only require enough running to assure that the proper distribution was attained. In practice, the matter is more difficult. It is hard to know how much is enough. A proposal distribution that jumps too far too often will result in most proposed new trees being rejected. A proposal distribution that moves too timidly may fail to get far enough to adequately explore tree space in the allotted time. At the moment the choice of a good proposal distribution involves the burning of incense, casting of chicken bones, use of magical incantations, and invoking the opinions of more prestigious colleagues.

Mau and Newton use a novel method of rearranging the tree. They note that at each node in a clocklike tree the order of the subtrees can be exchanged without changing the tree. They start by making an arbitrary choice at each interior node

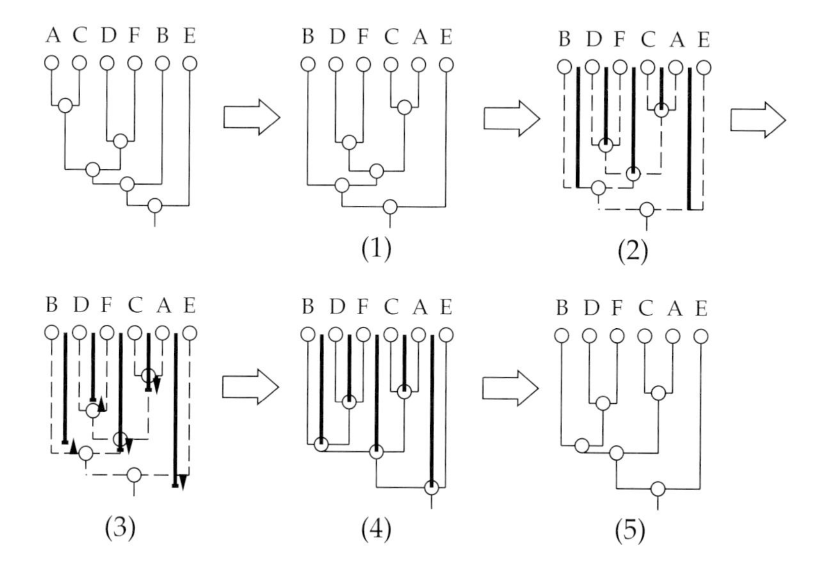

Figure 18.3: Steps in Mau and Newton's method of making proposals of new trees in their Bayesian MCMC method. The species (1) are reordered by rotating branches, (2) coalescence times between adjacent species are calculated, (3) these are modified, and (4, 5) a new tree is constructed with those coalescence times.

of which subtree will be on the left. They then consider the depth of the times of "coalescence" of each pair of adjacent tips. They alter these coalescence times by choosing a change for each from a distribution uniform about zero, except for very small coalescence times. In those cases they ensure that the coalescence time does not become negative by reflecting its value about zero. The process is illustrated in Figure 18.3. When two connected coalescence times pass one another as a result of the change, this causes a change in the tree topology. (When two nonconnected times pass one another, the labeled history changes.) Their scheme for perturbing the tree leads not only to changes of node times, but of tree topologies, including some changes that are larger than nearest-neighbor interchanges.

Mau and Newton's proposal method implicitly assumes that the prior on trees is uniform for all clocklike labeled histories. This is actually an impossible distribution, but we need to know whether that creates any difficulty. Mau, Newton, and Larget (1999) prove that this proposal mechanism is irreducible (i.e., it can reach any tree in the tree space from any other with enough steps). Larget and Simon (1999) extend the scheme to nonclocklike trees by altering both coalescence times and heights of tips. They do not make it clear how this works, and I have some doubts as to what prior on unrooted trees it assumes.

Both Larget and Simon (1999) and Li, Pearl, and Doss (2000) have a different proposal distribution that makes local modifications of the tree that can be nearest-neighbor interchanges. Larget and Simon also have a nonclocklike version of this.

In Mau and Newton (1997) and the papers following up on it, the specific calculations in their acceptance-rejection step are not explained. I assume that they are using the likelihood ratio of the trees after and before the proposed rearrangement, without any term for the prior. This implicitly assumes that the prior is flat. Li, Pearl, and Doss (2000) do use this procedure, and thus also assume a flat prior, but they constrain the root at a fixed time. They do not use a prior for the rate of substitution, but estimate it in a somewhat arbitrary way. As this quantity is the multiplier that determines the branch lengths given the times of the nodes, this leaves it uncertain what the prior on the tree is.

Yang and Rannala (1997) move among tree topologies, ones that do not have node times or branch lengths. Thus these are discrete, and there are a finite number of them. Starting with one topology, they then use a nearest-neighbor interchange at a randomly selected internal node to propose another topology. For each of these they sample a large number of speciation times from their conditional distribution. The likelihoods of the trees before and after rearrangement are computed approximately from these samples of trees, each of which has had its likelihood computed. The Metropolis algorithm is then used to accept or reject the proposed change of topology.

It is likely that there will be many more proposal mechanisms invented in the future. Choice among them is a matter of practicality, trying to achieve a mechanism that moves around the tree space enough but without having its proposals rejected too often.

Computing the likelihoods

In all of these methods we must compute the acceptance ratio R. This involves the ratio of the priors of the two trees and also the ratio of their likelihoods, as we can see in equation 18.8. For the methods mentioned above, the priors are taken to be uniformly distributed over all relevant trees, so their ratio is 1. In Chapter 16 we saw how to efficiently compute likelihoods, summing for each character over all possible assignments of states to interior nodes of the tree. In the cases in which proposal distributions modify only a local area of a tree (as in the cases that generate only nearest-neighbor rearrangements), we can take advantage of the way that the likelihood computation uses conditional likelihoods. This can allow the likelihood computation to be local, without any need to recompute most of the conditional likelihoods. We have already discussed a similar economy in Chapters 2 and 4, illustrating it in Figure 2.3.

In the method of Li, Pearl, and Doss (2000) there is an additional consideration. In specifying the phylogeny, they specify not only the topology and node times, but also the DNA sequences at the interior nodes. This enormously speeds the computation of the likelihood, which then is merely a product of the probabilities of the changes on each branch. They can also take advantage of the fact that their proposal mechanism involves only local rearrangements of the tree, so that the ratio of likelihoods before and after a change involves only terms from nearby branches. However, this simplification is purchased at a cost, because they must run their Metropolis sampler long enough to collect a sample of the possible assignments of DNA sequences to interior nodes. This can be a considerable burden.

There is no reason that the models of change used in the likelihood computations need to be restricted to the standard DNA models. Jow et al. (2002) use six- and seven-state RNA models, like those of Tillier (1994) and Muse (1995), that reflect pairing of sites in the secondary structure of the molecule. We may expect more elaborate models to be used as protein structure is more intensively investigated. For some of these models the likelihoods may be difficult to compute by integrating out the states of the interior nodes of the tree. A procedure like that of Li, Pearl, and Doss may then have to be adopted, with MCMC over both trees and interior node states.

Summarizing the posterior

Having a good sample of trees from the posterior, how are we to summarize the information they convey? It is tempting to simply ask how frequent each tree is. As the trees have node times that vary continuously, there is no answer to this question, as each tree is infinitely improbable. If we confine ourselves to asking about the tree topologies, or about the labeled histories, these have probabilities, but they may be hard to assess. If the data strongly constrain the trees, then we might find only a few trees accounting for most of the probability in the posterior. But if the data are fairly noisy, there might be millions of different trees in the posterior, and in the worst case no two of the trees we sample will be alike.

Thus there is good reason to be interested in methods of summarizing the posterior other than the empirical tree probabilities. One solution is to use clade probabilities. For clades of interest such as {Human, Chimp} we can sum the posterior probabilities of all trees that contain the clade. Then we can say that (for example) the posterior probability of that clade being on the tree is 0.87. This could be done for many different clades. The probabilities are not for independent events: If {Human, Gorilla} has a clade probability of 0.40 and {Orang, Gibbon} has a clade probability of 0.37, there might be no tree in the sample from the posterior that has both, or there might be as many as 0.37 of the trees that contained both. Whether there is a problem analogous to a "multiple-tests" problem if we look for the clade with the highest posterior probability has not been examined, as far as I know. Clade probabilities were used by Larget and Simon (1999) and are calculated in the Bayesian phylogeny inference program `MrBayes` by Huelsenbeck and Ronquist (2001).

An alternative way of summarizing the posterior was used by Li, Pearl, and Doss. Considering the cloud of sampled trees, they use a triples distance measure (Critchlow, Pearl, and Qian, 1996; see below, Chapter 30) to find the tree that is central to the cloud. They then take as their interval of trees the 90% of all trees that are closest to this central tree.

Of course, to obtain a valid result, we must have sampled the Markov chain for a long enough time to reach equilibrium, and to sample from it adequately. Little is known about how long a run is needed. Controversy exists (Suzuki, Glazko, and Nei, 2002; Wilcox et al., 2002) as to whether current programs sample for a long enough time.

Priors on trees

A serious issue in Bayesian inference is what kind of prior to use. If the prior is agreed by all to be a valid one, then there can be no controversy about using Bayesian inference. For example, in cases where a random mechanism such as Mendelian segregation provides us with prior probabilities of genotypes, all statisticians, whether Bayesians or not, will happily use Bayes' theorem and consider the posterior probabilities to be valid.

Priors on trees fall into three classes:

1. Priors from a birth-death process
2. Flat priors
3. Priors using an arbitrary distribution for branch lengths such as an exponential

Rannala and Yang (1996) used a birth-death process of speciation, which can also be followed (Yang and Rannala, 1997) by a random sampling from the resulting species to get the smaller number of species actually analyzed. They needed parameter values for birth and death rates. It would be possible to put prior distributions on the birth and death rates in order to do a full Bayesian inference. They chose instead to estimate the birth and death rates by maximizing the sum of all posterior probabilities. This is an "empirical Bayes" method.

Huelsenbeck and Ronquist's (2001) program `MrBayes` allows the user the specify priors on branch lengths, using either uniform or exponential distributions. This gives a prior different from a birth-death process.

As we have seen, the other methods assume one or another form of a flat prior. One of the issues that arises with flat priors is how far out from the origin they extend. For example, we might use a prior that gave equal probability to all labeled histories, with the time back to the root node drawn from a uniform distribution on the interval $(0, B]$. The times of the interior nodes could then be drawn uniformly between 0 and the time of that node, and assigned in a way consistent with the topology of the tree. (Exactly how the times are parameterized is less important than one might think, as Jacobian terms for the density of the prior will usually cancel out; in some more complicated cases a "Hastings term" is created, which also cancels out.)

Controversies over Bayesian inference

Although the use of Markov chain Monte Carlo methods has greatly increased interest in Bayesian methods, Bayesian methods remain somewhat controversial in statistics. The issues at first seem to be technical ones in statistics; in fact, they are really issues in the philosophy of science. Different positions on statistical inference are really different positions on how we should think about drawing inferences from scientific data. I will not try to cover these issues in any detail here. For philosophers' viewpoints supportive of Bayesian methods, see Howson and Urbach (1993) and Rosenkrantz (1977); for one critical of Bayesian inference, see Mayo (1996). For a likelihoodist view see Royall (1997). The likelihoodist monograph of Edwards (1972) was originally inspired by his work on inferring phylogenies.

All of the controversy is about the priors. The issue of whether we can say that there really is a prior (for example, whether we can put a probability on whether there are or are not Little Green Men on the planet Mars) leads us directly into philosophy, and beyond the scope of this book. Two other issues concerning the prior are worth note here: whether one person's prior can be another's, and whether flat priors lead to trouble in cases of unbounded quantities.

Universality of the prior

The researcher doing a Bayesian analysis of a data set applies a prior of their own choice. When the software chooses the prior, it is one chosen by the author of the software. Even when the software allows the user the choice, the issue arises whether this is the same prior that would have been chosen by the reader of the resulting paper. If it is, there is no problem, but if it differs by much from the reader's prior, the conclusion can be affected.

For example, suppose that we send a space probe to Mars and have it use a camera to look for Little Green Men. It sees none. Assume further that it would have had a $1/3$ chance of not seeing them if they really were there (and of course a 100% chance of not seeing them if they really weren't there). If my prior belief in Little Green Men would give odds of $4:1$ against their existence, we can use the odds ratio formulation to calculate that the posterior odds ratio for the existence of Little Green Men is $1/4 \times 1/3 = 1/12$. But if I publish this, it reflects my prior. If you were willing to give odds $4:1$ in favor of the existence of Little Green Men, your posterior odds ratio would have been $4/1 \times 1/3 = 4/3$, a very different result.

It might be argued that the correct thing to do in such a case is to publish the likelihood ratio $1/3$ and let the reader provide their own prior. This is the likelihoodist position. A Bayesian is defined, not by using a prior, but by being willing to use a controversial prior.

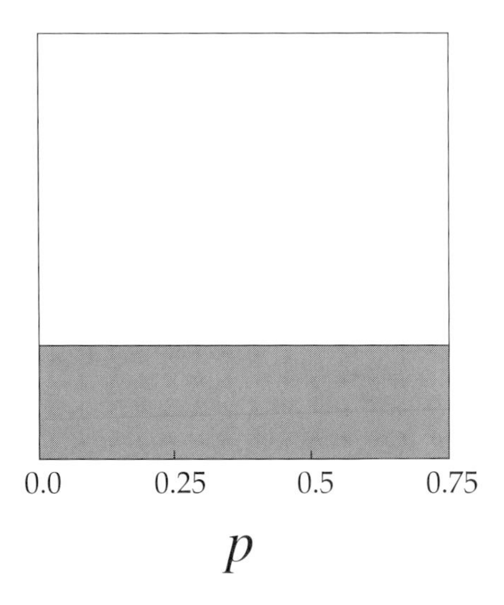

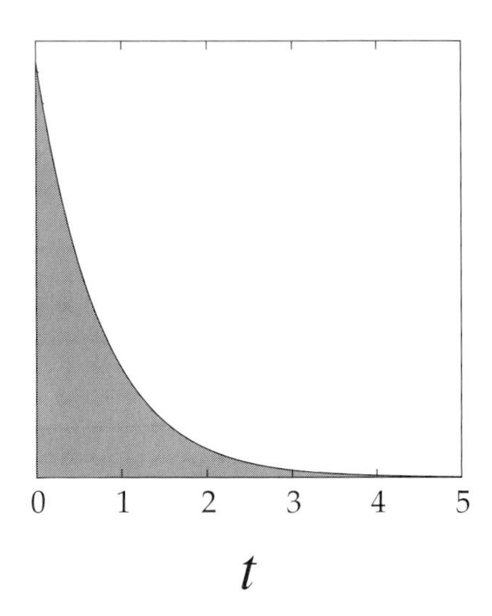

Figure 18.4: A flat prior on the scale of the net probability of change (left) and the prior on branch length to which this corresponds (right) under a Jukes-Cantor model of base change.

Flat priors and doubts about them

One way of reducing the controversy over priors is to use flat priors (sometimes called "uninformative" priors). It becomes important to know what scale we are on. To discuss this, we will use the simplest case of parametric inference of phylogenies. Imagine that we have only two species, and infer the tree connecting them. If we use a simple Jukes-Cantor model of DNA sequence evolution, the resulting tree is unrooted. There is then no issue of tree topology. Inference of the tree is in effect the same as estimating the distance between the two species.

Issues of scale. In our example, if we use a scale of branch length, this can go from 0 to ∞. Alternatively, we could use a scale of the net probability p of change at a site. This goes from 0 to 3/4. Figure 18.4 shows (on the left) a flat prior between 0 and 3/4. On the right is the prior on the branch length that this implies, using equation 11.17. It should be immediately apparent that this is far from a flat prior on branch length.

If we were doing likelihood inference, it would not matter which of these two quantities we used. Figure 18.5 shows the case in which we have observed 10 sites, of which 3 differ between the two sequences. On the left the likelihood is plotted as a function of the branch length t, on the right as a function of the net probability of change p. These look quite different, but the maximum likelihood estimates correspond exactly. The value of t (0.383119) at which the likelihood is

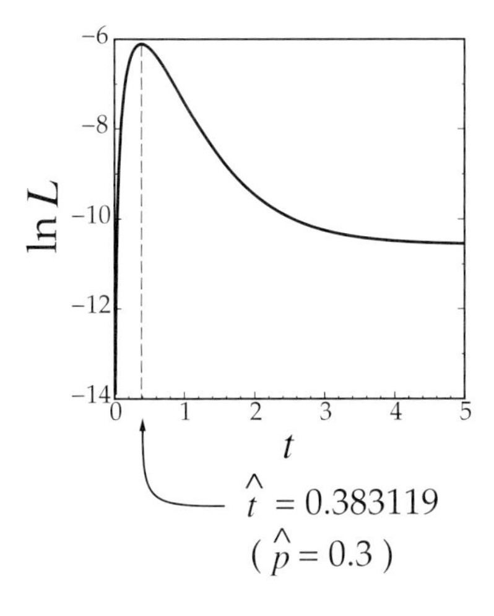

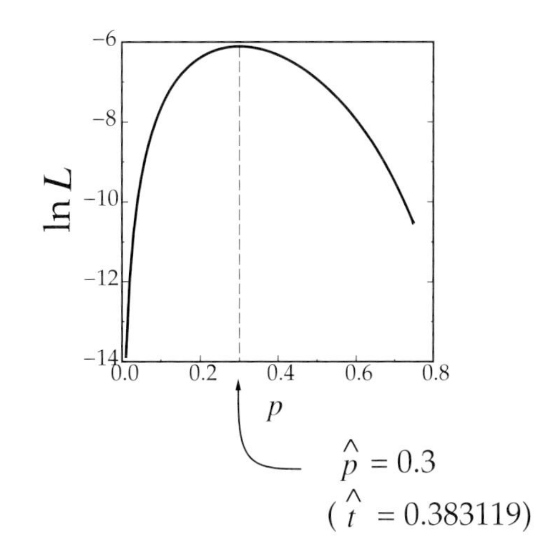

Figure 18.5: Log-likelihood curves for the case of two DNA sequences differing at 3 sites out of 10, under a Jukes-Cantor model. Left, plotted as a function of branch length t. Right, plotted as a function of the net probability of change p. The maxima are at precisely corresponding points on the two scales.

maximized is exactly the value of t that corresponds to the value of p (0.3) at which the likelihood is maximized. This property of the estimate being independent of the scale is called the functional property of maximum likelihood. It does not hold for Bayesian inference.

Flat priors on unbounded quantities. More problems are uncovered if we try to put a flat prior on a quantity that is unbounded. The problem is that if a prior is to be flat from 0 to ∞, the density will be zero everywhere. So we must have the prior be truncated somewhere. Figure 18.6 shows what happens if we use a flat prior on t that is arbitrarily truncated at $t = 5$. When plotted in terms of p, the prior is strongly slanted toward values of p near 3/4. Even worse, when we change the truncation to occur at (say) 100, the slanting on the p scale is even stronger.

It might be hoped that it would not matter much where a flat prior on t was truncated. It can matter, but this depends on how we are summarizing the posterior distribution. If we note only the single tree of highest posterior probability, this will also be the maximum likelihood estimate of the tree and will not depend on where the prior is truncated, as long as that is beyond the position of the ML estimate. If we instead make a Bayesian "credible interval" which contains (say) the middle 95% of the posterior distribution, this can depend critically on where the truncation point is.

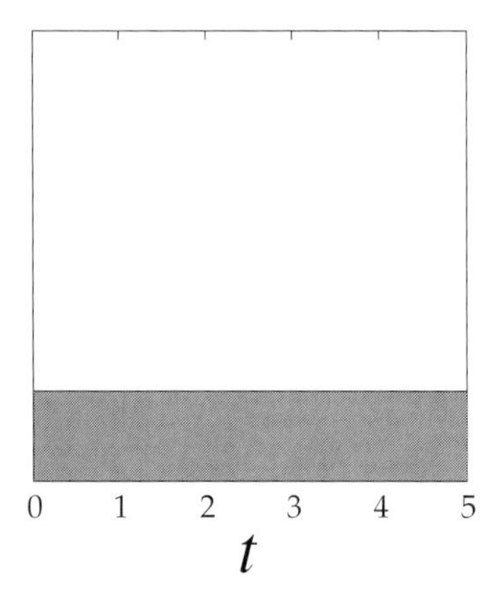

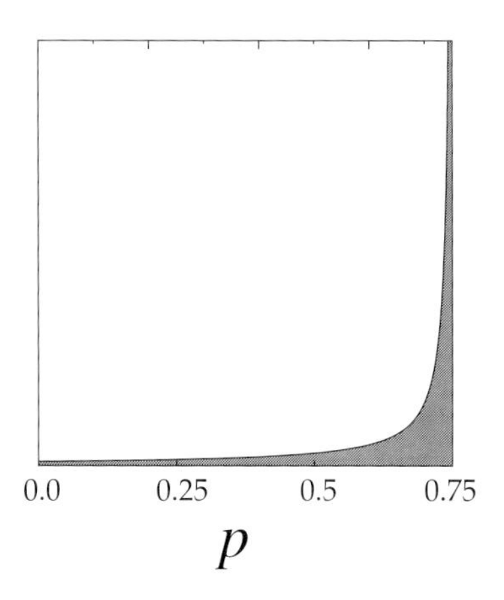

Figure 18.6: A flat prior on t between 0 and 5, and the prior on p which is equivalent to it. The larger the number at which the flat prior is truncated, the more strongly the prior on p is attracted to $p = 3/4$.

To see this in our Jukes-Cantor model example, note that the likelihood curve as a function of branch length t (see Figure 18.5) has a long tail that stretches infinitely far to the right. It has very low density, but, as it goes out to infinity, it contains an infinite area. If we try a prior that is truncated sufficiently far out, it will have most of the posterior be in that long, low tail. Figure 18.7 shows the upper and lower ends of the credible interval as a function of the truncation point. When it reaches about 700, the credible interval does not contain the maximum likelihood value 0.383119. This shows that flat priors on quantities like branch lengths can lead to strong truncation point effects.

The problem can be severe when credible intervals are defined in this way. But there are alternative ways of defining them that do not suffer from this difficulty. Instead of defining the interval by having 2.5% of the posterior probability lie beyond each end, we can make it as short as possible, by preferentially including the highest parts of the posterior density. The probabilities beyond each end of the interval are then asymmetric, but the mode of the distribution always lies within the interval, so in this case the interval always contains the maximum likelihood estimate.

Applications of Bayesian methods

Although it is still early days for Bayesian methods in phylogenetic inference, they have already been applied to a number of problems beside inferring the phylogeny itself:

- Huelsenbeck, Rannala, and Yang (1997) have tested hypotheses about rates of host switching and cospeciation in host/parasite systems. This is discussed further below in Chapter 31.

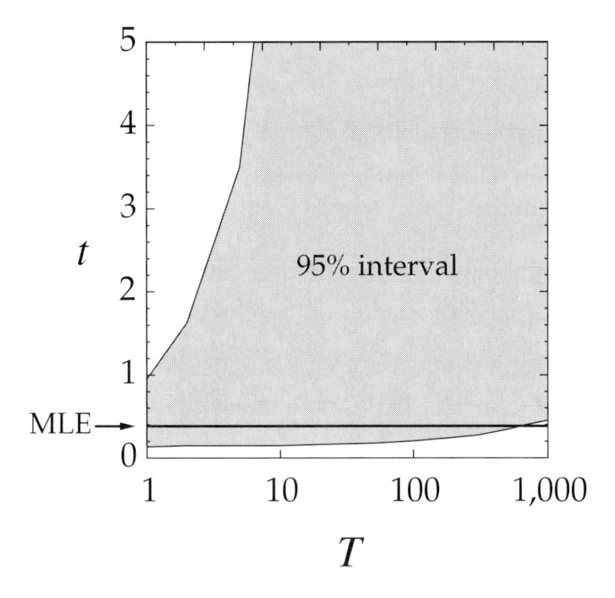

Figure 18.7: Effects of having a flat prior of branch length, in our example of a two-species phylogeny with 10 sites, when the flat prior is truncated at various large branch lengths. As the truncation point T becomes large, both the upper and the lower ends of the Bayesian 95% credible interval rise, so that ultimately that interval does not contain the maximum likelihood value 0.383119.

- Thorne, Kishino, and Painter (1998) used an autocorrelated prior distribution on rates of evolution among branches to sample from combinations of rates on different branches of a phylogeny in doing MCMC integration of likelihood of a phylogeny. This has been discussed above in Chapter 16.
- Huelsenbeck, Larget, and Swofford (2000) had a similar model for change of rates among branches, different in changing rates at discrete points on the tree. This has also been discussed above in Chapter 16.
- Huelsenbeck, Rannala, and Masly (2000) have used them to infer the rate change of states of a character (horned soldier castes in aphids) and the bias in the rate of gain of the character.
- Suchard, Weiss, and Sinsheimer (2001) have chosen among models of base change using Markov chain Monte Carlo methods on a space that includes both trees and models. We have mentioned this in Chapter 13.
- Huelsenbeck and Bollback (2001) have used a Bayesian framework to investigate the accuracy of inference of ancestral states. This has the complication that on some phylogenies a given ancestral node (such as the immedi-

ate common ancestor of chimpanzees and humans) may not exist. We have mentioned this work in Chapter 16.

- Metzler et al. (2001) placed priors on the parameters of a model that allows for deletion and insertion and samples alignments of a set of related sequences. This will be discussed below in Chapter 29.
- Huelsenbeck, Bollback, and Levine (2002) have inferred the position of the root of the tree, testing the relative merit of basing this on a clock, on an outgroup, or on nonreversibility of the model of DNA base change.
- Bollback (2002) used Bayesian inference to choose among models of DNA change in the light of the specific data set.

There is no reason to believe that this list will not expand until it covers all hypotheses of any interest.

Chapter 19

Testing models, trees, and clocks

In this chapter I will consider the ways in which likelihood ratios can be used to test hypotheses about phylogenies or about evolutionary models. I will also discuss the related issue of "interior branch tests" and the wider issue of tests of tree topologies. After considering testing of the molecular clock, I will also cover more direct approaches to statistical testing and confidence intervals, ones that do not use the asymptotic framework of likelihood ratio testing. Tests involving bootstrap and other resampling methods, tests involving invariants, and paired-sites tests will be covered in later chapters.

Likelihood and tests

Likelihood does not only allow us to make a point estimate of the phylogeny; it also gives us information about the uncertainty of our estimate. We can see in the odds-ratio form of the likelihood equations (equation 16.1) that the likelihood ratio affects the posterior probabilities of different hypotheses. Even when we do not have prior probabilities (and hence do not have posterior probabilities), there is a way of using the likelihood curve to test hypotheses and to make interval estimates.

From one data set to another, the likelihood surface moves back and forth around the true parameter values. Our estimates of them vary, and the height of the curve at the peak is greater than the height at the true estimates. We can test whether some suggested parameter values are too far down the likelihood surface to be the true values.

Asymptotically, with large amounts of data, the parameter values are estimated with great accuracy, so that as long as we are in this "asymptopia" we can consider only the values of the parameters near their true values. For well-behaved likelihood curves the log-likelihood will be close to quadratic in this region. (One can simply consider it as approximated by the first few terms of a Taylor series in that vicinity.) In these cases it is possible to show that the maximum likelihood estimate is asymptotically distributed as a multivariate normal distribution, with mean zero and variance given by the negative inverse of the matrix of curvatures of the expectation of the log-likelihood function. We will not attempt the proof here: it is straightforward but too tedious. It will be found in many theoretical statistics texts, such as Kendall and Stuart (1973, pp. 57ff.) and Rao (1973, pp. 364ff., p. 416).

Likelihood ratios near asymptopia

If only one parameter, θ, is being estimated, the asymptotic variance of the estimate of θ is given by the curvature of the log-likelihood:

$$\mathrm{Var}\left[\widehat{\theta}\right] \approx -\frac{1}{\left[\dfrac{d^2\mathbb{E}\,\log(L)}{d\theta^2}\right]} \tag{19.1}$$

With multiple parameters, the variance is replaced by a covariance matrix, the curvature by a matrix of curvatures, and the inverse by matrix inversion:

$$\mathrm{Var}\left[\widehat{\boldsymbol{\theta}}\right] = \mathbf{V} \approx -\mathbf{C}^{-1} \tag{19.2}$$

where $\mathbf{C}$ is the matrix of curvatures of the expected log-likelihood:

$$C_{ij} = \mathbb{E}\left[\frac{\partial^2 \log(L)}{\partial\theta_i\,\partial\theta_j}\right] \tag{19.3}$$

the θ_i being the individual parameters that we are estimating. The expected log-likelihood is not known in most cases; however, one can substitute the observed log-likelihood and make only a small error if one is in asymptopia.

These rules about the variances and covariances of parameters are closely connected with asymptotic distribution of the likelihood ratio. Consider the case of a single parameter θ. Asymptotically, with large amounts of data, the ML estimate $\widehat{\theta}$ is normally distributed around its true value θ_0, as we have already mentioned. Suppose that its variance, which is also the inverse of the curvature of the log-likelihood, is called v. Then, using this mean and variance, a standard normal variate is

$$\frac{\widehat{\theta} - \theta_0}{\sqrt{v}} \sim \mathcal{N}(0, 1) \tag{19.4}$$

The likelihood will (by definition) be highest at the estimate $\widehat{\theta}$ and somewhat lower at the true value of θ, which we will call θ_0. How much lower? It is relatively easy to see, since we know the negative of the curvature of the log-likelihood, $1/v$, and we are so close to the peak that we can consider the log-likelihood curve as locally quadratic. The log-likelihood is approximately its value at $\widehat{\theta}$, less a quadratic function with that curvature:

$$\ln L(\theta_0) = \ln L(\widehat{\theta}) - \frac{1}{2}\frac{(\theta_0 - \widehat{\theta})^2}{v} \tag{19.5}$$

Subtracting $\ln L(\theta_0)$ from both sides and using the normal distribution in equation 19.4, we can rearrange equation 19.5, and we immediately see that twice the difference in the log-likelihoods between the peak and the value at the true value θ_0 will be distributed as the square of a standard normal variate. This is a chi-square variate with 1 degree of freedom:

$$2\left[\ln L(\widehat{\theta}) - \ln L(\theta_0)\right] \sim \chi_1^2 \tag{19.6}$$

This result is a consequence of the fact that asymptotic variances of the parameters turn out to be obtainable from the curvatures of the log-likelihood surface. It follows that there is a simple rule for how far down the log-likelihood surface the true value of the parameter is expected to be. If θ_0 is in fact the true θ, it should not be too far down the curve. That is, maximizing over all θ should not improve the likelihood too much above the likelihood of the true θ. If we select some tail probability (probability of Type I error) such as 0.05, we can compute the left-hand side of equation 19.6 by doubling the difference of log-likelihoods between the peak and a proposed true value of θ. We see whether that exceeds 3.8414, which is the 95th percentile of the distribution of χ^2 with 1 degree of freedom.

Thus we have a test of the hypothesis that $\theta = \theta_0$, rejecting it when some other value of θ fits the data significantly better. Alternatively, we could use the same logic to make an interval estimate of θ. The acceptable values of θ are those whose log-likelihoods are within 1.9207 (= 3.8414/2) of the peak of the likelihood curve. (For the likelihoods themselves, this implies that they should be within a factor of 6.826 of the maximum.)

Multiple parameters

For multiple parameters, we can use the same logic to get the distribution of the likelihood ratio between the maximum likelihood estimates and the true parameter values. Suppose that there are p parameters (such as branch lengths on the phylogeny) that are being estimated from the data and their values are made into a vector $\boldsymbol{\theta} = (\theta_1, \theta_2, \ldots, \theta_p)$. We have seen that, in the asymptotic limit of large amounts of data, these will be distributed in a multivariate normal distribution with means equal to the true parameter values and covariance matrix $-\mathbf{C}^{-1}$, where $\mathbf{C}$ is the expectation of the curvatures of the log-likelihoods. We now go through a process very similar to the one-parameter case. The likelihood can be expanded around the true parameter values in a multivariate Taylor series. Since the slopes of the likelihood at the maximum likelihood value $\boldsymbol{\theta}$ are zero,

$$\ln L(\boldsymbol{\theta}_0) \approx \ln L(\boldsymbol{\theta}_0) - \frac{1}{2}(\boldsymbol{\theta}_0 - \boldsymbol{\theta})^T \mathbf{C}(\boldsymbol{\theta}_0 - \boldsymbol{\theta}) \quad (19.7)$$

We know that the maximum likelihood estimate $\boldsymbol{\theta}$ is multivariate normally distributed around the true value $\boldsymbol{\theta}_0$ with mean $\mathbf{0}$ and covariance matrix $-\mathbf{C}^{-1}$. It follows from the usual sorts of manipulations of normal distributions that

$$(\boldsymbol{\theta} - \boldsymbol{\theta}_0)^T \mathbf{C}(\boldsymbol{\theta} - \boldsymbol{\theta}_0) \sim \chi^2_p \quad (19.8)$$

as it is the sum of squares of p independent standardized normal variates. The result is a straightforward extension of equation 19.6:

$$2\left[\ln L(\widehat{\boldsymbol{\theta}}) - \ln L(\boldsymbol{\theta}_0)\right] \sim \chi^2_p \quad (19.9)$$

To test whether the null hypothesis that the parameter values are $\boldsymbol{\theta}_0$ is acceptable, we double the difference between the highest log-likelihood and the log-likelihood of this value, and we look this up in a χ^2 distribution with p degrees of freedom. Intervals can also be constructed in a straightforward way, although they will be ellipses in p dimensions.

Some parameters constrained, some not

If we have a total of p parameters, and under the null hypothesis q of them are constrained, there is a very similar rule. It turns out that we can double the difference in log-likelihoods between the two hypotheses, and then look up the resulting quantity in a table of the χ^2 distribution with q degrees of freedom:

$$2\left[\ln L(\widehat{\boldsymbol{\theta}}) - \ln L(\boldsymbol{\theta}_0)\right] \sim \chi^2_q \quad (19.10)$$

This includes all the other cases. (For example, equation 19.9 is simply the case in which $q = p$.) Thus the number of degrees of freedom of the χ^2 is the number of parameters that are constrained to get to the null hypothesis. This includes not only the cases in which q parameters have their values specified, but also cases in which q algebraic relationships between the parameters are proposed.

Conditions

The likelihood ratio test (LRT) requires some rather stringent conditions. Most important is that the null hypothesis should be in the interior of the space that contains the alternative hypotheses. If there are q parameters that have been constrained, there must be a possibility to vary in both directions in all q of them. Thus if we are testing hypotheses about a parameter θ that can takes values between 0 and 1, the null hypothesis that it is 0 does not allow for the usual form of the likelihood ratio test.

A modification is possible in such cases. In the one-parameter case, when the null hypothesis is at the end of the possible interval, as random sampling effects in the data cannot cause the parameter to vary in both directions, the probability of achieving a given likelihood ratio is only half as great as it would be if the null hypothesis were in the interior of the region. If the data were to try to recommend a value less than 0, we would estimate the parameter to be 0 in those cases. Thus if L_0 restricts one parameter to the end of its range, the distribution of the doubled log of the likelihood ratio has half of its mass at 0 and the other half in the usual chi-squared distribution. Thus in this case we should compute the likelihood ratio statistic, look it up in the usual chi-square table with one degree of freedom, and then halve the tail probability. This distribution has been described by Ota et al. (2000). They have also gone on to consider more complicated cases where more than one parameter are constrained to be at the limits of their range.

It is important to keep in mind that the LRT is only valid asymptotically. The proof of its validity requires that we be close enough to the true values of the parameters that we can approximate the likelihood curve as being shaped like a normal distribution. That is true with very large amounts of data. Statisticians are accustomed to using it with modest amounts of data, but it is well to remember that it is an approximation in all such cases.

Curvature or height?

If we do not have an infinite amount of data, there are two possible ways to proceed. One is to use the curvatures of the expected log-likelihood (as approximated by the curvatures of the observed log-likelihood) and assume that the estimates are normally distributed, with covariances defined by the inverse of the curvature matrix. The null hypothesis is then rejected if it lies too far out in the tails of that normal distribution.

The other method is simply to use the likelihood ratio. It is not at all obvious which of these methods will be better, for in the asymptotic case in which the derivations take place, both of them are the same. In fact, there is some reason to believe that the second procedure (the use of the likelihood ratio) will generally be more accurate.

Interval estimates

Statistical tests such as the likelihood ratio test can be used to make estimates of intervals that might contain the true values. As we have seen, this is done by finding the maximum likelihood estimate (of whatever you are estimating) and then finding all values that cannot be rejected compared to it, using the test. For the moment we will talk in terms of parameters of an evolutionary model, for as we will see, tree topologies are more difficult to deal with. Interval estimates are sets of values, all of which are acceptable given the data. As they go hand in hand with tests, we will discuss both simultaneously.

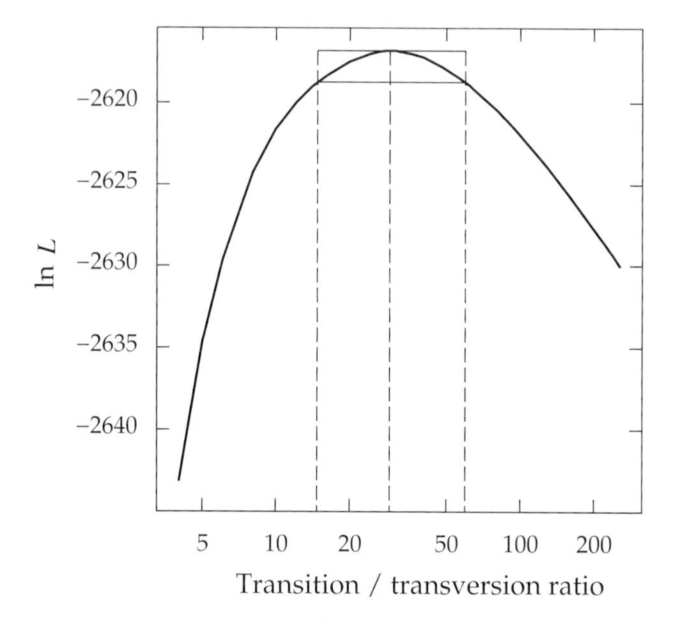

Figure 19.1: Likelihood curve for the transition/transversion ratio R in the F84 model, for a 14-species primate data set selected by Masami Hasegawa. For each value of R, the tree of highest likelihood is found and the resulting likelihood plotted. Note that the horizontal scale is logarithmic. The maximum likelihood estimate of R is approximately 29, and the likelihood interval for R is $(14.68, 59.7)$.

Testing assertions about parameters

As examples of likelihood ratio tests, let us consider testing assertions about the parameters of the model of evolution. For the moment, we defer consideration of testing different topologies, as that turns out to have some serious complications.

If we have DNA sequences and a model of change of the bases (such as the Tamura-Nei model), we can imagine testing various hypotheses about the parameters of the model. We will use a particular data set, 14 sequences each of 232 bases, assembled by Masami Hasegawa, as our example. Note that this data set omits sites that were first or second positions of codons, so it is not contiguous sequence. If we use the F84 model of base change, we have three parameters for the base frequencies (the fourth is determined by those three as the frequencies must add to 1), plus a transition/transversion ratio. Suppose that we try different transition/transversion ratios, plotting the likelihood against that ratio, which we call R.

Figure 19.1 shows the likelihood curve we get as we vary the transition/transversion ratio R. The peak, which is at –2616.86231, is near 29. As we are specifying one parameter, we are restricting one degree of freedom in looking at any one point on this curve. We can reject those that have their log-likelihood less than the maximum by more than 3.8414/2 = 1.9207. This defines an interval of acceptable values between 14.68 and 59.7. Given the limited amount of data, this rather wide confidence interval is reasonable. Although the computation searches for the best tree for each one of these points, the tree topology found is the same for all the values shown here. The computation can be speeded up by simply keeping the tree topology unchanged and searching only among different branch lengths. This would have to be checked occasionally by a full search among tree topologies. It may seem that the curvature of the log-likelihood could also have been used give a good approximation to this interval, as the curve is close to a quadratic. But note that the horizontal axis is the logarithm of R: In fact, the interval is quite asymmetric.

We could also vary other parameters such as the base frequencies. We could define the interval of acceptable base frequencies by finding all those combinations of base frequencies that achieved a log-likelihood no more than $\chi_3^2(0.95)/2 = 7.815/2 = 3.908$ below the maximum. Alternatively, we could have a Bayesian prior on models. Suchard, Weiss, and Sinsheimer (2001) have used a Bayesian MCMC method to choose among models of DNA substitution at the same time as they choose among trees.

Coins in a barrel

For some parameters, the matter is not quite so simple. In the preceding chapter we saw that likelihoods could be computed for multiple evolutionary rates, and that variation of rates over sites could be modeled by a hidden Markov model (HMM). Consider the case where there is no autocorrelation of rates, so that each site has its evolutionary rate drawn from a distribution of rates. There is the perplexing question of how many different rates to use. As we add rates one by one, each model contains the preceding model as a subcase, because by adding an additional rate that is the same as one of the previous rates we do not change the likelihood. Therefore, the likelihood can never decrease on adding a new rate — it will usually increase at least slightly. Each rate brings with it two new parameters: one rate and one expected fraction of times that this rate will be used. What should we do? Add rates until the increase is not significant? This is tempting, but it leaves the possibility that adding one more rate would again increase the likelihood substantially.

We can see the dilemma, and one type of solution to it, by considering a simpler case, that of Coins in a Barrel. We imagine that we have a large barrel filled with well-mixed coins. We draw coins from it, toss each 100 times, and then return them to the barrel. The barrel is large enough that we draw a new coin each time. The coins do not necessarily all have the same probability of heads. We want to fit hypotheses about the mixture of heads probabilities. If, for example, we wanted to fit the hypothesis that there were three kinds of coins, one having frequency f_1

and heads probability p_1, one frequency f_2 and heads probability p_2, and one with frequency f_3 and heads probability p_3, the probability of a sequence of N tosses that got M heads and $N-M$ tails would be

$$L = f_1 \binom{N}{M} p_1^M (1-p_1)^{N-M} + f_2 \binom{N}{M} p_2^M (1-p_2)^{N-M} + f_3 \binom{N}{M} p_3^M (1-p_3)^{N-M} \tag{19.11}$$

We can immediately see that, as with evolutionary rates, we will usually increase the likelihood each time we add another type of coin, and there is the puzzling issue of when to stop and consider the model to fit adequately.

As with evolutionary rates, the heads probabilities could be slightly different for each coin. Thus it would make more sense to imagine a distribution $f(p)$ of the heads probabilities. The likelihood for a set of tosses would then be

$$L = \int_0^1 f(p) \binom{N}{M} p^M (1-p)^{N-M} \, dp \tag{19.12}$$

At first sight, this problem appears even worse. There are uncountably infinitely many possible density functions, $f(p)$. But a closer examination shows that equation 19.12 depends on those density functions only through the first N moments of each (the kth moment is the expectation of p^k). If the functions $f(p)$ have N moments, with the jth moment being

$$M_j = \mathbb{E}[p^j] = \int_0^1 f(p) \, p^j \, dp \tag{19.13}$$

and if we expand the powers of $(1-p)$ in equation 19.12, it follows that the likelihood L is a linear combination of $M_1, M_2, \ldots, M_N$. Two functions f that have the same first N moments, but different values of M_{N+1}, will achieve the same likelihood L.

Thus, if we can find the first N moments of the distribution of p, we can maximize the likelihood. Finding a sequence of real numbers while ensuring that they are possible moments of a distribution is a well-known problem in statistics. The point that is relevant here is that only N quantities are involved — it is possible (though not easy) to find the likelihood of the best-fitting model. Note that the process is *not* the same as fitting a different heads probability for each coin tossed — that would estimate far too many parameters.

Evolutionary rates instead of coins

Turning to evolutionary rates, we find a case that behaves this way. For DNA likelihoods, if autocorrelation of rates among sites is not permitted, the likelihood is a function of the data through the observed pattern frequencies of patterns of nucleotides. For n species, there are 4^n possible patterns (`AAAA...A` through `TTTT...T`). Each of these has an expected frequency for each possible phylogeny.

Two combinations of rates will have distinct likelihoods only if they differ in the expected frequencies of some of these. So all combinations of rates that achieve the same expected frequencies of patterns have equally high likelihoods. The best we can do is to achieve the best set of pattern frequencies that can be predicted by a tree. This is not the same as the set of pattern frequencies that is exactly the same as the observed pattern frequencies; usually those will not be expected from any tree.

Most of the models of DNA change have the property that the transition probabilities are linear combinations of exponentials, as in Table 13.1 where the Tamura-Nei model has probabilities that (for branch lengths rt) are functions of $\exp[-(\alpha_R + \beta)rt]$, $\exp[-(\alpha_Y + \beta)rt]$ and $\exp(-\beta\, ut)$. This in turn means that, using equations like 16.10 and 16.11, we can write the pattern frequency of any given pattern as a sum of exponentials of sums of branch lengths

$$\text{Prob}\,(ACCCG) \;=\; \sum_i c_i \,\exp\Big(-\sum_j a_{ij}\, r\, v_j\Big) \tag{19.14}$$

In this equation, the first summation (i) is over all ways that states can be assigned to hypothetical ancestral nodes in the tree, and the second summation is over branches in the tree. For each branch in each reconstruction, the coefficient a_{ij} is either 0 or is one of $\alpha_R + \beta$, $\alpha_Y + \beta$, or β, depending on what kind of event has been reconstructed in that branch. The coefficient c_i also depends on the details of the reconstruction. It contains, for the Tamura-Nei model, only constant terms that do not depend on the branch lengths or evolutionary rates.

When we take expectations over the distribution of rates, the upshot is that the expected pattern frequencies depend on terms like $\mathbb{E}\,[\exp(-rx)]$, where r is the evolutionary rate and x is a weighted sum of some of the branch lengths. This expression is the Laplace transform of the distribution of evolutionary rates, evaluated at $-x$. There will be a large number of different values of x, but we can at least say that any two distributions of evolutionary rates that have the same values of their Laplace transforms at that large set of values will achieve the same likelihood. This again gives us the hint that the structure of the problem with varying evolutionary rates is similar to the structure of the coins-in-a-barrel problem. A derivation closely related to this is given for phylogenies by Kelly and Rice (1996). We break off here; no one has yet solved the problem of maximizing the likelihood over all possible rate distributions.

Choosing among nonnested hypotheses: AIC and BIC

As in the case of evolutionary rates, we often have sets of models with ever-greater numbers of parameters. We can test these by likelihood ratio tests, but it may be difficult to know which model to prefer. It will always be the case that a more general model will have a higher likelihood than a restricted subcase of that model. Choosing the model with highest likelihood may lead to one that is unnecessarily complex.

Statisticians have come up with methods that try to compromise goodness of fit with the complexity of the model. The most famous of these is the *Akaike information criterion* (AIC). It computes the expectation of the log-likelihood for a new data set of the same size as the current one. The argument is roughly as follows:

- The expectation of the log-likelihood is highest at the true value of the parameters.
- However, we can't compute it directly; we can only compute the log-likelihood for the current data.
- If there are p extra parameters in our model, the parameter estimate for the current data is at a point where the expectation of the log-likelihood is on average p units lower.
- With a large amount of data, the shape of the log-likelihood curve is about the same as the shape of the (unknown) expected log-likelihood curve, only displaced from the true values.

I realize that, put this way, the argument is hard to follow, but we do not have space here for a more exact derivation. If we take the negative of twice the log-likelihood of each hypothesis and penalize it by adding twice the number of parameters, so that for hypothesis i with p_i parameters,

$$\mathrm{AIC}_i = -2\ln L_i + 2p_i \tag{19.15}$$

we get quantities that can be compared among hypotheses. One prefers the hypothesis that has the lowest value of the AIC. We could equally well subtract p_i from the log-likelihood and maximize that.

An alternative to the AIC is the *Bayesian information criterion* (BIC). It differs by using a penalty that is dependent on the sample size n:

$$\mathrm{BIC}_i = -2\ln L_i + p_i \ln(n) \tag{19.16}$$

It is justified by a rather general Bayesian argument that places priors on hypotheses but does not place many requirements on those priors. For all cases with more than a few data points, the penalty for extra parameters in the model will be greater for the BIC than for the AIC.

Of these the AIC has been most frequently used. Hasegawa (1990), Kishino and Hasegawa (1990), Hasegawa et al. (1990), and Reeves (1992) discussed its use to select among models. Kishino and Hasegawa (1990) also used it to select among tree topologies. Completely resolved tree topologies all have the same number of parameters (their branch lengths); thus choosing the best AIC or BIC value is the same as choosing the tree of highest likelihood. However, when one tree has a multifurcation, it thereby has fewer parameters, so that it may be preferred if it does not lose much likelihood.

Ren, Tanaka, and Gojobori (1995; see also Tanaka et al., 1997, 1999) have presented a "model-based complexity" criterion. It is closely related to the minimum description length (MDL) criterion of J. Rissanen. In effect it adds to the log-likelihood a term depending on the number of internal nodes in the tree, and one depending on the number of parameters in the model of evolution. The net effect is very much like the BIC criterion. All of these criteria can be used to decide among models of evolution and whether multifurcating trees are to be preferred to bifurcating trees that further resolve them.

Posada and Crandall (2001) discuss the AIC and BIC criteria, which are available in their model-testing program (Posada and Crandall, 1998).

An example using the AIC criterion

To illustrate the use of the AIC criterion, I have run the 14-species primate mitochondrial DNA data set (which we used in Figure 19.1). Table 19.1 shows the log-likelihoods, the numbers of parameters, and the AIC criterion for a series of models of evolution. The Jukes-Cantor, Kimura two-parameter, and F84 models are evaluated, the latter two with transition/transversion ratio R either estimated or set to a value common for mammalian nuclear DNA. As this is not nuclear DNA, that value is essentially an arbitrary one. In this case, the AIC prefers the most complicated of these models. The set of likelihood ratio tests that can be performed on these hypotheses is shown in Figure 19.2. These form a lattice. Posada and Crandall (2001) have suggested using likelihood ratio tests with a Bonferroni correction for multiple tests. In this case there are seven arrows in the diagram. In all cases the unrooted tree topology is identical, a condition which is needed for the likelihood ratio test to be valid. Using the tail probability $0.05/7 = 0.00714$, we find that six of the arrows correspond to significant tests. Following these arrows, we arrive at the same conclusion about the best model that we would with the AIC criterion.

Table 19.1: An example of the use of the AIC criterion to choose a model of evolution. The 14-species primate mitochondrial data set is used, evaluated with PAUP* for a variety of models of evolution. The F84 model with estimated transition/transversion ratio $R = 28.95$ is the preferred model among those shown here.

Model	$\ln L$	Number of parameters	AIC
Jukes-Cantor	-3068.29186	25	6186.58
K2P, $R = 2.0$	-2953.15830	25	5956.32
K2P, $\widehat{R} = 1.889$	-2952.94264	26	5957.89
F81	-2935.25430	28	5926.51
F84, $R = 2.0$	-2680.32982	28	5416.66
F84, $\widehat{R} = 28.95$	-2616.3981	29	5290.80

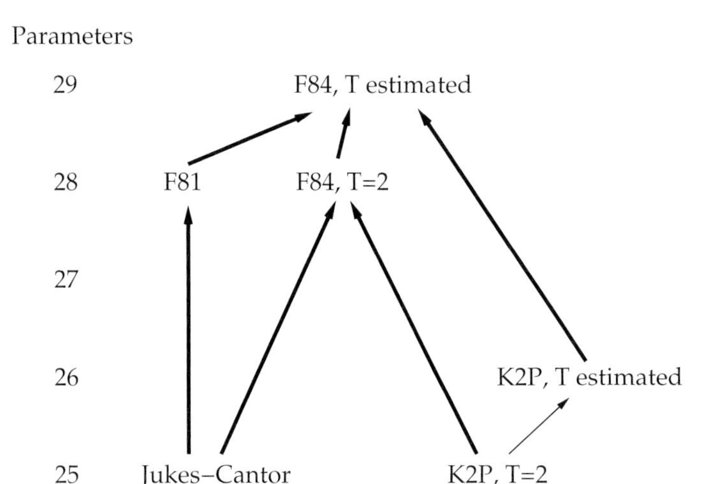

Figure 19.2: A lattice of likelihood ratio tests, for the cases shown in Table 19.1. The number of parameters in each hypothesis is given on the scale to the left, and the arrows represent likelihood ratio tests. The tests that come out strongly significant are shown as dark arrows; the one that is not significant is shown as a thin arrow.

Posada and Crandall (2001) compared movement in a likelihood ratio test lattice with use of the AIC and BIC criteria. In simulations in which the true model was known, the likelihood ratio test lattice did well and outperformed AIC and BIC. However, it is not always easy to see in which order to perform the likelihood ratio tests. In the example shown here, if one starts with K2P with $R = 2$, and first tests whether that parameter should be estimated, one might be tempted to abandon the effort to move further. A different route would lead to the top, including use of an estimated value of R. It is also not obvious what number of tests to use in computing the Bonferroni multiple-tests correction. If we start at the simplest hypotheses, we may stop after considering only some of the possible tests, so we would overcorrect if we corrected for the presence of all of them.

The problem of multiple topologies

We have so far avoided the issue of testing assertions about tree topologies, and for a reason. Naively, we could imagine taking two tree topologies, maximizing the likelihood for each, and comparing their likelihoods. How many parameters are there in each case? If we have an unrooted tree with n tips, there are $2n - 3$ branch lengths. These are the parameters (though there may be others as well that are parameters of the model of evolutionary change). One might want there to be parameters for the tree topology, but technically there are not. The tree topologies are discrete and cannot be considered to be parameters. The theory of the likelihood ratio is asymptotically correct with large amounts of data. In that case we are so close to the true tree that there is only one possible topology. Thus the theory gives us no hint as to how to test tree topologies.

LRTs and single branches

If the two trees that we compare are both fully resolved and unrooted, both have $2n-3$ degrees of freedom. Thus the difference in the number of degrees of freedom must be 0. We cannot do a likelihood ratio test between them, as neither will be a subcase of the other. Thus there is no easy way to do the likelihood ratio test between two trees.

One might think that we could do a conservative test by comparing each tree to an appropriate multifurcation, as I have myself suggested (Felsenstein, 1988b). This will not work. For example, suppose that we have one tree on which we see the groupings ((Human,Chimp),Gorilla). If we can reject the tree that is a trifurcation: (Human,Chimp,Gorilla), compared to ((Human,Chimp),Gorilla), does this mean that we can rule out other topologies such as ((Chimp,Gorilla),Human)? In fact, we cannot. The likelihood may actually have a peak within both of these bifurcating topologies, but be substantially lower at the trifurcation that is between them.

Figure 19.3 shows a simple numerical example of the presence of two peaks in adjacent tree topologies, for the case of a molecular clock. The example data has 64 sites and three species, A, B, and C. Of the sites, 37 are invariant, while the other 27 have nucleotide patterns xxy, xyx, and yxx in proportions $14:13:0$. There are no sites with pattern xyz. When analyzed with a Jukes-Cantor model, the results are as shown. Note that if we were to test the internal branch length t, we would find that the value 0 can be excluded, as it is more than 2.2 units of log-likelihood down from the peak. But there is another peak within the tree topology ((A,C),B), which is only 0.78 units lower than the highest peak.

We can test against the trifurcation with one degree of freedom, for the trifurcation restricts one branch length of the bifurcating topology to be 0. As the example shows, this does not exclude the possibility of alternative peaks, including fairly plausible ones. An example equivalent to this, and coming to the same conclusion, is given by Farris, Källersjö, and De Laet (2001).

If we cannot use the likelihood ratio test on the length of an internal branch to rule out nearby topologies, can we do a likelihood ratio test between them directly? Sadly, we cannot. Both bifurcating trees have the same number of branch lengths varying. The difference in their numbers of parameters is 0. We cannot do the test with 0 degrees of freedom. More to the point, the two topologies are not nested within each other. Neither of them is a subcase of the other. There is no way that branch lengths of ((Human,Chimp),Gorilla) can be restricted so as to obtain a tree of topology ((Chimp,Gorilla),Human). We could do this to get the trifurcation, but not to get an alternative bifurcation.

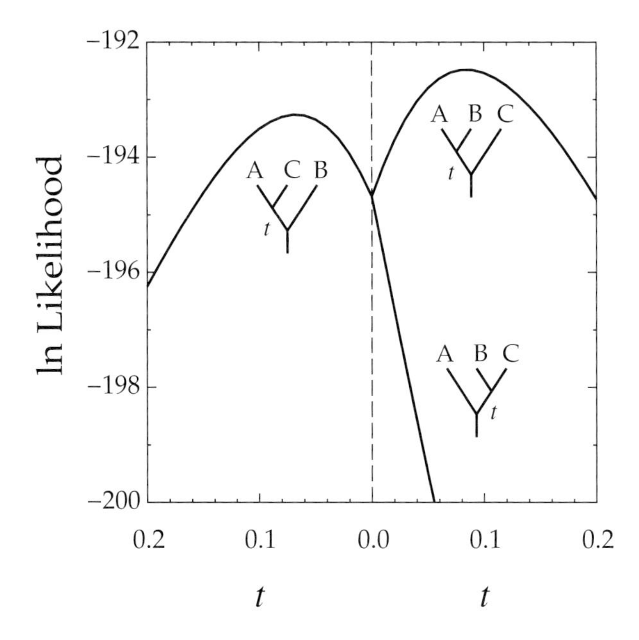

Figure 19.3: A numerical example of multiple peaks within different tree topologies, for a 64-site three-species case described in the text. Note that the scale of branch lengths runs backwards on the left-hand side of the graph. The log-likelihoods are plotted against the internal branch length t, for the case of a molecular clock. The log-likelihoods are profile log-likelihoods; that is, for each value of t, they show the highest log-likelihood that can be achieved by optimizing the value of the other branch length parameter.

Thus there is the open issue of how to make a likelihood ratio test of different tree topologies.

Interior branch tests

Another class of tests for tree topology are the *interior branch tests*. These were introduced by Nei, Stephens, and Saitou (1985). Other variations on interior branch tests are due to Li (1989), Tajima (1992), and Rzhetsky and Nei (1992). We have seen in Chapter 11 that for methods such as least squares, we can compute the variance on the estimate of the length of a branch in the interior of the tree, although at the expense of a fair amount of computation. In all of these papers, various approximations are made enabling quick computation of the variance of the length of an interior branch in the tree, without considering the topology of other parts of the tree. In these cases, the authors argue that when the length of a branch is significantly different from 0, we regard the reality of the branch as established.

Variances of branch lengths can be computed in distance matrix methods, and this has been done by Chakraborty (1977), who approximated sequence evolution by a Poisson process, assuming that the tree was obtained by a UPGMA clustering method. The likelihood approach mentioned above can also address the same questions. We have seen that we can plot likelihood against branch length. This allows all other branches to adjust their lengths in response to changes in the length of this branch. The curvature of the log-likelihood of the tree as a function of length of this branch allows us to compute an approximate standard deviation of the branch length. Alternatively, we can use the height of the log-likelihood rather than its curvature; this may be more accurate.

If we were to use this height to construct an equivalent of the interior branch test, we would accept the branch if the LRT against a branch length of 0 was significant (cf. Navidi, Churchill, and von Haeseler, 1992). In the numerical example I gave in Figure 19.3, the result would be misleading. Does this prove that interior branch tests are dubious? Although this case is disquieting, it is not strong evidence against interior branch tests. The data set was one concocted by hand, rather than occurring in nature or in a computer simulation. (A similar argument, also based on a concocted data set, is given by Farris et al., 1999.) Real data sets, especially real simulated data sets, would rarely have this strong signal for two of the three possible topologies, with lack of signal for the third. In general, the logic of the interior branch test will not be misleading, particularly if a branch length of 0 is strongly rejected.

Sitnikova, Rzhetsky, and Nei (1995) have compared the behavior of bootstrap tests and interior branch tests. In general, they found good concordance of their results: When an interior branch test recommended a branch strongly, so did the bootstrap. (Bootstrapping is covered in Chapter 20.) They argue for the superiority of the interior branch test, when one particular branch is of interest. Dopazo (1994) and Sitnikova (1996) used bootstrapping to calculate the variance of the length of an individual branch; Sitnikova found that this was more robust against variation in evolutionary rates among sites than was the standard interior branch test. She also presented corrections for the P value computed from the estimated branch length and its variance, to correct for the tree topology having been chosen on the basis of its having a positive estimate of that branch length.

Interior branch tests using parsimony

Although most of the literature on interior branch tests has used distance matrix methods and likelihood can also be used, there has been at least one attempt using parsimony. Sneath (1986) used parsimony reconstructions of the number of changes in branches of the tree. Taking an interior node of a bifurcating tree, he considered the three branches connecting to it. For each he calculated a standard deviation, assuming that the number of changes was drawn from a Poisson distribution, as it would be if these were observed changes and evolution were sufficiently clocklike. This enabled him to place a P value on the presence of the branch. When the length of one of the branches is not significantly different from zero, this is taken as an indication that rearrangement of the topology in that region of the tree is not ruled out.

Sneath's method is limited by its assumption that the parsimony method allows the reconstructed changes to be treated as observations. As with all the interior branch tests, it also treats branches in isolation without consideration of covariation in their statistical uncertainty.

A multiple-branch counterpart of interior branch tests

If a tree has b branches, each of which has a length, then when we test a proposed tree with given branch lengths, it restricts b parameters. Thus the trees (of that topology) that should be in the interval are ones whose log-likelihoods are less than the maximum by an amount that is half the size of a significant chi-square variable with b degrees of freedom. This is a more general counterpart of the interior branch test, in which one finds an interval that is supposed to contain the whole tree.

For example, if an unrooted tree has n tips, it has $2n - 3$ branches. For $n = 10$, that means that there are 17 branches. The value of a χ^2 variate that has tail probability $\alpha = 0.05$ with 17 degrees of freedom is 27.6. Thus all trees that come within $27.6/2 = 13.8$ of the log-likelihood of the maximum likelihood tree are within the interval.

This sounds interesting, but it has the same difficulty as the interior branch test — there may be trees of a different tree topology that have likelihoods in this range. The asymptotic theory that justifies this interval does not allow for multiple topologies, so that it is not clear how seriously to take the interval if it extends over multiple topologies. There is also the very real issue of how one finds and reports the trees that are in this interval. This is a serious computational task that has received too little attention as yet.

Testing the molecular clock

When tree topology is not an issue, we can consider how to test the hypothesis of a molecular clock. There are two forms of the clock hypothesis. One asserts that all lineages have the same rate of evolution, which does not change through time. A less restrictive version asserts that all lineages have the same rate of evolution, but do not prevent it from changing through time, as long as the changes apply simultaneously to all lineages. Here we will be primarily concerned with the latter. Unless we have sampled individuals at different times, we cannot distinguish between these two forms of the molecular clock. I briefly summarize here the large literature on testing the molecular clock.

Parsimony-based methods

The first attempt at a comprehensive statistical test of the molecular clock was made by Langley and Fitch (1973, 1974). Using protein sequences, they assigned amino acid replacements to branches in a known tree. They then used these as if they were known data, and estimated branch lengths by maximum likelihood. They then constructed a chi-square test of whether the number of substitutions on each branch were proportional to these clocklike branch lengths. In spite of this statistical sophistication, their test relies on the accuracy of the parsimony reconstruction, and is limited by it. A similar issue arises for the paper of Cutler (2000a), which extends Langley and Fitch's work to allow for an episodic clock that has bursts of substitution.

Distance-based methods

One can also use distances on trees fit by least squares, and use variance ratios to test for a clock. I gave such a test (Felsenstein, 1984) in which the increase in the sum of squares imposed by assuming a clock is to be compared with the sum of squares without a clock. This can be done if the two tree topologies are the same. The test is an F ratio with degrees of freedom $n - 1$ in the numerator and $(n - 2)(n - 3)/2$ in the denominator, when a triangular distance matrix is used. The difficulty with this test is that it assumes that the distances have independent noise, which will not be true if they are based on molecular sequences. Some correction for the covariances of the errors can be made, as we have seen in the discussion of generalized least squares in Chapter 11. In principle this could be used to make a more accurate distance-based test. But the computational burden would be heavy.

Likelihood-based methods

The likelihood ratio test can be used to test the molecular clock (Felsenstein, 1981b, 1988b). Suppose that we have estimated a phylogeny under a molecular clock and also one without it. Suppose further that these turn out to have the same unrooted tree topology. Restricting our search of topologies and branch lengths to those that reflect a molecular clock cannot result in a higher likelihood. If we have done the search properly, the search in the absence of a clock will consider all trees that have a clock as well as all trees that do not, so that it cannot find a worse tree. If we restrict our consideration to trees that are clocklike, we are searching in fewer dimensions. With n tips, the clocklike trees are specified by knowing the branch lengths down from the tips to each of the $n - 1$ internal nodes. The full tree has $2n - 3$ branch lengths, each of which can be varied, while a clocklike tree of the same topology has $n - 1$ lengths that can be independently varied. The test of a molecular clock thus has $(2n - 3) - (n - 1) = n - 2$ degrees of freedom.

If $n = 3$, there are 3 degrees of freedom in a nonclocklike tree, but only 2 in a clocklike tree. If the tree has topology $((A, B), C)$, the likelihood ratio test of

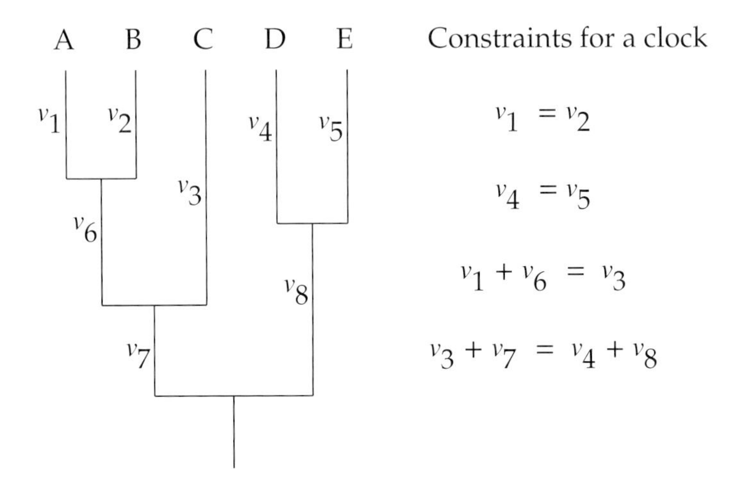

Figure 19.4: The constraints resulting from a molecular clock in a case with six species, in which there are 3 degrees of freedom constrained in enforcing a molecular clock. The bottommost constraint does not actually have an effect in restricting $v_7 + v_8$, so it is not counted.

the clock is in effect testing whether the branches leading to A and to B are the same length. That places one restriction on the three branch lengths, reducing the number of parameters free to vary to 2. The likelihood ratio test has one degree of freedom.

Figure 19.4 shows a tree with 6 species, and the branch length constraints that are involved in making it adhere to a molecular clock. Note that although there are four equations in Figure 19.4, the last one does not place a constraint on the value of one of the branch lengths in the unrooted tree.

The relative rate test

Sarich and Wilson (1973) introduced the idea of a *relative rate test*, testing whether the two ingroup branch lengths (such as the ones leading to A and to B in the above example) are equal. They did not develop it as a statistical test. This has been done by Wu and Li (1985). They considered the number of sites at which sequences A and C differ, the number of sites at which B and C differ, and they gave an approximation for the variance of the difference between the differences. They also gave similar expressions for the differences based only on transitions. An approximate normality of the difference of differences is justified in most cases, and their simulations showed that the variance approximations were appropriate. Tajima (1993) presented a simplified test that simply counted whether the patterns xyx and yxx occurred in significantly different numbers, where the species are in the order ABC. He also presented extensions that distinguished between transitions and transversions.

Muse and Weir (1992; see also Weir, 1990) addressed the problem in a likelihood framework, using the likelihood ratio test mentioned above, with one degree of freedom. They showed by simulation advantages of the likelihood ratio test when rates of transition and transversion differed greatly.

Extension to multiple species. The difficulty with the relative rate test is that when there are many species, one has many tests to perform if all triples are considered, and there will be considerable nonindependence of these tests. Tajima (1993) shows a table with many tests performed for a 6-species data set. When some of these are significant, it is hard to know how to combine them to infer where in the tree there are rate inequalities, and how to correct for multiple tests. Li and Bousquet (1992), Takezaki, Rzhetsky, and Nei (1995), and Steel, Cooper, and Penny (1996) showed ways to compute the variance of a relative rate test statistic, when the statistic compared the average distances of members of two clades to an outgroup. The latter authors also showed how to narrow confidence intervals on the timing of the root of the tree when there are multiple species in the data. Robinson et al. (1998) presented a method for weighting these averages to take the internal structure of the clades into account and presented simulations showing that this improved the statistical power. A full likelihood approach would be better yet, allowing all forms of nonindependence in the data to be taken into account. We have seen above that a likelihood ratio test can make a simultaneous test of the whole tree for clockness.

Discovering local violations of clockness. Bromham et al. (2000) showed by simulation that relative rate tests between triples of species have relatively poor power to detect lineages that are evolving at a different rate. We may be interested in tests that focus on particular parts of the tree or that explore which parts have the least clocklike behavior. It would in principle be possible to take any partition of the species into two or more groups and test the hypothesis that tips within each partition have equal height. Takezaki, Rzhetsky, and Nei (1995) suggested such an approach. For each nonroot interior node in a rooted bifurcating tree, they test whether the two clades descended from the node have equal rates of evolution. They use average distance calculations and use the structure within each clade to calculate covariances of the distances. This is not a likelihood approach but does take the structure of the tree partly into account. They suggest a sequence of such tests coming up from the root, with clades being eliminated from the tree if they have rates of change sufficiently different from the rest. Thus groups that were judged clocklike at one level could later be judged to have subclades that did not adhere to that clock. They eliminate groups; the tree that remains has its branch lengths re-estimated under the constraint of a clock. They call this a "linearized tree".

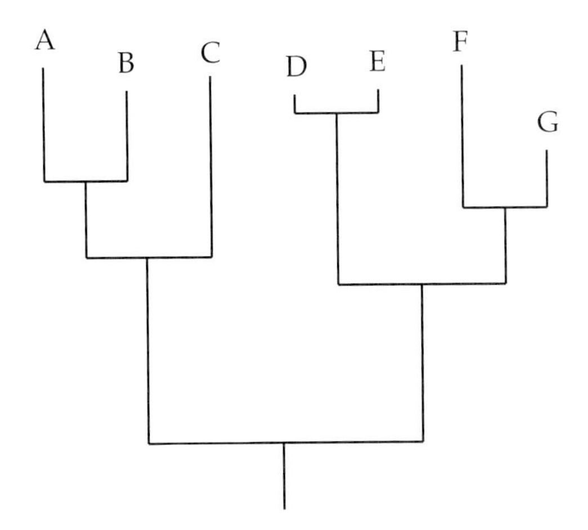

Figure 19.5: A tree with seven species, used as an example in the discussion of testing various parts of it for a molecular clock.

A similar sequence of tests could be done coming down from the tips, using likelihood ratio tests or other tests. As an example, in Figure 19.5 we could ask whether the sets of species {A, C, F}, {D, E}, and {B, G} represent three different heights, with A, C, and F all at the same height, D and E at another height, and B and G at still another. There is a large lattice of such tests, one for each partition of the set of 7 species into sets. Some of these are restrictions on others. For example, the test for these three sets is a further constraint on the test of the pair {D, E}. The least constrained tests are the one with all species in different subsets (no clock at all) and the one with all species in the same subset (one overall clock).

This framework raises all sorts of interesting questions about how one would search among all possibilities and do tests in a way that allowed for the large number of tests possible. But perhaps this approach is not worth pursuing. If we ask for these tests where the inequalities of rates in the tree would have to be, it becomes immediately apparent that many of the possibilities are not biologically sensible. If two of the subsets have their species connected by lineages that overlap, then those lineages will be evolving according to two different clocks at the same time. For example, the pairs {A, C} and {B, G} are connected by paths that overlap, and we cannot assume that those branches simultaneously have two different rates of evolution.

It becomes apparent that we should be interested only in subsets connected by nonoverlapping paths. One combination of tests that seems natural follows the hierarchical structure of the tree. Table 19.2 shows the sequence of tests we would do, and which ones are possible only after others show clockness. The tests are, in effect, one for each interior node of the tree, and each test is only meaningful if all the tests for nodes above it in that subtree are passed. Note that there is no test for

Table 19.2: Tests of clockness in different parts of the tree in Figure 19.5, showing for each which previous tests must have shown clockness for it to be meaningful.

Test number	Tests equal rate in set	Meaningful if pass test(s)
1	{A, B}	–
2	{A, B, C}	1
3	{D, E}	–
4	{F, G}	–
5	{D, E, F, G}	3 and 4

the bottommost node of the tree, as there is no outgroup allowing for relative rate tests for pairs of species in the different subtrees there.

The five tests account for the 5 degrees of freedom ($n - 2$) in the overall clock test. They can each be done by a likelihood ratio test that constrains the tips in that set to be at a common height. As there are multiple tests being done, a correction such as the Bonferroni correction would need to be done, to ensure that the overall probability of rejection of a clock in some region of the tree would be less than the desired value such as 0.05.

There is more to be done on this subject, but I must leave that as an exercise for the readers.

Using relaxed clocks. I have noted, in Chapter 16 that a number of authors have discussed reasons why the clock may be violated, and that some of them have put forward models of relaxed clocks. Likelihood analysis is difficult under relaxed clock models. One has to allow for bursts of evolution within sites, and also for the presence of these bursts to be correlated in different sites. This nonindependence of evolution in different sites is particularly troublesome. One approach that has been used is to assume that a parsimony method correctly reconstructs how many changes have occurred in each site in each branch of the tree. This is the approach taken by Langley and Fitch (1973, 1974) and later by Sanderson (1997, 2002) and by Cutler (2000b). These papers make statistical tests of constancy of the rate of change, using the parsimony reconstructions as if they were data.

A more tedious, but more accurate, approach is to use Markov chain Monte Carlo techniques. In effect, these average over various patterns of rates along the tree. The variation of rates is addressed by letting the MCMC algorithm assign rates to different parts of the tree, sampling from the patterns that are possible. We have discussed above (in Chapters 16 and 18) the papers of Thorne, Kishino, and Painter (1998), Huelsenbeck, Larget, and Swofford (2000), and Kishino, Thorne, and Bruno (2001) that use MCMC methods to carry out Bayesian analysis of models in which rates of evolution vary through time. These analyses allow more fully for the uncertainty of reconstruction of placement of substitutions. By inferring posterior distributions for the parameter that affects departure from a clock, these papers can test the molecular clock against a model that departs from a clock.

Simulation tests based on likelihood

We have seen that the standard Likelihood Ratio Test is often wrong or dubious, for two reasons. For tests of different tree topologies, the hypotheses are not nested properly and thus do not satisfy the assumptions of the LRT. Even when the hypotheses are nested properly, the use of a chi-square distribution may be unwarranted if the sample size is not large enough for the asymptotic approximations to be valid.

Goldman (1993) has proposed using computer simulation to find the distribution of the LRT statistic, using a method first introduced in other statistical contexts by Cox (1961). For the case of a relative rate test, it would work like this:

1. Infer the phylogeny with and without a clock, calling these trees $\widehat{T}_c$ and $\widehat{T}$ and their log-likelihoods ℓ_c and ℓ.
2. Use the null hypothesis that the tree is clocklike, by taking the tree to be our best clocklike estimate, $\widehat{T}_c$, and simulating a large number R of data sets using the same statistical model of evolution.
3. For each these simulated data sets, find $\widehat{T}_c$, $\widehat{T}$, and their likelihoods.
4. Consider the distribution of the differences of log-likelihood $\ell - \ell_c$ for all R replicates. Include as well this difference of the log-likelihoods computed from the original data.
5. If the original difference is in the upper 5% of these $R + 1$ differences, we may reject the hypothesis of clockness at the 95% level.

In effect, Goldman's test uses simulation to empirically tabulate the distribution of the log-likelihood ratio test statistic and uses that empirical tabulation instead of the chi-square distribution. If the chi-square distribution were actually adequate, then we would also expect Goldman's empirical distribution to be close to it. When we are far from asymptopia, one cannot prove that this simulation test makes optimal use of the data. Perhaps a test statistic different than $\ell - \ell_c$ would be better, but I would guess that this LRT statistic does about as well as can be done.

Huelsenbeck and Bull (1996) have suggested using a simulation test to determine whether two data sets suggest different trees for the same group.

In the next chapter, when we discuss the bootstrap method, we will see that these simulation tests are in effect an application of the method called the parametric bootstrap. As in that case, the simulation test relies heavily on the approximate correctness of the stochastic model of evolution. That reliance is shared by all uses of the Likelihood Ratio Test.

Further literature

For a more extensive discussion of the use of likelihood ratio tests of models and tree topologies, and of the use of computer simulation to generate the distribution of likelihood ratio test statistics, the reader is referred to the review article by Huelsenbeck and Crandall (1997). Posada and Crandall (2001) have argued on grounds of theory and simulation for use of a hierarchical lattice of models of evolution, with likelihood ratio tests used to select a model out of a complete set of models.

More exact tests and confidence intervals

Likelihood ratio tests are useful and powerful, but we may wish to derive tests or confidence intervals from first principles, in order to avoid assuming that we are in asymptopia. Cavender (1978) has been the pioneer of direct approaches to statistical testing. He took the parsimony statistic as the relevant one; he examined its distribution on all possible tree topologies in a four-species case without a molecular clock. As he was using the symmetric 0/1 model, it turns out that in the worst case, long branch attraction could cause as many as one-third of the characters to misleadingly favor the wrong tree topology. Cavender therefore argued that if the number of characters favoring a tree was significantly greater than 1/3, they could not be the result of long branch attraction. For nucleotide sequences, the analogous worst case has 3/16 of the characters favoring a particular wrong topology. Cavender's argument came at the same time as my own discovery of long branch attraction (Felsenstein, 1978b) and must be considered an independent discovery of that phenomenon.

We can take the primate data set of Hasegawa (which we have used earlier in this chapter) as an example. If we consider only the four species Human, Chimp, Gorilla, and Orang, we find that ((Human,Chimp),(Gorilla,Orang)) is the best-supported unrooted tree topology. Of the 232 characters, 24 have differences in the number of changes on the three tree topologies, and of those, 12 support this tree and 6 support each of the two alternative ones. The fraction 12/232 is far less than 3/16, so that in this case Cavender's test would not distinguish between the tree topologies.

We might hope to make Cavender's test more powerful by discarding invariant sites. In our example, this raises the fraction 3/16 to 3/15, which does not change the result much. If we are more daring and discard all sites that show no difference between the topologies (leaving us with only the "phylogenetically informative" sites), we pay a terrible price. The worst case would be one that has all of the phylogenetically informative sites backing the wrong tree topology. Thus we would not be able to reject any topologies using this approach.

Tests for three species with a clock

We can make the test more powerful by assuming a molecular clock. In the mitochondrial data example, we could assume that the clade of Human, Chimp, and Gorilla follows a molecular clock. I have investigated tests based on simple statistics in such cases (Felsenstein, 1985c). In those cases we can ignore all sites that do not vary, as well as those that show three different states in the three species. In the case of a clock, this leaves us with those sites that show patterns xxy, xyx, or yxx. For this example, there are 85 such sites, and these patterns occur 33, 29, and 23 times, respectively.

This is support for the ((Human,Chimp),Gorilla) tree. For the other two possible rooted bifurcating trees, the closest fit to these data would be obtained by assuming that the interior branch was as short as possible, so that the tree was in effect a trifurcation. In a trifurcation, all three patterns would be expected to occur equally often. I investigated two statistics. One was S, the difference between the number of steps favoring the best tree and the next best tree. For this case, we would have $S = 33 - 29 = 4$. The other statistic C, was simply the number of sites favoring the best tree (here $C = 33$).

A confidence interval is constructed by taking a statistic and, for each possible hypothesis, finding the set of values of the statistic that are in the extreme tail for that statistic. Thus, for the S statistic, we want to know for each possible tree what the probability is that S exceeds the observed value of 4. The confidence interval is obtained by "inverting the test," finding all those trees for which 4 is not an extreme value. A tree could be excluded from a 95% confidence interval if the value of S favored another tree by enough that a value that large in favor of a wrong tree would occur less than 5% of the time.

In this simplified case, I assumed that the worst case had a zero-length internal branch, so that there was a trifurcation (in fact, this can be proven). We need to know for $S = 4$ whether the probability of favoring one of the two wrong trees by four steps or more is less than 5%. In this simple case that can be done by enumerating all possible cases and working out by direct tabulation the probabilities of favoring the wrong tree by four steps or more. Taking the probabilities of the three outcomes to be equal, the probability that we get n_1, n_2, n_3 in the three classes is simply

$$\mathrm{Prob}\,(n_1, n_2, n_3) = \binom{n}{n_1\; n_2\; n_3} \left(\frac{1}{3}\right)^{n_1+n_2+n_3} = \frac{n!}{n_1!\; n_2!\; n_3!} \left(\frac{1}{3}\right)^{n} \tag{19.17}$$

Table 19.3 shows values of the S and C statistics that are significant at the 95% level.

Bremer support

The S statistic measures the difference between the number of changes for a tree with a particular group and the number without it. I did not put this statistic forward for general use in situations with more than three species and/or without a

Table 19.3: Values of the S and C statistics that are required to have the boundary of the 95% confidence limits in a three-species tree with a clock exclude the two other tree topologies.

Characters	S	C	Characters	S	C
4	4	4	21	7	12
5	5	5	22–23	7	13
6	4	6	24–26	7	14
7	5	7	27–28	7	15
8	4	6	29	7	16
9–10	5	7	30	8	16
11–12	5	8	31–33	8	17
13	5	9	34–35	8	18
14	6	9	36–38	8	19
15–16	6	10	39	8	20
17–19	6	11	40	9	20
20	6	12	50	9	24

clock. Bremer (1988) suggested an index (called by various authors *Bremer support*, the support index, or the decay index). He used it for all groups on a tree. For each group he asked what tree which did not have the group had the fewest changes of state. If the most parsimonious tree that had group ABC had 138 changes of state, and the most parsimonious tree that lacked that group had 143 changes, the Bremer support for that group would be $143 - 138 = 5$. Bremer support has no immediate statistical interpretation in the general case. With three species and a clock, the above table shows that the level of Bremer support that is significant depends on the number of characters that illuminate that trichotomy. As we see later in this section, we cannot specify what level of Bremer support is statistically significant in more general cases. Gatesy (2000) has suggested a measure of joint Bremer support for two branches of a tree.

Zander's conditional probability of reconstruction

The three-species clock calculation bears comparison with the test used in Zander's (2001) Bayesian calculation of posterior probabilities of different ways of resolving a trichotomy. Zander takes a tree inferred by parsimony and, for each internal branch, considers the two possible nearest-neighbor interchange (NNI) rearrangements of the tree at that branch. For each he reconstructs by parsimony the number of changes seen on the interior branch. Suppose that the original tree shows 60 changes in the interior branch, but only 37 and 30 in the corresponding interior branch when we do the two NNIs. He then assumes that these three numbers can be treated as observations, and uses a chi-square test with three classes to test these three numbers against equality. If they significantly depart from equality, he does a subsidiary test of the two smallest of the three numbers to see whether they are significantly unequal. In this example, the three numbers are significantly unequal and the inequality of 37 and 30 is not significant. He then uses the tail probability of the former chi-square as the probability that the tree is correct at this point. For these numbers, the chi-square value would be 11.64, with 2 degrees of freedom. A chi-square variate with 2 degrees of freedom would be smaller than or equal to this 0.997 of the time, and Zander takes the posterior probability of the original resolution to be 0.997.

We can compare Zander's test to the confidence limits in the three-species clocklike case. In the numerical example there would be 60, 37, and 30 characters supporting the three trees. With 127 characters in all, the values of C and S that are significant will be 54 and 14. In this case the values of C and S are 60 and 23, which is strongly significant, as Zander's test was as well. The probability of exceeding $C = 60$ is 0.00159, which is about half the tail probability that Zander gets. The two tests are thus similar, though not identical. Zander intends his test to be used in nonclocklike cases and for all interior branches of a tree. This cannot be justified, as the expectation of equal numbers of changes in the interior branches of all three trees will not hold in such cases, and in the extreme, long branch attraction may make his test declare the wrong tree significantly supported.

In cases where the three numbers of changes are small enough to make the chi-square calculation dubious, Zander takes the posterior probabilities of the trees to be proportional to the numbers of changes in the interior branches of the three trees. This leads to some curious results, such as the statement that when there is 1 step in one of them and 0 in the two others, the posterior probability for the best tree is 100%. This will not be similar to the confidence statements in the clocklike three-species case, and looks difficult to justify. I have already commented in Chapter 18 on the difficulties of interpretation of Zander's posterior probabilities as Bayesian posteriors.

More generalized confidence sets

Can this result be generalized? Only a little is known. It is not simple to generalize it to resolving a trifurcation in the interior of a larger clocklike tree. In that case the lineages leading from the trifurcation are of unequal length and then split into various numbers of lineages. The application of the S and C statistics is not simple, especially since many characters may vary within each of the lineages. Inequalities of the lengths of the lineages have a serious effect on the use of statistics like S and C. It is no longer true that, when there is a trifurcation, characters supporting the three possible tree topologies are equally frequent. If the lineages leading to A and to B were long, and that leading to C was short, we expect more characters in which A and B share the same derived state than we do for any other pair of lineages. In the extreme case in which there is long branch attraction, we will be back in Cavender's worst case. Williams and Goodman (1989) suggest that the three-species test using the C statistic is valid even when there is no molecular clock; I do not see why it would then be valid.

Even in the case of three species with a clock, we do not know that the S and C statistics are the most effective for delineating confidence sets. Some results can be obtained for the C statistic that at first seem to offer hope. Using the classic confidence region machinery due to Neyman and Pearson, one can examine testing of one hypothesis against another. Neyman and Pearson pointed out that if one wanted a confidence region with (say) a 95% probability, the most powerful test would exclude those regions that had the greatest ratio of the probability of that point (or density there) under the alternative hypothesis to that under the null hypothesis. Thus one wants to use the ratio of the likelihoods under the two hypotheses, excluding those regions that had the highest ratio.

In our three-species example with only the three character patterns used, this likelihood ratio for topology ((A,B),C) versus topology (B,(A,C)) is

$$\frac{\text{Prob}\,(n_1, n_2, n_3 \mid p_1, q_1)}{\text{Prob}\,(n_1, n_2, n_3 \mid p_2, q_2)} = \frac{p_1^{n_1}\; q_1^{n_2}\; q_1^{n_3}}{q_2^{n_1}\; p_2^{n_2}\; q_2^{n_3}} \tag{19.18}$$

This is simpler as a log-likelihood:

$$\ln\left(\frac{\text{Prob}\,(n_1, n_2, n_3 \mid p_1, q_1)}{\text{Prob}\,(n_1, n_2, n_3 \mid p_2, q_2)}\right) = n_1 \ln\left(\frac{p_1}{q_2}\right) + n_2 \ln\left(\frac{q_1}{p_2}\right) + n_3 \ln\left(\frac{q_1}{q_2}\right) \tag{19.19}$$

If the null hypothesis were the trifurcation $p_2 = q_2 = 1/3$, then the likelihood ratio depends only on $n_1 - (n_2 + n_3)$, which is $2n_1 - n$. Thus the best statistic to define the test and the confidence region is the compatibility statistic n_1. A 95% region is defined by taking the values of n_1 that are in the lower 95% of its distribution. It is most powerful against all alternative values of p_1. That most powerful test is then described as "uniformly most powerful."

Before we get intoxicated by this success, we have to remember that it is only for the trifurcation tested against one particular alternative topology. When the other topology is considered, it is reasonable to use the analogous statistic and to have the rejection regions guard equally against the two alternative trees. But the resulting test is then not uniformly most powerful, because for any one alternative tree, the most powerful test is one that does not take any account of the other possible alternative topology. Tests (and their associated confidence regions) that allow for both alternative topologies cannot be as powerful against one of them as a test specifically designed for it.

Equation 19.19 shows that even for a comparison of two specific tree topologies, there is no uniformly most powerful test or confidence region. For example, if we have data of $n_1 = 60$, $n_2 = 25$, and $n_3 = 15$, and if we compare a tree of topology ((A,B),C) with one of topology (B,(A,C)), if $p_1 = p_2$, the log-likelihood ratio is a function of $n_1 - n_2$. But if instead $p_2 = 1/3$, it is a function of only n_1, as we saw earlier. So the most powerful statistic differs, depending on which trees are being compared. Any confidence region we form using one statistic is necessarily a compromise based on different possible alternative trees.

All of this considers only three of the five base patterns. If we take all of them into account, the news cannot improve. And there the matter rests. There has been no recent work on confidence regions for trees beyond the work reported here. Though there is little reason to expect easy and powerful methods, the problem does bear more looking into.

Chapter 20

Bootstrap, jackknife, and permutation tests

As likelihood requires us to believe the probability model of evolution, it may underestimate the amount of uncertainty about the tree. It would be desirable to have a less parametric approach to testing phylogenies. Bootstrap, jackknife, and randomization tests are one way to be less dependent on a complete parametric model. They use empirical information about the variation from character to character in evolutionary processes. A second reason for using these resampling techniques is that they allow us to infer the variability of parameters in models that are too complex for easy calculation of their variances.

Bootstrap and jackknife tests on phylogenies started with the work of Mueller and Ayala (1982), who used a jackknife approach to estimating the variance of the length of a branch in a UPGMA phylogeny from gene frequency data. This was followed by my own paper on the bootstrap (1985b) and those of Penny and Hendy (1985, 1986), who used random partitioning of the characters into two halves.

The bootstrap and the jackknife

The jackknife and bootstrap are statistical techniques for empirically estimating the variability of an estimate. They differ, but are of the same family of techniques. The *jackknife*, which is the older of the two, involves dropping one observation at a time from one's sample, and calculating the estimate each time. The variability of the estimate is then inferred from the rather small variations that this causes, by an extrapolation. The *bootstrap* involves resampling from one's sample with replacement, and making a fictional sample of the same size. We start by giving a general explanation of the bootstrap, and then consider how it can be applied to phylogenies.

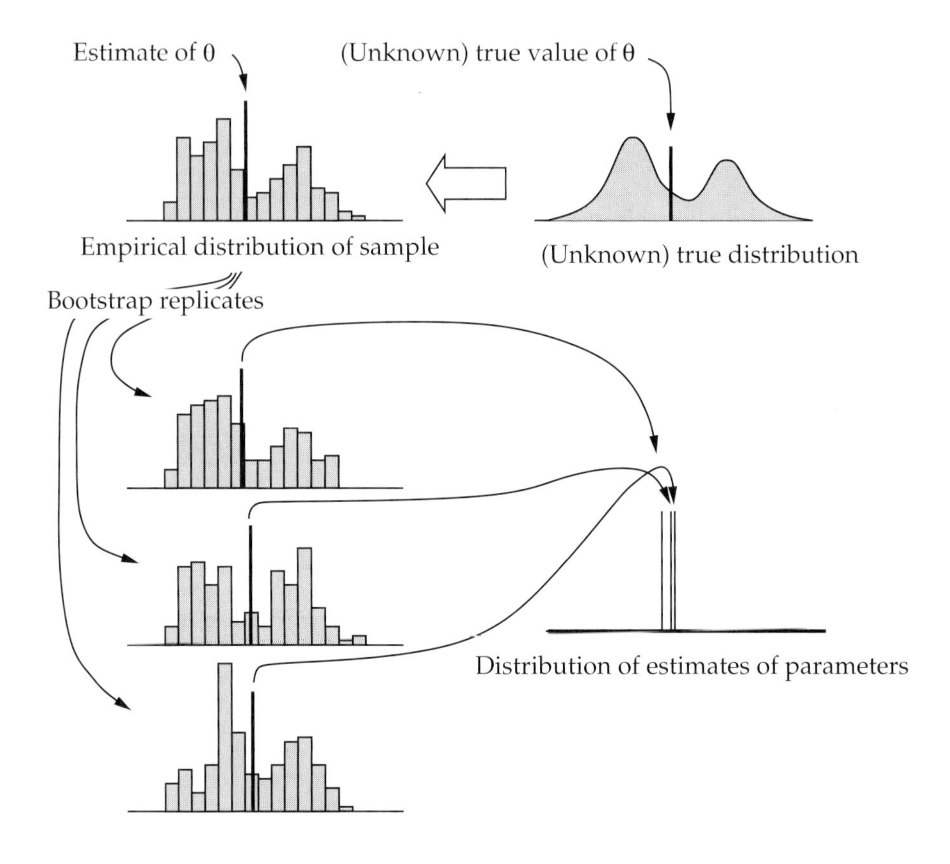

Figure 20.1: The bootstrap. The distribution of independent data items is taken as an estimate of the unknown true distribution. In this case the true distribution is a 60:40 mixture of two normal distributions, with means 7 and 14 and standard deviations both 1.5. By drawing samples of size n (in this case $n = 150$) from it and analyzing these, we can approximate the kinds of variation in our estimate that would be seen if we could draw new samples of that size from the unknown true distribution. The parameter estimated in this example is the population mean.

The bootstrap was invented by Bradley Efron (1979) as a general-purpose statistical tool analogous to the jackknife. Figure 20.1 shows a diagram of the method. Imagine that we have some data points $x_1, x_2, x_3, \ldots, x_n$ that are drawn independently from a distribution $F(\theta)$, that depends on an unknown parameter, θ. From them we are computing an estimate $t(x_1, x_2, \ldots, x_n)$ of the parameter θ. We would like to know the variability of the distribution of these estimates. If we knew the family of distributions from which F came, and if the estimator $t(\mathbf{x})$ were mathematically tractable, then we could know the distribution of estimates and how it depended on the true θ. For instance, when $F(\theta)$ is a normal distribution with mean θ and variance 1, and $t(\mathbf{x})$ is simply the sample mean, we know precisely what the distribution of estimates of θ is for every possible value of θ. (It is normal, with mean θ, and variance $1/n$.) That helps greatly in understanding what an estimate of θ implies.

However, we may not know the distribution F, or the estimator $t(\mathbf{x})$ may not be mathematically tractable. Efron's insight was that in this case, if the sample size n is sufficiently large, we can consider the empirical distribution of data in our sample (which we can call $\widehat{F}$) to estimate the true distribution F. Of course, the overall estimate of θ is not precisely correct, but the kinds of variation that the collection of values $x_1, x_2, \ldots, x_n$ display should be typical of the variation we would see in any large sample from the true distribution.

We would like to know what variation we would see in the estimate, $\widehat{\theta}$, if we drew new data sets of size n from the unknown distribution and analyzed them in the same way. The bootstrap infers this variation by using our current data set, by drawing new data sets not from F but from the empirical distribution $\widehat{F}$ in our data. Drawing a sample of size n from the empirical distribution is the same as drawing a sample of points $x_1^*, x_2^*, \ldots x_n^*$ from the existing data, drawing them independently, and sampling *with replacement*. If we instead sampled n points without replacement, we would simply end up drawing each point once, and we would back get our original data, although the points would be in a different order. This would not result in a different estimate of θ. But drawing with replacement means that points in the original data may be sampled different numbers of times. Some may be sampled twice, some once, some not at all (and some larger numbers of times). The numbers of times each one is drawn, $m_1, m_2, \ldots, m_n$ is a sample from a multinomial distribution with n classes that have equal probabilities of being drawn.

This sample $\mathbf{x}^*$ is called a *bootstrap replicate*. Each such replicate can be analyzed using the estimator t to get $\widehat{\theta}^* = t(\mathbf{x}^*)$. To get a picture of the variation of the estimates θ, we draw many different bootstrap replicates and infer θ from each one. The amount and kinds of variation in the resulting cloud of estimates of θ is then taken to be typical of the kinds of variation we would see if we could somehow sample new data sets from the unknown distribution F. For many well-behaved distributions and many well-behaved estimators $t(\mathbf{x})$ there are theorems assuring us that this picture of the variability of $\widehat{\theta}$ will be accurate, if n is large and if a large number of bootstrap replicates are taken.

Bootstrapping and phylogenies

To use the bootstrap to assess the uncertainty of our estimate of the phylogeny, the data should be a series of independently sampled points. We typically have,

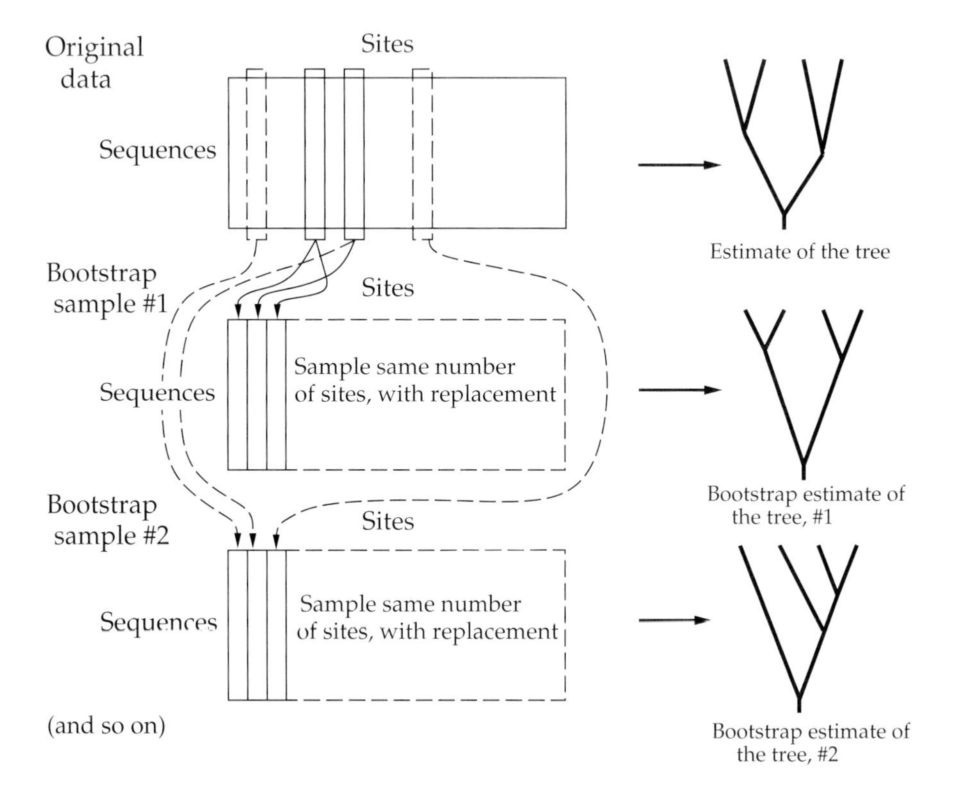

Figure 20.2: The bootstrap for phylogenies. Sites (or characters) are drawn independently, with replacement, with the order of species within each column of the data matrix remaining the same. Data sets with n characters are drawn, and each is analyzed to infer the phylogeny. The resulting sample of phylogenies should show approximately the same variation as a sample obtained by collecting n new sites for each tree.

instead, a matrix of species × characters. We cannot consider the species to be independent samples — they come instead as tips on an unknown phylogeny, some closely related to each other. In fact, the whole point of our analysis is to discover this structure. The characters (or sites) are a better candidate for being independent samples. If different characters evolve independently on the same phylogeny, they will satisfy the independence assumptions of the bootstrap, since the outcome of evolution at each character cannot be predicted from that in neighboring characters. Of course, evolutionary outcomes and processes in different characters may be related, in which case the independence assumption is incorrect. We return this subject later in this chapter.

To apply the bootstrap, we sample whole characters from the set of n characters, with replacement, and do so n times. The result is a data matrix with the same number of species and the same number of characters as in our original data matrix. Some of the original characters may have been sampled several times, others left out entirely. Figure 20.2 shows the process. For each data matrix we use our favorite phylogeny method to infer the phylogeny. The method may be a parsimony, distance matrix, or likelihood method. If a distance matrix method is used, the resampling occurs on the original character data (or sequences) before the distance matrix is computed. We end up with a collection of different estimates of the phylogeny. Some methods may give us more than one estimate of the phylogeny (parsimony methods, for example, often find multiple trees that are tied for best). In such cases we can consider that if 10 tied estimates are found for one bootstrap replicate, we consider each to be one-tenth of a tree, so that the results from that bootstrap replicate are not overemphasized when the trees are combined.

The delete-half jackknife

Other resampling methods are possible, and may have approximately equivalent behavior. The *delete-half jackknife* (e.g., Wu, 1986; Felsenstein, 1985b) is one, which has many of the same properties as the bootstrap. It involves sampling, not n times with replacement, but $n/2$ times without replacement. Thus we are taking a random half of the characters. Actually, if there are r parameters being estimated for each sample, we are supposed to take a random fraction $(n + r - 1)/2$ of the characters. For largish n this will not make much difference, and it is hard to know what the value of r is for a phylogeny. The matter needs a closer examination.

One way to put the bootstrap and the delete-half jackknife into a common context is to consider them as randomly reweighting the data. Drawing a bootstrap sample is equivalent to putting new weights on the original data, with the weight on character i being the number of times, m_i, that it is sampled in the bootstrap. As noted above, the weights m_i have a multinomial distribution, with n trials and equal probabilities for all n characters. It is not hard to show that the mean weight of a character is then 1, and the variance of the weight is $1 - 1/n$. Their coefficient of variation (the ratio of the standard deviation to the mean) is then $\sqrt{1 - 1/n}$, which is nearly 1.

A jackknife that deletes a fraction f of the characters can be thought of as weighting the deleted characters 0 and the included characters 1. This implies a mean weight per character of $1 - f$ and a variance of $f(1 - f)$. The coefficient of variation is then $\sqrt{f/(1 - f)}$. When $f = 1/2$ we have a coefficient of variation of 1. It can be shown that any random weighting scheme that achieves the same coefficient of variation will also approximate the bootstrap.

It is not clear whether the delete-half jackknife has any substantial advantages over the bootstrap.

Farris et al. (1996) have advocated using a delete-$1/e$ jackknife together with a parsimony estimate of the phylogeny (their "Parsimony Jackknife"). $1/e$ is

0.36788, so this amounts to deleting substantially fewer characters, so that groups will appear to have more support than they would under a delete-half jackknife (or a bootstrap). We can evaluate this method by checking its behavior in a case where exact computations can be done. Suppose that we have 100 characters, 10 of which back group I, and 8 of which back group II, these two groups being incompatible. The other 82 characters do not discriminate between the two alternatives. We can calculate by exact enumeration of outcomes, calculating the probability of each, that a bootstrap sample will favor the first group 0.63836 of the time, with a tie 0.08461 of the time. It seems fairest to count the resampling as favoring the first group half of the time when there is a tie. This will be $0.6386 + (0.08461)/2 = 0.68066$ of the time. If we do a delete-half jackknife, the corresponding number is 0.67555, while in a delete-$1/e$ jackknife that samples 63 characters it is 0.72402. Thus the delete-half jackknife gets results much more consistent with the bootstrap.

Farris et al. chose delete-$1/e$ based on the behavior when all the support is for group I. If two characters support group I and none group II, then the probability of favoring group I is 0.86738 for the bootstrap, 0.75253 for the delete-half jackknife, and 0.86545 for the delete-$1/e$ jackknife. However, this match between the delete-$1/e$ jackknife and the bootstrap vanishes quickly as more characters favor group II. With just a few of them, the delete-half jackknife becomes closer to the bootstrap. Of course, if the bootstrap is to be the standard, this speaks in favor of using it instead.

The bootstrap and jackknife for phylogenies

Once we use the bootstrap (or the jackknife) to resample characters, we will have a cloud of trees, the results of estimating the phylogeny for each bootstrap or jackknife replicate. In the simple case of estimating a real-valued parameter, we can make a histogram of the estimates. How are we to do this with phylogenies? They have discrete topologies, but continuous branch lengths. We could use the bootstrap to make a histogram of branch lengths, but only if the branch in question existed in all of our estimates of the phylogeny. We might then, for example, make an interval estimate of the branch length by finding the upper 95% of the branch length histogram, so that we could infer a lower limit on the branch length.

If this lower limit were positive, we would then be asserting the existence of that branch. But suppose that the branch is missing in some of the bootstrap (or jackknife) estimates of the phylogenies. It seems reasonable to assume that those cases can be lumped with ones that have a zero branch length for this branch. If we do that, then we can assign the probability P to the branch if a fraction P of the bootstrap (or jackknife) replicates have the branch present. In cases where there are several tied trees in a bootstrap (or jackknife) estimate, some with the branch and some without, we can count each one as conferring fractional support for the branch. An alternative, and equivalent, way of looking at this is to imagine an indicator variable that is 1 if the branch exists in the bootstrap (or jackknife) esti-

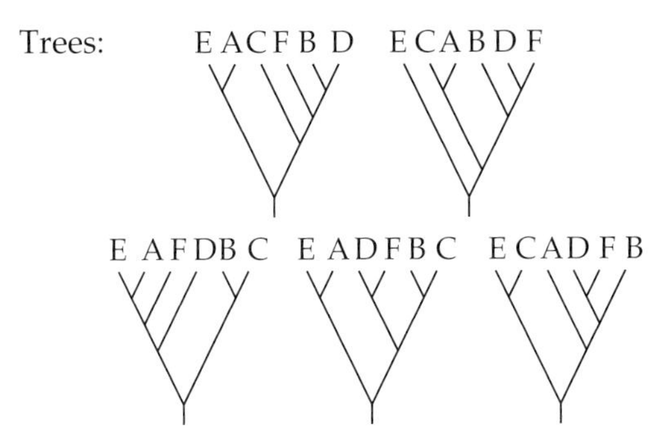

Number of times each partition of species is found:

AE | BCDF	3
ACE | BDF	3
ACEF | BD	1
AC | BDEF	1
AEF | BCD	1
ADEF | BC	2
ABDF | EC	1
ABCE | DF	3

Majority−rule consensus tree of the unrooted trees:

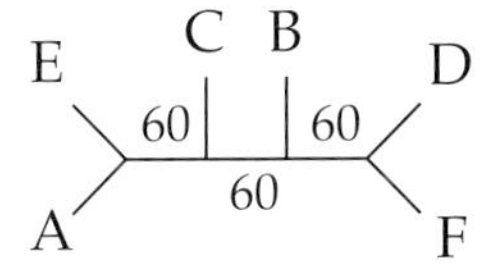

Figure 20.3: A set of five trees and their majority-rule consensus tree, with the percentage of support for each interior branch shown. Note that the majority-rule consensus tree is not identical to any of the five trees. Although shown here as if rooted, the trees are considered unrooted in the computation.

mate, and 0 if it does not exist. If we make a histogram of this indicator variable, and make an interval estimate for it by finding the upper 95% of this histogram, then when the branch appears more than 95% of the time, the upper 95% confidence interval contains only cases in which the branch is present, and so we can place a P value of 0.95 or greater on the hypothesis that the branch is present.

To implement this method, we must scan through the bootstrap or jackknife estimates of the trees, tabulating how often each branch occurs. We are only interested in ones that occur a large fraction of the time. If we have many branches that are of interest to us, keeping track of all of this is a tedious task. Fortunately, there is a consensus tree method that helps with this. Margush and McMorris (1981) defined the M_ℓ family of consensus tree methods. One of these is the *majority-rule consensus tree.* A consensus tree is, as we shall see in more detail in Chapter 30, a tree that summarizes a series of trees. Margush and McMorris's majority-rule consensus tree is simply a tree that consists of those groups that occur in a majority of the trees. It may not be obvious that these will form a tree. In fact, they will. If two groups each occur in more than 50% of the trees, then there must be at least one tree that has both of them. If two groups are compatible, then they are either disjoint, or identical, or one must be contained within the other. Suppose that we make up for each group a 0/1 character, which has 1s for each species that is in the group, and 0s otherwise. The compatible groups will then all have compatible characters. The pairwise compatibility theorem that we saw in Chapter 8 then guarantees that all these groups can be placed on the same tree.

The majority-rule consensus tree is found by tabulating all groups that occur on all trees and retaining those that occur on a majority of the trees. When we use it on the bootstrap estimates of the tree, the result is a single tree. All of the groups that appear on it are present in more than 50% of the bootstrap estimates. A simple extension of the majority-rule consensus tree is to note, next to each group, in what fraction of bootstrap replicates it has appeared. We can quickly see which groups have strong support, and which weak support. Figure 20.3 shows five trees and the resulting list of partitions of the species, as well as the majority-rule consensus tree.

The P value for each branch is intended to give an estimate of the amount of support the branch has. As we shall see below, this number turns out to be biased, underestimating the value of P when it is large.

The multiple-tests problem

One problem with the use of these P values is that we may not know in advance which group interests us. If we instead look for the most strongly supported group on the tree and then report its value of P, we have a "multiple-tests problem" (Felsenstein, 1985b). If there were actually no significant evidence for the existence of any of the groups, then P values on the branches would be drawn from a uniform distribution, with 5% of them expected to fall above 0.95. So one out of every 20 branches of a tree would be expected to reach the "significance" level of 0.95.

One way to correct for this is to use the well-known *Bonferroni correction.* In this case that simply amounts to dividing the desired tail probability (say 0.05) by the number of tests. Thus if we want to know for which value of P the most significant out of n tests has only a 5% chance of reaching that value, when the null hypothesis (of no significant structure) is true, we should require our groups to attain a support of $P = 1 - 0.05/n$. Thus with (say) 15 groups in a tree, the P value required for significance would be taken to be $1 - 0.05/15 = 0.99666$. This is a conservative procedure and allows for us to find the most significantly supported group out of n, even when the support for different groups is not quite independent.

Independence of characters

The most telling criticism of the bootstrap for phylogenies is that the assumptions of independence of the characters may not be met (Felsenstein, 1985b). The easiest way to see what effect this has is to imagine a case in which pairs of characters are identical. In other words, in collecting characters, we have inadvertently collected two characters that are so closely correlated that they are effectively providing the same information about evolution. We have done this so often that each character has, somewhere in our data, an identical partner.

A little consideration will show that the proper method of bootstrapping would be to draw once for each identical pair, as we then have $n/2$ independent characters, not n. The proper bootstrapping technique would be to draw $n/2$ times, each time drawing one character. If instead we draw n times, we will be sampling too often, the variation between bootstrap samples will be too small, and the trees they generate will be too similar. There will appear to be more corroborating evidence for groups on the tree than there really is.

Less complete correlation between characters is more realistic. It will cause similar problems — the appearance of too much evidence for groups on the tree. Unfortunately, there is usually no easy way to know how much correlation there is between characters, and thus no easy way to choose the number of characters to draw in a bootstrap sample. In certain cases, such as molecular sequences, one may be able to assume that the correlation of characters occurs mostly between nearby sites in the sequence. For example, we might have correlations that are mostly between sites that are within five nucleotides of each other.

Künsch (1989) has proposed a *block bootstrap* method that can cope with that correlation. He suggests drawing, not single sites, but blocks of B sites, the starting position for each block being drawn at random. Instead of drawing n individual sites, he draws n/B blocks of B sites, so that the bootstrap sample ends up consisting of n sites. Künsch shows that this corrects for autocorrelations along the sequence that are no longer than B sites. If the distance between correlated sites averages five sites, then $B = 10$ would seem to be a good choice. If we are mistaken and there is actually no autocorrelation, block-bootstrapping has the happy property of being a correct method anyway.

Note that in the imaginary example above, where pairs of characters have perfect correlation, if these pairs were adjacent characters, the data set would consist of $n/2$ adjacent pairs. One could use Künsch's method with, say, $B = 4$ in such a case.

Identical distribution — a problem?

In drawing a statistical sample, one commonly assumes that the draws are independent and identically distributed (i.i.d.). This is also the assumption of the bootstrap. We have seen that nonindependence is a potentially serious difficulty for the bootstrap, particularly if the dependent characters are not adjacent. Is failure to be identically distributed an equally difficult problem? I don't think so.

It is evident that the evolutionary processes in different characters (and in different sites in a molecule) can differ substantially. The differences in evolutionary rate from site to site in molecules are one example. Given that, is there any way to use the bootstrap? The approach I have proposed in such cases (Felsenstein, 1985b) is to consider the characters as samples from a larger pool of characters. Suppose that rates are assigned independently to sites in a molecule, so that each site has a rate randomly drawn from a distribution of rates. The characters have randomly assigned rate of evolution, and then the outcome of evolution is the result of a random process running at that rate. To get the data for a character, we draw a rate from the pool of rates, then evolve the character independently at that rate. In that case, the outcomes at the characters are still i.i.d., even though their rates of evolution differ.

In that original paper, I may have created unnecessary difficulties by saying that the bootstrap assumes that "each character is ... a random sample from a distribution of all possible configurations of characters," and by describing the systematist as sampling from "a pool of different kinds of characters." Others (Carpenter, 1992; see also Sanderson, 1995) have rejected this argument by disagreeing with the notion that characters are drawn from the universe of all possible characters. Although the notion of there being such a universe is indeed dubious, it is not actually necessary to the argument. All we need to assume is that the characters are drawn independently from *some* universe of characters, from some pool of characters.

In both molecules and morphology we may have characters that occur in blocks, such as data sets that have 10 skull characters followed by 10 limb characters, or molecules that have a fast region followed by a slow region. The issue that these data sets raise is not identical distribution, but independence. If we could consider successive characters as independently drawn, having a mix of rates of evolution, or a mix of body regions, would not endanger the bootstrap. The existence of these blocks of characters calls into question the assertion of independence, but the heterogeneity of evolutionary processes in the different characters is not the problem.

Invariant characters and resampling methods

The bootstrap and related resampling methods have also been argued to be sensitive to the number of invariant characters included in the data set. Suppose that we are using a method of phylogenetic inference, such as parsimony, that is not affected by the presence of characters that show no variation. Will we get substantially different bootstrap values by omitting the invariant characters from the analysis? It has been repeatedly argued (Faith and Cranston, 1991; Carpenter, 1992; Kluge and Wolf, 1993; Farris et al., 1996; Carpenter, 1996) that the bootstrap

Table 20.1: The probability of a character being omitted from a bootstrap sample, for different numbers of characters (N) in the data set.

N	$(1-1/N)^N$	N	$(1-1/N)^N$	N	$(1-1/N)^N$
1	0	14	0.35434	60	0.36479
2	0.25	15	0.35526	70	0.36524
3	0.29630	16	0.35607	80	0.36557
4	0.31641	17	0.35679	90	0.36583
5	0.32768	18	0.35742	100	0.36603
6	0.33490	19	0.35798	150	0.36665
7	0.33992	20	0.35849	200	0.36696
8	0.34361	25	0.36040	250	0.36714
9	0.34644	30	0.36166	300	0.36727
10	0.34868	35	0.36256	400	0.36742
11	0.35049	40	0.36323	500	0.36751
12	0.35200	45	0.36375	1000	0.36770
13	0.35326	50	0.36417	∞	0.36788

will give substantially different results without the invariant characters. Harshman (1994) has argued that it will not.

Consider a single character that does show variation in the data set. How often will it appear in the bootstrap replicates? If there are N characters in all, it will be chosen with probability $1/N$ each time a character is sampled. Thus it will be omitted $1-1/N$ of the time for each character sampled. The chance that it will be omitted entirely is thus (Harshman, 1994) $(1-1/N)^N$.

Adding M invariant characters to a data set changes this probability by increasing the value of N. Harshman argues that this quantity is very close to being constant at $e^{-1} = 0.36788$, no matter what the value of M. Farris et al. (1996) argue that it is not constant, that its complement (the probability of the character being included) "decreases as N increases." Table 20.1 shows the probabilities of the character being omitted.

The values *do* increase (and the probabilities of inclusion decrease), but not by much: They reach 90% of their ultimate value at $N = 6$, and 99% of the ultimate value at about $N = 50$. We can conclude, with Harshman, that the inclusion or exclusion of invariant characters will have little effect on the support given any particular group by the bootstrap method. The delete-half jackknife will behave similarly.

Of course, if the method for inferring phylogenies assumes that all characters are present (as do distance and likelihood methods), then we cannot drop invariant characters without doing serious violence to the trees.

Biases in bootstrap and jackknife probabilities

For years after the introduction of the bootstrap method for phylogenies, people had complained that the P values that the bootstrap method provided seemed too pessimistic. When they were noticeably lower than 95%, there still seemed to be a very high chance that the groups were real. Zharkikh and Li (1992; Li and Zharkikh, 1994) carefully examined the statistical properties of this inference and showed that the support was indeed underestimated. Hillis and Bull (1993) carried out a large simulation study that reached the same conclusion. They argued that a P value as small as 70% might indicate a significantly supported group. Felsenstein and Kishino (1993) have agreed that the bias is present, but they argued that it is not due to the bootstrap sampling itself, but instead to the use of a P value to describe the presence or absence of particular clades. Efron, Halloran, and Holmes (1996) argued that there was not always a bias downwards; they are correct, but for high values of P the bias is almost entirely in that direction. Newton (1996) has verified the validity of the bootstrap for discrete entities such as tree topologies, and has also verified that there is this bias.

P values in a simple normal case

To show that this bias is not due to the bootstrap, we argued that it would appear even in cases where there was no bootstrapping. For example, suppose that we draw n points from a normal distribution whose standard deviation is known to be 1, but whose mean is unknown. We are interested in whether the mean is positive or negative. This is analogous to asking whether a branch is present or absent, with the value of the mean playing the same role as the branch length. Our estimate of the mean will be the empirical mean of the sample, $\bar{x}$.

To obtain a level of significance for the proposition that the true mean is positive, we consider that the sample mean is normally distributed around the true mean with variance $1/n$. The conventional way of constructing P values is to use pivotal statistics. Thus we have the difference between the true mean, μ and the sample mean $\bar{x}$, which is $\bar{x} - \mu$. That difference has a normal distribution with mean 0 and variance $1/n$. It follows that when we multiply it by $\sqrt{n}$ it will become a quantity with mean 0 and variance 1. The probability that this quantity is greater than some particular value is then easily computed from tables of the normal distribution. We can then say, for example, that there is a 95% probability that $\sqrt{n}\,(\bar{x} - \mu)$ is greater than 1.64485. This can be turned into a statement assigning a level of significance to the statement that $\mu > 0$. For example, if $n = 10$ and $\bar{x} = 0.7$, we know that $\sqrt{10}\,(0.7 - \mu)$ has a normal distribution with mean 0 and variance 1. The probability that $\mu < 0$ is then the probability that a standard normal deviate lies below $3.162 \times (-0.7) = -2.214$, which is about 0.014. Then the probability that $\mu > 0$ is thus approximately 0.986.

We will have to return and ask what this really means. It seems entirely too neat (and so it is). But for the moment it tells us how to assign a P value to the

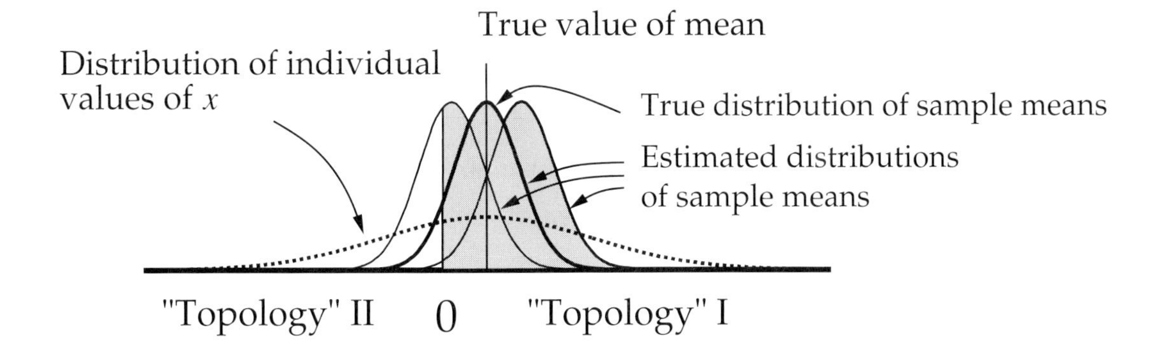

Figure 20.4: An example of assigning values of P to regions of a space that resemble tree topologies. We draw a sample of points from a true distribution (dashed curve) and there is a resulting distribution of the mean of n such points (density function with darkest line). The two other density functions show what we might infer this density function to be if the mean came out a bit closer to, or a bit farther away from 0. In each case the P value assigned is given by the shaded area of the curve.

statement that $\mu > 0$. We consider the distribution of $\bar{x}$, which in this case we know. For each observed value of $\bar{x}$ we ask how many standard deviations away from it 0 is. The area of the standard normal distribution for the appropriate tail then gives us P. Figure 20.4 illustrates this process. It shows the distribution from which the individual data points are drawn (the dashed curve), and the regions above and below 0, which are the two "topologies." The density function with the darkest curve is the true distribution of $\bar{x}$. The actual value of $\bar{x}$ could come from anywhere in this distribution. Three possible outcomes are shown — having it come out equal to the true mean μ, having it come out somewhat higher, and somewhat lower. For each one we will make a different estimate $\widehat{\mu} = \bar{x}$, and consider a different estimated density function. The P values we will get in each case are the fractions of those distributions that are above 0 (the shaded areas of the curves).

The correct P value to assign is the tail area of the true distribution of $\bar{x}$, which tells us the probability that our samples will get the true "tree topology." The actual P values vary around this, and it is immediately apparent that they do not vary symmetrically. When $\bar{x}$ is too close to 0, they drop substantially. When it is too far from 0, in an event that is equally likely to occur, the P rises by a much smaller amount.

The result is that there is a bias in P. When P should be (say) 0.95, the value we get is on average smaller than 0.95, leading to statements that are on average too conservative. Figure 20.5 shows the average P values as a function of the true

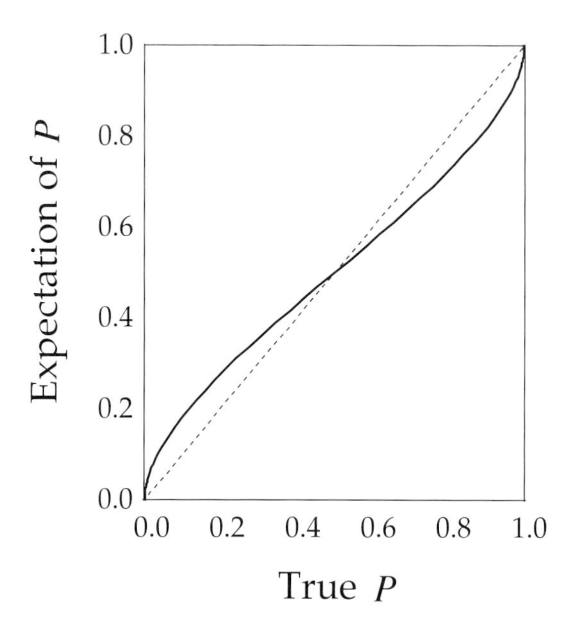

Figure 20.5: The expected value of P for the hypothesis that $\mu > 0$ in the case of n points drawn from a normal distribution with expectation μ and variance 1 (as in Figure 20.4). The expectation of P is plotted as a function of the true probability that a sample will have $\bar{x} > 0$.

P values, which are easily computed for this example (Felsenstein and Kishino, 1993). The bias of P is apparent. It is always toward 0.5, which, for the large values we are interested in, means that the P's are on average too conservative.

In Figure 20.4 we can also see that when the "true" value is P, the estimate P_e will be greater than P half of the time, and less than P the other half of the time. It is less obvious, but also true, that the estimate P_e will be greater than 0.5 a fraction P of the time. Thus when the true $P = 0.95$, the estimated P will exceed 0.95 half of the time, and the fraction of times that P_e will exceed 0.5 is 95%.

One of the sources of the conservatism of the estimated P values is that we are taking statements about the "branch length" μ and reducing them to statements only about the "tree topologies" $\mu > 0$ and $\mu < 0$. If the observed mean turns out to be above $1.95996/\sqrt{n}$ (the one-tailed 95% point of a normal distribution), we will conclude that the confidence set is entirely of "topology" I. When it is below $-1.95996/\sqrt{n}$, we conclude the opposite, that it is entirely of "topology" II. Anywhere in between, we will conclude that both "topologies" are possible.

If the true value of μ were (say) infinitesimally less than 0, so that the "topology" was II, but just barely so, we would draw the wrong conclusion 5% of the time, as that is how often we would get an observed mean that exceeded $1.95996/\sqrt{n}$. The other 95% of the time the confidence set would include the correct "topology." For any more negative value of μ, the probability of type I error (falsely rejecting the true "topology") is less than 5%, often considerably so. For example, when $n = 10$ and $\mu = -0.1$, the value $1.95996/\sqrt{10} = 0.61979$ is 2.27619 standard deviations away from that true mean. Thus we get the false conclusion that the "topology" is I only about 0.0114 of the time, as this is the fraction of a normal distribution that lies beyond 2.27619.

These results are true for the analogy of tree topologies with regions of positive and negative values of a normally distributed quantity. Will similar behavior be seen for actual tree topologies? This is not known, but I suspect that topologies will behave very similarly.

This analogy leads us to one interpretation of the bootstrap P value. If we see a group that occurs a fraction P of the time, we can say that the probability that it would have obtained this much support if it were not actually present on the true tree is less than $1 - P$. Thus a group that obtains a P value of precisely 95% will be expected to obtain that much support, when it is not actually present, less than 5% of the time. We must, however, note that the proof of this conservative interpretation has not yet been made for the case of phylogenies.

Methods of reducing the bias

The bias of the P value becomes even greater when consider that we are in a space of trees and consider the multiple topologies near each tree. As we will see, the effect is to increase the bias. Four methods have been proposed to correct for this bias. I will describe each briefly, and then suggest some connections between them.

- **The complete and partial bootstrap.** Zharkikh and Li (1995) developed a method which at the time seemed strange. It looks much less strange now that it has been joined by other methods and the connections between them become more apparent. Zharkikh and Li considered a case where there were K different character patterns, each backing a different tree topology. Using normal approximations and simulations, they showed that the bias of the bootstrap P value grew greater as K got much larger than 2. They went on to derive the *complete and partial bootstrap method* to correct these P values. We do not know what the relevant value of K is for a space of tree topologies. But they noted that for the case of K classes, if we do two bootstrap analyses with different numbers of characters resampled, we can estimate the effective values of K and of the probability of the correct class, and then use it to correct the bias. Suppose that we draw a regular bootstrap sample and obtain $P = P^*$. We also do partial bootstraps, in which we sample only $1/r$ as many characters (thus, if $r = 3$, we resample a number of times only one-third the number of characters). Call the fraction of these smaller resamplings that support that particular outcome P_r^*. Zharkikh and Li then were able to compute from the values of P^* and P_r^* what was the effective value of K, and use that to correct the bootstrap P value.

- **The method of Efron, Halloran, and Holmes.** Efron, Halloran, and Holmes (1996) applied a correction due to Efron (1987) to get a less biased P value for presence of a group in a phylogeny. They first bootstrap the data and infer trees. I have noted above that bootstrapping can be regarded as a reweighting of characters, where each of the original characters has a weight corresponding to the number of times it occurs. Thus if character i occurs n_i times, this would be the same as having it have weight n_i. They now take the samples that do not show the particular group, such as {Human, Chimp}. For each of these they try to adjust the weights back toward equality, so as to arrive at a set of weights that results in the group just barely being absent. One searches for the fraction f that determines weights $f + (1 - f)n_i$, such that these weights just barely result in the absence of the group. Efron, Halloran, and Holmes point out that this can be done by a simple "line search." The data set with these weights is a least favorable case, one that lacks the group but comes as close as possible to the original data set. They now bootstrap from these reweighted data sets. If the weights are w_i, the bootstrap draws character i with probability w_i. Analyzing this second level of bootstrap samples, they see what fraction of the resulting trees contain the group. After computing a constant a from the weights for each of these reweighted data sets, they then use a formula from Efron (1987) to calculate a bias-corrected P value.

- **The iterated bootstrap.** Rodrigo (1993) adapted methods invented by P. Hall and R. Beran in the statistical literature to propose the *iterated bootstrap*. He uses no less than three levels of bootstrapping. First one takes the usual R bootstrap replicates and estimates the tree for each. Then for each of these bootstrap sampled data sets, one bootstraps R more times from it, so that one has done $R+R^2$ bootstrap samples in all. Not content with this, one goes one more level, to make a triple bootstrap with a total of $R + R^2 + R^3$ replicates. We assume that our interest is in some particular group (such as {Human, Chimp}), and we want to discover what fraction of times P it should appear in the bootstrap estimates to make its appearance give us 95% confidence in its existence.

  We would ideally like to know the true tree, sample more data sets generated on it, and see how often we rejected the group when it was present on the true tree. This we cannot do: If we knew the true tree, we would not even bother to ask the remaining questions. The iterated bootstrap takes the R bootstrap estimates of the tree as true, and for each takes the R^2 second-level bootstrap samples to approximate the variation of data generated on such trees. Then the third level of sampling is used to find out, for each of these R^2 data sets, whether the group in question would be judged to have significant support. This is done for the first-level trees that have the group and for the first-level trees that do not. These are used to approximate the proba-

Table 20.2: P value for the 50% partial bootstrap at which the corrected P value does not reach 0.95, for the Zharkikh and Li method and the AU method.

	ZL	AU		ZL	AU
P^* for complete bootstrap	$P > 0.95$ when P_r^* less than		P^* for complete bootstrap	$P > 0.95$ when P_r^* less than	
0.99	0.9602	0.9704	0.86	0.7499	0.7137
0.98	0.9354	0.9445	0.84	0.7257	0.6820
0.97	0.9143	0.9213	0.82	0.7024	0.6515
0.96	0.8952	0.8989	0.80	0.6799	0.6222
0.95	0.8776	0.8776	0.75	0.6265	0.5532
0.94	0.8611	0.8571	0.70	0.5765	0.4899
0.93	0.8454	0.8374	0.65	0.5288	0.4314
0.92	0.8303	0.8182	0.60	0.4833	0.3772
0.91	0.8159	0.7997	0.55	0.4394	0.3270
0.90	0.8019	0.7818	0.50	0.3969	0.2804
0.88	0.7752	0.7468	0.40	0.3152	0.1976

bility that a group that is not present will be significantly supported, and the probabilities that a group that is present will be significantly supported.

- **The AU method of Shimodaira.** Shimodaira (2002) has developed a method similar to Zharkikh and Li's complete and partial bootstrap. It uses a series of bootstraps of different sizes. One might be the original bootstrap, another might sample $n/2$ sites, and another $2n$ sites (which is perfectly possible since sampling is with replacement). By fitting curves through the resulting P values, he obtains constants needed for a correction formula. Shimodaira and Hasegawa (2001) have described a computer program to do this.

The correction formulas for three of the methods look generically similar, which suggests that the methods are related. They have much in common. The first two explore the shape of the region of data space that lead to inferring the group. The partial bootstrap (used in the Zharkikh and Li method and in the AU method) has us spread out more widely from the original data set and see what this does to the probability of inferring the presence of the group. The Efron-Halloran-Holmes (EHH) method moves to the nearest edge of the region and uses the bootstrap to ask about the geometry of the region there. They argue that they are in effect asking about the convexity of the region in that neighborhood. Shimodaira discusses the matter in more detail and points out the close relationship of his method with these two methods. It is less easy to see that the iterated bootstrap also does something similar, as it works more empirically without any explicit geometry.

The methods differ in computational effort. The iterated bootstrap can be quite tiresome, as it replaces bootstrap R replicates with more then R^3 replicates. This would replace 100 replicates by more than a million replicates. The method of Efron, Halloran, and Holmes takes a fraction of the bootstrap replicates, reweights their characters, and then resamples from these. This will be tedious but not nearly as burdensome as the iterated bootstrap. The complete and partial bootstrap method and the AU method are the least difficult because they can be carried out with as few as two bootstrap samplings. However, those samplings may need a large number of replicates to obtain sufficiently accurate P values. Shimodaira presents computer simulation results comparing the ZL, AU, and EHH methods, and finds that AU is most accurate.

We can make tables to carry out both of these methods with two bootstraps. Suppose that we have a complete bootstrap plus a partial bootstrap that samples half as many characters. Call their observed bootstrap P values P^* and P_r^*, respectively. Table 20.2 shows for each method which values of P_r^* are small enough to allow the bias-corrected P to reach 0.95 for a number of different P values for the complete bootstrap.

The drug testing analogy

In Hillis and Bull's (1993) simulations, they asked what fraction of the time a group that had 95% bootstrap support would be on the true tree. They found that groups that had as little as 70% support had a 95% chance of being true. This was the outcome of a simulation in which they took randomly branching trees and evolved characters along them.

Will this prove to be a general result? If so, then we might hope for general rules allowing us to correct the P values and interpret the result as a probability that the group is correct. The following Bayesian analogy shows that there is some reason for doubting this. Suppose that we are carrying out product tests for a pharmaceutical company, testing whether their drugs cure a particular disease. We do a blind test of the proposition that the drug is ineffective, and come up with a tail probability α for the test. Some of the time we reject this null hypothesis. Consider a group of proposed drugs that have each achieved $\alpha = 0.05$. What fraction of them actually work?

It depends heavily on who selected the drugs. They are submitted to us by the drug development branch of the company. If that branch is highly competent, they will submit to us mostly drugs that work. In that case many of them will reach the $\alpha = 0.05$ threshold, and the probability that a drug that reaches $\alpha = 0.05$ actually works is then very high, probably much higher than 0.95. On the other hand, if the drug development branch is not competent, then the drugs they submit for testing will mostly be ineffective. Few drugs will reach the 0.05 threshold, and when one does, it will have a small chance of being one that actually works, being more likely to be one of the 1 drugs in 20 that accidentally appears to work.

Hillis and Bull (1993) had, in effect, a fairly competent drug development laboratory. They used computer simulation on randomly branching trees. If there is a moderate amount of evolution on the branches of the tree, and a large number of characters, the groups recovered will tend to have a large probability of being correct. If, however, there is too much change between nodes on the tree, the groups recovered will reflect mostly random noise, and have a good chance of being incorrect.

We can use the normal distribution analogy to show this phenomenon. Suppose that μ itself is drawn from a normal distribution with mean 0 and variance σ^2. We know that, for n characters, a group reaches $P = 0.95$ when its sample mean $\bar{x}$ is $1.95996/\sqrt{n}$. If we take n data points from a normal distribution with variance 1, whose mean is itself normally distributed with mean 0 and variance σ^2, that mean, $\bar{x}$, will come from a normal distribution with mean 0 and variance $\sigma^2 + 1/n$. We can now ask about the conditional distribution of the true μ given the observed $\bar{x}$. This too is normal. It has mean $b_{\mu.\bar{x}}\,\bar{x}$, where $b_{\mu.\bar{x}}$ is the regression of μ on $\bar{x}$. That regression is the fraction of the total variance $\sigma^2 + 1/n$ which comes from the variation of μ, namely

$$b_{\mu.\bar{x}} = \frac{\sigma^2}{\sigma^2 + \frac{1}{n}} \tag{20.1}$$

The variance of μ given $\bar{x}$ is also easy to obtain. It is the residual variance in μ after the variance due to regression is taken out, which can be calculated to be

$$\sigma^2 - b^2_{\mu.\bar{x}}\left(\sigma^2 + \frac{1}{n}\right) \tag{20.2}$$

Using equation 20.1, this variance is easily shown to be

$$\text{Var}\ (\mu \mid \bar{x}) = \frac{\sigma^2}{n\sigma^2 + 1} \tag{20.3}$$

Thus given a group that has significance level P, we can calculate the probability that it truly has $\mu > 0$. All we need to do is (1) find the standard normal deviate that has area P below it, (2) multiply this by $\sqrt{\sigma^2 + 1/n}$ to get the corresponding value of $\bar{x}$, (3) multiply that by the regression coefficient $\sigma^2/\left(\sigma^2 + \frac{1}{n}\right)$ to get the mean of μ, (4) calculate how many standard deviations this is from 0 when the variance is given by equation 20.2, and (5) work out what fraction of that conditional distribution of μ's lies above that point. Note that 0 lies in the left tail of this distribution of μ's, and thus we are asking about the area above that point.

This has been done by Felsenstein and Kishino (1993). Figure 20.6 shows the results, with the probability that $\mu > 0$ plotted against P. The result depends on the value of $n\sigma^2$, and these values are indicated next to the curves. When $n\sigma^2 = 0.1$, in effect there is very little genuine signal (the drug development group

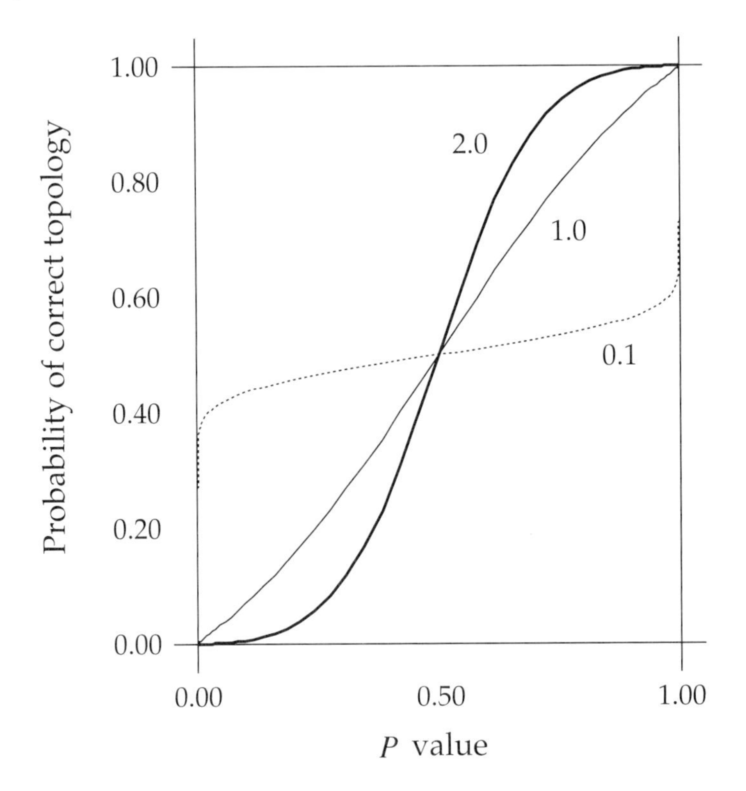

Figure 20.6: The probability that $\mu > 0$ when we draw n points from a normal distribution whose expectation μ is itself normally distributed with mean 0 and variance σ^2. The probability is plotted as a function of the P value for the observed mean. The value of $n\sigma^2$ is shown next to each curve.

is sending drugs that are generally ineffective). Even when a test reaches $P = 0.95$, the probability is not much greater than 50% that the true mean is above 0. When $n\sigma^2$ is 1, the curve is nearly a straight line, and when a test reaches $P = 0.95$, it has a bit more than 95% chance that the true mean is above 0. Hillis and Bull's (1993) results looked more like the case $n\sigma^2 = 2$, as they found that when $P = 0.70$, the group appeared on the true tree about 95% of the time.

These results are for the normal distribution analogy. What use can a user of the bootstrap make of them? Until further simulation testing on phylogenies is done, one has to be cautious. We do not know whether Hillis and Bull's rule of thumb is general. We do not know whether other cases are similar in the parameters that correspond to $n\sigma^2$. Note that with more information (larger n) the bootstrap becomes more conservative. One way to get a feel (but no more than that) for the conservatism of the bootstrap would be to look at all the P values on the tree.

If they are all large, this indicates that n is large, and we may then cautiously conclude that P values much less than 95% may indicate groups that have a high probability of being true. But if the P values are mostly small, then n is not large and we must be much more cautious in concluding that they indicate that a group is true.

Berry and Gascuel (1996) have argued that if correctness of trees is judged by the symmetric difference metric (which will be explained in Chapter 30), and if we count Type I and Type II errors as equally serious, the best value of P to use to resolve the tree partially would be $P = 0.5$. Their argument relies on a particular form of the relationship between the measured P value and the probability of the grouping being correct, one which makes this probability 0.5 when $P = 0.5$. It seems unlikely that this is true in general, so that their proposed rule needs further examination.

Alternatives to P values

Another difficulty with P values on groups is that one "rogue" species that is of uncertain placement can disrupt the signal in a majority-rule consensus tree. If the group ABCDEF occurs in most trees, but half of the time with species G in it and half of the time without, the majority-rule consensus tree may not contain either ABCDEF or ABCDEFG. The majority-rule method does not give a group credit for a partial appearance, or for appearance only in a larger group. Sanderson (1989) has suggested coping with this by setting a number n of extra individuals allowed into a group. Thus, if n is 2, we note that ABCDEF is present whenever a group containing those species and no more than 2 others is present. In the example above, ABCDEF would be given high support when $n = 1$, as then ABCDEFG would count towards it being present.

Wilkinson (1996) proposed another method: computing a reduced majority-rule consensus tree which shows trees of groups that are present among the bootstrap estimates of the trees, when we drop various species from consideration. Thus, dropping species G, we find ABCEDF present a large fraction of the time. He did not present efficient algorithms for finding the set of reduced majority-rule consensus trees. He notes that they require us to specify the desired tradeoff between number of species dropped and strength of support for groups. Algorithms to find these trees efficiently are still lacking.

In both cases some of the problems from noise are reduced by asking a somewhat looser question. Computational issues aside, the question that must be faced is whether this looser question is meaningful enough. Is it helpful to know that the group {Human, Chimp} occurs often if some additional species are allowed in the group, if the broader group turns out to be {Human, Chimp, Mouse}?

Brown (1994a) suggests other questions: Does a group appear significantly more frequently than another, and does a group appear significantly more often than 50%? I cannot see that these are useful: With enough bootstrap replicates a group that appears 51% of the time will be declared to appear significantly more often than 50%. But does this mean that its appearance on the true tree is supported? I suspect not.

Probabilities of trees

An alternative to the puzzle of how to describe support for groups is to simply take the distribution of trees and measure support for the different tree topologies. If we have a modest number of species we may be able to look at all possible trees. With 5 species and unrooted trees, there are 15 bifurcating tree topologies, and we can count how often each of them occurs among the bootstrap estimates of the topology. One way of constructing a confidence interval on trees is then to take the most frequent topologies until their probabilities add up to at least 95%. As the number of species increases, it will be less and less practical to do this. The number of possible phylogenies increases greatly, and it will soon become rare that two bootstrap replicates will lead us to estimate the same tree topology. We then end up with two classes of tree topologies—those that occurred once, and those that did not occur. We might order the ones that occurred once according to their goodness-of-fit to the original data (as judged by likelihood, parsimony, or whatever criterion we are using). The real problem is that we are then not concentrating our attention on the trees that contain a group of interest, so that we lose power in evaluating such a group.

Tree probabilities estimated from a bootstrap are used in Lake's (1995) "bootstrapper's gambit" method. There each bootstrap sample has its quartets analyzed, and if these all are compatible, a tree is constructed from them. When the tree probabilities are calculated, their interpretation is marred by the omission of all bootstrap samples that have incompatibilities among their quartets. Lake's tree probabilities must therefore be regarded as upper limits on the actual values.

Using tree distances

In Chapter 30 distance measures between trees will be described, in particular the symmetric difference metric. Penny and Hendy (1985, 1986) used this difference, together with the jackknife, to discover how far from the true tree we are. They randomly sampled a fraction of all characters, and constructed a tree from this resampled data. They calculated the mean distance between the trees from different samplings. They could show that, as the fraction of characters that were sampled increased, the trees became closer to each other. Plotting the decline of distance between trees against the number of characters sampled allowed them to infer how much sequence data was necessary to infer the true tree accurately.

Miller (2003) has used a similar plot (although using distance from a reference tree rather than distance between different sampled data sets). Like Penny and Hendy, his interest is in distances between trees, in order to understand the accuracy of the whole tree.

Jackknifing species

Early on, Lanyon (1985) suggested using a jackknife across species, removing one species at a time from the tree to see what effect this had on the estimate of the relationships of the remaining species. It is not easy to see what statistical meaning this jackknifing of species will have. Species are not independent and identically distributed — they come to us on some phylogeny, where they are highly clustered. This has been a major barrier to any attempt to make a statistical interpretation of jackknifing or bootstrapping species instead of characters.

Parametric bootstrapping

In the bootstrap, the resampling of the data set is intended to mimic the variability that we would get if we could sample more data sets from the underlying true distribution. In effect, that would be what we would get if we could simulate data sets on the true tree using the true model. The data sets we get from bootstrapping would be similar in the kinds of variability they contained. As we have seen in the discussion of biases, the trees they yield vary around the estimate that the original data set gives rather than around the true tree.

On the assumption that our estimate of the tree is somewhere near the true tree and that our model is somewhere near the true model, we could also imagine using our estimate and making new data sets on it by computer simulation. We would hope that these data sets also contain the same kinds of variability as would

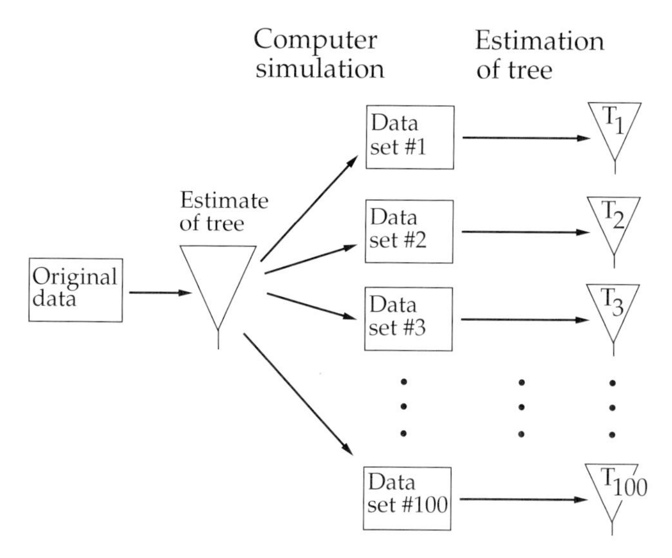

Figure 20.7: The parametric bootstrap. The data sets are obtained by simulation on our best estimate of the tree rather than by resampling columns of the original data matrix.

data sets from the true tree. The data sets could be treated in much the same way as are bootstrap samples. This method is the *parametric bootstrap*. The technique was introduced by Efron (1985). It was introduced for phylogenies by a number of people (Felsenstein, 1988b; Goldman, 1993; Adell and Dopazo, 1994; Huelsenbeck, Hillis, and Jones, 1996).

The closeness of the relationship between parametric bootstrapping and the ordinary bootstrap has led to the latter being referred to as the nonparametric bootstrap. With a single variable this is particularly apparent. Sampling from the original data is the same as sampling from an empirical histogram of data points. This histogram is regarded as an estimate, hopefully a close one, of the original distribution from which the data were drawn. Parametric bootstrapping replaces this histogram with a distribution from a parametric family, with the parameters being those that would be inferred from the data.

Figure 20.7 diagrams the process of using the parametric bootstrap with R replicates:

1. A single best estimate of the tree is made from the data set.
2. R (in this case, 100) computer simulations are then used to produce R data sets of the same size from this tree.
3. Each of these simulated data sets is used to infer the tree, using the same method used on the original data set.
4. The resulting trees are then analyzed in the same way as in the ordinary (nonparametric) bootstrap, such as by making a majority-rule consensus tree and P values for branches in the tree.

Advantages and disadvantages of the parametric bootstrap

Parametric bootstrapping can be used as a general replacement for nonparametric bootstrapping. For small data sets, it will have the advantage that it can sample from the desired distribution, even when sampling columns of the data matrix might leave many kinds of variation in the data unrepresented. The main concern is its close reliance on the correctness of the statistical model of evolution. When the model is correct, the type of variation that we will get between different bootstrap sample data sets will closely mimic the type of variation that we will get between the simulated data sets. It will not matter much whether we use parametric or nonparametric bootstrapping. But when the model is not correct, they will behave differently. The sampling of columns of the data matrix in ordinary nonparametric bootstrapping will reflect the variation in the correct model, while the simulation in parametric bootstrapping will reflect the variation in our incorrect model. In this situation, the ordinary (nonparametric) bootstrap will have the advantage. The more trust we have in the adequacy of our model, the more we will be willing to use instead the parametric bootstrap.

Permutation tests

An alternative to resampling is to reorder one's data. Permutation tests are standard methods in nonparametric statistics. For example, if we have two samples, one with 34 points and one with 43 points, that are supposed to be drawn independently from the same distribution, we can do a nonparametric version of a t-test by computing the difference in their means. Rather than assume we know the distribution from which they came, we can simply reshuffle the points many times. Suppose that we take all 77 values and shuffle them into a random order. Take the first 34 as being in sample 1, the second 43 as being in sample 2. Compute the difference in their means.

If we continue shuffling into random orders, and each time compute the difference of means, we get a large sample from the distribution of means under the null hypothesis that the two samples are from the same distribution. If we draw (say) 999 such samples, we can take these differences of means, and consider also the actual difference of means. Of these 1,000 numbers, if the actual difference lies in the top 25 or the bottom 25, we can reject the null hypothesis with $\alpha = 0.05$. Under the null hypothesis, all 1,000 values are from the same distribution, and the probability of being in these tails is 0.05.

Notice that the samples are not precisely from the full distribution because they always involve the same 67 numbers. Notice also that there are only a finite number of possible outcomes. There are only $77!/(34!\,43!)$ possible outcomes, but this is a satisfyingly large number, over 8.1×10^{21}. The power of the test is also dependent on the intelligent choice of a statistic. If the underlying distribution is one that generates samples whose means are dominated by a few extreme values, this test would not be particularly sensible.

A number of permutation strategies have been suggested:

Permuting species within characters

Archie (1989) and Faith and Cranston (1991) have suggested a permutation test for the presence of taxonomic structure in a data set. It is often called the *permutation tail probability test* (PTP). They take each character (column) in the data matrix and shuffle its values, reassigning them to species at random. All of the columns are shuffled independently of each other. The hope is that this will produce data sets that have no phylogeny but have numbers and distributions of states that are typical of the data. The distribution of goodness-of-fit measures such as likelihood or parsimony score among these permuted data sets are then compared to the value from the original data. If the actual value lies far enough into the tail (in the direction of higher likelihood or lower parsimony score), then there is significant taxonomic structure in the data. Källersjö et al. (1992) suggest some approximate strategies for more rapidly sampling trees and approximately computing the tail probability, based on Chebyshev's inequality and an exponential approximation to the distribution of tree lengths.

There are two difficulties with the PTP test. One is that structure may be detected for relatively trivial reasons. Suppose that two species are sibling species and that these are nearly identical. This may be enough to cause the test to be significant. It is true that there is then relatedness among species being detected, but it is only this rather obvious relationship between sibling species, and it does not mean that other larger-scale relationships are being detected. For reasons similar to this, Slowinski and Crother (1998) argue that the PTP test too readily detects significant structure. Simulation tests of the PTP method have disagreed whether or not its probability of type I error is too high (Peres-Neto and Marques, 2000; Wilkinson et al., 2002).

A second, and more serious problem was pointed out by Thorne (in Swofford et al., 1996). A tree with only a single internal node, with all lineages branching from it in a great multifurcation, can show significant structure in the test, if the branches are of substantially unequal lengths. An example of such a case is given in that paper. One possible response to this is that such a case does have structure. If one lineage is much longer than the others, and if the tree is unrooted and we regard it as the outgroup, then the other species can be regarded as forming a group distinct from the outgroup. Källersjö et al. (1992) have suggested that, rather than using parsimony score to characterize the degree of monophyly, we use a total support criterion, which is the sum over all branches of the Bremer support values. This will be zero if there is no unambiguous support for any monophyletic group. Farris et al. (1994a) gave an example data set where there was no such unambiguous support for any monophyletic group, but where the PTP test using parsimony score is significant. The example shows definite structure — for example, placing the species in a linear order — so that it is possible to argue that the behavior of the PTP test is appropriate. There has been debate back and forth over these examples (Carpenter, 1992; Faith, 1992; Källersjö et al., 1992; Trueman, 1993; Faith and Ballard, 1994; Farris et al., 1994a; Farris, 1995; Trueman, 1996; Carpenter, Goloboff, and Farris, 1998). The debate revolves around what the null hypothesis and alternative hypotheses of the PTP test really are. Some of these concerns have been raised on philosophical grounds (Goloboff, 1991; Bryant, 1992), but the matter will be more readily resolved in a statistical context. This needs more examination, so that we can understand what are the assumptions and behaviors of the test.

In an effort to concentrate the test's attention on hierarchical structure, Alroy (1994) has suggested using the PTP permutation strategy but computing different statistics, based on the number of pairs of characters that are compatible.

A variation of the PTP test is suggested by Faith and Cranston (1991). The tree topology is held constant and the data permuted, evaluating each permutation on that topology. This is held to test whether that tree has support greater than random. As this way the tree cannot adapt to the data, the test is quite likely to reject randomness. Brown (1994a) suggests using the permutation while examining whether a particular group appears significantly more often among bootstrap estimates than among estimates from the permuted data. As the presence of almost any internal structure within the group can cause that to happen, this seems unlikely to be a useful question.

Topology-dependent permutation of species. Faith (1991) developed a version of this permutation method designed to test whether a specific branch, which divides the species into two groups, is supported. He permutes the data as above, but instead of computing the total parsimony score for each data set, he computes the Bremer support for a given split of the species into two groups. For each data set we must compute the difference between the best tree and the best tree that does not contain this split. Faith argues for randomization that does not include the outgroup species. Swofford et al. (1996) disagree and prefer randomization of each character over all species. Faith's randomization test, the *topology-dependent permutation tail probability* (T-PTP) test, is designed to test whether there is nonrandom support for that particular split.

There has also been some uncertainty as to what should be done with the outgroups in the randomization process. Trueman (1996) argued that exclusion of the outgroup species from the randomization process was appropriate, and suggested ways of making the test more conservative.

As with the PTP test, there has been much discussion of the usefulness and validity of this test (Faith, 1992; Trueman, 1993; Farris et al., 1994a; Faith and Ballard, 1994; Trueman, 1996; Faith and Trueman, 1996; Carpenter, Goloboff, and Farris, 1998). I note here one particular criticism. Swofford et al. (1996) concentrated on whether other, irrelevant structure in the data could cause the test to reject randomness too often. (Farris, 1995 had earlier made an equivalent suggestion.) Swofford et al. simulated evolution on a tree of topology (O, (A, B, (C, D))), and found that the group (B, C, D) was supported too often. The presence of group (C, D) thus made the randomization procedure inappropriate, as it often broke up this group. The T-PTP test may thus have a null hypothesis of no structure anywhere in the tree, which dilutes its focus on the monophyly of the one group. Faith and Trueman (1996) have argued that this criticism is invalid, being based on the wrong choice of null hypothesis.

Permuting characters

Many phylogeny inference methods are insensitive to the order of the characters (the exception is likelihood or Bayesian methods that allow for autocorrelation of rates among sites). It might thus seem uninteresting to permute the order of the characters in a data set. But when there are two data sets, we might wish to know whether they are inferring noticeably different trees. If the data sets have, respectively, n_1 and n_2 characters, this can be addressed by a permutation test. We take all n_1+n_2 characters, and allocate them randomly into two data sets of size n_1 and n_2. This is most easily done by permuting the order of the $n_1 + n_2$ characters, and taking the first n_1 to be the first data set and the second n_2 to be the second data set.

The test is carried out by doing this R times, and measuring for all of these replications some aspect of the difference between the phylogenies inferred from the two data sets. We add the original data set into the picture and see whether the difference between their phylogenies is in the top 5% of these $R + 1$ numbers. Permutation tests like this are standard in statistics; in systematics they go back at least to the paper of Rohlf (1965), who did not compare phylogenies but measured the correlation between distances inferred from both data sets. Penny and Hendy (1985) used random divisions of a data set into halves to measure the average distance between the resulting trees, and from that get an idea of how accurately the tree was being estimated. The permutation test of whether the trees from two data sets are significantly different was introduced by Farris et al. (1994b) as the *incongruence length difference* (ILD) test, and independently by Swofford (1995) as the *partition homogeneity test*. It is most often known by the former name.

For the ILD family of tests, one computes for each replicate (and for the original two data sets) a measure of the extent to which the two data sets result in different trees. This can be done for parsimony, distance matrix, or likelihood methods. For parsimony, suppose that $T(D)$ is the tree estimate from data set D, and $N(D, T(D))$ is the number of changes required to evolve data set D on that tree. If the data sets are D_1 and D_2, and if when combined they are the larger data set D, then the suggestion of Farris et al. (1994b; Farris et al., 1995b) is to use the measure of Mickevich and Farris (1981), which is $N(D, T(D)) - N(D_1, T(D_1)) - N(D_2, T(D_2))$, a number that cannot be negative. (I leave it to the reader to discover why.) Swofford (1995) notes that the first term is unnecessary as it is the same in all permutations of a data set.

There are other possible measures. For example, one could use a tree distance (for which see Chapter 30) to measure how dissimilar the two trees are. Generalizations using the criteria for distance or likelihood methods are also straightforward, as long as one takes into account that higher is better in likelihood. In any of these cases one tests whether the measure of difference in outcome for the actual data sets is in the top 5% of the distribution, where the other R replicates are generated by permutation. If it is significantly extreme, this is an indication that the two data sets have significantly different signal.

ILD tests have been fairly widely used to analyze real data sets.

The ambiguity in these permutation tests is exactly what a significant result implies. Trees can be different in topology and/or in branch length. Simulations by Dolphin et al. (2000), Dowton and Austin (2002), Darlu and Lecointre (2002), and Barker and Lutzoni (2002) found that inequalities of rates of evolution in different data sets, using the same tree, could cause an elevated rate of rejection of the null hypothesis. This suggests caution in concluding that two data sets imply different trees.

Skewness of tree length distribution

A technique that is not really a permutation test, but which should be discussed along with them, is the skewness test of Hillis (1991; see also Fitch, 1979, 1984), which is discussed more extensively by Huelsenbeck (1991). This looks at the numbers of changes on all possible tree topologies, using parsimony. There is judged to be phylogenetic signal in the data if the distribution is significantly skewed. The rationale for this is that a few trees of much lower score than the others will create negative skewness. It can be computationally burdensome to examine all possible topologies, when the number of species is not small. The burden can be largely avoided by instead sampling randomly from the distribution of all possible topologies.

This method has been criticized by Källersjö et al. (1992), who gave a data set on which it did not behave properly. The fact that skewness is affected by all parts of the tree distribution, and does not concentrate its attention on the better trees, means that it may be of limited power in detecting phylogenetic signal.

Chapter 21

Paired-sites tests

A relative of likelihood ratio tests and bootstrapping is the class of tests that I will call *paired-sites tests*. The exact relationship of these tests to bootstraps and likelihood ratio tests is not obvious, but there is one. Paired-sites tests were first developed by Alan Templeton (1983a, b) for restriction site data. His test was rather complex, owing to his grouping the sites according to which restriction enzyme cuts them. A simplified version of the test was developed by Allan Wilson (in Prager and Wilson, 1988; also analyzed by Felsenstein, 1985c), called the *winning sites test*. Kishino and Hasegawa (1989) introduced a form of paired-sites test appropriate for maximum likelihood trees. It has come to be known as the KH test or the KHT test.

The basic idea of paired-sites tests is that we can compare two trees for either their parsimony or likelihood scores. The expected log-likelihood of a tree is the average log-likelihood we would get per site as the number of sites grows without limit. If evolution in different sites is independent (as we will assume in this chapter), then if two trees have equal expected log-likelihoods, the differences in log-likelihood at each site will be drawn independently from some distribution whose expectation is zero. If we do a statistical test of whether the mean of these differences is zero, we are then also testing whether there is significant statistical evidence that one tree is better than another. A similar principle holds for parsimony.

There have been a number of forms of the test proposed. They include:

- **The winning sites test.** For each site, score which tree is better, so that each site is assigned either a + or a – (or is assigned a 0 if both trees are tied at that site). Use a binomial distribution to test whether the fraction of + versus — is significantly different from 1/2.

- **The z test.** I suggested (in documentation for my PHYLIP program package in 1993) assuming that the differences are normally distributed. We can then estimate the variance of the differences of the scores at each site, from this calculate the variance of the sum of differences, and from that calculate by how many standard deviations the sum of differences departs from 0. We use the table of the normal distribution to calculate the probability that the mean would differ by this much or more from 0. A similar normal approximation was made by Kishino, Miyata, and Hasegawa (1990).

- **The t test.** Swofford et al. (1996) suggested using a t test instead of a z test. Thus the test reduces to a standard test of whether means from two normal samples are equal. Of course, we intend to test sums of the differences rather than means, but this turns out to be the same test, as these differ by a multiple of n (the number of sites), and so do their variances.

- **The Wilcoxon signed ranks test.** Templeton's (1983a, b) original method used the Wilcoxon signed ranks test, which replaces the absolute values of the differences by their ranks, then re-applies their signs. The sum of these values for whichever sign is less numerous is used as the test statistic. In practice, above 16 sites this test becomes a z test, with mean and variance known from the distribution of ranks.

- **The RELL test.** Kishino and Hasegawa (1989) developed a suggestion of mine (cf. Kishino, Miyata, and Hasegawa, 1990) to use bootstrap sampling to infer the distribution of the sum of differences of scores and see whether 0 lay in the tails of the distribution. This is perhaps the most accurate test.

An example

To give the reader some idea of how these tests differ, let us carry them out on a data set. Consider the 7-species 232-site mitochondrial DNA data set selected by Hasegawa, Kishino, and Yano (1985) from the 896-site data set of Brown et al. (1982), with five great apes and two other mammals (Mouse and Bovine). Suppose that we want to compare the two tree topologies shown in Figure 21.1, using maximum likelihood with an F84 model of evolution with transition/transversion rate of 2.0 and equal rates of change at all sites. We infer the branch lengths for each tree (they turn out to be the ones shown in the figure). The two trees differ primarily in the position of the Chimpanzee. Figure 21.2 shows part of a table of the log-likelihoods of sites for these two trees. The trees differ by 3.19 in their total log-likelihood. Figure 21.3 shows a histogram of the differences in log-likelihood at all 232 sites. These have a noticeably nonnormal distribution. The histogram shows them grouped in classes of width 0.05.

If we carry out the tests listed previously, here is what we find:

- **The winning sites test.** Of the 232 sites, 160 show tree I to have higher log-likelihood, 72 show tree II to have higher log-likelihood. The binomial probability that, out of 232 coin tosses, 160 or more would come out heads,

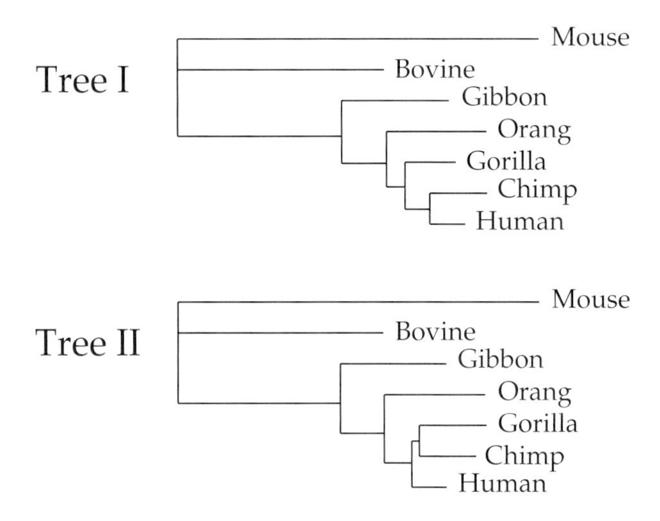

Figure 21.1: The two trees that are used as an example for the paired-sites methods. They differ primarily in the position of the Chimp. The trees are inferred as unrooted trees.

or that 160 or more would come out tails, is very small, about 3.279×10^{-9}. If we instead carry out a chi-square test on the observed numbers 160 : 72 with expectations 116 : 116, we get a similar result, that $\chi^2 = 33.37931$ with one degree of freedom, so that the probability of a value this large or larger is

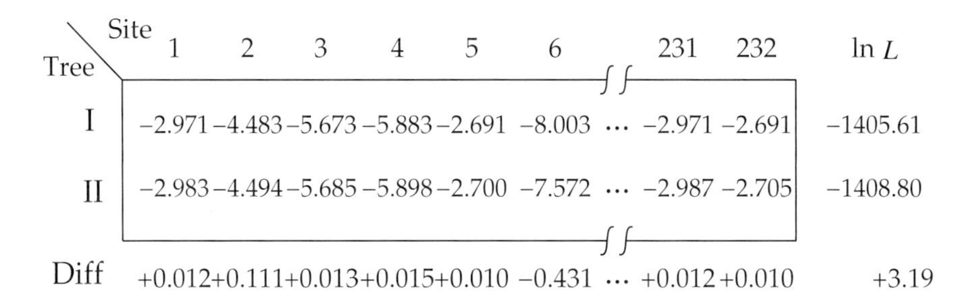

Tree \ Site	1	2	3	4	5	6		231	232	ln L
I	−2.971	−4.483	−5.673	−5.883	−2.691	−8.003	···	−2.971	−2.691	−1405.61
II	−2.983	−4.494	−5.685	−5.898	−2.700	−7.572	···	−2.987	−2.705	−1408.80
Diff	+0.012	+0.111	+0.013	+0.015	+0.010	−0.431	···	+0.012	+0.010	+3.19

Figure 21.2: Partial table of log-likelihoods at different sites for the trees of Figure 21.1.

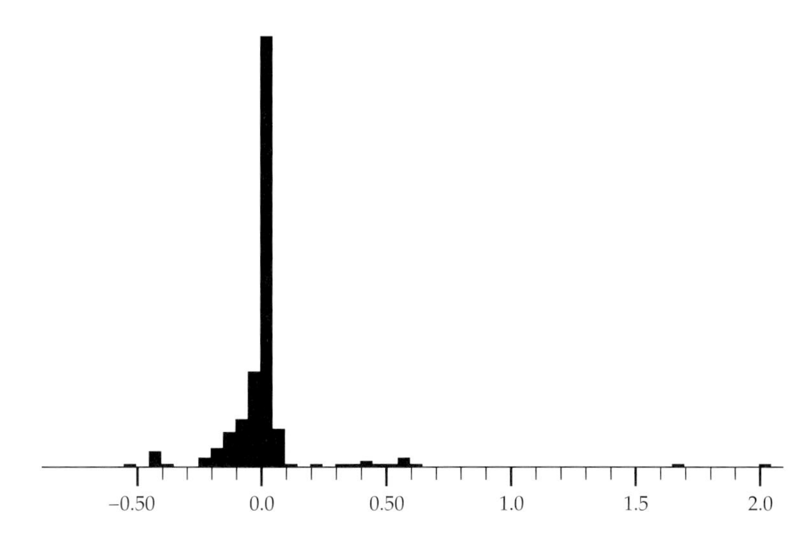

Figure 21.3: Histogram of log-likelihood differences between the two trees in Figure 21.1, with values grouped in classes of width 0.05.

3.791×10^{-9}. The winning sites test thus speaks strongly for the significance of the difference between the two trees.

- **The z test.** The difference in the sum of log-likelihoods is 3.18907. The variance of differences in log-likelihoods is 0.0487867, so that the variance of the sum of differences is 232 times as great, or 11.31852. The standard deviation of the sum of log-likelihoods is thus 3.364302. The z statistic is the ratio of the difference to its standard deviation, so that $z = 3.18907/3.364302 = 0.948104$. This statistic is assumed to be normally distributed. The probability that a normal variable is farther than 0.948104 away from zero (in either direction) is 0.343077, which speaks strongly against the significance of the difference between the two trees.

- **The t test.** As there are 231 degrees of freedom, a paired t test turns out to be essentially the same as a z test, thus giving the same result. Above 30 degrees of freedom, t tests are approximated by doing z tests.

- **The Wilcoxon signed ranks test.** If we take the differences, rank their absolute values, and then add up the ranks of those that are negative, we get a rank sum of 8,573. With samples of size greater than 16 sites, the Wilcoxon test uses a normal approximation. The rank sum with n sites has expectation $\frac{1}{4}n(n+1)$, and its standard deviation will be $\sqrt{n(n+1)(2n+1)/24}$. As $n = 232$, the value of 8,573 is drawn from a distribution expected to be approximately normal with mean 13,514 and standard deviation 1,023.39386. Thus it is 4.82805 standard deviations below its expected value (i.e., there are too few negative values). The Wilcoxon test then amounts to a two-tailed test from the normal distribution. The probability that the value is more than 4.82805 standard deviations away from its expectation is 0.000001378765.

- **The RELL test.** We draw a large number (here, 10,000) of bootstrap samples of sites. For each we ask whether the sum of differences of log-likelihoods is positive or negative. In this case out of 10,000 samples, 8,326 had a positive sum. This would imply a one-sided tail probability of approximately 0.1674 so that the two-sided tail probability is double this, or 0.3348. This is quite close to the result with the z test.

Which of these tests is correct? My own favorite is the RELL test, and the z and t tests seem to be giving the same answer. The winning sites test and the Wilcoxon test give a different answer. A look at the histogram of log-likelihood differences (Figure 21.3 shows that the distribution is not normal and not symmetrical. The two classes closest to 0 are very unequal in size, with far more values above 0 than below. The winning sites test is impressed by this disparity. Further out along the axis, we see a smaller number of sites that make a much larger positive or negative contribution to the log-likelihood. The RELL, z and t tests respond mostly to the signal from those sites, and as there are far fewer of them, they give much less significance to the difference of log-likelihoods between the trees. The Wilcoxon signed ranks test is somewhere in between, not responding fully to the sizes of the sites that have a large effect.

An examination of the sites with large differences of log-likelihoods shows that all that have a log-likelihood difference greater than 0.5 in absolute value show a difference in parsimony score between the two trees. These are the sites that reconstruct changes of state within the {Human, Chimp, Gorilla} clade. Of 15 sites that show a log-likelihood difference greater than 0.2, 11 of them have parsimony score differences between the two trees. Of 13 sites that show parsimony score differences, 12 of them have log-likelihood differences greater then 0.2. The RELL, z-test, and t-test results are influenced mostly by these sites.

There are interesting exceptions, such as site 130, in which all species have A except Human and Orang, which have G. There is no difference in parsimony score between the two trees at this site, but the log-likelihood difference is 0.41608, favoring a {Chimp, Gorilla} clade. This presumably reflects the low frequency of G in the sequences, which leads a likelihood method to prefer to explain the G as having arisen once in the lineage leading to the {Human, Chimp, Gorilla, Orang} clade, and then having been lost once in the ancestor of Chimp and Gorilla. This is an interesting example of way likelihood uses base frequency information in ways that have no parallel in parsimony methods.

The remaining sites show smaller log-likelihood differences because they are reconstructed as changing in other areas of the tree. Small branch-length differences between the trees then influence whether they seem to favor one tree or the other. There are many more of these sites that favor Tree I than Tree II, and thus the winning sites test and the Wilcoxon test back Tree I much more strongly than do the other tests.

I tend to be impressed by the evidence from the few sites of large effect, and unimpressed by the disparity of signs between sites of small effect, so I end up feeling that the RELL, z, and t tests are giving the most meaningful results. You may feel differently.

Like the bootstrap, the KHT class of tests compares likelihood differences among tree topologies to the empirical variation in log-likelihoods. This gives it some robustness against small violations of the models. Hasegawa and Kishino (1994) gave simulation results showing the RELL method to arrive at probabilities for tree topologies very similar to those from a full bootstrap sampling. Emerson, Ibrahim, and Hewitt (2001) found in a simulation test that the KHT test coped better with violation of the evolutionary model than did others.

Multiple trees

In many cases (including the tests I built into my own phylogeny program package), when there are many trees to test, confidence intervals on the acceptable trees have been constructed by comparing each tree to the best tree, and accepting all trees that cannot be rejected by the KHT test. Shimodaira and Hasegawa (1999) and Goldman, Anderson, and Rodrigo (2000) have pointed out that this is an incorrect way of doing multiple tests, and will accept too few trees. When only two trees are being compared, but one is chosen because it is the overall maximum likelihood tree, one can correct the test by treating it as a one-sided test instead of a two-sided test.

It might be thought that the problem is the number tests being done, and that a simple multiple-tests correction such as the Bonferroni correction can be used to solve the problem. That it is not this simple is seen by considering a t test to place confidence limits on the mean of a normal distribution. We can carry out the t test of the observed sample mean against any proposed other value, in the usual way. Each proposed value is compared with the observed mean, and a t test carried out. Those proposed means that are not rejected by this test form the confidence interval for the mean. Note that in doing this we make no correction for the number of proposed values we try. There could be one, or there could be a million — the test for any one value is not affected. There should be no correction for the number of tests because the results of all such tests respond to fluctuations in the data in a highly correlated way. In the phylogenies case this is not true — there needs to be a correction for all the different ways the data can vary, ways that support different trees.

The SH test

Shimodaira and Hasegawa (1999) have described a resampling method that approximately corrects for testing multiple trees. They suggest that we

1. Make R bootstrap samples of the N sites. For each compute the total log-likelihood. (This is most conveniently done by RELL sampling where we add up sitewise log-likelihoods without re-estimating branch lengths or other parameters.)
2. For each tree, subtract from the sum of the resampled log-likelihoods its mean across all R bootstrap samples. This "centering" has the effect of adjusting all trees so their resampled log-likelihoods have the same expectation. Thus if the total log-likelihood of the ith tree in the jth bootstrap sample is $\tilde{\ell}_{ij}$, compute the centered value for it as

$$\tilde{R}_{ij} = \tilde{\ell}_{ij} - \frac{1}{R}\sum_{k=1}^{R}\tilde{\ell}_{ik} \qquad (21.1)$$

3. For the jth bootstrap replicate, compute for the ith tree how far that centered value is below the maximum across all trees for that replicate:

$$\tilde{S}_{ij} = \left(\max_{k} \tilde{R}_{kj}\right) - \tilde{R}_{ij} \qquad (21.2)$$

4. For each tree i, the tail probability is then taken to be the fraction of the bootstrap replicates in which $\tilde{S}_{ij}$ is less than the actual difference between the maximum likelihood and the log-likelihood L_i of that tree.

The trees with tail probabilities above our target value (say, 0.05) cannot be rejected. In effect, the resampling constructs a "least favorable" case in which the trees show the same patterns of covariation of site evidence as in the actual data but don't differ in overall log-likelihood. Resampling mimics data sets arising in such a case, and it asks how often a particular tree falls as far below the maximum across these trees as it was observed to do in reality.

An alternative to the SH bootstrap resampling procedure has been suggested by Shimodaira (1998). It computes a covariance matrix of log-likelihoods from the table of log-likelihoods per site for each tree. From that one can estimate the covariances of sums of log-likelihoods. Sampling from a normal distribution with means zero and this covariance matrix, we can get values of the likelihood more quickly than in the RELL sampling mentioned above. This is analogous to the normal distribution approximation of Kishino, Miyata, and Hasegawa (1990).

One limitation of the SH test is that it assumes that all of the proposed trees are possibly equal in likelihood and resamples under that "least favorable" assumption. This means that if we have 10 reasonable trees to evaluate, but out of defensiveness we add another 90 implausible trees to the analysis, they make the test more conservative. The presence of the implausible trees makes it harder to reject any of the 10 plausible trees, by giving the resampling more opportunities to result in a large log-likelihood difference. Shimodaira and Hasegawa (1999) warn against including too many trees in the analysis, as it will dilute the power.

Other multiple-comparison tests

Bar-Hen and Kishino (2000) have another alternative based on the multivariate normal distribution. They compute covariances of the log-likelihoods per site (these are simply $1/n$ of the log-likelihoods, so have similar covariances). They make use of the asymptotic multivariate normality of the log-likelihoods per site, and base some multiple-comparisons statistics on that. If $l(\theta_i)$ is the log-likelihood per site for the ith tree, then the sum of squares of differences of it between all pairs of trees is approximately a multiple of a chi-square variate with approximate degrees of freedom:

$$\sum_i \sum_{j>i} \left[l(\theta_i) - l(\theta_j)\right]^2 = \frac{1}{2}\left(\left[\sum_i l(\theta_i)\right]^2 - \sum_i l(\theta_i)^2\right) \sim a\,\chi_b^2 \qquad (21.3)$$

where the multiple a and the effective degrees-of-freedom b can be computed from the trace of the variance-covariance matrix $\mathbf{A}$ of log-likelihoods per site, and from the trace of its square:

$$\begin{aligned} a &= \frac{\mathrm{tr}(\mathbf{A}^2)}{\mathrm{tr}(\mathbf{A})} \\ b &= \frac{[\mathrm{tr}(\mathbf{A})]^2}{\mathrm{tr}(\mathbf{A}^2)} \end{aligned} \qquad (21.4)$$

This gives them a test of the hypothesis that all members of the set of trees have indistinguishable likelihoods. When there are only two trees, the test is identical to the z test mentioned above. They also give a test of whether a tree is different in log-likelihood than a set of others.

Their tests assume that the trees are designated as of interest in advance. It is not inevitable that the set of trees of interest after the fact will be those with the highest likelihoods, but this will often be true. It would be of great interest to have a sequential testing procedure that could work its way along a series of trees in order of decreasing log-likelihoods, deciding when to reject all trees beyond that point.

Goldman, Anderson, and Rodrigo (2000) point out a test first mentioned by Swofford et al. (1996), which they call the SOWH test. For the ith proposed tree, it uses parametric bootstrapping, with tree i taken to be the true tree. Data sets generated by simulation on that tree, and for each the maximum likelihood tree is found, as well as the likelihood of tree i. The distribution of the differences of log-likelihoods between these is found, and the same quantity is calculated from the original data. It is significant if the value is in the tail of the distribution. Note that by using tree i as the true tree in the simulation, we automatically are in the "least favorable" case. Goldman, Anderson, and Rodrigo (2000) give some variations on the SOWH test as well. These tests are more computationally burdensome than the SH test. Buckley (2002) makes some comparisons of the behavior of the SOWH test, the SH test, and Bayesian approaches using data sets in which the true tree is considered well-known. He finds that the SOWH test and Bayesian methods place too much credence in the correctness of the assumed model of base change, while the SH test is more conservative. Like the nonparametric bootstrap, it uses empirical variation between sites rather than generating it from a model, as parametric bootstrap methods and Bayesian methods do.

Testing other parameters

The discussion so far may make it seem that paired-sites tests are useful primarily to compare tree topologies. Certainly they are very useful for that, but they can compare any aspect of trees or of models. We can compare sitewise likelihoods with different lengths of a branch and the same tree topology, or we can compare different transition/transversion rates. We can compare clocks with no clock.

Perspective

The generality and robustness of the paired-sites tests guarantees that they will continue to be of great interest. If they lose a bit of power, they make up for this in robustness. They are open to criticism for relying heavily on the assumption that there is no correlation of evolutionary rates or outcomes among nearby sites. The SOWH test could be made to cope with autocorrelations—the others assume that it is absent. It is clear that the testing of multiple trees is an area still under active development, so that we can expect to see more convenient and relevant test procedures in the future.

Chapter 22

Invariants

Invariants provide a very different way of looking at the inference of phylogenies. Usually we start with possible phylogenies. For each we ask what patterns of data the phylogeny predicts, and then we try to choose the phylogeny that makes the best overall prediction. We do so by searching in tree space and inferring branch lengths. Invariants work in the other direction — looking at specific patterns in the data to see whether they show relationships that are predicted on particular tree topologies. They escape from the need to infer branch lengths and may also allow us to rapidly eliminate many tree topologies.

Invariants methods start with the same probabilistic models as do distance matrix and likelihood methods. If these models assume independent evolution at each character (given the tree), then they predict particular pattern frequencies for each tree. For example, for DNA sequences, a phylogeny with eight species gives particular predicted frequencies to all $4^8 = 2^{16} = 65,536$ possible nucleotide patterns, from *AAAAAAAA* through *AAAAAAAC* to *TTTTTTTT*. If we consider all the eight-species trees that have one given bifurcating tree topology and we vary all $2 \times 8 - 3 = 13$ branch lengths, then each of them will predict the values in a vector of 65,536 expected pattern frequencies. Although the vectors of pattern frequencies are in a space with 65,535 dimensions (one is lost as the frequencies have to add up to 1), the collection of vectors that are predicted using the model and the tree will only have 13 dimensions, as each combination of the 13 branch lengths predicts a different vector of 65,536 quantities. These expected frequencies thus are arranged in a subspace that is only 13-dimensional.

A simple counting of degrees of freedom will lead us to expect that the expected pattern frequencies must obey many constraints, as they only vary in a 13-dimensional subspace. Every time we specify an equation that the expected pattern frequencies must obey, we reduce the dimensionality by one. For example, if we insist that all the expected patterns f_i satisfy an equation such as

$$f_{AAAAAAAA}\, f_{CCCCCCCC} = f_{GGGGGGGG}\, f_{TTTTTTTT} \qquad (22.1)$$

we find that we are constraining ourselves to a subspace one dimension smaller. To reach a 13-dimensional subspace, we would expect there would be $65,535 - 13 = 65,522$ such constraints. There should be 65,522 equations we could write down in the expected pattern frequencies, all of which would be satisfied for all trees of that topology. If we express all these equations in the form

$$F\left(f_{AAAAAAAA},\ f_{AAAAAAAC},\ \ldots,\ f_{TTTTTTTT}\right) = 0 \tag{22.2}$$

then the expressions are called the *invariants*. They are zero for all the expected pattern frequencies, no matter what the branch lengths.

Some of these equations might be satisfied for all tree topologies; others would be satisfied for some tree topologies but not all. These latter are called the *phylogenetic invariants* (Felsenstein, 1991). If we know the phylogenetic invariants for each possible tree topology, we can tabulate the pattern frequencies, and we can then look at the observed pattern frequencies to see what the values of the phylogenetic invariants are. If the observed pattern frequencies are close to their expectations, by seeing which invariants come close to their expected values we should be able to find the tree topology.

There are two catches: knowing what the phylogenetic invariants are for each possible tree topology, and allowing for the random variation of the observed pattern frequencies around their expected values. The development and terminology used in this chapter will follow a paper of mine (Felsenstein, 1991) that tried to list all invariants for a case with four species and a Jukes-Cantor model of base change, and that named some of them the phylogenetic, symmetry, and clock invariants.

We will frequently use counts of the degrees of freedom. Evans and Zhou (1998) have proven for a relevant class of symmetric models of base change that the number of invariants is in fact equal to the number of degrees of freedom left after the parameters are estimated: that is, the difference between the number of possible base patterns and the number of parameters estimated. It wasn't obvious that this had to be true: The notion of degrees of freedom comes from linear models, whereas invariants are often nonlinear. Hagedorn (2000) gave a more general proof that the degrees of freedom gives the correct count.

In this chapter I will discuss four topics. For simple three- and four-species trees and a Jukes-Cantor model, we will attempt to find all invariants. In the process we will see examples of the best-known classes of invariants. As we do so, for an eight-species tree I will try to count the number of each of these kinds of invariants, showing that they account for most but not all of the degrees of freedom. I will also discuss papers that present general machinery that can be used to find all invariants in some cases. Finally, I will try to explain why invariants are potentially important, although not at present very useful.

Table 22.1: Different pattern types for eight species, the number of types in each category, then number of patterns of each type, and the resulting number of invariants for each category.

Category of types	Example of a type	Pattern types of that category	Patterns of that type	Total invariants that result
8x	xxxxxxxx	1	4	3
7x, 1y	xxxxxxxy	8	12	88
6x, 2y	xxxxxxyy	28	12	308
5x, 3y	xxxxxyyy	56	12	616
4x, 4y	xxxxyyyy	35	12	385
6x, 1y, 1z	xxxxxxyz	28	24	644
5x, 2y, 1z	xxxxxyyz	168	24	3,864
4x, 3y, 1z	xxxxyyyz	280	24	6,440
4x, 2y, 2z	xxxxyyzz	210	24	4,830
3x, 3y, 2z	xxxyyyzz	280	24	6,440
5x, 1y, 1z, 1w	xxxxxyzw	56	24	1,288
4x, 2y, 1z, 1w	xxxxyyzw	420	24	9,660
3x, 3y, 1z, 1w	xxxyyyzw	280	24	6,440
3x, 2y, 2z, 1w	xxxyyzzw	840	24	19,320
2x, 2y, 2z, 2w	xxyyzzww	105	24	2,415
	Totals:	2,795		62,741

Symmetry invariants

Some of the invariants are simply consequences of the symmetry of the model of base change, and will not be phylogenetic invariants. These I have called *symmetry invariants*. For example, if the model of base change is the Jukes-Cantor model, the four bases can be exchanged without altering the pattern frequency. If we have eight species, the expected frequency of the pattern *ACAATTAA* should be the same as the expected frequency of *CGCCAACC*, a pattern that is obtained by replacing *A*, *C* and *T* with *C*, *G* and *A*, respectively. Thus one of the invariants will be:

$$f_{ACAATTAA} - f_{CGCCAACC} = 0 \tag{22.3}$$

There are other such replacements that would also lead to invariants. In fact, this pattern should have the same frequency as any pattern of the form xyxxzzxx, where x, y and z are any three distinct nucleotides. Thus for pattern type xyxxzzxx, there will be $4 \times 3 \times 2 = 24$ different patterns, all of which will have equal expected pattern frequencies. That implies that there are 23 invariants, one of which is in the equation above.

Table 22.2: Pattern types for four species with a Jukes-Cantor model, showing the number of symmetry invariants that result.

Category of types	Example of a type	Pattern types of that category	Patterns of that type	Total invariants that result
4x	xxxx	1	4	3
3x, 1y	xxxy	4	12	44
2x, 2y	xxyy	3	12	33
2x, 1y, 1z	xxyz	6	24	138
1x 1y 1z 1w	xyzw	1	24	23
	Totals:	15		241

Table 22.1 shows the pattern types for the case of eight species, an example of each, how many such patterns would be possible, and how many invariants each would give rise to. Note the rather obscure terminology: A pattern would be, say, *AACAAAGT*, its pattern type would be `xxyxxxzw`, and that would be a member of the category of pattern types that had 5x's, 1y, 1z, and 1w.

Of the 65,536 total degrees of freedom, one is lost to the trivial requirement that the expected frequencies add up to 1, and 62,741 of the remainder are symmetry invariants, leaving only 2,794 that could help us discriminate among phylogenies.

In the four-species Jukes-Cantor case we have 256 patterns, and lose one degree of freedom because they sum to 1. Table 22.2 counts the symmetry invariants — there are 241 of them, leaving us with only $255 - 241 = 14$ degrees of freedom.

We can use the symmetry invariants to test the symmetry of the model of base change. For example, one could take a pattern type such as `xyxzyxxw` and test whether the 24 different patterns of this type are equally frequent. This could be done by a chi-square test, except that in practice many of the patterns might have a low expected frequency, rendering the chi-square approximation dubious. The test could be done in that case by computer simulation of the distribution of the chi-square statistic. There may even be some cases in which all of the patterns of a given type are absent, rendering the test of their equality moot.

Three-species invariants

If we inquire how many invariants are present in cases with few species, this turns out to find some that can be used in larger cases. The simplest possible case is a one-species "tree." That, of course, is just a single DNA sequence. The symmetry invariants amount to a test of whether the sequence has equal numbers of As, Cs, Gs, and Ts. The case of two species is a tree with one branch, connecting the two. The pattern types are `xx` and `xy`. There are 16 degrees of freedom. One disappears because the pattern frequencies must add to 1. There are $3 + 11 = 14$ symmetry invariants. That leaves one degree of freedom. This is soaked up by estimation of the branch length between the species (in effect, as we have seen, estimation of the distance between them). There are then no additional invariants.

The smallest case that has real nonsymmetry invariants is three species. There are $4 \times 4 \times 4 = 64$ total degrees of freedom. The table of pattern types is:

Pattern type	Number	Symmetry invariants
xxx	4	3
xxy	12	11
xyx	12	11
xyy	12	11
xyz	24	23
Total	64	59

Thus there are 5 degrees of freedom ($64 - 59$) available that are not symmetry invariants. One is lost owing to the requirement that the pattern frequencies add to 1. Three more are lost by estimation of the lengths of the three branches in the unrooted three-species tree. That leaves one degree of freedom unaccounted for.

In fact, it does correspond to a nonsymmetry invariant. We can see this by writing the equations for the five pattern frequencies. In the Jukes-Cantor case, which we consider, the probability of net change on a branch of length v_i can be written as (see equation 11.17)

$$p_i = \frac{3}{4}\left(1 - e^{-\frac{4}{3}v_i}\right) \tag{22.4}$$

A bit of careful consideration (Felsenstein, 1991) will then show that we can write the expected pattern frequencies as

$$\begin{aligned}
P_{xxx} &= (1-p_1)(1-p_2)(1-p_3) + p_1p_2p_3/9 \\
P_{xxy} &= (1-p_1)(1-p_2)p_3 + 1/3\, p_1p_2(1-p_3) + 2/9\, p_1p_2p_3 \\
P_{xyx} &= (1-p_1)p_2(1-p_3) + 1/3\, p_1(1-p_2)p_3 + 2/9\, p_1p_2p_3 \\
P_{xyy} &= p_1(1-p_2)(1-p_3) + 1/3\,(1-p_1)p_2p_3 + 2/9\, p_1p_2p_3 \\
P_{xyz} &= 2/3\, p_1p_2(1-p_3) + 2/3\, p_1(1-p_2)p_3 \\
&\quad + 2/3\,(1-p_1)p_2p_3 + 2/9\, p_1p_2p_3
\end{aligned} \tag{22.5}$$

These five equations must add up to 1 (that is, the fifth equation can be obtained by adding the others and then subtracting that sum from 1). Thus there are in effect four equations in three unknowns. This means that there is one relationship that must hold between the P's for the equations to always yield the same p's no matter which three of the four equations we choose to solve.

Each equation is linear in each of the p_i. So in principle we can solve the equations by solving the first one for p_1, then substituting that expression into the second equation. We then have an equation with p_2 and p_3 but no longer any p_1. This second equation can then be solved for p_2, and the resulting expression substituted into the third equation, which now has only p_3. Solving for that, we then have expressions in the P's for p_1, p_2, and p_3. These can be substituted into the fourth equation, yielding the equation that is the constraint on the P's that is needed to make the whole system self-consistent. That equation tells us the invariant.

This is possible with computer algebra systems, but there are easier ways. In the Appendix to my paper (Felsenstein, 1991) I show a derivation that is more straightforward, if idiosyncratic. The result is the invariant

$$\begin{aligned} &3\,(2P_{xxx} - 2P_{xxy} - 2P_{xyx} - 2P_{xyy} + 1)^2 \\ &- [4(P_{xxx} + P_{xxy}) - 1]\,[4(P_{xxx} + P_{xyx}) - 1]\,[4(P_{xxx} + P_{xyy}) - 1] \;=\; 0 \end{aligned} \tag{22.6}$$

It is rather difficult to give any simple intuitive explanation of this invariant. It seems to be testing whether the substitutions in different branch lengths are independent, whether having changed away from a base you are more likely to return to it. If one of the three branch lengths (let's say, v_2) is zero, then it turns out to be simply a test of whether the pattern xyx is, as expected, half as frequent as xyz.

Note that the three-species cubic invariants are not, and cannot be, a phylogenetic invariant. There is only one possible unrooted tree topology. In cases with more species, each triple of them has a three-species invariant. We can compute them by taking those three species. In our eight-species example, there is one three-species invariant for each of the $8 \times 7 \times 6/(1 \times 2 \times 3) = 56$ ways that we could choose a triple from the eight species. These invariants are all independent, as they are each based on a three-species marginal distribution of the full eight-species distribution of nucleotides. It is rather easy to show that there are pairs of sets of the full eight-species expected pattern frequencies that have the same marginals for species 1, 2, and 3, but different marginals for (say) species 2, 3, and 4. Even knowing all but one of the three-species marginal distributions does not allow us to predict the remaining three-species marginal. Thus all 56 of these three-species invariants are independent. We have already reduced the number of invariants that could be phylogenetic invariants to 2,795. These 56 invariants can be removed from that number, as they are definitely not phylogenetic invariants since they hold for all tree topologies. Thus we have 2,739 degrees of freedom left.

For the four-species Jukes-Cantor model there will be four three-species invariants, one for each possible three-tuple of species. This leaves us with but 10 degrees of freedom.

Lake's linear invariants

One of the two original papers that founded work on invariants was the paper of Lake (1987), in which the technique is called "evolutionary parsimony." Lake considered a more general model of evolutionary change than the Jukes-Cantor model. He considered the probabilities of sites that have, among four particular species, two with a purine and two with a pyrimidine. The two purines may or may not be the same, and the two pyrimidines may or may not be the same. If we use the IUB ambiguity code and let R stand for purine and Y for pyrimidine, Lake asks about the patterns RRYY, YYRR, RYRY, YRYR, RYYR, and YRRY. These naturally divide into three pairs of patterns, according to whether the purine/pyrimidine distinction unites species A and B, A and C, or A and D. Consider one of these, say, the pair RRYY and YYRR which unite species A with B.

Lake considered the probabilities of two events:

1. The two purines are the same and so are the two pyrimidines, or the two purines differ and so do the two pyrimidines
2. The two purines differ but the two pyrimidines are the same, or else the two pyrimidines differ but the two purines are the same.

A useful notation of his is to denote the base in the first species as 1. If it is a purine, call the other purine 2. If it is a pyrimidine call the other pyrimidine 2. Now find the next species that does not have either base 1 or base 2. Call its base 3, and the remaining one 4.

Considering the purine/pyrimidine patterns RRYY and YYRR, these could be 1133, 1134, 1233, or 1234. Lake considers the total frequency of patterns 1133 and 1234, and compares this to the total frequency of patterns 1134 and 1233. He shows that for the trees ((A,C),(B,D)) and ((A,D),(B,C)), the expected frequencies of these two classes of patterns are equal. In other words:

$$P_{1133} + P_{1234} \;=\; P_{1134} + P_{1233} \tag{22.7}$$

This is true when the central branch of the four-species tree separates the two purines from each other and the two pyrimidines from each other.

Lake's invariants distinguish strongly between the purines and the pyrimidines, but they work even when there is a Jukes-Cantor model. In terms of commonly used models, they are valid for the Kimura two-parameter model, of which the Jukes-Cantor model is a special case. In terms of the pattern classes that we defined for the Jukes-Cantor model, 1133 is the portion of the pattern xxyy in which x and y are one purine and one pyrimidine (in either order). This will be 2/3. Likewise, 1134 is 1/3 of xxyz, 1233 is 1/3 of xyzz, and 1234 is 1/3 of xyzw. Thus the Lake invariants for the tree ((A,B),(C,D)) will be, in that notation,

$$\frac{2}{3}P_{xyxy} \;+\; \frac{1}{3}P_{xyzw} \;-\; \frac{1}{3}P_{xyxz} \;-\; \frac{1}{3}P_{xyzy} \;=\; 0 \tag{22.8}$$

and

$$\frac{2}{3}P_{xyyx} \;+\; \frac{1}{3}P_{xyzw} \;-\; \frac{1}{3}P_{xyzx} \;-\; \frac{1}{3}P_{xyyz} \;=\; 0 \tag{22.9}$$

In the eight-species case that we have been using as our example, there will be two Lake invariants for each quartet of species. As there will be $8 \times 7 \times 6 \times 5/(1 \times 2 \times 3 \times 4) = 70$ four-species quartets, there will be 140 Lake invariants. These will be phylogenetic invariants. The number of other invariants as yet unaccounted for is now down to $2,739 - 140 = 2,599$.

In the four-species case there are two Lake invariants for each tree topology, leaving us with eight degrees of freedom still unaccounted for.

Cavender's quadratic invariants

The other original paper on invariants was that of Cavender and Felsenstein (1987). Cavender discovered two quadratic invariants in the case of two states (my own contribution to that paper came long after he had discovered them). These are known as the K and L invariants. Although they are presented as rather arbitrary quadratic expressions, both of them can be derived in straightforward ways.

The K *invariants*

The *K invariant* is a consequence of the Buneman four-point metric (FPM) condition (Drolet and Sankoff, 1990). This was discussed in Chapter 13. It simply says that, for the four-species tree ((a,b),(c,d)), that if v_{ij} is the total branch length between species i and species j,

$$v_{ac} + v_{bd} \ = \ v_{bc} + v_{ad} \tag{22.10}$$

That this is true is easily seen: The expressions on both sides of the equation are equal to the sum of all five branch lengths, plus the length of the interior branch.

In the Jukes-Cantor case, the expected fraction of difference between two species (say, the D_{ac}) is a simple function of the branch length between them. Calling that branch length v_{ac},

$$D_{ac} \ = \ \frac{3}{4}\left(1 - e^{-\frac{4}{3}v_{ac}}\right) \tag{22.11}$$

(as we saw in equation 11.17). Solving for v in terms of D, we get

$$v_{ac} \ = \ -\frac{3}{4}\ln\left(1 - \frac{4}{3}D_{ac}\right) \tag{22.12}$$

(which we also saw in equation 11.18). We can plug these into Buneman's equation 22.10, and the factors of $-\frac{3}{4}$ will disappear. We have

$$\ln\left(1 - \frac{4}{3}D_{ac}\right) + \ln\left(1 - \frac{4}{3}D_{bd}\right) \ = \ \ln\left(1 - \frac{4}{3}D_{ad}\right) + \ln\left(1 - \frac{4}{3}D_{bc}\right) \tag{22.13}$$

from which we can exponentiate both sides to eliminate the logarithms and turn the sums into products to get

$$\left(1 - \frac{4}{3}D_{ac}\right)\left(1 - \frac{4}{3}D_{bd}\right) \ = \ \left(1 - \frac{4}{3}D_{ad}\right)\left(1 - \frac{4}{3}D_{bc}\right) \tag{22.14}$$

This simplifies to

$$D_{ac} + D_{bd} - D_{bc} - D_{ad} + \frac{4}{3} D_{ac} D_{bd} - \frac{4}{3} D_{bc} D_{ad} \ = \ 0 \tag{22.15}$$

The D's can be written in terms of the pattern probabilities. The expected difference D_{ac} will, for example, be the sum of the probabilities of all patterns that differ in the symbols for the first and third species. The equation is:

$$\begin{aligned} D_{ac} \quad = \quad & P_{xxyx} + P_{xxyy} + P_{xyyx} + P_{xyyy} + P_{xxyz} \\ & + P_{xyyz} + P_{xyzx} + P_{xyzy} + P_{xyzz} + P_{xyzw} \end{aligned} \tag{22.16}$$

There are similar equations for the other differences:

$$\begin{aligned} D_{ad} \quad = \quad & P_{xxxy} + P_{xxyy} + P_{xyxy} + P_{xyyy} + P_{xxyz} \\ & + P_{xyxz} + P_{xyyz} + P_{xyzy} + P_{xyzz} + P_{xyzw} \\ D_{bc} \quad = \quad & P_{xxyx} + P_{xxyy} + P_{xyxx} + P_{xyxy} + P_{xxyz} \\ & + P_{xyxz} + P_{xyzx} + P_{xyzy} + P_{xyzz} + P_{xyzw} \\ D_{bd} \quad = \quad & P_{xxxy} + P_{xxyy} + P_{xyxx} + P_{xyyx} + P_{xxyz} \\ & + P_{xyxz} + P_{xyyz} + P_{xyzx} + P_{xyzz} + P_{xyzw} \end{aligned} \tag{22.17}$$

We have given the derivation in terms of four-state differences. Cavender (Cavender and Felsenstein, 1987) gave it for a two-state model. Though it won't be obvious, the two invariants are the same (with replacement of $\frac{4}{3}$ by 2). One can see this by noting that when we take the purines as one state and the pyrimidines as the other, the two-state differences we obtain are 2/3 the size of the four-state differences. Noting that, it is easy to show that the four-state invariant becomes the two-state invariant.

Every quartet of species in the data will have one of these K invariants. However these are not all independent of each other. Note that the K invariants depend on the data only through the pairwise differences. For n species there are $n(n-1)/2$ of these, and they have $2n-3$ branch-length parameters predicting them. Thus we cannot have more than $(n-3)(n-2)/2$ independent quantities among them. For example, in our eight-species example, there cannot be more than 15 of them. By contrast there are $8 \times 7 \times 6 \times 5/(2 \times 3 \times 4) = 70$ quartets possible. So many of the K invariants must depend on each other. Taking out the 15 independent invariants, the number of degrees of freedom left is $2,599 - 15 = 2,584$.

For the four-species Jukes-Cantor model, there is for each tree topology one K invariant, as given above. Now we are down to seven degrees of freedom.

The L *invariants*

Cavender's L invariants are another set of quadratic invariants. They have a particularly nice interpretation, one which also makes them easy to generalize. Consider a symmetric model of DNA change (such as the Jukes-Cantor DNA model, or the two-state model that Cavender originally used). As Cavender and I pointed out (Cavender and Felsenstein, 1987), when the tree is ((A,B),(C,D)) whether or not species A and species B are identical depends on events that have occurred on the branch between them. Likewise, whether or not C and D are identical depends on the events in the branches between them. As these two sets of branches do not overlap, this implies that the identity of these two pairs of species is independent. Then the probability of both happening is simply the product of the probabilities of each:

$$\text{Prob}\ [(A = B)\ \&\ (C = D)] \;=\; \text{Prob}\ [(A = B)]\ \ \text{Prob}\ [(C = D)] \tag{22.18}$$

The probability of both being equal is, for the four-species Jukes-Cantor case, $P_{xxxx} + P_{xxyy}$. The probability that $(A = B)$ is $1 - D_{ab}$, and the probability that $(C = D)$ is $1 - D_{cd}$. The expressions for these are analogous to equation 22.16:

$$\begin{aligned} D_{ab} \;=\;& P_{xyxx} + P_{xyxy} + P_{xyyx} + P_{xyyy} + P_{xyxz} \\ &+ P_{xyyz} + P_{xyzx} + P_{xyzz} + P_{xyzy} + P_{xyzw} \end{aligned} \tag{22.19}$$

and

$$\begin{aligned} D_{cd} \;=\;& P_{xxxy} + P_{xyxy} + P_{xyyx} + P_{xxyx} + P_{xyxz} \\ &+ P_{xxyz} + P_{xyyz} + P_{xyzx} + P_{xyzy} + P_{xyzw} \end{aligned} \tag{22.20}$$

Substituting all these into 22.18 results in a quadratic equation in the P's, which yields the L invariant. It is not given here.

The L invariant thus is a simple expression of the independence of similarity in different parts of the tree. As such it can be easily tested statistically: One would simply construct a 2×2 contingency table, with individual sites counted to test whether $(A = B)$ and $(C = D)$ were independent across sites.

One might assume that the same independence would hold even when the model of DNA change is asymmetric. In fact, it will not hold in that case. Suppose that the tree is rather small, composed of short branches. Suppose further that one base (say G) is far more frequent than the other three. Knowing that $(A = B)$ is then usually the knowledge that both of these species share the frequent state. As the central branch of the tree is short, this also informs us that species C and D are most likely also in state G. Thus the two events $(A = B)$ and $(C = D)$ are far from independent. In the symmetric case knowing that A and B have the same state does not give us any information that helps us know whether C and D share the same state.

Generalization of Cavender's L *invariants*

Sankoff (1990) has shown that Cavender's L invariants can be generalized to larger numbers of species rather straightforwardly. For example, in the five-species tree in Figure 22.1 we could ask whether the patterns in the pair of species *a* and *b* differ, and also whether the differences in the triple of species *c*, *d*, and *e* differ. In this tree

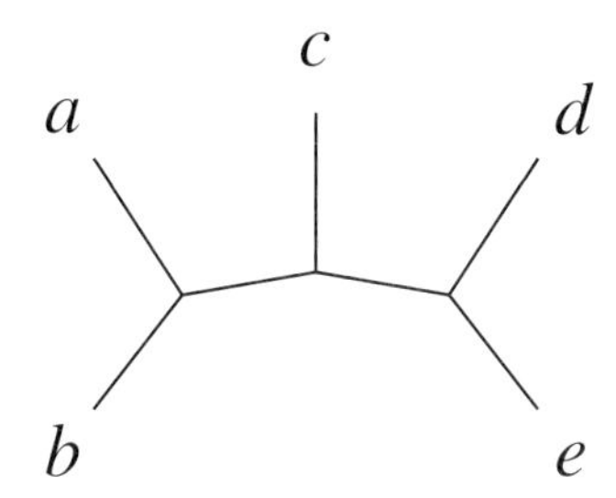

Figure 22.1: A five-species tree used as an example.

topology, with a symmetric model of base substitution such as the Jukes-Cantor model, the differences between these two sets of species should be independent.

To be more precise, the pair of species a and b could have either of two patterns, $\texttt{xx}$ and $\texttt{xy}$. The triple of species cde could have any one of five patterns ($\texttt{xxx}$, $\texttt{xxy}$, $\texttt{xyx}$, $\texttt{xyy}$, and $\texttt{xyz}$). As these two sets of species can be isolated from each other by dissolving a branch, in a tree of this topology the pattern that is seen in each of these sets of species depends only on events within the set of branches that connects them. We could imagine testing this postulated independence of the patterns by making up a 2×5 contingency table. It would have $(2-1) \times (5-1) = 4$ degrees of freedom.

In the same tree, there is another internal branch that can be dissolved, and it too has four degrees of freedom, implying four invariants. Sankoff (1990) has pointed out that there are advantages to paying attention to the invariants that test important conflicting hypotheses about the structure of the tree, and not spending effort on most of the invariants. He also notes that in trees of six or more species, one can dissolve more than one internal branch at a time and test the simultaneous independence of three (or more) sets of species.

We have been discussing a hypothetical eight-species tree. How many L invariants would it have? This may depend on the tree topology. For example, with the eight-species tree topology of Figure 22.2, there are five internal branches that could be dissolved. Making the L invariants for each of these branches (dissolving only that one branch), we see that the contingency tables would be 2×6, 3×5, 4×4, 2×6, and 6×2 in size, and would, respectively, have 5, 8, 9, 5, and 5 invariants. I am assuming without proof that these would all be independent. They account for 32 degrees of freedom, leaving us with $2,584 - 32 = 2,552$ degrees of freedom.

We can also dissolve two branches at a time as long as they do not leave us with any one-species sets. There are five ways to do this, leading to the partitions $\{ab \,|\, cd \,|\, efgh\}$, $\{ab \,|\, cdef \,|\, gh\}$, $\{abc \,|\, def \,|\, gh\}$, $\{abc \,|\, dgh \,|\, ef\}$, and $\{abcd \,|\, ef \,|\, gh\}$. For each of these partitions, the number of independent invariants should be the product over its sets of one less than the size of the set. Thus they should have

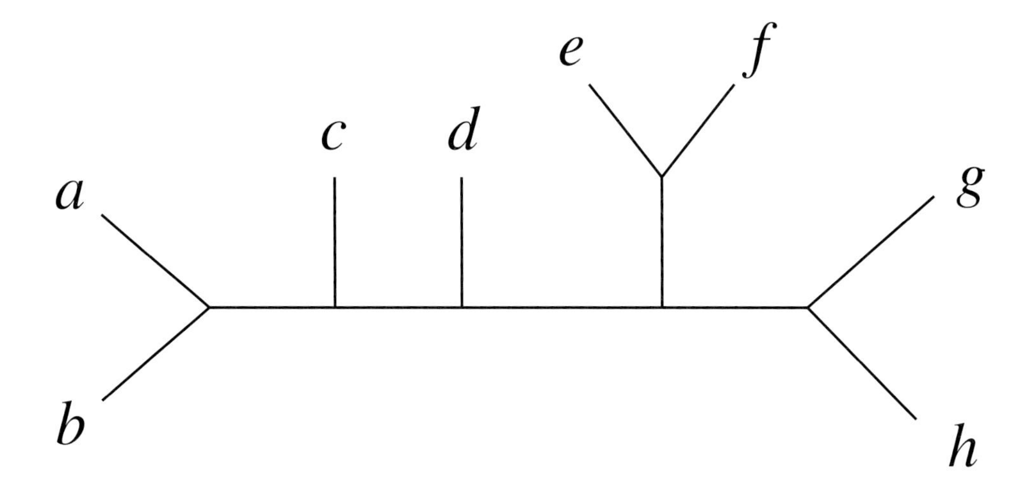

Figure 22.2: An eight-species tree used as an example in counting the number of generalized L invariants

$1 \times 1 \times 3, 1 \times 3 \times 1, 2 \times 2 \times 1, 2 \times 2 \times 1$, and $3 \times 1 \times 1$. This is a total of 17 invariants, leaving us with 2,535 degrees of freedom.

In addition, there is one way that the species can be partitioned into four sets, with each set having more than one species. It is $\{ab|cd|ef|gh\}$, which has one invariant. Sankoff also points out that we can also make invariants for less than the full set of species. For example, dropping species a, we have a seven-species tree. It has 24 degrees of freedom for L invariants. Dropping the other species one at a time, we find that three of the other seven-species trees have 24 L invariants, and four of them have 26 L invariants. Thus the seven-species trees have 200 L invariants in all.

Dropping two species at a time, we find that of the 28 ways we could drop two species, 6 have 10 L invariants, and 22 have 11 L invariants, for a total of 302 degrees of freedom. The 56 ways of dropping three species at a time yield five-species trees, each of which have four L invariants, for a total of 224 degrees of freedom. There are 70 ways of dropping four species at a time, yielding one L invariant each, for a total of 70.

The result is that, if all of these L invariants are independent, we now have reduced the number of degrees of freedom down to $2,535 - 200 - 302 - 224 - 70 = 1,739$. This argument has not shown that all of these degrees of freedom are independent, so the reduction of degrees of freedom may be less than this.

In the four-species Jukes-Cantor model, for each tree topology there is one L invariant, leaving us with six degrees of freedom. As there are five branch lengths to estimate, this leaves us with one degree of freedom still unaccounted for.

Drolet and Sankoff's k-state quadratic invariants

Drolet and Sankoff (1990) generalized the Cavender K and L invariants to a k-state symmetric Jukes-Cantor-like model of change. They prove for $k = 2$, $k = 3$, and $k = 4$, and conjecture for higher values of k that the Cavender K invariant is

$$\begin{aligned}
&\left[k\left(P_{xxxx} + P_{xyxy} + P_{xyxz} + P_{xyzy}\right) - 1\right] \\
&\quad \times \left[k\left(P_{xxxx} + P_{xyxy} + P_{xyyy} + P_{xxyx} + P_{xyxx} + P_{xxxy}\right) - 1\right] \\
&+ k\left(P_{xyyy} + P_{xxyx} - P_{xyxz}\right) \times k\left(P_{xyxx} + P_{xxxy} - P_{xyzy}\right) \\
&- \left[k\left(P_{xxxx} + P_{xyyx} + P_{xyzx} + P_{xyyz}\right) - 1\right] \\
&\quad \times \left[k\left(P_{xxxx} + P_{xyyx} + P_{xyyy} + P_{xxxy} + P_{xxyx} + P_{xyxx}\right) - 1\right] \\
&- k\left(P_{xyyy} + P_{xxxy} - P_{xyzx}\right) \times k\left(P_{xxyx} + P_{xyxx} - P_{xyyz}\right) = 0
\end{aligned} \qquad (22.21)$$

We will not try to count how many degrees of freedom this accounts for. In the four-species case it might be thought to account for the last remaining degree of freedom, but Ferretti and Sankoff (1993) have shown that this is not so. It is not an independent invariant in that case — they show that it can be written as a linear combination of the other invariants. For the four-species case, after taking out five degrees of freedom for the branch lengths, we are left with one degree of freedom unaccounted for, so there must be another invariant, or unknown form. For the eight-species, case, we have at least 1,739 degrees of freedom unaccounted for.

Clock invariants

We have been discussing invariants for trees that are unrooted and have no assumption of a molecular clock. If we impose the molecular clock, that places constraints on the branch lengths. That in turn means that there are more degrees of freedom since there are fewer branch length variables. Thus there are some more invariants. For example, in our hypothetical eight-species case there are $8 \times 2 - 3 = 13$ branch lengths in the tree, but when a clock is imposed, there are only seven degrees of freedom used to specify the tree. These are the times of the seven ancestral nodes. There then must be six constraints on the pattern frequencies that come from the assumption of the clock.

Finding them is easy. Consider the eight-species clocklike tree in Figure 22.3. We need only look at the six interior nodes other than the bottommost one. These have been numbered for easier reference. In this tree the species happen to be in alphabetic order left to right, which makes discussion easier.

If there were only three species, *a*, *b*, and *c* then there would be one clock constraint: that the pattern xyx should have the same expected frequency as yxx, which is more properly called xyy. We can apply the same condition to any triple of species in the tree. That appears to be far too many conditions. In fact, many of these triples test redundant information. The total number of degrees of freedom

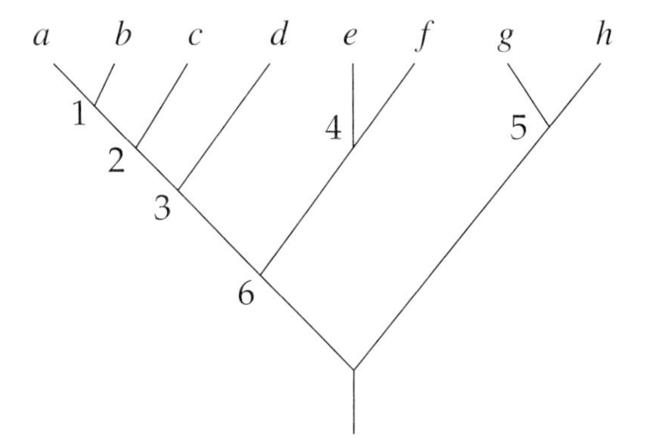

Figure 22.3: Eight-species clocklike tree used as an example in the discussion of clock invariants. The internal nodes of the tree have been assigned numbers.

accounted for by clockness with n species is $n - 2$. For eight species, that is, seven degrees of freedom. At the moment there is not a convenient way of writing down all seven of the linear clock invariants.

General methods for finding invariants

The methods used above of finding invariants are informal, and not easy to replicate in other cases. There are formal ways of finding all invariants.

Fourier transform methods

A quite general method of finding invariants is the Fourier transform approach originated by Evans and Speed (1993) and discussed further by Evans and Zhou (1998). It applies to the case in which the model of base change is the Kimura 3ST model (see Table 17.5 in Chapter 17, which includes the K2P model and the Jukes-Cantor model as special cases).

Evans and Zhou's algorithm starts with a rooted tree with m tips and a total of n nodes, counting the interior nodes and root as well. Their algorithm examines all but one of the 4^m patterns of bases. For each it uses a rule based on the multiplication table in a well-known group called a Klein 4-group to assign states to interior nodes of the tree. These assignments are in turn used to set coefficients of 1 or 0 in $3n$ vectors, each of length $4^m - 1$. Standard matrix algebra computational methods are then used to find all the coefficients of the basis vectors in the null space of this set of vectors (that is, to find a minimal set of vectors that each are orthogonal to all of the $3n$ vectors and can generate all vectors orthogonal to them as linear combinations). They then give a simple recipe for turning the coefficients of these basis vectors into polynomials. These are all of the invariants.

They also show how the same can be done for the K2P model and the Jukes-Cantor model, and also for cases in which the state distribution at the root of the tree is not assumed to be in equilibrium under the base substitution model. For our four-species Jukes-Cantor model, their method constructs five equations in 255 variables. The null space for these will consist of 250 vectors. The coefficients of these define 250 invariants, and we also have the trivial sum invariant, which declares the sum of the expected frequencies of all patterns to be 1.

Evans and Zhou's machinery establishes that in these cases the count of invariants is always equal to the degrees-of-freedom count, which we have been using. Evans and Speed (1993) had conjectured that the degrees of freedom count held in general; Evans and Zhou confirmed this for the Kimura 3ST model and its submodels.

Gröbner bases and other general methods

Ferretti and Sankoff (1996) and Kim (2000) have noted that finding the invariants for a tree is the problem of finding a "Gröbner basis" of a set of polynomials in algebraic geometry. This is discussed by Hagedorn and Landweber (2000) and by Kim (2000). Hagedorn (2000), for a fairly general framework, proves that the number of invariants is in fact given by the degrees-of-freedom count. The difficulty with using the computational machinery for the Gröbner basis is that, although cut-and-dried, it involves quite heavy computation even for cases as small as four species. By contrast, the Evans and Zhou machinery is much less onerous. The case they treat is also the one that the Hadamard machinery (covered in Chapter 17) handles. In fact, there is a close relationship between these works. The zeros in the Hadamard conjugation correspond precisely to the phylogenetic invariants. For the Kimura 3ST model, the invariants can be found readily from the expressions that must then be zero.

The most general method so far described is due to Allman and Rhodes (2003). Their methods find all invariants for the case with general transition probability matrices on all branches of a rooted tree. It requires only matrix algebra, and is both more general than the Hadamard methods and easier computationally than Gröbner basis methods.

Expressions for all the 3ST invariants

Steel et al. (1993) use this approach to find all invariants for any number of species for the Kimura 3ST model. Their expressions avoid any need for heavy computations. They treated the case of n species with a rooted tree with arbitrary base frequencies at the root. I have concentrated here on the more interesting case in which the base frequencies at the root are equal, as if they resulted from evolution under the Kimura 3ST model. Their results could be simplified for this case, and also for the K2P and Jukes-Cantor models.

Finding all invariants empirically

A quite novel and simple method of finding invariants was invented by Ferretti and Sankoff (1993, 1995). It takes the phylogeny and the substitution model, and picks random values of the parameters (such as the branch lengths). For each set of parameters, they compute the expected frequencies of all nucleotide patterns. They do this many times. Suppose that in the jth set of parameters the frequency of the ith pattern is P_{ij}. The points in the 4^n-dimensional space of pattern frequencies then lie in a subspace. We need to find equations in the pattern frequencies that define this subspace. For example, suppose we look for linear equations. If the invariant we seek is

$$\sum_{i=1}^{4^n} a_i \, P_{ij} \;=\; 0 \tag{22.22}$$

then this should hold for all the sets of parameters. If we take 4^n sets of parameters we then have a set of equations

$$\sum_{i=1}^{4^n} a_i \, P_{ij} \;=\; 0, \qquad j = 1, 2, \ldots, 4^n \tag{22.23}$$

as we know the P_{ij}, we can solve this set of equations for the a_i, (once we standardize one of the a_i to 1). Similarly, if we searched for a quadratic formula, we get a similar set of equations with both linear terms P_{ij} and quadratic terms $P_{ij}P_{kj}$. Those quadratic terms are known, and the equations are all linear in the unknown coefficients a_i and b_{ik}. So these can always be found by solving linear equations. One can go on to find cubic and quartic invariants in the same "empirical" way. Then number of equations can get large for higher-order invariants; a more serious problem, which Ferretti and Sankoff did not entirely solve, is simplifying the resulting set of invariants.

All linear invariants

There is a literature on finding all linear invariants, which I will mention only briefly. The difficulty is that these are of limited interest. In the symmetric models of base change, most of the linear invariants test the symmetry of the model; in the four-species Jukes-Cantor case, only two of the linear invariants, the Lake invariants, are phylogenetic invariants. Testing for symmetry is of interest, but can be done straightforwardly as described above in the section on symmetry invariants. Papers on finding linear invariants include those of Cavender (1989, 1991), Fu and Li (1992), Fu (1995), and Steel and Fu (1995).

Hendy and Penny (1996) consider models with and without a molecular clock. The invariants they find are of particular interest as they should include all the clock invariants.

Special cases and extensions

I have concentrated on the symmetric models of base change, with the base frequencies at the root of the tree being equal, as these are the cases of greatest interest. Many of the papers mentioned have also dealt with one or another asymmetric case. These are of interest, but no one model of asymmetric change has been subject to a comprehensive treatment. For example, Lake (1987) treated a case in which the tree is rooted and no assumption of equilibrium base composition is made at the root. His model of base change was fairly asymmetric, with only certain symmetry assumptions made. Nevertheless, the part of his work that connects to the other papers in the invariants literature is its specialization to the Kimura K2P model. Ferretti, Lang, and Sankoff (1994) gave a detailed discussion of the problem of finding invariants in asymmetric models of base substitution. The issue of how invariants can be extended to allow for correlation of evolution in adjacent sites is discussed by Steel et al. (1993).

Invariants and evolutionary rates

Invariants are defined for a particular evolutionary model, usually one with the same rate of evolution at all sites. What happens to them if the rate of evolution differs from site to site? Cavender and Felsenstein (1987) and Lake (1987) considered this problem, and both came to the conclusion that the quadratic invariants were no longer valid in such a case. However, Lake pointed out that the linear invariants continued to be invariants of a model with different evolutionary rates at each site, and even for models in which the rates differ in each character in each branch of the tree.

This would seem to make Lake's linear invariants appropriate for use when rates may vary among sites, whereas quadratic invariants would not work. The difficulty with practical use of the Lake invariants is that they account for only a small fraction of the information in the data set. They are computed using only those sites that have two of their bases purines and two pyrimidines, ignoring the information from all other sites. Consequently, computer simulation tests have shown that other methods have much greater power to discriminate among phylogenies (Huelsenbeck and Hillis, 1993; Hillis, Huelsenbeck, and Cunningham, 1994).

Testing invariants

We have seen that linear invariants can be tested using a simple chi-square test of a contingency table. Some quadratic invariants (such as the Cavender L invariants) can also be tested using contingency tables. It would also be possible to test other invariants for data sets with many sites, by using asymptotic formulas for their variance generated by the well-known "delta method." In principle, if we test all invariants, we are testing whether the model and tree fit the data. It should be equivalent (at least asymptotically) to testing whether all the nucleotide patterns are occurring at their expected frequencies. Navidi, Churchill, and von Haeseler (1991) have suggested doing such testing. However, in practice, such tests almost always reject the model. The tree may be strongly supported by the data, but the combination of the model and the tree is strongly rejected. This rejection results mostly from violation of some of the symmetry assumptions. Thus we might have strong evidence for the tree, but find that since AAAAA is very unequal in frequency to CCCCC, the fit of model plus tree is poor. Thus, at the moment, simultaneous statistical tests of all invariants are not of much use.

What use are invariants?

If invariants typically have problems with rate variation, and typically do not all fit their expected values, what use are they? They have found some use in comparative genomics, where Sankoff and Blanchette (1999) have used invariants with a Jukes-Cantor-like model of the appearance and disappearance of breakpoints of chromosome rearrangements. Sinsheimer, Lake, and Little (1997) have integrated use of Lake's linear invariants with a parsimony method, and they report that the combination helps counteract the low power of use of the invariants alone.

But these uses aside, invariants are worth attention, not for what they do for us now, but what they might lead to in the future. They are a very different way of considering tree topologies and branch lengths. Instead of crawling about in a tree space, trying to find the tree of best fit, they have us look at relationships of pattern probabilities in a space of pattern frequencies, and build up our inferences of the tree in that space. For the cases in which both invariants and the Hadamard conjugation apply, this is essentially the same as looking at which partitions show support in the Hadamard transform analysis. Both invariants and the Hadamard conjugation lead to interesting mathematics, and both give us a view of phylogenetic inference from a new direction. That alone would be enough to justify continued development of these interesting methods.

Chapter 23

Brownian motion and gene frequencies

Many morphological characters that are used for phylogenetic inference are not discrete, but in their original form they are quantitative characters measured on a continuous scale. Although in practice they are often discretized before being used, it is of interest to know how to treat the original data without making them discrete. Gene frequencies are another type of continuous data. The original phylogeny methods of Edwards and Cavalli-Sforza (1964) were developed to analyze them. Edwards and Cavalli-Sforza's method was to approximate the process by Brownian motion. I will describe Brownian motion, and how likelihoods are computed and maximized for a phylogeny using it. This involves much use of the density function of the normal distribution. Then I will consider how well the approximation describes gene frequencies, and (in the next chapter) how well it describes quantitative characters. At the end of this chapter there will also be some consideration of parsimony methods for gene frequencies.

Brownian motion

Robert Brown (1773–1858) was a systematic botanist and microscopist. In fact, he was the first to distinguish the gymnosperms from the angiosperms, the first to distinguish the monocots from the dicots, and it was he who discovered and named the cell nucleus. In 1827, he observed that pollen grains in solution jiggled ceaselessly. The explanation, in terms of random molecular impacts, came much later, and led on to Einstein's work on the subject. Ford (1992) gives a good account of Brown's work on Brownian motion.

Mathematicians, notably Norbert Wiener, have constructed a stochastic process that is intended to approximate Brownian motion. In it, a particle takes a great many small steps on an axis. Each step is independent of the others, and each displaces the particle by a random amount. The mean displacement is zero, and the variance of the displacements is the same no matter where the particle is on the axis. If the variance of one displacement is s^2 and there are n displacements, the net displacement is the sum of the individual steps. Its variance is the variance of this sum, and as the individual steps are independent, the variance of the sum is the sum of the variances ns^2.

Wiener's mathematical process of Brownian motion is the limit as we take the variance s^2 smaller and smaller, and the number of steps n larger and larger, such that their product remains constant. The process is thus infinitely jiggly, unlike the physical process of Brownian motion, which has discrete impacts. There are many fascinating facts about the mathematical process of Brownian motion, mostly connected with its infinite jiggliness. For our purposes we do not need to consider them, but need only know what the distribution is of the net displacement after time t.

The displacements are independent and are summed, so that their variances can be added up. Thus if σ^2 is the variance that is expected to accumulate per unit time, then the variance of the net displacement after time t will be $\sigma^2 t$. The fact that the total displacement along a branch is the sum of a large (in fact, infinite) number of independent quantities implies that it is normally distributed. Thus when we approximate an evolutionary process by saying that it follows a Brownian motion with variance σ^2 per unit time, we know that the net change along a branch of length t is drawn from a normal distribution with mean 0 and variance $\sigma^2 t$. The displacements in different branches of a tree are independent, because the small steps of which they are composed are all independent.

We now have to show what this implies for evolution. There were in fact already some interesting connections between Robert Brown and evolution. Charles Darwin knew Robert Brown and asked his opinion on what microscopes to take with him on his voyage on the *Beagle*. Brown's connection with evolution goes further, and did not even stop with his death. His death made available a meeting time slot before the Linnean Society: It was filled by the famous session at which Darwin's and Wallace's theories of natural selection were first presented. Brown would probably have been astounded, and perhaps dismayed, to see his observation of Brownian motion become the basis of Edwards and Cavalli-Sforza's paper on the numerical inference of phylogenies.

Likelihood for a phylogeny

If we have a measurable character that we know evolves according to a Brownian motion process, we can compute the likelihood for any given tree rather easily, owing to the normality of the distributions of changes in individual branches, and the independence of the change in each branch. This was done by Edwards and Cavalli-Sforza (1964; see also Cavalli-Sforza and Edwards, 1967), who showed how to compute the likelihood for a given tree on which the values of the characters at each interior node were known. This turns out to give a likelihood with unpleasant singularities.

David Gomberg (1968), in an unpublished manuscript that nevertheless influenced subsequent workers, formulated the problem in Bayesian terms, with prior probabilities on trees provided by a random process of branching of lineages. He used the result that the joint distribution of the tips of the tree was multivariate normal. He did not have a very practical computational method, but understood the structure of the problem well.

Figure 23.1 shows a tree, with the observed values of the character at the tips, and true unknown values at the interior nodes. The change of the character in each branch, which is the difference between the states at each end of the branch, is shown. The branch lengths are given as the v_i.

The Brownian motion process, as we are using it, has the property that its changes in different lineages are independent. As with the other stochastic processes that we have used for phylogenies, sister lineages start at the same point, but their changes are independent. The character value at the end of lineage 5 in Figure 23.1 for example, is the sum of four terms:

$$x_5 = x_0 + (x_{12} - x_0) + (x_{11} - x_{12}) + (x_5 - x_{11}) \tag{23.1}$$

Except for x_0, which is a constant, the terms for changes on the right side are all independent and drawn from normal distributions. This guarantees that their sum is also normally distributed. Its mean is x_0 and its variance is the sum of the variances of the individual terms, so that

$$\text{Var}\ (x_5) = \sigma^2 v_{12} + \sigma^2 v_{11} + \sigma^2 v_5 \tag{23.2}$$

where σ^2 is the variance that accumulates in the Brownian motion per unit branch length.

Similarly, the character value for lineage 7 in Figure 23.1 is the sum of normally distributed terms:

$$x_7 = x_0 + (x_{12} - x_0) + (x_{11} - x_{12}) + (x_{10} - x_{11}) + (x_7 - x_{10}) \tag{23.3}$$

and of course

$$\text{Var}\ (x_7) = \sigma^2 v_{12} + \sigma^2 v_{11} + \sigma^2 v_{10} + \sigma^2 v_7 \tag{23.4}$$

The values x_5 and x_7 are not independent of each other, because the expressions on the right sides of equations 23.1 and 23.3 share some terms. It is easy to show that the covariance of two such random sums is simply the variance contributed by the terms that they share; in this case it is

$$\text{Cov}\ (x_5, x_7) = \text{Var}\ [(x_{12} - x_0) + (x_{11} - x_{12})] = \sigma^2 v_{12} + \sigma^2 v_{11} \tag{23.5}$$

One can do this for all pairs of tips on the tree, with the analogous result. They are jointly multivariate normally distributed, all with the expectation x_0, and with

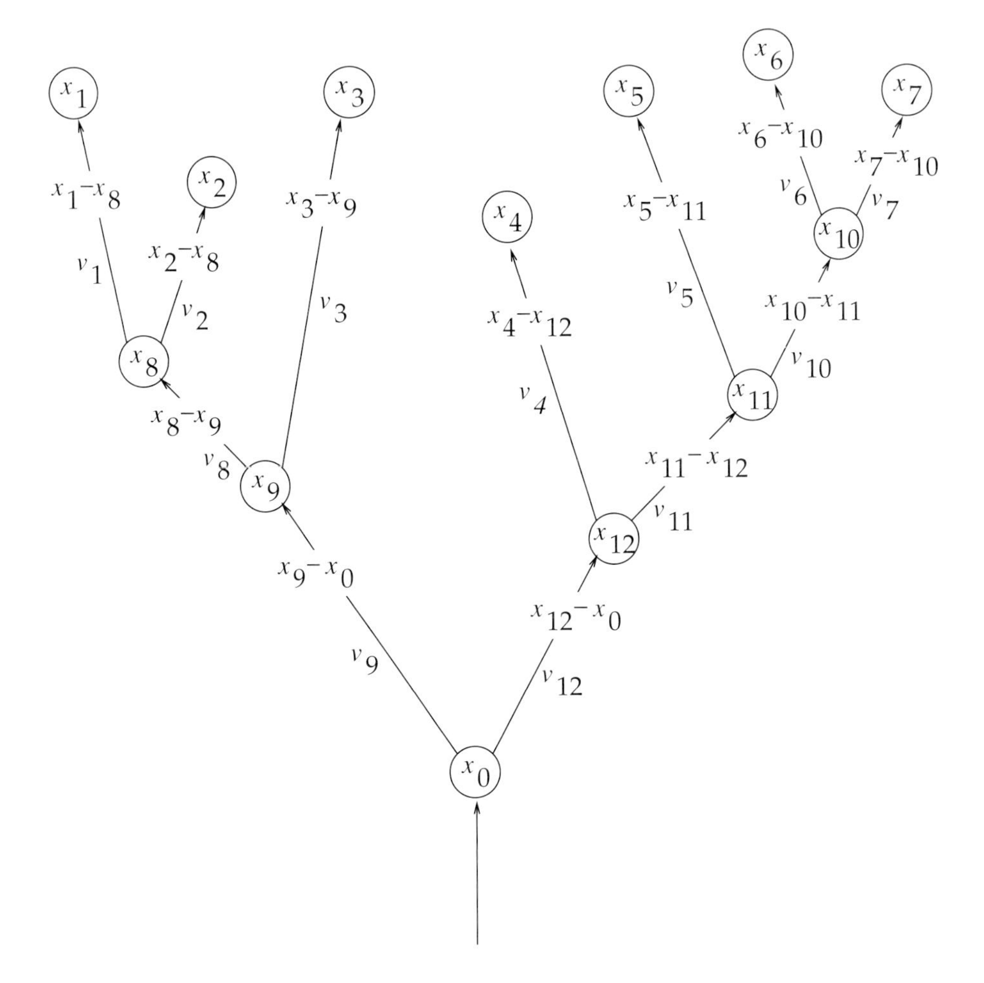

Figure 23.1: A tree with a character that has evolved by Brownian motion. The values of the character are shown at each tip, and the unknown true values at each interior node, including the starting value at the root. Next to each branch is shown the branch length v, as well as the net change of the character along that branch. Note that these true values may differ from any estimate that we might make of them.

covariances that are the variance contributed by the branches that are shared by the path from the root up to the most recent common ancestor of these two tips. In matrix terms, if $\mathcal{N}$ indicates the (multivariate) normal distribution,

$$\mathbf{x} \sim \mathcal{N}\left(x_0\mathbf{1},\ \sigma^2\mathbf{T}\right) \tag{23.6}$$

where $\mathbf{1}$ is a column vector of 1s, and $\mathbf{T}$ is the matrix whose (i, j) element is the branch length shared by the paths from the root up to tips i and j. As an illustration, here is the $\mathbf{T}$ matrix for the tree in Figure 23.1:

$$\begin{bmatrix} v_1+v_8+v_9 & v_8+v_9 & v_9 & 0 & 0 & 0 & 0 \\ v_8+v_9 & v_2+v_8+v_9 & v_9 & 0 & 0 & 0 & 0 \\ v_9 & v_9 & v_3+v_9 & 0 & 0 & 0 & 0 \\ 0 & 0 & 0 & v_4+v_{12} & v_{12} & v_{12} & v_{12} \\ 0 & 0 & 0 & v_{12} & v_5+v_{11}+v_{12} & v_{11}+v_{12} & v_{11}+v_{12} \\ 0 & 0 & 0 & v_{12} & v_{11}+v_{12} & v_6+v_{10}+v_{11}+v_{12} & v_{10}+v_{11}+v_{12} \\ 0 & 0 & 0 & v_{12} & v_{11}+v_{12} & v_{10}+v_{11}+v_{12} & v_7+v_{10}+v_{11}+v_{12} \end{bmatrix} \quad (23.7)$$

This matrix has a structure strongly reminiscent of the tree (and not surprisingly so, given the way we constructed it). It is of the form

$$\left[\begin{array}{c|c|c|c|c|c|c} a & b & c & 0 & 0 & 0 & 0 \\ \cline{1-2} b & d & c & 0 & 0 & 0 & 0 \\ \cline{1-3} c & c & e & 0 & 0 & 0 & 0 \\ \hline 0 & 0 & 0 & f & g & g & g \\ \cline{4-7} 0 & 0 & 0 & g & h & i & i \\ \cline{5-7} 0 & 0 & 0 & g & i & j & k \\ \cline{6-7} 0 & 0 & 0 & g & i & k & l \end{array}\right] \quad (23.8)$$

Figure 23.2 shows a realization of Brownian motion of one character, on a five-species tree. The open circles indicate where the forks are — these tend to get obscured by all the criss-crossing of curves just after them. In fact, the tree has the same topology as Figure 23.2. The character shows how noisy the relationship between any one character and the tree is. It takes multiple characters to make a decent estimate, even when the Brownian motion process applies exactly.

What likelihood to compute?

Although we still have not shown how to do the likelihood computations or search for the best tree, so far everything seems straightforward. This section will show that it is not. We will look closely at the two-species case, where the algebra is simplest. Nevertheless, a detailed mathematical derivation will be needed. The results are startling — maximum likelihood estimation does not work, and needs fixing.

Imagine that we have collected values of p different characters in two species. Denote the value of character i in species 1 by x_{1i}, and the value of that character in species 2 by x_{2i}. We will assume that the expected variance per unit of branch length in character i is σ_i^2. In most analyses one standardizes the characters so that $\sigma_i^2 = 1$, but we will not need to do that here. There is only one possible tree topology, with two branch lengths, v_1 and v_2. We also need to use the value of the character i at the root, which we call x_{0i}.

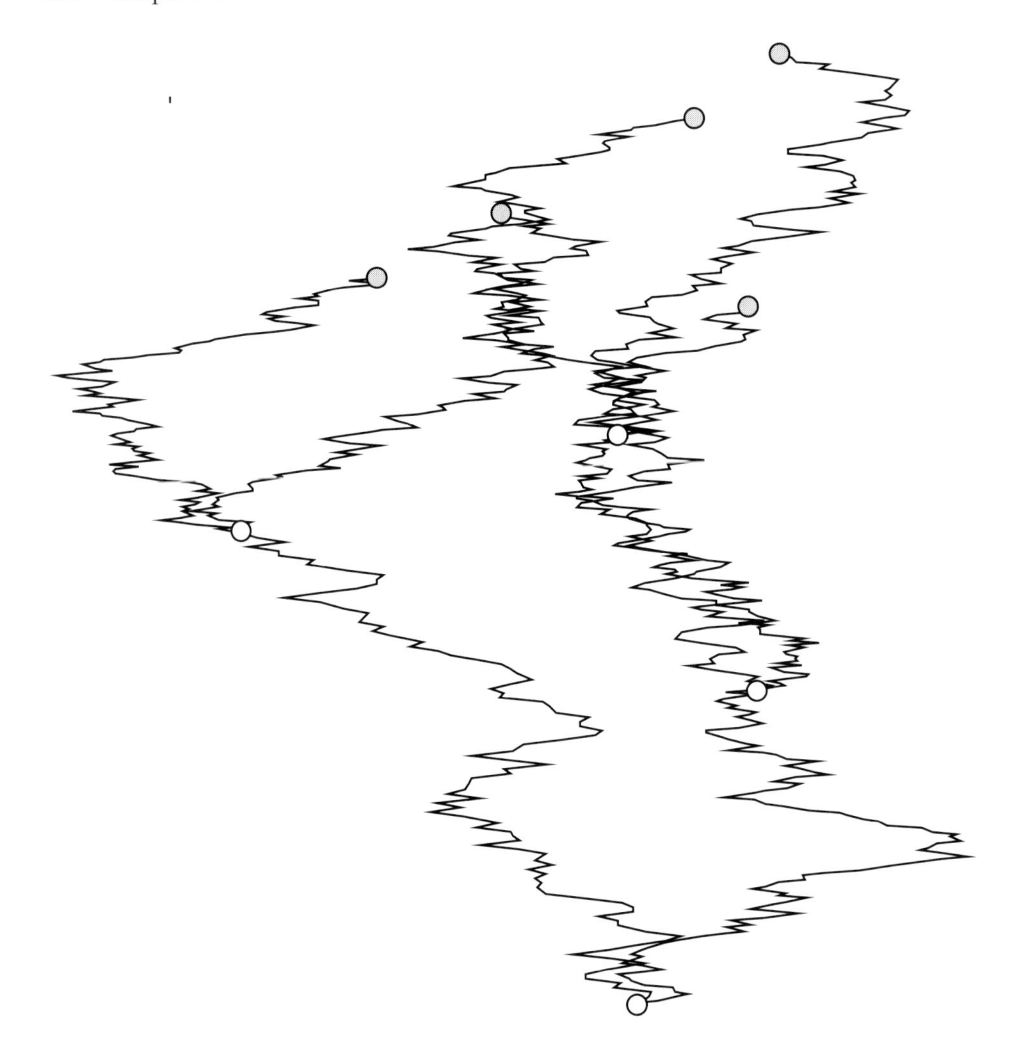

Figure 23.2: A Brownian motion process in one character on a five-species tree. Open circles show the locations of the forks. The tree has the same topology as Figure 23.1.

The likelihood is the product of terms for each character, since the characters evolve independently on the tree. It is also the product of the probabilities of the changes in each of the two lineages. In lineage 1, character i changes from x_{0i} to x_{1i}, the probability of this change being the density of a normal distribution whose variate is $x_{1i} - x_{0i}$ and whose expectation is 0 with variance $\sigma_i^2 v_1$. There is a similar term for the second lineage. Since the probability density for a normal variate x whose expectation is μ and whose variance is σ^2 is

$$f\left(x;\mu,\sigma^2\right) \;=\; \frac{1}{\sigma\sqrt{2\pi}}\,\exp\left[-\frac{(x-\mu)^2}{2\sigma^2}\right] \tag{23.9}$$

and since μ is 0 and σ^2 is $\sigma_i^2 v_1$, the probability density at the observed values is then the product of p pairs of these normal densities:

$$L \;=\; \prod_{i=1}^{p} \frac{1}{(2\pi)\sigma_i^2\sqrt{v_1 v_2}}\,\exp\left(-\frac{1}{2\sigma_i^2}\left[\frac{\left(x_{1i}-x_{0i}\right)^2}{v_1}+\frac{\left(x_{2i}-x_{0i}\right)^2}{v_2}\right]\right) \tag{23.10}$$

This has parameters v_1, v_2, and the x_{0i}. We are going to try to find values that maximize it.

We start with the x_{0i}. Holding v_1 and v_2 constant, we search for the value of x_{0i} that maximizes L. Since x_{0i} appears in only the ith term in the product in equation 23.10, we can simply find the value that maximizes that term. The relevant part of the term is

$$Q \;=\; \frac{\left(x_{1i}-x_{0i}\right)^2}{v_1}+\frac{\left(x_{2i}-x_{0i}\right)^2}{v_2} \tag{23.11}$$

which is to be minimized. To find the minimum of this with respect to x_{0i}, we take the derivative of Q and equate it to zero:

$$\frac{dQ}{dx_{0i}} \;=\; -2\frac{(x_{1i}-x_{0i})}{v_1}-2\frac{(x_{2i}-x_{0i})}{v_2} \;=\; 0 \tag{23.12}$$

The result is a simple linear equation in x_{0i} whose solution is

$$\widehat{x}_{0i} \;=\; \frac{\frac{1}{v_1}x_{1i}+\frac{1}{v_2}x_{2i}}{\frac{1}{v_1}+\frac{1}{v_2}} \tag{23.13}$$

So far, so good. The values look very reasonable: The starting point x_{0i} for evolution in the ith character is simply a weighted average of the tip values for that character, with weights that are the inverses of the variances for the two tips.

Now we need to substitute the maximum likelihood estimates of all the x_{0i} into our expression for the likelihood. Substituting the estimates into Q in equation 23.11, we find after collecting terms that

$$Q \;=\; \frac{\left(x_{1i}-x_{2i}\right)^2}{v_1+v_2} \tag{23.14}$$

Substituting that into the full likelihood (equation 23.10), we can move the product across characters into the exponent, where it becomes a sum:

$$\begin{aligned} L \;&=\; \prod_{i=1}^{p} \tfrac{1}{(2\pi)\sigma_i^2\sqrt{v_1 v_2}}\,\exp\left(-\tfrac{1}{2\sigma_i^2}\left[\tfrac{(x_{1i}-x_{2i})^2}{v_1+v_2}\right]\right) \\ &=\; (2\pi)^{-p}\left(v_1 v_2\right)^{-\frac{1}{2}p}\left(\prod_{i=1}^{p}\sigma_i^{-2}\right)\,\exp\left(-\tfrac{1}{2}\sum_{i=1}^{p}\tfrac{(x_{1i}-x_{2i})^2}{\sigma_i^2(v_1+v_2)}\right) \end{aligned} \tag{23.15}$$

Now we have our likelihood in terms of two parameters, v_1 and v_2 (as we are treating the σ_i^2 as known, and have set the x_{0i} to their maximum likelihood values). It is usually easier to maximize the logarithm of the likelihood rather than the likelihood itself, and one obtains the same estimates of the parameters. Taking the logarithm of L in equation 23.15,

$$\ln L = -p \ln(2\pi) - 2\sum_{i=1}^{p} \ln \sigma_i - \frac{1}{2} p \ln(v_1 v_2) - \frac{1}{2}\sum_{i=1}^{p} \frac{(x_{1i} - x_{2i})^2}{\sigma_i^2 (v_1 + v_2)} \tag{23.16}$$

All that remains is to maximize this with respect to v_1 and v_2. To simplify this, we replace the variables v_1 and v_2 by, respectively, $F v_T$ and $(1 - F) v_T$, so that v_T is the total $v_1 + v_2$ and F is the fraction of it that is due to v_1. Making a one-to-one transformation of variables like this should not affect the maximum likelihood: after finding the best values of v_T and F, we can easily find which values of v_1 and v_2 those imply.

To simplify equation 23.16 we will also replace the sum of squares of standardized differences in the x's by D^2:

$$D^2 = \sum_{i=1}^{p} \left(\frac{x_{1i} - x_{2i}}{\sigma_i} \right)^2 \tag{23.17}$$

This is, in effect, a squared distance measure between tips 1 and 2 in the coordinates of the phenotype space, with each axis being standardized by its Brownian motion variance σ_i^2. The result is

$$\ln L = K - \frac{1}{2} p \ln \left[F v_T (1 - F) v_T \right] - \frac{1}{2} \frac{D^2}{v_T} \tag{23.18}$$

Here K is just a placeholder for the terms in equation 23.17 that do not have F or v_T in them. We can rewrite equation 23.18 a bit more simply as

$$\ln L = K - \frac{1}{2} p \ln F - \frac{1}{2} p \ln(1 - F) - p \ln v_T - \frac{1}{2} \frac{D^2}{v_T} \tag{23.19}$$

Now we are ready for the horrible shock. We could maximize this with respect to v_T easily enough, and we do that later. But if we look at the way the likelihood depends on F, we see something strange. As F decreases towards zero, $-\ln F$ goes to infinity. And if instead it increases toward 1, $-\ln(1 - F)$ goes to infinity. There is a point ($F = 0.5$, in fact) where the slope of the log-likelihood with respect to F is zero, but that is a *minimum* rather than a maximum! By taking F or $1 - F$ to zero, we can make the likelihood infinitely large. This corresponds to allowing v_1 to go to zero, or else allowing v_2 to go to zero.

In fact, Cavalli-Sforza and Edwards (1967) found an exactly analogous behavior in their pioneering studies of maximum likelihood phylogeny estimation, and it blocked their use of the method. They estimated the phenotypes not only at the root, but also at all interior nodes on the tree. They found that if any interior branch of the tree was taken to be of zero length, then the inferred phenotypes at the nodes at the two ends of that branch would become equal, and the likelihood of the tree would become infinite.

This bizarre behavior has a simple explanation. Remember that the likelihood we are computing is not the probability of the observed data, but the probability *density* at the observed data. By allowing the root to coincide with one of the data points (which is what happens if we take v_1 or v_2 to zero) we are making the probability of the changes on that branch be 1. For any other value of v_1 and v_2 we have an infinitesimally small probability of this change, as the probability density is multiplied by an infinitesimal, dx. By allowing (say) v_1 to go to 0, we in effect multiply the likelihood by $1/dx$, which is infinite. The result is an infinitely strong attraction to shortening a branch to length 0. Another way of thinking about this is that by allowing a branch length to shrink to zero, we are permitting the method to drop a parameter from the estimation problem.

Assuming a clock

There are several possible ways to solve this problem. One is to constrain the branch lengths. For example, we could decree the presence of an evolutionary clock and insist that $v_1 = v_2$. That would fix F at $1/2$. This was the approach taken by Elizabeth Thompson in a brilliant but difficult monograph (1975) for the more general case of n populations. It eliminates the singularity and the infiniteness of the likelihood. In our case, where $n = 2$, the likelihood becomes

$$\ln L = K' - p \ln v_T - \frac{1}{2} \frac{D^2}{v_T} \tag{23.20}$$

(the constant K' having absorbed the now-uninteresting terms in F). This is readily solved by differentiating with respect to v_T to get

$$\widehat{v}_T = D^2/(2p) \tag{23.21}$$

which of course means that

$$\widehat{v}_1 = \widehat{v}_2 = D^2/(4p) \tag{23.22}$$

This would be an entirely satisfactory solution, although it leaves us unable to handle nonclocklike cases, and also is a strongly biased answer. In fact, the answer is only half as big as it ought to be. We can detect this by computing the expectation of D^2 and putting it into equation 23.22. Note that in character i,

$$x_{1i} - x_{2i} = (x_{1i} - x_{0i}) - (x_{2i} - x_{0i}) \tag{23.23}$$

and that the two terms on the right side are independent; they have expectation zero and variances $\sigma_i^2 v_1$ and $\sigma_i^2 v_2$. Then the expectation

$$\mathbb{E}\left[\frac{(x_{1i} - x_{2i})^2}{\sigma_i^2}\right] = v_1 + v_2 \tag{23.24}$$

So that the expected value of D^2 is

$$\mathbb{E}\left[D^2\right] = p\,(v_1 + v_2) \tag{23.25}$$

As a result, the expectation of $\widehat{v}_1$ and of $\widehat{v}_2$, from equation 23.21, is

$$\mathbb{E}\left[\widehat{v}_1\right] = \mathbb{E}\left[\widehat{v}_2\right] = \frac{1}{2}v_1 = \frac{1}{2}v_2 \tag{23.26}$$

This is a bias by a factor of two. Nor is this bias confined to the case of two populations. In use of Thompson's n-population results, one frequently finds multiway splits at the base of the tree, often fourway splits. These are symptoms of a tendency to collapse small branches near the base of the tree.

In fact, this bias is an old and familiar one. In the case of two populations, if all the σ_i^2 happen to equal 1, the estimation of v_1 (or, of course, v_2) is exactly equivalent to estimating the variance of a set of distributions that all have the same true variance but different means. It is well known in statistics that if we estimate the variance v_1 of a sample of two points from its maximum likelihood estimator, we compute

$$\widehat{v}_1 = \frac{(x_1 - \bar{x})^2 + (x_2 - \bar{x})^2}{2} \tag{23.27}$$

This estimate is biased, because $\bar{x}$ is an estimate rather than the true mean. It is common to use instead the unbiased estimator, which requires a denominator of $n - 1$ rather than n, to correct for $\bar{x}$ being too close to the observed values x_1 and x_2. In the case $n = 2$, the factor $(n - 1)/n$, which is the bias in the estimate of the variance, is $1/2$. In the case of p characters, we average over p characters a series of estimators of v_1 that are each biased by being too small. The number of data points provides a spurious sense of accuracy.

Another way of thinking of this bias is that it arises from having to estimate so many parameters other than v_1 and v_2. There is one original phenotype per character to be estimated, namely the quantities x_{0i}. The number of parameters to be estimated rises proportionally to the number of characters examined. The ratio of data points to parameters thus does not rise indefinitely, but approaches an asymptote. It is in cases like these that the so-called infinitely many parameters problem arises; frequently it causes serious problems for likelihood.

Thompson's (1975) solution is to assume an evolutionary clock and to press ahead in the face of any concerns about biases near the root of the tree, considering bias an irrelevant criterion. However, there is another possible approach that avoids the issue of bias and does not need to assume a clock.

The REML approach

In my own papers (Felsenstein, 1973c, 1981a) on the problem I have adopted a different approach. It avoids the bias and the assumption of a clock, but at a different cost, that of omitting a bit of the data. It is a part of the data that seems not to be relevant anyway. The approach is to use only the differences between the populations and to omit any information about where they actually are in the phenotype space. Thus if we have a set of populations that, for the first character, had phenotypes x_1, x_2, x_3, ... x_n, we would use only their differences. That means we would know that x_2 was $x_2 - x_1$ above the first population, that population 3 was $x_3 - x_1$ above the first population, and so on. We discard all information about where x_1 is on the phenotype scale. Note that we are *not* reducing all information to a set of pairwise differences between populations, as we keep track of the differences in each character separately.

Offhand, this would not seem to be too severe a limitation. It should be immediately obvious that taking a set of populations and increasing the phenotypes in all of them by the same constant K should have no effect on our estimate of the tree, only on the estimate of the starting phenotype x_{0i}. What we are doing is to retain the information on relative positions on the phenotype scale but to discard the information about absolute positions. Then we will do maximum likelihood estimation of the phylogeny from the differences. This procedure, applied to mixed model analysis of variance, is the *REML* procedure of Patterson and Thompson (1971). The acronym "REML" has been variously translated as "restricted maximum likelihood," "reduced maximum likelihood," and "residual maximum likelihood." It is most generally described as taking the contrasts among the observations, and then doing maximum likelihood estimation as if we were given only those contrasts and not the original data. We have dropped the overall mean of the characters, so that we know only their relative displacement from each other, and not where they are located on the character axis. REML drops a small, and seemingly irrelevant, part of the information, so there seems some chance that it would be almost as powerful as maximum likelihood. We will see that it eliminates the problem of the bias.

In the two-species example that we have been using, the REML procedure consists of reducing the information for each character to the difference $x_1 - x_2$, thus discarding the information about where on the phenotype scale x_1 actually is. The expectation of the difference $x_1 - x_2$ is of course 0, and the variance is $\sigma_i^2\,(v_1 + v_2)$. The likelihood is then from equation (23.15) the product

$$L = \prod_{i=1}^{p} \frac{1}{\sqrt{2\pi}\sigma_i\sqrt{v_1+v_2}} \exp\left(-\frac{1}{2\sigma_i^2}\frac{(x_{1i}-x_{2i})^2}{v_1+v_2}\right) \tag{23.28}$$

If we take logarithms in the usual way, this becomes

$$\ln L = K - \frac{p}{2}\ln(v_1+v_2) + \frac{1}{2\,(v_1+v_2)}\sum_{i=1}^{n}\left(\frac{x_{i1}-x_{i2}}{\sigma_i}\right)^2 \tag{23.29}$$

where K contains all the uninteresting terms that do not have any of the $v's$. The summation on the right side is just the standardized distance D^2 as before, so this becomes

$$\ln L \;=\; K \;-\; \frac{p}{2}\ln\left(v_1 + v_2\right) \;+\; \frac{D^2}{2\left(v_1 + v_2\right)} \tag{23.30}$$

We cannot estimate v_1 and v_2 separately in this case, as they always appear as their sum v_T. Putting it in,

$$\ln L \;=\; K - \frac{p}{2}\ln\left(v_T\right) + \frac{D^2}{2v_T} \tag{23.31}$$

and when this is differentiated with respect to v_T and the derivative set equal to 0, we get the maximum likelihood estimate

$$\widehat{v}_T \;=\; D^2/p \tag{23.32}$$

Note that this differs from equation 23.21 by the all-important factor of two. In fact, unlike the estimate of equation 23.21, it is not biased at all, having expectation $v_1 + v_2$. We obtained this lack of bias by dropping any attempt at estimating the starting point x_0, but this does not seem too great a sacrifice to make.

The result is that we can estimate, in a reasonable manner, the unrooted phylogeny and its branch lengths. We cannot infer the position of the root on that phylogeny, or in the phenotype space. The derivation here has been for only two populations ($n = 2$), but the method applies to larger numbers as well. In matrix terms, if we define $\mathbf{C}$ as the $(n-1) \times n$ matrix of contrasts between the populations, then the joint distribution of the contrasts will be a normal distribution with means 0 and covariances calculated using the matrix of contrasts $\mathbf{C}$ and the original covariance matrix of populations $\mathbf{V}$:

$$\mathbf{y} \;=\; \mathbf{C}\,\mathbf{x} \;\sim\; \mathcal{N}\left(\mathbf{0},\; \mathbf{C}\,\mathbf{V}\,\mathbf{C}^T\right) \tag{23.33}$$

Note that this is the equation for the distribution of a vector of values, the contrasts $\mathbf{y}$ between populations. However it is for a single character. The variance of the Brownian motion for that character, σ_i^2, is contained in the covariance matrix $\mathbf{V}$.

Multiple characters and Kronecker products

A more complete set of data will have multiple characters, as we have been assuming in our discussion of the two-population case above. For now, let us consider the case in which the characters evolve independently on the tree. If x_{ij} is the value of the ith character in the jth population, it seems sensible to take the same set of contrasts for each character. The result would be a set of y_{ij}. For each character we have a vector of $n-1$ values, one for each of the $n-1$ contrasts.

We can use matrix notation to express this by taking the vectors $x_1, x_2, \ldots$ and stacking them on top of each other. The values of all n characters in population 1 are on top. Below them are the values of all n characters in population 2, and so on. The resulting $np \times 1$ vector has a normal distribution. We are assuming for the moment that the characters are uncorrelated with each other, so the covariance matrix of these characters will consist of an $p \times p$ array of blocks, each block being $n \times n$ in size. The covariance matrix looks like this:

$$\begin{bmatrix} \mathbf{V}^{(1)} & \mathbf{0} & \dots & \mathbf{0} \\ \mathbf{0} & \mathbf{V}^{(2)} & & \mathbf{0} \\ \vdots & & \ddots & \vdots \\ \mathbf{0} & \mathbf{0} & \dots & \mathbf{V}^{(n)} \end{bmatrix} \tag{23.34}$$

From equation (23.6), the covariance matrices $\mathbf{V}^{(i)}$ are $\sigma_i^2\mathbf{T}$, where σ_i^2 is the Brownian motion variance in character i. Thus if we were to make up a matrix

$$\mathbf{A} = \begin{bmatrix} \sigma_1^2 & 0 & \dots & 0 \\ 0 & \sigma_2^2 & & 0 \\ \vdots & & \ddots & \vdots \\ 0 & 0 & \dots & \sigma_p^2 \end{bmatrix} \tag{23.35}$$

one could say that the overall covariance matrix consisted of each entry in $\mathbf{A}$ multiplied by the tree covariance matrix $\mathbf{T}$. This kind of a product is called a *Kronecker product*. In the Kronecker product $\mathbf{A} \otimes \mathbf{T}$, the elements of $\mathbf{A}$ are each replaced by the entire matrix $\mathbf{T}$, with every term in that $\mathbf{T}$ being multiplied by that a_{ij}.

Thus we can write the overall distribution of $\mathbf{x}$ as

$$\mathbf{x} \sim \mathcal{N}(\mathbf{0},\ \mathbf{A} \otimes \mathbf{T}) \tag{23.36}$$

We will see in the next chapter that this expression generalizes easily to the case of nonindependent characters.

When we take the same $n-1$ contrasts for each of these p characters, the matrix of contrasts applied to the stacked-up vector of values $\mathbf{x}$ is

$$\begin{bmatrix} \mathbf{C} & \mathbf{0} & \dots & \mathbf{0} \\ \mathbf{0} & \mathbf{C} & & \mathbf{0} \\ \vdots & & \ddots & \vdots \\ \mathbf{0} & \mathbf{0} & \dots & \mathbf{C} \end{bmatrix} \tag{23.37}$$

and this can also be written as the Kronecker product of the $p \times p$ identity matrix and $\mathbf{C}$, so that it is $\mathbf{I} \otimes \mathbf{C}$, which is an $(n-1)p \times (n-1)p$ matrix. We will call the resulting stacked-up vector of contrasts $\mathbf{y}$. We saw that for one character the covariance matrix of the contrasts was $\mathbf{CVC}^T$. For p characters it is

$$(\mathbf{I} \otimes \mathbf{C})\,(\mathbf{A} \otimes \mathbf{T})\,(\mathbf{I} \otimes \mathbf{C})^T \tag{23.38}$$

This can be simplified using two rules for Kronecker products: The transpose of $\mathbf{A} \otimes \mathbf{B}$ is $\mathbf{A}^T \otimes \mathbf{B}^T$, and the product of $(\mathbf{A} \otimes \mathbf{B})$ and $(\mathbf{D} \otimes \mathbf{E})$ is $(\mathbf{AD} \otimes \mathbf{BE})$. Applying these we see that the distribution of the stacked vector $\mathbf{y}$ of contrasts of the p characters is

$$\mathcal{N}\left(\mathbf{0},\ \mathbf{A} \otimes \mathbf{C}\ \mathbf{T}\ \mathbf{C}^T\right) \tag{23.39}$$

There are many possible sets of contrasts that could be taken, but in the next section we will see one that is particularly convenient, as it results in contrasts that are statistically independent, and easy to compute from the form of the tree. We will see that they also simplify a great deal the task of computing the likelihood of the tree.

Pruning the likelihood

If we have a set of n species (or populations) that have been measured for p characters, and these are assumed to have undergone evolution by Brownian motion, we can use the contrasts to simplify the calculation of the likelihood of a phylogeny. In fact, we can read the contrasts off the tree, and compute the likelihood, all at the same time.

Consider as an example the tree in Figure 23.1. The method for computing contrasts and likelihoods will start by finding two adjacent tips, that is, two tips that are sister species. On that tree there are two such pairs, (1,2) and (6,7). We will choose (1, 2). We can make a contrast in each character between tips 1 and 2. In character i it will be $x_{1i} - x_{2i}$. The variance of this contrast will be (by equations 23.6 and 23.7)

$$\begin{aligned} \text{Var}\left[x_{1i} - x_{2i}\right] &= \text{Var}\left[x_{1i}\right] + \text{Var}\left[x_{2i}\right] - 2\text{Cov}\left[x_{1i}, x_{2i}\right] \\ &= \sigma_i^2\left(v_1 + v_8 + v_9\right) + \sigma_i^2\left(v_2 + v_8 + v_9\right) - 2\sigma_i^2\left(v_8 + v_9\right) \\ &= \sigma_i^2\left(v_1 + v_2\right) \end{aligned} \tag{23.40}$$

This will be a general result: When a contrast is taken among two adjacent tips, the variance of the contrast will be the sum of lengths of the branches between them, multiplied by the variance of the Brownian motion.

We can also imagine computing a weighted average of the character in the two tips, choosing the weight so that the average is independent of the contrast. Suppose that the weighted average is $fx_{1i} + (1-f)x_{2i}$. What value of f should we use? We want the covariance of the weighted average with the contrast between 1 and 2 to be zero:

$$\begin{aligned} &\text{Cov}\left[x_{1i} - x_{2i},\ fx_{1i} + (1-f)x_{2i}\right] \\ &= f\ \text{Cov}\left[x_{1i}, x_{1i}\right] - (1-f)\ \text{Cov}\left[x_{2i}, x_{2i}\right]\ +\ (1-2f)\text{Cov}\left[x_{1i}, x_{2i}\right] \\ &= f\ \text{Var}\left[x_{1i}\right]\ -\ (1-f)\ \text{Var}\left[x_{2i}\right]\ +\ (1-2f)\ \text{Cov}\left[x_{1i}, x_{2i}\right] \\ &= \sigma_i^2\left[f\left(v_1 + v_8 + v_9\right)\ -\ (1-f)\left(v_2 + v_8 + v_9\right)\ +\ (1-2f)\left(v_8 + v_9\right)\right] \\ &= \sigma_i^2\left(fv_1 - (1-f)v_2\right) \end{aligned} \tag{23.41}$$

which will be zero when $f : 1 - f = v_2 : v_1$. The result is

$$f = v_2/(v_1 + v_2) \tag{23.42}$$

The interesting fact about the weighted average is that we can consider it to be the "phenotype" of the interior node 8, at least insofar as that can be inferred from its descendants. If we calculate

$$x'_{8i} = \frac{v_2}{v_1 + v_2} x_1 + \frac{v_1}{v_1 + v_2} x_2 \tag{23.43}$$

we can ask about the variance of this fictional value, and its covariance with the other tips. The covariance is easy. We already know that for any other tip j,

$$\text{Cov}\,[x_{ji}, x_{1i}] = \text{Cov}\,[x_{ji}, x_{2i}] \tag{23.44}$$

so that

$$\begin{aligned} \text{Cov}\,[x_{8i}, x_{ji}] &= \text{Cov}\,[f x_{1i} + (1-f) x_{2i},\; x_{ji}] \\ &= f\,\text{Cov}\,[x_{1i}, x_{ji}] + (1-f)\,\text{Cov}\,[x_{2i}, x_{ji}] \\ &= \text{Cov}\,[x_{1i}, x_{ji}] = \text{Cov}\,[x_{2i}, x_{ji}] \end{aligned} \tag{23.45}$$

So the covariances of the new "tip," 8, with the remaining tips are the same as the covariances of 1 and of 2 with them.

We also need to know the variance of the new "tip." This can be computed as:

$$\begin{aligned} \text{Var}\,[x_{8i}] &= \text{Var}\,[f x_{1i} + (1-f) x_{2i}] \\ &= f^2\,\text{Var}\,[x_{1i}] + (1-f)^2\,\text{Var}\,[x_{2i}] + 2f(1-f)\,\text{Cov}\,[x_{1i}, x_{2i}] \\ &= \sigma_i^2 \left[f^2 (v_1 + v_8 + v_9) + (1-f)^2 (v_2 + v_8 + v_9) + 2f(1-f)(v_8 + v_9) \right] \\ &= \sigma_i^2 \left[\left(v_2^2 v_1 + v_1^2 v_2 \right) / \left(v_1 + v_2 \right)^2 + v_8 + v_9 \right] \\ &= \sigma_i^2 \left[v_1 v_2 / \left(v_1 + v_2 \right) + v_8 + v_9 \right] \end{aligned} \tag{23.46}$$

In other words, we compute the x_{8i} as a weighted average of the two tips above it, the tips being weighted inversely by their values of v_i. The resulting fictional tip has the same covariances with other tips as the two immediate descendants had. However, it has some extra variance, as if the branch leading to node 8 had been lengthened by $v_1 v_2/(v_1 + v_2)$, which is twice the harmonic mean of v_1 and v_2. We thus have a new branch length, which may be denoted v'_8.

The result of the contrast between 1 and 2, is to decompose the tree into two trees. One has only a single tip and change $x_{1i} - x_{2i}$ from the root to that tip (or it can be considered to have two tips with phenotypes x_1 and x_2). The other has $n-1$ tips. We have carried out a linear transformation of the original tip phenotypes, so that the result is still a multivariate normal distribution. The n values that we had are now separated into one that is independent of all the others. So the likelihood is the product of the likelihoods of these two parts of the data, in effect the product of the likelihood of the two new trees. This process is shown in Figure 23.3. We have not mentioned the Jacobian of the transformation. We will have the density of the multivariate normal distribution be the product of the two density functions only if the Jacobian of the transformation happens to be 1. Fortunately this is true.

We can continue the process, of course. We can take the new tree that has $n-1$ tips, find two of them that are sister lineages, and form another contrast. Now the original tree has been decomposed into three trees, the two contrasts and a tree of $n-2$ tips. The likelihood is the product of their likelihoods. We continue until there are $n-1$ contrasts. The original tree has now been decomposed into n trees. The last of these trees has a single tip, and its phenotype values are weighted averages of all the original tip phenotypes. It is possible to compute the ordinary likelihood by taking the product of the likelihoods of all n trees. However, the REML likelihood is also computable, by taking the products of the likelihoods of the $n-1$ contrasts, and ignoring the likelihood of this weighted average phenotype.

This process rapidly computes the REML likelihood of the tree, in an amount of effort proportional to the product of n and p. In effect it diagonalizes the covariance matrix. Owing to the special structure of that matrix, this can be done rapidly, in a way related to the structure of the tree. The computation of the independent contrasts and the computation of the likelihood in this way was introduced by me (Felsenstein, 1968, 1973a, 1981c). I call it "pruning," in analogy to the method of "peeling" for computation of likelihoods along pedigrees in human genetics (Hilden, 1970; Elston and Stewart, 1971; Heuch and Li, 1972).

Maximizing the likelihood

The pruning process also gives us a natural way to optimize branch lengths of a given tree topology. The method is entirely analogous to the one we have already seen for molecular sequence likelihoods (Chapter 16). It will turn out, on close examination, that the likelihood for REML treatment of the Brownian motion model depends only on the unrooted tree, not on the placement of the root. (I leave this as an exercise for the reader.) This is true for the REML likelihood but not for the ML likelihood, as the Brownian motion model is not reversible. Given the unrootedness of the tree in Figure 23.1, for example, the likelihood depends on branch lengths v_9 and v_{12} only through their sum, $v_9 + v_{12}$. Therefore, for our purposes, we can consider this as one branch, which we may call branch 0. Its length, $v_9 + v_{12}$, will be called v_0.

We can use pruning to compute the conditional likelihoods at both ends of branch 0, that is, at nodes 9 and 12. Note also that this means that we also compute augmentations to the lengths of branches 9 and 12. Call these δ_9 and δ_{12}. Now we

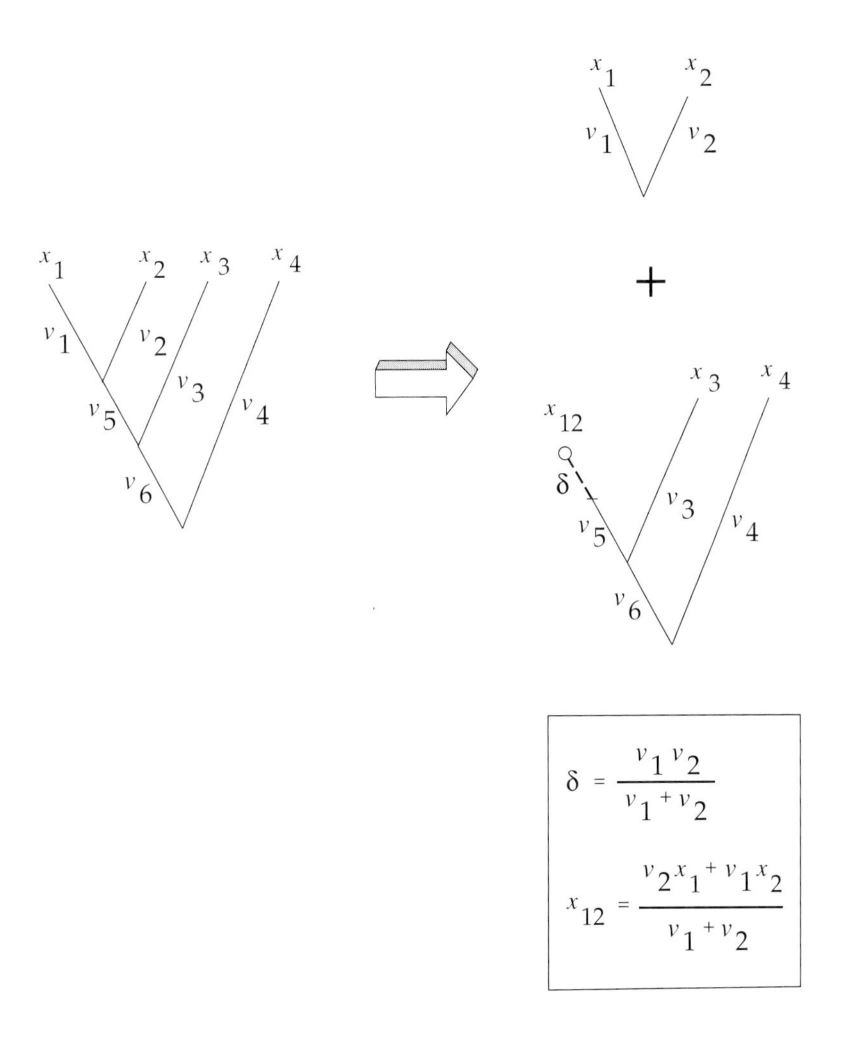

Figure 23.3: "Pruning" REML likelihoods on a Brownian motion tree by decomposing the tree into a two-species tree and one with $n - 1$ species. The likelihood is the product of the likelihoods of these two trees. The pair of sister species that are pruned off are replaced by a single tip whose phenotypes are weighted averages of the phenotypes in the two species (this is done for all characters). The weighted average and the extra branch length that must be added to the new tip are shown in the equations in the box. The process can be continued until the tree is decomposed into $n - 1$ two-species trees plus a single tip.

have a two-species tree with one branch, namely branch 0. Using equations 23.32 and 23.17, we can easily compute the REML estimate of this branch length. We can then get the estimate of v_0 by subtracting $\delta_9 + \delta_{12}$. On occasion, the resulting branch length is negative. If so, it is not hard to show that the best estimate of v_0 is zero.

As the tree is effectively unrooted, we can relocate the root anywhere without affecting its likelihood. It is therefore possible to carry out this procedure on any branch. We relocate the root to that branch, prune the likelihoods down to both ends of the branch, and obtain a REML estimate of the branch length. In each case, we are optimizing the REML likelihood with respect to the length of that one branch. As we did in the ML estimation for molecular sequences, we can move around the tree, successively maximizing the likelihood with respect to the length of one branch after another. If we repeatedly visit all branches, the process should converge. Technically, all we know is that the likelihood converges, as it is a monotone sequence bounded above (by the value of the maximum of the likelihood). When the likelihood stops changing, it is quite probable that we are at a maximum of the likelihood function. Strictly speaking, we have not proved this, but I have never seen a real case in which this was not true. Note that we find a maximum but not necessarily the highest maximum. This complication is unlikely to occur in real data.

We thus have an effective method for finding a maximum of the likelihood for a given tree topology, though it is not inevitable that it always finds the highest likelihood in that tree topology. As with all other cases in which we maximize a criterion over all tree topologies, we can make use of the heuristic search methods outlined in Chapter 4. The problem of searching among tree topologies is just as difficult in this case as in all the others; in this case, the presence of maximum likelihood does not result in any analytical proofs that provide improvements over standard rearrangement methods.

Inferring ancestral states

We have seen in Chapter 16 that one can use likelihood to infer the states of ancestors, for models with discrete states. This can be done as well for characters evolving by Brownian motion. If we wish to infer the state at an interior node i of a tree, we can reroot the tree at that node and then use the pruning algorithm to obtain the estimate of the state x_i at that node, as well as the increment δ_i of the variance there. We can infer the ancestral state to be x_i, and the variance of this estimate is δ_i.

This is equivalent to the procedure of Schluter et al. (1997). Note that the procedure uses ML, not REML, since REML dispenses with the position on the scale, leaving us unable to infer ancestral states. Maximum likelihood inference in the Brownian motion case has the ancestral state as one of its parameters. In this it differs from the reversible models of evolution used for molecular sequences. If we want to infer ancestral states at more than one interior node, these cannot all be parameters of the model at the same time. One way to avoid these issues is to ask whether the objective is to infer the ancestral states. Usually it is not: The objective is to make an inference about correlation of traits with each other or with an environment. The ancestral state is not data — it is an inference that can be treated as data only by ignoring the uncertainty of the inference. The comparative methods that will be discussed in Chapter 25 are a sounder basis for such inferences than is treating ancestral states as if they were directly observed.

Squared-change parsimony

One can also find that combination of values at the interior nodes of the tree that makes the largest contribution to the likelihood. For this the method of *squared-change parsimony* is useful. It searches for the combination of values that minimizes the sum of squares of changes along the tree, where the squares are weighted inversely by the length of that branch:

$$Q = \sum_{i=1}^{\text{branches}} \frac{(x_i - x_{i'})^2}{v_i} \tag{23.47}$$

(where i and i' are the nodes at the opposite end of the branch). The unweighted case of this criterion (with all the v_i equal) was introduced by Huey and Bennett (1987). It was suggested to them by me, in hopes that it would find a reasonable combination of ancestral phenotype values. A simple iterative algorithm exists to obtain the x_i for any tree and set of values at the tips. It can be shown that if node i is connected to nodes j, k, and ℓ by branches of length v_j, v_k, and v_ℓ, then the most parsimonious value of x_i is the weighted average

$$x_i = \left(\frac{x_j}{v_j} + \frac{x_k}{v_k} + \frac{x_\ell}{v_\ell}\right) \Big/ \left(\frac{1}{v_j} + \frac{1}{v_k} + \frac{1}{v_\ell}\right) \tag{23.48}$$

which weights the value of each neighbor inversely by the branch length to it. If we compute this iteratively for one node after another, then each pass through the tree will cause the values to converge closer to the most parsimonious values. For the case in which the v_i are all equal, this is simply an unweighted average of the neighbors' values.

Wayne Maddison (1991) stated the general weighted case (as above) and gave a dynamic programming algorithm to find the most parsimonious combination of values. McArdle and Rodrigo (1994) showed how to do the same computation in matrix form. Maddison also proved that the squared-change parsimony solution is more than that — it is also the combination of values that makes the largest contribution to the likelihood if there is a Brownian motion model. This makes it analogous to the criterion of Yang, Kumar, and Nei (1995), which sought the combination of discrete ancestral states that made the largest contribution to the likelihood. Maddison's algorithm is the continuous counterpart to the dynamic programming algorithm of Pupko et al. (2000). Generalized least squares methods can also be used to infer ancestral states (Martins and Hansen, 1997). Martins (1999) shows computer simulation studies indicating that squared-change parsimony, maximum likelihood, and generalized least squares all yield closely similar estimates of ancestral states.

For a broader review of ancestral state estimation by likelihood and parsimony methods, see Cunningham, Omland, and Oakley (1998). Webster and Purvis (2002) have made an interesting empirical test of several methods with foraminiferan data where the ancestral states can also be observed directly. They found that a likelihood method based on the Ornstein-Uhlenbeck process (which will be described in the next chapter) did the best job of predicting ancestral states.

Gene frequencies and Brownian motion

The model that Edwards and Cavalli-Sforza used for gene frequencies involved approximating genetic drift by Brownian motion. This they had to do because genetic drift itself is not sufficiently tractable. It is not easy enough to calculate transition probabilities between different gene frequencies over the length of a branch of a phylogeny. For example, the distribution showing where the gene frequency is expected to be in a population of constant size N after t generations, given that it starts at 0.6, cannot be calculated exactly under the classical Wright-Fisher model of theoretical population genetics without multiplying large transition matrices. Even in the approximation of genetic drift by a diffusion process, this density function cannot be computed without using a power series of Gegenbaur polynomials.

Edwards and Cavalli-Sforza therefore assumed instead that the gene frequencies wandered on an infinite scale by Brownian motion. This is a good approximation for small divergence times, which is what they were considering. With Brownian motion, the gene frequency change is the sum of a huge number of very small changes. The change after t units of time has expectation zero, but a variance $\sigma^2 t$. The variance that is expected to accumulate is the same in all parts of the scale. In this the process differs from gene frequency change, which necessarily has variance that dies away to 0 as the gene frequency reaches the ends of its scale, at frequencies of 0 or of 1. Recall that the variance of gene frequency change in one generation in a two-allele case is the binomial variance $p(1-p)/(2N)$.

We might hope to make gene frequency changes better approximated by Brownian motion by changing the scale on which the diffusion occurs. To some extent this is possible. A simple example is the case of two alleles. The classical arc-sine square root transform method of equalizing variances when there are binomial distributions attempts this scale change. Suppose that we change from the gene frequency scale p to

$$y = \sin^{-1}\left(\sqrt{p}\right) \tag{23.49}$$

A bit of algebra finds that

$$\frac{dy}{dp} = \frac{1}{2\sqrt{p}\sqrt{1-p}} \tag{23.50}$$

The standard "delta method" of approximating the variance of a transformed variable says that for small values of $\text{Var}(p)$,

$$\text{Var}(y) \sim \left(\frac{dy}{dp}\right)^2 \text{Var}(p) \tag{23.51}$$

which on substituting from equation 23.50 leads to the one-generation variance of change

$$\text{Var}(y) \sim \frac{1}{4p(1-p)} \frac{p(1-p)}{2N} = \frac{1}{8N} \tag{23.52}$$

We seem to have found that by using not the original gene frequencies, but their arc-sine square root transform, we have approximately equal variances at all points on the new scale. Does this mean that the stochastic process on the new scale is a Brownian motion? Alas, no such luck. We have ignored the mean of the process. It was zero on the original scale, but it is not on the new scale. We can use a Taylor series (in effect, the delta method again) to compute it. Expanding it to the second-derivative term,

$$\mathbb{E}(y) \approx \mathbb{E}\left(p_0 + (p - p_0)\left[\frac{dy}{dp}\right]_{p=p_0} + \frac{1}{2}(p-p_0)^2 \left[\frac{d^2y}{dp^2}\right]_{p=p_0}\right) \tag{23.53}$$

We then substitute in the derivatives of y with respect to p. Taking the second derivative using equation 23.53:

$$\mathbb{E}(y) \approx y_0 + \mathbb{E}\left[(p-p_0)\right] \frac{1}{2\sqrt{p_0}\sqrt{1-p_0}} + \frac{1}{2}\mathbb{E}\left[(p-p_0)^2\right] \frac{2p_0 - 1}{4[p_0(1-p_0)]^{3/2}} \tag{23.54}$$

Since the expectation of $p - p_0$ in binomial sampling is 0, and the expectation of its square is the binomial variance $p(1-p)/(2N)$,

$$\mathbb{E}(y) \approx y_0 + \frac{1}{8N} \frac{2p_0 - 1}{\sqrt{p_0(1-p_0)}} \tag{23.55}$$

The mean change is thus not zero, as it is in Brownian motion and as it was on the original gene frequency scale. Instead it has a directional force that pushes the gene frequency away from 1/2 and toward 0 or 1. The process on the new scale may be a diffusion process with constant variance, but it is not Brownian motion. Thus any treatment of gene frequency evolution by Brownian motion is of necessity an approximation.

Using approximate Brownian motion

In spite of the approximateness of the Brownian motion model, one can go ahead and use it. Cavalli-Sforza and Edwards (1967) used the arc-sine square root transformation of one allele's frequency for two-allele loci. For multiple alleles the issue is more complex. One method proposed by Edwards (1971) replaces a k-allele locus with k variables. If the gene frequencies in a population are $p_1, p_2, \ldots, p_k$, the new coordinates are

$$y_i = \frac{2(\sqrt{p_i})}{1 + \sum_{i=1}^{k} \sqrt{p_i/k}} - \frac{1}{\sqrt{k}}, \qquad i = 1, 2, \ldots, k \tag{23.56}$$

This transformation does exaggerate differences somewhat when the alleles are not equally frequent. The new variables all lie in a plane, and thus the k of them change in only $k-1$ dimensions. To obtain variables that change independently one can take any set of orthogonal contrasts of the y_i (Thompson, 1975, p. 26). Thompson (1972) has investigated Edwards's transform further and has shown that it does an optimal job of approximating genetic drift by Brownian motion. The transform is centered at the point where all alleles are equally frequent. It would be even better to center it near the centroid of the frequencies in all observed populations. This would amount to the transform

$$y_i = \frac{2(\sqrt{p_i})}{1 + \sum_{i=1}^{k} \sqrt{p_i \bar{p}_i}} - \sqrt{\bar{p}_i}, \qquad i = 1, 2, \ldots, k \tag{23.57}$$

where $\bar{p}_i$ is the mean gene frequency of allele i across populations.

Distances from gene frequencies

Rather than using likelihood, one can compute genetic distances based on gene frequencies. These have a large, complex, and remarkably confusing literature, largely because there has been no agreement on a common task for the distances to perform or common standards for the distances to be judged by. The authors often sound as if they are competing for the prize for "most mathematically beautiful distance measure." When we are inferring phylogenies the task is clearer and the criteria more straightforward. I have made (Felsenstein, 1985d) an examination of bias of these genetic distances in the case where the divergence of species is due entirely to genetic drift.

For the case of pure genetic drift, suppose that the gene frequencies in two populations at allele j of locus i are x_{ij} and y_{ij}, with a total of n loci. One popular distance is the chord distance of Cavalli-Sforza and Edwards:

$$D_{CSE} = 4 \sum_i \left(1 - \sum_j \sqrt{x_{ij} y_{ij}}\right) \tag{23.58}$$

Another is the distance of Reynolds, Weir, and Cockerham (1983):

$$D_{RWC} = \frac{\sum_i \left(\sum_j (x_{ij} - y_{ij})^2 \right)}{\sum_i \left(1 - \sum_j x_{ij} y_{ij} \right)} \tag{23.59}$$

These have been given here in a form that is expected to rise linearly with elapsed time when two populations have been separated and diverge through genetic drift.

When there is mutation, the genetic distance of Nei (1972) is by far the most widely used:

$$D_N = -\ln \left[\sum_i \sum_j x_{ij} y_{ij} \Big/ \left(\sum_i \sum_j x_{ij}^2 \right)^{1/2} \left(\sum_i \sum_j y_{ij}^2 \right)^{1/2} \right] \tag{23.60}$$

Nei's distance assumes that all loci have neutral mutation at the same rate, with each mutant to a new allele. Nei's distance does a better job when there is mutation, but is still liable to rise nonlinearly with time when mutation rates vary among loci. When there is no mutation, Nei's distance is not particularly well-behaved (Felsenstein, 1985c) but this should be no surprise as in this case it is being used when its model is inapplicable.

Use of an appropriate distance measure with distance matrix methods is a reasonable competitor to likelihood using Brownian motion models, particularly when there is mutation.

A more exact likelihood method

Nielsen et al. (1998) give a likelihood method, allowing for coalescent phenomena within species, in the absence of mutation. We will explain this further in Chapter 28 where we cover coalescent effects in trees of closely related species. Their likelihood calculation can compute likelihoods for trees of populations within a species or trees of species. It infers not only a tree but a set of effective population sizes, one for each branch. As the effective population size is used in the form $t/(2N_e)$, the likelihood does not actually depend on the time and population size separately. The approach of Nielsen et al., unlike the Brownian motion approximation, is almost exact. It requires Markov chain Monte Carlo methods to evaluate the likelihood.

Gene frequency parsimony

In addition to likelihood and distance methods, there are also parsimony methods for gene frequencies. Edwards and Cavalli-Sforza (1964) introduced the first of these, which minimized the sum of the distances between adjacent nodes in the tree, using ordinary Euclidean distance. Thompson (1973) developed further methods for finding this tree. Swofford and Berlocher (1987) introduced methods minimizing the length of branches using the sum of absolute values of differences of gene frequency (the Manhattan metric, which we mention further in Chapter 24). They used well-known linear programming methods to infer the gene frequencies at each locus at the interior nodes of the tree. They later (Berlocher and Swofford, 1997; see also Wiens, 1995) proposed an approximate optimality criterion that was easier to compute, and they argued that it tended to choose the same trees as the original criterion.

There are also methods that code alleles only present and absent, using discrete-characters parsimony without taking allele frequencies into account (Mickevich and Johnson, 1976; Mickevich and Mitter, 1981, 1983). These lose much information by not taking gene frequency changes into account and are also overly sensitive to the use of small population samples, which may omit an allele that is actually present. They have been criticized by Swofford and Berlocher (1987), among others.

Wiens (2000) has examined the performance of different methods of taking gene frequencies into account. He comes to the conclusion that using maximum likelihood with the Brownian motion approximation does at least as well as other methods, even when gene frequency changes are affected by mutation in addition to genetic drift. As comforting as this might be, I suspect that taking mutation into account in the statistical model would be preferable in such cases.

Chapter 24

Quantitative characters

Brownian motion models have also been used as approximate models of the evolution of quantitative characters. *Quantitative characters* are those that are described by values on a continuous numerical scale. They are also called "continuous characters" and sometimes "polygenic characters," the latter being on the presumption that their variation is controlled by many loci.

The theory of the quantitative genetics of these characters has made use of a model, which traces back to Fisher (1918) and even before, that has the characters as the sum of effects contributed by a number of loci, these effects being independent of each other and of environmental contributions to the trait. By using a simple sum, we are ruling out all interaction among the loci. Frequently the model is taken further, and it is assumed that the number of loci contributing to the variation of the character is large and the individual contribution of each locus is small. In that case the sum will be normally distributed.

The theory of quantitative genetics, under these conditions, allows the computation of within-population covariances among relatives, and the response to directional selection. All of the relevant information about the loci and their gene frequencies affects this theory through only a few quantities: the population mean and the additive, dominance, and environmental components of variance. The theory can also be extended without difficulty to multiple characters, taking into account by genetic covariances the pleiotropy of the loci, the extent to which the same alleles affect multiple characters. The result is an enormously powerful and supple theory that links observable quantities by a few parameters (μ, V_A, V_D, V_E) and avoids having to know the detailed genomics of the characters. We can infer the variance components from covariances among relatives, and we can predict selection response from that, without actually having to know how many loci are affecting the trait, where they are in the genome, or what are the effects of their alleles. Falconer and MacKay (1996) and Lynch and Walsh (1998) are good places to start to explore the large literature of quantitative genetics.

As remarkable and powerful as this theory is, its limitations are many. It is basically a short-term theory. Quantities such as the additive genetic variance V_A depend on the gene frequencies at all of the relevant loci, and they change as these gene frequencies change. The response to selection then also changes, and so the relationship between the measurable quantities is not reliable over the long run.

Neutral models of quantitative characters

Brownian motion comes closest to being an adequate model of change of quantitative characters when their mutations are selectively neutral. A number of quantitative geneticists (Lande 1976; Chakraborty and Nei, 1982; Lynch and Hill, 1986) have constructed models of evolutionary change in which neutral mutations occur at multiple loci, acting additively on a phenotype. The implications of such a model for the process of character change is most easily seen by making it as simple as possible. Suppose that there are n unlinked loci, whose alleles have numerical values that are added to get the genotypic contribution to a character. A normally distributed environmental effect is then added, independently for each individual. In other words, there are no dominance or epistatic effects, no genotype-environmental interaction, no genotype-environmental covariance.

We assume that the environmental effect has mean 0 and variance σ_E^2. We will also assume a mutational process at each locus that is the same, so that the n loci are fundamentally equal in their contribution to the variance. Each allele has a probability μ per generation of mutating. The mutation adds a quantity ε to the numerical value of the allele, where ε is normally distributed with mean 0 and variance σ_m^2. Note that assuming that the mean is 0 has major implications. Even if the allele currently has a large positive value, mutation is no more likely to make it smaller. There are thus no limits to the extent of phenotypic evolution.

In an idealized population of newborn diploid organisms, suppose that the genotypic mean (the totals of the values of the alleles) has variance V_A among the newborn individuals. If the next thing that happens is genetic drift caused by random survival to adulthood, with N individuals surviving, the adult generation will be a sample of N individuals drawn at random from the distribution of newborns. It is not hard to show that the mean of such a sample will differ from the parental mean by a random amount whose mean is 0 and whose variance is V_A/N. But what happens to the variance V_A as a result of this sampling?

In the sampling process of genetic drift, it is, on average, reduced. If the genotypic mean (the sum of allelic effects) for adult individual i is g_i, the deviation of a random one of the survivors (say individual 1) from the mean of the survivors is

$$\begin{aligned} g_1 - \bar{g} &= g_1 - \left(\frac{g_1+g_2+\ldots+g_N}{N}\right) \\ &= \left(1 - \frac{1}{N}\right) g_1 - \frac{1}{N} g_2 - \frac{1}{N} g_3 - \ldots - \frac{1}{N} g_N \end{aligned} \tag{24.1}$$

As the g_i are independently drawn from a distribution whose variance is V_A, the variance of this weighted sum is that variance times the sum of squares of the coefficients:

$$\text{Var}\,(g_1 - \bar{g}) \;=\; \left[\left(1 - \frac{1}{N}\right)^2 + \frac{1}{N^2} + \frac{1}{N^2} + \ldots + \frac{1}{N^2}\right] V_A \;=\; \left(1 - \frac{1}{N}\right) V_A \qquad (24.2)$$

We have not specified what the distribution of the g_i is. The adults will be a random sample of N from the newborns, with a variance that is, by the above equation, on average $1 - \frac{1}{N}$ times as great. It is possible to show that, when we derive $2N$ gametes from these survivors and combine them at random, the genetic variance V_A is reduced, not by $\frac{1}{N}$, but by $\frac{1}{2N}$, once one takes within-individual genetic variability into account. Thus, on average, a fraction $\frac{1}{2N}$ of the variance disappears with each generation of genetic drift. However, this varies randomly, and some generations may see an increase of variance.

After they reach adulthood, the individuals will be assumed to reproduce, producing infinitely many offspring who make up the newborns of the next generation. The distribution of the genotypic means is no longer normal, but it is easy to show that its expectation and variance do not change. Each allele in the newborns may or may not have suffered a mutation. It will have added to it a random quantity, which is 0 with probability $1 - \mu$, and with probability μ is drawn from a normal distribution with mean 0 and variance σ_m^2. It is easy to show that the variance of the amount added to the allele is then $\mu\sigma_m^2$. If there are n such loci, each with two copies of the gene, the total amount added to the genotypic mean of the individual will be the overall mutational variance

$$V_M \;=\; 2n\,\mu\,\sigma_m^2 \qquad (24.3)$$

and, as there are infinitely many newborns, the variance among them will be increased by exactly this amount.

If we stop at this point, survey the newborns, and compare them to the previous generation's newborns, their mean will have changed by an amount which is random but has mean 0 and variance V_A/N. Their variance will have an expectation

$$\mathbb{E}\left[V_A^{(t+1)}\right] \;=\; V_A^{(t)}\left(1 - \frac{1}{2N}\right) + V_M \qquad (24.4)$$

Taking expectations in generation t, we can equate $\mathbb{E}[V_A]$ in the two successive generations and find that at equilibrium we can solve equation 24.4 to get

$$\mathbb{E}\,[V_A] \;=\; 2NV_M \qquad (24.5)$$

This is the amount of genetic variance that we expect at an equilibrium between mutation and genetic drift. This will not be the value of V_A, only its expectation. The actual value of V_A will fluctuate around it, sometimes higher and sometimes lower. In neutral phenotypic evolution, this process occurs essentially independently at each locus. The more loci (the higher the value of n), the more the fluctuations in V_A will tend to average out across loci. For large values of n it will be found that V_A stays nearly constant; for small values it will fluctuate. Similarly, for large N the process of genetic drift proceeds slowly and the amount of genetic variance maintained at equilibrium will be nearly constant.

The genetic variance thus fluctuates around an equilibrium. But the population mean does not. Each generation it changes by a random amount whose mean is 0 and whose variance, as we have seen, is V_A/N. If $V_A \approx 2NV_M$, then the variance of the change in population mean is

$$\text{Var}\left[\Delta\bar{g}\right] \;=\; \left(2NV_M\right)/N \;=\; 2V_M \tag{24.6}$$

This equation was first derived by Clayton and Robertson (1955). The conditions for it to hold are discussed by Turelli, Gillespie, and Lande (1988). It is strikingly like Kimura's famous result for neutral molecular evolution: The long-term change of the phenotype depends only on the mutational processes, and the population size cancels out! In each generation the population takes a step in random walk, a step whose variance is the total mutational variance V_M. In actuality, these steps are not independent. The variance of the step may be higher in one generation than in another, as the standing genetic variance V_A in the populations fluctuates up and down around its expectation. The number of loci n and the population size N affect how much fluctuation there is in values of V_A from generation to generation. Note that although the distribution of genotypic means will be approximately normal, it will not be exactly normal.

Thus the mean of the population undergoes a random walk which, added up over generations, is not far from a Brownian motion. The standing genetic variance in the population will be, on average, $2N$ times the variance of change per generation. The approximation of the process by Brownian motion is fairly robust, but only if we can continue to assume that mutation is not affected by how far the process has wandered. If there is a tendency for mutations to push the allele effects back towards some particular value, the process will be approximated better by the Ornstein-Uhlenbeck process, of which we hear more soon. It is the random walk of an elastically bound particle, which is tethered to a post by an elastic band. It does not wander infinitely far from the post in the way that a Brownian motion does. To know which kind of model is more appropriate, we need to think hard about the process of mutation.

The model we have discussed is for one character. The same result holds for multiple characters — the covariance matrix of the characters in the population will be N times the covariances of their mutational changes, and the covariance matrix of mutational changes will be the covariances of change through time. The same approximations and qualifications apply.

Changes due to natural selection

Neutral models of quantative characters have a limited range of application. They may work well near the species level, but large changes, such as we might see over longer periods, are much more likely to be constrained or encouraged by natural selection. In fact, it is also possible to model responses natural selection by Brownian motion. To do so it is helpful to start with the classical quantitative genetic theory of selection response, which is due to R. A. Fisher and Sewall Wright. As made widely known among animal breeders by Lush (1937), their major result was that the change per generation due to selection (Δz) was the product of the heritability (h^2) times the selection differential S:

$$\Delta z \;=\; h^2\,S \tag{24.7}$$

The heritability is the fraction of phenotypic variance (V_P) that is due to the additive genetic component (V_A). The selection differential is the difference between the means of the selected individuals and the herd from which they are chosen. The same equation applies to natural selection. The question is, what the selection differential will be. If there is a fitness curve $w(x)$ that relates fitness to phenotype, then it turns out that the difference between the means before and after selection is the phenotypic variance times the slope of the log of mean fitness with respect to mean population phenotype

$$S \;=\; V_P\,\frac{d\log\left(\bar{w}\right)}{d\bar{x}} \tag{24.8}$$

(Lande, 1976). Alternatively, we can combine this with equation 24.7 and note that

$$\Delta z \;=\; (V_A/V_P)\;V_P\,\frac{d\log\left(\bar{w}\right)}{d\bar{x}} \;=\; V_A\,\frac{d\log\left(\bar{w}\right)}{d\bar{x}} \tag{24.9}$$

For the simple case in which fitness is an exponentially rising function $\exp(\alpha P)$ of phenotype, the derivative of log mean fitness ($\bar{w}$) will be the same as the derivative of log fitness. (It is different in other cases, the more different the more the fitness function departs from an exponential curve.) There is also a multivariate version of all of these equations, in which V_A and V_P are replaced by additive-genetic and phenotypic covariances among the characters, the change Δz by a vector of changes, and the gradient of log fitness by a corresponding vector of derivatives.

Selective correlation

The point of all of this is that the change in any one generation is a product of terms, one reflecting the additive genetic variance, and the other the effect of the character on fitness. For multiple characters these are products of a matrix and a vector. If we consider a lineage changing through time, the wanderings of the mean will depend on the additive genetic covariances and also on the vagaries of the way fitness depends on phenotype. Characters can evolve in a correlated way for either or both of two reasons:

1. The characters are genetically correlated, owing to the same alleles having effects on both.
2. The selection pressures on the characters are correlated.

This last has been called "selective correlation" by Stebbins (1950). For example, suppose that a mammalian lineage enters a cold environment. Its limbs grow shorter, its body grows larger, and its color grows darker. These are (respectively) Allen's, Bergmann's, and Glogler's rules. They result, not from any correlation in the genetic basis of these traits, but because selection tends to favor all three in the same circumstances. Thus both genetic correlations and selective correlations must be understood in order to predict the correlations of changes of characters through time.

Covariances of multiple characters in multiple lineages

In the previous chapter, we saw that for multiple characters the Brownian motion model could be written with the covariance matrix of multiple species and multiple characters being a Kronecker product of a matrix representing the tree and a matrix reflecting the different variances of the different characters. That model assumed that all characters were evolving independently, although with possibly unequal variances of change through time. In fact, the equation continues to hold when characters can covary. If $\mathbf{A}$ is the covariance matrix of changes in the characters per unit time (or branch length) in a single lineage, the covariance of those characters after t units of branch length will simply be $t\mathbf{A}$. If two lineages evolve together for t_k units of branch length and then separate, and if we observe character i in one lineage and character j in the other, the covariances of the changes in these two characters will simply be $t_k\mathbf{A}$, as after the lineages separate there can be no further covariance. The upshot is that when we consider a vector that has the values of all p characters "stacked" atop other such vectors for all n species, the covariances of this vector of np numbers are the Kronecker product

$$\text{Cov}\left[\mathbf{x}\right] = \mathbf{T} \otimes \mathbf{A} \tag{24.10}$$

In effect this means that for populations i and j, whose common ancestor is at branch length t_{ij} above the root, the covariance of character k in population i with character ℓ in population j is

$$\text{Cov}\left[x_{ik}, x_{j\ell}\right] = t_{ij}\, a_{k\ell} \tag{24.11}$$

The model is still a multivariate normal model; if we knew the covariances of change in characters in evolution, we could use it to infer the tree, which is reflected in the matrix $\mathbf{T}$. Note that $\mathbf{A}$ will reflect both genetic covariances and covariances of selection pressures.

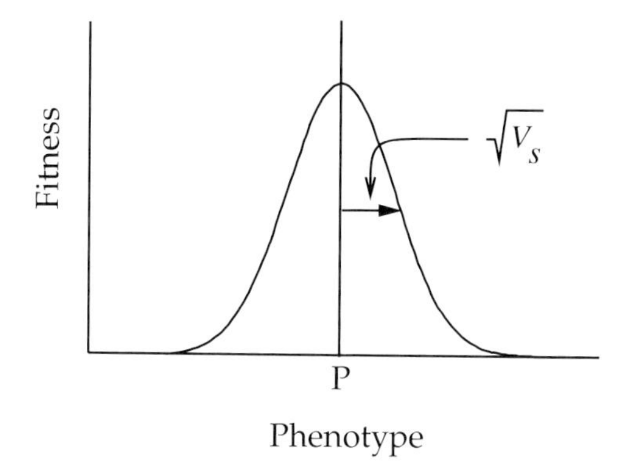

Figure 24.1: The Gaussian optimum selection curve that is the most mathematically tractable scheme of selection for an optimum. The optimum phenotype is at the peak, and the distance out to the inflection point (arrow) is $\sqrt{V_s}$. The larger that value, the weaker is the selection.

Selection for an optimum

Many characters are selected for closeness to an optimum value. Fitness may fall off as one departs from the optimum. The simplest mathematically tractable model of this is to take the fitness curve to have a Gaussian shape, with fitness 1 at the peak. If the optimum value is p and the mean is m, fitness is taken to be

$$w(x) = \exp\left[-\frac{(x-p)^2}{2V_s}\right] \tag{24.12}$$

The "variance" term V_s is here simply a quantity inversely related to the strength of selection. When very wide departures from the optimum are tolerated, V_s is large, but when selection is strong, V_s is small. Figure 24.1 shows the shape of this adaptive peak. There is also a multivariate version of this fitness curve, with mean and phenotypic optimum vectors and the quantity V_s replaced by a matrix of coefficients.

It is particularly easy to derive the change in phenotype when the selection is of the form of equation 24.12. If the distribution of phenotypes is Gaussian, with mean m and variance V_P, some tedious but straightforward completing of squares will show that after the selection, the mean of the phenotypes of survivors becomes

$$m' = \left(\frac{m}{V_P} + \frac{p}{V_s}\right) \Big/ \left(\frac{1}{V_P} + \frac{1}{V_s}\right) = \frac{V_s\, m + V_P\, p}{V_s + V_P} \tag{24.13}$$

and therefore the change in phenotype among them is:

$$m' - m = \frac{V_s\, m + V_P\, p}{V_s + V_P} - m = \frac{V_P}{V_s + V_P}\,(p - m) \tag{24.14}$$

We can then say that the change that selection makes in phenotype within a generation is a fraction of the difference between the optimum and the mean.

The response to this selection is roughly the product of the right side of equation 24.14 with the heritability V_A/V_P, so that it is

$$m' - m = \frac{V_A}{V_s + V_P} (p - m) \tag{24.15}$$

Multivariate versions of this exist as well, with m and p replaced by vectors; V_A, V_P, and V_s by matrices; and the inverse replaced by matrix inversion.

Brownian motion and selection

Equation 24.15 implies that m moves toward the optimum, to the extent that there is selection, and that there is additive genetic variance. If we wait long enough and the optimum stays at the same value, then the phenotypic mean should approach the optimum and go to the top of the fitness peak. If the peak keeps moving, however, the mean phenotype constantly chases the peak. In the long run, the phenotype goes where the peak goes, so its long-term movement will be a rough image of the movement of the peak. We can invoke a Brownian motion model for the phenotype, but only by making the assumption that the optimum phenotype wanders in an approximately Brownian matter. The mean phenotype will then wander after it, though it will not actually carry out a Brownian motion, but instead a smoother process that is roughly similar in its gross features. It will be smoother because the phenotype will change the direction of its movement only when the peak passes the phenotype. As long as the phenotype is on one side of the peak, it will continue to move in that direction. Thus the short-term properties of the phenotype's movement will show much less noise than does Brownian motion.

We can add to this picture a genetic drift process as well; it will cause the population mean to wander about the peak even when the peak does not move. The effect would be to cause short-term change to be less smooth. The population mean will still track the movement of the peak in the long run.

Correcting for correlations

If we want to use Brownian motion models and make maximum likelihood inference of phylogenies, we need to ensure that there are no covariances among the characters, and that they are expected to have equal variances of change through time. These are the assumptions of the likelihoods for inferring phylogenies under Brownian motion. They will typically require that the characters be transformed to remove covariances and to standardize the Brownian motions of the new characters.

In the case of neutral evolution, we would be able to do this only if we knew the covariance matrix of mutational changes. However, the standing covariation

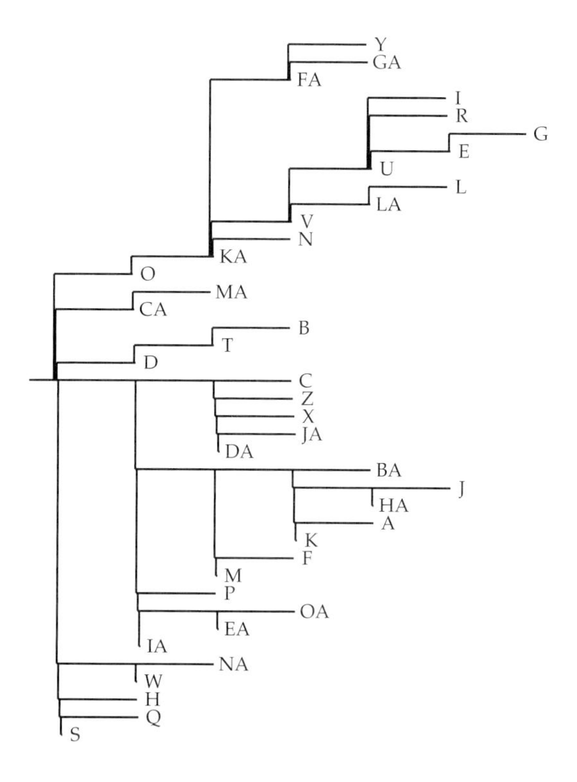

Figure 24.2: A phylogeny of 40 species with a punctuated equilibrium model. At each speciation event, the daughter lineage undergoes a normally distributed change, while the parent lineage does not change. This is reflected by having a branch length of 1.0 in the daughter lineages. In this figure, a small branch length has been assigned to the parent lineages, so that the tree topology is more easily seen. The labels (A–Z and AA–OA) are arbitrary and are randomly assigned.

in the population is N times this, so it can be used to estimate the mutational variance and remove the correlations. This does, however, require that we use the additive genetic covariances, whereas we may have observed only the phenotypic covariances.

In the case of natural selection, the situation is even worse. We still need to know the additive genetic covariances. It is possible to estimate these by breeding experiments, though this information is not usually available in studies of morphological evolution. But we need more: We need to know the selective covariances. It is not obvious how we are to infer these, absent a mechanistic model of

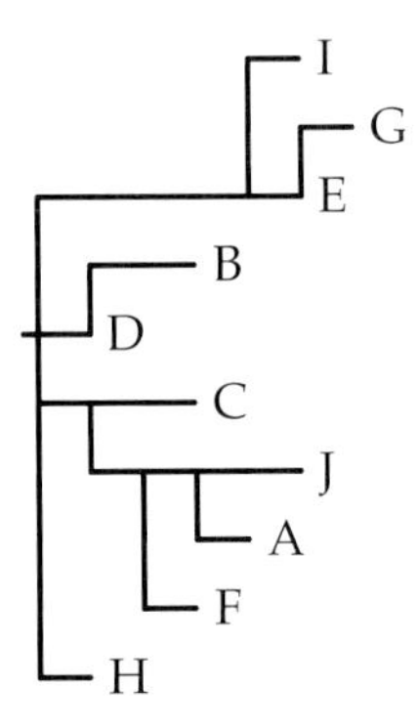

Figure 24.3: A random subtree of 10 species chosen from the tree of the previous figure. The shape of the tree is less obviously punctuational, with only two speciations (those involving E and D) having an unchanged parent species visible.

natural selection on the traits and much information about the changes of environment.

I have discussed these dilemmas (Felsenstein, 1973a, 1981c, 1988a, 2002) and concluded that we cannot at present correct for character covariation. We may instead want to use molecular sequences to infer the phylogenies and then use the changes of the characters along the phylogenies to infer the covariances of their evolutionary changes. We will see in the next chapter that this can be done once an underlying Brownian motion is assumed.

Punctuational models

The models above have been strictly gradualist. Change continues along a branch of a phylogeny, and there is no special process of change when a speciation occurs. We may wonder what a punctuational model would look like. It would show stasis (no change) within branches, and change only at the time that a branch originates. If we see all the species that ever existed, with each origin of a new lineage reflected by a fork in the tree, then change should occur only at forks. For most punctuational models, only one of the two lineages emerging from each fork changes, the one that is the daughter lineage. One way to reflect this is to assign daughter lineages all the same length (say 1.0), and parent lineages all length 0. Brownian motion on such a phylogeny will then reasonably well reflect the punctuated equilibrium model. Figure 24.2 shows a phylogeny produced by a stochastic process of birth events, with daughter lineages displaced from the parent lineages and with the assignment of branch lengths that would come from a punctuated equilibrium model. If we saw this tree, we could readily discern that it was the result of punctuational processes.

In fact, Figure 24.2 is not realistic, because we do not sample all of the species that ever arose. Figure 24.3 shows the subtree consisting of only the randomly selected species A–J. It shows a considerably less punctuated character, with nonzero branch lengths on both sides of most splits, and the species more nearly equidistant from the root than in Figure 24.2. As the fraction of all species that we are examining becomes smaller, either as a result of extinction occurring, or as a result of sampling of species by the biologist, the tree will look less and less obviously punctuational.

Change of characters through time will be far from a Brownian motion process in the punctuational case, as the character shows no change at all most of the time. As the branches come to reflect multiple punctuational events (as in Figure 24.2) the net change will be more nearly like a Brownian motion, as it will reflect superposition of multiple normally distributed changes. Interestingly, if we express branch lengths in units of time, punctuational models will depart from Brownian motion. But if we express them instead in terms of the total variance that accumulates as a result of the lineage being the daughter lineage in a series of speciations, the net change will be normally distributed and there will be no departure from Brownian motion. This can be done because being a daughter lineage is an experience shared by all characters.

Inferring phylogenies and correlations

We have seen that if we can correct for the variances and covariances of evolutionary change in the characters, we can make a maximum likelihood estimate of the phylogeny. We shall see in the next chapter that, given a phylogeny, we can make a maximum likelihood estimate of the covariances. Is there any hope of making a joint maximum likelihood estimate of both? It seem that there is, though one may need a large number of species to make it work well.

The phylogeny with n species has $2n - 3$ branch lengths to estimate, plus, of course, the tree topology, which does not count as parameters. The character covariation for p characters involves $p(p+1)/2$ parameters. Since branch lengths in the tree are expressed in terms of expected accumulated variance, these two sets of parameters must have one parameter in common. Thus there are a total of $(2n-4) + p(p+1)/2$ parameters to estimate. Is this too many? No, if there are enough species. With n species and p characters, we can estimate $(n-1)p$ parameters, when we do REML inference that discards the mean of each character. Thus with (say) five characters and five species, there are 23 parameters to estimate and only 20 degrees of freedom of the data. But with eight species there are 29 parameters to estimate and 35 degrees of freedom in the data. In general, when the number of species is more than twice the number of characters, there are enough data to think of estimating both the tree and the character covariation.

For the moment we defer doing this until the next chapter.

Chasing a common optimum

When a number of species all are being selected towards a common optimum, we can model their motion by the *Ornstein-Uhlenbeck (OU) process,* which is Brownian motion with a force continually pulling towards a point. If we have a single character changing according to an OU process, with variance V_ε accumulating per unit time, and a returning force pushing toward an optimum at a per unit time, the transition probability density will be a normal distribution with mean $x_0 \exp(-at)$ and variance

$$\text{Var}\left[x_t \mid x_0, t\right] = \frac{1}{2a}\left(1 - e^{-2at}\right) V_\varepsilon \tag{24.16}$$

Starting at $x_0 = 0$, a group will end up in a multivariate normal distribution, and the model will become Brownian motion if we transform time appropriately. However, this is not actually worth pursuing, as with multiple characters the transformations of time for different characters, subject to different forces returning them toward their optima, are different. A number of people (Lande, 1976; Felsenstein, 1988a; Hansen and Martins, 1996) have proposed using the Ornstein-Uhlenbeck model for evolutionary inferences. Webster and Purvis (2002) found that a single-character Ornstein-Uhlenbeck model did a superior job of inferring ancestral states of quantitative characters, when used on foraminiferan data where fossils made it possible to check against observed ancestral states.

For multiple characters it would require not only phylogenies but enough species to permit the estimation of the optimum values and also the transformation to a space in which covariation around these optima would be independent. This is a tall order, and methods based on this attractive model have not yet been made practical.

The character-coding "problem"

It is often considered desirable to carry out parsimony analysis of continuous quantitative characters. As many parsimony computer programs assume discrete characters, methods have been sought to convert continuous characters to discrete ones, to enable their analysis by discrete parsimony methods. A variety of methods have been proposed. These include gap coding (Mickevich and Johnson, 1976), segment coding (Colless, 1980), and generalized gap coding (Archie, 1985). An early review covering some additional methods was done by Archie (1985); more recent references and further methods will be found in the papers by Thiele (1993) and Wiens (2001).

Gap coding places boundaries between states at parts of the numerical scale where there are gaps between species. If we have histograms of values of the quantitative character, whenever these overlap the two species are coded as being in the same discrete state. Figure 24.4 shows a hypothetical example resulting in three states.

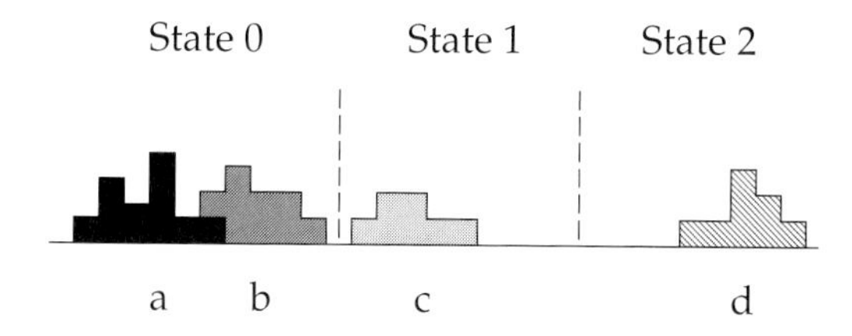

Figure 24.4: An example of gap coding. The discrete character has three states, the boundaries of these states being the gaps in the distribution of the quantitative character.

Generalized gap coding and segment coding do not simply use the gaps but take into account how large these gaps are. *Generalized gap coding* finds the same gaps as gap coding, but creates additional gaps as well. The resulting states cover ranges of values, which can be overlapping. Additive binary coding (see Chapter 7) is used, with each of the new states coded present or absent in a series of new binary factors. *Segment coding* divides the range of species means into as many states as possible, placing each species into the state that results. This has the effect that a gap is not simply coded as a change of state, independent of how large it

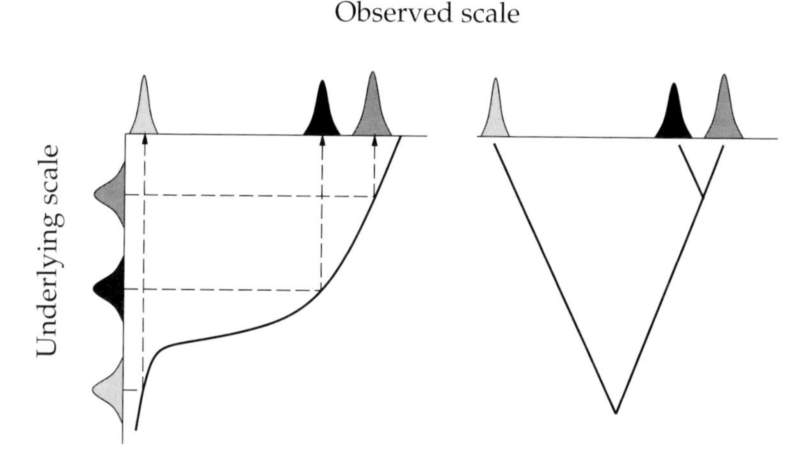

Figure 24.5: Two interpretations of a gap between species histograms on a quantitative scale. On the left, as a result of a nonlinear mapping from an underlying character to the visible character, it is a region in which species change rapidly and from which they tend to move. On the right, it is simply a region separating values that are on opposite sides of a deep fork in the tree.

is. In Figure 24.4, segment coding and Thiele's (1993) "step-matrix gap weighting" might require one or more additional states to cover the large gap. The four species might then have states 0, 0, 1, and 3, allowing us to take the size of the large gap into account to some extent.

There is an implicit assumption in gap coding that gaps are not historical artifacts, but are regions of the scale that evolution traverses with difficulty and perhaps does not like to remain in. Figure 24.5 shows two interpretations of a gap: on the left, that it represents a region of the scale in which a species does not remain long, and on the right, that it is a region across which there happens to have been more time to evolve. Consideration of other characters could in principle distinguish between these.

However, it is not at all obvious that we *should* reduce quantitative characters to a discrete coding. The methods of Thiele (1993) and Wiens (2001), as well as the earlier segment coding (Colless, 1980) are, in effect, using discrete-state parsimony programs to approximate the results of remaining on the original scale and using parsimony. We have seen that there are statistical methods for likelihood analysis of quantitative characters. There are also, as we will see in the next section, parsimony methods that remain on the original scale. Too hastily moving to a discrete scale may lead us to ignore correlations between characters; it would be better to make some attempt to remove these. When methods exist for using the original scale (or a reasonable transformation of it), it would seem best to "just say no."

Continuous-character parsimony methods

Manhattan metric parsimony

If we are not converting continuous characters into discrete characters, but using their values on the continuous scale, in addition to likelihood methods there are also parsimony methods available. In fact, some of the early parsimony methods allowed values to be on a continuous scale. Kluge and Farris (1969) and Farris (1970) assumed that the character values lay on an ordered scale, on which they could be either continuous or discrete. They gave algorithms for reconstructing states on the interior of the tree and for counting the total change required in that character on that tree. This they give as

$$d(A, B) = \sum_i |X(A, i) - X(B, i)| \tag{24.17}$$

where $X(A, i)$ is the value of character i in node A on the tree. The sum of absolute differences is defined both for continuous characters and for discrete characters that are points on the continuous scale. The values at the interior nodes are easy to calculate, as they are the medians of the values of the nodes to which that node connects (Kluge and Farris, 1969). Of course, one may need to iterate the taking of medians, as the nearby nodes may themselves be interior nodes. Maddison and

Slatkin (1990) have shown that the assignment of values to interior nodes in the tree under this iterated median rule results in values that do not depend on the spacing of the species values on the scale, but only on their ranking.

We can consider the species to be points in a continuum, with their character values for the ith character being the ith coordinate of the points. Thus a species with five characters, whose values are 1, 2.5, 0, 3.4, and 6 occupies the point (1, 2.5, 0, 3.4, 6) in five-dimensional Euclidean space. The parsimonious tree is a Steiner tree in this space (see Chapter 5), with the measure of distance between points A and B being given by equation 24.17. This is well-known mathematically as the L_1 metric, and is often called the *Manhattan metric*, as it measures the distance between points in the regular grid of streets where one is not allowed to go diagonally. It is approximated by the street map of Manhattan, New York (provided one ignores the East Side Highway, Broadway, the roads in Central Park, and nearly everything below 14th Street).

Other parsimony methods

Another parsimony method, squared-change parsimony, was mentioned in the previous chapter. It has been used mostly to infer ancestral states under a Brownian motion model, rather than to choose among phylogenies. The original parsimony method of Edwards and Cavalli-Sforza (1964) summed the distances rather than the squared distances between nodes on the tree. It has not been applied to quantitative characters, but only to gene frequencies.

Threshold models

Many morphological characters come to us as discrete characters (such as presence or absence of a ridge on a scale, when that ridge may have been observed but not measured). Quantitative character models can be used to analyze such traits. The literature on this goes back to Sewall Wright (1934), who used a "threshold model" to analyze the inheritance of different numbers of digits on the hind foot of guinea pigs. It was further developed by Falconer (1965) as a model of disease in human populations, particularly diabetes. The model assumes an underlying quantitative character, called "liability," which cannot be directly observed. That character is assumed to be normally distributed. There is a developmental threshold, and if the liability exceeds the threshold, the observed character has state 1 rather than state 0.

In the genetics literature, the model has been used with inbred strains or pedigrees. It remains to adapt it to phylogenies. I have discussed doing so (Felsenstein, 1988a, 2002), though without presenting a practical method. The model would assume a quantitative character evolving by Brownian motion, with the observed character being determined by whether the character in an individual exceeded the threshold. Note that the model predicts not only change between species but polymorphism within species as well. If a species had its mean liability 1 standard

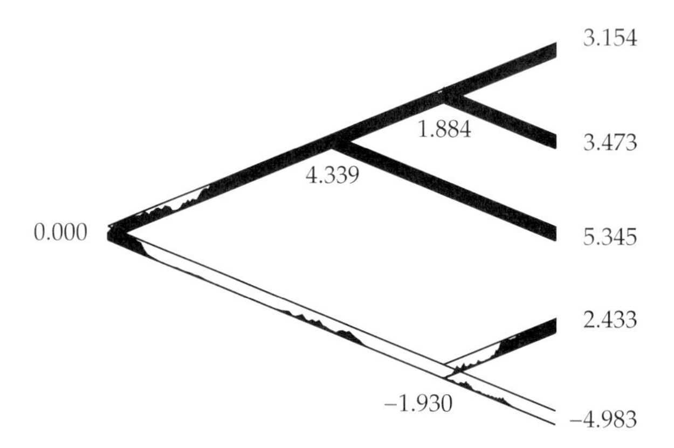

Figure 24.6: A simulation of evolution in a threshold character, with threshold 0 and standard deviation of liability 1. The mean population liability is shown at each node of the tree. The branches are shaded to show (in black) the fraction of the population that are of state 1. Four of the species end up with most individuals having state 1.

deviation below the threshold, we would predict that 0.8413 of the individuals in the population will have state 0, and 0.1587 of them will have state 1. Another biologically appealing feature of the model is that it predicts that after state 1 arises in a population, it can more easily be lost in the near future than later. When the population has just become mostly of state 1, it is fairly close to the threshold value, and further wanderings might soon carry it back across the threshold. But later on, it may well have wandered further from the threshold into the zone for state 1, and then it will be less likely that we will soon see reversion to state 1.

Figure 24.6 shows a simulation of evolution of a 0/1 threshold character on a simple phylogeny. The threshold is assumed to be at 0, and the standard deviation of the liability is assumed to be 1. In this simulation I arbitrarily started the liability at a value of 0. At each node in the tree we see the mean liability, and the branches of the tree show the fraction of the population (in black) that have state 1.

The liability is on an arbitrary scale, so we may as well choose the threshold value to be 0 (in two-state cases) and choose its standard deviation to be 1. Wright (1934) modeled a meristic character with many states (and even with partial states), using multiple thresholds.

If we have observed one individual at each tip, the likelihood for a given phylogeny under the threshold will be the total probability of that combination of individuals. Thus for five species, the first four of which are 1 and the last 0,

$$
\begin{aligned}
L \quad = \quad & \int_0^\infty \int_0^\infty \int_0^\infty \int_0^\infty \int_{-\infty}^0 \\
& \qquad \text{Prob}\,(x_1 > 0,\ x_2 > 0,\ x_3 > 0,\ x_4 > 0,\ x_5 \leq 0 \mid \mu, T) \qquad (24.18) \\
& \qquad\qquad dx_1\ dx_2\ dx_3\ dx_4\ dx_5
\end{aligned}
$$

where the tree T is specified with branch lengths and the probability of the x_i is the usual Gaussian density with covariances depending on the branch lengths. This is an integral of a corner in a multivariate normal density. It also depends on the starting liability μ.

The model has the additional great advantage that it can easily be extended to allow for correlations among discrete traits, simply by assuming that the underlying Brownian motions are correlated. For p characters we then need only $p(p-1)/2$ additional parameters for their correlations, and there is the hope of inferring them, given enough species. But there is one major disadvantage. Integrals of corners with multivariate normal densities have no analytical formula. Such integrals are difficult to compute by conventional numerical integration — it looks like they will yield only to Markov chain Monte Carlo methods. Note that our model is essentially a hidden Markov model as well, with the liability being the hidden state.

Chapter 25

Comparative methods

Comparative methods use the distribution of traits across species to make inferences about the effect on their evolution of other traits or of environments. Gallons of ink have been spilled over the years by biologists writing about the importance of the comparative method, but only in the last 25 years have they understood that use of the comparative method requires phylogenies and statistical methods that use them. Figures 25.1 and 25.3 show how, when we do not use phylogenies, a seemingly straightforward analysis of individual species can create an artifactual signal.

An example with discrete states

Figure 25.1 shows a phylogeny with 10 species and two 0/1 characters. On the phylogeny both characters happen to change once, in the same branch. When the species are taken as individual sample points and a contingency table is created, it appears to support a strong correlation between the states of the character. Yet the phylogeny makes it clear that this is an illusion. If all branches on the phylogeny were of equal length, it would be safer to make a contingency table of branches. If the branch where both characters changed was longer than the other branches, the coincidence would be even less surprising.

The contingency table on the left side of the figure shows the apparent tight correlation between the two characters. If we accepted the species as being independently distributed, we could use a Fisher's exact test and we would find that the probability of a correlation this close or tighter is very small, $1/210 = 0.0047619$. This would seem to establish that the two characters have evolved in a closely correlated manner. But if we made the calculation this way, we would be deceiving ourselves. The species cannot be regarded as independent outcomes of evolution. They are related, having evolved on a phylogeny.

A more relevant way of looking at the evolution of these characters is to consider on which branches of the phylogeny the characters changed. The figure re-

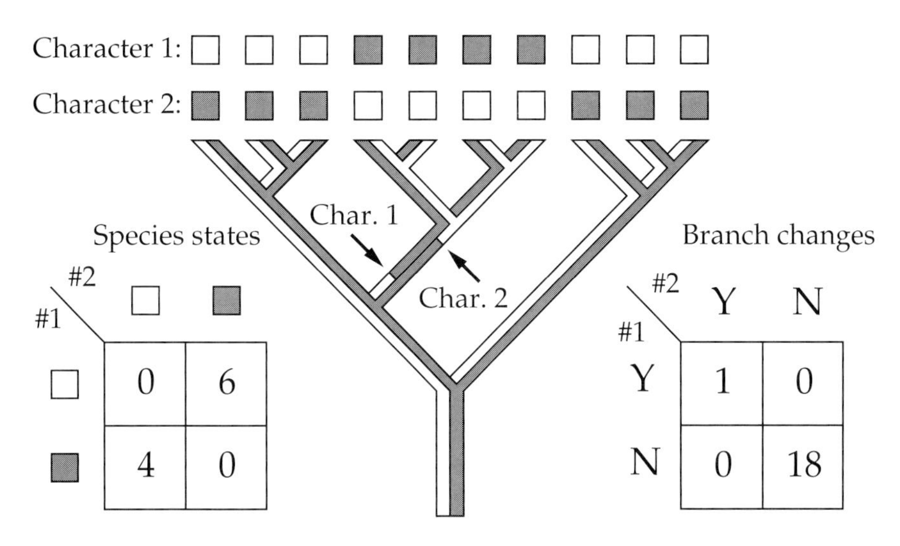

Figure 25.1: An example of two discrete two-state characters evolving along a phylogeny so that their distributions among the species are correlated. Each branch of the phylogeny consists of two regions side by side, for characters 1 and 2. Shading shows which state each is in. The states of the two characters are shown at the tips of the tree by empty or shaded boxes.

constructs where each character changed. Both changes happened on the same branch of the phylogeny. Not counting the root, there are 19 branches on this phylogeny (18 if we remove the root and unite the two most basal branches). The probability that both characters would change on the same branch would be 1/18 = 0.05555. This fails to be significant; it is far less extreme than the previous probability.

An example with continuous characters

Figure 25.2 shows a tree with 22 species, which has two clades of 11 species each. The vertical scale indicates the branch length. There is no structure within each of the clades, but they have been separated for some time. Figure 25.3 shows two characters that have evolved by Brownian motion along this phylogeny, with no correlation between the evolution of the characters. The 22 species are plotted as points, and in the left plot there seems to be a correlation between the two characters — a larger value of one goes with a larger value of the other. But on the right side we have shaded in the points according to which clade they come from (the shadings are the same as in Figure 25.2).

Looking at the shaded points, we can see that within each of the clades there is no particular sign that the two characters are correlated. All of the correlation

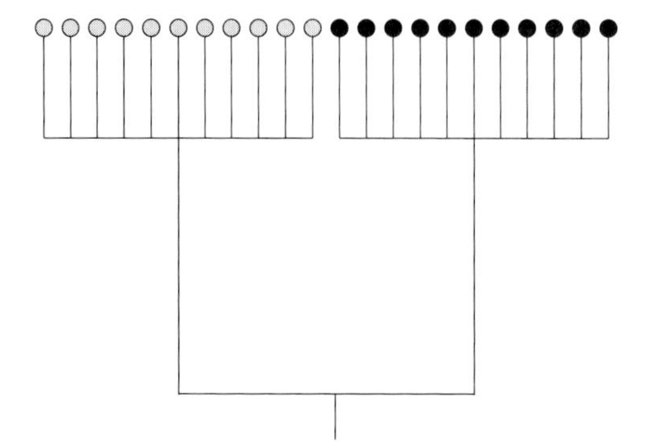

Figure 25.2: A tree with 22 species and two clades, which was used to evolve the characters shown in Figure 25.3.

arises from the difference between the clades — from the changes of the two characters along the two large internal branches of the tree. Those two branches are effectively only one branch, as we have no outgroup to give us separate information about the root of the tree. So the correlation in a plot of 22 points comes from 11 of them being displaced upwards and to the right, compared to the other 11. This is due to a single event, a burst of change in the branch separating the two

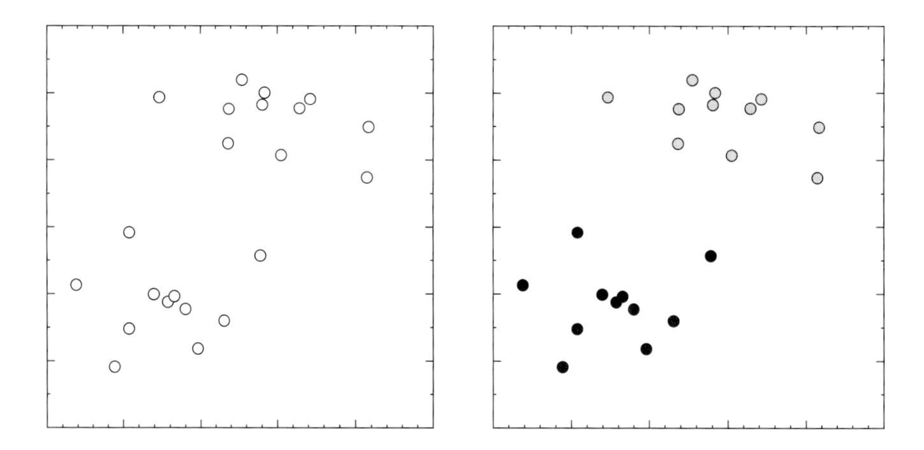

Figure 25.3: Two characters that evolved in an uncorrelated fashion on the tree of Figure 25.2. There appears to be a correlation, but when the points are shaded to show which clade they came from, it becomes apparent that this is an artifact of the phylogeny, as within each of the two clades there is no correlation. The scales on the axes are omitted, as they are arbitrary.

clades. The number of points (22) is a misleading indication of the number of independent sources of variation.

The whole point of phylogenies is that species did not evolve independently, but that historical events affected large groups of species. Only by untangling the correlation of the species can we untangle the correlations of the characters.

The contrasts method

That there was a problem in the statistical analysis of the comparative method was recognized by a number of evolutionary biologists in the late 1970s (cf. Clutton-Brock and Harvey, 1977, 1979; Harvey and Mace, 1982; Ridley, 1983). Methods based on analysis of variance were attempted (Harvey and Mace, 1982), and Ridley (1983) discussed how to analyze data on discrete characters. We return to Ridley's method and other discrete characters methods later; for the moment we concentrate on the analysis of continuous character data, for which the answer is found in the contrasts method.

The *contrasts method* was introduced by me (Felsenstein, 1985a), using techniques developed for computational efficiency when computing likelihoods on phylogenies where the characters have evolved by Brownian motion. In Chapter 23 we have seen that, for any character evolving by Brownian motion along a phylogeny, we can find a series of contrasts between the character values at the tips that are statistically independent. Figure 25.4 shows a phylogeny, together with the contrasts it implies. Below the figure are the contrasts that are derived from this phylogeny, together with their variances.

The contrasts are expressed by their coefficients, shown in the figure. When these are used, the resulting numbers may be called contrast scores. The contrast scores may be taken as new variables. They express all the variation between species, leaving out only the grand mean of the character. We can also divide each contrast score by the square root of its variance, so that (in our example):

$$\begin{aligned} Y_1 &= y_1/\sqrt{0.4} \\ Y_2 &= y_2/\sqrt{0.975} \\ Y_3 &= y_3/\sqrt{0.2} \\ Y_4 &= y_4/\sqrt{1.11666} \end{aligned} \qquad (25.1)$$

and we obtain standardized contrast scores.

Assuming that the Brownian motion model is correct and so is the tree, the contrasts all have the same expected mean and variance. Note that

1. The contrasts all have expectation zero.

2. They assume that we know the mean phenotype of the populations at the tips of the trees precisely.

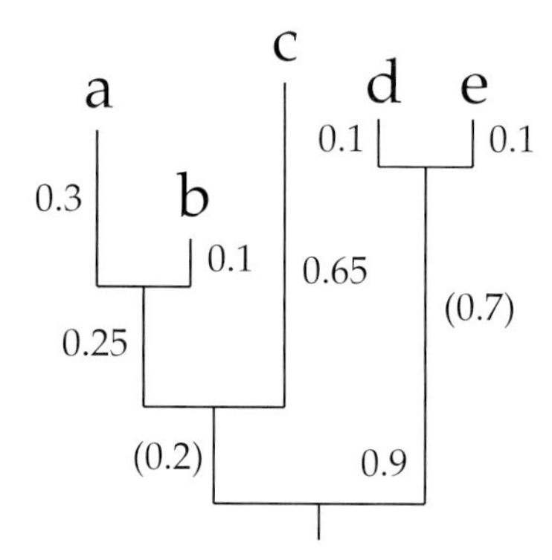

		Contrast									Variance proportional to
y_1	$=$	x_a	$-$	x_b							0.4
y_2	$=$	$\frac{1}{4}\,x_a$	$+$	$\frac{3}{4}\,x_b$	$-$	x_c					0.975
y_3	$=$							x_d	$-$	x_e	0.2
y_4	$=$	$\frac{1}{6}\,x_a$	$+$	$\frac{1}{2}\,x_b$	$+$	$\frac{1}{3}\,x_c$	$-$	$\frac{1}{2}\,x_d$	$-$	$\frac{1}{2}\,x_e$	1.11666

Figure 25.4: An example phylogeny and the independent contrasts that it implies under a model of evolution by Brownian motion. The branch passing through the bottommost node has total length 0.9, as two branches of length 0.2 and 0.7 separate these two clades.

3. If the tree's branch lengths are all multiplied by the same constant (say, 2.347) the contrasts will still all be independent, have zero means, and all have the same variance; their variances will all be divided by 2.347.

Figure 25.5 shows the contrasts for the numerical example of Figures 25.2 and 25.3. The points for the contrasts within the two clades have the corresponding shadings, and the single contrast between the clades is unshaded.

Correlations between characters

If we have two or more characters, we can apply the contrasts method to each simultaneously. The contrast formulas will be the same in each character. We have seen in equation 23.39 that when there are multiple characters undergoing correlated Brownian motion, a set of contrasts can be found that are independent. The covariances of the characters in these contrasts will be proportional to the covariances ($\mathbf{A}$) of evolutionary change among the characters.

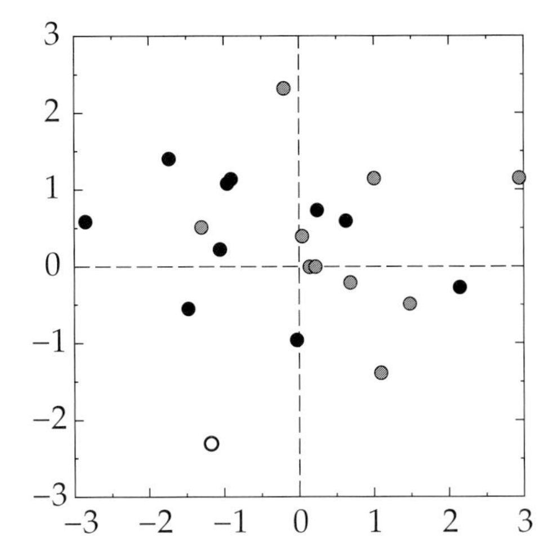

Figure 25.5: The contrasts for the two characters in the example of Figures 25.2 and 25.3. The 10 contrasts within each of the two clades are shaded correspondingly, and the single contrast between the two clades is unshaded. The scales are in standard deviation units.

As an example, consider the tree of Figure 25.4 with three characters. We have five species and three characters, so that we have 15 species means. If we take the four contrasts in each of the characters, for contrast y_1 we will have three values, one for each of the characters. The covariance matrix of changes in the three characters along the tree (**A**) will then also be the covariance we expect to see among the three values of y_1. There will be a similar covariance among the three values of y_2, with the same covariance matrix. But the y_1 values and the y_2 values will be independent.

Thus we can simply take the four contrast scores for each of the three characters, and feed them into a standard multivariate statistics package. They give us four samples with which to estimate the covariances among the characters. We can do the usual kinds of multivariate statistical analysis on them, such as using principal components to find the combination of characters that shows the most change along the tree.

When the tree is not completely known

The contrasts method requires that we know the phylogeny, including its branch lengths. Frequently, some regions of the tree have unresolved structure, as they are considered to be poorly known. These are often described as "soft polytomies," as they are not positively known to be polytomous, but describe our ignorance instead. A number of suggestions have been made as to how to deal with these:

- I suggested (1988a) that we bootstrap-sample the original data that inferred the tree, inferring a tree from each bootstrap sample. The comparative method would be performed on each bootstrap sample. To place a confidence limit on an inferred parameter such as a slope, one would make a histogram of its values from all these analyses, and reject a value such as 0 only if it fell in the appropriate tail of the histogram.

- Grafen (1989), Harvey and Pagel (1991), and Pagel (1992) proposed methods of dealing with polyfurcations in the tree, methods that go beyond the contrasts computation. Purvis and Garland (1993) suggested reducing the degrees of freedom to count for the polyfurcation. Garland and Díaz-Uriarte (1999) presented simulation results arguing that this worked well. Rohlf (2001) pointed out that in their method if the tree were a complete multifurcation, there would be zero degrees of freedom in spite of the presence of data.

- Martins (1996a) proposed that when a tree topology is completely unknown, one simulate trees at random for these species, then analyze the data on each and take the mean of the inferred quantity as the best estimate. Housworth and Martins (2001) suggested ways of simulating random parts of trees when only a portion of the tree is unknown. Abouheif (1998) suggested that randomly generated bifurcating trees will give the same result as having only a single multifurcation. In this he was supported by the simulations of Symonds (2002). It would seem possible to treat this analytically. The approach of Martins and Housworth (2002) would seem relevant.

If polyfurcations in the tree are taken to be real ("hard polytomies"), the correct way of dealing with them is to resolve them into bifurcations by adding zero-length branches in any reasonable way. It can be shown (Rohlf, 2001) that how this is done does not affect the values of regressions and correlations calculated from ordinary multivariate analyses.

Inferring change in a branch

McPeek (1995) has discussed using the contrasts method to infer the amount of change along a particular branch. If we take the tree as unrooted, then prune the phenotype values down to both ends of the branch, we will obtain values such as x'_1 and x'_2, with extra variances δ_1 and δ_2 added to the two ends. If the branch has original length v_3, a simple regression argument shows that the actual change of the character along that branch is distributed normally around a mean $x'_1 - x_2'$ multiplied by $v_3/(v_3 + \delta_1 + \delta_2)$. The variance of the distribution is $v_3(\delta_1 + \delta_2)/(v_3 + \delta_1 + \delta_2)$. McPeek points out that reconstructed changes on different branches can be correlated (in fact, all of them are). However, his formulas for the reconstructed changes differ from the ones I have given here. His formulas give nonzero reconstructed change even when $v_3 = 0$, while the above formulas correctly infer that change is zero in such a case.

Sampling error

Ricklefs and Starck (1996, p.169), in the midst of a skeptical review of applicability of contrast methods, make an insightful point about sampling error. If we do not have the actual character mean for each species, because we have only a finite sample of individuals, this adds an extra source of error beyond the randomness of Brownian motion. This sampling error will depend on sample sizes, but not on branch lengths of the tree. If a contrast is taken between two species that are neighbors on the tree, the contrast will be divided by a small quantity, the branch length separating those two tips. This assumes that the variance of the difference between them is substantially smaller than if they had been farther apart on the tree. But if sampling error is an important source of variance, the variance of the contrast between neighbors is then underestimated. Ricklefs and Starck find that contrasts of closely related species are often outliers in their regressions. This is presumably an artifact of sampling error (some of which could be due to measurement error).

The solution to this problem is to take the sampling error explicitly into account. Riska (1991) discussed the need to include sampling error in our models. We must add a variance component due to variation between individuals within a population, and allow for its variance when inferring covariances between characters. Each individual's measurement is distributed normally, with a term added to the model for within-species variation. We have seen that if all the variation is due to evolutionary change of population means, the covariance matrix of multiple characters in multiple species is $\mathbf{T} \otimes \mathbf{A}$. When there are also sampling effects, suppose that we draw multiple individuals from species i. The covariances between characters k and ℓ will then have an extra component $\sigma_{k\ell}$ if they are measured in the same individual. This means that an extra component of variance (which we call e_{kl}) is added to covariances of characters in the same individual.

The model can be written concisely by taking the individuals as the unit. We can imagine a phylogeny like Figure 25.6 connecting all the individuals in the study. The tree that is shown is basically the same as in Figure 25.4, except that extra branches have been added for the individuals sampled for each species. In this model, the covariances added by these branches are not derived from the covariances of changes in phenotype in evolution, but from the covariances of characters within species. The effect of this is that the covariance matrix can be written as

$$\mathbf{T} \otimes \mathbf{A} + \mathbf{I} \otimes \mathbf{E} \tag{25.2}$$

where $\mathbf{T}$ is a matrix for a tree that has tip i replicated n_i times, where n_i is the sample size for species i on that tree. The extra branches that are added have length 0. The new term in this covariance matrix, $\mathbf{I} \otimes \mathbf{E}$, adds covariances of different charac-

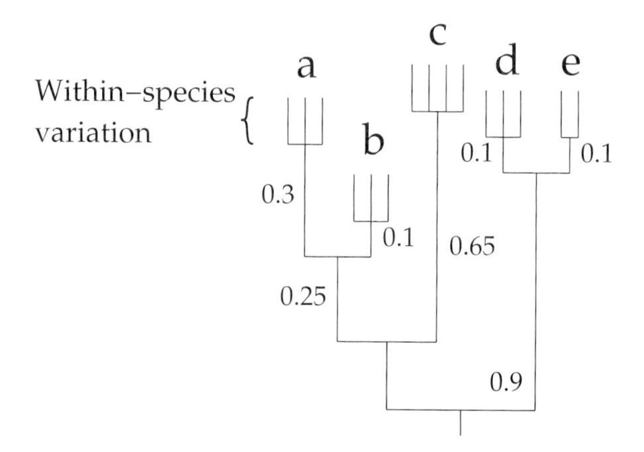

Figure 25.6: The same tree as in Figure 25.2, but with extra branches for each species, one per member of the sample for that species.

ters within each individual, which are the within-species phenotypic covariances of characters. The distribution of phenotypes in individuals is

$$\boldsymbol{x} \sim \mathcal{N}(\mathbf{1} \otimes \boldsymbol{\mu},\ \mathbf{T} \otimes \mathbf{A} + \mathbf{I} \otimes \mathbf{E}) \tag{25.3}$$

where $\mathbf{x}$ is a vector that not only stacks up the different species, but within them the members of the sample and within those their different characters, and where $\mathbf{T}$ is a tree with a branch of length 0 for each sampled individual.

We cannot simply regard these extra branches, one for each individual in the samples, as involving the same differences as those that accumulate in evolution. It is true that contrasts among these individual branches can be taken, and they are independent of each other. If a species has n_i individuals in its sample, it is quite easy to construct $n_i - 1$ contrasts among the individuals, all of which are independent. However, the character covariances in those contrasts are the covariances $\mathbf{E}$, the within-species phenotypic covariances, and these may be quite different from the covariances $\mathbf{A}$ of between-species phenotypic change. If we take all the within-species contrasts, this leaves at each species its arithmetic mean phenotype, as seems appropriate.

You might think that there is a neat separation between one set of contrasts that measures within-species phenotypic variation and another that measures between-species changes. But when we set out to take between-species contrasts, we should remember that the species means reflect not only the random changes between species, but also our sampling error due to having only a limited number of individuals when there is within-species variation. For example, if species a has a sample size of 3, then when the within-species variance of a character is σ^2, this creates an extra variability of $\sigma^2/3$ in the species mean beyond the variability of evolutionary change.

It will not necessarily be true that the covariation of characters among species due to evolutionary change is proportional to the covariation within species due to phenotypic variation and finiteness of the samples. Two characters could covary positively within species, and yet change in a negatively correlated way between species.

One case of this model has already been treated in the literature. Lynch (1991) used a model where, in addition to its phylogenetic covariances, each species has a component due to recent change that is specific to that species and has its own covariances. This gives the same model as if we had sampled only one individual per species. Using an REML method from quantitative genetics, Lynch was able to provide iterative equations for estimating the different variance components.

I have (Felsenstein, *in prep.*) provided an alternative method of computation for the model with sampling error, using contrasts that are not standardized in the usual way, but are orthonormal contrasts (the sum of squares of their coefficients sum to 1). As we saw with the variance of a single character, the finiteness of the sample sizes implies that a small portion of the within-species phenotypic covariances becomes added to the between-species covariances. The computation must be iterative, converging on REML estimates of the within- and between-species covariances. We can no longer simply take the contrasts of all characters and feed them into a standard multivariate statistics package, but we can maximize likelihood with respect to the general model and with respect to a restrictive model (such as a model that has no phylogenetic covariance between two characters), and do a likelihood ratio test in the usual way.

The effect of allowing for sampling error and within-species phenotypic variation is to discount many of the contrasts between closely-related species. If we have a pair of species that are close relatives, ordinary contrasts analysis inflates the contrast between them, dividing it by the square root of the branch length separating them. But if there is also sampling error, most of the contrast may come from that. When the sampling error is properly taken into account, the contrast between that pair of species may contribute little to the inference of phylogenetic covariances and correlations. This makes intuitive sense.

When Lynch's method was tested (E. Martins and M. Lynch, personal communication; Felsenstein, *in prep.*) it was found that estimates of phylogenetic correlation between traits tend to be too extreme. This simply reflects the small effective sample sizes; if we have 20 species, but many fewer clusters of closely related ones, the correlation coefficients are estimated mostly from the between-cluster contrasts, which are fewer. Correlations are usually biased to be too extreme, and this bias is serious when sample size is small. For example, when sample size is 2 we know that the correlations that can be observed are almost always +1 or –1. The extreme values found in simulations reflect the fact that the between-species observations are equivalent to a rather small number of independent points.

Correction for sampling error is still rare in comparative studies; hopefully it will become more common. It should place a premium on having a sufficiently great diversity of species in the study.

The standard regression and other variations

Grafen (1989, 1992) has given a multivariate statistical framework that is an alternative to the contrasts method; he calls it the *standard regression*. In the case in which the tree is fully bifurcating, careful consideration of his method will show that it is simply an alternative computational scheme to contrasts and will always give the same results. His methods use matrix calculations instead of recursive derivation of contrasts. The computation of the contrasts is, in effect, a way of obtaining eigenvalues and eigenvectors of the covariance matrix of species.

Generalized least squares

Martins (1994) presents an alternative generalized least-squares framework that is not exactly the same as REML, but that will approximate it in most cases. Rohlf (2001) argues that these two methods are actually equivalent. Paradis and Claude (2002) suggest using generalized estimating equations (GEE) rather than the generalized least squares (GLS) framework of Martins. Their method is more general than GLS but reduces to it in the Brownian motion case. The chief utility of these methods is that when we have other evolutionary mechanisms, they can be modified to approximate them. Martins, Diniz-Filho, and Housworth (2002) have examined by computer simulation how robust some of these methods are to variations in the evolutionary model.

Phylogenetic autocorrelation

Cheverud, Dow, and Leutenegger (1985) used a method developed to correct for geographic structure in data, adapting it to phylogenies (see also Gittleman and Kot, 1990). Their "phylogenetic autocorrelation" method has been tested against contrasts methods in simulations, with indifferent results (Martins, 1996b). It has been criticized by Rohlf (2001) as having some assumptions that conflict with any possible evolutionary mechanisms. Chief among these is the way error enters the model. Differences between closely related species are expected to be as variable as differences between distantly related ones; this is inconsistent with mechanisms such as Brownian motion.

Transformations of time

Gittleman and Kot (1990) added to their method a transformation of the time scale. Suspecting that the Brownian motion model might not accurately represent the evolutionary process, they allowed the time scale to be nonlinearly transformed. Actually, their scale was not so much time as it was a distance that represented taxonomic dissimilarity. By adding a parameter α to make the covariances proportional to this distance raised to the αth power, they allow the data to dictate how taxonomic groupings are reflected in covariances. The phylogenetic autocorrelation method also has a constant ρ that controls how much phylogeny influences similarity.

Pagel (1994), in his method for discrete characters, used a different approach in which the length of each branch of the tree is raised to a power, with the power being estimated.

Should we use the phylogeny at all?

The most drastic modification of the contrast method is to discard it. As phylogenetic comparative methods have first disrupted comparative work, and then become compulsory, this has led to questioning of their celebrity status. A particularly interesting exchange in the pages of *Journal of Ecology* raises the issue of whether seeing an effect of phylogeny proves that the effect is not due to natural selection (Westoby, Leishman, and Lord, 1995a, b, c; Harvey, Read, and Nee, 1995a, b; Ackerly and Donoghue, 1995). The paper of Ricklefs and Starck (1996) is another interesting critique.

Freckleton, Harvey, and Pagel (2002) suggest that a parameter defined by Pagel (1999a) can be used to test whether there are any phylogenetic effects at all. The covariance matrix between species is taken to be λ times the matrix predicted from the phylogeny, plus $(1 - \lambda)$ times a matrix in which all species are independent. By doing likelihood ratio tests on λ one can test whether there is any sign of an effect of phylogeny. This is essentially the same as a comparable test in the work of Lynch (1991) in which the model included both a phylogenetic effect and an individual species effect.

Paired-lineage tests

An alternative to the contrasts method is to look at pairs of species, chosen from the tree in such a way that the paths between them do not overlap. Looking at pairs of related species is an old method that dates back to at least the work of Salisbury (1942). I have extended it (Felsenstein, 1985a) to allow pairs that are connected by nonintersecting paths on the tree. Figure 25.7 shows the choice of such pairs. We can make a sign test of whether two characters change in the same direction (+) or opposite directions (–) between members of the pairs. If there is no correlation in the evolution of the two characters, our expectation is that these two outcomes would be equally frequent. So a simple sign test suffices.

Figure 25.7 shows a tree with one of the ways that the species can be divided into pairs. The paths between the species are shown by the dark lines. In general there are multiple ways that species can be grouped into pairs so that the paths between the members of the pairs do not overlap. On this tree there is only one grouping that satisfies this condition and finds four pairs, and it is shown here.

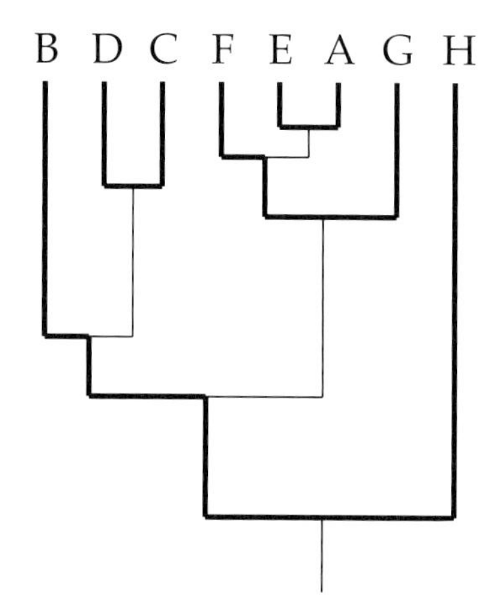

Figure 25.7: Pairs of species chosen so that the paths between the two members of each pair do not overlap. The paths between the pairs of species are indicated by the dark lines.

The paired-lineages test is appealingly simple, but it does lose some statistical information. A set of 19 species will have eight such pairs of lineages, but it will have 18 contrasts in the contrast method. An additional issue is that some pairs may have long paths between them. Depending on the exact scheme of evolution these pairs could fail to show a correlation. Of course, the same could be said for contrasts methods using a contrast with long branch lengths between the groups.

Ackerly (2000) presents simulation results on different sampling strategies used with different methods of analysis, including paired-lineage tests. He argues that choosing pairs of species so as to maximize the differences between species in one character can bias the regression of another character on that one.

Discrete characters

The contrasts method does not help us if we have discrete characters. Both parsimony and likelihood methods have been suggested to test whether characters are correlated in their evolution.

Ridley's method

Mark Ridley (1983), in probably the earliest method for treating comparative method data statistically, suggested mapping the two traits onto a tree using parsimony. Ridley's method is not entirely clear in his book. Wayne Maddison (1990) expresses some uncertainty as to what it is; Harvey and Pagel (1991) argue that it involves investigating the independence of the states in two characters at the upper ends of all branches. Sillén-Tullberg (1988) interprets Ridley's method as involving examining whether the origins of the derived state in one character occur more often with one state of the other character than with the other state. However, she does not present a statistical test. Still another interpretation is that Ridley intends examining all branches and scoring each character as to whether it has changed or not in that branch. A contingency table can then be used to test association (as was done at the beginning of this chapter).

Concentrated-changes tests

Wayne Maddison (1990) has proposed a statistical test with discrete 0/1 characters. Depending on one's interpretation of Ridley's method, it may be an implementation of it. It asks whether the chance of a change in one character (call it character #1) is affected by the state of another character (call it character #2). He would have us start by reconstructing the regions of a tree that have one state or the other in character #2. Then he reconstructs the locations of changes in character #1. Suppose that there are seen to be seven occurrences of state 1 in character #1, and that these represent five changes of state in the tree, four of which are in regions with character state 1 in the other character. Is this number (4) a surprisingly large number?

Maddison uses a recursive algorithm to compute the probability of having four or more changes, out of five in total, in regions having character state 1 in character #2. He computes the number of different ways that one could have five changes of state of character #1 on the tree, and the fraction of those in which there are four or more changes in the regions that have state 1 in character #2. This fraction is taken as the probability, under the null hypothesis of no concentration of character #1's changes.

Sillén-Tullberg (1993) presents an alternative to Maddison's concentrated-changes test. Like Maddison, she makes parsimony reconstructions, including assignment of states to interior nodes in the tree. She looks at those branches that have state 0 of character #1 at their base and makes a contingency table. She classifies them according to whether or not there is a change to state 1, and whether or not the branch has state 0 or state 1 for character #2. One ambiguity in this test is when character #2 changes in a branch; it is not obvious which state of that character to assign to the branch. If multiple, equally parsimonious reconstructions are possible in either character, this can also cause trouble.

There are other possible ways to compute a probability for a concentrated-changes test. One could, for example, randomize the occurrences of state 1 in character #1 among all the tips on the tree, and count in how many of those cases there were more than four changes in the relevant regions. This would give a different result, but it would not be correct, because it would implicitly assume that even sister species were not correlated in their states. Maddison's test does a better job of allowing neighboring species on the tree to have correlated states. However, it does restrict its attention to those outcomes with a certain number of total changes in character #1. If we had a stochastic model for the evolution of that character, we could have randomized over outcomes of evolution. However, in the absence of that kind of model, Maddison's quantity is probably as good as we can do.

In addition to this issue, this concentrated-changes test is completely dependent on the accuracy of the reconstruction of state changes by parsimony, as Wayne Maddison (1990) acknowledges. It assumes that we can know, without error, where each character's changes of state were. In doing so, it fails to take into account our full uncertainty as to how characters have changed.

A paired-lineages test

Read and Nee (1995) have proposed a test with discrete binary characters which is a natural counterpart to the continuous-characters paired-lineages test. Taking care to keep lineages from overlapping, the test takes pairs of species that differ at both characters. A simple sign test can then be done to check whether the character correlations are more in one direction than the other. A difficulty with this test is that there may be too few pairs of lineages that differ in both characters to allow for much statistical power. Wayne Maddison (2000) has discussed algorithms for choosing these pairs in as effective a way as possible.

Methods using likelihood

Pagel (1994) made the first likelihood-based comparative method for discrete characters. He assumed that two characters each have two states, 0 and 1, and that there is a simple stochastic process of change back and forth between them. There were two rates of change, so that the probability of change in a tiny interval of time of length dt would be $\alpha\, dt$ if the character were in state 0, and $\beta\, dt$ if it were in state 1. If the two characters were changing independently, for any given tree with branch lengths, we could compute the likelihoods for each character on the tree and take their product as the overall likelihood. The algorithms are essentially identical to those in Chapter 16, except that the number of states is two instead of four.

However, if the two characters are changing in a correlated fashion, the state of each can affect the probability of change of the other. In effect, we then have four states, for the four different combinations of states at the two characters. If both characters are in state 0, the combination is counted as being in state 00. There are also states 01, 10, and 11. If we assume that in any tiny interval of time only one of the characters can change, the matrix of rates of change of the character combinations is as shown in Table 25.1.

The notation in the table, which differs from Pagel's, has rates α and β of forward and backward change at the first character, and γ and δ of change at the second character. However, they differ according to the state of the other character. When character 2 is 0, the rates of change at character 1 are α_0 and β_0. But when character 2 is 1, they are α_1 and β_1. Similarly, the rates γ and δ for character 2 are subscripted according to the state of character 1.

Table 25.1: Rates of change between combinations of states in Pagel's (1994) discrete-character comparative method.

From : / To :	00	01	10	11
00	--	γ_0	α_0	0
01	δ_0	--	0	α_1
10	β_0	0	--	γ_1
11	0	β_1	δ_1	--

I will not go into the computational details, but it is possible to compute transition probabilities and equilibrium frequencies for the four states for this model. Using them, we can compute the likelihood for a tree (perhaps one obtained by molecular methods). More tediously, one can maximize this likelihood over the values of the parameters.

Some hypotheses of interest are restrictions of the values of these parameters. If the subscripting of α, β, γ, and δ does not affect their values, then the two characters are evolving independently. This is the set of constraints

$$\begin{aligned} \alpha_0 &= \alpha_1 \\ \beta_0 &= \beta_1 \\ \gamma_0 &= \gamma_1 \\ \delta_0 &= \delta_1 \end{aligned} \tag{25.4}$$

If we maximize the likelihood while maintaining these constraints, we are restricting four of the eight parameters. It is possible to do a likelihood ratio test comparing this likelihood to the likelihood with unrestricted parameter values, and this has 4 degrees of freedom. It is also possible to test, in a similar fashion, whether character 1 is unaffected by character 2, and whether character 2 is unaffected by character 1. These each involve restricting two of the parameters (α and β or γ and δ). Pagel noted that one can also test individual ones of these four parameters for being unaffected by the state of the other character, and each of these tests has 1 degree of freedom.

Another possibility is that one character affects the rate of evolution of the other, but not the equilibrium frequency of its two states. If $\alpha_0/\beta_0 = \alpha_1/\beta_1$, then character 1 will have the same equilibrium frequency no matter what the state of character 2. There is a similar condition for character 2. Each of these restricts 1 degree of freedom. A likelihood ratio test of these assertions could be done individually or simultaneously. The latter test has 2 degrees of freedom.

Pagel's framework provides a straightforward test of the independence of change in two characters. It can be extended to more elaborate forms of dependence. If one is provided with an unrooted tree, it may be necessary to constrain the two-character model of change to be reversible, so as to avoid having to know where the root is. This can be done by adding only one constraint, which is that

$$\frac{\gamma_0 \beta_0}{\delta_0 \alpha_0} = \frac{\gamma_1 \beta_1}{\delta_1 \alpha_1} \tag{25.5}$$

Advantages of the likelihood approach

Pagel's discrete characters comparative method has some important advantages over Maddison's and Ridley's parsimony-based approaches. It takes branch lengths into account, which the methods of Maddison and Ridley cannot (as Maddison pointed out in introducing his method). When some of the branches in the tree are quite short, it will automatically adjust for this and not assume that there could be large amounts of change in those parts of the tree. When a branch is very long, it will treat the evidence from the groups connected to its opposite ends as relatively independent. Another disadvantage of the parsimony-based approaches is that they reconstruct the placement of changes and then treat those events as observations. Any time we find ourselves using something that is an estimate, and using it as a definite observation, we should be suspicious. In such cases the errors that may arise from the uncertainty of the reconstruction are not taken into account in the analysis.

All of these methods suffer from a common limitation—they use a rather naive model of character change. Populations do not make instantaneous changes from one state to another. Having arrived at state 1, they may be more likely to revert to state 0 soon, but less likely later. None of these are taken into account in parsimony or likelihood uses of a simple model of change between two states. We have seen in Chapter 24 that a model with underlying quantitative characters and a threshold can come closer to reality. It allows for polymorphism within species and for different probabilities of reversion soon after a change and later. It is also easier to generalize to covariances of change among multiple characters than the 0/1 stochastic model. It will be important to develop a comparative method that uses the threshold model.

Molecular applications

Molecular evolutionists frequently want to know whether substitutions at different sites are correlated. This might help reconstruct protein structure so as to place correlated sites near each other in space. The great difficulty with using discrete-characters comparative methods for this is that there are 20 amino acids, and many possible pairs of sites to examine, so that there are far too many possible parameters. Any successful application of these methods would necessarily involve constraining parameters and examining sets of pairs of amino acids that could all be neighbors. Studies of correlated evolution of amino acid positions in a protein have often used nonphylogenetic measures of correlation. Wollenberg and Atchley (2000) have done so, but have used a parametric bootstrapping approach to simulate data on phylogenies to see how much of the correlations could be coming from the phylogeny. We have already seen, in Chapter 13, models of change that maintain pairing of sites in RNA structures.

Chapter 26

Coalescent trees

What happens to phylogenies when we consider individual copies of genes within populations? Trees do in fact exist, but are no longer trees of species. We are seeing the pattern of molecular evolution when it gets down below the species level. Copies of genes can be related by a tree, but different loci in the same individual are related by different trees, and the trees can change even within a gene. To make it clear how these trees form, let us consider a small population, mating at random in an idealized fashion. Figure 26.1 shows the genealogy of gene copies at a single locus in a random-mating population of 10 individuals.

The genealogy that is shown in Figure 26.1 differs from ordinary genealogies in that it shows connections between gene copies, rather than between individuals. Each line goes from a gene up to a gene that is descended from it. The mating system is that of an idealized Wright-Fisher model, commonly used in theoretical evolutionary genetics to investigate the effects of genetic drift. According to that model, each gene at a locus comes from a randomly chosen parent, copied from one of its two genes at random. The population is thus in effect monoecious, and selfing occasionally occurs when the two genes in an offspring happen to be descended from the same parent. This may seem biologically unrealistic, but in evolutionary genetics the effects of other mating systems are usually taken into account by computing an effective population number N_e and putting it in place of the actual population number. This has been extensively investigated and is found to work surprisingly well.

The genealogy in Figure 26.1 is the result of a computer simulation of 11 generations of descent in a Wright-Fisher model with 10 individuals. It is almost impossible to comprehend. In an effort to make it easier to look at, we can erase the circles that indicate individuals (Figure 26.2). The result is still too tangled to convey much. If we abandon any attempt to put genes from the same individual near each other, we can swap gene copies left-to-right and untangle the genealogy. The result is shown in Figure 26.3. No lines cross. The figure resembles a branching

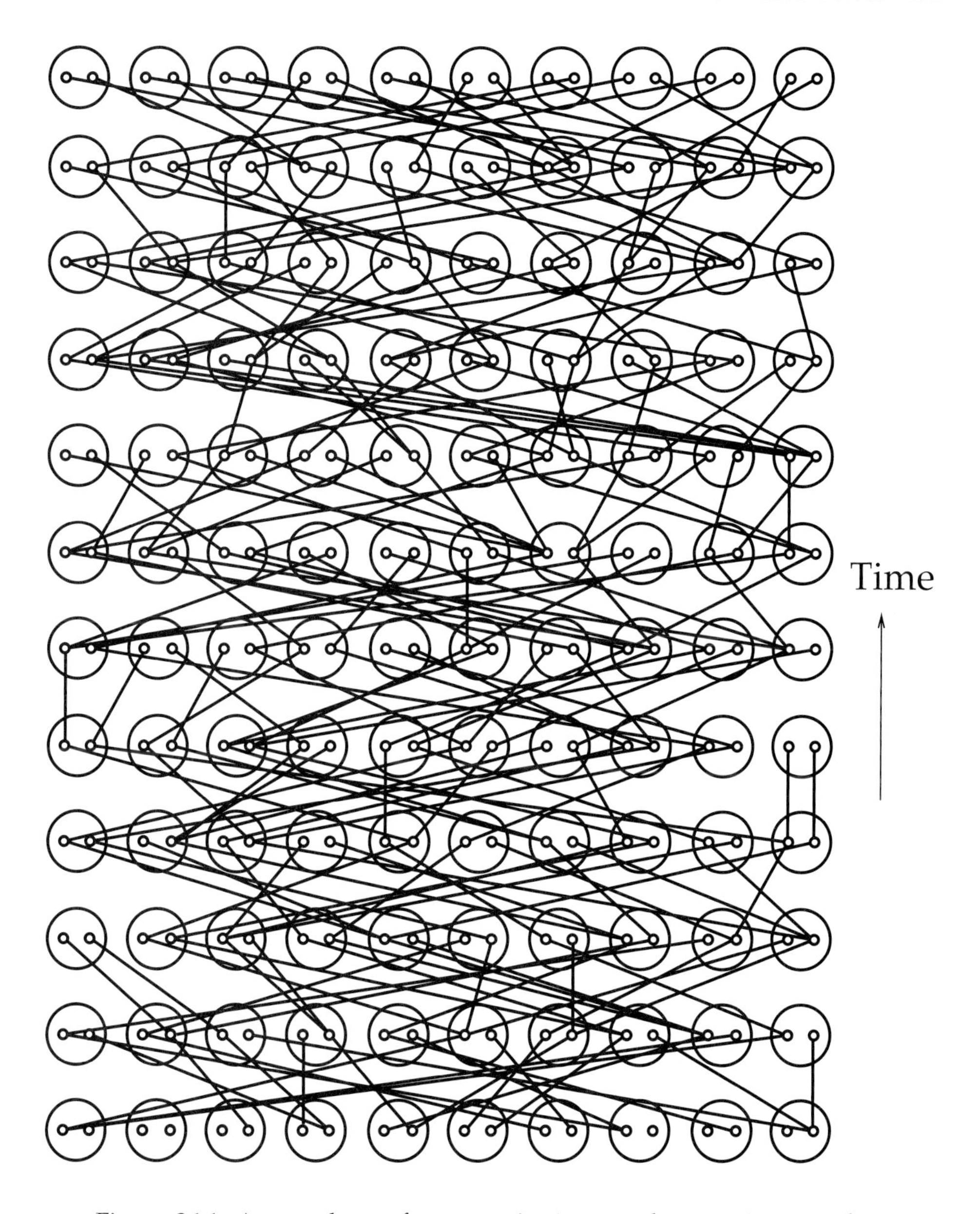

Figure 26.1: A genealogy of gene copies in a random-mating population of size 10, for 11 generations. Lines connect genes to their descendant copies in offspring. The model of reproduction is a Wright-Fisher model. Large circles are individuals, small ones are gene copies.

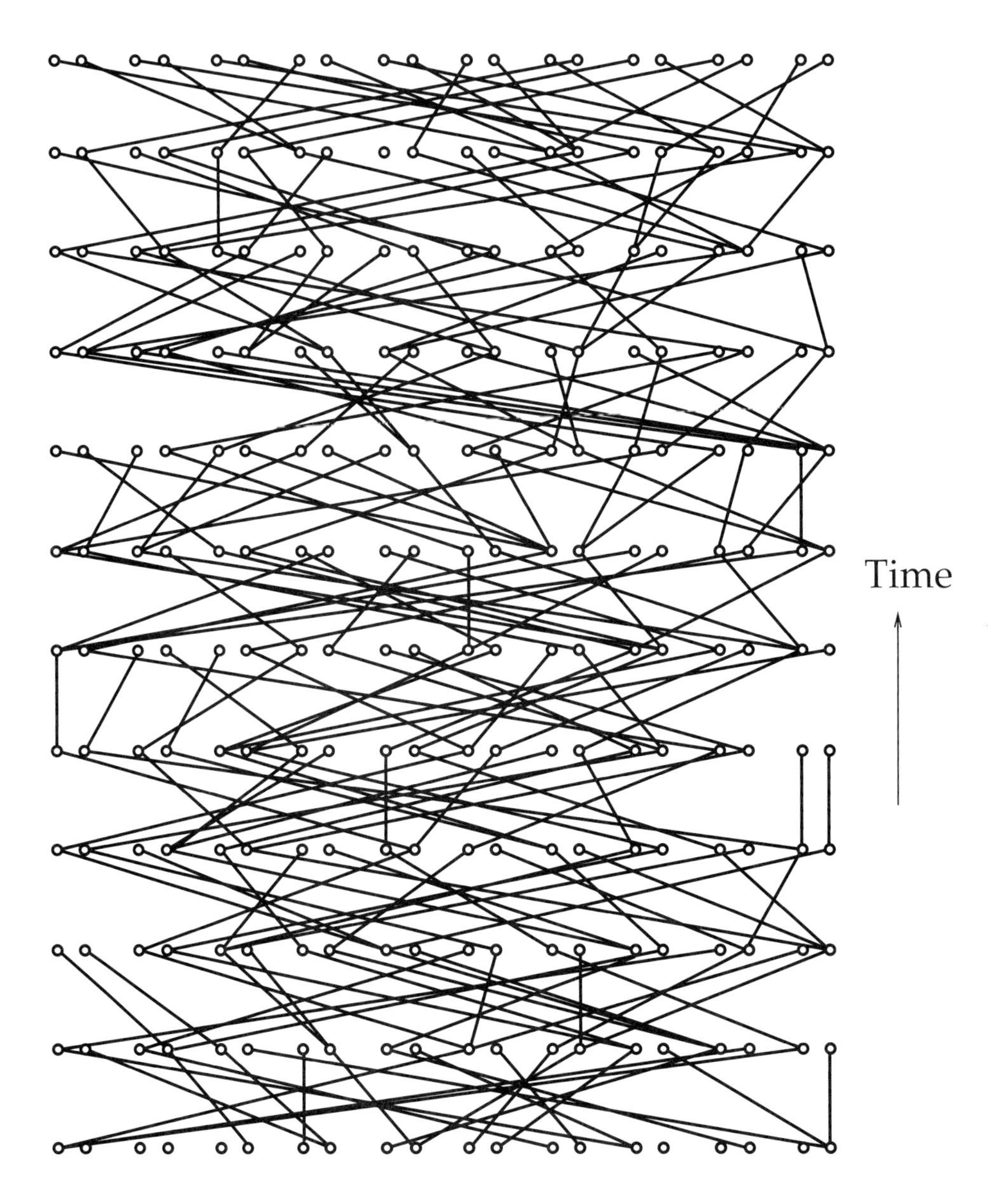

Figure 26.2: The same genealogy of genes as in Figure 26.1, with the individuals erased.

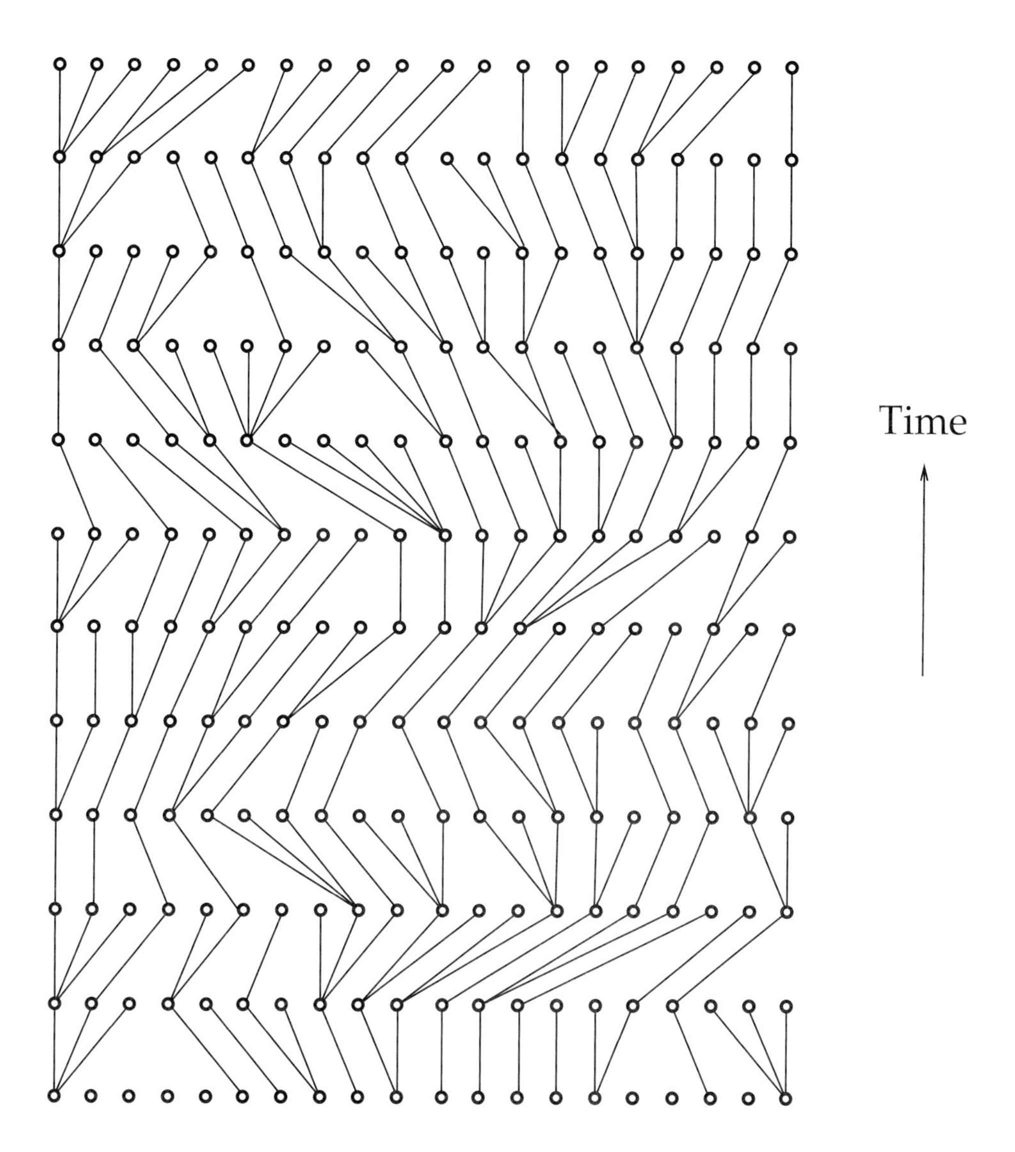

Figure 26.3: The same genealogy of genes as in Figure 26.2, with lines swapped left-to-right to untangle it, removing all crossed lines.

river system, with small tributaries at the top feeding ever-larger rivers that flow toward the bottom. In fact, of all the genes in the top generation, the first six are all descended from the leftmost gene in the bottom generation, and the next 14 are all descended from the gene number 10 in that generation. All other copies present in the bottom generation did not leave descendants by the time of the top generation.

Usually we are actually considering, not the entire genealogy of genes in a population, but the genealogy of a sample from the population. Figure 26.4 shows the genealogy of a particular sample of three copies from the current (top) generation in the genealogy of the previous figures. The members of the sample are related by a genealogical tree.

Kingman's coalescent

The structure of these trees of gene copies that form in random-mating populations was greatly illuminated by the probabilist J. F. C. Kingman (1982a, b, c). Kingman's result is an approximation, but such a good one that few evolutionary geneticists have tried to investigate the exact structure of such trees (nor will I). Kingman's results are generalizations of a result for two copies that was obtained by the famous evolutionary geneticist Sewall Wright (1931). Wright noted that in a finite population of size N, which is monoecious and has selfing allowed, the probability that two gene copies come from the same copy in the preceding generation is $1/(2N)$. In each generation there is the same probability. The distribution of the number of generations until the two copies finally have a common ancestor is thus exactly the same as the distribution of the number of times one must toss a coin until "heads" is obtained, where the probability of "heads" is $1/(2N)$ on each toss.

That distribution is called a geometric distribution. It has mean $2N$. It is very well approximated, as Wright was aware, by an exponential distribution that also has mean $2N$. Kingman's result is the extension of this result to a population with k copies of the gene. Going back in time, there will be a number of generations until two or more of these k copies have a common ancestor. Rather than following Kingman's algebra in detail, we can use a result from my own paper (1971) on genetic drift with multiple alleles. We compute the probability that no two of the k alleles in the current generation came from the same copy in the preceding generation, i.e., that all of them came from distinct copies.

The first copy came from some copy in the preceding generation. The second has probability $1 - 1/(2N)$ of coming from a different one. Given that (so that two copies in the preceding generation are now represented), the chance that the third copy came from a copy different from both of these is $1 - 2/(2N)$. Given that, the fourth copy has probability $1 - 3/(2N)$ of coming from a different copy from all of these. Continuing in this fashion, the probability that all k of them came from different copies is

$$G_{kk} = \left(1 - \frac{1}{2N}\right)\left(1 - \frac{2}{2N}\right)\left(1 - \frac{3}{2N}\right)\cdots\left(1 - \frac{k-1}{2N}\right) \qquad (26.1)$$

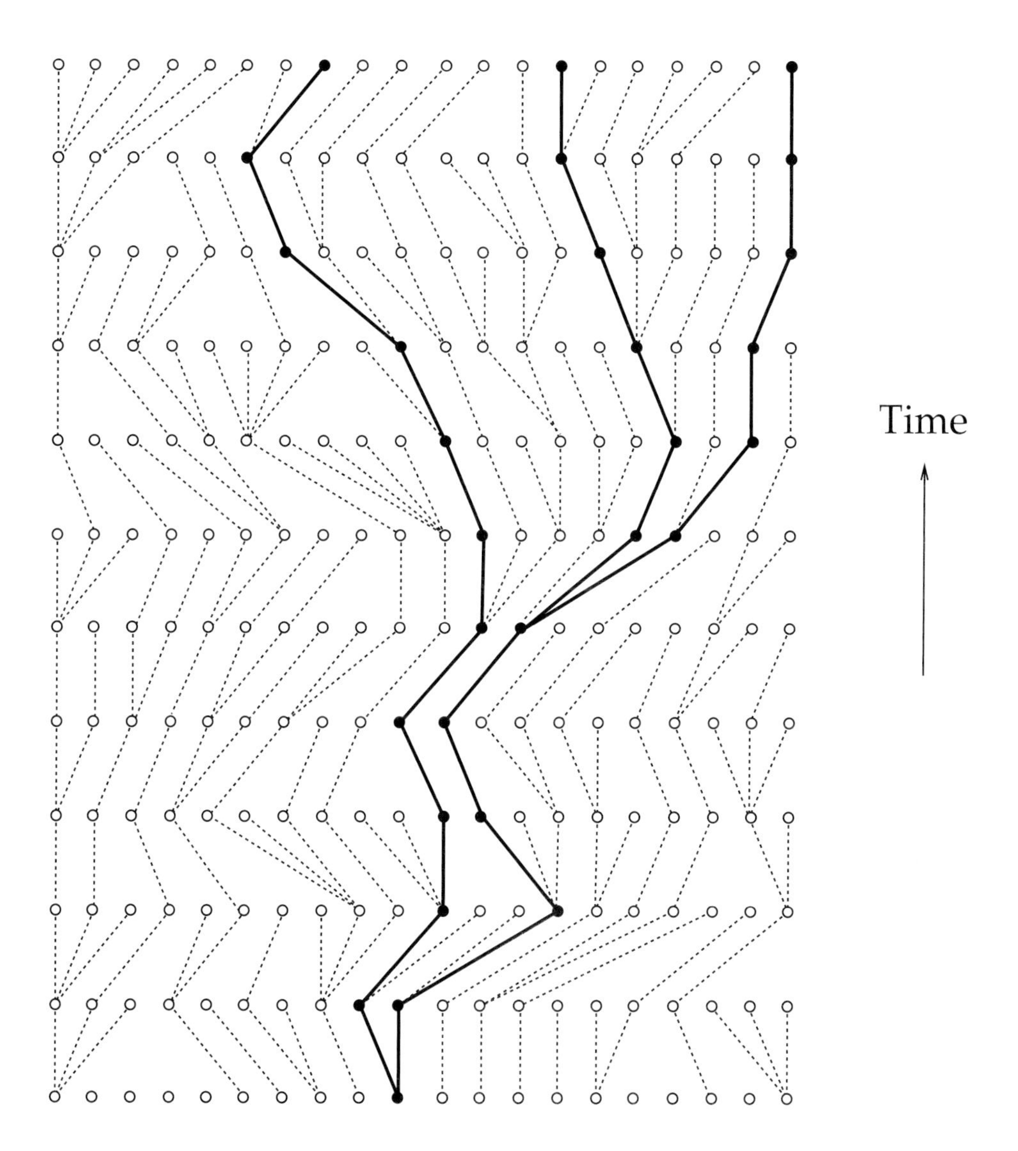

Figure 26.4: A genealogy of three gene copies sampled from the final generation of the preceding figures.

The right side can be multiplied out, and yields

$$G_{kk} = 1 - [1 + 2 + 3 + \ldots + (k-1)]/(2N) + \text{terms in } \frac{1}{N^2} \quad (26.2)$$

The sum of the integers from 1 to $k-1$ is well-known to be $k(k-1)/2$. Kingman's results amount to showing that ignoring the terms in $1/N^2$ is a good approximation. This will be true as long as the quantity $k(k-1)$ is much smaller than the population size N, which is usually the case. The events that are envisaged in the first terms on the right side of equation 26.2 are those in which precisely two of the genes are copies of the same parent gene. So Kingman's approximation in effect says that events in which three or more lineages collide are rare compared to ones in which two lineages collide.

We can then say that, to good approximation, in each generation a coin is tossed that has probability

$$1 - G_{kk} \approx \frac{k(k-1)}{4N} \quad (26.3)$$

of "heads." The number of tosses (generations) that are needed to get a "heads" is geometrically distributed, with mean being the reciprocal of the "heads" probability. Calling this time u_k, we have its expectation as

$$\mathbb{E}(u_k) = \frac{4N}{k(k-1)} \quad (26.4)$$

The time is also well-approximated by an exponential distribution with the same expectation.

From the process it should also be obvious which lineages are the ones that collide—a random pair. Thus Kingman's recipe for constructing a genealogical tree of k gene copies is simply:

1. Go back a number of generations drawn from an exponential distribution with expectation $4N/[k(k-1)]$.
2. Combine two randomly chosen lineages.
3. Decrease k by 1.
4. If $k = 1$, stop. Otherwise, go to step 1.

The resulting stochastic process was called by Kingman the *n-coalescent*. The name has stuck (though without the n): Genealogical trees of ancestry of multiple gene copies are widely known as coalescents. We should keep in mind that Kingman's coalescent is an approximation, in which it is impossible for three lineages to collide simultaneously. But as long as $k(k-1) \ll N$, it is a very good approximation. The remarkable thing about Kingman's coalescent is that it describes the genealogy of a sample of k genes, without making it necessary to know the genealogy of the rest of the population. This can be a great economy.

As the number of copies grows smaller, the expectation of the time for them to coalesce grows longer. The expected total time for k copies to coalesce is readily computed. Note that $1/(k(k-1)) = 1/(k-1) - 1/k$, so that

$$\begin{aligned} & \frac{4N}{k(k-1)} + \frac{4N}{(k-1)(k-2)} + \frac{4N}{(k-2)(k-3)} + \ldots + \frac{4N}{2} \\ = \; & 4N\left(\frac{1}{k-1} - \frac{1}{k} + \frac{1}{k-2} - \frac{1}{k-1} + \frac{1}{k-3} - \frac{1}{k-2} + \ldots + \frac{1}{1} - \frac{1}{2}\right) \\ = \; & 4N\left(1 - \frac{1}{k}\right) \end{aligned} \tag{26.5}$$

The results are a bit surprising. When there are many copies it takes on average about $4N$ generations for all of their ancestral lineages to coalesce! But when there are two copies, it takes on average $2N$ generations. That implies that a bit more than half of the depth of a coalescent tree is spent waiting for the last two copies to coalesce. More generally, $(1 - 1/n)/(1 - 1/k)$ of the time is spent waiting for the last n copies to coalesce. So with 100 copies in all, $0.9/0.99$ or 0.90909 of the time is spent waiting for the last 10 copies to coalesce. Only 9% of the time is spent on the first 90 coalescent events! (Of course, we mean "first" in the sense of going backwards in time.) One gets the picture that lineages coalesce rather rapidly at first and then the process gradually slows down.

These figures are based on expectations, and as expectations of ratios are not quite the same things as ratios of expectations, they may be a bit off, but in this case they are a reliable guide to what coalescent trees look like.

One might also ask how unbalanced these random trees of lineages are. Farris (1976) and Slowinski and Guyer (1989) considered that the basal split of a random tree with k tips could have any number of lineages from 1 through $k-1$ on the left-hand side. They showed that all $k-1$ of these values are in fact equiprobable. This also gives us useful information about the effect of adding one lineage to a tree. If we have 100 lineages and add one lineage, what is the probability that it will connect to this tree below the pre-existing root? Their result shows that the probability is only 2/100 that the root of the 101-lineage tree separates one lineage from the rest. And even if it does, the chance is only 1/101 that this single lineage is the new one that we added. Thus the chance that the new lineage establishes a new root below the pre-existing one is only $2/(101 \times 100) = 0.000198$.

Figure 26.5 shows nine realizations of a coalescent with 20 gene copies, all drawn to the same scale. This will show both the pattern of increasing lengths of time for coalescence to occur as the number of lineages decreases, and the enormous variability around that implied by the exponential distributions involved. It also shows a reasonable agreement with the Farris-Slowinski-Guyer uniform distribution of numbers of lineages on each side of the bottom split.

Figure 26.6 shows the tendency for the first few lineages to have in their ancestry the long lines at the bottom of the tree. It shows a sample of 50 gene copies.

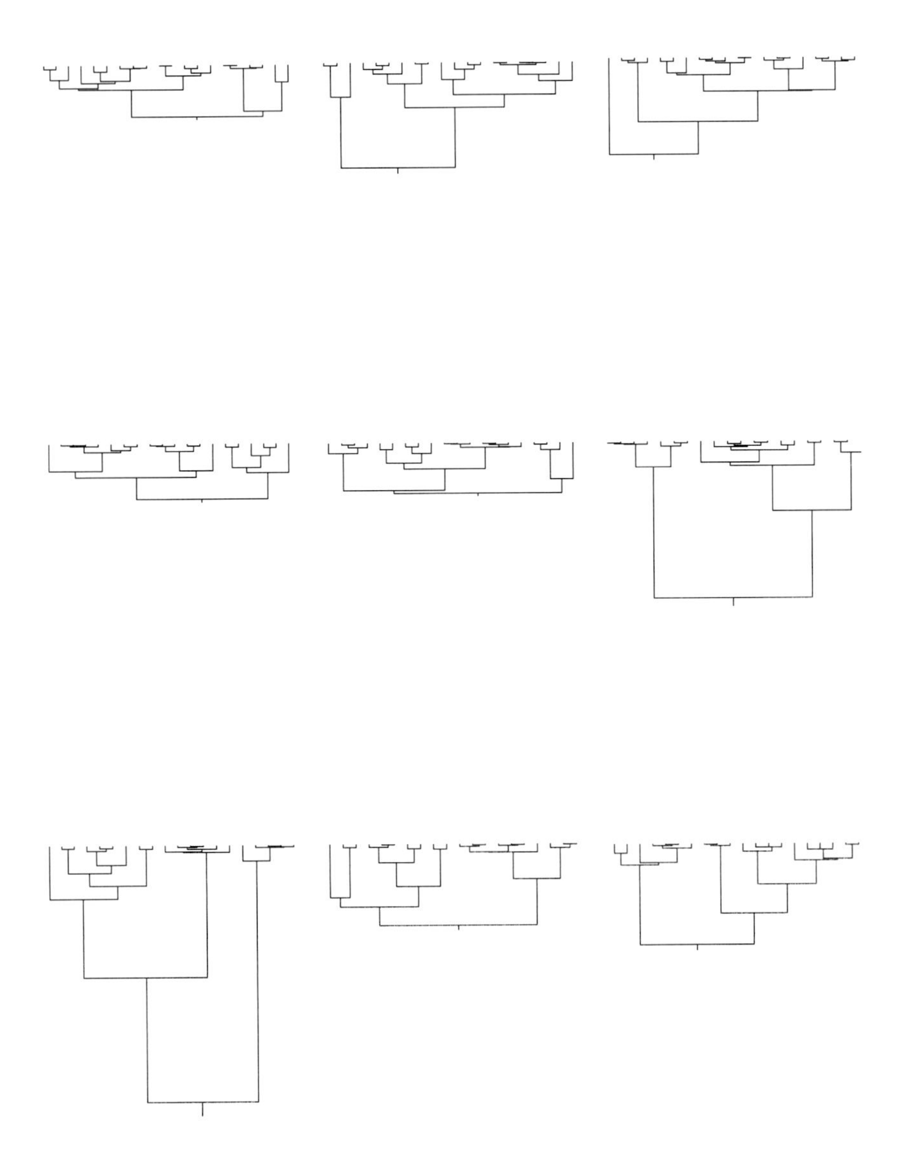

Figure 26.5: Nine outcomes of the coalescent process with 20 gene copies, drawn to the same scale.

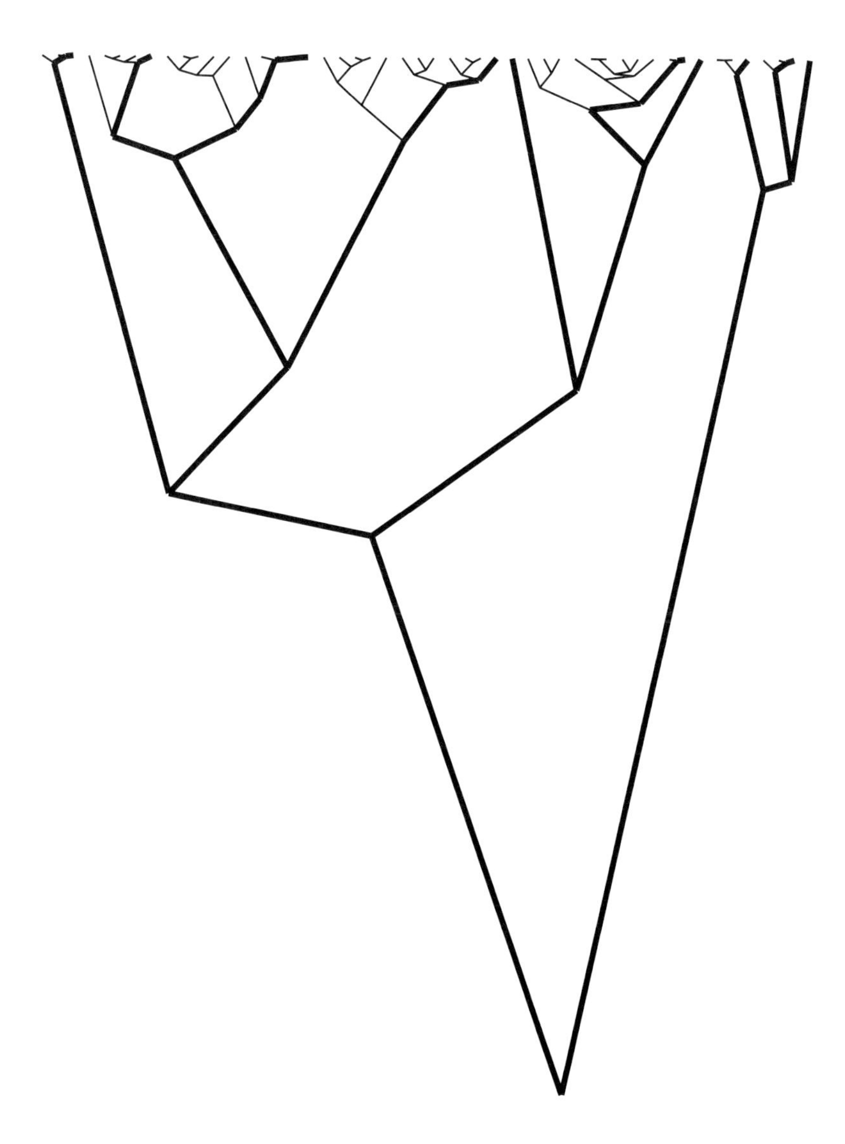

Figure 26.6: A sample genealogy of 50 gene copies, with the ancestry of a random 10 of them indicated by bold lines. Note that adding 40 more gene copies to the sample discloses no new lines in the bottom part of the diagram.

The ancestry of a random subsample of 10 of them is indicated by making the lines bolder. Adding the 40 to the 10 actually adds no new lines to the bottom part of the tree: They tend to connect to the existing lines by short branches and rarely add much length to the tree.

Bugs in a box—an analogy

We can make a physical analogy (if a somewhat fanciful one) by considering a box containing hyperactive, indiscriminate, voracious, and insatiable bugs. We put k bugs into the box. They run about without paying any attention to where they are going. Occasionally two bugs collide. When they do, one instantly eats the other. Being insatiable, it then resumes running as quickly as before. It is obvious what will happen. The number of bugs in the box gradually falls from k to $k-1$, to $k-2$, as the bugs coalesce, until finally only one bug is left.

The analogy is actually fairly precise. The number of pairs of bugs that can collide is $k(k-1)/2$. If there are $2N$ "places" in the box that can be occupied, the probability of a collision will be proportional to $k(k-1)/4N$. The size of the population corresponds to the size of the box. A box with twice as many "places" will slow the coalescence process down by a factor of two. So a simpleminded physical analysis of the bugs-in-a-box process will have the Kingman coalescent distribution as the probability distribution of its outcomes.

Effect of varying population size

We have been assuming that effective population size does not change through time. In reality it will, and we will also want to make inferences about its changes. Working backwards in time, when we get (back) to the point where the effective population size is $N(t)$, we will find there that the instantaneous rate of coalescence of k lineages is $k(k-1)/4N(t)$. If the population size $N(t)$ is half the value that it has today, these coalescences will happen twice as fast as they do now. The effect is to make it appear that time is passing twice as fast. This suggests a simple time transformation that allows us to find the distribution of coalescence times in the case in which the effective population size is $N(t)$ at time t ago.

Suppose that we imagine a time scale in which time passes at a rate proportional to $N(0)/N(t)$, where $N(0)$ is the effective population size now. Let us call this fictional time scale τ, where

$$d\tau = \frac{N(0)}{N(t)}\, dt \tag{26.6}$$

The total amount of this fictional time that elapses going back from the present to time t ago will then be the integral

$$\tau = \int d\tau = \int \frac{N(0)}{N(t)}\, dt \tag{26.7}$$

Whatever the course of population size change, as long as its inverse can be integrated, we can use equation 26.7 to derive the formula for the fictional time.

The usefulness of this new time scale comes from the fact that the rate of coalescence of k lineages on it is then $k(k-1)/4N(0)$ per unit of (fictional) time. On the fictional time scale, the original Kingman coalescent is valid. Thus if we have a function $N(t)$ whose inverse can be integrated, we can generate a coalescent for that population size history simply by generating a coalescent from the original Kingman process, then converting the times of coalescence from fictional to real time by evaluating t's , by solving for t in terms of τ in equation 26.7.

The simplest interesting example is exponential growth. Suppose that the population has been growing exponentially at rate g per generation. Then if time is measured going backwards from the present,

$$N(t) \;=\; N(0)\, e^{-gt} \tag{26.8}$$

From equations 26.8 and 26.7 we find that

$$\tau \;=\; \frac{1}{g}\left(e^{gt}-1\right) \tag{26.9}$$

and we can solve for t:

$$t \;=\; \frac{1}{g}\ln(1+g\tau) \tag{26.10}$$

To generate a coalescent outcome in a population that has been growing exponentially, we first generate a coalescent in a population of fixed size $N(0)$. Then we take the times of coalescence (which are fictional times τ) and use equation 26.10 to obtain the real times.

This time transformation was first given by Kingman (1982c) and was used by Slatkin and Hudson (1991) and by Griffiths and Tavaré (1994a).

Done this way, we never need to use the actual distribution of coalescence times. In fact, the density function for the time u back until coalescence, when there are k copies at present, is

$$\text{Prob}\,(t \leq u < t+dt) \;=\; \frac{k(k-1)}{4N_e}\exp\left[-\frac{k(k-1)}{4N_e}\frac{\left(e^{gt}-1\right)}{g}+gt\right]\,dt \tag{26.11}$$

This can be obtained by passing the exponential density of the ordinary coalescent through the time transformation of equation 26.10. It was given by Kingman (1982c) and by Slatkin and Hudson (1991). The density function for a particular coalescent tree may be obtained from this straightforwardly by taking a product of these densities.

Migration

If we have more than one population, with migration between them, the analytical theory becomes very difficult. But simulation of trees from the prior distribution

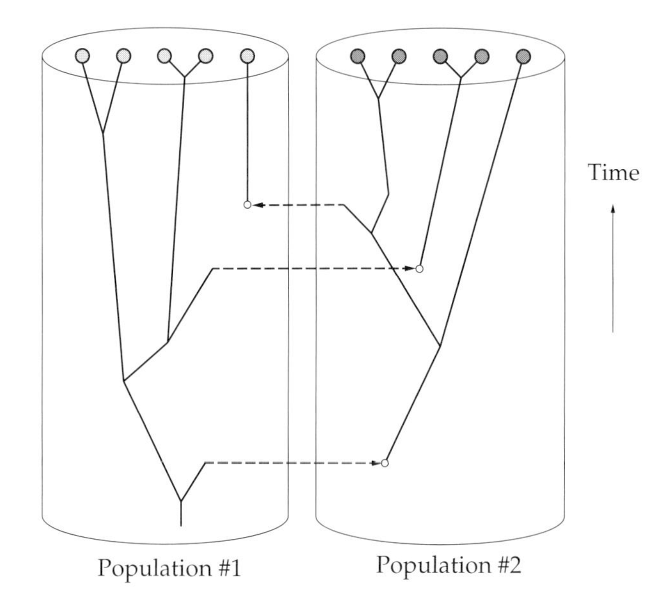

Figure 26.7: A coalescent tree with migration between two populations (represented by the two cylinders). The migration events are indicated by the dashed arrows that lead to small clear circles.

is not difficult, nor is the calculation of the density function of a given tree. As we shall see below, these are the only calculations we will need when doing likelihood analysis of data.

Imagine two populations, each of effective size N. There is a constant rate of migration between them. If there are k_1 copies in population #1 and k_2 in population #2, the rate of coalescence per unit time (going backwards, as usual) is $k_1(k_1-1)/(4N)+k_2(k_2-1)/(4N)$. Another event that can occur is for a lineage to migrate. This has probability m per copy, so that the total rate of migration events is k_1m+k_2m. So the total rate of all events is the sum of these,

$$\lambda \;=\; k_1(k_1-1)/(4N)+k_2(k_2-1)/(4N)+k_1m+k_2m \tag{26.12}$$

The time back until the next event is exponentially distributed, with the rate of occurrence per unit time equal to λ. Once one of these events has occurred, we need to know which. The probability of a coalescence in the first population is the fraction of λ that is contributed by its first term, and similarly for the other possible events. The lineages that coalesce or the lineages that migrate are chosen from the possibilities at random.

Figure 26.7 shows the events in a coalescent tree connecting gene copies in two populations. Migration events are the clear circles. We can generate such a multi-population coalescent by starting with the numbers of copies in each population

(k_1 and k_2 in the two-population case), and generating an exponential variable with the rate λ being specified by equation 26.12. Then we know when the first event, going backward, is. We choose what kind of event it is by choosing from the four possibilities (coalescence in two populations, and two directions of migration). These will occur with probabilities proportional to the four terms on the right side of equation 26.12. If the event is a coalescence, we then choose at random two lineages in the proper population. If it is a migration, we choose one lineage in the proper population to be the immigrant.

It will turn out that to do likelihood computations on coalescents, we need to be able to compute the probability density for a given coalescent, with the times of the migration events known. This is straightforward. We work our way down the coalescent, encountering successive events. In each interval between events, we have current values of the numbers k_1 and k_2 of lineages in each population. We then can use equation 26.12 to compute the current value of λ. That interval in the tree then gives rise to a term $\exp(-\lambda t)$. The event at the bottom of the interval gives rise to a term $m_{ij}\,dt$ if it is a migration arriving in population i from population j, or $1/2N_i\,dt$ if it is a coalescence in population i.

The resulting expression for the probability density of a tree G is the product over intervals, the ith of which is t_i generations long:

$$\text{Prob}(G) \;=\; \prod_i \left[e^{\left[-\sum_j k_{ji}(k_{ji}-1)/(4N_j) - \sum_{j,\ell,j\neq\ell} k_{\ell i} m_{\ell j}\right] t_i} \left(\prod_{j\ell} m_{j\ell}^{\delta_{j\ell,i}}\right) \prod_j \left(\frac{1}{2N_j}\right)^{\varepsilon_{ij}} \right] \tag{26.13}$$

(omitting the dt terms). In this expression the number of lineages in population j, k_{ji}, may change with each interval, $\delta_{j\ell,i}$ is an indicator variable that is 1 if the event at the bottom of interval i is a migration from population ℓ to population j, and 0 otherwise, and ε_{ij} is an indicator variable that is 1 if event i is a coalescence in population j. If the tree G has its branch lengths in mutational units (so that an average site has probability 1 of mutating per unit time), then equation 26.13 becomes instead:

$$\text{Prob}(G) \;=\; \prod_i e^{\left(-\sum_j k_{ji}(k_{ji}-1)/(4N_j\mu) - \sum_{j\ell,j\neq\ell} k_{ji} m_{j\ell}/\mu\right) t_i} \left(\prod_{ij\ell} (m_{j\ell,i}/\mu)^{\delta_{j\ell,i}}\right) \prod_{ij} \left(\frac{1}{2N_j\mu}\right)^{\varepsilon_{ji}} \tag{26.14}$$

Note that the migration events each give rise to a single factor of m, not k_{ji} of them, and the coalescence events give rise to $1/(2N)$, not $k_{ji}(k_{ji}-1)/(4N)$. This reflects the probabilities of choosing the particular pairs of lineages that are going to migrate or coalesce.

Although we can retrospectively calculate the probability density of any particular tree, you may be surprised to find out that much simpler quantities are impossible to calculate prospectively. For example, Takahata (1988) derived the mean time to coalescence of a pair of genes in two populations that exchange migrants, but he was unable to derive the distribution of coalescence times for the pair. In one population the corresponding distribution is a simple exponential, and this has been known since Sewall Wright's work of 1931. If we observe one copy in each of the two populations, there is an exponentially long wait with mean $1/(2m)$ generations for one of the genes to migrate, following which there is an exponentially long wait with mean $1/[1/(2m) + 2N]$ for the pair to coalesce or for migration to occur again, and so on, until finally they coalesce. Thus the overall distribution is a sum of a geometrically distributed number of sums of pairs of exponentials. This does not have a simple density function. Recently, its Laplace transform has been derived (Bahlo and Griffiths, 2001), and that can be be used to compute the distribution numerically. And that's the simplest possible case.

Fortunately, for likelihood inference with geographically separated populations, we do not need to solve any of these prospective problems, but can simply compute the density function of a given genealogy with given migration events, retrospectively.

Effect of recombination

We have been assuming so far that there is no recombination in any of the lineages. If there were, then we could not sustain the assumption that each gene copy comes from a single gene copy in its parent. A recombination event within the gene would instead mean that part of the copy came from the one of the parent's two copies and the rest from the other. Figure 26.8 shows what a single recombination does to a coalescent tree with three lineages. The arrow shows where there was a recombination between sites 138 and 139 of this gene. There are really two coalescent trees here, one for sites 1–138, and the other for sites 139–204.

As one moves along the genome, one will have a series of sites that all have the same coalescent tree, with no recombination anywhere in that tree. But suddenly there will be a recombination somewhere in the tree. It will cause a particular kind of rearrangement of the tree (breaking of one lineage and its attachment elsewhere). That tree will then hold for a while as one moves along the genome, and then there will be another breakage and reattachment. After some distance along the genome, the tree has changed to be totally different.

How far will this be? Your intuition may tell you that it must be many map units, perhaps enough to have 30% recombination. If so, your intuition is very wrong. The distance is quite short. We can get a good feel for it by asking how far one has to go to expect one recombination event on an average lineage that leads down from a tip to the root of the tree. The length of such a lineage will be about $4N$ generations. If the recombination fraction between two points in the genome is r, the average number of recombinations between them on this lineage is $4Nr$.

Figure 26.8: A coalescent tree with a single recombination event. The recombination is between sites 138 and 139 at the point indicated by the horizontal arrow. The result is a pair of related coalescent trees.

The condition we are looking for is then $4Nr \gg 1$, which is $r \gg 1/(4N)$, How far along the genome we have to go to see the tree change substantially thus depends on the population size.

In humans, one expects about one recombination event every 10^8 bases. If we take human effective population size in the (distant) past to have been about 10^4, then $4Nr = 1$ will hold when $r = 2.5 \times 10^{-5}$. That works out to be about 2,500 bases. This is a surprisingly small number. Actually, recombination is not evenly spread along the genome, but is patchily distributed, so the number 2,500 is probably too small. But if human population size has instead been closer to 10^5, the figure of 2,500 bases changes to only 250 bases! This calculation is due to Robertson and Hill (1983).

Interestingly, the conditions for change of the coalescent tree are actually the same as those for genetic drift not to form a substantial amount of linkage disequilibrium. Thus trees and D's are closely related, being in some sense just different ways of describing the same situation. A perfectly compatible (and hence perfectly treelike) set of sites is also one that shows complete linkage disequilibrium.

We must then think of the coalescents in a genome as each holding for only a small region of the genome, and that there may be a million different coalescent trees that characterize the ancestry of our genome. Virtually every locus has its own Eve or Adam, at widely varying times and places. We not only have as our ancestors mitochondrial Eve and Y-chromosome Adam (who did not know each other), but also hemoglobin Sam and cytochrome Frieda, and a great many others as well. Once we are inside a species, we have lost by genetic recombination the single branching genealogy that exists between species. But as we will see, the study of coalescent trees provides an important framework for thinking about evolutionary genetics and estimating the parameters of its processes.

The effect of recombination on coalescents in creating loops in the genealogies was first discussed by Hudson (1983; Hudson and Kaplan, 1985). The random process of forming a coalescent from a present-day sample of k haplotypes has considerable similarity to the process for migration. Going backward from the k haplotypes, there is a constant risk $k(k-1)/4N$ per unit time of coalescence. If the entirety of the k haplotypes can recombine, if each is L nucleotides in length, and if there is r recombination per base, their total map length is $k(L-1)r$. Thus, going back in time, we have two kinds of events, with a total rate $k(k-1)/4N+k(L-1)r$. As in the migration case, we draw from an exponential distribution with this rate, and then use the relative sizes of the two terms to decide randomly which of the two kinds of events occur.

In the example of Figure 26.8, $k = 3$ and $L = 204$. If (say) $N = 10^6$ and $r = 10^{-8}$, we have rates $3\times 2/(4\times 10^6) = 0.0000015$ of coalescence, and $3\times 203\times 10^{-8} = 0.00000609$ of recombination. We draw the time back to an event, then choose whether it is to be a coalescence or a recombination, then choose who coalesces or where the recombination event is. In the tree shown, the first event (going backward) is a recombination, between sites 138 and 139. Going further back there are four haplotypes. The rate of coalescence is straightforward: $4 \times 3/(4 \times 10^6) = 0.000003$. But we must be careful about the rate of recombination. We need only follow recombinations that occur in an interval that ends up in the sample. The leftmost and rightmost haplotypes have 203 intervals between bases that can have a recombination. But the second haplotype has only $138 - 1 = 137$ bases at which recombination affects the ancestry of the sampled haplotypes. The third has $204 - 139 = 65$ bases. The total number of intervals between sites in which we want to keep track of recombinations is then $406 + 137 + 65 = 608$ bases. The total rate of coalescence and recombination is then $0.000003 + 0.00000608$. In general, on each haplotype, there will be a leftmost site that is ancestral to one in our sample, and a rightmost site that is, and we want to pay attention to recombinations between these, even ones in interior regions that may happen not to be ancestral to sites in our sample. Proceeding backward in this way, we can simulate the genealogy of a recombining coalescent. We need only go far enough back that every site has coalesced— before that, all coalescences of lineages are irrelevant.

An alternative way of simulating a recombining coalescent is to draw an ordinary coalescent for the first site. If it has total branch length (say) 10^7, then we can ask how far down the chromosome we have to go to encounter a recombination somewhere on it. We might find that with the rate $r = 10^{-8}$ we would draw (say) 138.6. This would put the recombination between bases 138 and 139. Choosing a random place on the tree for the recombination, we break the branch there, erase the branch below it down to the next coalescence, and then simulate the ancestry below the place where the recombination occurred. In any time interval in which there are ℓ other lineages in existence, the chance of coalescing with each is $1/(2N)$ per unit time, for a total rate of coalescence of $\ell/(2N)$. Once the lineage reattaches by coalescence, we recompute the length of the tree and again draw how far along the chromosome the next recombination is. This continues until the last base is passed. This method of simulating a series of coalescent trees, each applying to a range of bases, was introduced by Wiuf and Hein (1999).

Coalescents and natural selection

In all the discussion so far, it has been assumed that all genetic variation is neutral, so that each copy of a locus in the population has an equal chance of surviving. When that is not true, it is hard to see what the genealogy of a sample will be. If the sample contains 12 copies of an advantageous allele, and eight copies of a less favored one, it is likely that the advantageous allele has been increasing in frequency. Its copies may all be descended from a single mutation, and if so, they may all have a common ancestor more recently than do the other samples.

If we have one site in a locus that is under selection, and it has two alleles, we can imagine that we knew the gene frequencies of the alleles at all times in the past. The gene copies in the population fall into two classes according to whether they have allele A_1 or A_2. We can consider the two classes of copies as if they were separate subpopulations. The A_1 copies coalesce within their subpopulation, whose size is determined by the past gene frequencies of A_1, and similarly for A_2. Kaplan, Darden, and Hudson (1988) discussed this process. They considered particularly the case in which strong balancing selection keeps the two alleles at constant gene frequencies. In that case, it is easy to generate the genealogy, as the mathematics is the same as with two populations connected by migration. Mutation of the A_1 and A_2 alleles to each other moves copies between the two subpopulations and thus plays the role of migration. Hudson and Kaplan (1988) extended the method to allow for recombination between neutral sites and the selected site, and Kaplan, Hudson, and Iizuka (1991) extended it further to allow for geographic structure, with the possibility of different equilibrium gene frequencies in each population. Kaplan, Hudson, and Langley (1989) used a similar approach to approximate the case where favorable mutations linked to a locus affect the genealogy. The resulting rapid changes in gene frequency of alleles are called "hitchhiking," "selective sweeps," or "periodic selection."

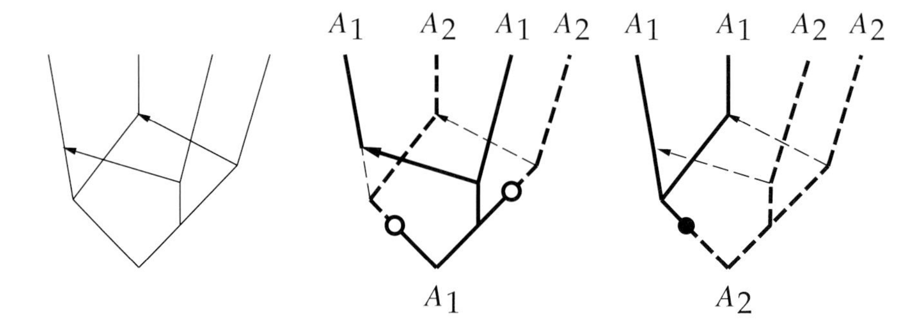

Figure 26.9: An example of an ancestral selection graph, with two selection events (left). In the center, the lineages that are used if the ancestral allele at the selected site is A_1 are shown in bold. On the right, those that are used if the ancestral allele is A_2 are shown in bold. Mutational events are shown as circles, and lineages where the allele is A_2 are dashed.

Neuhauser and Krone's method

Until recently, there were no good ways of drawing samples from the coalescent with moderately strong natural selection. The best one could do was simulate the ancestry of the whole population and then retrospectively extract the coalescent from a detailed population genealogy. This is so difficult that it was almost never done. Neuhauser and Krone (1997; Krone and Neuhauser, 1997) have made a major breakthrough that enables the simulation of coalescents with selection.

I will not attempt to derive their method, but will simply show the algorithm. Suppose we have two alleles at the selected site, as before A_1 and A_2. The fitnesses of the three diploid genotypes will be taken to be $1+2s : 1+s : 1$, and the population size to be N. There will be a mutation rate μ from A_1 to A_2 and mutations at rate ν in the opposite direction. As with the Kingman coalescent, their process is a diffusion approximation. We go backward in time from our sample, with two sorts of events happening: coalescence and a special selection event. The special selection event happens with rate $4Ns$ on each lineage. When such an event happens, there is a fork in the tree, with two lineages below that point. One is designated as the incoming lineage. This diagram is called the *ancestral selection graph*.

Figure 26.9 shows such lineages, with the incoming lineages ending in arrowheads. To simulate a sample, we start from the root, and choose its state from the equilibrium distribution under this mutation-selection model. We assign the state of the selected locus at the base of the tree (either A_1 or A_2). Then we work our way up, putting in mutations appropriately. Note that the unequal rates of mutation in the two alleles may make the locations of the mutations dependent on which allele is present. As the alleles evolve up the lineages, at the selection events there is a simple rule that decides which arrow leading into the event is the real lineage. If the incoming arrow carries an A_1, it prevails; otherwise, it does not. The middle and right parts of Figure 26.9 show this process with two selection events each. The two parts correspond to different choices of ancestral allele. The bold lines (either solid or dashed) in the figure indicate the lineages that end up supplying genetic material to the sample.

The ancestral selection graph differs from the ordinary coalescent in that the shape of the graph depends on the alleles at one of the sites. In the usual coalescent we can determine the shape before filling in the alleles. The ancestral selection graph will reflect some of the processes that we might hope to see—that when an A_1 lineage is present among A_2 lineages, it will tend to be more successful. The ancestral selection graph is derived from the diffusion approximation to gene frequency change. Any two finite-population models that have the same diffusion approximation will have the same ancestral selection process and will be indistinguishable when using them for analysis. Neuhauser (1999) has extended the analysis to models in which there is heterozygote advantage or frequency-dependent selection, using ancestral selection graphs with two incoming lineages in each selection event. Slade (2001a, b) has developed methods for simulating population samples using the ancestral selection graph,

We will see in the next chapter that the ancestral selection graph does not fit easily into some methods of likelihood inference for population samples. Nevertheless it is a remarkable advance in our ability to treat natural selection by coalescent methods.

Chapter 27

Likelihood calculations on coalescents

How can we make statistical inferences using coalescents? A basic point is that we are not much interested in the coalescent genealogy itself. Every locus in an organism may have a different coalescent tree, so that the individual trees are of little interest. Knowing that, at the particular locus we are investigating, Freddie Field Mouse is the closest relative in our sample to Frieda Field Mouse is not going to be a very riveting fact, particularly given that the next locus may give a completely different result. In addition, our estimates of each coalescent will usually be very poor, as few sites may be varying in that region of the genome.

The individual genealogies are drawn from a population of genealogies, whose parameters reflect population sizes, growth rates, migration rates, and recombination fractions; these parameters are of great interest to population biologists. We would like to estimate these and make statements about confidence intervals.

The basic equation

The straightforward way to estimate the parameters is to compute their likelihoods, averaging over the uncertainty in the genealogy. In principle, this is easy (but only in principle). A simple application of conditional probabilities shows that if we consider all possible genealogies, denoting a typical one of them by G^*, the likelihood given data D for a parameter α is (Felsenstein, 1988b)

$$L = \sum_{G^*} \text{Prob}\,(G^*|\boldsymbol{\alpha})\ \text{Prob}\,(D|G^*, \mu) \tag{27.1}$$

The parameter α could be a single number, or a collection of numbers. The important requirement is that they be the parameters that control the distribution of genealogies, G^*. Note that the G^* are not simply the tree topologies; they include the branch lengths.

In equation 27.1 the coalescent parameters $\boldsymbol{\alpha}$ are separated from the mutational parameter μ. The assumption is that the genetic data D is produced by neutral mutation acting along a genealogy G^*. This is a somewhat restrictive assumption: For example, it rules out natural selection as the cause of any of the variation in the data, as the presence of the selection would affect G^*, by affecting the probabilities of birth and death.

The branch lengths in G^* are most naturally scaled in generations. However, when G^* is inferred from genetic data D, it is usual for the branch lengths to have units like neutral mutations per site. In that case, the parameter μ disappears from the second term on the right side of equation 27.1: The probability of the data can be computed directly from the rescaled genealogy, which we will call G. For example, knowing that a branch length is 0.014 tells us that the probability of change of a given site on that branch is (nearly) 0.014, differing from it only by considerations of multiple change at the same site.

The mutational parameters disappear from the right term when we change from G^* to G, but they reappear in the other term. Note that $\text{Prob}\,(G^*|\boldsymbol{\alpha})$ is simply the density function of the coalescent, the function which we computed for several different cases in the previous chapter. In those computations, the branch lengths t were scaled in generations. They occur in the formulas not by themselves, but multiplied by the strengths of evolutionary forces. In equation 26.13, for example, the i-th coalescence interval t_i appears only in the products $k_{ji}(k_{ji}-1)t_i/(4N_j)$ and $k_i m_{j\ell} t_i$. When we change the time scales to make them reflect mutations per site rather than generations, these quantities become instead $k_{ji}(k_{ji}-1)t_i/(4N_j\mu)$ and $k_i(m_{j\ell}/\mu)$. In addition, terms that were once $1/(2N)$ and m become $1/(2N\mu)$ and m/μ. We saw this in equation 26.14.

The upshot is that we cannot actually estimate parameters such as N, μ, and $m_{j\ell}$ by themselves. They instead appear in the formulas only in the compound parameters $4N_j\mu$ or $m_{j\ell}/\mu$. So doubling the m's and μ, and at the same time halving the N's, does not change these compound parameters and does not change the likelihood at all. Similar considerations apply for cases with population growth (one gets $4Ng$) or recombination ($4Nr$). If we call the vector of these compound parameters $\boldsymbol{\beta}$, we can rewrite equation 27.1 as:

$$L = \sum_G \text{Prob}\,(G|\boldsymbol{\beta})\ \text{Prob}\,(D|G) \tag{27.2}$$

This is the form in which we will use the formula. We will replace the products and ratios of parameters such as $4N_e\mu$ each by a single quantity (in that case, Θ).

Using accurate genealogies—a reverie

We might wonder how to make use of this formula in the simple case in which there are so many molecular sites that the genealogy G can be estimated with total precision. I have addressed the simplest case (Felsenstein, 1992a). In this case, all of the likelihood is contributed by one gene tree G, with $\mathrm{Prob}\,(D|G)$ being zero for all other trees. Then we can remove the summation from equation 27.2. If we let $\mathrm{Prob}\,(D|G)$ be 1 for G and 0 for all others, equation 27.2 reduces to

$$L = \mathrm{Prob}\,(G|\boldsymbol{\beta}) \tag{27.3}$$

In the simplest case, of a single isolated population of constant size, the likelihood is the product of $n-1$ terms, one for each coalescence event. If we express branch lengths in terms of expected neutral mutations per site, the product of terms for $n-1, n-2, \cdots, 2$ lineages gives the probability density

$$\mathrm{Prob}\,(G \mid \Theta) = \prod_{k=2}^{n} \left(\frac{k(k-1)}{\Theta}\right) \exp\left(\frac{-k(k-1)u_k}{\Theta}\right) \, du_2 \, du_3 \cdots du_n \tag{27.4}$$

where u_k is the interval between the coalescence event that produces k lineages and the one that produces $k-1$ lineages. This can be rewritten by grouping together the exponentials and summing their exponents, to get

$$\mathrm{Prob}\,(G \mid \Theta) = \frac{n!(n-1)!}{(\Theta)^{n-1}} \exp\left(-\sum_{k=2}^{n} \frac{k(k-1)u_k}{\Theta}\right) \, du_2 \, du_3 \cdots du_n \tag{27.5}$$

The log-likelihood is the logarithm of the density:

$$\begin{aligned} \ln L &= \ln \, \mathrm{Prob}\,(G|\Theta) \\ &= \ln(n!) + \ln((n-1)!) - (n-1)\ln(\Theta) - \sum_{k=2}^{n} \frac{k(k-1)u_k}{\Theta} \end{aligned} \tag{27.6}$$

To estimate Θ we simply take the derivative of equation 27.6 with respect to Θ, equate it to zero, and solve for Θ. The result is (Felsenstein, 1992a)

$$\frac{d \ln L}{d\,\Theta} = -\frac{(n-1)}{\Theta} + \frac{1}{(\Theta)^2} \sum_{k=2}^{n} k(k-1)u_k \tag{27.7}$$

When the right side of equation 27.7 is equated to zero, the resulting estimate is

$$\widehat{\Theta} = \frac{\sum_{k=2}^{n} k(k-1)u_k}{n-1} \tag{27.8}$$

It is also not hard to show, from the variances of the u_k, that the variance of this is $(\Theta)^2/(n-1)$, so that its coefficient of variation is $1/\sqrt{n-1}$.

Thus if we could only observe the coalescent tree with total accuracy, the expression for the maximum likelihood estimate of Θ would be quite simple. Even then, there would be a substantial uncertainty in the estimate, as reflected by the coefficient of variation, which even for 50 sequences would be as large as 0.1428. Note that this means that part of the variation comes from the random variation in the coalescent tree itself, not just from the mutational variability that makes it hard for us to infer it.

Two random sampling methods

To cope with the uncertainty about what the coalescent tree actually is, two groups have proposed Monte Carlo integration methods. The objective is to do the integrals in equation 27.2 by drawing a large random sample of trees. This approach, obtaining the approximate value of an integral by evaluating the function at a large random sample of points, is a standard one. It was first intensively used in designing the first hydrogen bombs; now it is being put to a more constructive use. In this case there is not just one integral but a very large number of them.

Griffiths and Tavaré (1994a, b, c) were the first to use this approach to evaluate likelihoods involving coalescents [see also Griffiths (1989)]. I will start, however, with an approach from my own group (Kuhner, Yamato, and Felsenstein, 1995), which is easier to describe. Then I will explain the work of Griffiths and Tavaré in the same terms.

A Metropolis-Hastings method

Kuhner, Yamato, and Felsenstein (1995) described a method of wandering through tree space that achieves a random sample from a relevant distribution. Suppose that we have chosen a value of $\Theta = 4N_e\mu$ that seems in the vicinity of the proper estimate. Call this value Θ_0. Imagine drawing a large random sample of trees from a distribution which is proportional to the quantity $\text{Prob}\,(G|\Theta_0)\ \text{Prob}\,(D|G)$. We will show below that, given such a random sample, we can estimate the likelihood $L(\Theta)$ for any nearby value of Θ.

To achieve this sample, we use the Metropolis algorithm, which was previously described in Chapter 18. This works by taking an initial tree, altering it somehow, and then either accepting or rejecting the alteration. Metropolis et al. (1953) showed how to do the acceptance and rejection to achieve the proper distribution. Hastings (1970) gave the modification that is needed to cope with biased distributions of proposed changes. Their methods are general for a wide variety of problems, not just likelihoods for coalescents. They form the basis of the widely used optimization method called *simulated annealing*, though we will not be using that method.

The *Metropolis-Hastings method* considers a tree G and a proposed change to a tree G'. Suppose that the probability of G' being proposed in this case is $Q(G'|G)$, but the probability of G being proposed if the tree is initially G' is $Q(G|G')$. The

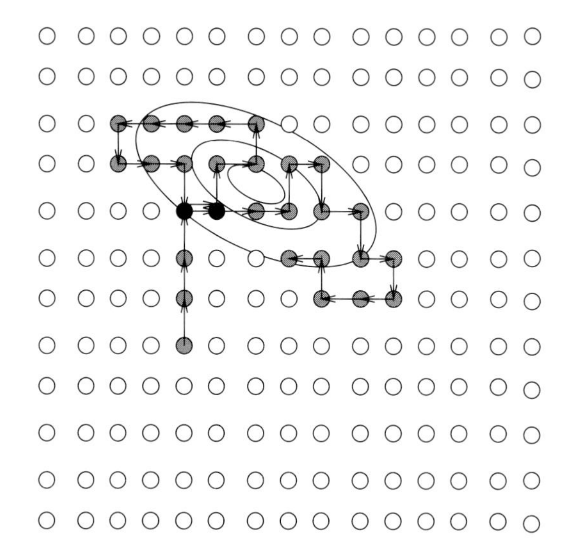

Figure 27.1: The Metropolis-Hastings sampler wandering in a space of coalescent trees, represented here as two-dimensional.

method decides whether to accept a tree by drawing a uniformly distributed random variable from the interval $[0, 1]$ and accepting G' if that random variable R satisfies

$$R < \frac{Q(G|G')}{Q(G'|G)} \frac{\text{Prob}\,(G'|\Theta_0)\ \text{Prob}\,(D|G')}{\text{Prob}\,(G|\Theta_0)\ \text{Prob}\,(D|G)} \tag{27.9}$$

It thus helps us accept a tree if it has a higher data likelihood $\text{Prob}\,(D|G')$, or if it has a higher probability $\text{Prob}\,(G'|\Theta_0)$ under the Kingman prior. It also helps us accept it if it is more likely that G will be proposed when we start at G' than vice versa. This latter helps counteract the biases of the proposal method.

Note, however, that the method may not always accept a tree if it improves the expression $\text{Prob}\,(G|\Theta_0)\ \text{Prob}\,(D|G)$, if the Q's differ enough. Most noticeably, a tree G' will sometimes be accepted even though it lowers the expression $\text{Prob}\,(G|\Theta_0)\ \text{Prob}\,(D|G)$. Thus the method will tend to wander towards "peaks" in the desired distribution , but will continually also wander down off of those peaks and also explore valleys. It can be proven that if this Metropolis-Hastings sampler is run long enough, it will explore every part of the space of trees in proportion to its contribution to the integral 27.2. Figure 27.1 shows a cartoon of this process, in an imaginary case where the tree space is only two-dimensional. It is actually much more complex than that.

Kuhner, Yamato, and Felsenstein (1995) used this sampler to estimate Θ. They used a method of locally rearranging trees, resulting in a slightly different tree topology or locally different branch lengths, to propose new trees G'. The two key problems with this approach are: (1) whether the sampler can be run long enough to obtain a reasonably representative sample, and (2) whether we can use the sample to calculate likelihoods for values of Θ other than Θ_0.

The latter problem is solved by the "importance sampling" formula. Suppose that we are computing the average of a quantity $g(x)$ over a distribution whose density is $f(x)\,dx$. We could, of course, do this by drawing points at random from the distribution $f(x)\,dx$ and averaging $g(x)$ for those points. But suppose that we instead draw points from a somewhat different distribution, $f_0(x)\,dx$. Since

$$\begin{aligned} \mathbb{E}_f\left[g(x)\right] &= \int f(x)\, g(x)\, dx \\ &= \int f_0(x)\, \frac{f(x)}{f_0(x)}\, g(x)\, dx \\ &= \mathbb{E}_{f_0}\left[\frac{f(x)}{f_0(x)} g(x)\right] \end{aligned} \tag{27.10}$$

it follows that we can also compute it by drawing points from the distribution whose density is $f_0(x)\,dx$ and averaging the quantity $[f(x)/f_0(x)]g(x)$ across those points.

In the present case we are trying to integrate the quantity $f(x) = \text{Prob}(G|\Theta)\ \text{Prob}\,(D|G)$. We set $g(x)$ to 1 and $f_0(x)$ is $\text{Prob}\,(G|\Theta_0)\ \text{Prob}\,(D|G)$. The sampler is drawing points from f_0, and we need to average over those points the ratio f/f_0. The term $\text{Prob}\,(D|G)$ cancels out, and we end up with

$$\frac{L(\Theta)}{L(\Theta_0)} = \frac{1}{n}\sum_{i=1}^{n} \frac{\text{Prob}\,(G|\Theta)}{\text{Prob}\,(G|\Theta_0)} \tag{27.11}$$

In theory the sample of points has to be infinitely large, but in practice the sample $G_1, G_2, \ldots, G_n$ can be sufficiently large when it has a few hundred points, one drawn every 100 proposed trees. However, the formula in equation 27.11 is then valid only for values of Θ that were not too far from Θ_0. Kuhner, Yamato, and Felsenstein coped with that limitation by running a Metropolis-Hastings sampler for a modest number of steps, computing $L(\Theta)/L(\Theta_0)$, and maximizing it with respect to Θ. This new value of Θ then replaced the the previous Θ_0 and the process continued. By running 10 of these "short chains" and two long ones, Kuhner, Yamato, and Felsenstein were able to use equation 27.11 to find the optimal value of Θ.

This approach to inferring parameters such as Θ is computationally tedious, but is also straightforward and does not depend critically on the details of the model of evolution. It can be done for any situation in which we can calculate the probability $\text{Prob}\,(D|G)$ of the data on the tree, and in which we can also calculate the coalescent probability $\text{Prob}\,(G|\boldsymbol{\beta})$. Figure 27.2 shows the result of a run with 10 short chains and two long chains of the Metropolis-Hastings sampler on a simulated data set if we assume an isolated population of constant size, with 50

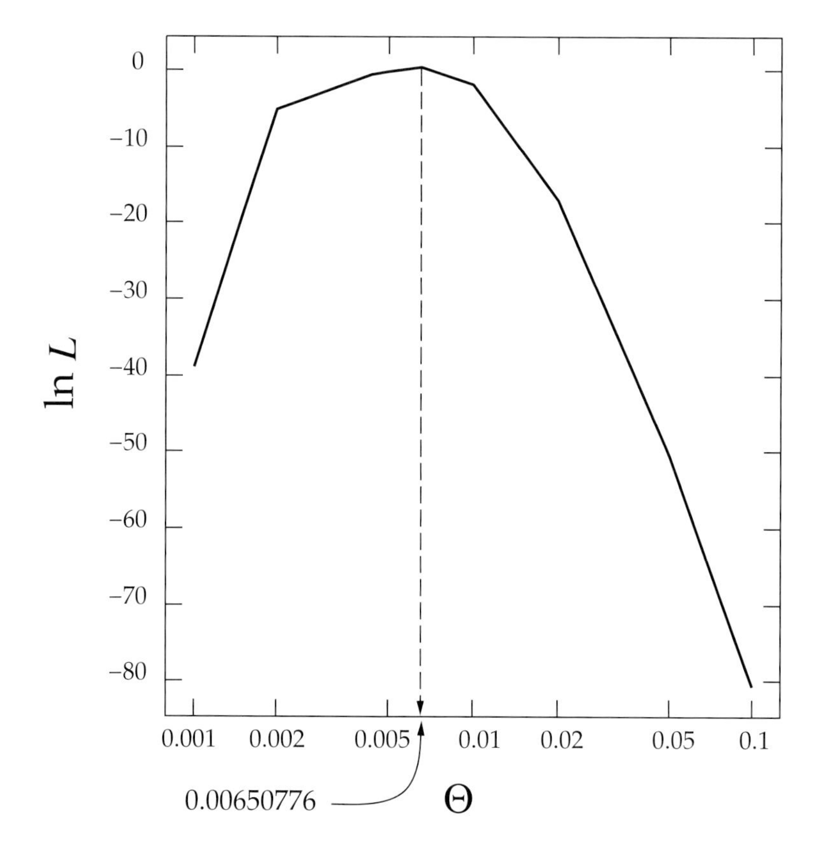

Figure 27.2: A typical likelihood curve inferred for a simulated data set, using the Metropolis-Hastings sampler. The true Θ is 0.01, the estimate from 50 sequences of 500 sites each is in this case 0.00650776.

sequences each of 500 sites. The true value of Θ is 0.01. In this case the estimate turns out to be 0.00650776.

Griffiths and Tavaré's method

The Metropolis-Hastings method has an unknown computational burden, because we are not sure how long to run the Markov chain to collect a given number of reasonably independent samples from the distribution. Griffiths and Tavaré (Griffiths, 1989; Griffiths and Tavaré, 1994a, b, c) invented a method that is considerably faster, especially as it samples independent points. Their method was really the first application of Monte Carlo integration to coalescent likelihoods; we have delayed explaining it because it is easier to explain once the other method is understood. The objective of their method is the same as that of the Metropolis-Hastings method: the computation of the likelihood curve. The models of character evolution and of coalescence are also the same. Thus there is no controversy over this objective.

Although their papers explain the method differently from they way in which we will explain it here, we can think of their method as computing the integral in equation 27.2. This justification for their method was introduced by us (Felsenstein et al., 1999). There is one important difference between their method and the previous MCMC method. Instead of the genealogy G, which is a tree with a topology and branching times, they use (in effect) a different structure, which shows the topology and the mutational events, but does not have times of branching. Let us call this history of events H. In computing a quantity like $\text{Prob}\,(D|G)$, we needed node times so as to allow us to compute the probabilities of different numbers of mutations and to average over them. If the mutational events (and the coalescences) are specified, then the times become irrelevant. For example, we can calculate the probability of arriving at sequence s_2 given a start at sequence s_1 if we know the time separating them, or we can calculate it if we know exactly how many mutational events occur between them, and at exactly which sites.

The basic equation 27.2 then becomes, trivially,

$$L = \sum_{H} \text{Prob}\,(H|\boldsymbol{\beta})\ \text{Prob}\,(D|H) \tag{27.12}$$

However, the calculation of the terms in equation 27.2 is now different and, in fact, much easier. Assume that we have described the history H of events by specifying not only the order of coalescent events, and which lineages coalesce, but that this ordered sequence of events also specifies the precise order of mutational events, the site at which each one occurs, and the result of each mutation. Once this sequence of events is specified, it either leads to the observed data set D, or it does not. If it does, the probability $\text{Prob}\,(D|H) = 1$. If not, this probability is zero. Thus the sum in equation 27.12 reduces to a sum of $\text{Prob}\,(H|\boldsymbol{\beta})$ over all those H's that lead to the observed data. There are a vast number of these H's.

Each H is specified as a sequence of events, in order. Figure 27.3 shows one sequence of events leading to a set of four molecular sequences. Note that the earliest event is that the sequence at the root of the tree is one particular sequence (`CATTAACGTCG`). Kingman was able to calculate the probability of a genealogy G as a product of terms, one for each coalescence interval. Griffiths and Tavaré are able to calculate the probability of one of their histories H as a product of terms, one for each event. Starting at the present and going back in time, there are (given the observed sequences) in Figure 27.3 two sequences that are identical, sequences 1 and 3. These might coalesce. The probability of this particular coalescence is $1/(2N_e)$ per generation. Note that although there are five other pairs of lineages that might have coalesced, the observed sequences rule out the most recent event involving the coalescence of any but lineages 1 and 3.

For simplicity, let us assume that the model of mutation is the completely symmetric Jukes-Cantor model. Then the probability of a mutation is μ per site per

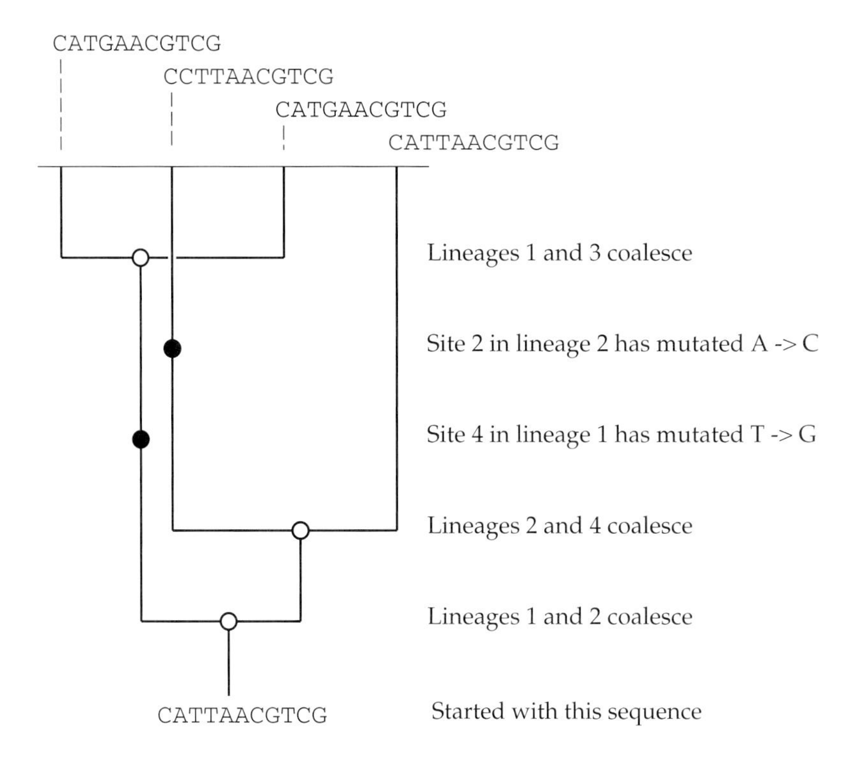

Figure 27.3: A sequence of events, leading to four molecular sequences.

generation. Note that we are computing the probability unconditioned on the present data. All four of the sequences might have mutated. There being 11 loci in four lineages in this example, each of which might mutate with rate μ per generation, the total rate of events that might have been the most recent event is $6/(2N_e) + 44\mu$. Thus the probability of the event that did happen, the coalescence, is

$$\frac{\frac{1}{2N_e}}{\left(\frac{6}{2N_e} + 44\mu\right)} = \frac{1}{6 + 22\Theta} \tag{27.13}$$

The immediately preceding event is a mutation in site 2 from A to C. This has probability $\mu/3$ per generation of occurring. There are at that point three lineages, so that there are three pairs of lineages. Thus the probability of coalescence is $3/(2N_e)$. Therefore, the probability of this event, given that some event occurs, is

$$\frac{\frac{\mu}{3}}{\left(\frac{3}{2N_e} + 3 \times 11\mu\right)} = \frac{\Theta}{18 + 99\Theta} \tag{27.14}$$

The successive events, as we work backward down the tree, are in fact independent. We can continue in this fashion, working out the probability of each of the events in the history H until we get back to the bottom of the tree. We take the product of the probabilities. Then we need to multiply this product by the probability that the initial sequence is CATTAACGTCG. Under the Jukes-Cantor model this is simply $1/4^{11}$. The result is that the probability of the event history H is

$$\begin{aligned} & \left(\frac{\frac{1}{2N_e}}{\frac{6}{2N_e} + 4 \times 11\mu}\right)\left(\frac{\frac{\mu}{3}}{\frac{3}{2N_e} + 3 \times 11\mu}\right)\left(\frac{\frac{\mu}{3}}{\frac{3}{2N_e} + 3 \times 11\mu}\right) \\ & \quad \times \left(\frac{\frac{1}{2N_e}}{\frac{3}{2N_e} + 3 \times 11\mu}\right)\left(\frac{\frac{1}{2N_e}}{\frac{1}{2N_e} + 2 \times 11\mu}\right)\left(\frac{1}{4^{11}}\right) \\ = & \left(\frac{1}{6+22\Theta}\right)\left(\frac{\Theta}{18+99\Theta}\right)\left(\frac{\Theta}{18+99\Theta}\right)\left(\frac{2}{6+33\Theta}\right)\left(\frac{1}{1+11\Theta}\right)\left(\frac{1}{4^{11}}\right) \end{aligned} \tag{27.15}$$

If we could enumerate all possible histories H, we could compute these expressions for each, and add them to obtain the likelihood in equation 27.12. There are infinitely many of them in the DNA sequence case, which is our example. Nevertheless, such a summation is possible in other (infinite-sites) mutation models, for very small sample sizes and very small numbers of sites. To cope with the size of the summation in all other cases, Griffiths and Tavaré (1994a, b, c) introduced a technique of sampling from the set of possible histories. As with genealogies, the problem is that most of the possible histories contribute vanishingly small amounts to the likelihood. Griffith and Tavaré therefore used importance sampling to concentrate the sampling on the histories of interest. They used the probabilities of occurrence of events at each stage, concentrating the sampling proportional to those probabilities. The events they allow are those that are compatible with the data. For example, in the reconstruction of the history in Figure 27.3 the most recent event, given the data, could have been a coalescence of lineages 1 and 3 or any one of 44×3 different mutational events. All possible mutational events could be compatible with the data, because the finite-sites DNA model allows any change to be reversed later. The coalescence has probability $1/(2N_e)$, and each of the mutations probability $\mu/3$. Griffiths and Tavaré sample from among these events in proportion to these numbers. Thus, the probability that the coalescence is chosen (as it was to make the history in Figure 27.3) is

$$\frac{\frac{1}{2N_e}}{\left(\frac{1}{2N_e} + 44\mu\right)} = \frac{2}{2 + 44\Theta} \tag{27.16}$$

while for each of the possible mutations, the probability of sampling it is (by a similar argument)

$$\frac{\frac{\mu}{3}}{\left(\frac{1}{2N_e} + 44\mu\right)} = \frac{\Theta}{6 + 132\Theta} \tag{27.17}$$

This sampling is continued until a complete history is sampled. The probability that it contributes can then be computed by taking the product of the sampling probabilities, as in equation 27.15.

However, their sampling is concentrated on some histories, and therefore we need a correction term for this importance sampling. The method that Griffiths and Tavaré use is logically equivalent to the importance sampling formula. In this case we have a series of stages of sampling, in the ith of which there are choices dictated by terms whose probabilities are proportional to $1/(2N_e)$, $\mu/3$, etc. Let us call these terms, in the jth choice for the ith possible history, $a_{ij1}(\Theta)$, $a_{ij2}(\Theta)$, ..., $a_{ijn_{ij}}(\Theta)$. Let the probability for the one that actually occurs at stage j of history i be $b_{ij}(\Theta)$. The probability of our taking this choice is

$$\frac{b_{ij}(\Theta)}{\sum_k a_{ijk}(\Theta)} \tag{27.18}$$

The probability that we will choose some particular history, H_i, is simply the product of this probability over all the choices $j = 1, \ldots, n_i$ that it entails. So

$$\prod_j \left[\frac{b_{ij}(\Theta)}{\sum\limits_k a_{ijk}(\Theta)} \right] = \frac{\prod\limits_j b_{ij}(\Theta)}{\prod\limits_j \left[\sum\limits_k a_{ijk}(\Theta) \right]} \tag{27.19}$$

The summations and products in equation 27.19 are as long as they need to be in each history, and this issue need not concern us.

Thus we have replaced sampling from all possible H's by sampling from a more restricted distribution. It only draws histories H that are compatible with the data, so that $\text{Prob}(D|H)$ is always 1. The discrete probability distribution that it draws from has probability given by equation 27.19 of drawing a particular history H_i.

Notice that the basic equation that underlies the method of Griffiths and Tavaré (1994a, b, c) is equation 27.12. This can be thought of as the weighted sum, over all possible event histories, of $\text{Prob}(D|H)$. The weight is the probability of that event history, given β. Let us call this prior distribution f and refer to it as the *Griffiths prior*. It is like the *Kingman prior*, but differs in that the events it describes include coalescences and mutations but do not contain any times for the coalescences. The Kingman prior is a distribution on possible genealogies that describe no mutations but do specify the times of the coalescences.

The likelihood we seek to compute is simply the expectation of $\text{Prob}(D|H)$ under the Griffiths prior. The Griffiths-Tavaré sampling procedure does not sample from the Griffiths prior (f) but concentrates the sampling on those histories that are compatible with the data. To correct for this different distribution (which we call g), we need to make an importance sampling correction. The importance sampling correction for an expectation is, as we saw in equation 27.10,

$$\mathbb{E}_f[Y] = \mathbb{E}_g\left[\frac{f}{g}Y\right] \tag{27.20}$$

where we want to know the expectation of the quantity Y over samples from a distribution whose density is f, but we are actually drawing from one whose density is g.

The likelihood can then be calculated from this importance sampling formula (27.10),

$$L(\Theta) = \mathbb{E}_f[\,\mathrm{Prob}\,(D|H)] = \mathbb{E}_g\left[\frac{f}{g}\,\mathrm{Prob}\,(D|H)\right] \tag{27.21}$$

but the trees drawn from distribution g always have $\mathrm{Prob}\,(D|H) = 1$, so we can drop that term from the expression. Substituting expression 27.19 for the probability of history H_i under distributions f and g, we get

$$L(\Theta) = \mathbb{E}_g\left[\frac{\left(\frac{\Pi_j b_{ij}(\Theta)}{\Pi_j\left(\sum_k a_{ijk}(\Theta)\right)}\right)}{\left(\frac{\Pi_j b_{ij}(\Theta_0)}{\Pi_j\left(\sum_k a_{ijk}(\Theta_0)\right)}\right)}\right] = \mathbb{E}_g\left[\prod_j\left(\frac{b_{ij}(\Theta)}{b_{ij}(\Theta_0)}\frac{\sum_i a_{ijk}(\Theta_0)}{\sum_i a_{ijk}(\Theta)}\right)\right] \tag{27.22}$$

This expectation is approximated by sampling many trees and averaging this quantity.

I have not used the notation that Griffiths and Tavaré did, which is very precise but not transparent. Mine is rougher but easier to discuss. I also have presented their method as one of sampling histories, to make its relationship to the Metropolis-Hastings sampling method clearer. Griffiths and Tavaré view their method differently, as a recursive method for computing the exact probability of a data set, with the sampling method simply a way of choosing paths through the recursion. The two views seem very different, but do not lead to different methods or formulas.

The remarkable thing about the Griffiths-Tavaré sampling method is that the histories H that it samples are independent. We need not worry about how long it will take the process to wander to another part of the tree space, as we must with the Metropolis-Hastings sampler. Each sample is an independent draw from the distribution. In addition, the process is much faster than the Metropolis-Hastings sampler. There are no terms $\mathrm{Prob}\,(D|G)$ to compute. In the Metropolis-Hastings sampler, we must compute that term for each proposed tree. That term is the likelihood for the tree, which is much slower to compute than the Kingman prior $\mathrm{Prob}\,(G|\Theta)$. In practice, much of the effort can be avoided by reusing parts of this tree likelihood that are shared between successive trees. But the Griffiths-Tavaré sampling avoids the data likelihood calculation altogether, achieving much faster speeds.

It would seem to be so much more efficient that there would be no sense using the Metropolis-Hastings sampler at all. But there is one limitation of Griffiths and Tavaré's sampling that is not shared by the Metropolis-Hastings sampler. Note that it does not sample from the posterior distribution $\text{Prob}(H|\beta)\ \text{Prob}(D|H)$. In the case of nucleotide sequences, the Griffiths-Tavaré sampler can waste large amounts of time on histories H that make little contribution to this distribution. Griffiths and Tavaré (1994c) have pointed out this problem. They noted that further developments may be necessary to concentrate their sampler on the more productive histories.

Stephens and Donnelly (2000) have modified the importance sampling terms of the Griffiths-Tavaré method to concentrate the sampling more on the relevant histories. Their method, which tries to predispose the choice of mutational events so as to converge the sequences more, achieves a substantial speedup. Slatkin (2002) has described another method of independent sampling, not based directly on the Griffiths-Tavaré method, which could compete with these methods.

Bayesian methods

Although both of the sampling approaches have been described as methods for computing likelihood curves, the Metropolis-Hastings approach can also be used for Bayesian inference. If we have a prior density for the parameters, $\text{Prob}(\Theta)$, the Metropolis-Hastings method can simply sample from a density proportional to the posterior, one proportional to $\text{Prob}(\Theta)\ \text{Prob}(G\,|\,\Theta)\ \text{Prob}(D\,|\,G)$. The sampling will then wander through not only the genealogy space (the values of G), but also the parameter space (the values of Θ). The acceptance ratio will then have an extra term (as we saw in equation 18.8 when Bayesian MCMC was introduced). Bayesian MCMC methods allow changes in either G or Θ, or in both. One reasonable strategy is to change G some of the time and the rest of the time change Θ. In the former case, the term $\text{Prob}(\Theta)$ cancels out of the acceptance ratio; in the latter case, $\text{Prob}(D\,|\,G)$ cancels out. The use of Bayesian MCMC in the coalescent framework has been pioneered by Wilson and Balding (1998) and by Beaumont (1999).

MCMC for a variety of coalescent models

There has developed a substantial literature using sampling methods to infer parameters in a variety of models in evolutionary genetics. Although these methods are not yet widely used, they will be. There is every reason to expect them to become the standard methods of inference in evolutionary genetics. Only if some progress is unexpectedly made in finding formulas for the integrals over genealogies would we be able to supersede the sampling approaches. Some of the models used are:

- Models with population growth. The paper of Griffiths and Tavaré (1994a) allowed exponential growth, estimating the scaled rate of growth as well as $4N_e\mu$. Kuhner, Yamato, and Felsenstein (1998) used the same model of growth with their Metropolis-Hastings approach. (Another such approach is that of Beaumont, 1999.) They noted that the estimation of growth rate is strongly biased toward high growth rate, though with a likelihood curve usually broad enough not to exclude the actual growth rate.

- Models with migration. Beerli and Felsenstein (1999, 2001) applied a Metropolis-Hastings sampler to inference of population sizes and migration rates (each scaled by mutation rates) in two-population and in n-population models. The corresponding method using the Griffiths-Tavaré sampler was developed by Bahlo and Griffiths (2000).

- Models with recombination. The models so far have been for nonrecombining loci. As we saw in the previous chapter, recombination within the region leads to a genealogy with loops in it (or more properly, a collection of different treelike genealogies, differing for different sets of sites). Griffiths and Marjoram (1997) have used the Griffiths-Tavaré sampler to compute likelihoods with recombination. The Metropolis-Hastings approach to models with recombination is given by Kuhner, Yamato, and Felsenstein (2000).

- Models with diploidy. The preceding models assumed that haplotypes were collected. When the genotypes are diploid, there is the problem of resolving the haplotypes. One cannot actually resolve them—even a maximum likelihood inference of the haplotypes is not correct. What is needed is a way to sum over all possible haplotype resolutions. This can be done with MCMC methods, if we add an additional class of events that are resolutions of the diploid genotypes into haplotypes. We can wander through the space of these resolutions as we wander through the space of trees. In fact, one must move through both at the same time, as a change of a haplotype resolution raises the need for the placement of the haplotype on the genealogical tree to change as well. Analysis of diploid genotypes by these methods has been investigated by Kuhner and Felsenstein (2000). With diploidy and recombination dealt with, likelihood approaches are becoming available that will enable the estimation of population parameters and genetic parameters from nuclear genes, which will add greatly to the information available from mitochondrial and Y-chromosome sequences.

- Models with SNPs. We have seen that likelihoods can be computed for phylogenies for molecular sequences, restriction sites, and microsatellites. SNP (single nucleotide polymorphisms) will play a large role as well within species. These can be analyzed by using ascertainment corrections similar to those we discussed in Chapter 15 for restriction sites. The ascertainment corrections needed for SNPs have been outlined by Nielsen (2000) and Kuhner et al. (2000).

- Inference of ages of divergence or mutation. There has been much recent concern with inferring the age of mutations in humans, and the time of the most recent common ancestor of a sample. Relevant papers include those of Tavaré et al. (1997), Slatkin and Rannala (1997, 2000), Griffiths and Tavaré (1999), Stephens (2000), and Nielsen (2002). The time of coalescence and the age of an allele are not parameters of the model. Although normally it would make little sense to infer them, they can provide clues as to when the model may be violated, if their values are outside an expected range.

Models involving two or more species are covered in the next chapter. I will not try to cover this rapidly expanding area more closely here. Sampling approaches will have a major impact on genomics, as the work on recombination leads naturally to sampling-based likelihood analysis of linkage disequilibrium mapping. For approximate approaches see the papers of Graham and Thompson (1998), Rannala and Reeve (2001), Garner and Slatkin (2002), Larribe, Lessard, and Schork (2002), and Morris, Whittaker, and Balding (2002). Inference methods involving natural selection will also appear in the near future.

Single-tree methods

I have concentrated here on sampling methods, as these compute the full likelihood (although approximately). However they are computationally tedious and can be difficult to understand. We have seen that if we could somehow know the coalescent tree exactly, we could base inference on it and avoid many of these complications. Given the importance of phylogeny inference above the species level, it is tempting to infer the tree, treat that inference as if it were an observation, and then proceed using that single tree.

Avise (1989; see also Avise et al., 1987) have taken this approach, though without a detailed statistical methodology. A more statistical approach was taken by Templeton (1998), using the nested clade analysis tree reconstruction methods introduced earlier by Templeton et al. (1988). Although well-defined enough to be implemented by computer programs (Clement, Posada, Crandall, 2000; Posada, Crandall, and Templeton, 2000), these methods do not attempt to take into account the uncertainty of the estimate of the tree, and there has been little study of their statistical properties. A notable exception is the paper by Knowles and Maddison (2002). Although the need to use manual steps in the analysis limited the number of replications they could make, they found that the single-tree approach was problematic.

Slatkin and Maddison's method

Slatkin and Maddison (1989) used a single-tree method to infer rates of migration, using as their statistic the parsimony count of migration events on a coalescent tree that was assumed correct. A notable feature of their work was a simulation-based correction for the biases inherent in this method.

Fu's method

Another statistically-based single-tree method is due to Yun-Xin Fu (1994). His is also a phylogenetic method, but needs only one tree. In effect, it estimates the best tree it can for the data, then makes a correction for the uncertainty of the tree. Fu obtained the correction by simulation. One need only use his formula to side-step the messy sampling of alternative trees. The accuracy of his method has not yet been carefully compared to the tree-sampling methods of Kuhner, Yamato, and Felsenstein or of Griffiths and Tavaré. When it is, one suspects that it will do well when the tree is relatively well-known (as in the case of a large number of sites but a modest number of samples) but will perform less well for fewer sites or larger population samples.

Summary-statistic methods

For these problems, even before single-tree methods were developed, methods that summarized the statistical information in single statistics were widely used. The most famous is Sewall Wright's F_{ST} statistic for geographic differentiation of populations, which has achieved a near-cult status. One of the best-investigated statistically is the number of segregating sites.

Watterson's method

The simplest method of all is that of Watterson (1975). He derived a simple formula for the estimation of θ (mutation rate per locus) for an infinite-sites model. That model allows mutation to occur only once per site in a population, so that there is never any ambiguity as to how many mutations have occurred. Watterson noted that in a coalescence interval where k lineages are about to coalesce to give $k-1$, the expected number of sites that mutate in a locus of L sites will be $L\mu k$ times the expected coalescence time, where $L\mu$ is the mutation rate per locus. In Watterson's model, L is assumed to be very large, so that mutations never occur more than once per site. The coalescence time is $4N_e/[k(k-1)]$, so that the expected number of mutants during the coalescence interval is $4N_eL\mu/(k-1)$ or $\theta/(k-1)$. Summing over all $n-1$ coalescence intervals, we get the expected number of segregating sites, $\mathbb{E}(S)$,

$$\mathbb{E}[S] = \theta \sum_{k=2}^{n} \frac{1}{k-1} = \theta \sum_{k=1}^{n-1} \frac{1}{k} \tag{27.23}$$

Equating the observed S to its expectation and solving for θ, we get an estimator of θ:

$$\widehat{\theta} = \frac{S}{\sum\limits_{k=1}^{n-1} \frac{1}{k}} \tag{27.24}$$

Using equation 27.23 we can easily show that this estimator is unbiased:

$$\mathbb{E}\left[\widehat{\theta}\right] = \mathbb{E}\left[\frac{S}{\sum_{k=1}^{n-1} \frac{1}{k}}\right] = \frac{\mathbb{E}[S]}{\sum_{k=1}^{n-1} \frac{1}{k}} = \frac{\theta \sum_{k=1}^{n-1} \frac{1}{k}}{\sum_{k=1}^{n-1} \frac{1}{k}} = \theta \qquad (27.25)$$

The variance of $\widehat{\theta}$ is also easily computed by considering the variances of the coalescence times and the means of the variances of the numbers of mutants per coalescence. In this way one can compute Watterson's (1975) variance formula.

When it is applied to sequences of finite length, Watterson's estimator will underestimate θ, owing to the occurrence of multiple mutations at some sites. I have argued (Felsenstein, 1992a) that Watterson's estimator will be inefficient when the number of sites is large and the number of samples is also large. Fu and Li (1993) have used conservative approximations to argue that Watterson's estimator, although less efficient than maximum likelihood, is not as bad as I implied, once finite numbers of sites are allowed. Simulation studies (Felsenstein, in prep.) show that Watterson's estimator, although not perfectly efficient, is remarkably good. It remains to be seen whether estimates based on it can be extended to more complex situations, as the tree-sampling methods can.

Other summary-statistic methods

Examination of summary-statistic methods carries us beyond the scope of this book, as they do not even attempt to use coalescent trees. Nevertheless, we should note some of the statistically better-justified methods using summary statistics. Wakeley (1998) extended Watterson's estimator to a symmetric migration model. Weiss and von Haeseler (1998) used a simulation technique to compute likelihoods for scaled population size and growth rate, when the data were represented only by the mean pairwise difference and the number of segregating sites. Beaumont, Zhang, and Balding (2002) have taken a similar approach using a Bayesian framework.

Testing for recombination

There is an extensive literature on methods testing whether a set of sequences seems to have undergone recombination and reconstructing where the recombinations may have been. Many of the approaches use single-tree or summary-statistic approaches.

Sneath, Sackin, and Ambler (1975) suggested using compatibility matrices among sites to detect groups of adjacent compatible sites; this was implemented by Jakobsen, Wilson, and Easteal (1997; Jakobsen and Easteal, 1996). Similarly, Sawyer (1989) based a statistical test on runs of adjacent substitutions that might indicate where the tree had changed. This has some advantages over the earlier runs test of Stephens (1985). Hudson and Kaplan (1985) used pairwise compatibility among sites to infer the level of recombination.

Hein (1993) inferred recombination events by parsimony, penalizing recombinations the same as base substitutions. Maynard Smith and Smith (1998) use parsimony reconstructions of the tree and test whether the number of cases of homoplasy exceeds that expected. Grassly and Holmes (1997) used a sliding window, likelihoods, and a parametric bootstrap distribution to detect changes in phylogeny along a sequence. McGuire, Wright, and Prentice (2000) discussed a Bayesian hidden Markov model approach to identifying where in a sequence the phylogeny changes.

Drouin et al. (1999) give an interesting comparison of the behavior of many of these methods on a large data set. The MCMC likelihood approaches using outlined above allow a test for the presence of recombination, by comparing likelihoods with and without allowing it. This should be more powerful than single-tree or summary-statistic methods.

Chapter 28

Coalescents and species trees

As the study of molecular evolution nears the species level, coalescent phenomena become noticeable. There is no longer a simple correspondence between trees of copies of genes and phylogenies of species. Figure 28.1 shows why coalescents can cause inconsistency between the species tree and the tree of gene copies. In the figure there is one copy of the gene sampled from each species. The copies from A and B have lineages that coexist in their ancestor, the interior branch of the tree which is delimited by dashed lines. If it happens that these two lineages do not coalesce with each other by the bottom of that branch, all three lineages will then find themselves in the same ancestral species. Two-thirds of the time, the two that coalesce first (going backwards in time) will not be the lineages from A and B. The result is a genealogy whose topology is different from the phylogeny of the species. In the figure, the result is that the copies from A and C are more closely related to each other than either is to the copy from B. If we were looking at molecular sequences, we might see this relationship.

The cause of the lack of coalescence is, of course, large population size in the common ancestor of species A and B. If this population size is large, we might expect such discrepancies to arise fairly often, though there is a chance that when the lineages ancestral to A and B were in the common ancestor of all three species, they would happen to coalesce with each other. In that case, the topology of the coalescent genealogy would be consistent with the phylogeny of species, though the divergence time between A and B would be exaggerated.

A number of molecular evolutionists noticed that discrepancies between coalescent genealogies and species trees might occur. Gillespie and Langley (1979) noted that divergence times between gene copies in different species would be greater than the divergence time between the species, owing to coalescent phenomena. Tajima (1983) and Hudson (1983) gave probabilities that the coalescent genealogy would be incompatible with the species tree in a three-species case. Tajima's formulas are for mitochondrial DNA, but they are easily modified to

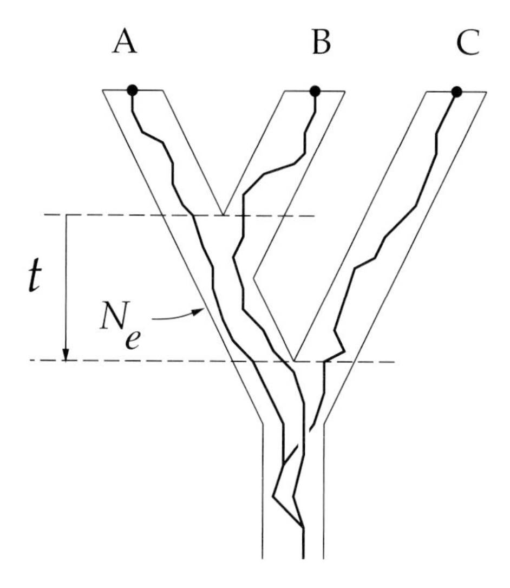

Figure 28.1: A three-species phylogeny (light lines) and a genealogy of gene copies that is inconsistent with it (darker lines). The time span of the interior branch (t) and the effective population size in that branch (N_e) determine the probability of getting a gene tree that is inconsistent with the species tree.

apply to nuclear genes. Pamilo and Nei (1988) have extended this computation, considering the effect of adding more independent loci and more gene copies sampled from the populations.

We can get a picture of how population sizes affect the probability of the coalescent tree being incompatible with the phylogeny of the species by looking at Hudson's and Tajima's calculation (the latter modified to apply to diploids). Figure 28.1 shows a three-species tree and one of the ways that the coalescent genealogy can be inconsistent with it. The incorrect topology will be obtained if both of the following hold true:

- The gene lineages from A and B do not coalesce before one gets back to the common ancestor with C. The probability of this is

$$e^{-\frac{t}{2N_e}}$$

- After one gets back to the common ancestor of all three, the two that coalesce first as one goes farther back are not A and B. The probability of this is 2/3.

These two possibilities are shown in the two trees in Figure 28.2. The net probability of having a tree incompatible with the species phylogeny is

$$\frac{2}{3}\,e^{-\frac{t}{2N_e}} \tag{28.1}$$

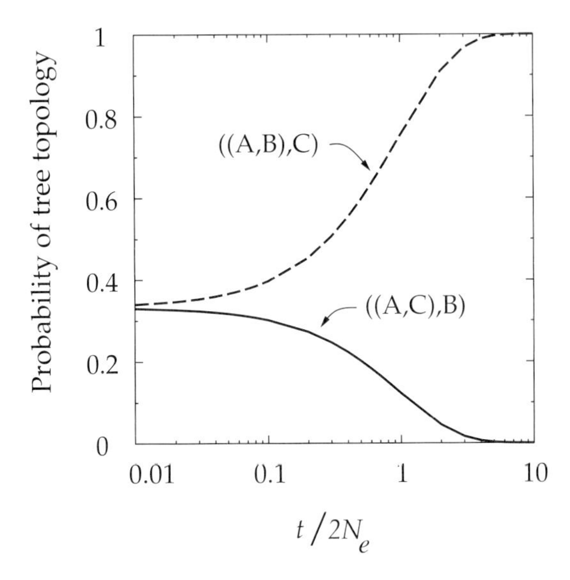

Figure 28.2: Probability, in a three-species tree like Figure 28.1, that the topology of the coalescent is one of the two incorrect tree topologies (solid curve) and that it is the correct tree topology (dashed curve). These are given as a function of the ratio of branch length in generations (t) to twice the effective population size of the interior branch of the tree ($2N_e$).

trees. The figure shows the probabilities in this case of the coalescent being one of the two wrong tree topologies, and the probability of it being the right tree topology, as a function of $t/(2N_e)$. As the number of generations becomes larger than $2N_e$, the probability of a wrong topology declines to zero. Pamilo and Nei (1988) have given probabilities of various coalescent topologies in cases with 3, 4, or 5 species. Takahata (1989) has extended their work to larger sample sizes for three species. Rosenberg (2002) has discussed the difference between the concordance criteria used by these authors and suggested an alternative definition of concordance appropriate for use with samples of more than one gene copy per species. For pairs of species, each with a sample of sequences, there is particular interest in the case where the gene tree is consistent with the species trees, so that the lineages for each species coalesce before we get back to their common ancestor. Hudson and Coyne (2002) have investigated conditions for reciprocal monophyly, and discussed the implications of these conditions for use of coalescent trees in species definitions. An interesting case where discordance of coalescent trees and species trees must be considered is pointed out by Shedlock, Milinkovitch, and Okada (2000) who discuss the importance of coalescent phenomena as a source of discrepancy in using SINE element insertion to delimit monophyletic groups.

Methods of inferring the species phylogeny

Given that there are possible conflicts between the species tree and the coalescent genealogy, what methods can be used to infer the species tree? If there is only one locus and one gene sampled per species, then there is nothing we can do but accept the coalescent genealogy as being the gene tree. But with multiple copies per species, and especially with multiple loci, we can hope to iron out the discrepancies. A variety of methods have been proposed to make inferences about the species tree. I will describe them before focusing on likelihood and Bayesian methods.

Takahata and Nei (1985) discussed how one could infer species divergence times from the divergence of gene copies on the coalescent genealogies. Nei and Li (1979) had proposed that divergence time between two species x and y be estimated using

$$d = d_{xy} - \frac{1}{2}d_x - \frac{1}{2}d_y \tag{28.2}$$

This assumes that the effective sizes of the two species and of their ancestor were the same. Takahata and Nei showed that the variance of this estimator is quite large when the divergence time is much less than $2N_e$, the effective size of a species. They gave a general formula for the variance of the estimator as a function of the number of copies m and n sampled from the two species. They found that while variance could be reduced by taking m and n large, it would be much more helpful to use multiple loci. Wakeley and Hey (1997) give other estimators based on the number of polymorphic sites that are shared.

The variance from locus to locus is of course precisely due to the variation of the coalescence times. Even with enough sites to enable precise knowledge of the coalescent genealogy for a single locus, the variation of its coalescence times has not been subdued. To do so, we need multiple loci that have independent coalescents. Baum (1992) and Doyle (1992) suggested combining trees estimated from different loci by coding each tree as a single multistate character. The nonadditive binary coding method, which was explained in Chapter 7, was to be used. That would make each branch of a coalescent tree be represented by a single binary character. They suggested using these together with the morphological characters in an overall parsimony analysis.

Implicitly, their method assumes that each coalescent genealogy is known precisely—they give no way to take its uncertainty into account. However, in most respects their procedure is quite conservative because it does not give credit to the coalescents for the amount of data that went into them. If the effective population sizes N_e are small, we may want to give more weight to coalescent trees that are based on long stretches of DNA, or to ones whose rate of evolution makes them more likely to be accurate.

Lyons-Weiler and Milinkovitch (1997) have suggested using Lyons-Weiler's RASA method (for which see Chapter 12) to test for difference of topology in different gene trees. Chen and Li (2001) used parsimony to reconstruct trees for 53

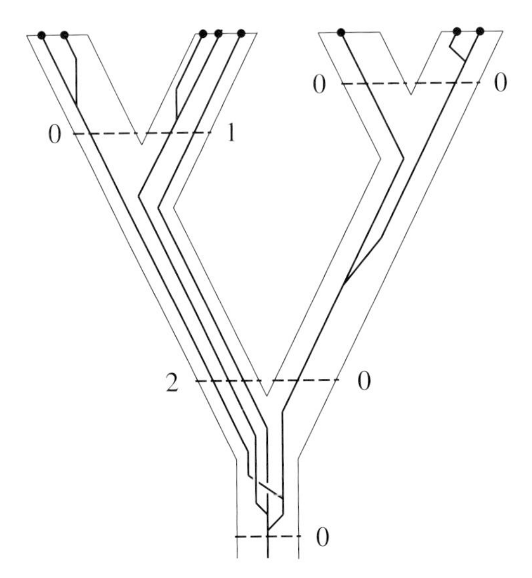

Figure 28.3: An example of the count of uncoalesced lineages in gene tree parsimony. The count is made at the dashed lines, which are at the bottom of each branch. The number of events is the number of gene lineages reaching the bottom of the branch, less 1. In this tree, the total number of events is 3.

autosomal regions for hominids. They were able to infer $t/(2N_e)$ for the ancestor of human and chimp from the fraction of these regions in which humans were closest to gorillas.

Reconciled tree parsimony approaches

Page and Charleston (1997), Wayne Maddison (1997) and Slowinski, Knight, and Rooney (1997) have pointed out that the parsimony approach of Goodman et al. (1979) can be applied to coalescent genealogies. The event that is, in the context of a gene family, interpreted as a duplication, is in this case considered to be a deep coalescence. One must ban losses, as they have no natural counterpart when coalescence is the source of discrepancy between the gene genealogy and the species tree. As explained by Slowinski and Page (1999), one should count, not the number of duplications, but the number of branches in which lineages fail to coalesce. They use the number of lineages not coalescing in the branch, less 1, as the number of events in their parsimony method. Thus if four lineages are in the top of the branch, with three present at its bottom, that counts as 3 – 1 = 2 events. They call this method *gene tree parsimony*.

Figure 28.3 shows an example of a count of uncoalesced lineages in gene tree parsimony. The dashed line at the bottom end of each branch shows the census of "extra" lineages that have not coalesced by the time the bottom of the branch is reached. There are a total of three of these lack-of-coalescence events. This parsimony method is easily combined with parsimony of molecular sites or morphological characters. Doing that, one takes into account both the number of sites in each gene and the discrepancies between the coalescent genealogies and the species tree. However, the weighting of failures to coalesce in this method is rather arbitrary.

We may recall (from Chapter 9) that the natural formula for the weight on an event in a parsimony analysis is $-\ln[p(t)]$, where $p(t)$ is the probability of that event happening in a branch of length t. If we have k lineages that do not coalesce in a branch of length t, the probability of noncoalescence is (from equation 26.3)

$$e^{-\frac{k(k-1)\,t}{4N_e}} \tag{28.3}$$

so that the weight should be $k(k-1)t/(4N_e)$. Like the true weights for all parsimony methods, this depends on the branch length; even if it did not, it would still be k times as great as the weight assigned in gene tree parsimony. When there are k lineages at the top of a branch, and all but j of them coalesce before the bottom, the probabilities are more complicated and the weights even less obvious. When the branch is very short, it is coalescence rather than lack of coalescence that is the rare event that should be counted.

Likelihood

A more comprehensive method would, of course, use likelihood or Bayesian methods. Estimation of ancestral population sizes has spurred the development of a likelihood framework for coalescent genealogies in multiple species. Takahata (1986) had used means and variances of divergence between two species across a number of loci to infer the population size in their ancestor. Takahata, Satta, and Klein (1995) used maximum likelihood for two or three species, under the assumption that one could observe all substitutions between each pair of species. In effect, this assumes an infinite-sites model of change. They used one gene copy per locus per species and applied their method to human ancestry. Subsequently, they have (Li, Satta, and Takahata, 1999) also applied it to the *Drosophila* species data of Hey and Kliman (1993). Yang (1997a) extended this method to make an infinite-sites analysis of ancestral population sizes, taking variation of evolutionary rate from site to site into account. Edwards and Beerli (2000) have extended Yang's method for two species so as not to require the infinite-sites model. These studies pioneered the application of likelihood methods to multiple species and multiple loci. They all assumed one lineage per species, and some analyzed only two species at a time. Yang (2002) gave likelihood and Bayesian approaches that allowed three species with one copy per locus per species, and did not assume an infinite-sites model. Wall (2003) considered four species, with one copy per locus per species, with the assumption that we can assign all substitutions to branches on the tree (which is close to assuming an infinite-sites model). He used coalescent simulations to approximate summation over all genealogies. The simulations allowed recombination to occur, so that Wall's approach allows for intragenic recombination. Nielsen et al. (1998) made a likelihood analysis for the case in which, with two or three species, we have samples of genes where all polymorphism is assumed to be pre-existing and where no mutation occurs thereafter. This case had been the subject of a different likelihood analysis when single copies are sampled from each species (Wu, 1991; Hudson, 1992).

Each of these studies carried us nearer the goal of an analysis that was fully statistical, allowed use of standard DNA models, analyzed an arbitrary number of species, and allowed for sampling of more than one gene copy within each species. Wayne Maddison (1997) noted that since evolutionary genetics provided formulas for the probability of coalescence of lineages, a likelihood approach to fitting coalescent genealogies to species trees was possible. Nielsen (1998) has come closest to the goal of a general treatment, allowing in his analysis for more than one sampled gene copy per locus per species. Using the infinite-sites model and the case of two species, he used Griffiths-Tavaré sampling methods to calculate the likelihood for samples from the two species. The parameters are the divergence time, mutation rate, and the effective population sizes of the two populations, plus an effective population size of the population ancestral to both of them. Figure 28.4 shows his model.

An interesting extension of this model was given by Nielsen and Wakeley (2001). It allows migration between the two populations after they separate. The AIC criterion was suggested to choose between the model with and without migration. The model assumes that the population size is constant within each branch, which is obviously an oversimplification. If we wanted to fit a model with a population bottleneck at the start of a lineage, that would be an additional complication, but such complications could easily be accommodated in this framework. In Nielsen's model the mutation rate is not an independent parameter, as one can scale time by it, so that t and the N_i would actually correspond to parameters μt and $\Theta_i = 4N_i\mu$.

It is straightforward to generalize Nielsen's likelihood method to an arbitrary phylogeny, arbitrary sample sizes per species, and an arbitrary model of molecular change. Suppose that T is the species tree, G the coalescent genealogy, and D the data. The likelihood can then be written as a function of T, the vector of population sizes $\mathbf{N} = (N_1, N_2, \ldots, N_n)$, and a mutation rate μ as:

$$L = \text{Prob}(D \mid T, \mathbf{N}, \mu) = \sum_G \text{Prob}(D \mid G, \mu)\ \text{Prob}(G \mid T, \mathbf{N}) \tag{28.4}$$

If the times on T and G are scaled in mutational units (so that one unit of time is the expected time until a base mutates), we can drop the argument μ and replace the vector $\mathbf{N}$ by the vector $\boldsymbol{\Theta} = (4N_1\mu,\ 4N_2\mu,\ \ldots,\ 4N_{2n-1}\mu)$:

$$L = \text{Prob}(D \mid T, \boldsymbol{\Theta}) = \sum_G \text{Prob}(D \mid G)\ \text{Prob}(G \mid T, \boldsymbol{\Theta}) \tag{28.5}$$

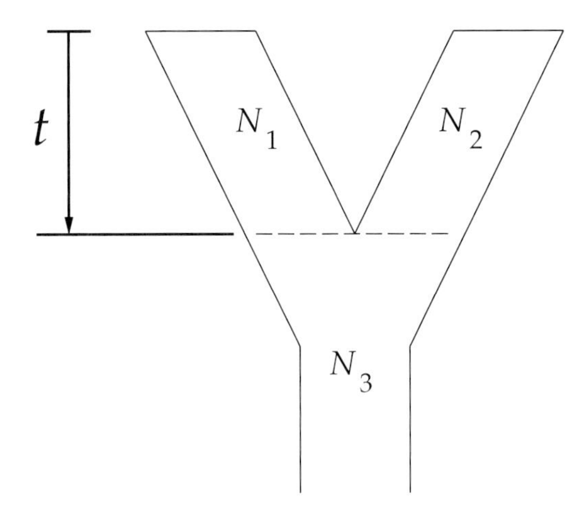

Figure 28.4: Nielsen's (1998) model. Each branch on the tree has a separate effective population size, and the divergence time is also a parameter of the model.

This equation is for a single locus. For multiple loci we have a separate coalescent genealogy G_i for each locus but common values of T and $\mathbf{\Theta}$. If D_i are the data for locus i. the counterpart of equation 28.4 is

$$L = \prod_i \text{Prob}\,(D_i \,|\, T, \mathbf{\Theta}) = \prod_i \sum_G \text{Prob}\,(D_i \,|\, G)\;\text{Prob}\,(G \,|\, T, \mathbf{\Theta}) \tag{28.6}$$

The overall likelihood is simply the product across loci of the likelihoods given in equation 28.5. Bayesian inference can also be carried out, with a prior on $\mathbf{\Theta}$ or perhaps on both T and $\mathbf{\Theta}$, multiplying that product of likelihoods. In principle, these methods can also be extended to allow for intragenic recombination as well.

Markov chain Monte Carlo methods would be used to carry out the summation, as we saw in Chapter 27. Within each branch the coalescence is governed by the branch's value of $\Theta_i = 4N_i\mu$. At the top of an interior branch of the tree, lineages arrive from the two immediate descendants. These then coalesce going down that branch, under the control of its value of $\Theta_j = 4N_j\mu$. Markov chain Monte Carlo computation of the likelihood is possible, as are estimates of the Θ_i.

It is also possible to search for the maximum likelihood species tree $\widehat{T}$, using methods such as the Metropolis-Hastings sampler. One would update the estimates of the genealogy for each locus separately. In updating estimates of the species tree, one might hold the gene tree constant while changing the species tree, and see whether the new species tree was rejected for having too low a likelihood. This general approach will underly most future statistical inference on the fit of coalescent trees to species trees.

Chapter 29

Alignment, gene families, and genomics

In principle, all parts of the genome are descended from one much smaller ancestral piece of nucleic acid. All genes, and all genomes, have originated from this ancestral genome by gene duplication, loss of parts of the resulting genomes, insertions, and rearrangements of various sorts. Chromosome complements of many different species have been known for many decades, and genetic maps of some of these organisms have been available more recently. Information about different genes that appear to have been descended from one ancestral gene has been collected since the 1950s, when the protein sequences of the globin family became known. As complete genomes have been sequenced, both gene family and genome structure information have become available in those species, starting with organelle genomes, continuing to prokaryote genomes, and now to eukaryotic genomes.

Putting all of this information into a phylogenetic context requires models of deletion, insertion, gene duplication, and genome rearrangement. Specialized methods are required to gain an understanding of the events in gene family and genome evolution, and to use this to assist in inferring phylogenies.

In this chapter I will start with sequence alignment, which involves inferring where deletions and insertions may have taken place within genes. I then move on to the analysis of gene families, where I take into account gene duplication and loss but not the exact placement of the loci in the genome. I then continue to the making of inferences using the order of genes in the genome, taking explicit account of the genetic map. Less work has been done on this topic, and there has been no serious attempt to combine these problems. This will happen inevitably, and fairly soon.

Table 29.1: A multiple sequence alignment of five nucleotide sequences, with gaps at positions 3–5 and 8–9.

Alpha	`ACCGAAT--ATTAGGCTC`
Beta	`AC---AT--AGTGGGATC`
Gamma	`AA---AGGCATTAGGATC`
Delta	`GA---AGGCATTAGCATC`
Epsilon	`CACGAAGGCATTGGGCTC`

Alignment

Almost as soon as algorithms for the alignment of pairs of sequences were available, the issue of how they relate to phylogenies was raised. Alignment is really an inference of the descent of all bases (or amino acids) at a given position from a common ancestor. As computational molecular biologists started to develop methods for alignment of multiple sequences, they were making inferences about evolution, usually without acknowledging the fact. But the issue had already been raised in 1973, by Sankoff, Morel, and Cedergren. Using 5S RNA sequences, they attempted to find the phylogeny and the alignment, which, taken together, involved the smallest total penalty from base substitutions, insertions, and deletions. The papers by Sankoff (1975) and Sankoff and Rousseau (1975), which we have already discussed in the section on counting changes in parsimony, were actually made general enough to encompass insertions and deletions as well. Sankoff and Cedergren (1983) expanded on the relationships between phylogenies and multiple sequence alignment. These pioneering papers were ignored for almost two decades, until the success of alignment programs such as `ClustalV` revived interest in the connection between phylogenies and alignment.

Why phylogenies are important

It may seem that we can assess how many gaps (insertions or deletions) are required in a multiple sequence alignment, simply by examining a table of the alignment. Table 29.1 shows a hypothetical multiple sequence alignment. There appear to be two gaps present, one at positions 3–5 and the other at positions 8–9. However, distribution of the gaps relative to each other shows that the two gaps are incompatible with each other. All four combinations of gapped and ungapped regions exist in the data, so that it is not possible for there to be a phylogeny on which the two gaps can be explained with only one insertion (or deletion) event each. Figure 29.1 shows one possible phylogeny for these sequences: It requires two insertion events and one deletion event. Note that if the tree were rerooted, the number of events would not change, but the interpretation of which ones were insertions and which ones deletions would change. In this tree the hash marks are base substitutions: 10 of them are required as well.

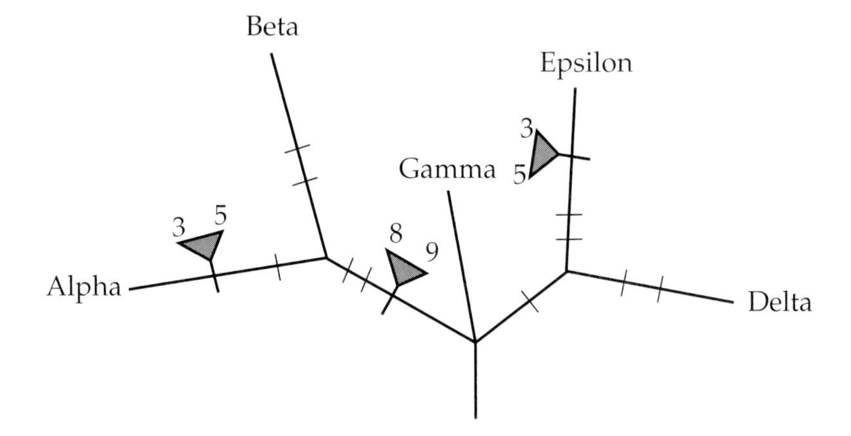

Figure 29.1: A phylogeny with a reconstruction of the insertion, deletion, and substitution events needed to evolve the sequences in the multiple sequence alignment in Table 29.1. The deletions and insertions are indicated by the triangles, which show which positions are involved. The substitutions are shown by the hash marks on the lineages. In this case, two insertion events and one deletion event are required.

Parsimony method

Sankoff, Morel, and Cedergren (1973) carried out multiple sequence alignment and inference of the phylogeny simultaneously, by a parsimony method. They used the penalty function for alignment of two sequences. This has penalties for base substitution, insertion, and deletion. They inferred sequences at the interior nodes of the tree, computed the alignment penalty for each branch of the tree, and summed these. The resulting number was the total penalty for the tree, and they chose the tree that minimized this quantity,

In their method, with each tree there is a combination of interior nodes that achieves the lowest penalty for that tree. Upon choosing the best tree, one automatically finds a multiple sequence alignment (or perhaps several tied alignments) at the same time.

The computational burden involved is great. In principle, for each tree of n species with sequence length about L, the computation can be done by a dynamic programming algorithm with effort approximately L^n. This is impractical for all but a few species. Sankoff, Morel, and Cedergren suggested an approximate alternative. Starting with initial guesses at the sequences in the interior nodes, we compute improved guesses by evaluating one node at a time. For a bifurcating tree, we update the sequence at an interior node by considering its three neighbors, using dynamic programming in a three-dimensional array.

I briefly outline the algorithm here. Each node in that array represents a partial sequence. Thus if the three neighboring sequences have lengths L_1, L_2, and L_3, the array is of size $(L_1 + 1) \times (L_2 + 1) \times (L_3 + 1)$. The points in the array are designated as (i, j, k), where i takes values from 0 to L_1, and similarly for j and k. The number a_{ijk} contains the score for the best alignment of the first i positions of neighbor #1, the first j positions of neighbor #2, and the first k positions of neighbor #3. As with alignment of two sequences, each possible path through the array corresponds to one three-species alignment. The algorithm computes the values for a_{ijk} from those of $a_{i-1,j,k}$, $a_{i,j-1,k}$, $a_{i,j,k-1}$, $a_{i-1,j-1,k}$, $a_{i-1,j,k-1}$, $a_{i,j-1,k-1}$, and $a_{i-1,j-1,k-1}$. The value of a_{ijk} is taken as the minimum of the penalties for each of these seven routes arriving at point (i, j, k). For the route from $a_{i-1,j,k-1}$, for example, there is a gap added at position j in sequence #2, and an alignment of position i of sequence #1 with position k of sequence #3. The cost of this route will be $a_{i-1,j,k-1} + c + d_{i,k}$, where c is the cost of a gap of one base, and $d_{i,k}$ is the cost for substituting the base at position i of sequence #1 with the base at position k of sequence #3. This latter will be 0 if these two bases are identical. The dynamic programming algorithm evaluates all seven possible routes that arrive at (i, j, k) and chooses the one with the smallest cost. After point (L_1, L_2, L_3) is reached, one can determine the route of smallest overall cost by backtracking through the array.

This computation involves effort $(L_1 + 1)(L_2 + 1)(L_3 + 1)$, which can be substantial. For example, if the three sequences are each of length 100, this dynamic programming algorithm requires $101^3 \approx 10^6$ steps. This is done for each interior node of the tree in turn. At each step, its three neighbors are used to infer the optimal sequence at the interior node. Each time, the total penalty of the tree cannot increase, as the worst that the three-species dynamic programming algorithm can do is to find the same interior node sequence as before. Of course, there can be ties among multiple sequences as the best interior node sequence. One could simply take one of the tied sequences, but a better choice would be to use the "sequence graph" of Hein (1989). It summarizes all the tied alignments in one graph.

The method is easily extended to allow for gap events that involve more than one base, provided that the penalties are linear functions of the number of bases involved. The array then needs three quantities in each cell instead of one, much as is done when two sequences are aligned.

The interior nodes of the tree are updated until no further improvement occurs. Rearrangements of the tree topology can be tried, each one to be evaluated by the same iterated updating of interior node sequences or sequence graphs. When a tree is rearranged, many of the previous interior node sequences will be good first guesses for this iteration.

Sankoff, Morel, and Cedergren (1973) had only been able to do their calculation by using the most powerful computers available to them; their work did not result in a practical computer program that could be widely disseminated. Hein (1989, 1990) produced a program, TREEALIGN, that carried out a relative of the Sankoff-

Morel-Cedergren algorithm, but with some tree-building steps using distance matrix methods rather than evaluating trees strictly by parsimony. More recently increases in computer speed have permitted some more precise implementations of the Sankoff-Morel-Cedergren approach (Wheeler and Gladstein, 1994).

An active literature has also arisen in computer science on the complexity of the computation when we seek the optimal tree and alignment simultaneously. Wang and Jiang (1993) proved that the problem was in a class called "MAX SNP hard" which does not have polynomial algorithms that approximate the result closely. Wareham (1995) simplified the proof. Jiang, Wang, and Lawler (1994; see also Wang, Jiang, and Lawler, 1996) gave an approximation algorithm that could be proven to come within a factor of 2 of achieving the minimum cost. Wang and Gusfield (1997) improved the approximation algorithm. Generally, the heuristic approximations mentioned earlier do much better than coming within a factor of 2, but they lack theoretical proofs of how well they do. The problem had to be at least NP-hard, because it contained as a subcase the problem of finding the best phylogeny using parsimony on base substitution. Further progress on approximation of the problem has been made by Schwikowski and Vingron (1997a, b). Gonnet, Korostensky, and Benner (2000) have argued that a circular permutation of species can be used to put bounds on the score achievable in any phylogeny.

Approximations and progressive alignment

Hogeweg and Hesper (1984; see also Konings, Hogeweg, and Hesper, 1987) and Feng and Doolittle (1987) have introduced the strategy of *progressive alignment*. In this approach a "guide tree" is inferred by doing pairwise alignments among the sequences, and then using the alignment penalties for the pairs as a distance measure. Using this tree, neighboring sequences on the tree are then aligned with each other. As one proceeds down the tree, these alignments are combined using some rather arbitrary rules. Progressive alignment is a more approximate method than the original Sankoff-Morel-Cedergren method, but programs based on it have performed well and have been widely used in multiple sequence alignment. Chief among these is the `ClustalW` program (Higgins and Sharp, 1988; Thompson, Higgins, and Gibson, 1994; Higgins, Thompson, and Gibson, 1996).

The success of programs based on progressive alignment has drawn attention to the interaction of trees and alignments, and ultimately to methods using less arbitrary algorithms. The difficulty with progressive alignment is that, once a decision has been made that aligns sequences of some of the species, this alignment is never reconsidered in light of decisions made for other species. Thus if the tree contains both mammals and birds, and if at the root of the mammals we have the alignment

```
gorilla  AGGTT
horse    AG-TT
panda    AG-TT
```

while at the root of the birds the algorithm has chosen

```
penguin   A-GTT
chicken   A-GTT
ostrich   AGGTT
```

when these are put together at the common ancestor of both groups, the alignment is likely to be

```
gorilla   AGGTT
horse     AG-TT
panda     AG-TT
penguin   A-GTT
chicken   A-GTT
ostrich   AGGTT
```

rather than allowing the subsequent information to force reconsideration of the placement of the gap. One's eyeball immediately suggests reconsideration, aligning all the gaps in the same column and all the Gs on the same side of them. A Sankoff-Morel-Cedergren parsimony approach would reconsider this, but progressive alignment is more approximate and does not.

Another approximate approach was introduced by Vingron and von Haeseler (1997). They use a sequential addition strategy to add sequences to a multiple sequence alignment. The method maintains a tree. As each sequence is added to the tree, the lineage leading to it branches off of an interior branch of the tree. This location divides the previous tree in two. Branch lengths are computed, using the average alignment scores between the new sequence and the sequences in the two parts of the tree, as well as the average alignment score between sequences in the two parts. The location that adds the smallest amount to the total length of the tree is chosen. Note that this differs from the Sankoff-Morel-Cedergren method, in that sequences are not reconstructed at interior nodes. Vingron and von Haeseler point out that their method is more closely related to the "sum of pairs" alignment criterion, a nonphylogenetic method. Nevertheless their method is capable of reconsidering a previous alignment once a new sequence is added.

An even rougher approximation is the "optimization alignment" of Wheeler (1996). He treats the observed sequences at the tips of the tree as the only possible states, uses the alignment scores between them as the penalties for change between them, and then assigns states to ancestors by Sankoff optimization. This rules out any other sequences existing at interior nodes of the tree. It is very rough, but fast; interestingly, Wang and Jiang (1993) show that a very similar procedure approximates the true tree score within a factor of 2.

Probabilistic models

The Sankoff-Morel-Cedergren approach uses parsimony: It penalizes a gap or a base substitution without regard to the length of the branch on which it occurs, and the method considers only the reconstructions of events that achieve the smallest penalty. Obviously, it would be desirable to use a probabilistic model and make inferences based on it.

Bishop and Thompson's method

The pioneering paper on statistical inference using probabilistic models of alignment is by Bishop and Thompson (1986). They used a model in which bases could substitute, but also single bases could be inserted after any base, or at the beginning of the sequence. Single bases could also be deleted. Calculation of transition probabilities in such a model is difficult, so they dealt only with cases in which two sequences were diverged for a small enough time (or branch length) that multiple gaps could not be superimposed. They were able to calculate the likelihood for different branch lengths by a dynamic programming calculation. Having the maximum likelihood estimate of the branch length, they could then find (by backtracking through the dynamic programming array) the alignment that made the largest contribution to the likelihood. Note that this is slightly different from inferring the maximum likelihood alignment. The rest of the papers in the statistical alignment literature have followed their lead and done the same calculation.

The minimum message length method

Allison and Yee (1990) have used a different framework, *minimum message length* (MML), with a model of evolution similar to that of Bishop and Thompson. They allow substitutions and insertions or deletions of one base. For a given pair of sequences, they take the probabilities of insertion and deletion to be unknown and infer these from the data. Like Bishop and Thompson, they compute the probability of getting one sequence from another by summing over all possible alignments of the two sequences using a dynamic programming algorithm. I will describe a similar algorithm in more detail in the section below for the TKF model. Minimum message length estimation is a framework based on algorithmic information theory. It gives results that are extremely close to maximum likelihood. At times it seems to be a recasting of maximum likelihood, though it can actually be shown to be slightly different.

Thus we can regard the MML method for inferring parameters of a probabilistic model of insertion, deletion, and base substitution as being an extension of Bishop and Thompson's approach, as it is almost that. Yee and Allison (1993) explained it further, with simulations showing that it converged on the correct parameter values. Allison, Wallace, and Yee (1992) extended the results to allow for more complex models in which there are three or five parameters. Allison and Wallace (1994) made the further extension to more than two species. This allowed the method to infer net probabilities of insertion, deletion, and base substitution on all branches of a phylogeny. The method did not assume a single model that operated on all branches, but allowed the model parameters to be arbitrarily different on each branch of the phylogeny.

This MML approach achieves an interesting level of generality, but is lacking in two respects. There is no model that allows insertions to be made at a site after others have been made earlier in the same branch. In short, there is no actual time scale: The model simply infers parameters for the probabilities of insertions, deletions, and base changes that result. This also makes it difficult to assume the same process on all branches. For these reasons we will concentrate on explaining the TKF model, which does not have these limitations, but the importance of this MML model in extending the Bishop and Thompson framework ought not be overlooked.

The TKF model

Jeff Thorne's *TKF model* (Thorne, Kishino, and Felsenstein, 1991) is the first model to allow gap events to overlay one another along a branch of a tree. It involves a clever bookkeeping method that prevents double-counting of possible events that could lead from one sequence to another. In the TKF model, each base is accompanied by a "link" representing the covalent bond connecting it to the next base in the sequence. A special link called the "immortal link" precedes the first base. In the model there is a rate of base substitution (which we can set to 1 per unit time) with an ordinary reversible model of base substitution. In addition, each link has a rate λ at which a new base will be inserted at that link. When it is, a new link is inserted as well, to the right of the new base. There is as well a rate at which bases are deleted. Each has a constant risk μ per unit time of being deleted, and when it is, the link to its right is deleted as well. Note that all this means that the immortal link, which is to the left of the sequence, cannot be deleted.

This model implies that the stretch of sequence has its length follow a birth-death process with immigration, with the birth rate and the immigration rate equal. The probability per unit time of adding a base to the sequence is $(n+1)\lambda$, where the current length is n, and the probability of losing a base is $n\mu$ per unit time. The sequence cannot be lost forever, as even when it is of length 0 there is a rate at which new bases are inserted. The new bases inserted are assumed to be drawn with the equilibrium frequencies of the base substitution model. The birth-death process with immigration is one which is well-known in theory of stochastic processes. If $\mu > \lambda$, the equilibrium distribution of lengths will be a geometric distribution with mean μ/λ. The equilibrium base composition of these sequences will simply be that expected from the mutation model.

This model has the great advantage that its transition probabilities can be calculated. Figure 29.2 shows a case that I will use to illustrate this. Under the model, it is not hard to show that the length of the sequence will be drawn from a geometric distribution with expectation $\mu/(\mu-\lambda)$. We would like to be able to com-

```
- C - - G T T A A - G G
C C T T - - - A - A C -

- C G T - - T A A G - G
C C - - T T - A - A C -
```

Figure 29.2: Two alignments explaining the evolution of one sequence into another. The different placements of the gap symbols correspond, under Thorne's TKF model, to different sequences of events that led from the top sequence to the bottom one. There are many other possible sequences of events as well.

pute the transition probability from one sequence to another as part of likelihood (or Bayesian) inference of the divergence time and the parameters. If we have an alignment showing which links are homologous, as we do in the figure, we can calculate the probability that one sequence will arise from the other. For the events in the top part of Figure 29.2, going from the top sequence to the bottom sequence, they are:

- A *C* is inserted after the immortal link.
- A *C* remains unchanged without insertion or deletion.
- Two *T*s are inserted to its right.
- A *G* is deleted.
- Two *T*s are deleted.
- An *A* remains unchanged.
- An *A* is deleted.
- Before that *A* is deleted, an *A* is inserted to its right.
- A *G* changes into a *C* without any insertion or deletion.
- A *G* is deleted.

Note that this particular combination of events, although it accounts for the change of sequences, is not the most plausible one. It would seem more likely that the two Ts in the middle of the sequence remained unchanged and that the G in front of them was deleted.

Note also that the TKF model makes a particular interpretation of the alignment. If we see the alignment

```
T - A G
T A - G
```

on the usual interpretation of alignments, this does not seem to be much different from

```
T A - G
T - A G
```

But in the TKF model these are different events. When a base is inserted, its arises from the base to the left of it in the alignment. Thus in the first alignment, the A in the second sequence arises from the T that is to the left of it. In the second alignment, the A arises from the A in the first sequence. Under the TKF model these are separate events with separate possibilities, and to compute the sum of all these probabilities we need to compute both of them. Figure 29.3 shows the three kinds of events that are possible under the TKF model.

The transition probability from one sequence to the other is computed by summing over all possible sequences of events. As usual with sequence alignments, there are a very large number of these. One great advantage of the TKF model is that it allows us to sum up the probability of getting one sequence from another, using a dynamic programming algorithm. Although there is not enough space here to go through all the details, we can sketch the method. The probability that a link ends up having n links as its descendants is given by one of the three quantities $p_n(t)$, $p'_n(t)$, and $p''_n(t)$, depending on whether the link is a mortal link that itself survives, is a mortal link that does not itself survive, or is the immortal link. The probabilities of all of these can be computed from well-known results for birth-death processes. With the one-base insertion scheme in the TKF model, it turns out that $p_n(t) = [\beta(t)\lambda]^n p_1(t)$, with this being true for all three quantities with the same value of $\beta(t)$. Thus the three functions of n all are geometric series for $n > 1$, all declining at the same rate.

Consider the first m bases A_m of the first sequence, and the first n bases B_n of the second sequence. Define three likelihoods, conditional on different events at the end of the sequences:

$$\begin{aligned} L^{(0)}_{mn}(\lambda,\mu,t) &= \text{Prob}\,(A_m, B_n \mid \text{rightmost link of } A_m \text{ has 0 descendants}) \\ L^{(1)}_{mn}(\lambda,\mu,t) &= \text{Prob}\,(A_m, B_n \mid \text{rightmost link of } A_m \text{ has 1 descendant}) \\ L^{(2)}_{mn}(\lambda,\mu,t) &= \text{Prob}\,(A_m, B_n \mid \text{rightmost link of } A_m \text{ has} \geq \text{ 2 descendants}) \end{aligned} \tag{29.1}$$

These are joint probabilities of the two subsequences, given the parameters. It can then be shown that we can compute each of these from the three subsequences that are shorter by one base. For example, $L^{(0)}_{mn}$ can be computed from $L^{(0)}_{m-1,n}$, $L^{(1)}_{m-1,n}$ and $L^{(2)}_{m-1,n}$. This is possible because if we have a gap opposite the last nucleotide in sequence A_m, this could represent the extension of a gap (0), the initiation of a gap after an aligned nucleotide (1), or the initiation of a gap after insertion of one or more bases in B_{n-1} (2). The probabilities of insertion and deletion in the single-base insertion and deletion model make these expressions cover all possibilities for computing $L^{(0)}_{mn}$.

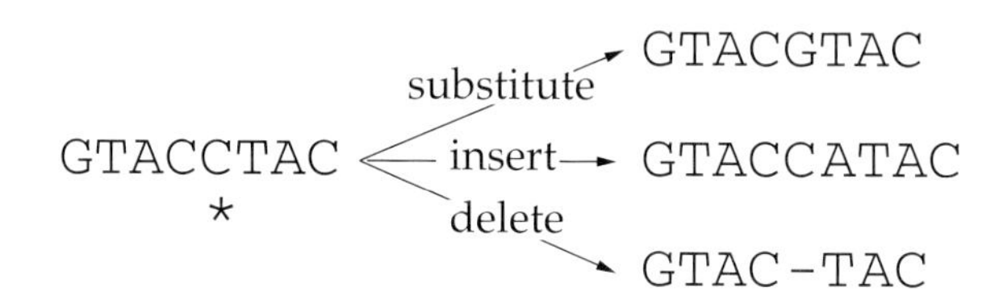

Figure 29.3: The three types of events possible under the TKF 1991 model. They are base substitution, insertion of a single base to the right of the link, and deletion of a single base and its link. The star indicates the base at which these events happen, and the consequences are shown to the right.

As we have a dynamic programming algorithm for computing the likelihood for given parameters t, λ, and μ, we can use it to infer those parameters by maximum likelihood, in the usual way.

Hein et al. (2000) have made improvements in the calculation, including simplification of the recursions and "corner-cutting" approximations. These speed up the calculation greatly. They have also discussed goodness-of-fit testing for this model. Thorne and Churchill (1995) have used an EM algorithm for inferring parameters. They have also sampled from different alignments in proportion to their contribution to the likelihood to indicate the range of possible alignments. A similar sampling was used in the MML framework by Allison and Wallace (1994). Metzler et al. (2001) have assumed priors on the parameters of the TKF model, and have used Bayesian MCMC methods to sample the posterior distribution of alignments and parameter values.

Multibase insertions and deletions

The one-base model of substitution that Thorne, Kishino, and Felsenstein (1991) used has the great advantage of allowing computation of the likelihood by a dynamic programming algorithm. But it has the great disadvantage of omitting multibase insertions and deletions. As actual insertions and deletions of more than one base at a time are common, this is an obvious defect. If not remedied, for a long gap it would lead us to infer many deletions instead of one long deletion, and would thereby overestimate the probability of deletion (and similarly for insertion).

Thorne, Kishino, and Felsenstein (1992) attempted to remedy this by assuming that it was not individual bases, but instead multibase segments that were inserted and deleted. There could be a distribution of lengths of these segments. This model has an obvious limitation: Once a segment has been inserted, if it is deleted thereafter, the whole segment must be deleted. There is no event that can delete part of a segment once it has been inserted. Thus overlapping gaps cannot be handled properly. Metzler (2003) has used a version of the TKF model which he calls the *fragment insertion and deletion* (FID) model. It takes the insertion and deletion rates to be equal. It is not clear that equality is tenable, as the resulting model of sequence-length variation then has no equilibrium distribution, and reversibility of the process is not guaranteed.

Tree HMMs

Another approach to gaps uses the *tree HMMs* of Mitchison (1999; see also Mitchison and Durbin, 1995). These are arrays of pointers that specify which bases in a sequence will be deleted. Figure 29.4 shows one of these arrays. At each site, it has two states, "match" (M) and "delete" (D). For each state there is a pointer that indicates whether its successor state is a D or an M. At the start of the sequence there is also such a pointer. At the M states, there are bases (or amino acids, if this is a protein). These change according to one of the usual substitution models. The pointers also change direction according to simple Markov process models with two states (along and across) as one moves through time. The sequence at any given time is determined by the pointers and by the bases that are associated with the M states. In the figure, the sequence shown is (start)MMDDM(end). This might correspond to the sequence `GA--T`. In evolution, the bases and the pointers change according to Markov processes. If the second M were to have its pointer change, the states would then be MMMMM and the sequence might then be `GACTT`.

This model has the advantage that it can have deletions that are more than one base long. Even so, it suffers from some lack of realism, because when a group of adjacent bases is deleted, the bases retain information (in the M states) about the base sequence, and if they are inserted again, there will be some memory of the original base sequence. Thus when two adjacent As are deleted, if bases are reinserted there soon after, they are quite likely to turn out to be As. Holmes and Bruno (2001) point out that it is also possible that when a series of bases is reinserted there may be "memory" of an internal gap that was once there and that now returns with them.

Tree HMMs do not have the convenient mathematical properties that enable the TKF 1991 single-base insertion and deletion model to be used in a dynamic programming algorithm. This is needed to sum the likelihood over all alignments. Mitchison (1999) uses a Bayesian MCMC method to sum over all alignments instead. If TKF models could be extended to multibase insertions and deletions, they would have the advantage. Otherwise, MCMC methods would be needed and tree HMMs would then be strong competition.

Trees

Multiple-sequence alignments cannot be properly inferred without taking into account the phylogeny of the sequences. Thorne and Kishino (1992) have used the model of deletion and insertion of segments for all pairs of sequences, computing a maximum likelihood estimate of divergence time $\widehat{t}$ for each pair of species. The other parameters are estimated for each pair. The median of the estimates of each

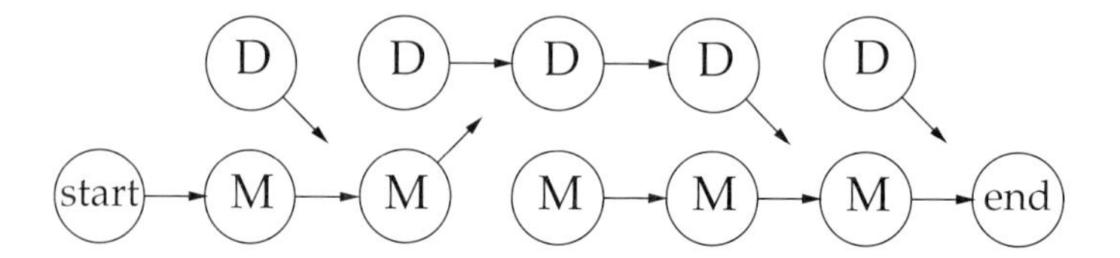

Figure 29.4: Mitchison's tree-HMM, which has two states, "match" (M) and "delete" (D), at each site. Each state has a pointer that designates its successor, and each of these pointers is reoriented by having its own simple two-state stochastic process as one moves along the tree. The effect is to introduce and remove deletions that may be more than one base long.

parameter is calculated and these are used in a reanalysis of the pairwise values of $\hat{t}$. These can be treated as distances in a distance matrix method. They use bootstrap sampling of the resulting distances, but as they are perturbing each distance independently, they cannot take the covariances of distances into account. They thus describe their bootstrap values as being quite conservative, as those correlations would cause less disruption to the tree if taken into account.

Another approximate approach, discussed in some detail by McGuire, Denham, and Balding (2001) is to simplify the insertion/deletion models by treating the "gap" symbol as a fifth base in the sequence. This allows the insertion/deletion processes to run separately at each site, and it greatly simplifies likelihood computations. However, this approach has some unrealistic properties. If three neighboring bases are deleted (by each changing to the gap state), there is then a "memory" of the original size of the sequence in this region. Insertion in the region cannot then put in more than three bases.

Some efforts have been made to compute likelihoods for trees under the TKF model with single-base insertion and deletion. Steel and Hein (2001) worked out algorithms for computing the likelihood on a completely unresolved "star" phylogeny. Subsequently, Miklós (2002) gave an improvement that speeds up this calculation. Hein (2001) gave algorithms for a bifurcating tree that would be useful for up to about seven or eight species. Holmes and Bruno (2001) have used methods related to those of Hein et al. (2000) to reconstruct alignments on a given binary tree, including reconstruction of the sequences at interior nodes and their alignment to the observed sequences. They also allowed sampling from the distribution of interior node sequences (and hence of alignments) proportional to their contribution to the likelihood. They describe this as Bayesian inference, though as their parameters are not given a prior, there is some question as to whether it should be thought of this way.

These are important steps in the development of a reasonably realistic likelihood-based multiple sequence alignment method. They still lack a good way of taking multibase insertions and deletions, including overlapping ones, into account. This is one major task for the immediate future (another is allowing for regions of the sequence to be more or less conserved). If the proper transition probabilities cannot be calculated, it will be necessary to use MCMC methods that avoid them, by having the individual events all represented in the tree and sampled. Such an approach can make use of complicated, and even bizarre models, but it may be difficult to make them "mix" adequately, requiring long runs to explore the space of possible combinations of events.

The statistical alignment literature is currently quite active. If these problems can be solved, the use of probability models of evolution in multiple sequence alignment will rapidly move from its present genteel obscurity and become central to all alignment inferences.

Inferring the alignment

The reader may have been under the impression that I was discussing how to infer the alignment. However, all of the probabilistic methods mentioned have summed over all alignments. The likelihoods are maximized with respect to the unknown parameters and the divergence times, and the alignment disappears from the inference by being summed out, because it is not a parameter of the model. The probabilistic methods have thus eliminated the alignment from consideration. After the maximum for the parameters is found, we may want to know which alignment makes the largest contribution to the likelihood. However, it must be kept in mind that even when an alignment maximizes this contribution, it may account for a very small fraction of the likelihood (or, in the case of Bayesian inference, a very small fraction of the posterior probability). Both the TKF 1991 model and tree HMMs can be made to find the alignment that makes the maximum contribution. I have mentioned above papers by Allison and Wallace (1994), Thorne and Churchill (1995), Metzler et al. (2001), and Holmes and Bruno (2001), which sample from the distribution of possible alignments. Seeing some of this variation may be an effective antidote to overreliance on the single best-fitting alignment.

In the case of likelihood, it is somewhat unclear whether we ought to refer to the inference of the alignment as maximum likelihood estimation, since the alignment is not, strictly speaking, a parameter.

Gene families

Whole loci can duplicate or be lost as well as individual bases or amino acids. Consideration of the history of gene duplication involves inferences about gene families. The analysis of families of genes goes back to some of the earliest work on molecular evolution. Among the first protein sequences available were globins, and multiple globin genes were sequenced in humans. Zuckerkandl and Pauling (1962) considered differences between globins. As some of the comparisons they made were between members of a gene family within a species, these involved gene trees rather than phylogenies.

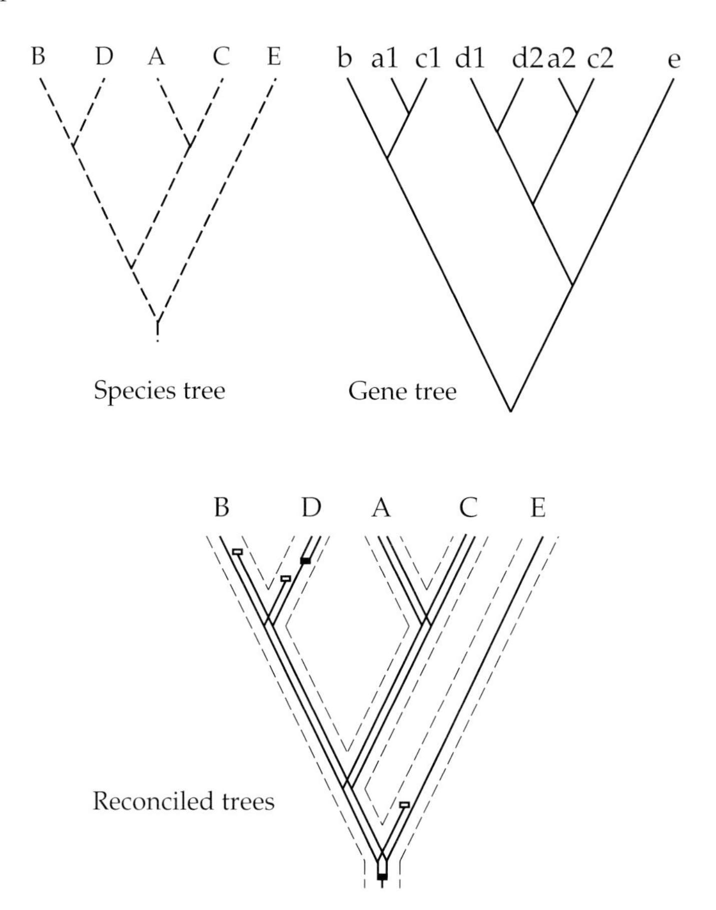

Figure 29.5: A species tree (upper left), a gene tree (upper right), and the two trees reconciled (bottom, species tree in dashed lines). Gene duplications are shown as filled boxes, losses as open boxes. There are two duplications and three losses.

Reconciled trees

Goodman et al. (1979) were first to investigate algorithms for reconciling gene trees with species phylogenies. They proposed finding that combination of species tree and gene tree that minimized the total of the numbers of substitutions, gene duplications, and gene losses. The issue is illustrated here in Figure 29.5. Page and Charleston (1997) have discussed reconciled trees, assuming that the gene tree and the species tree are both known. They use a tree-mapping algorithm of Page (1994b) to count the number of duplications and losses necessary to reconcile the trees. Like Goodman et al., they envisage searching for the species tree that re-

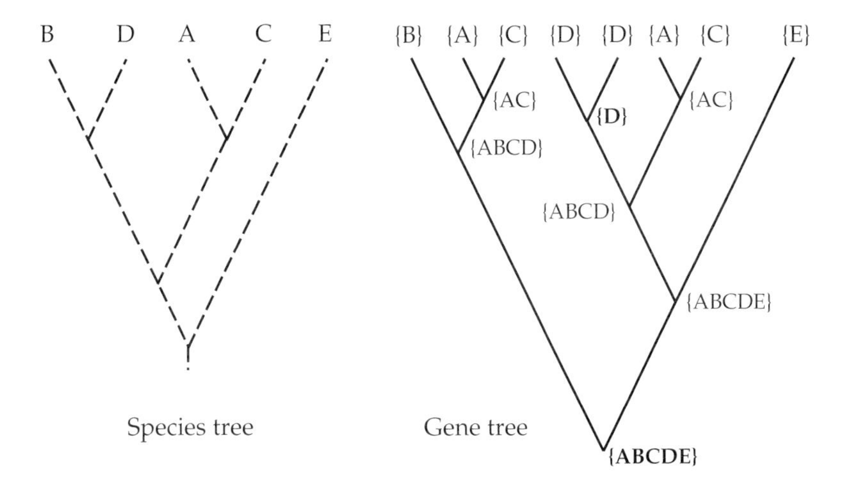

Figure 29.6: The same gene and species trees as in Figure 29.5, with the construction of sets at the interior nodes of the gene tree shown. Each is the set of species descended from the latest common ancestor in the species tree of the union of the two sets above that node. A gene duplication is inferred if a node has a set equal to the sets in either of its two immediate descendants. (These cases are boldfaced.)

quires the fewest total duplications and losses to account for a set of gene trees observed on these same species. Mirkin, Muchnik, and Smith (1995) gave a different algorithm to compute the same quantity. Zhang (1997) and Eulenstein and Vingron (1998) have proven that the two algorithms give the same result. Guigó, Muchnik, and Smith (1996) have discussed methods for attributing duplications at individual loci to a few duplications of the entire genome.

An advantage of the original approach of Goodman et al. is that it allowed for uncertainty in the gene tree, by adding to the criterion the number of changes in the sequences and then optimizing over the shapes of both species trees and gene trees. What we have called losses, they referred to as gene expression events, as they saw them as being either deletions or losses of detectable gene expression.

Reconstructing duplications

The reconciled-tree approach requires us to be able to count the number of duplications and losses (or just the number of duplications) for a given rooted gene tree and a given rooted species tree. The algorithm for bifurcating rooted gene and species trees is relatively simple in its conception. Figure 29.6 shows the process for the case of Figure 29.5. We start by replacing the tip labels on the gene tree by a set containing the species that the gene copy appears in. Then we work down the gene tree (a postorder tree traversal). At each node we construct a set that is the union of the sets in the two nodes above it. Thus the upper-leftmost set has {AC}, as {A} and {C} are above it.

These sets are then augmented by reference to the species tree. The latest common ancestor of all of the species in the set is found, and the list of species descended from that ancestor. For {AC} we do find a node on the species tree that is the latest common ancestor of A and C. It has only these two species descended from it. So the set is not augmented by adding any additional species. But for the set immediately below it, we initially have {ABC}. The latest common ancestor of these species also has D descended from it. The set thus becomes {ABCD}. We continue down in this way until we fill in sets on all interior nodes of the gene tree.

Gene duplications are then inferred for any node that has a set that is the same as one or both of the sets in its immediate ancestors. In this example, there are two such cases. At the bottommost node, {ABCDE} is identical to the set in its immediate right descendant. Further up, an interior node has set {D}, and so do both of its immediate descendants. We infer that these two forks in the gene tree are duplications, as we can see in Figure 29.5.

The algorithmics involved in doing this efficiently have led to a number of papers. Page (1994b) gave an algorithm whose time would be of order n^3, where n is the number of genes in the gene tree. Zhang (1997) and Chen, Durand, and Farach-Colton (2000) improved this to be proportional to n. Zmasek and Eddy (2001) presented a much simpler algorithm that can be slower in the worst case, but is usually faster than these methods. Hallett and Lagergren (2000) showed that the search for the most parsimonious species tree based on duplications or duplications and losses requires only polynomial time, if the maximum number of duplications needed is bounded by some number k.

Reconstructing the number of losses is also straightforward, given the sets at the interior nodes (cf. Ma, Li, and Zhang, 2000). It is interesting that the reconstruction of the number of duplications is the same whether we are trying to minimize the number of duplications or the number of duplications plus losses.

Rooting unrooted trees

The algorithms for reconstructing the numbers of duplications and losses use rooted gene trees and rooted species trees. Unless a molecular clock is used to root them, gene trees will typically be unrooted. To find the best rooted species tree, one can root the gene tree in all possible places and, for each such rooting, evaluate possible species trees. If there are many gene trees in the same analysis, the number of possible combinations of their rootings will be too large to make this method practical. Chen, Durand, and Farach-Colton (2000) discuss the algorithmics involved.

Duplications have been used to root trees that have no useable outgroups. The most notable of these is the Universal Tree of life, for which a rooting between the Bacteria and the {Archaea, Eukarya} clade has been proposed by Gogarten et al. (1989) and by Iwabe et al. (1989). Figure 29.7 shows three rooted trees for the three

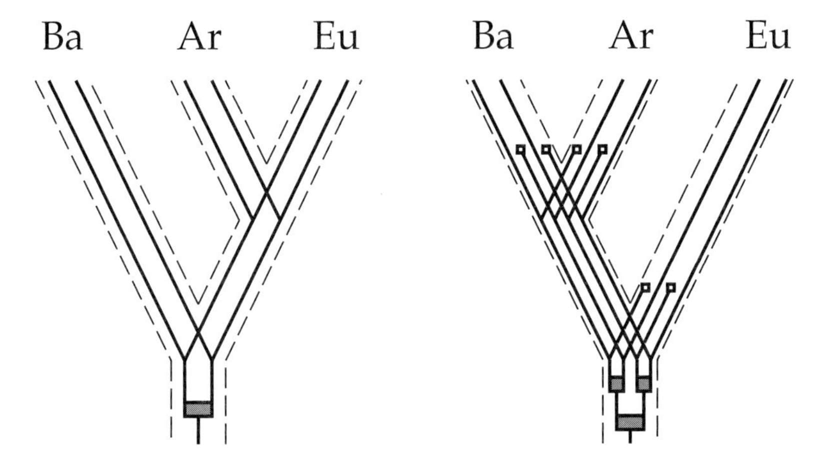

Figure 29.7: Two rooted trees for the domains Bacteria (Ba), Archaea (Ar), and Eukarya (Eu), showing Iwabe et al.'s (1989) gene tree for the EF-TU/1a and EF-G2 family of proteins reconciled with each. Shaded boxes represent duplications, open boxes represent losses.

domains, with the tree for the gene family of EF-TU/1a and EF-G2 reconciled with each. The first tree requires one duplication. The second requires three duplications and six losses. This speaks strongly in favor of the first tree. The tree that unites Bacteria and Eukarya has the same reconstruction as the tree on the right, but with the Ar and Eu labels exchanged. The rootedness of the trees inferred from duplications comes from the assumed irreversible nature of a duplication event. Genes can be lost, but we do not allow events that truly reverse the duplication, by merging two genes. Thus "the outgroup is rocks"[1] because we assume that the ancestral state is the absence of a duplication. This is equivalent to outgroup-rooting with the state of the outgroup—rocks—taken to be the absence of the duplication.

To make the search for the species tree easier, Simmons, Bailey, and Nixon (2000) have introduced a coding method *uninode coding*, for use in the presence of gene duplications. They start with a gene tree and infer the locations on it of the duplication events. Then they take each pair of duplicate genes as separate loci, which they do not align to each other or to the original copy. They infer the sequence of the locus at the moment of duplication. This sequence is assigned to the original locus after the duplication and is also assigned to the duplicated loci before the duplication. Adding one more character, a 0/1 character indicating the presence of the duplication, they use parsimony to reconstruct the species phylogeny. A difficulty with this method is the need to reconstruct a single un-

[1] A phrase I uttered in a discussion at the Molecular Evolution course at the Marine Biological Laboratory in Woods Hole, Massachusetts, in the mid-1990s. It was meant seriously but is usually assumed to be a joke.

ambiguous sequence at the moment of duplication. It is not obvious that this reconstruction can always be done.

The difficulty in doing such an analysis is illustrated by the ambiguities connected with the words "paralogy" and "orthology." When a gene duplicates, the two copies are paralogs and should not be aligned. But each is orthologous to the original copy, and if it is present, both should be aligned to it. The obvious impossibility of following these dictates underlines the need for methods that involve reconciliation of the gene and species trees.

A likelihood analysis

Lindsey Dubb (*in prep.*) has made a likelihood analysis of a model with base substitution, gene duplication, and loss. The gene tree exists within the lineages of the species tree, as in Figure 29.5. The tips and forks of the species tree have times, and a birth-death process is assumed, to account for the duplication and loss events. This has the usual rate parameters λ and μ. The individual members of the gene family are duplicated or lost at constant rates, independently of one another. This birth-death process model of gene duplication has been proposed by Nei and Hughes (1992). It has the limitation that it has no surveillance by natural selection to ensure that some functional locus in the family remains in the genome, or to ensure that too many functional loci are not produced. But its mathematical tractability will ensure it a continuing central role in models of gene duplication and loss.

In addition, the sequences are assumed to change by a standard DNA model. Dubb computes the likelihood of the species tree by integrating over all possible gene trees:

$$L \;=\; \text{Prob}\,(\text{Data}\,|\,\lambda,\mu) \;=\; \int_G \text{Prob}\,(\text{Data}\,|\,G)\;\; \text{Prob}\,(G\,|\,\lambda,\mu)\; dG \qquad (29.2)$$

where G is the gene tree. The two terms inside the integral are easily worked out: The first is the usual likelihood for DNA sequences on a tree, and the other is the probability computed by the birth-death process for this particular gene tree. It is more complex, involving a calculation summing over all possible number of loci that could have existed at each interior node of the tree. That includes loci all of whose descendants were subsequently lost.

The integration requires use of Markov chain Monte Carlo methods, in this case the Metropolis-Hastings method, which we saw in Chapter 27. The result is a likelihood surface for the parameters λ and μ. Although the method considers many gene trees, it does not deliver a single estimate of the gene tree. At first glance, this may seem unsatisfactory, but it is λ and μ that are the parameters of the model, not the gene tree. Estimation of these rates, which are the rates of duplication and loss, can be very noisy unless they can be assumed to hold for multiple gene families and information accumulated over them. It will also be possible to test whether these rates are different between different categories of genes.

If one uses statistical methods like Dubb's that integrate over our uncertainty about the gene tree and estimate birth and death parameters, it will be possible to do more complete analyses of duplication and loss rates. Current methods (cf. Lynch and Conery, 2000) examine only duplicates within species; they cannot take into account the nonindependence of comparisons in different species when the duplication has occurred in their ancestor.

The greatest strength of these statistical methods is that they can adequately take into account the uncertainty of the gene tree, which parsimony-based reconciled tree methods will not. Of tree-based approaches to gene family evolution, parsimony-based approaches are better-known at this point, but as with many other topics in this book, it is the probabilistic statistical methods such as likelihood that hold the promise for future work. In the interim, one could at least use resampling methods such as the bootstrap, and analyze the cloud of resulting estimates of the gene tree. Even this is not done at present. Users of reconciled tree methods may have to break themselves of the habit of first inferring the gene tree and then treating this inference as if it were data. This is closely similar to the unfortunate way single-tree methods treat coalescent trees.

Comparative genomics

The models of gene duplication and loss have one major limitation. They implicitly assume that each gene family's duplications and losses happen independently. As the loci examined become denser in the genetic map, deletions and duplications will be seen to involve more than one gene locus at a time. Models that allow each locus to be duplicated or lost separately will be increasingly inadequate as the number of gene families considered increases.

So far there is no model or method that involves duplications and deletions in a map as well as change in the DNA sequences. There is a literature involving rearrangement of genetic maps, and even efforts to use these models to infer phylogenies. Most of the work has come from the laboratory of David Sankoff.

Tandemly repeated genes

One step in the direction of genomic analysis has been taken in the literature on tandemly repeated genes. Fitch (1977) was the first to notice that duplication by unequal crossing-over in a tandem gene family would constrain the topology of the resulting gene tree. He derived rules allowing a tree to be checked to see whether it was consistent with the unequal crossing-over mechanism. The problem was independently rediscovered by Elemento, Gascuel, and Lefranc (2002; see also Benson and Dong, 1999, and Tang, Waterman, and Yooseph, 2002). Gascuel et al. (2003) counted the number of possible trees consistent with the mechanism.

As interesting as these papers are, they analyze only the duplicates within a single species. They are just barely concerned with genomics, since they analyze only a small piece of the genome.

Inversions

One of the first models of evolution of a genomic map was rearrangement of a linear or circular genetic map using only inversions. Watterson et al. (1982) were the first to pose the problem of how many inversions it takes to get from one map to another (they considered the case of a circular map). Kececioglu and Sankoff (1995) developed some approximation algorithms and conjectured that the problem was NP-hard. Caprara (1999) proved that it is. However, a slight change in the problem makes it much more tractable. The inversions ("reversals") problem assumes that we characterize each map by the list of its markers, in order. If we assume that we also have the orientation of each marker (for example, that we know, for each, which end is the start of the protein) then we have the problem of signed reversals. As complete sequencing of genomes is done, it yields precisely this kind of evidence. Hannenhalli and Pevzner (1995, 1999) found a polynomial-time algorithm to compute the minimum number of inversions (reversals) needed. Berman and Hannenhalli (1996) and Kaplan, Shamir, and Tarjan (1997) have given algorithms for computing the minimum number of inversions in time linear in the number of markers (see also Bader, Moret, and Yan, 2001).

Inversions in trees

With the problem of counting changes between two ends of a branch proven NP-complete, it is even more difficult to infer the states of common ancestors and to count the number of changes needed on a given phylogeny. Hannenhalli et al. (1995) and Sankoff, Sundaram, and Kececioglu (1996) gave some algorithms for finding gene orders at interior nodes in a small tree. Algorithms like these can form the basis of a heuristic parsimony algorithm. Bourque and Pevzner (2002) have applied the reversal distances to searching for a most parsimonious set of reversals explaining a multispecies data set.

Inversions, transpositions, and translocations

Sankoff (1992) investigated the broader question of how many inversions, transpositions, and translocations are needed to transform one genome into another. Transpositions delete a block from one chromosome and insert it elsewhere, while translocations exchange the terminal sections of two chromosomes. Hannenhalli (1995) showed for the case of signed genomes that the minimum number of translocations necessary could be computed in polynomial time. Blanchette, Kunisawa, and Sankoff (1996) presented a branch-and-bound algorithm to compare two genomes by minimizing a weighted combination of the numbers of inversions, transpositions, and translocations. Bourque and Pevzner (2002) described a faster algorithm that proposes inversions, transpositions, fissions, and fusions that bring neighboring genomes in a tree closer to one another. These are used to update interior node genomes in the tree, as part of a broader search among tree topologies. Wu and Gu (2003) described another heuristic parsimony method.

Sankoff (1999) has discussed the difficulties of taking gene duplication into account in models of rearrangement of genomes. If events that cause duplications are not taken into account, families of duplicated genes can be represented by a single exemplar locus. He discusses the algorithmics of choosing which copy is the best exemplar.

Breakpoint and neighbor-coding approximations

As the full parsimony problem is computationally difficult, approximations are needed. Blanchette, Bourque, and Sankoff (1997) suggested the use of a count of the number of breakpoints in transforming genomes along branches of a tree. A breakpoint is a pair of adjacent markers in one genome that are not adjacent in another. Figure 29.8 shows a set of four genomes of 10 loci, generated on a phylogeny by four inversions, together with the table of presence and absence of breakpoints. The loci are numbered 1–10, based on their order in genome A, as we might number them if we knew only the genomes A–D. Note that a breakpoint such as 1 : 2 is considered to be the same as –2 : –1.

Solving for the tree with the fewest breakpoints is not the same as optimizing a weighted combination of rearrangements, but it preserves some of that information and is faster. The number of breakpoints between two genomes can be rapidly computed for a pair of genomes, but the problem is not as simple for a phylogeny. Blanchette, Bourque, and Sankoff introduced a heuristic method involving solving multiple traveling salesman problems to infer breakpoint patterns at interior nodes of the tree, until there is no change in the number of breakpoint changes in the tree. The number is then used to choose among phylogenies in an exhaustive search of trees. Cosner et al. (2000; Moret et al., 2001) have used the binary presence/absence coding for breakpoints. They then used standard parsimony heuristics to search for the best tree. Gallut, Barriel, and Vignes (2000, Gallut and Barriel, 2002) have used a modified breakpoint coding with states that represent a marker with its two neighbors, assuming unordered parsimony change between these states. In inferring hypothetical ancestors within the tree, they retain only those combinations of states that would yield a full genome.

Synteny

Another, even rougher, approximation is to score only whether or not genes are present on the same chromosome, without regard to where they actually are on it. Nadeau and Sankoff (1998) review work on *synteny*. Ferretti, Nadeau, and Sankoff (1996) developed a method for estimating the minimum number of events—reciprocal translocations and Robertsonian fissions or fusions—needed to change one synteny pattern into another. DasGupta et al. (1997) showed that

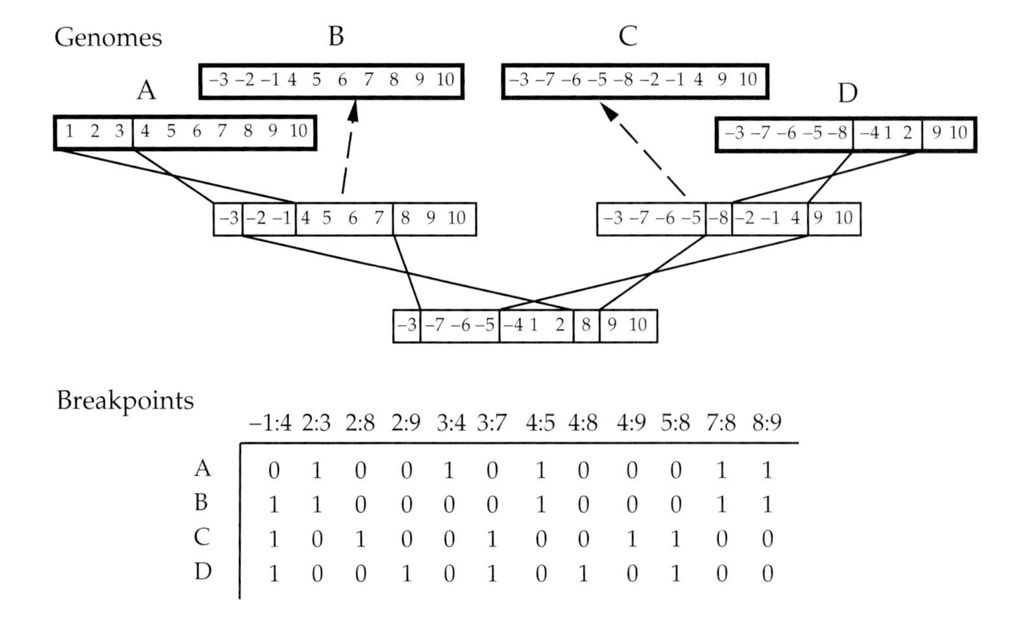

	−1:4	2:3	2:8	2:9	3:4	3:7	4:5	4:8	4:9	5:8	7:8	8:9
A	0	1	0	0	1	0	1	0	0	0	1	1
B	1	1	0	0	0	0	1	0	0	0	1	1
C	1	0	1	0	0	1	0	0	1	1	0	0
D	1	0	0	1	0	1	0	1	0	1	0	0

Figure 29.8: Four 10-locus genomes generated from a common ancestor by inversions (which are shown) and the table of breakpoints that corresponds to the observed genomes (the ones with darker borders and names A–D). The breakpoints table shows for each pair of markers in which genomes they will be found to be adjacent. Pairs that are always adjacent are omitted. Numbering of markers is relative to genome A.

this problem is NP-hard. In general, synteny alone is less and less used, as more detailed maps showing locations of loci become ever more widely available.

Probabilistic models

However, all the methods above are parsimony methods. They are subject to the usual worries about statistical good behavior. Sankoff and Goldstein (1989) were the first to make a parametric model of rearrangement of a map (using only inversions and not allowing deletion or duplication). Sankoff and Blanchette (1999) have made a start on statistical analysis of these models by using only the breakpoints information and computing phylogenetic invariants for a model of breakpoints independently arising and disappearing in a model of unsigned inversions ("reversals").

It seems likely that to make enough progress on semi-realistic models of genome rearrangement, statistical inference can be carried on only by using Markov chain Monte Carlo methods. These could incorporate moderately realistic (and therefore moderately ugly) models. Larget, Simon, and Kadane (2002) have made the first MCMC model of genomic change, for mitochondrial gene order that is assumed to change by inversions. Their Bayesian MCMC reconsidered both gene orders and phylogenies.

It is probable that parsimony methods will do better in the case of gene orders than they do for aligned molecular sequences, because the chance of parallel change or reversion is much lower in the gene order case. The extra computational difficulties in doing parsimony on gene order come from the large size of the space of possible states. That in turn makes parallelism and reversal less probable than in molecular sequences. Parsimony may behave better on gene orders than on nucleotide sequences, so that there is less need to do a full probabilistic treatment.

Genome signature methods

One cannot leave the topic of comparative genomics without treating *genome signature methods*, though it is hard to know where they fit in. In fact, this uncertainty is an indication of their importance. These methods originated in the work of Gibbs et al. (1971) on alignment-free methods for detecting similarity between protein sequences. They tabulated, for each protein, the frequency of pairs of adjacent pairs of amino acids. Thus if Serine (S) is followed by Tryptophane (W) twice in one protein, they placed a 2 in the 20×20 table of amino acid pairs. They then used a clustering method to make trees of protein sequences.

This did not involve any alignment step. If one protein had two SW pairs, and so did another, but these did not occur at corresponding positions, these two proteins would nevertheless appear similar. There is thus some necessary loss of information, as we must forego the information provided by the alignment.

Blaisdell (1986) used adjacent pair and triple frequencies to construct a phylogeny, his measure of difference of sequences being derived from the significance of a chi-square test of homogeneity when the sequences were combined. From a simulation, he inferred a relationship (Blaisdell, 1989a) between his measure and alignment mismatch scores, and he tested the effectiveness of these alignment-free methods compared to methods that used alignments (Blaisdell, 1989b).

Karlin and Ladunga (1994) used dinucleotide relative abundances

$$\rho_{XY} = f_{XY}/(f_X f_Y) \tag{29.3}$$

computed from the base frequencies f_X and the dinucleotide frequencies f_{XY}, which correct for differences in base composition between genomes. They also developed a distance between arrays of dinucleotide relative abundances. Karlin, Ladunga, and Blaisdell (1994) also developed an analogous relative abundance for trinucleotides.

In ordinary models of random change of nucleotides, we would expect the dinucleotide relative abundances to fluctuate randomly, and we would expect related species to share some of those fluctuations. There would be phylogenetic signal, but there would also be a large loss of power owing to not using alignment information. It is when the model of evolution is not one of the standard ones that these genome signature methods could prove valuable. They may indicate departures from conventional models, involving genomically significant events such as spread of transposons. Thus the genome signature methods could serve as a valuable exploratory tool where conventional models would fail. The uncertainty about where they fit in this chapter comes from their lack of reliance on the processes that we normally assume to be present. It is precisely because they do not rely on detailed assumptions about base substitution processes or genome rearrangement processes that they are of interest.

Chapter 30

Consensus trees and distances between trees

The development of phylogenetic inference led quickly to situations in which a researcher had two or more trees for the same group, often from different types of data. The issue then arose as to how to put those together to get one overall estimate of the tree. Alternatively, how can one describe the extent of difference between the trees? These tasks are, respectively, the computation of consensus trees and of distances between trees. We consider them in turn and then return to the issue of whether one ought to use these methods. The article by Swofford (1991) contains an excellent introduction to this subject; Bryant (2003) has a good survey of consensus tree methods from a more mathematical viewpoint.

We have already introduced the majority-rule consensus tree in Chapter 20, but we will introduce it again here, in a somewhat more generalized treatment.

Consensus trees

Consensus trees are trees that summarize, as nearly as possible, the information contained in a set of trees whose tips are all the same species. Figure 30.1 shows a set of trees whose consensus we want to compute.

Note the differences between the trees. The first tree differs from the second and third trees. Those two trees are identical to each other. The differences between the first tree and the others is entirely in the group BDEF.

For any consensus tree method or tree distance it is important to remember whether the trees are to be regarded as rooted or unrooted. With an unrooted consensus tree method and two trees that differ only in the placement of their root. the two trees are considered to be identical. When the consensus tree method is rooted, they are considered to be in conflict.

Figure 30.1: A set of trees over the same species, which will be used as the example for consensus methods and tree distances in this chapter.

Strict consensus

The *strict consensus* method (Rohlf, 1982) is the simplest. It constructs the tree that contains all groups that occur on all trees. "Groups" means monophyletic groups if the trees are regarded as rooted. If they are regarded as unrooted, each branch is regarded as creating a partition of the species, and instead of monophyletic groups we consider the presence or absence of the partitions. The first tree in Figure 30.1 has a branch that separates group BEDF from the rest of the tree; if this were an unrooted tree, this branch would be considered to define the partition {ACG | BDEF}.

In the example of Figure 30.1, the trees can be considered to be rooted. (They might have been rooted by using G as an outgroup.) Figure 30.2 shows their strict consensus tree. Three of the groups are present in the strict consensus tree.

However, the strict consensus is too strict for most purposes. Consider, for example, the two trees in Figure 30.3. They are nearly identical, differing only in species A being moved. Yet there is no monophyletic group (or, considered as

Figure 30.2: Strict consensus of the trees in Figure 30.1.

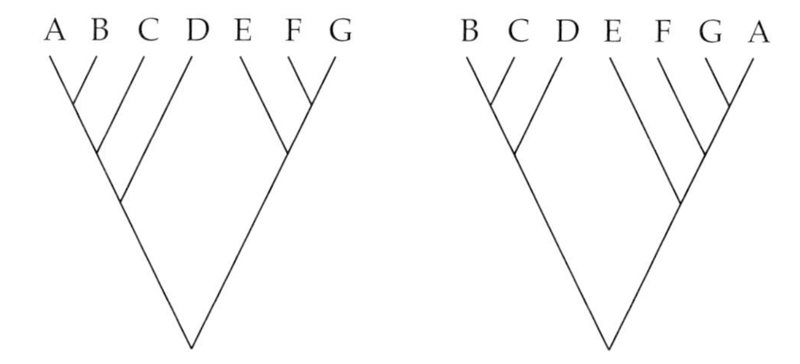

Figure 30.3: Example of the limitations of strict consensus trees. These two trees differ only in the placement of species A and are otherwise quite similar. Yet their strict consensus tree is completely unresolved, because none of the sets of species that have B, C, or D on the right tree have species A and all of the sets of species on the right tree that have E, F, or G also have species A in them.

unrooted trees, partition of the species) that is present on both trees—the strict consensus is a totally unresolved tree. There has therefore been an effort to find methods that retain some of the information about common structure without being as rigid as the strict consensus method.

Majority-rule consensus

We have already been introduced to the majority-rule consensus tree in Chapter 20. You will recall that it is the tree consisting of those groups that are present in a majority of the trees whose consensus is being taken. Figure 30.4 shows the majority-rule consensus tree for the example of Figure 30.1. In fact, there is an entire family of consensus tree methods that includes the majority-rule consensus and the strict consensus. These are the M_ℓ consensus trees defined by Margush and McMorris (1981). The parameter ℓ is a percentage that ranges from 50% to 100%. The corresponding consensus tree method constructs a tree that contains all those groups that occur more than that percentage of the time. (If the percentage is 100, we take those that occur 100% of the time.) It will be clear that M_{50} is the majority-rule consensus tree, and M_{100} the strict consensus tree. M_{80}, for example, is the tree containing all groups that occur more than 80% of the time. All of these methods avoid putting two groups that might conflict on the tree, because they all ensure that among the input trees there is at least one that contains both groups, so that the two groups cannot conflict.

In the example in Chapter 20, the majority-rule consensus tree was fully resolved, but was not the same as any of the five fully resolved trees for which it was their consensus. It contained three groups, but none of the input trees had all of those three groups. In the present example, there are five groups, and the

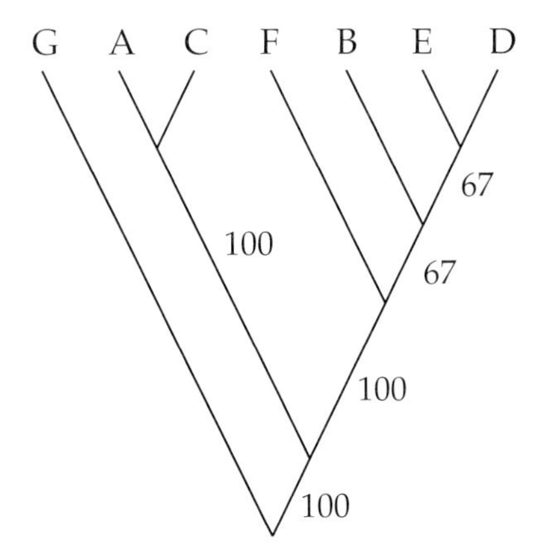

Figure 30.4: The majority-rule consensus tree for the example of Figure 30.1.

rightmost two trees of Figure 30.1 are trees that contain all of them. In Figure 30.4, I have put numbers next to each of the branches, indicating which fraction of the time each branch is present among the input trees. The ones that have 67 are present two-thirds of the time. We can immediately see what the M_ℓ family will look like: When $\ell \leq 66$ the M_ℓ consensus tree will be identical to the majority-rule consensus tree. For ℓ above 66, the M_ℓ consensus tree will be identical to the strict consensus tree. It will always be the case that the members of the M_ℓ family for higher values of ℓ can be constructed by collapsing branches.

Adams consensus tree

The *Adams consensus tree* (Adams, 1972, 1986) is less strict than the strict consensus, while not relying as majority-rule consensus does on "taking a vote." If we add an extra copy of one of our trees, the majority-rule consensus result may change, while the Adams consensus will not. This makes Adams consensus of great interest, though, as we shall see, it has one major limitation. The basic idea of the Adams consensus tree is to find in all of the trees all three-taxon statements. As was mentioned in Chapter 12, three-taxon statements are triples of three species that show two of the species to be more closely related than is the third. The Adams consensus tree is formed by finding all those three-taxon statements that are not contradicted by any of the trees, and then forming a tree from them. For example, in our example of three trees in Figure 30.1, the triple of species ((D,E),B) appears in that order in all three trees. In other words, if we were to eliminate all species except those three, all three trees would have the same three-species tree for these three species. The Adams consensus tree will contain the group (D,E), showing that those two species are nested relative to species B. However, the triple of species BEF does not have any nested topology that is shared by all three trees.

Figure 30.5: The Adams consensus tree for the example of Figure 30.1.

In the first tree, the three-taxon statement for these species would be (B,(E,F)) and in the remaining two trees it would be (F,(B,E)). Thus F, B, and E are represented in the Adams consensus tree as being a trifurcation. If one of the input trees had a trifurcation for these three species, this would be counted as not conflicting with any three-taxon statement.

Figure 30.5 shows the Adams consensus tree for this case. The group DE shows up in the Adams consensus tree, but does not appear in the strict consensus tree, as in the first tree it is not present as a monophyletic group. The group BED shows up in the majority-rule consensus tree, as it is present in two-thirds of the trees. It does not show up in the Adams consensus tree, as neither the three-taxon statement (F,(B,D)) nor the three-taxon statement (F,(B,E)) is present in all of the trees whose consensus is being taken.

You may have detected the limitation of the Adams consensus tree already. Three-taxon statements have no meaning unless we have a rooted tree. The Adams consensus tree is defined only for rooted trees, and there is no counterpart for unrooted trees. One might think that we could define an unrooted Adams consensus tree method by rooting the tree at a particular species (say, species #1), constructing the Adams consensus tree of the resulting rooted trees, and then considering the resulting consensus tree as an unrooted tree. This does not work because the result turns out to depend on which species is taken as the root! We may obtain different unrooted consensus trees when using species #1 as the root than we do when using species #2 as the root.

A dismaying result

We might wonder whether there is some way to define a consensus method parallel to the Adams consensus method, but for unrooted trees. Instead of requiring that we make a consensus of all of the three-taxon statements, we would want to make a consensus of quartet statements. Thus if all of the trees have the quartet ((I,J)(K,L)) in them, this quartet should also appear in the consensus tree. Steel,

Dress, and Böcker (2000) have proven that there cannot be such a method. They make three requirements for any reasonable consensus tree method. One is that relabeling of the species at the tip of the tree should yield the same result, with the species appropriately labeled. The second is that the result should not depend on the order in which the trees are input. The third is the requirement that a quartet that appears in all trees also appear in the consensus tree.

They are then able to show a simple example in which the third requirement cannot be satisfied if the first two are assumed to hold! It is the two unrooted trees ((a,b),(c,d),(e,f)) and ((b,c),(d,e),(f,a)). There therefore cannot exist a satisfactory analogue to the Adams consensus tree method for unrooted trees. There does not seem to be much that we can do about it except grumble.

Consensus using branch lengths

There has been less attention to formulating consensus tree methods that use the branch lengths of the tree as well as the topology. One generalization of majority-rule consensus trees would be to take the mean or the median of all the lengths of a particular branch. But what to do when the branch is not present? I suggest that we count each branch as being of length 0 when it is not present, and otherwise use its length. The length assigned to this branch could then be the median of these quantities. If the branch is always present, that means that the length assigned it in the consensus tree will simply be its median length. If it is absent some of the time, those trees count as having 0 for this branch length. Thus if there are 11 trees, four of which do not have the branch, and the others have it present with lengths 0.1, 0.2, 0.2, 0.3, 0.34, 0.4, and 0.5, we need to take the median of the 11 quantities 0, 0, 0, 0, 0.1, 0.2, 0.2, 0.3, 0.34, 0.4, and 0.5. This is 0.2. The branch will then be listed as present in the consensus tree, with length 0.2.

If there are an even number of trees, exactly half of which have the branch present, we must be careful. In order to avoid making a "tree" with contradictory branches in it, we must count the branch as absent. Following the rule that any branch whose median length is 0 is absent, we would then have to redefine the median as the smallest value that has half the branch lengths less than or equal to it. To be consistent with this, if we have (say) four branches of length 0.1, 0.2, 0.3, and 0.4, the median will be taken to be, not 0.25, but 0.2.

This consensus tree method could be called the median branch length (MBL) consensus method. If we consider only the resulting tree topology, it will always be the same as the majority-rule consensus tree. Note that this method tends to shorten branches when they are absent some of the time in the input trees. It is not self-evident whether this is a good thing to do. If not, we could take the average branch length only over all those that have the group present.

Other consensus tree methods

There are other consensus tree methods. Some are listed here; for more information about them see also the excellent survey by Bryant (2003).

- The *combinable component consensus* method (Bremer, 1990). This is similar to the strict consensus. It includes all groups that occur on at least one tree and are not contradicted on any of the trees. Thus it can allow some groups that would be rejected by the strict consensus. Two groups contradict each other when they overlap, but neither is included in the other. If all trees are fully resolved, combinable component consensus is the same as strict consensus. The presence of an unresolved region in a tree does not contradict structure within it. Thus the unresolved group ABC on one tree is not contradictory to the group AB on another tree. In that case, the combinable component consensus method can retain the group AB, if it is not contradicted in any other tree. But a strict consensus would reject AB because it was not present in all trees. The combinable component consensus is also sometimes called the *loose consensus* or the *semi-strict consensus*.

- *Nelson consensus* (Nelson, 1979) finds the largest clique of groups that are all compatible with each other. Two groups are compatible if, in the above sense, they do not contradict each other. Page (1989) has distinguished between Nelson consensus and strict consensus, with which it has sometimes been confused. He has also (Page, 1990) modified Nelson's algorithm to remove some difficulties. We can follow Bryant (2003) in calling the resulting method *Nelson-Page consensus*. Nelson-Page consensus will come close to being majority-rule consensus. When a fully resolved majority-rule consensus tree exists, this will also be the Nelson-Page consensus tree. Swofford (1991) gives a careful discussion of the literature on Nelson-Page consensus. Another consensus method closely related to Nelson-Page consensus is the "asymmetric median tree" of Phillips and Warnow (1996).

- Consensus trees based on rooted trees with branch lengths. A number of methods have been developed that use rooted trees and remove groups whose branches do not rise far enough from the root (Neumann, 1983; Stinebrickner, 1984, 1986).

- Consensus trees based on path distances. Lapointe and Cucumel (1997) have introduced the *average consensus* method. It takes unrooted trees with branch lengths, constructs a set of predicted distances from each, and then finds a tree that fits these best by least squares, where the sum of squares is computed separately for each set of predicted distances and then added up. I have already mentioned this in Chapter 12 as it is a consensus supertree method as well. Buneman (1971) suggested another distance-based consensus tree method, based on averaging the path-length distances over all trees, then fitting a tree to these averages.

- The *MRP consensus*. The representation of trees by binary characters used by Baum (1992) and Ragan (1992b) can be used to find the tree that is most parsimonious using these as a data set. This too has been mentioned in Chapter

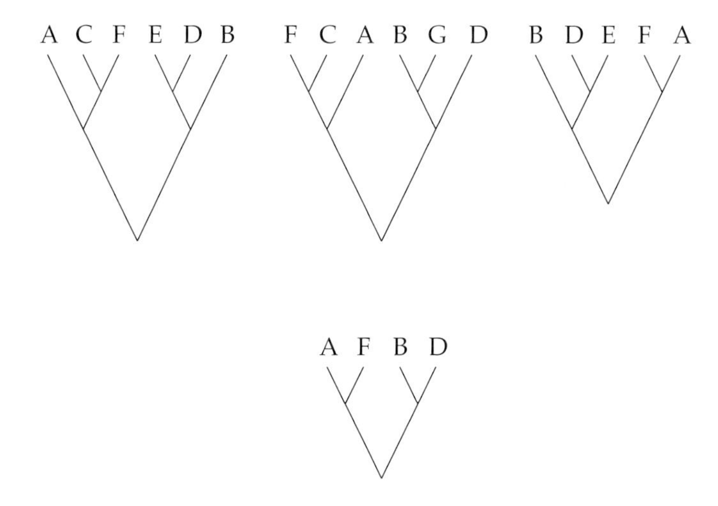

Figure 30.6: Three trees (top) and a consensus subtree (bottom).

> 12 as a consensus supertree method. The method of deriving binary characters from trees was first introduced by Kluge and Farris (1969) for character state trees and by Farris (1973a) for phylogenies.

References to a number of other consensus methods will be found in the review by Bryant (2003).

Consensus subtrees

A counterpart to supertrees is to take a set of trees, all of which have the same species on them, and to drop some, hopefully a small number, until we get a set of subtrees that are all the same. Figure 30.6 shows this process done for a set of three trees. Gordon (1980) suggested deleting the smallest number of objects that would result in all trees being the same. Steel and Warnow (1993) gave an algorithm for two trees whose speed was quadratic in the size of the trees. Faster algorithms to find strict consensus subtrees were given by Amir and Keselman (1997) and Henzinger, King, and Warnow (1999).

Swofford (1991) gives an example suggesting that a strict consensus may not be the most meaningful objective. Wilkinson (1994) has suggested methods of computing subtrees corresponding to Adams consensus trees, and he has also (Wilkinson, 1996) suggested obtaining majority-rule reduced consensus trees.

Distances between trees

In addition to knowing what common structure is implied by a set of trees, we may alternatively be interested in measuring how different they are. A number of methods for measuring difference between trees have been proposed.

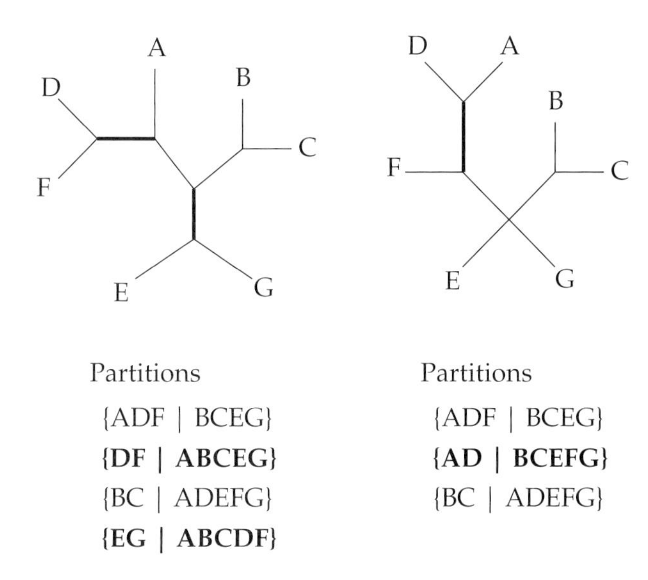

Figure 30.7: Two trees whose symmetric difference is computed. The branches that differ between the tree are highlighted with darker lines. The tables of partitions for each tree are shown, and the partitions that differ between the trees are also highlighted in bold. As there are three of these, the symmetric difference is 3.

The symmetric difference

Bourque (1978; see also Robinson and Foulds, 1981) defined a distance between a pair of trees based on the number of branches that differ between the trees. This has become known as the *symmetric difference* or the *partition metric*. If we have two unrooted trees and ignore their branch lengths, each can be considered as a set of branches, and each branch divides the species into a partition with two sets, one connected to each end of the branch. For each tree, make a list of the partitions it implies. The symmetric difference is simply a count of how many partitions there are, among these lists, that are not shared with the other tree. An example is shown in Figure 30.7. The two trees shown there have some branches that are not shared between them. These are shown by darker lines in the figure. The symmetric difference is easy to compute, but it is highly sensitive to all differences between trees. Penny, Foulds, and Hendy (1982) sampled randomly generated trees for 11 species. They found that 81% of pairs between them had the maximum possible distance when the symmetric difference was used. Some of the pairs that achieved the maximum distance may have had partial similarities, but this was not enough to reduce their difference. An example is the pair of trees in Figure 30.3. In spite of having much structure in common, the trees will share no partitions and thus will achieve the maximum possible value of the symmetric difference.

Although I have described the symmetric difference as computed for unrooted trees, it can also be defined for rooted trees. In that case, each branch defines a set of species, those connected to its upper end. The symmetric difference is then the number of sets that differ between the two trees.

The quartets distance

A distance measure that is more sensitive to partial similarities of structure between trees is the *quartets distance* of Estabrook, McMorris, and Meacham (1985). They actually defined four different dissimilarity measures, based on looking at all possible quartets of species. If we use as our example Figure 30.3, which showed the limitations of strict consensus, we will find that the two trees there each have 35 possible quartets of species ($7 \times 6 \times 5 \times 4/1 \times 2 \times 3 \times 4$). Of these, all the 15 quartets that do not involve species A are resolved identically in the two trees. For example, the quartet BDEF has unrooted tree topology ((B,D),(E,F)) in both trees. However all 20 quartets that involve species A differ in topology. One of these is the quartet ABCG, which has unrooted tree topology ((A,B),(C,G)) in the first tree and ((A,G),(B,C)) in the second.

In their paper Estabrook, McMorris, and Meacham defined different dissimilarity measures depending on how unresolved quartets were handled in trees that were not entirely bifurcating. In the fully resolved case considered here, they would not differ: All of the measures would be 20/35 = 0.5714. It would be possible to define a distance that was simply the number of quartets differing in topology, which would be 20 in this example, but it seems more meaningful to scale it by the total number of quartets. Note that this distance does have some sensitivity to the similarity of the two trees: They are not 100% different but only 57% different.

Estabrook (1992) further developed these measures to indicate for particular species to what extent their placement on the two trees differed. Critchlow, Pearl, and Qian (1996) developed a similar distance for rooted trees, which used triples (three-taxon statements) rather than quartets. The quartets distance at first looks difficult to compute, since the number of quartets in a tree is proportional to the fourth power of the number of species. However, Brodal, Fagerberg, and Pedersen (2001) have discovered how to compute it in a time proportional to $n(\ln n)^2$, which is almost linear in the number of species.

The nearest-neighbor interchange distance

Waterman and Smith (1978) suggested another measure of distance between trees. They proposed that we use the number of nearest-neighbor interchange (NNI) rearrangements that are needed to go from one tree to the other. We discussed these rearrangements in Chapter 4. As the rearrangements can be precisely reversed to go the other way, it would not matter which of the two trees we chose as the starting point. For the trees in Figure 30.3, we can get from one to the other in only four NNI rearrangements, each time moving species A. The NNI distance between these trees is then 4. This distance has the advantage of not seeing those two trees as greatly different, so that it does not have the same problem as the symmetric difference.

The difficulty with the NNI distance is that, for large trees that are very different, it is impossibly hard to compute. Li and Zhang (1999) have shown that computing it for unrooted bifurcating trees is NP-complete. Interestingly, the problem seems related to one in quantum theory (Fack, Lievens, and Van der Jeugt, 1999).

Allen and Steel (2001) have investigated extensions of the NNI approach to SPR and TBR rearrangements (recall that these were defined in Chapter 4). They have shown that a distance based on TBR rearrangements is actually much easier to compute than one based on NNIs.

The path-length-difference metric

Penny, Watson, and Steel (1993) suggest a distance that measures, for each pair of species, the number of branches that separate them on the tree. (This will be 1 greater than the number of nodes that separate them.) The table of branch numbers is computed for each tree. The distance between trees is the square root of the sum of squares of the differences between these numbers. This distance is related to earlier suggestions by Farris (1969b), Williams and Clifford (1971), and Phipps (1971).

Distances using branch lengths

All of these distances use the tree topologies but not the branch lengths. Two distances have been defined that use the branch lengths. They are both related to the symmetric difference. Robinson and Foulds (1979) defined one, and Mary Kuhner and I (Kuhner and Felsenstein, 1994) defined the other. Ours was a squared distance called the *branch score*. Both start with a list of all possible partitions of the species. To each partition assign a number that is 0 if the partition is absent in the tree, or is the branch length if the branch is present. The Robinson-Foulds distance is the sum of absolute values of the differences between these two lists of numbers. The branch score is the sum of squares of differences between these numbers for two trees. If the same branch is present in both trees, it will contribute the absolute value of the difference, or the square of the difference, between the two branch lengths. If the branch is absent from one tree, it will contribute its length in the other tree (or the square of its length). If absent from both, it will contribute zero, so that we actually need only consider those partitions that are present in one tree or the other. Note that we include all partitions present on either tree, not just the ones corresponding to internal branches.

Figure 30.8 shows the calculation for the two trees previously used in Figure 30.7. It shows a table of all partitions and the branch lengths corresponding to each one. Partitions for external branches of the tree are included. Where a partition corresponds to a branch that is not present on the tree, the branch length is

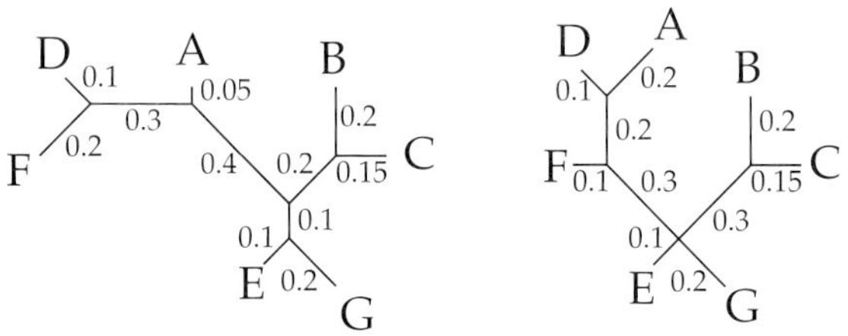

Partitions	Branch lengths	
{AD | BCEFG}	none	0.2
{ADF | BCEG}	0.4	0.3
{BC | ADEFG}	0.2	0.3
{DF | ABCEG}	0.3	none
{EG | ABCDF}	0.1	none
{A | BCDEFG}	0.05	0.2
{B | ACDEFG}	0.2	0.2
{C | ABDEFG}	0.15	0.15
{D | ABCEFG}	0.1	0.1
{E | ABCDFG}	0.1	0.1
{F | ABCDEG}	0.2	0.1
{G | ABCDEF}	0.2	0.2

Figure 30.8: Two trees (with the same topology as in Figure 30.7) with branch lengths, showing the list of branch lengths needed in the computation of the branch score or the Robinson-Foulds distance.

indicated by "none." The Robinson-Foulds distance is the sum of the absolute values of the differences between the two columns in that figure. The branch score is the sum of the squares of the differences between the two columns of branch lengths in the figure. In both cases we count "none" as zero. The result for the Robinson-Foulds distance is (ignoring all pairs of branch lengths that are identical in the two lists)

$$\begin{aligned} &|0-0.2| + |0.4-0.3| + |0.2-0.3| + |0.3-0| \\ &+|0.1-0|+|0.05-0.2| + |0.2-0.1| = 1.05 \end{aligned} \tag{30.1}$$

and for the branch score it is

$$\begin{aligned} &(0-0.2)^2 + (0.4-0.3)^2 + (0.2-0.3)^2 + (0.3-0)^2 \\ &+ (0.1-0)^2 + (0.05-0.2)^2 + (0.2-0.1)^2 = 0.1925 \end{aligned} \tag{30.2}$$

The relationship of the branch score to the symmetric difference should be apparent from the figure. The symmetric difference is the sum of the number of rows in the table that have "none" in one of their columns (here it would be 3). In fact, if all branch lengths of both trees are 1.0, the symmetric difference is equal to both the Robinson-Foulds distance and the branch score. There is no such simple relationship to the path length difference metric of Penny, Watson, and Steel. Differences in interior branches, through which more paths between species pass, affect the path-length-difference metric more than do differences between terminal branches. The Robinson-Foulds distance and the branch score treat both kinds of branches equally.

Since we want to compute a distance, it seems natural to take the square root of the branch score and call that the distance. Let us call that the *branch-length distance* (BLD). It will equal the ordinary Euclidean distance in the space defined by the branch lengths if the tree topologies do not differ. The Robinson-Foulds distance does not need square-rooting.

Are these distances truly distances?

One might stop to ask whether the "distances" we have been defining satisfy the mathematical requirements of being called a distance. Those are the three conditions for being a metric: that the distance from an object to itself is zero, that the distance from A to B is the same as the distance from B to A, and that the Triangle Inequality holds true. It requires that the distance directly from A to B is never greater than the distance from A to C plus the distance from C to B. In other words, the direct route is never longer than an indirect route.

The first two requirements are satisfied by all the distances we have mentioned. The only issue is satisfying the Triangle Inequality. All of the distances we have mentioned can be computed from a list of numbers for each tree. For the symmetric difference, this is a list that has all possible partitions of the species, with the entries in it being either 1 or 0, depending on whether that partition is or is not present in the tree. If we have such a list for each tree, the symmetric difference is the sum of the sum of absolute values of the differences between the lists. Any such formula can easily be shown to satisfy the Triangle Inequality. So the symmetric difference is a metric, and a mathematician will not object to calling it a "distance."

We can do a similar proof for the quartets distance, since each tree can be characterized by a list of numbers, three for each possible quartet. For each quartet we have three possible ways that it can be resolved. ACDF can be either ((A,C),(D,F)) or ((A,D),(C,F)) or ((A,F),(C,D)). We make a list that has three quantities for each possible quartet, and have it contain a 1 when that quartet is resolved in that way in the tree, and 0 otherwise. Then we find that the quartets distance is the sum of absolute values of differences between the lists for the two trees. As in the case of the symmetric difference, this establishes that the quartets distance satisfies the Triangle Inequality.

For the nearest-neighbor distance, we can simply note that we can imagine the graph of all possible trees, with connections whenever one tree can be obtained from another by a nearest-neighbor interchange. The NNI distance is the distance in this graph. It is easy to show that it must then satisfy all the requirements of being a metric, including the Triangle Inequality.

The branch score is a bit more problematic. It is defined as the sum of squares of differences between two lists of numbers, one for each tree. The branch-length distance derived from it is easily shown to be a metric. (It is also not hard to show that the branch score itself is a metric, but we prefer to have the BLD be the quantity that is called a distance, since for trees of the same topology it is the ordinary Euclidean distance in a space whose coordinates are the branch lengths.) One can also show that the sum of absolute values of differences of branch lengths is a metric. Maybe this should be called the branch length absolute difference (BLAD).

Consensus trees and distances

Consensus trees and distances between trees seem to be somehow related. In certain cases this relationship can be made explicit. We could imagine finding a tree that lay in the center of a cloud of trees, in the sense that its total distance to all of them was as small as possible. This is called a *median tree*. We could think of doing this for any of these distances. In the case of the symmetric difference, we can identify the median tree. It turns out to be essentially the majority-rule consensus tree. We have to say "essentially," because there is one important qualification. Barthélemy and McMorris (1986) show that if the number of trees is odd, the median tree is simply the majority-rule consensus tree. If the number of trees is even, it is possible for there to be two median trees, each having some groups that occur in exactly 50% of the trees. These would not, strictly speaking, be majority-rule consensus trees, as 50% is not quite a majority.

This relationship between median trees and majority-rule consensus trees may seem subtle, but it is easily derived. One need only consider that for each partition in any tree in the tree space, the contribution that partition makes to the distance to all other trees is equal to the number of trees in which it does not occur. I leave the rest of the proof to you as an exercise.

Trees significantly the same? different?

It is common for biologists to ask whether two trees can be shown to be significantly similar, or significantly different. When this is done directly from the trees without recourse to data, it is problematic. It is possible to ask whether two trees show more similarity than would two trees drawn at random. There is the issue of what random distribution of trees is relevant. Random distributions can arise in a number of ways, such as by random branching of a lineage, or by random choice from the list of all possible trees. Steel and Penny (1993) show results for the means and variances of the symmetric difference, the quartets distance, and the path-length differences for trees drawn randomly from a number of such distributions. They find that the asymptotic distributions are Poisson distributions, and they also investigate them by simulation.

One can use such results to test whether two trees are closer than two random trees would be. But this may not be what you really want to know. If your data set contains even a single pair of sibling species that are always adjacent in the tree, this may be enough to cause the trees to be significantly closer than random. There is no guarantee that the signal comes diffusely from the whole tree—it may instead be responding to a feature that is in common but is of little interest to the biologist. Thus these tests will usually be of limited interest.

The question whether two trees are different is even less exciting—because it is usually vacuous. If I ask you whether two friends of yours are different, the answer is always that they are. They differ in some respect, however small. Likewise, two trees that are not identical are different—and that is all that means. Whether they are significantly different is more complex, and requires a statistical model for the variation of the features of the tree. This statistical model is lacking if we ask only about the trees and do not consider the data that generated them. So, yes, those trees are different, and no, it isn't a very good question to ask.

What do consensus trees and tree distances tell us?

Before using consensus trees or tree distances, it is worth asking what they do and do not tell us. They treat each tree equally, and they treat each feature of each tree equally. This may or may not be appropriate. They do not necessarily tell us whether a feature of a tree is well-supported by the data, nor do they tell us whether we care about the feature. Suppose that we have studied 10 different loci in the great apes, and nine of them give strong support to a clade consisting of humans and chimpanzees. One of the loci gives weak support to an alternative clade (say, chimps and gorillas), but comes close to supporting the human-chimp clade. Nevertheless, that locus has as its best estimate a tree with the chimp-gorilla clade. If we were making a strict consensus tree of the trees inferred from the 10 loci, we would get no resolution of the human-chimp-gorilla trichotomy.

This shows us both of the difficulties. The consensus tree does not take into account the strength of the evidence supporting the groups in each tree. It simply counts them as present or absent, without asking how present or how absent they are. The consensus tree also fails to take into account our great interest in the human-chimp-gorilla trichotomy, preferring to babble on and on about what is happening among the gibbons. One might object that requiring a strict consensus is too strong, but the same problems occur with less stringent consensus methods. If, of our 10 loci, four supported a human-chimp clade very strongly, and six supported chimp-gorilla, but only slightly, then using a majority-rule consensus tree would give a misleading result. It would simply take the vote (6 to 4) without considering how strongly held the opinions were. The six loci might be short and evolving at a high rate, with lots of contradictory phylogenetic signal. The four might be large loci with lower rates of evolution that gave clear results.

Tree distances have similar problems. They can show large differences between trees that agree in a crucial area. We can have trees that all show the human-chimp clade, but they might have large distances from each other as a result of differences in the placement of gibbons.

One application of consensus trees that escapes these problems is their use in summarizing the results of bootstraps or jackknifes. (Somehow it seems wrong to call these "jackknives," so I have not done so.) The frequency with which a partition (or clade) appears among the bootstrap replicates or the jackknife replicates is then a direct reflection of the strength of data supporting the grouping, and consensus trees are then entirely appropriate. This suggests that if each data set were represented by a cloud of bootstrap or jackknife tree estimates, there might be interesting consensus tree methods that summarized the different studies with a common tree. If a partition of the species was strongly supported by bootstrap or jackknife estimates for one or more data sets and only weakly contradicted by other data sets, we would want the consensus method to include it. A consensus method that simply combined all trees would be inappropriate here, as it would lose the information about how strongly supported a grouping was in any one study.

Sets of trees that simply reflect analysis of different data sets, or use of different phylogeny methods, or trees tied for best under a single method do not have the same meaning, and their consensus may be more problematic.

The total evidence debate

The validity of using consensus trees is at the heart of a vigorous debate in systematics. Opponents of using them advocate instead the *total evidence* approach. With a multiple-locus data set, instead of inferring separate trees for each locus and then making a consensus tree, they would put all loci into a common data set, in effect concatenating their sequences end to end. They then analyze these using parsimony, to find the most parsimonious tree or trees.

This has the great advantage of taking into account the different amounts of sequence in different loci and of combining the evidence in a single tree that does not depend on an arbitrary choice of consensus tree method. As advocates of total evidence use parsimony methods, they also can incorporate discretely coded morphological characters, including fossil data. The total evidence approach originated in the molecules-versus-morphology literature of the 1980s. It has been most forcefully advocated by Kluge (1989; Eernisse and Kluge, 1993; Kluge, 1998).

The case against the total evidence approach is made by Bull et al. (1993), by De Queiroz, Donoghue, and Kim (1995), and by Miyamoto and Fitch (1995). If different loci have substantially different rates of change, combining them into one data set obscures evidence that indicates that one locus should be treated differently from another. In the limiting case in which all loci have low (though possibly unequal) rates of evolution, parsimony does a reasonable job of combining evidence. In Chapter 9, we saw that in this limit parsimony and likelihood infer the same tree, and the optimal weights for changes become equal in sites that change at different rates. However, when we are not in this limiting case, parsimony will do less well. When two trees differ in parsimony score by a single step, this may conflict much less with the evidence for a locus that has a high rate of change than it does for a locus that has a low rate of change. In that situation, the overall parsimony score exaggerates the evidence coming from the loci with high evolutionary rates. The use of consensus trees is a response to this problem.

The consensus approach may involve predefined subsets of data, or may involve testing whether data sets should be divided into partitions whose trees need to be analyzed separately. This type of test has been advocated by Bull et al. (1993), Rodrigo et al. (1993), and De Queiroz (1993). It frequently involves use of the ILD test, which was described in Chapter 20. There is a large and complex literature on this, which is reviewed by De Queiroz, Donoghue, and Kim (1995); Huelsenbeck, Bull; and Cunningham (1996), and Page (1996).

The last word has not been said in this controversy, so readers should also look for more recent papers that cite these studies.

A modest proposal

In certain cases it may be possible to have the best of both approaches and to circumvent the issue of testing combinability of data. Suppose that the issue is variation in rate of evolution from locus to locus. For variation of rate of evolution from site to site, we have seen in Chapter 9 that, in a parsimony method, this argues for unequal weighting of changes. More changes in a site lead us to consider it a site with a high rate of change, and this in turn will lead us to weight it less heavily. In Chapter 7 we saw successive and nonsuccessive weighting methods that give less weight to changes in a site the more of them there are.

The same will hold for loci. If we have evolutionary rates that vary not only from site to site, but also from locus to locus, we should use the number of changes at a locus to help decide what the overall rate of change at that locus is. The more changes there are per site at a locus, the lower the weight of each individual change should be. We could construct a weighting method that gave changes at site j of locus i a weight $w(n_{ij}, N_i)$ that depended on both the number of changes n_{ij} at that site and the number of changes N_i at that whole locus. Without going into details, I will simply say that this function can be chosen to reflect the degree of variability of rates among loci and among sites within loci. The upshot is a weighted parsimony method that will automatically de-emphasize information from loci that have large numbers of changes. This will be a total evidence approach, but in effect it will also come close to what a consensus method would do. It counts not simply changes in individual sites but also evidence from whole loci.

One need not confine this method to using parsimony. Likelihood methods can use hidden Markov models (HMMs) to take variation of evolutionary rates from site to site into account (as we have seen in Chapter 16). Why not do the same for loci? We could have rates of evolution varying among loci and assume a distribution of these per-locus rates. The overall likelihood at a locus would then be an integral over the per-locus rates, weighted by the probability density $f(r)$ of these rates, where the rates were treated as multipliers affecting the branch lengths:

$$L^{(i)} = \int_0^\infty f(r) \; \mathrm{Prob}\,(D^{(i)} \,|\, T, r) \; dr \qquad (30.3)$$

I have recently (Felsenstein, 2001b) discussed variation of evolutionary rates over loci, using a population-genetic model to relate different strengths of selection at different loci to their evolutionary rates.

If such a model were used, it could be described as a total evidence approach (likelihood). Perhaps this should be abbreviated TEAL. It would not have the formation of consensus trees as part of the method, but if the per-locus evolutionary rates varied greatly, it would combine the evidence in ways that came close to use of a consensus method, while still taking the total evidence into account.

In the above discussion, we have been assuming that the source of heterogeneity among results from different loci is high evolutionary rates at some loci. Another possible source of difficulty is discrepancy between species trees and coalescent "gene trees." This was dealt with in Chapter 29. It requires somewhat different methods. Similarly, horizontal gene transfer and paralogy in gene families can cause discrepancies and again require different methods for analysis.

Chapter 31

Biogeography, hosts, and parasites

Two different bodies of work have arisen that examine the correspondence between two trees, one in biogeography and the other in parasitology. They are closely related, so I will treat them together here. In biogeography, the book by Nelson and Platnick (1981) attempted a quantitative treatment of vicariance biogeography. In parasitology, the paper by Brooks (1981) placed the problem of correspondence of host and parasite phylogenies (Fahrenholz's Rule) in a parsimony context.

The similarity of the two problems is not accidental. Vicariance biogeography explains the distribution of organisms by assuming that an ancestral area has successively been subdivided by geological, climatological, or vegetation changes, and that these divisions have brought about allopatric speciation of the organisms. We may be fitting a phylogeny of the organisms to a particular "area cladogram" that reflects this subdivision, or we may be searching among possible area cladograms to find the one that fits best. In parasitology, the hosts are the "geography" on which the parasites are distributed, and the speciation of the hosts brings about allopatric speciation of their parasites. In both cases, exceptions to an exact similarity of the two trees can result from extinctions, undiscovered species, speciation of the parasite (or group) without host speciation (or without vicariance), and by migration across geographic barriers or invasion of one host from another. One difference between the two problems is that in parasitological studies there is usually independent evidence about the host's phylogeny, and we do not need to use the parasite phylogeny to infer the host phylogeny.

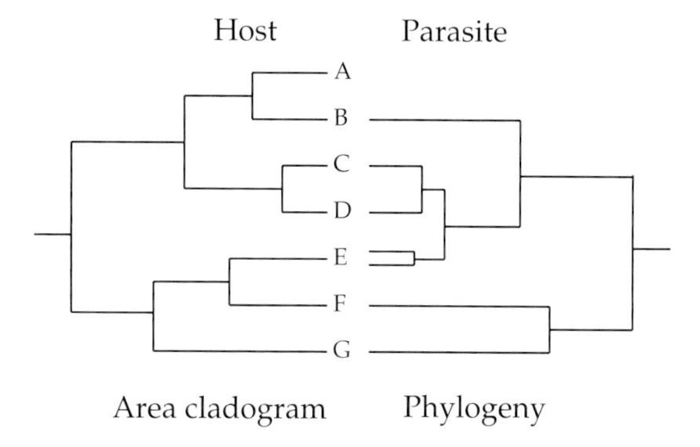

Figure 31.1: Two trees as an example for the biogeography and host/parasite cases. The left tree is either the area cladogram, which reflects the successive subdivision of a geographic region, leftmost fork first, or it is the phylogeny of the hosts. The right tree is the species phylogeny in the biogeography case (below), or the parasite phylogeny in the host/parasite case (above).

Component compatibility

Nelson and Platnick (1981) suggested inferring area cladograms in vicariance biogeography by looking for repeated groups of areas in cladograms of different groups, where each individual species had been replaced by the name of the area or areas in which it is found to occur. Consider the trees in Figure 31.1. In the biogeography case, we interpret the left-hand tree as the area cladogram, reflecting

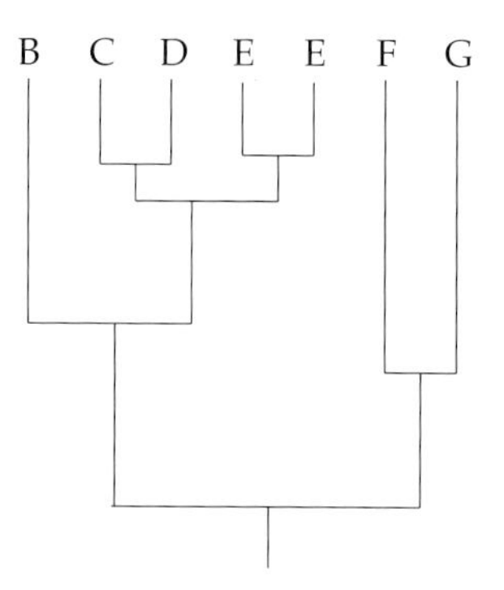

Figure 31.2: The tree of species from our example, with the names of the regions in which each species occurs placed at the tips instead of the species names.

vicariance events that successively sudivide the region (the deepest fork being the first subdivision). If we label the tips on the area cladogram A through G and then take the right tree and place the area in which each species occurs at the tip instead of the species name, we get the tree in Figure 31.2. The tree is the estimate we would make of the area cladogram (except that E would be assumed to be undivided). It is not entirely correct. It leaves out A, and puts E in the wrong subdivisions of the area. We might imagine having more than one group for which we have a phylogeny, distributed in the regions A–G. How can we compromise the information from these phylogenies to infer the area cladogram? Nelson and Platnick (1981) stress the finding of repeated components in the phylogenies. Their method is closely related to the consensus method advocated by Nelson (1979; see further comments by Page, 1988). As I have mentioned in Chapter 30, this method has been discussed by Page (1989), who argues that it is a consensus method somewhat similar to majority-rule consensus.

Methods analogous to this are used less often in parasitology, where one does not often use the distribution of parasites to infer the host phylogeny.

Brooks parsimony

Brooks (1981) introduced the first parsimony algorithm in parasitology, which has become known as Brooks parsimony (developed further in Brooks, 1990; see also Brooks, van Veller, and McLennan, 2001). He suggests taking the parasite tree,

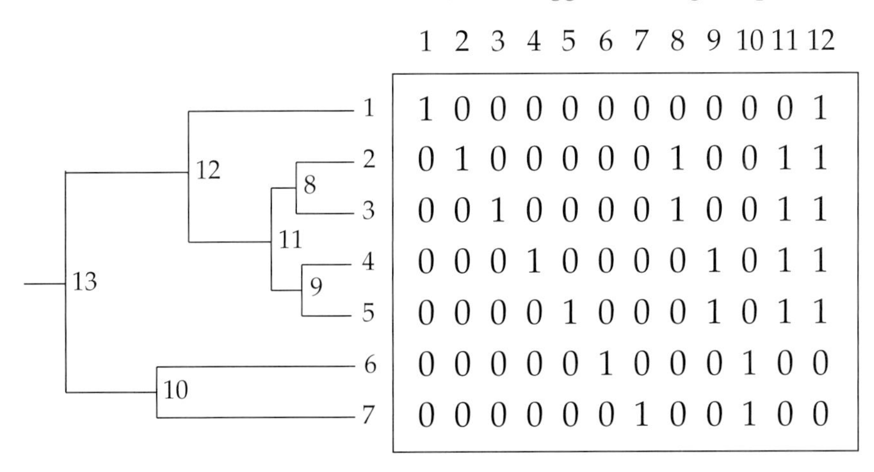

Figure 31.3: The parasite tree in the example, and a set of binary characters that are derived from treating it as a character state tree and using the recoding method of Kluge and Farris (1969). Numbering of nodes and tips corresponds to the number of the binary character.

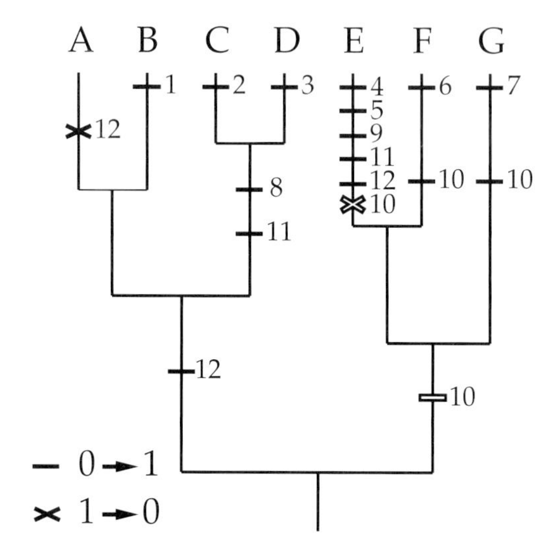

Figure 31.4: Reconstruction of the parasite binary characters from Figure 31.3 on the host phylogeny. The loss of character 12 is implied by the absence of a parasite from host A. For character 10, two different placements of the changes are shown, one with filled events and one with empty events.

treating it as a character state tree, and making up a set of nonadditive binary characters for it. This recoding has been discussed above in Chapter 7. The characters for the parasite tree are then reconstructed on the host tree. It is assumed that the ancestral state of each of these binary characters is 0, which is reasonable since sufficiently far back in time there was only a single ancestral parasite lineage. Figure 31.3 shows the binary characters corresponding to the parasite tree.

Figure 31.4 shows the reconstruction of the parasite characters on the host phylogeny. I have added a loss of character 12 on the lineage leading to A. For character 10 there are two possible placements of the changes, one with two gains (solid bars) and one with a gain and a loss (the empty box and the empty cross). In addition to this, the reconstruction shows states 11 and 12 arising twice. The parasite species ancestral to species 4 and 5 was one of the species arising when state 11 arose. If we imagine that it invaded the lineage to host E, which then lost the parasite species it had, this would account for the reconstructed states.

The Brooks parsimony reconstruction does not specify exactly what happened when there is a parallelism like this. If the lineages on which state 11 arises are in existence simultaneously, a host switch could account for the parallelism. However, the trees used in the analysis do not have branch lengths or times. At a minimum, a labeled history (as described in Chapter 3) would be required in order to know whether a host shift could have happened. Otherwise, some more complex

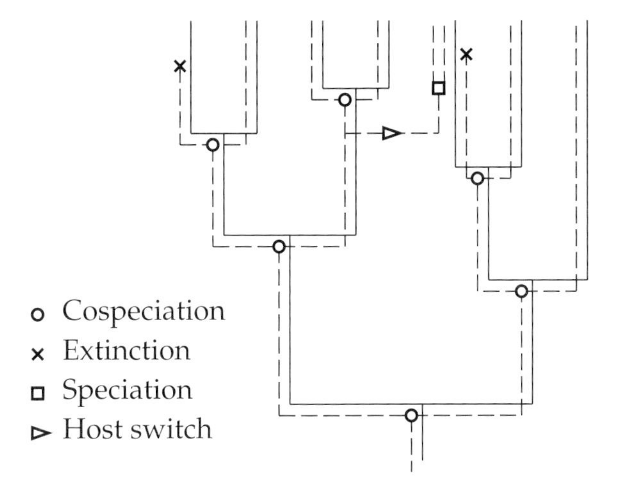

Figure 31.5: Reconstructed history of our example parasite tree (dashed lines) on the host tree, using cospeciation, extinction, host switching, and speciation events.

set of events would be required involving early presence of the parasite and later multiple losses. The events counted in Brooks parsimony analysis are not exactly independent, and their numbers are not always easily interpretable.

Brooks parsimony analysis can also be done in biogeography. Coding trees of individual groups of organisms, all of whom live in the same geographical areas, one can infer an area cladogram that takes the phylogenies of all these organisms into account. In the host/parasite problem, Brooks parsimony has led directly to more precise parsimony approaches. Dowling (2002) has tested Brooks parsimony against Page's (1994b) parsimony method in a simulation test. Neither reconstructed events perfectly, but Brooks's method performed better.

Event-based parsimony methods

A number of papers by Roderic Page, Fredrik Ronquist, and Michael Charleston (Ronquist and Nylin, 1990; Page 1994a; Page and Charleston, 1997; Ronquist, 1994, 1996a, 1997, 1998b; Charleston, 1998) have constructed parsimony methods that represent actual biological events. For a detailed review and comparison of the methods, the review by Ronquist (2002) will be helpful.

Figure 31.5 shows a series of events needed to reconcile the parasite tree with the host tree in our example. Figure 31.6 shows a similar series of events needed when our example is instead interpreted as vicariance biogeography.

The events that are involved are listed below, with the names that might be used for them in the host/parasite and vicariance biogeography cases:

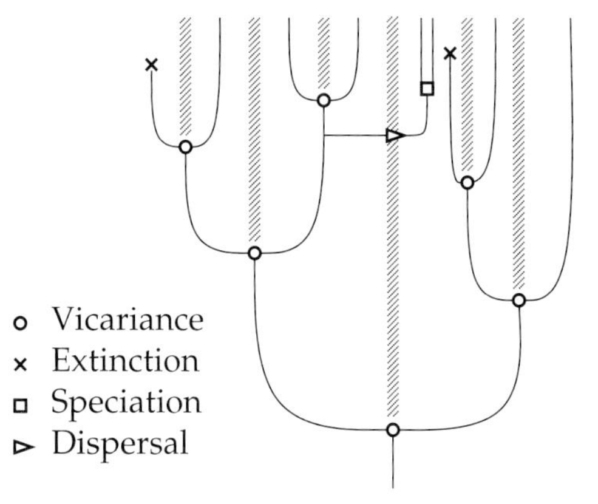

Figure 31.6: Reconstructed history of our example organism tree in the case of vicariance biogeography, using vicariant speciation, extinction, dispersal, and speciation events. The shaded bars show the vicariance events implied by the area cladogram.

- *Cospeciation* or *vicariant speciation*, in which a host speciation or the subdivision of a geographic region is accompanied by allopatric speciation.
- *Extinction* of a parasite species or a species in one geographic region.
- *Speciation* of a parasite on one host, or of a species within one geographic region.
- *Partial host switching* or *dispersal* in which the parasite colonizes a new host or a species invades a different geographic region.

Ronquist (2002) argues that two other kinds of events are unnecessary, though they have sometimes been invoked. These are complete host switching, in which a parasite invades a new host while simultaneously becoming extinct in the old host, and sorting, in which a host speciation or a vicariance event leads to the species surviving in only one of the resulting hosts or resulting regions. It should be noted that both of these patterns are biologically unlikely. In the rare event that one is seen to happen, it would be proper to consider it as the result of two events. In the case of sorting, the events would be a partial host switch followed by an extinction, or a dispersal followed by an extinction. In the case of complete host switching, the events would be invasion of a different host followed by extinction in the old host.

These methods all assume that once a host speciates, or a region is subdivided, that the populations that inhabit them must become different species. This "one host per parasite" assumption is widely used. The corresponding "one parasite per host" assumption is not usually made, because it would allow a partial host switch only when the host's parasite had already gone extinct, or when the invading parasite forced an extinction of the previous parasite. The counterparts to these scenarios in biogeography seem equally artificial.

We have seen in Chapter 9 that a statistical framework would suggest that changes be weighted more heavily in a parsimony method the rarer the event is expected to be. Of the four types of events given above, one, cospeciation, is expected to be frequent. Charleston (1998) assigns cospeciation a negative cost. Ronquist (2002) disagrees with this assignment because it seems to carry the method outside of the framework of parsimony; he assigns cospeciation zero cost. In the limit, when the rates of the other events are very small, the weights of those events should be equal.

There is not space here to go into the algorithms employed to count changes in these parsimony methods. Ronquist (1996a) and Charleston (1998) give algorithms, both quite complicated. They are not identical—Ronquist (2002) argues that Charleston's "jungles" algorithm does not handle some situations correctly. A glance at either will show why I have not tried to explain them here.

Relation to tree reconciliation

In either the biogeography or the host/parasite case, we are trying to reconstruct events so as to reconcile one tree with another. We have seen in the discussion of gene families in Chapter 29 that Page (1994b) and Mirkin, Muchnik, and Smith (1995) put forward tree reconciliation methods that count gene duplications and gene losses, and try to reconcile the trees so that the total number of these are minimized. Figure 29.5 shows that gene duplications create the same pattern as speciation on one host or in one geographic region, and gene losses have the same result as extinctions. Thus tree reconciliation by parsimony methods is a special case of the parsimony methods used in biogeography or parasitology. The gene family methods have no event corresponding to a host switch or a dispersal. If they did, it would in effect model horizontal gene transfer.

Randomization tests

Legendre, Desdevises, and Bazin (2002) have described a randomization test that uses matrices representing the host and the parasite trees, and one that indicates which host each parasite is on. Some least-squares statistics are computed that would be correct if the trees and host/parasite connections could be thought of as random variables. They are not, but the actual test is done by a randomization procedure. Legendre, Desdevises, and Bazin describe two tests. One sees whether the fit of host to parasite tree is better than random, where the null hypothesis is that the parasites are assigned to their hosts at random. The second test is whether the change in fit between parasite and host trees is significantly better when a particular parasite/host connection is removed. The fit is judged against a distribution of the same statistic computed from random associations of host and parasite.

The first test is of the hypothesis of no nonrandom association of parasite phylogeny with host phylogeny. The second test is more focused, but it is less obvious that the distribution with which it is compared is relevant, as there is no association of host and parasite phylogeny in that null hypothesis.

Johnson, Drown, and Clayton (2001) took a rather different approach. They used the ILD (or partition homogeneity) congruence test (Farris et al., 1994b; Swofford, 1995; see Chapter 20) to see whether there is evidence that the host and parasite trees differ. They removed troublesome species until the two trees were not significantly different. Those common subtrees were then held constant while adding the offending hosts and parasites back into their respective trees at most parsimonious locations. This, they argue, avoids postulating unnecessary events for which evidence is weak. It is not completely clear to me what their method is for reconstructing the events after the species are added back.

Statistical inference

Model-based statistical methods have also begun to be introduced. For the host/parasite problem, Huelsenbeck, Rannala, and Yang (1997) used a model in which there is a host-switching rate, but no speciation events within hosts. They assumed that there could be only one parasite per host, and that when a parasite switched to a new host, the parasite previously present on that host would be made extinct. They used the symmetric difference metric described in Chapter 30 and asked by simulation what rate of host switching led to the observed symmetric difference between host and parasite phylogenies. They also developed a likelihood ratio test for the hypothesis that identical host and parasite phylogenies had identical divergence times, and a parametric bootstrap likelihood ratio test for whether the host and parasite phylogenies do or do not have identical topologies.

A fuller statistical analysis was made by Huelsenbeck, Rannala, and Larget (2000). They used Bayesian inference by Markov chain Monte Carlo with the model used by Huelsenbeck, Rannala, and Yang (1997) in which there are no speciation events or independent extinctions of parasites. They could obtain a posterior distribution of rates of partial host switching, which allowed tests on this rate. As in the previous analysis, they could allow for uncertainty about both the host and the parasite phylogenies. Posterior plots of the placements of speciation events, the number of partial host switches, and the pairs of lineages involved in the host switches gave a particularly clear picture of the range of possibilities. They also extended their model to have the probability of a partial host switch depend inversely on the distance along the host tree between the two hosts.

As noted by Huelsenbeck, Rannala, and Larget, their model is not as rich as the parsimony models described above. They do not allow independent extinction of a parasite on a host or speciation of a parasite in the absence of host speciation. Both of these are needed to treat cases in which there is not a perfect one-to-one correspondence between hosts and parasites.

They also suggest that their method would be useful in biogeography. The model seems on less sure ground there, as it would not allow for effects due to the different sizes of regions, or their geographic adjacency. We are still a bit short of satisfactory probabilistic models in the host/parasite and biogeographic cases.

Chapter 32

Phylogenies and paleontology

Fossils have had a special place in reconstructions of evolutionary history. The issue that arises with numerical and statistical methods is whether, or how, to give them a special status. It might be thought obvious that they give us a snapshot of the ancestors of present-day species. That is not necessarily so, as can be seen in Figure 32.1. It shows the true phylogeny of a small group (indicated by the cylindrical species lineages and the oval specimens present at each of five geological strata. If we observe only the specimens shaded black (moderate numbers at the present but only two per geological stratum), then the true phylogeny of the observed species is shown on the right side of the figure. Note that of the eight fossils, only three of them are actually from species ancestral to the present-day species (these are numbers 2, 3, and 6). The others are relatives of the ancestors, but are not themselves ancestors.

As the fraction of past species in a group that have been observed in the fossil record declines, the fraction of them that will be ancestors of the present-day species will also decline. It is this picture, together with knowing that the fraction of species available to us in the fossil record is very small, that motivates phylogenetic systematists, who insist that we cannot infer ancestor-descendant relationships. In other words, it is dangerous to interpret fossils as ancestors.

The converse is also true: If the fossil record of a group has been searched thoroughly enough, then we should not only be allowed to interpret fossils as ancestors, we should be encouraged to do so. Anthropologists have searched long and hard to find any species 2 million years ago that could be ancestral to modern humans. Since *Homo erectus* fits the bill morphologically, it can be assumed to be our ancestor, since if there were another candidate we would presumably have found it by now. This is not allowed in the strict interpretation of phylogenetic systematics, which, insisting that one cannot infer ancestor-descendant relationships, places *Homo erectus* on its own branch, a lineage parallel to ourselves [cf. British Museum (Natural History), 1980].

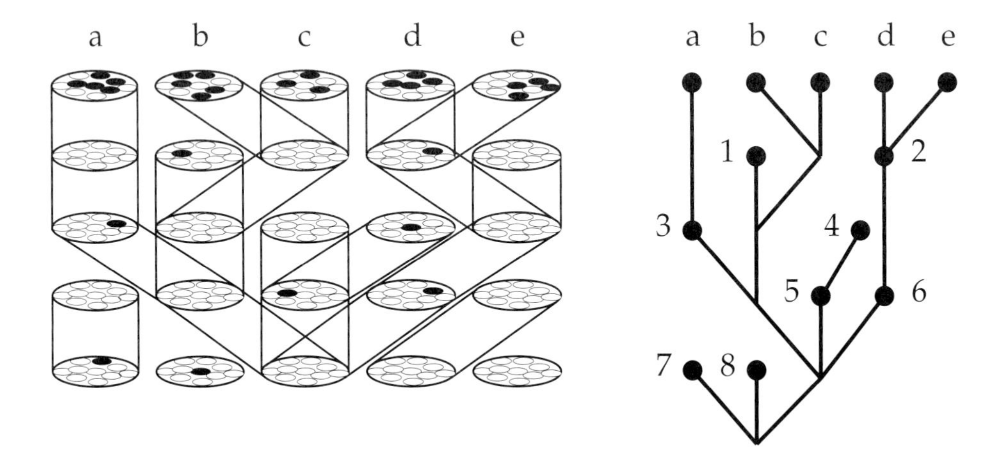

Figure 32.1: How the place of fossils in the phylogeny can be affected by how complete the fossil record is thought to be: An example.

There have been a number of numerical and statistical approaches to making these inferences, including stratophenetics, stratocladistics, and "stratolikelihood." I will describe these briefly here, with emphasis on the criteria they use rather than the detailed algorithmics. The field is a quite active one—I must skip over numerous numerical methods for assessing geological species ranges and fossilization rates, concentrating only on the methods that use or involve a phylogeny.

Stratigraphic indices

One approach to taking the fit of a phylogeny to stratigraphic information is to compute an index that describes how well the phylogeny fits the stratigraphy. There are a number of proposed indices, including the Spearman rank correlation measure (Norell and Novacek, 1992), the *relative completeness index* of Benton (1994), the *stratigraphic consistency index* (Huelsenbeck, 1994), the *retention index for the stratigraphic character* of Clyde and Fisher (1997), the *Manhattan stratigraphic measure* (Siddall, 1998a; Pol and Norell, 2001), the *gap excess ratio* (Wills, 1999), and the character-based measure of Angielczyk (2002). There has been a certain amount of controversy over which of these measures is best (cf. Benton, Hitchin, and Wills, 1999; Finarelli and Clyde, 2002).

I will not describe these measures or the tests of which are best, preferring to concentrate on methods that have some explicit means of compromising the phylogeny and the stratigraphy.

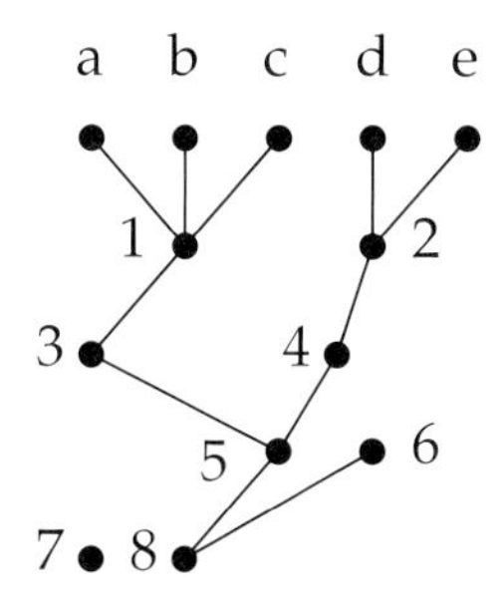

Figure 32.2: An example of stratophenetic linking, using specimens from Figure 32.1. The horizontal axis, which was arbitrary in that figure, plays the role of a phenotypic axis here.

Stratophenetics

Gingerich, (1979a, b, 1992) put forward the method of *stratophenetics*. In spite of its name, it is intended to infer phylogenies. (The "phenetics" is there because it uses a clustering step.) It assumes, as do all these methods, that we have observed specimens in a series of geological strata (one sometimes being the present). Gingerich proposes that the specimens within each stratum be clustered into groups that are postulated to be species. After the species clusters are found and the strata arranged in temporal order, "then a species in a chosen level can be linked to other species in adjacent levels based on overall similarity" (Gingerich, 1979a). Elsewhere Gingerich describes this (1979b) as "phenetic linking of closely similar species samples in each horizon with those in adjacent horizons to form a minimal spanning tree of evolutionary lineages."

Care would be needed. One does not want to allow two different species at one stratigraphic level to be inferred to have the same descendant, but one does want to sometimes allow one species to have two different descendants. In addition, no criterion is stated allowing a species to be linked with a descendant species more than one stratum away. Thus it is unclear whether the method would ever infer that an ancestral species had been missed (though the examples given by Gingerich do seem to have some such events). The lineages are not in fact inferred correctly in this example (though perhaps only because I connected the dots in Figure 32.2 by eyeball rather than by an algorithm). Gingerich (1979b) does say that "the approach requires a relatively dense and continuous fossil record." It thus seems to have an implicit assumption that the fossil record in each stratum is nearly complete.

Stratophenetics cannot cope with all data sets, simply because it is not very well-defined, with no well-specified procedures for dealing with difficulties. It has mostly been implemented by "eyeball." Nevertheless, it has stimulated further work on well-defined numerical methods for phylogenies with fossil data.

Table 32.1: The simple data set with 0/1 characters, extended by adding four fossil species and a number showing which stratum each species is in (1 is earliest, 5 latest).

	Characters						
Species	1	2	3	4	5	6	Stratum
Alpha	1	0	0	1	1	0	5
Beta	0	0	1	0	0	0	5
Gamma	1	1	0	0	0	0	5
Delta	1	1	0	1	1	1	5
Epsilon	0	0	1	1	1	0	5
f1	1	1	0	1	0	0	4
f2	0	0	1	0	0	0	3
f3	1	0	0	0	0	0	2
f4	0	0	0	0	0	0	1

Stratocladistics

Stratocladistics is a relatively well-defined approach that tries to use the stratigraphical information in a parsimony method. Fisher (1991, 1992) suggested that, in addition to the count of the number of changes of state that the characters require, we compute a "stratigraphic parsimony debt." This counts the number of times that a lineage crosses a stratum without a fossil being observed. This is to be added to the number of changes to compute the overall score of a phylogeny. Table 32.1 shows our example data set from Table 1.1, extended by adding four fossil taxa. The extra column on the right side of the table shows which stratum each species is from. This can be thought of as a character with unidirectional change among its states: $1 \rightarrow 2 \rightarrow 3 \rightarrow 4 \rightarrow 5$ with no reversals permitted.

Figure 32.3 shows two trees evaluated by Fisher's criterion. The left-hand tree has 10 changes of character state for the ordinary characters, but it requires four instances in which a lineage crosses a stratum without leaving a fossil. These are indicated by dashed lines. The total score of the tree is thus 14. The right tree has 11 changes of the ordinary characters, but requires only two instances of a lineage crossing a stratum without a corresponding fossil being found. Thus it has an overall score of 13; it would be one of the trees tied for best according to stratocladistics.

Stratocladistics can be implemented using some existing parsimony methods. In our example each stratum-crossing by a lineage creates one extra change of the stratum character. With five states present, we would expect at least four changes of state in the stratum numbers. For the left-hand tree in Figure 32.3 one can optimize the stratum numbers on the tree; we find that they require eight changes of

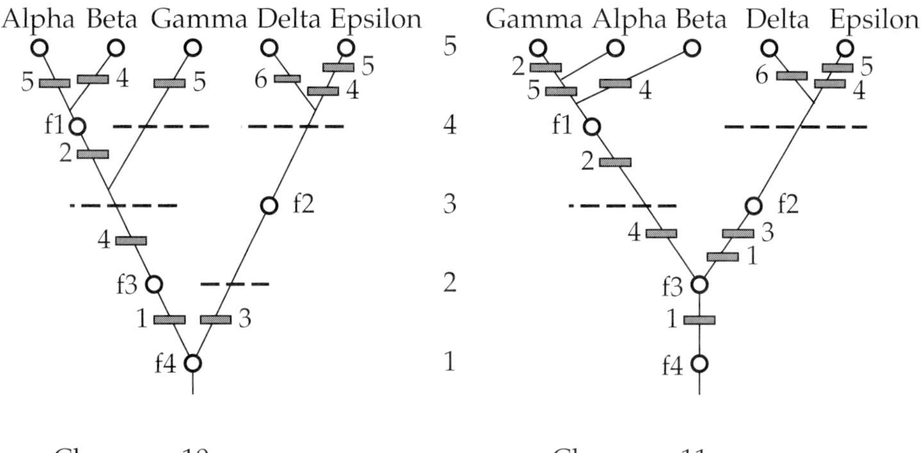

Changes: 10
Stratum crossings: 4
Total score: 14

Changes: 11
Stratum crossings: 2
Total score: 13

Figure 32.3: Two phylogenies evaluated for their fit using stratocladistics. Each fossil species is shown either as an ancestor or as a tip. The tree on the right shows more change in the characters, but this is more than offset by having fewer cases where lineages cross strata without a fossil species having been found (dashed lines).

state. We thus have four extra changes of state. In programs that allow mixtures of different kinds of parsimony in different characters, stratocladistic analysis is easily implemented — one need only specify that the stratum character has a linear, unidirectional character state tree with known ancestral state. The tree that will be favored will be the one preferred by stratocladistics, and by subtracting from the score of that tree 1 less than the number of strata, one can compute the score according to stratocladistics.

Stratocladistics is less likely than stratophenetics to assume that a fossil must be an ancestor. If a fossil species contains a derived state that does not appear in a later species, then it might be reconstructed as being an offshoot of the main lineage. Figure 32.4 shows two interpretations of the fossil when there is a two-state character in which the fossil has state 1 while preceding and later species all have state 0.

We have already seen in Chapter 9 that there is a statistical justification for parsimony when changes are expected to be rare, and that when they are sufficiently rare, weighted parsimony methods should approach equal weighting of changes. Stratocladistics weights changes and stratum-crossings equally. It thus implicitly

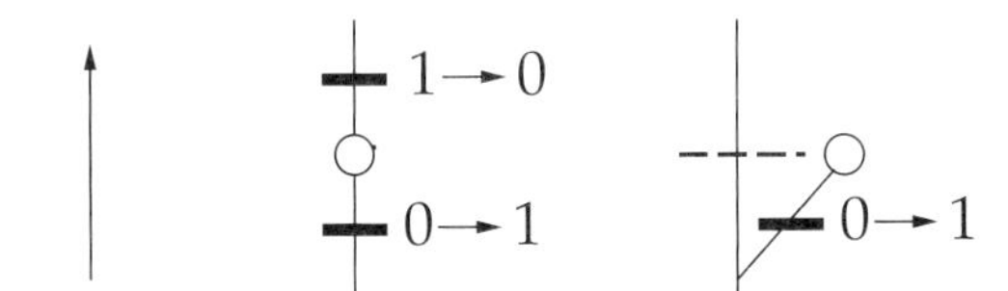

Figure 32.4: Two interpretations of the fossil species (circle) that contains one derived state (1) not present later. In the left interpretation, there are two changes of state (black bars). In the right interpretation, there is only one change of state, but one extra instance of a lineage crossing a stratum without there being a corresponding fossil species (dashed line). Stratocladistics would consider these possibilities equally well-supported.

assumes that it is a rare event for a lineage to cross a stratum without leaving a fossil species. Like stratophenetics, it thus assumes a relatively complete fossil record, as it considers examples of incompleteness as rare surprises.

Fisher (1992) acknowledged one arbitrary aspect of stratocladistics, which Huelsenbeck and Rannala (1997) also pointed out. This is the issue of how finely we subdivide the strata. If we divide the fossil record into more strata, this can result in increasing the weight given to the stratum characters. If the situation in Figure 32.4 were considered to have one stratum, the two scenarios are considered to be equally good. But if the fossil-bearing stratum were divided into two, with the fossil species present in both, then the right scenario has one character change plus two stratum crossings, and the interpretation of the fossil as an ancestor is preferred. As there is an implicit assumption of a relatively complete fossil record, the method then tries harder to reduce the number of missing fossils.

Controversies

The assumption, common to stratophenetics and stratocladistics, that failure to observe a fossil is a rare event, has generated a certain amount of controversy. The less adequate the fossil record of a group, the less well-justified this assumption will be. The cladistic parsimony school has tended to argue that one should use the character data and not the stratigraphic data in inferring the phylogeny. The argument may appear to be about the "adequacy" of the fossil record—in reality it is about whether lack of fossilization can be considered to be rare. This can be seen in the online debate from 1998 at `http://www.nature.com/nature/debates/fossil/` .

A not-quite-likelihood method

Wagner (1998) has proposed a "likelihood approach" to compromise fit of the tree to the data with fit of the tree to the stratigraphy. It is not a full likelihood method, but assumes that parsimony is the relevant statistic summarizing the fit of the data to the tree. For a given proposed tree, he computes a probability of the stratigraphic information implied, using a model of sampling of fossils. This is to be multiplied by the probability of the data given the tree. However, for the latter, he computes by a computer simulation the probability that the tree will result in a parsimony value equal to that observed. This is not equivalent to computing the probability of obtaining the observed data given the tree, which would be needed by a true likelihood method. Wagner's method does achieve a compromise between stratigraphic "debt" and parsimony score, and thus can be regarded as a statistically inspired variant of stratocladistics.

Stratolikelihood

If the rate of fossilization is relevant, why not develop a method that can cope with a variety of different rates? Huelsenbeck and Rannala (1997; see also Huelsenbeck and Rannala, 2000) have provided a maximum likelihood method for inferring phylogenies using a statistical model of the availability of fossils. They use a simple model of fossilization in which there is a constant rate λ of occurrence of fossils per lineage. If one of the lineages in a phylogeny starts at time t_f and continues to time t_l, the expected number of fossils observed will be $\lambda\,(t_l - t_f)$. The number of fossils observed will come from a Poisson distribution with this mean, so that the probability that the lineage has n fossils observed is

$$\text{Prob}\,(n \mid \lambda, t_f, t_l) \;=\; e^{-\lambda(t_l - t_f)}\;[\lambda(t_f - t_l)]^n\,/n! \tag{32.1}$$

However, note that we have three parameters, λ, t_f, and t_l. The observed times of the fossils $x_1, x_2, \ldots, x_n$ are informative about the t's. The x_i come to us ordered, so the joint probability of the x's given the t's is $n!/(t_l - t_f)^n$. Multiplying these to get the joint probability of the observations, we get

$$\text{Prob}\,(n \mid \lambda, t_f, t_l) \;=\; e^{-\lambda(t_l - t_f)}\;\lambda^n \tag{32.2}$$

The logarithm of this probability is

$$\ln\,\text{Prob}\,(n \mid \lambda, t_f, t_l) \;=\; -\lambda(t_l - t_f) \;+\; n\ln(\lambda) \tag{32.3}$$

The events of fossilization and discovery of the fossils are independent between different lineages. Thus the overall probability, given that each lineage i has n_i fossils and stretches from time $t_f^{(i)}$ to $t_l^{(i)}$, is

$$\ln\,\text{Prob}\,(\mathbf{n} \mid \lambda, \mathbf{t}) \;=\; -\lambda\,T + \sum_i n_i \ln\lambda \tag{32.4}$$

where T is the total length of the tree. If N is the sum of the numbers of fossils,

$$\ln \text{Prob}(\mathbf{n} \mid \lambda, \mathbf{t}) = -\lambda T + N \ln(\lambda) \tag{32.5}$$

If we have different proposed phylogenies that have different numbers n_i of fossils on their branches, and different times $t_f^{(i)}$ and $t_l^{(i)}$ for the beginnings and ends of their lineages, we can use equation 32.5 as the effect of the fossil part of the data on the overall likelihood of these trees. This assumes we know λ. Alternatively, we can infer λ for each tree. Taking equation 32.5 and differentiating it with respect to λ, equating the result to zero, and solving for λ, we get

$$\widehat{\lambda} = N/T \tag{32.6}$$

where N is the total number of observations of fossil species, and T is the total length of the proposed tree. Substituting this into equation 32.4, we get

$$\ln \text{Prob}(\mathbf{n} \mid \widehat{\lambda}, \mathbf{t}) = -N + N \ln N - N \ln(T) \tag{32.7}$$

If we consider different phylogenies, each with assignments of fossils to branches and times for beginning and end of each branch, this term will penalize phylogenies that are long.

This derivation is not the one given by Huelsenbeck and Rannala. They obtain a slightly different result by considering the first and last times o_f and o_l that fossils are observed in each ancestral lineage. They do not include a term for the density of the other x's given o_f and o_l. This would have made their result identical to the present result.

Making a full likelihood method

The terms above are only those dealing with occurrence of fossils. A complete treatment would also include probabilities (or probability densities) for the evolution of the characters in the present-day and the fossil species. We have seen in Chapter 24 the difficulties that must be faced in coming up with such a model. A full likelihood treatment of fossil and neontological data is in the future. Huelsenbeck and Rannala consider the terms they derive as components of such a model; they do not pretend to have a complete method at this stage. The part of the model that treats character change cannot be omitted: If it is, there is no way to know whether a given fossil is or is not a candidate to lie on a particular lineage.

More realistic fossilization models

Huelsenbeck and Rannala (1997) acknowledge that this simple Poisson process model of fossilization is naive. There are many more complications to take into account, including different rates of fossilization in different strata. They regard their model as a starting point, and suggest that it could be extended and that likelihood ratio tests of different fossil preservation models could be done using it.

It is not hard to alter the model to allow for discrete strata, in each of which a lineage has a probability p of being observed from one or more fossils (Foote and Raup, 1996). The results are similar to the above ones, including the possibility of inferring p by maximum likelihood for each tree. Greater realism could be achieved by allowing for it to be easier to find fossils in some strata than others, and having some strata exist in larger areas than others. One could estimate fossilization probabilities separately for each stratum, though there is the interesting issue of whether this is too many parameters and whether it would not be better to have a hidden Markov model assign fossilization probabilities to strata. We could also think of using data on how many individuals of each lineage in each stratum are seen. The present model simply counts the lineage as represented or not, without regard to how many times it is represented.

Fossils within species: Sequential sampling

Molecular data are often collected from present-day and fossil samples. You might think that this is limited by the poor survivability of fossil DNA, but, if so, you are not thinking of the right cases. In some viral cases molecular evolution is so rapid that substantial amounts of change occur in a few years. The most famous of these is of course HIV evolution, where rates of evolution are about a million times faster than in more normal organisms. Samples taken from the same patient a few months apart must then be treated by taking their sampling time into account. For these RNA viruses, the issue of sampling time becomes particularly important.

For a population of viruses evolving within a host, there is also a prior distribution of trees. In fact, it is the coalescent, which we have discussed above in Chapters 26 and 27. We are within a single species and should keep that in mind. Figure 32.5 shows a series of samples of virus taken at different times in the course of an imaginary infection. The coalescent assumes that there are a great many individuals in the population, as is certainly appropriate for virus particles in an infection. It is tempting to assume that the samples from 90 days ago are ancestral to the samples today, but this temptation must be resisted. The lineages reaching back to 90 days ago are found in genomes different from the ones that were sampled. Going back 90 days, we simply find that the genomes sampled then add new lineages to the coalescent. The resulting lineages then gradually coalesce with each other as we go further back.

A similar issue arises when we sample ancient DNA within the human species. It is not wise to assume that a gene from a modern Egyptian is the direct descendant of one from a mummy. The modern lineage almost certainly existed in someone else at that time. If the population then were even as small as 10,000 individuals, it would take about 800,000 years for the two lineages to coalesce!

Sequential sampling can be incorporated into likelihood analysis of coalescents (Felsenstein et al., 1999; Rodrigo and Felsenstein, 1999). It is necessary to keep the

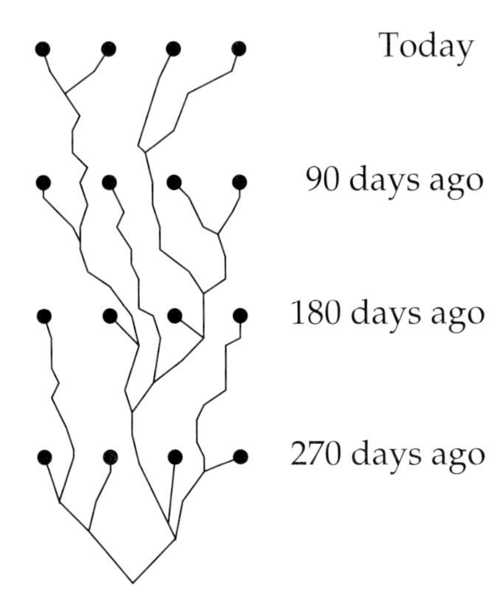

Figure 32.5: Hypothetical coalescent of virus genomes within a patient, showing samples of virus taken at a series of times. The black circles are the sampled genomes.

time scale and the branch-length scales separate. Suppose that we have k lineages, an effective population size of N_e, a mutation rate of μ per generation, and a generation time τ. The number of generations back to the most recent coalescence is exponentially distributed with expectation $4N_e/[k(k-1)]$. In time units it is exponentially distributed with expectation $4N_e\tau/[k(k-1)]$. The density of time until coalescence is

$$f(u_k \mid N_e, \tau) = \left[\frac{k(k-1)}{4N_e\tau}\right] \exp\left[-\frac{k(k-1)}{4N_e\tau}\, u_k\right] \tag{32.8}$$

which depends on $4N_e\tau$. The probability of the sequences given the genealogy depends on the tree topology and on the branch lengths: These are the ratio of mutation rate per generation and time per generation, which is μ/τ.

The overall likelihood is, as in equation 27.1,

$$L = \sum_G g(G \,|\, N_e\tau)\ \text{Prob}\,(D \,|\, G, \mu/\tau) \tag{32.9}$$

where g is the density function of the coalescent with serial sampling. This is like the usual coalescent, with one wrinkle. In each successive interval, we multiply by the probability that nothing happens for u_k generations, times the probability of whatever happens at the bottom of the interval. If the interval has a coalescent at its bottom, the density is as given in equation 32.8 above. But the interval could have a sampling time at its bottom. At that time, there is no coalescence event,

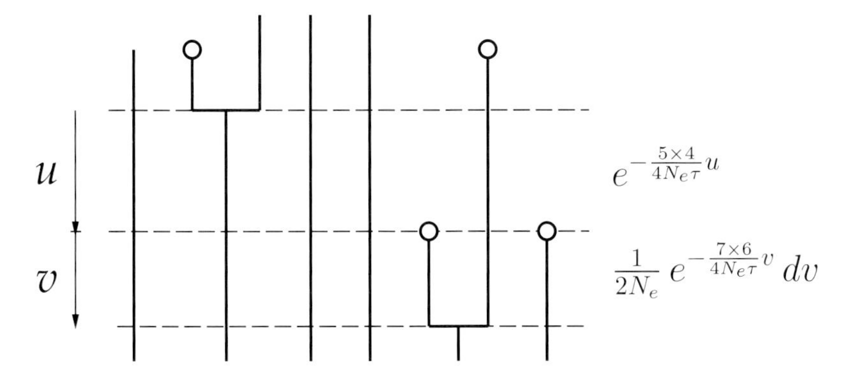

Figure 32.6: Stages in the calculation of the probability density of a coalescent, taking sequential sampling into account. The top term is the probability of noncoalescence of six lineages until they reach the next time of sampling. The term below it is the probability density of the time until two of the resulting lineages coalesce.

but the addition of n_t new lineages to the coalescence, as the sample size of the samples at time t is n_t. The probability of this event is 1, since we are given the schedule of samples. Thus the probability for that interval is

$$\text{Prob}\ (u_k \mid N_e\tau) \ = \ \exp\left[\ -\ \frac{k(k-1)}{4N_e\tau}\ u_k\right] \tag{32.10}$$

The function g is a product of intervals above a coalescence, and intervals above a sampling. The first have probability densities, the second probabilities, and their product is g. Figure 32.6 shows two of the terms that need to be calculated, one for non-coalescence until a sampling time is reached, the other for the time to coalescence. All of the terms are of one or the other type.

The likelihood depends upon the two parameters $4N_e\tau$ and μ/τ. Using Markov chain Monte Carlo methods, we can infer them separately. Their product is $4N_e\mu$. If we know the generation length τ, it can be used to make estimates of N_e and of μ. If we know μ, it can be used to infer $4N_e$ and τ. Similarly, if we know N_e, it can be used to infer τ and μ. But we cannot infer all three quantities together. Still, with knowledge of one of the three quantities, we can infer the other two. The existence of dated samples with known times gives us the ability to estimate one more parameter.

Between species

At present there is considerable uncertainty whether molecular sequences can be reliably sampled from more than 100,000 years ago. If they can, we will find ourselves facing many of the same issues as in sequential sampling in coalescents.

If we are working with samples from multiple species, the coalescent machinery does not apply, but we do continue to face the issue of branch length versus time. The amount of branch length that accumulates in an interval t is μ/τ per unit time. We may know the time at which each sample was taken, but unless we know the mutation rate per generation, μ, and the generation length, τ, we cannot turn time into branch length. Maximum likelihood estimation will then depend on estimating the compound parameter μ/τ that converts time to branch length. Similar issues arise with quantitative characters, where there will be a parameter for the rate of Brownian motion per unit time.

Chapter 33

Tests based on tree shape

As the numbers of species in studies of phylogenies has grown, so has the interest in using the shapes of these trees to test hypothesis about evolution. For example, if a few lineages give rise to most of the descendant species because they have acquired an important adaptation, this should be visible in the asymmetry of the resulting tree. If rates of speciation and extinction vary through time, this may be visible in the lengths of branches in the tree. If speciation or extinction is correlated with some particular phenotype, then its distribution on the phylogeny may be correlated with species richness of the clades.

Although stochastic processes were used to model phylogenies almost 80 years ago (Yule, 1924), most of the interest in using them to make inferences about processes of speciation and extinction has been during the last 15 years. Some of the questions that have been asked are:

- How can we estimate the speciation and extinction rates from a phylogeny?
- Have these rates changed through time?
- Have they changed in certain clades?
- Can we test whether they are correlated with particular characters?

Most methods have examined the shape of the phylogeny, without regard to the branch lengths or times, or they have examined the times of branching without regard to which lineages are doing the branching. Let us start with a basic result about tree topologies and then proceed to results about the time of branching.

Using the topology only

Much of the theoretical work has concentrated on the case in which branch lengths are not used. As a null hypothesis, the case of a randomly branching tree is important. Much of the work on this case has been motivated by the observation that phylogenies inferred by many methods tend to be more asymmetric than expected under this model. For a review of the extensive empirical literature on this and the rather unsatisfactory explanations offered so far, you can consult the excellent reviews by Mooers and Heard (1997) and by Aldous (2001).

Random processes generating trees by random splitting were introduced by Yule (1924) as a model for almost the same problem: the sizes of genera. The *Yule process* is the simplest pure-birth process, in which particles reproduce with a constant probability of giving birth per particle per unit time.

Harding (1971) examined the shapes of trees arising out of the Yule process. It turns out (Thompson, 1975) that his results are also valid for trees arising from a more general process in which there are random births and also random deaths. Harding's results for unlabeled rooted bifurcating shapes can most easily be explained using a result for the sizes of the clades separated by the root. Let us first look into that.

Imbalance at the root

If we have a rooted bifurcating tree with n unlabeled tips, and we look at the two clades that arise from the bottom fork, these divide the species into two subsets. Their sizes could be any of the possibilities $1 : n-1$, $2 : n-2$, $3 : n-3$, $\ldots$, $n-1 : 1$. If we generate trees by random branching, there is a remarkable fact about the probabilities of these sizes. They all have equal probability. There are $n - 1$ of them, so each has probability $1/(n - 1)$.

This was published in the biological literature by Farris (1976) and Slowinski and Guyer (1989). It is a consequence of the probability distribution of a well-known process, the *Polya urn model* (see, for example Feller, 1971, section VI.12). The Polya urn model is of an urn that contains two colors of balls, say, red and green. We start with r red and g green balls. At each step we draw one ball at random and look at its color. If it is (say) red, we toss it back and add also another red ball. If it is green, we toss it back and add one green ball. There are many good reasons to be interested in this process, but we will be concerned with it only as a model of speciating lineages. Drawing a ball may be considered as selecting a lineage to speciate. Returning it with another copy of itself models speciation. Polya showed what the distribution of numbers of red and green balls would be after s steps, starting from an urn with one red and one green ball. It is the flat rectangular distribution just mentioned.

I will not try to derive this result from scratch, but will show that it is the distribution by an argument using induction. We will show that it is true for $n = 2$, and then show that if it is true for $n - 1$, it is also true for n. This allows us to establish that the result is generally true for all n by induction. For $n = 2$, we have only one possibility, a $1 : 1$ split, which has probability 1, which is of course equal to $1/(n - 1)$. That is the starting point for the induction. Suppose that it is true for an urn that ends up with $n - 1$ balls, for which each possibility has probability $1/(n - 2)$. At the the next step, we could get numbers $k : n - k$ of red and green balls in either of two ways:

1. We could draw a red ball from an urn that already has $k - 1$ red balls out of $n - 1$ balls. This has probability $(k - 1)/(n - 1)$. The probability that an urn of $n - 1$ balls has $k - 1$ red balls is, as we have seen, $1/(n - 2)$.

2. We could draw a green ball from an urn that already has k red and $(n-1)-k$ green balls. This has probability $(n - 1 - k)/(n - 1)$. The probability that the urn has k red balls is, as we have seen, $1/(n - 2)$.

Putting these probabilities together, the total probability that an urn of n balls has k red balls is

$$\frac{1}{n-2}\,\frac{k-1}{n-1} + \frac{1}{n-2}\,\frac{(n-1)-k}{n-1} = \frac{1}{n-1} \tag{33.1}$$

This establishes the general result by induction. Starting with two lineages, when we reach a total of n species, we will have a probability $1/(n - 1)$ that there are k of them descended from the left-hand lineage. As this probability does not depend on k, the distribution of numbers of descendants of the left-hand lineage is uniform across all possible values. Thus, for example, if there are a total of eight species, the numbers descended from the left-hand lineage at the root are equally likely to have the values 1, 2, 3, 4, 5, 6, and 7.

We will use this result by ordering the lineages according to their numbers of descendants. If we see numbers of descendants 5:2, we will instead call this 2:5. Thus the probability of $k : n - k$ will be not $1/(n - 1)$ but $2/(n - 1)$, as we will also be including the possibility $n - k : k$. The exception to this is when $k = n - k$, in which case the two possibilities are the same, and their probability is $1/(n - 1)$.

Harding's probabilities of tree shapes

Harding (1971) has used this result to compute probabilities of tree shapes. Suppose that we have a tree T of n species. At its base is a fork that leads to two subtrees, T_1 and T_2, of n_1 and n_2 species, respectively. Harding notes that the probability of the shape of tree T is simply the probability of an $n_1 : n_2$ split at the base of the tree, times the probabilities of the shapes of the two subtrees. Once it has been decided that the bottommost fork separates n_1 from n_2 lineages, this is independent of what particular shape occurs within each of these subtrees. The bottommost fork gives rise to two lineages. We know only that they will ultimately give rise to n_1 and n_2 lineages, respectively. Given that knowledge, the shapes of the two subtrees are computed using the same rules as for the overall tree.

The effect of this independence is that we can calculate the probability of a bifurcating, rooted tree shape by simply computing the product of the probabilities of the splits at each fork. Figure 33.1 shows the calculation for a tree of 10 species. The overall probability of getting this tree shape from a process of randomly branching lineages is

$$\frac{2}{9} \times \frac{2}{5} \times \frac{2}{3} \times 1 \times 1 \times 1 \times \frac{1}{3} \times 1 \times 1 = \frac{8}{405}$$

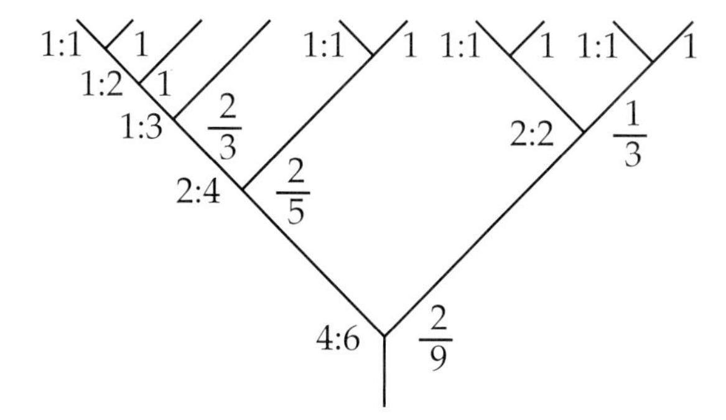

Figure 33.1: Computing the probability of the tree shape of a bifurcating rooted tree with 10 species. For each fork the numbers of lineages arising from the two lineages is shown (to the left of the fork), along with the probability of this pair of numbers (to its right). The overall probability of the tree shape is the product of the probabilities.

Stone and Repka (1998) derive a closed-form formula for this probability, arguing that this will be easier to compute. I have my doubts that their formula will be easier than this method of computation in practice.

A less obvious consequence of Harding's probabilities is that each labeled history of the species is equally likely. This is true because if we follow the n lineages back in time, the dual process of random branching is random coalescence. Brown (1994b) (see also Steel and McKenzie, 2001) derives a related expression for the number of labeled histories that correspond to a given tree topology.

Tests from shapes

Even though Harding's method allows us to compute probabilities of tree shapes quickly, it is not obvious how to derive a test from that. We would like to take all the rooted, bifurcating tree shapes, and place them in a natural order according to how asymmetrical they are. Having done that, we could ask whether an observed tree shape is significantly asymmetrical, say in the most extreme 5% of the ordering.

The difficulty lies in knowing what tradeoff to make between the number of lineages and the asymmetry. Is a 2 : 8 split more extreme than a 1 : 5 split, or less? That issue in turn can only be settled by consideration of what the alternative to random branching is. If at each split one lineage is more likely to further speciate than is the other, this will tend to generate asymmetrical trees. But unless we have specified more precisely what this alternative process is, we can't know how strong the difference is expected to be.

Measures of overall asymmetry

A number of people have suggested measures of overall asymmetry of rooted bifurcating trees:

- Sackin (1972) suggested using either σ_N^2, the variance of N_i across tips in the tree, or $\bar{N}$ its mean, where N_i is the number of nodes in the tree below tip i. For the tree in Figure 33.1 the variance is 0.65. Kirkpatrick and Slatkin (1993) gave an expression for the expectation of $\bar{N}$, and Rogers (1996) gave recursive expressions for its higher moments.

- Colless (1982) defined a measure C. If at an internal node in the tree the numbers of descendants of the two lineages are r_i and s_i, where we order them so that $r_i \geq s_i$, C is the sum over all internal nodes of $2(r_i - s_i)$, divided by the maximum possible value of this sum, $(n-1)(n-2)$. For the tree in Figure 33.1 this is 14/72. Heard (1992) and Rogers (1994) presented methods for computing the mean, variance, and the distribution of C.

- Shao and Sokal (1990) suggest the measure B_1, which takes each interior node in the tree (other than the root), and sums the reciprocals of the numbers of nodes below it in the tree. Thus the tree in Figure 33.1 has $B_1 = 2 \times 1 + 4 \times (1/2) + 1/3 + 1/4 = 4.58333$. Shao and Sokal (1990) also suggested that the numbers N_i for the n tips be used to compute another measure:

$$B_2 = \sum_{i=1}^{n} N_i / 2^{N_i}$$

  For the tree in Figure 33.1 this is 3.1875.

- Kirkpatrick and Slatkin (1993) suggest measuring the average number of nodes between the tips of a tree and its base (counting all nontip nodes below each tip). They show that this measure, $\bar{N}$ has an expectation of $2\sum_{i=2}^{n} 1/i$. The same result had earlier been obtained by computer scientists (Lynch, 1965; Mahmoud, 1992, p. 72). For the tree in Figure 33.1 this is 3.5.

- Hey (1992) suggested a statistic that asks for each interval between speciations whether the second speciation is in one of the lineages descended from the first. He computed the probability of the observed sequence of successive and nonsuccessive speciations, and compared it to a the probabilities of a large sample of these probabilities from simulated randomly branching trees.

- Kirkpatrick and Slatkin (1993) use the "left-light root ranking" of Furnas (1984) to make another measure. Furnas ordered trees from least symmetrical to most symmetrical, using a recursive numbering scheme that will not be explained here. For the tree in Figure 33.1, $R = 88$ out of a possible total of 98.

All of these measures are to some extent arbitrary. For almost none of them are their distributions under random branching of lineages known, though for some ($\bar{N}$ and C) their expectations can be calculated, and for C the variance and distribution can be calculated numerically.

Choosing a powerful test

Kirkpatrick and Slatkin (1993) set out to investigate six of these seven measures to find which might be the most powerful one to use to detect inequalities of rates of speciation and extinction that would make the trees asymmetrical. Their null hypothesis was random branching of lineages. To investigate the statistical behavior of these measures one must specify the alternative hypothesis. There are many ways one can do this (for example, one could imagine a quantitative character that evolved according to a Brownian motion process, and have the speciation and extinction rates depend on it). Kirkpatrick and Slatkin imagined a process in which at each split of the lineage, the rates of speciation of its daughter lineages were multiplied by $2x/(x+1)$ and $2/(x+1)$, respectively. This leaves the expected rate of speciation unchanged for the whole clade, with the two daughter lineages having a ratio of speciation rates of $x : 1$. This was done at each speciation event in the tree. Thus if $x = 3$ and the rate of speciation of the ancestral lineage is 1, after the first speciation the rates are 3/2 and 1/2. If the left-hand lineage speciates next, its descendant lineages will have rates of speciation 9/4 and 3/4, and so on.

Kirkpatrick and Slatkin used computer simulation to find the distributions of the six statistics under random branching of lineages for different numbers of species. They found the two-sided 95% confidence intervals for each of the measures. They then investigated, for their asymmetric speciation process, the probability of exceeding the 95% limit in the direction of asymmetry. Four of the six measures (all except R and B_2) did reasonably well, with B_1 performing best, and with Colless's measure C a close second.

As there are many other possible alternative hypotheses, this study does not exhaust the issue. It might even be possible to make a likelihood ratio test between well-specified null and alternative processes, once these could be decided upon. Losos and Adler (1995) have suggested that if speciation is not an instantaneous process, this might reduce the occurrence of successive speciations in a lineage, making the resulting tree more rather than less symmetrical (see also the reconsideration of this by Chan and Moore, 1999).

Tests using times

There has been considerable interest in recent years in using the timing of branching in phylogenies to make inferences about speciation and extinction. Although there has been some interest (see Hey, 1992) in using the lengths of branches, most work has been concentrated on the times of branching, which is more accessible to mathematical treatment.

Nee, May, and Harvey (1994) have introduced inference about the birth and death rates of lineages from plots of the number of lineages through time, without using the asymmetry of the tree. The objective is to evaluate the speciation and extinction rates, and hypotheses about their change through time, rather than to ask whether different lineages have different rates of speciation. If this is the question, Sanderson and Bharathan (1994) agree with Thompson (1975) that in simple models with random speciation and extinction, and with contemporary species, the details of the tree topology (more properly, of the labeled history) are irrelevant to testing of hypotheses about temporal variation of the rates of these events. The converse is not true: Testing whether lineages differ in their speciation and extinction rates is best done using information about both labeled history and speciation times. I will use a model quite similar to that of Nee, May, and Harvey, with some differences, to explain how likelihoods can be calculated for a simple model of speciation and extinction.

A standard stochastic process is the *birth and death process* or *birth-death process*, which I have mentioned earlier in this chapter. It imagines lineages that have a constant rate λ at which they each give rise to new lineages, and a constant rate μ at which they each die. The events are stochastic with these rates. If there are n lineages, in a very small interval of length in time dt, the probability that there is a lineage that gives birth (speciates) is $n\lambda\, dt$. The probability that there is a lineage that dies (goes extinct) in that time is $n\,\mu\, dt$.

The transition probabilities for this stochastic process are well-known (derived by Kendall, 1948). The probability that a single lineage has at least one descendant after time t is

$$s(t) \;=\; \text{Prob}\,(n > 0 \mid t) \;=\; \frac{(\lambda-\mu)e^{(\lambda-\mu)t}}{\lambda\, e^{(\lambda-\mu)t} - \mu} \tag{33.2}$$

To use this result to determine the expected distribution of times between successive branchings in a tree, we must realize that when we use a phylogeny of present-day organisms to look at the rate of speciation events, we see only those speciations that have led to both lineages having descendants in the present. This picture is the one that Nee, May, and Harvey (1994) call the *reconstructed evolutionary process*. Figure 33.2 shows the distinction between the full tree of lineages and the one we can see looking backward to the ancestors of the surviving species.

Lineage plots

Nee, May, and Harvey (1994; Harvey, May, and Nee, 1994) make a plot of the logarithm of the number of lineages against time, for both the reconstructed lineages and all lineages. Using a birth-death process, they obtain the curve showing the expected numbers of lineages through time, counting all lineages, and also the curve counting only reconstructed lineages. A birth-death process that starts with a single lineage has the expected number of descendants rise (or fall) exponentially: It is $\exp[(\lambda-\mu)t]$. If we consider an interval of time from 0 to T, and look

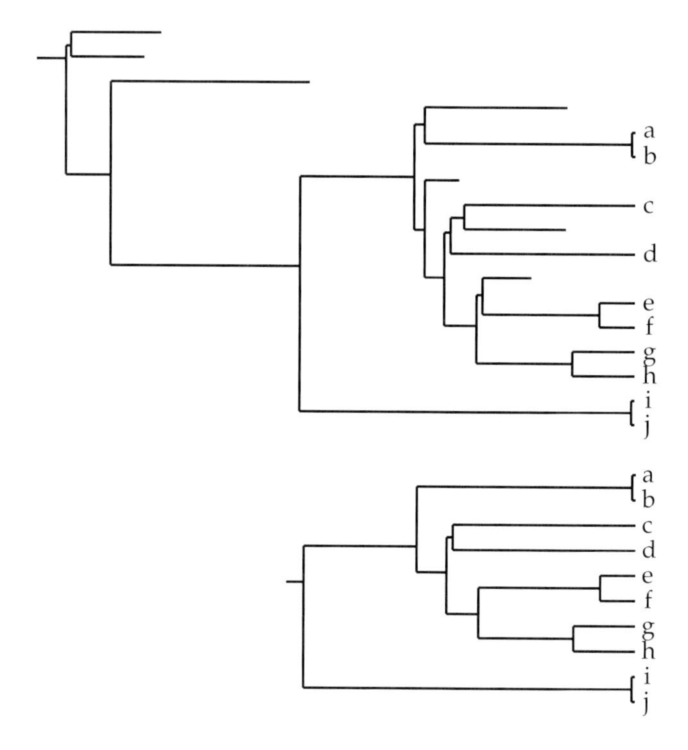

Figure 33.2: Results of a birth-death process generating phylogenies. Upper: The full process including lines that become extinct before the end of the period of time. Lower: The reconstructed process that shows only the lines that lead to surviving species.

at the number of lineages at time t, the fraction of these that have surviving descendants at time T is $s(T-t)$, using the function in equation 33.2. These are the lineages that would be seen in a reconstruction from species sampled at time T.

Nee, May, and Harvey argue that we must realize that we will have sampled only from cases in which at least one species survived to T. This is a fraction $s(T)$ of all cases. All others have no survivors. If we average the number of lineages at time t over only those cases where the number of survivors at T is positive, it is easy to show that the expected number of lineages is $\exp[(\lambda-\mu)t]/s(T)$. Of these a fraction $s(T-t)$ are expected to reach time T. Thus the expected number of lineages we reconstruct, given that we look only at cases where some lineages survive to the present, and given that we see only those lineages that have survived, is

$$\mathbb{E}[N_R(t,T)] = \frac{e^{(\lambda-\mu)t}\, s(T-t)}{s(T)} \tag{33.3}$$

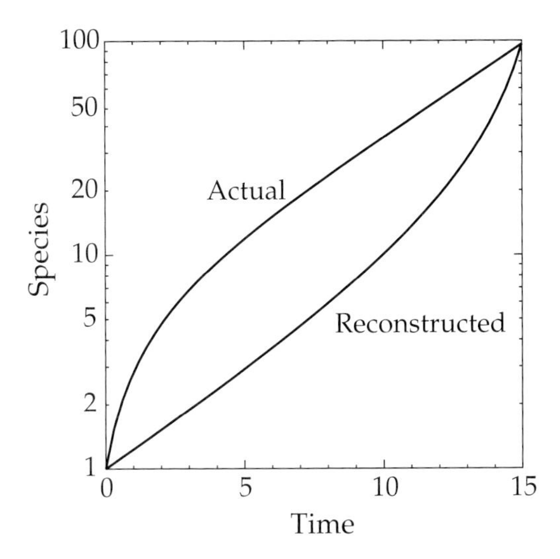

Figure 33.3: Plot of the logarithm of the number of lineages expected to exist (top curve) and of the number expected to be reconstructed (bottom curve) when $\lambda = 1.0$ and $\mu = 0.8$, over an interval of 15 units of time, starting with one lineage.

They also compute the expected number of actual lineages, counting the ones that will and will not be reconstructed, but restricting attention to the cases in which at least one lineage survives to the end. This they give as

$$\mathbb{E}[N(t,T)] = \frac{e^{(\lambda-\mu)t}}{s(T)} - \frac{s(t) - s(T)}{s(t)s(T-t)} \tag{33.4}$$

(see also the calculations by Kubo and Iwasa, 1995). I have made a numerical calculation to check this equation, and it seems to be correct.

Figure 33.3 shows an example of these two curves, plotted on a logarithmic scale. Their basic point is clear—that even if we were able to see only lineages that had at least one surviving descendant in the present, the rates of increase of numbers of lineages in different periods would give us enough information to estimate both speciation and extinction rates.

Likelihood formulas

Nee, May, and Harvey (1994) give a likelihood function for parameters λ and μ, presuming that we have observed the times of the forks in the tree of reconstructed lineages. Their expression is derived from one given by Thompson (1975, pp. 54–58), differing only in that theirs is conditioned on knowing the time of the earliest fork.

Both the lineage plots of Nee, May, and Harvey (1994) and the likelihood functions they give are conditional on having a tree whose first split is a time T ago. We may be interested in what the density function of trees would be, conditioned only on having n descendants alive in the present. The important result for computing this was obtained by Thompson (1975). Her derivation is a rather complex tour de force—I will give only a simplified rationalization here, one similar to the argument of Nee, May, and Harvey. I will compute a somewhat different quantity than either of these. A derivation similar to this one was given by Rannala (1997).

Suppose that we have a birth and death process with birth rate λ and death rate μ. We want to compute the probability density of obtaining a tree with n tip species and particular times of splitting. We measure time backward from the present. We want to know the probability density of obtaining an n-species tree with splitting times $t_1, t_2, \ldots, t_{n-1}$, where $T > t_1 > t_2 > \ldots > t_{n-1}$. If we know the function $p_1(t)$, which is the probability that a lineage has exactly one descendant after t units of time, the probability if we have $n = 1$ will simply be $p_1(T)$. This probability includes the possibilities that there have been some lineages branching off from this lineage, but that all those went extinct.

If we now alter the problem and ask for the probability (density) that we end up with two species, with the split between them at time t_1, this will be that same probability, but with the alteration that in the small interval $(t_1 + dt,\, t_1)$ there is an event with probability $\lambda\, dt\, p_1(t_1)$, the probability of a birth in this interval of a lineage that has exactly one descendant t_1 units of time later. Now if we alter it again to have a second split, on one or the other of the two lineages, at time t_2, this multiplies the probability density by $\lambda\, dt\, 2p_1(t_2)$. Continuing in this way, we get the probability density of a tree that has n species at the end, and has the ith species at time t_i:

$$\text{Prob}\,(n, t_1, t_2, \ldots, t_{n-1} \mid \lambda, \mu, T) \;=\; p_1(T)\, \lambda^{n-1}\, (n-1)! \prod_{i=1}^{n-1} p_1(t_i) \tag{33.5}$$

Note that these probabilities take into account all possible ways in which other speciation events could be arranged that do not lead to any further survivors in the present.

We would like to compute the probability density of the speciation times t_i conditional on ending up with n species. For this we need to divide the probability by its integral over all $t_1 \geq t_2 \geq \ldots \geq t_{n-1}$. Fortunately, this is not as hard to compute as might seem. The quantity being integrated is a product of the $p_1(t_i)$, and it follows that the integral is the same no matter what order we constrain the t_i to be in. As there are $(n-1)!$ orders, the integral for any one of them will be $1/(n-1)!$ of the integral over all possible orders, so that the resulting probability density is

$$\text{Prob}\,(t_1, t_2, \ldots, t_n \mid n, \lambda, \mu, T) \;=\; \prod_{i=1}^{n-1} \left[\frac{p_1(t_i)\; dt_i}{\int_0^T p_1(u)\; du} \right] \tag{33.6}$$

In this equation we have the usual birth-death process result (Kendall, 1948)

$$p_1(t) = \frac{(\lambda - \mu)^2 e^{(\lambda-\mu)t}}{\left[\lambda e^{(\lambda-\mu)t} - \mu\right]^2} \tag{33.7}$$

The expression on the right side of equation 33.6 can be used as the likelihood, to estimate λ and μ, much as Nee, May, and Harvey did. One advantage that it has over their approach is that we can consider its limit as $T \to \infty$. Nee, May, and Harvey, and also Thompson, took as the starting point the age of the first split. We have taken it before that. As Nee, May, and Harvey conditioned only on survival of the birth-death process, if λ exceeds μ, when we start from far in the past we expect a huge number of lineages. In the present case, we condition on getting n lineages, and this allows the probability density to converge as T increases, whether or not $\lambda > \mu$.

Other likelihood approaches

For the special case of the Yule process, which has speciation but no extinction, Hey (1992) and Sanderson and Bharathan (1993) have given expressions for the likelihood. These are completely consistent with the equations given here. Sanderson and Donoghue (1994) use these likelihood expressions to test whether different clades in a phylogeny have different rates of speciation.

Hey has also considered a model in which there are N lineages, and when one of them speciates, another goes extinct at the same moment. This maintains a constant number of species. It is equivalent to the "Moran model" of theoretical population genetics in which one individual dies and one is born in each event. One nice property of Hey's model is that a smaller sample of species $n \ll N$ has its reconstructed genealogy drawn from the distribution of a coalescent. Thus analysis of a sample of species rather than all extant species is possible. This is not possible in the birth and death process in general. Yang and Rannala (1997) have modeled sampling by having each lineage have an extra last-minute high probability of extinction. It is less than obvious to me that this is the correct way to model sampling, unless the systematist draws their samples in this way. It is not equivalent to drawing a fixed number n out of a larger number of B of species.

Other statistical approaches

The likelihood formula obtained above can replace some other statistical approaches. Wollenberg, Arnold, and Avise (1996) assumed a birth-death process and compared empirical distributions of speciation times to ones generated by simulation. Their method implicitly assumed that the birth rate equaled the death rate (Paradis, 1998b). Paradis (1997, 1998a, b) has developed likelihood methods using equations from survival analysis. Going backward in time, lineages decrease in number in a close parallel to deaths of patients under treatment (except that the patients die, they do not merge into each other). However, there is no event in survival analysis that is analogous to extinction. Given the reversal of time, that would be like the sudden appearance of a patient. Paradis's approximations amount to assuming that there is a pure birth process, extinction being unimportant by comparison (Pybus and Harvey, 2000). Kubo and Iwasa (1995) have used nonlinear curve fitting to infer parameters in a birth-death process. Their method can cope with both speciation and extinction, although it does not use the covariances that are expected between different points on the curve.

A time transformation

The result for n tips is not the same as the one Nee, May, and Harvey use. They note that the model of Hey (1992) that has a strict density-dependent limit on the number of lineages does lead to the coalescent, and they suggest using the coalescent to model the distribution of phylogenies. The present model does not result in the coalescent, though it comes very close if $\lambda = \mu$. It does lead to one interesting transformation that simplifies Rannala's (1997) suggested procedure for randomly sampling trees. We can take an imaginary time scale τ that satisfies

$$\frac{d\tau}{dt} = p_1(t) \tag{33.8}$$

Integrating this using equation 33.7, we get

$$\tau = \frac{e^{(\lambda-\mu)t} - 1}{\lambda e^{(\lambda-\mu)t} - \mu} \tag{33.9}$$

(This also provides the integrals for the denominators in equation 33.6.)

It is straightforward to show using equation 33.6 that this fictional time will be uniformly distributed between 0 and the value that is obtained for $t = T$. When $T \to \infty$, it is uniformly distributed between 0 and $1/\lambda$ if $\lambda > \mu$, and between 0 and $1/\mu$ if $\mu > \lambda$. When we have a n-species tree sampled from the probability density 33.6, its speciation times will be the order statistics of a sample of $n-1$ points from the appropriate uniform distribution. If we want to see whether a tree could be the reconstructed tree from the birth-death process with particular values λ and μ, we can compute the values of τ and see whether they appear to be uniformly distributed on the appropriate interval.

Alternatively, if we want to sample a reconstructed tree of n species, we can sample $n - 1$ points from the appropriate uniform distribution, order them, and then use the back-transformation

$$t = \frac{1}{\lambda - \mu} \ln\left(\frac{1 - \mu\tau}{1 - \lambda\tau}\right) \tag{33.10}$$

The labeled history can then be obtained by working backwards down the tree, choosing pairs of lineages randomly to coalesce at these times.

This result is equivalent to one suggested by Rannala (1997). It is somewhat different from the coalescent distribution suggested by Nee et al. (1995). They do not present a derivation from the birth-death process of their coalescent result but argue that it will prove useful in practice. Pybus and Harvey (2000) approach this problem differently. They use a statistic that would be exactly normally distributed if there were a pure birth process, and for cases in which there is death as well, they obtain its distribution by simulation.

The present result allows for both birth and death of lineages. It is not a coalescent and, unfortunately, does not share with it the property that it can predict the form of the tree reconstructed from a random sample of the species.

Characters and key innovations

A major question of interest has been whether clades that have acquired a particular phenotype—a putative "key innovation"—have thereby been enabled to spread more rapidly. Slowinski and Guyer (1993) have discussed testing of two sister clades, one of which has a putative key innovation. They use the uniform distribution of clade sizes under random branching to compute P values, and they suggest combining these across pairs of clades using Fisher's method of combining significance tests.

A likelihood-based test for this problem is presented by Sanderson and Donoghue (1994). They allow up to four different rates of speciation in a tree that has two clades, their ancestral lineage, and an outgroup. They compute a likelihood for each combination of rates by integrating over times of divergence of the two clades. Their model does not allow for extinction, and they do not have any penalty for a higher number of parameters in the model.

Work remaining

Lineage plots use branch-length information to infer the rates of speciation and extinction, and to test hypotheses about their changes through time. Tests using asymmetry of tree topology test whether some lineages have greater chance of speciating than others. What is still lacking is a testing and estimation framework that allows topologies and branch lengths to be used to answer both of these questions. For example, if a lineage has a greater rate of speciation than another, this should be reflected not only in its having more descendants but in shorter branch lengths between speciation events. At present we are lacking the following:

- Any method that uses both of these kinds of information (topologies and branch lengths)
- Any framework that takes into account the uncertainty of branch lengths and topologies
- Any method that makes use of a model of evolution of phenotypes of the species, to help see which traits might have contributed to the success of a clade

- Any method that uses a quantitative model of the evolution of the rate of speciation and/or the rate of extinction.

Clearly this literature is in its early days.

Chapter 34

Drawing trees

I am frequently asked for advice on how to "draw a tree" for some group from some data. Almost always, by "draw" the biologist means to infer the tree, and the issue of how to draw a picture of it is not really being raised. But drawing trees is worthy of discussion. There has been almost no literature on the subject, aside from descriptions of options in computer programs.

It might be thought not to be topic in science, but rather a matter of aesthetics. But how to draw diagrams of trees that convey the information most effectively is a legitimate concern of science. There is no scientific research on this topic, so this chapter will be mostly concerned with describing some possibilities, giving examples, and speculating as to how well they convey the information. Of course, trees can be drawn by hand, and programs such as TreeTool have been written enabling the user to move nodes manually. But my concern here will be how to draw a tree automatically, using an algorithm.

I will use a single tree as my example throughout this chapter. It will be described in the Newick tree format, which we have already used in this book. I will describe this format more in the next chapter. The tree I will use is

(((((((A:4,B:4):6,C:5):8,D:6):3,E:21):10,((F:4,G:12):14,H:8):13):13,
((I:5,J:2):30,(K:11,L:11):2):17):4,M:56);

There are two distinguishable cases, drawing a rooted tree, and drawing an unrooted tree. In both cases I will leave the left-right order of tips unchanged. It is worth noting that by reordering tips, you can change the viewer's impression of the closeness of relationships. For example, in drawings of the above tree, tip E looks close to F, simply because they are adjacent in the left-right order of tips. But we could equally well flip the order of branches in the tree so that A was immediately before L, and E and F were far apart. And yet these are really the same tree! A little judicious flipping may create a Great Chain of Being marching nicely along the sequence of names, even though the tree supports no such thing.

Issues in drawing rooted trees

In the examples shown here, the trees will "grow" from left to right. They might equally well grow from bottom to top, or from top to bottom. Growing from right to left is rarer, confined mostly to diagrams in which two trees abut head to head, in order to compare them. When we describe the coordinates of the interior nodes, x will be the horizontal distance and y the vertical distance. The horizontal (x) distances will be proportional to the branch lengths. (As we will see, this is not the only possible way of conveying branch lengths.)

Placement of interior nodes

Figure 34.1 shows the tree drawn with four different methods of placement of the interior nodes.

The horizontal axes are, as just mentioned, proportional to the branch lengths. The vertical coordinates of the tips are established by making a pass through the tree, from left to right and from top to bottom. (In the next chapter I will describe this more precisely as a postorder tree traversal.) As each tip is encountered, it is assigned a value of y, these being equally spaced. There are other possibilities, but these have their own difficulties.

The four methods shown of assigning the y coordinate of the interior nodes make use of the x and y coordinates of the descendants of the node, plus the branch lengths of the branches leading to the immediate descendants. I will assume here that the scale of the x axis is in branch-length units. The methods are, in order on Figure 34.1:

Intermediate: The y coordinate is halfway between the y coordinates of the first immediate descendant and the last immediate descendant. If there are only two immediate descendants, it is halfway between these. Thus if there are k immediate descendants,

$$y = (y_1 + y_k)/2 \tag{34.1}$$

Centered: The y coordinate is centered between the values of all descendant tips. Thus if there are K tips descended from the node, and these happen to be numbered $1, 2, \ldots, K$ in order,

$$y = (y_1 + y_K)/2 \tag{34.2}$$

Weighted: The y coordinate is the weighted average of the y coordinates of the first and last immediate descendants. The weighting is inversely proportional to the branch length leading from the node to each of these: If there

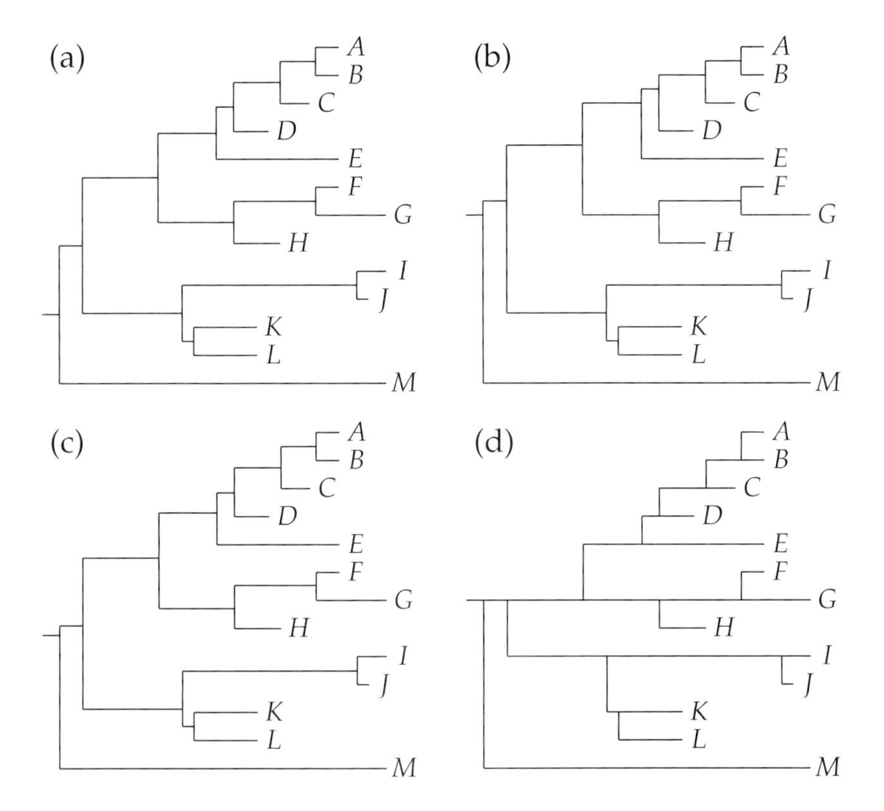

Figure 34.1: Four drawings of the same tree, using different methods of placement of the interior nodes. These are, respectively, (a) intermediate, (b) centered, (c) weighted, and (d) inner.

are k immediate descendants whose y coordinates are y_1 through y_k, and the first and last of these have branch lengths v_1 and v_k leading to them, then

$$y = \frac{(1/v_1)y_1 + (1/v_k)y_k}{(1/v_1) + (1/v_k)} \tag{34.3}$$

We could equally well define a version of the weighted method that weights all immediate descendants, rather than just the first and last ones.

Inner: The y coordinates of all tips descended from the node are considered. If these happen to be tips 1 through K, the one whose y value is the median of their y values is taken. If there are an even number of such tips, two will be tied to be the median. The one closer to the median of the tips on the whole tree is taken. The inner method ensures that each horizontal line on the tree has a tip at its right end.

You will immediately see that the effect of these systems is noticeably different. There must be many more methods; these are the ones that occurred to me when writing my program for drawing rooted trees. One other method, v-shaped, is not shown here, as it was designed to be used with a different method of drawing the lines. It will be described below.

Shapes of lineages

In these drawings, the lineages are each a vertical line followed by a horizontal line, so that the lineage has a right-angled turn in it. This works well with many of these node placement methods to keep lines from crossing. It also conveys the branch lengths very accurately, provided the viewer can concentrate on the x axis.

Figure 34.2 shows five other ways of drawing the lineages. The methods used to draw lineages are (in order, left to right and top to bottom):

Straight lines, with the node positions weighted. Note the crossed lines, an unfortunate effect that cannot always be avoided.

Straight lines, with the node positions determined so that the tree is v-shaped. Again, some lines are crossed. V-shaped trees are achieved by setting each node's y coordinate to be on a straight line between the node ancestral to it and one of its descendants. If the node is above its ancestor, its uppermost descendant is used. If it is below its ancestor, the lowermost ancestor is used. This is done iteratively until y coordinates of nodes cease changing.

Quarter ellipses. The lineage is one-quarter of an ellipse, going first vertically and finally horizontally.

Recurved quarter ellipses. The lineage is two quarter ellipses, the first beginning horizontally and becoming vertical, and the second going first vertically and then horizontally. The first ellipse (the one closer to the root of the tree) is half as large as the second one.

European. This imitates a style of cladogram popular among European researchers in the 1980s. The lineages are straight lines going diagonally for one-third of their horizontal extent, then going horizontally thereafter.

Circular tree. Here the tree is drawn with polar coordinates around the center of the page. What were formerly vertical lines become arcs of a circle. What were formerly horizontal lines now radiate outwards from the center of the circle. This system was devised by David Swofford and David Maddison.

Note that with many of these methods, one cannot always avoid having lines cross. I have deliberately used an example tree that shows this. The only way of avoiding crossing is to have the tips not be evenly spaced along the y axis.

With the straight and curved branches, the branch lengths may be misleading to many viewers. The lengths of the branches are proportional to the horizontal

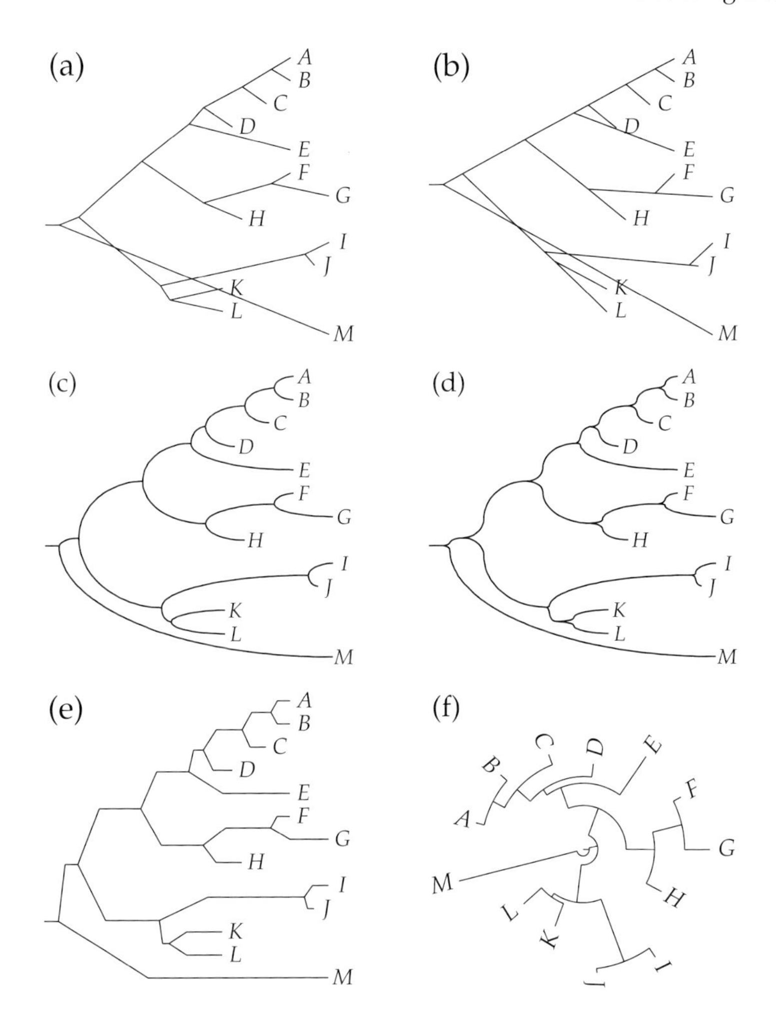

Figure 34.2: Methods of drawing lineages. These are, left to right and top to bottom, (a) diagonal line, (b) diagonal line with v-shaped node placement, (c) quarter ellipse, (d) recurved quarter ellipses, (e) European, and (f) circular.

extent of the branch. But our eyes are naturally attracted to the length of the line (or curve). For example, in the European tree the lineage ancestral to KL appears much longer than the lineage ancestral to J. Both are equal lengths on the original tree, but the lineage that leads to the common ancestor of K and L moves much farther vertically. In the case of recurved quarter ellipses, it is also hard to see where the lineages start diverging.

Unrooted trees

Unrooted trees pose even more challenges. Assuming that we want to preserve branch-length information, we can think of the tree as a collection of rods of given lengths, tied together at their ends. They cannot be stretched or compressed, but they can be rotated. Drawing the tree reduces essentially to the problem of determining the angles at all the connections between rods. Three classes of method will be described here. The problem is a special case (for trees) of the general problem of drawing graphs. An annotated bibliography of graph drawing algorithms (Di Battista et al., 1994) will serve as an introduction to the general problem.

The equal-angle algorithm

This method was invented by Christopher Meacham when he wrote a tree plotting program for my package. One starts from any internal node on the tree. The total number of species on the tree is counted, and the circle (total angle 2π radians or 360°) is divided into equal angles, one for each species. In our 13-species example, this will allocate 0.48332 radians (27.69°) per species. At the starting node, there will be three or more subtrees attached. Each is assigned a total angle equal to the share of the number of species that it contains. In our example tree, there is a node to which subtrees of size 8, 4, and 1 species are attached (these are species A–H, I–L, and M). These three subtrees would then be assigned angles 3.8666, 1.9333, and 0.48332 radians. We set up sectors of angles of these sizes around the node, and draw a line for that branch out into the middle of each sector, of length appropriate to the branch length.

As we reach each node we take the angle and divide it into parts, one for each subtree that is attached to the end of that branch (ignoring the parts of the tree attached to the node from which we came). Thus the eight-species subtree has a branch of length 13, and the node there connects to two subtrees, one with five species and one with three. We set up the sector for the eight-species subtree, now centering it at the end of the branch of length 13, in such a way that that branch points to the bisector of the angle. Then we divide the angle into parts proportional to the numbers of species in each subtree.

Thus the eight-species subtree starts with an angle of 3.8666 radians, moves out a distance 13 up the middle of that sector. It has a node there, which connects to two subtrees, with five and three species, respectively. The 3.8666 radian sector is set up at the node, with the branch of length 13 pointing at its bisector. This angle is then subdivided into two sectors, one of 2.4166 and the other of 1.4500 radians. Into each a branch is drawn, bisecting the angle, and of the appropriate length (in this example these would be of lengths 10 and 13).

The process continues, setting up an angle, dividing it into sectors for each subtree connected further on to that node, and drawing branches out into each sector, bisecting it.

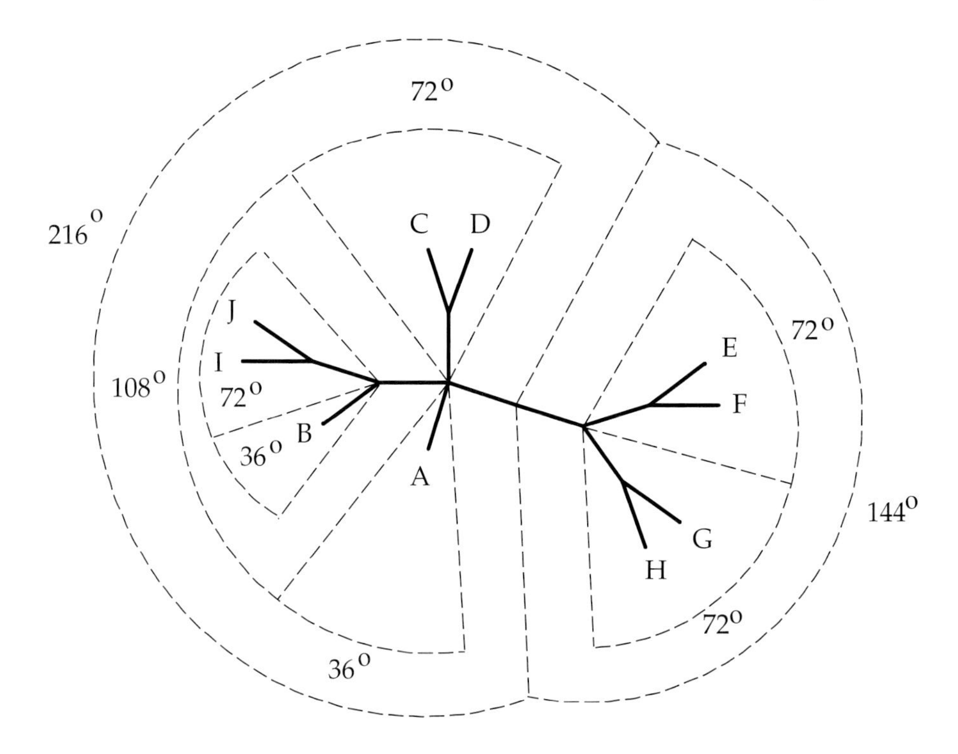

Figure 34.3: The equal-angle algorithm on a tree of 10 species. Angles are shown in degrees. The starting point is the midpoint of the branch that separates species EFGH from the others.

Figure 34.3 shows one such tree of 10 species, the angles (in degrees) allocated to each branch. An interesting, and not immediately obvious, fact about the algorithm is that it does not matter where you start on the tree. You can start at any node (even a tip) or even at any point on any branch. The resulting tree will differ at most by a simple rigid rotation and translation of the whole drawing, as a result of a different starting point. I will not prove this but leave it to you as an exercise. It is also easy to prove that the algorithm will never cause branches of the tree to cross. I am assuming that any branches of negative length will be drawn as if of positive length.

Figure 34.4 shows our 13-species example plotted by the equal-angle algorithm. This points up one of the limitations of this algorithm. The tree shows tips rather crowded together (A and B, F and G, for example), while at the same time there are large empty areas. These come from the fact that angles are allocated without regard to how far out the subtree is as a result of the length of the branch leading to it. In the example tree, subtree IJ sits on the end of a long branch, but is not allocated any extra angle that would allow I to swing further away from

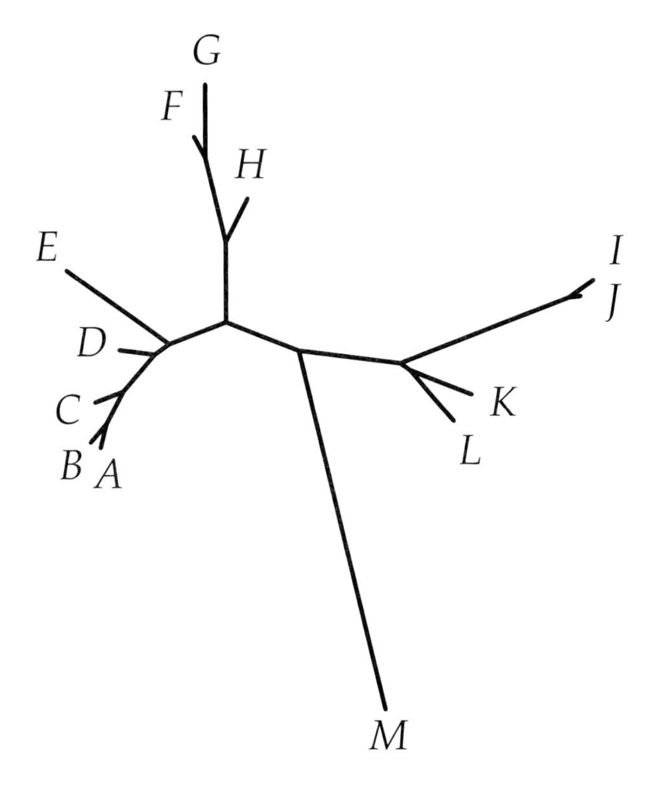

Figure 34.4: The example tree drawn by the equal-angle algorithm.

J, even though there is plenty of space to do so. To allow for this, we would need to know more about how long the branches within the subtree were.

n-Body algorithms

In an attempt to allow branches to swing away from each other and out into empty space, I have tried modeling the tree as a system of particles held together by rigid, but rotatable, rods. At each node, both at tips and interior nodes, a particle with an electric charge could be imagined to exist. If all the charges were of equal size and sign, they would repel each other, causing the tree to settle into some final configuration. This seemed like a good algorithm. But in testing it, a problem came to light. If the charges were at the ends of the branches, when there were long branches in different parts of the tree, one branch could actually pass through another! This can happen if one end of a branch approaches the middle of another. The end does not repel the middle of the other branch, and if the ends of that branch are far enough away it can be pushed (by forces from other parts of the tree) through the branch.

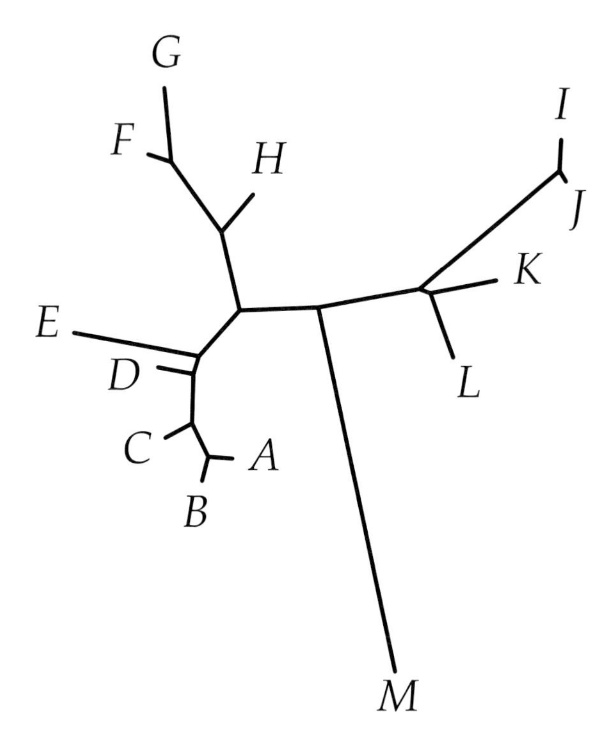

Figure 34.5: The example drawn by an n-body algorithm.

To avoid this, it has proven necessary to spread the electrostatic charge out along branches, so that the end of a branch will be repelled if it approaches the middle of another branch. At each moment, it is necessary to calculate the forces exerted by each branch on each other, and the rotations that would result per unit time. The branches are then moved a small distance, and then the forces must be recalculated.

The n-body problem in celestial mechanics is well-known to be a very difficult one. Taking too large a step size can cause accumulation of errors in the positions of the particles. Fortunately, it's not rocket science in our case. We don't care about the exact trajectories, just about the final locations of the particles. Figure 34.5 shows the result of applying an n-body algorithm to our example. The tree is quite a reasonable one. The branches leading to D and E get a bit too close to others for comfort, but the rest do a good job of using the previously empty space. On examples with much larger numbers of species, I did see cases where the n-body algorithm swung one subtree around behind another, which sometimes makes for less visual separation than one wants.

The main problem with the n-body algorithm is the high computational burden. Updating positions of subtrees and forces continually is computationally dif-

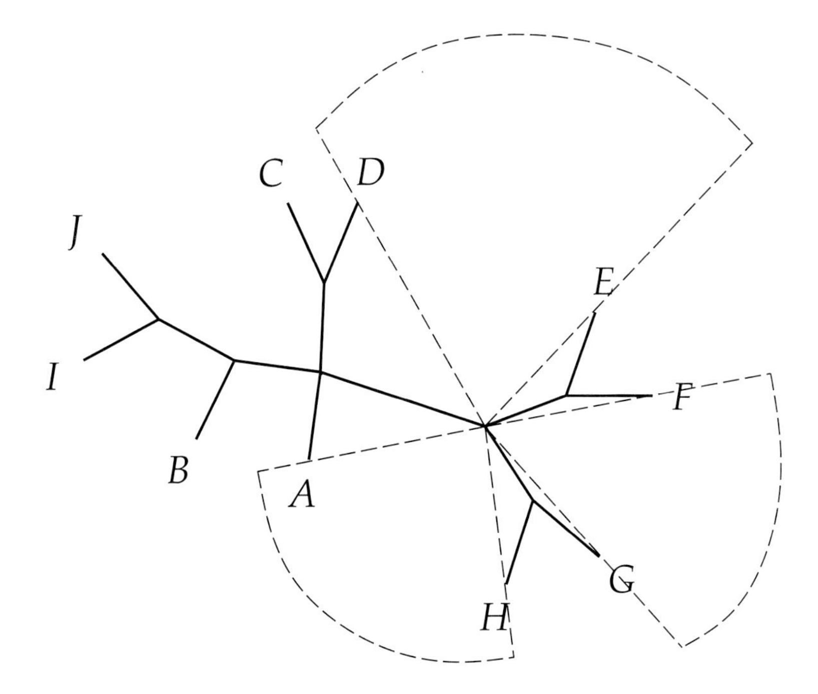

Figure 34.6: The angles of "daylight" between subtrees, calculated from one node of a tree.

ficult. I have therefore not given you many details of how my particular versions of n-body algorithms work. It turns out that there is a better, and quicker, method.

The equal-daylight algorithm

To avoid the heavy computation involved in the n-body algorithms, I reluctantly decided to try to find some rougher approach that would at least swing some subtrees out into the empty zones that haunt the equal-angle algorithm. A simple method suggested itself, one which equalizes the sizes of angular gaps between subtrees. When this was implemented, to my astonishment the trees were not just better than the equal-angle algorithm, they were outstanding.

The equal-daylight algorithm uses a starting tree (for which the equal angle algorithm tree is fine). For each internal node in the tree, we look at all subtrees connected to it (there will be three or more of these). We imagine lines from that node to each member of each subtree. For each subtree we find the rightmost and the leftmost such line. From these we can work out the angle between each subtree and the next. These angles tell us how much "daylight" there is between subtrees. We then change the angles with which each subtree connects to the node, so as to equalize the angles between the subtrees.

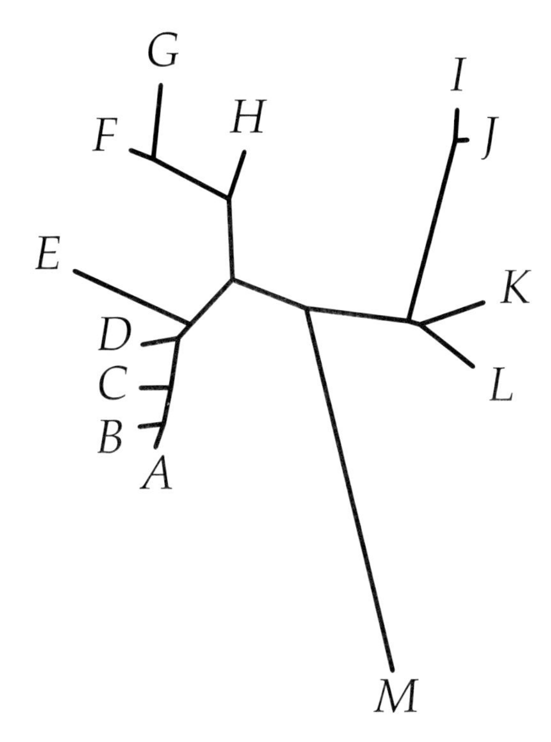

Figure 34.7: The equal-daylight algorithm applied to the example tree.

Thus if we find three subtrees, one shading the angles from $10°$ to $40°$, one from $90°$ to $150°$, and the third from $210°$ to $280°$, the angles between them are $50°$, $60°$, and $90°$. These daylight angles total $200°$. If they were made equal, they would all be $66\frac{2}{3}°$. To equalize them, we could rotate the second subtree by $16\frac{2}{3}$ more degrees, and the third by $16\frac{2}{3} + 6\frac{2}{3}$ more degrees.

We do this for each internal node in turn, and then make further passes through the tree, until the all of the daylight angles approach equality. It should be immediately apparent that this cannot cause any branches to cross, as long as we start with positive daylight angles. Figure 34.6 shows the daylight angles viewed from one internal node of a tree.

There are some problems that can arise in carrying out the equal-daylight algorithm. If the subtree angles sum to more than a full circle, it will be impossible to find a positive angle between subtrees, and the subtree angles will overlap. Fortunately, this cannot arise if the equal-angle algorithm is used for the starting tree. Figure 34.7 shows the result of applying the equal-daylight algorithm to the example tree. The results are quite good, showing better visual separation of nearby lineages than does the n-body algorithm. In general the equal-daylight algorithm runs considerably faster than the n-body algorithm, and shows the structure of the tree more clearly.

We might expect all parts of the tree to be seen particularly clearly with this algorithm. From each node, looking outwards, we can see a clear path out of the tree between each subtree, and good visual separation between the subtrees. Likewise, looking inwards from the outside, there will be straight paths in to each node from outside the tree. Subtrees will never move behind other subtrees, obscuring their visibility.

The equal-daylight algorithm is, for now, the best method of drawing unrooted trees. However, no one has defined general criteria for judging the adequacy of a tree drawing, so there is no way to prove that this method is optimal.

Challenges

There remains the task of placing tree drawing in a more scientific context by defining optimality criteria that make psychophysical sense. I have also omitted a number of problems. How best to orient the labels at the tips? How best to arrange the tree so that branch lengths and bootstrap percentages appear clearly on interior branches? This latter problem may have no solution: When a large number of small interior branches lie near each other, there may simply be no room to show branch-length values. With large trees containing many small branches, it may be important to provide views of the tree that have some subtrees collapsed and represented by triangles, so that only part of the tree is visible at any one time. One can collapse and expand these triangles interactively, allowing the user to explore the tree. A similar approach uses a fish-eye lens transformation to view structure in one part of the tree (cf. Lamping and Rao, 1994). By moving the center of the view, one can expand some parts of the tree while compressing others.

Chapter 35

Phylogeny software

Throughout this book I have been avoiding many of the details of algorithms and their computer implementation. This chapter is intended to provide some of this information. I will explain some details of computer representation of trees, both within programs and in files, and the process of traversing trees. Finally, I will provide a list of widely used packages for inferring phylogenies.

Trees, records, and pointers

The implementation of computer programs for handling phylogenies seems difficult, owing largely to the complicated and tedious bookkeeping involved. Trees are not as difficult to manipulate as might seem, because modern computer languages have records and pointers, plus the ability to do recursive function calls. Using these, we can move rapidly through a tree, updating information or changing its structure. Records are regions of computer memory containing several variables, that can be referred to by a single name (in Pascal they are "records," in C "structures," and in C++ and Java, "classes"). Variables that refer to records are called pointers.

For example, we could imagine a node of a tree, which we would want to have pointers to its left and right descendants, one to its ancestor, and a variable that indicated whether or not the node was a tip. Of course many other pieces of information might be desirable too, but we are describing only the minimum necessary to use when moving about the tree. Figure 35.1 diagrams the structure of part of the tree ((A,B),C). The record is a box, its variables are names within the box, and their values are shown by drawing an arrow to the other record they might point to. In this case `leftdesc` and `rightdesc` are the pointers that point to the left and right descendants, and `ancestor` points to the ancestor. `tip` shows by being 1 or 0 whether the record is or is not for a tip. Pointers that do not point to anything are shown as arrows with empty heads. There is a pointer in the program data called "root."

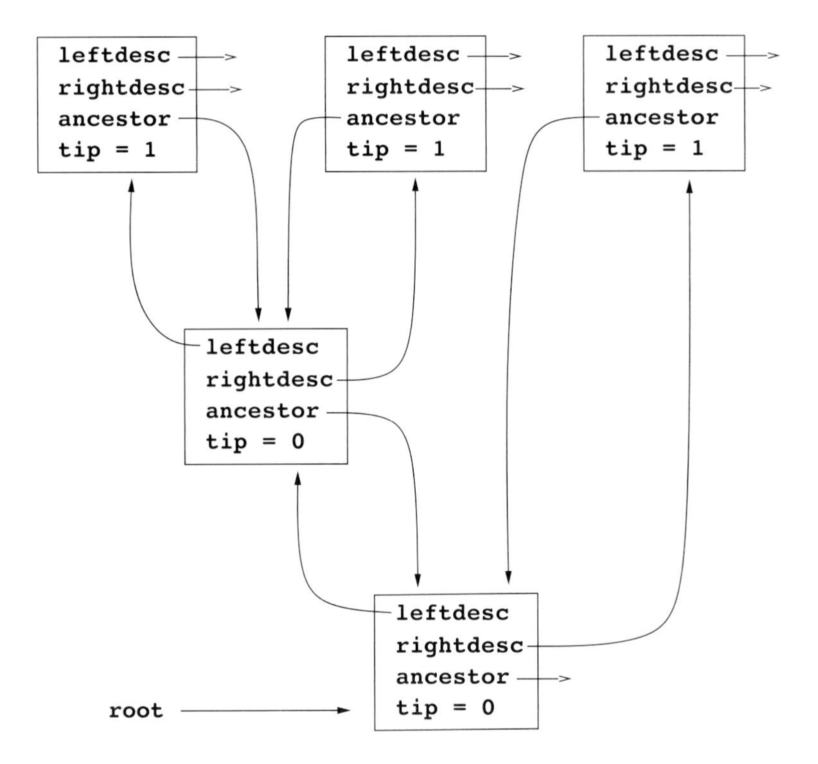

Figure 35.1: A record and pointer structure representing a simple rooted tree of form ((A,B),C). Each branch is represented by a pair of pointers, one pointing each direction along it.

This scheme is mentioned by Knuth (1973, Section 2.3.3) as a "triply linked tree."

Declaring records

In writing a program to set up and manipulate trees composed of records and pointers, we must inform the program, early on, of the structure of these records. Here is the declaration we might use in C:

```
typedef struct node {
  node *leftdesc, *rightdesc, *ancestor;
  int tip;
};
```

Here it is in C++:

```
class node
{
  public:
    node* leftdesc;
    node* rightdesc;
    node* ancestor;
    bool tip;
  ...
};
```

and here it is in Java:

```
class Node
{
  Node leftDesc;
  Node rightDesc;
  Node ancestor;
  boolean tip;
}
```

Each declares what variables will be in the records. In the case of C++ and Java we have actually declared, not simply records, but classes, which are entities that can contain functions as well as data.

Once these declarations are available, the program can actually set up instances of the records. The program would do this as it added lineages to a tree, either during a process of sequential addition, or while reading a representation of the tree from a file. I will not attempt to give the actual code that is used to create a record, make the pointers point to other records, and make the tip variable have the correct value.

Traversing the tree

The great advantage of this method of representing a tree is the ease of writing the code that moves about the tree. The main trick is using recursive function calls. Suppose we want to move through the tree, carrying out for each record `p` a function `f(p)`. This is known as *tree traversal*. We start at the bottommost fork in the tree. Two major types of tree traversal are preorder and postorder tree traversal. A *preorder tree traversal* carries out the function `f(p)` at each node before it visits its descendants. A *postorder tree traversal* carries it out at a node only after it visits its descendants.

A tree traversal is defined in terms of itself. A postorder tree traversal from node `p` does the following:

1. Traverses p's left descendant (if any).
2. Traverses p's right descendant (if any).
3. Carries out function f(p) at p.

Basically, one is instructing the program to (1) go up the left subtree and do all this with it, (2) go up the right subtree and do this with it, and then (3) carry out f(p) here. The effect of all this is that the program passes through the tree, going up the leftmost route until a tip is reached, then backtracking. On the way back down, as each node is reached for the last time, function f(p) is executed. So each node does not have f(p) done until this has been done for all its descendants.

The code for doing all this is startlingly simple. It uses recursive function calls. Functions are allowed to call themselves. Here is the code in C (and C++):

```
postorder (node* p) {
  if (!p->tip) {
    postorder (p->leftdesc);
    postorder (p->rightdesc);
  }
  f(p);
};
```

and here it is in Java:

```
public class PostOrder
{
  void Traverse(Node n)
  {
    if(!n.tip)
    {
      Traverse(n.leftDesc);
      Traverse(n.rightDesc);
    }
    F(n);
  }
}
```

To get the process started, one calls function traverse or Traverse on the rootmost fork. It then calls itself on the left descendant of that node, and that invocation of the function calls itself on the leftmost descendant of that node, and so on. Sooner or later a tip is reached, where there are no descendants. The test for the tip variable detects this and does not do further traversal calls. The function carries out function f(p) or F(p) at that tip. It then automatically goes back to being in the previous instance of the traversal function. There, it has just finished calling itself on the left descendant, and it now calls itself on the right descendant. It is automatically backtracking and exploring all parts of the tree. If you follow

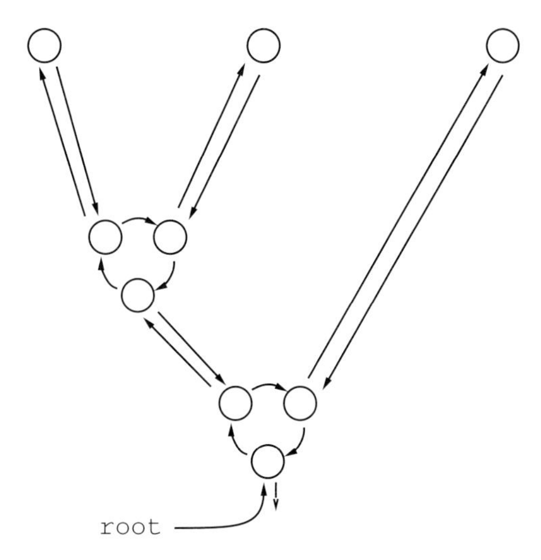

Figure 35.2: The same tree as in the previous figure, using the unrooted data structure that has rings of records at nodes. Each record has two pointers, a `next` pointer that goes around the ring, and an `out` pointer that points to the record at the opposite end of a branch.

the logic of the recursive calls carefully, you will see that they visit every node, and they call the function on each before calling it on any of its ancestors.

This is the easy way to implement going down the tree, carrying out a function on each node. It is not necessarily the fastest way. Back in the FORTRAN era, I used to represent a tree as a two-dimensional array. Each row represented one node, which had a number. It had entries in the row for the number of its left descendant, for the number of its right descendant, and for the number of its ancestor. Each of these was 0 if it pointed to nothing. The rows of the array were kept in order, so that each node's row occurred before that of its ancestor. To go down the tree carrying out a function, one just went along the array, carrying out the function at each node. This was quite fast. The trick was that, when one rearranged the tree, it was necessary to renumber and reorder some of the interior nodes. This was painful and error-prone.

Unrooted tree data structures

The system of records and pointers used above is well-adapted to strictly bifurcating rooted trees, but becomes cumbersome if there are multifurcations, or if the tree is to be regarded as unrooted. With multifurcations possible, we would have to replace `leftdesc` and `rightdesc` by an array of pointers, or by a linked list of pointers. If we want to reroot the tree, the above scheme makes it tiresome— we have to go through the tree swapping values of pointers in an appropriate way.

There is a simple tree representation that solves both of these problems. We represent each node in the tree by a circular ring of records, each with two pointers. One is the `next` pointer that points along the ring, and the other is the `out` pointer that points outwards along the branch to the nearest record in the next node. Figure 35.2 shows the same small tree using this data structure. The records are the small circles. Each has two pointers, respectively, the curved arrow `next` and the straight arrow `out`.

The great advantages of this structure are that it accommodates any degree of multifurcation, and if one wants to reroot the tree, this is done rather easily, with most of the nodes between the old and new locations of the root not changing at all. It is the structure used in my own program package PHYLIP, which originated its use in 1982 in the programs `DNAML` and `CONTML`.

Tree file formats

The representation of trees in readable format in files can be done in many ways. The most widely used is the Newick format. This format was designed by a committee (consisting of F. James Rohlf, David Swofford, Christopher Meacham, James Archie, Wayne Maddison, William H. E. Day, and me). It was convened informally by me at the 1986 meetings of the Society for the Study of Evolution in Durham, New Hampshire (it was not an activity formally organized by the Society). The second session of the committee meeting was held at Newick's lobster restaurant in nearby Dover, and the tree format that was produced has been named in honor of the restaurant, because we enjoyed the meal.

We have seen many examples of this format already, so I will describe it cursorily. Each internal node in a rooted tree is represented by a pair of matched parentheses. Within it is a list of the nodes immediately descended from this internal node. Ones that are themselves internal nodes are again in matched parentheses. Ones that are tips are represented by their names. Any node can be followed by a colon together with a number that is the branch length of the branch below that node. (Although we have not used this feature, any internal node can also be preceded by its name.) A tree ends in a semicolon.

Thus the simple tree in the figures in this chapter can be represented by

((A:0.102,B:0.23):0.06,C:0.4);

if its tips had names A, B, and C, and these lengths were assigned to the branches.

The Newick format is described further on its web page:
`http://evolution.gs.washington.edu/phylip/newicktree.html`
It is based on the famous relationship between trees and ways of putting parentheses into algebraic expressions noted by Cayley (1857).

The Newick format has been successful in spite of being nonunique for unrooted trees, which can be represented only as rooted trees (and there are many rooted trees corresponding to each unrooted tree). It also enforces a left-right order on a rooted tree, which may not need to have one.

It is about to be superseded by a wave of XML representations of trees. There is no standards committee for these—they will be standardized the way player piano rolls were a century ago, when every piano needed to be able to play the rolls of the most popular brand.[1] My own preference is that the XML representation of our small example tree be:

```
<phylogeny>
  <clade>
    <clade length="0.06">
      <clade length="0.102"><name>A</name></clade>
      <clade length="0.23"><name>B</name></clade>
    </clade>
    <clade length="0.4"><name>C</name></clade>
  </clade>
</phylogeny>
```

but the issue is not decided yet.

Widely used phylogeny programs and packages

There are now hundreds of phylogeny programs, and the number continues to grow. I have been attempting to keep track of them in a set of web pages at: `http://evolution.gs.washington.edu/phylip/software.html`
I will try to keep these pages available and current as long as I can. If you are reading this years from now and this link is not active, you may be able to find them, or a descendant of them, by searching for "Phylogeny Programs" in a search engine. Here I will list only a limited number of widely used programs and packages.

PHYLIP My own package, one of the earliest in wide distribution. It has been maintained since 1980, first as Pascal source code, then after 1993 as C source code, and with executables for major operating systems. It is available free from `http://evolution.gs.washington.edu/phylip.html`. The package contains programs to infer phylogenies by parsimony, distance matrix, and maximum likelihood methods, as well as programs for consensus trees, distances between trees, editing of trees, invariants, and comparative methods. There are programs for data types including nucleic acid sequences, protein sequences, gene frequencies, restriction sites and restriction fragments, discrete characters, and continuous characters.

PAUP* This is a comprehensive program with parsimony, likelihood, and distance matrix methods. It has many more features and methods than PHYLIP.

[1]Thanks to Chris Meacham for the analogy.

It competes with PHYLIP to be responsible for the most trees published. It is due to David Swofford and is distributed by Sinauer Associates of Sunderland, Massachusetts (also the publishers of this book). They sell executable versions for Windows, MacOS, and Linux or Unix systems. A PAUP* web page is at `http://paup.csit.fsu.edu`, where there are also links to its web page at Sinauer Associates.

MacClade An interactive MacOS program by Wayne and David Maddison. It can rearrange trees and watch the changes in the fit of the trees to data as judged by parsimony. MacClade has a great many features including a spreadsheet data editor and many descriptive statistics. It is particularly designed to export and import data to and from PAUP*. MacClade is available for purchase from Sinauer Associates. It is described in a web page at `http://phylogeny.arizona.edu/macclade/macclade.html`, where there are links to its Sinauer Associates web pages as well.

MEGA A Windows and DOS program by Sudhir Kumar, (written together with Koichiro Tamura and Masatoshi Nei while Kumar was a student in Nei's lab). It can carry out parsimony and distance matrix methods for DNA sequence data. An executable for Windows can be downloaded from `http://www.megasoftware.net`

PAML A package of programs by Ziheng Yang to carry out likelihood analysis of DNA and protein sequence data. PAML is particularly strong in the options for coping with variability of rates of evolution from site to site, though it is not designed to search particularly effectively for the best tree. It is available as C source code and as MacOS and Windows executables from its web site at `http://abacus.gene.ucl.ac.uk/software/paml.html`

TREE-PUZZLE This package by Korbinian Strimmer, Arndt von Haeseler, Martin Vingron, and Heiko Schmidt can carry out likelihood methods for DNA and protein data, searching by the strategy of "quartet puzzling," which Strimmer and von Haeseler invented. It can also compute distances. It superimposes trees estimated from many quartets of species. TREE-PUZZLE is available for Unix, MacOS, or Windows from its web site at `http://www.tree-puzzle.de/`

DAMBE A package by Xuhua Xia, it is a general-purpose package for DNA and protein sequence phylogenies. It can read and convert a number of file formats, and has many features for descriptive statistics, and can compute a number of commonly used distance matrix measures and infer phylogenies by parsimony, distance, or likelihood methods, including bootstrapping and jackknifing. There are a number of kinds of statistical tests of trees available. It can also display phylogenies. DAMBE includes a copy of the ClustalW alignment program as well;

DAMBE is distributed free as Windows executables from its web site at `http://aix1.uottawa.ca/~xxia/software/software.htm`

NONA A very fast parsimony program by Pablo Goloboff, capable of some relevant forms of weighted parsimony. It can handle either DNA sequence data or discrete characters. It is available as shareware from `http://www.cladistics.com/aboutNona.htm`. After a 30-day free trial, NONA must be purchased separately by sending a check to one of the addresses given on the web page.

TNT This program by Pablo Goloboff, J. S. Farris, and Kevin Nixon, is for searching large data sets for most parsimonious trees. TNT is described as faster than other methods, though not faster than NONA for small to medium data sets. Its distribution status is somewhat uncertain at present (see the web site `http://www.zmuc.dk/public/phylogeny/tnt`). It seems to be in limited beta release, but not yet freely available.

References

Abouheif, E. 1998. Random trees and the comparative method: A cautionary tale. *Evolution* **52:** 1197–1204.

Ackerly, D. D. and M. J. Donoghue. 1995. Phylogeny and ecology reconsidered. *Journal of Ecology* **83:** 730–733.

Ackerly, D. D. 2000. Taxon sampling, correlated evolution, and independent contrasts. *Evolution* **54:** 1480–1492.

Adachi, J. and M. Hasegawa. 1996. Model of amino acid substitution and applications to mitochondrial protein evolution. *Journal of Molecular Evolution* **42:** 459–468.

Adachi, J., P. J. Waddell, W. Martin, and M. Hasegawa. 2000. Plastid genome phylogeny and a model of amino acid substitution for proteins encoded by chloroplast DNA. *Journal of Molecular Evolution* **50:** 348–358.

Adams, E. N., III. 1972. Consensus techniques and the comparison of taxonomic trees. *Systematic Zoology* **21:** 390–397.

Adams, E. N., III. 1986. N-trees as nestings: complexity, similarity, and consensus. *Journal of Classification* **3:** 299–317.

Adell, J. C. and J. Dopazo. 1994. Monte Carlo simulation in phylogenies: An application to test the constancy of evolutionary rates. *Journal of Molecular Evolution* **38:** 305–309.

Agarwala, R. and D. Fernández-Baca. 1994: A polynomial time algorithm for the phylogeny problem when the number of states is fixed. *SIAM Journal on Computing* **23:** 1216–1224.

Aho, A. V., Y. Sagiv, T. G. Szymanski, and J. D. Ullman. 1981. Inferring a tree from lowest common ancestors with an application to the optimization of relational expressions. *SIAM Journal on Computing* **10:** 405–421.

Albert, V. A., B. D. Mishler, and M. W. Chase. 1992. Character-state weighting for restriction site data in phylogenetic reconstruction, with an example from chloroplast DNA. pp. 369–403 in *Molecular Systematics of Plants,* ed. P. S. Soltis, D. E. Soltis, and J. J. Doyle. Chapman and Hall, London.

Aldous, D. J. 2001. Stochastic models and descriptive statistics for phylogenetic trees, from Yule to today. *Statistical Science* **16:** 23–34.

Allen, B. L. and M. Steel. 2001. Subtree transfer operations and their induced metrics on evolutionary trees. *Annals of Combinatorics* **5:** 1–15.

Allison, L. and C. N. Yee. 1990. Minimum message length encoding and the comparison of macromolecules. *Bulletin of Mathematical Biology* **52:** 431–453.

Allison, L., C. S. Wallace, and C. N. Yee. 1992. Finite-state models in the alignment of macromolecules. *Journal of Molecular Evolution* **35:** 77–89.

Allison, L. and C. S. Wallace. 1994. The posterior probability distribution of alignments and its application to parameter estimation of evolutionary trees and to optimization of multiple alignments. *Journal of Molecular Evolution* **39:** 418–430.

Allman, E. S. and J. A. Rhodes. 2003. Phylogenetic invariants for the general Markov model of sequences mutation. *Mathematical Biosciences*, in press.

Alroy, J. 1994. Four permutation tests for the presence of phylogenetic structure. *Systematic Biology* **43:** 430–437.

Altschul, S. F. 1991. Amino acid substitution matrices from an information theoretic perspective. *Journal of Molecular Biology* **219:** 555–565.

Ambainis, A., M. Desper, M. Farach, and S. Kannan. 1997. Nearly tight bounds on the learnability of evolution. pp. 524–533 in *Proceedings of the 38th Annual Symposium on Foundations of Computer Science (FOCS '97), October 19–22, 1997, Miami Beach, Florida.* IEEE Computer Society Press, Los Alamitos, California.

Amir, A. and D. Keselman. 1997. Maximum agreement subtree in a set of evolutionary trees: Metrics and efficient algorithms. *SIAM Journal on Computing* **26:** 1656–1669.

Angielczyk, K. D. 2002. A character-based method for measuring the fit of a cladogram to the fossil record. *Systematic Biology* **51:** 137–152.

Archie, J. W. 1985. Methods for coding variable morphological features for numerical taxonomic analysis. *Systematic Zoology* **34:** 326–345.

Archie, J. W. 1989. A randomization test for phylogenetic information in systematic data. *Systematic Zoology* **38:** 239–252.

Atteson, K. 1999. The performance of neighbor-joining methods of phylogenetic reconstruction. *Algorithmica* **25:** 251–278.

Avise, J. C., J. Arnold, R. M. Ball, E. Bermingham, T. Lamb, J. E. Neigel, C. A. Reeb and N. C. Saunders. 1987. Intraspecific phylogeography—the mitochondrial-DNA bridge between population-genetics and systematics. *Annual Review of Ecology and Systematics* **18:** 489–522.

Avise, J. C. 1989. Gene trees and organismal histories—a phylogenetic approach to population biology. *Evolution* **43:** 1192–1208.

Backeljau, T., L. De Bruyn, H. De Wolf, K. Jordaens, S. Van Dongen, R. Verhagen, and B. Winnepenninckx. 1995. Random amplified polymorphic DNA (RAPD) and parsimony methods. *Cladistics* **11:** 119–130.

Backeljau, T., L. De Bruyn, H. De Wolf, K. Jordaens, S. Van Dongen, and B. Winnepenninckx. 1996. Multiple UPGMA and neighbor-joining trees and the performance of some computer packages. *Molecular Biology Evolution* **13:** 309–313.

Bader, D. A., B. M. E. Moret, and M. Yan. 2001. A linear-time algorithm for computing inversion distance between signed permutations with an experimental study. *Journal of Computational Biology* **8:** 483–491.

Bahlo, M. and R. C. Griffiths. 2000. Inference from gene trees in a subdivided population. *Theoretical Population Biology* **57:** 79–95.

Bahlo, M. and R. C. Griffiths. 2001. Coalescence time for two genes from a subdivided population. *Journal of Mathematical Biology* **43:** 397–410.

Bandelt, H.-J. and A. Dress. 1986. Reconstructing the shape of a tree from observed dissimilarity data. *Advances in Applied Mathematics* **7:** 309–343.

Bandelt, H.-J. and A. W. Dress. 1992a. A canonical decomposition theory for metrics on a finite set. *Advances in Mathematics* **92:** 47–105.

Bandelt, H.-J. and A. W. Dress. 1992b. Split decomposition: A new and useful approach to phylogenetic analysis of distance data. *Molecular Phylogenetics and Evolution* **1:** 242–252.

Bandelt, H.-J., P. Forster, B. C. Sykes, and M. B. Richards. 1995. Mitochondrial portraits of human populations. *Genetics* **141:** 743–753.

Bar-Hen, A. and H. Kishino. 2000. Comparing the likelihood functions of phylogenetic trees. *Annals of the Institute of Statistical Mathematics* **52:** 43–56.

Barker, F. K. and F. M. Lutzoni. 2002. The utility of the incongruence length difference test. *Systematic Biology* **51:** 625–637.

Barry, D. and J. A. Hartigan. 1987. Statistical analysis of hominoid molecular evolution. *Statistical Science* **2:** 191–210.

Bartcher, R. 1966. *FORTRAN IV program for estimation of cladistic relationships using the IBM 7040.* Computer Contribution 6. State Geological Survey. University of Kansas, Lawrence, Kansas.

Barthélemy, J. P. and F. R. McMorris. 1986. The median procedure for n-trees. *Journal of Classification* **3:** 329–334.

Bastert, O., D. Rockmore, P. F. Stadler, and G. Tinhofer. 2002. Landscapes on spaces of trees. *Applied Mathematics and Computation* **131:** 439–459.

Baum, B. R. 1992. Combining trees as a way of combining data sets for phylogenetic inference, and the desirability of combining gene trees. *Taxon* **41:** 3–10.

Baum, L. E. and T. Petrie. 1966. Statistical inference for probabilistic functions of finite state Markov chains. *Annals of Mathematical Statistics* **37:** 1554–1563.

Baum, L. E., T. Petrie, G. Soules, and N. Weiss. 1970. A maximization technique occurring in the statistical analysis of probabilistic functions of Markov chains. *Annals of Mathematical Statistics* **41:** 164–171.

Beatty, J. and W. L. Fink. 1979. Review of "Simplicity," by Elliott Sober. *Systematic Zoology* **28:** 643–651.

Beaumont, M. A. 1999. Detecting population expansion and decline using microsatellites. *Genetics* **153:** 2013–2029.

Beaumont, M. A., W. Zhang, and D. J. Balding. 2002. Approximate Bayesian computation in population genetics. *Genetics* **162:** 2025–2035.

Beerli, P. and J. Felsenstein. 1999. Maximum-likelihood estimation of migration rates and effective population numbers in two populations using a coalescent approach. *Genetics* **152:** 763–773.

Beerli, P. and J. Felsenstein. 2001. Maximum likelihood estimation of a migration matrix and effective population sizes in n subpopulations by using a coalescent approach. *Proceedings of the National Academy of Sciences, USA* **98:** 4563–4568.

Benham, C., S. Kannan, M. Paterson, and T. Warnow. 1996. Hen's teeth and whale's feet: Generalized character compatibility. *Journal of Computational Biology* **2:** 527–536.

Benner, S. A., M. A. Cohen, and G. H. Gonnet. 1994. Amino acid substitution during functionally constrained divergent evolution of protein sequences. *Protein Engineering* **7:** 1323–1332.

Benson, G. and L. Dong. 1999. Reconstructing the duplication history of a tandem repeat. pp. 44–53 in *Proceedings of the Sixth International Conference on Intelligent Systems for Molecular Biology,* ed. J, Glasgow, T. Littlejohn, F. Major, R. Lathrop, D. Sankoff, and C. Sensen. AAAI Press, Menlo Park, California.

Benton, M. J. 1994. Palaeontological data, and identifying mass extinctions. *Trends in Ecology and Evolution* **9:** 181–185.

Benton, M. J., R. Hitchin, and M. A. Wills. 1999. Assessing congruence between cladistic and stratigraphic data. *Systematic Biology* **48:** 581–596.

Berlocher, S. H. and D. L. Swofford. 1997. Searching for phylogenetic trees under the frequency parsimony criterion: An approximation using generalized parsimony. *Systematic Biology* **46:** 211–215.

Berman, P. and S. Hannenhalli. 1996. Fast sorting by reversal. pp. 168–185 in *Combinatorial Pattern Matching, 7th Annual Symposium, CPM 96, Laguna Beach, California, USA, June 10–12, 1996, Proceedings.* ed. D. Hirschberg and G. Myers. Lecture Notes in Computer Science No. 1075, Springer-Verlag, New York.

Berry, V. and O. Gascuel. 1996. On the interpretation of bootstrap trees: Appropriate threshold of clade selection and induced gain. *Molecular Biology and Evolution* **13:** 999–1011.

Berry, V. and O. Gascuel. 2000. Inferring evolutionary trees with strong combinatorial evidence. *Theoretical Computer Science* **240:** 271–298.

Beyer, W., M. Stein, T. Smith, and S. Ulam. 1974. A molecular sequence metric and evolutionary trees. *Mathematical Biosciences* **19:** 9–25.

Bickel, D. R. and B. J. West. 1998. Molecular evolution model as a fractal renewal point process in agreement with the dispersion of substitutions in mammalian genes. *Journal of Molecular Evolution* **47:** 551–556.

Billera, L. J., S. P. Holmes, and K. Vogtmann. 2001. Geometry of the space of phylogenetic trees. *Advances in Applied Mathematics* **27:** 733–767.

Bininda-Emonds, O. R. P., J. L. Gittleman, and M. A. Steel. 2002. The (super) tree of life: Procedures, problems, and prospects. *Annual Review of Ecology and Systematics* **33:** 265–289.

Bishop, M. J. and E. A. Thompson. 1986. Maximum likelihood alignment of DNA sequences. *Journal of Molecular Biology* **190:** 159–165.

Blaisdell, B. E. 1986. A measure of the similarity of sets of sequences not requiring sequence alignment. *Proceedings of the National Academy of Sciences, USA* **83:** 5155–5159.

Blaisdell, B. E. 1989a. Average values of a dissimilarity measure not requiring sequence alignment are twice the averages of conventional mismatch counts requiring sequence alignment for a computer-generated model system. *Journal of Molecular Evolution* **29:** 538–547.

Blaisdell, B. E. 1989b. Effectiveness of measures requiring and not requiring prior sequence alignment for estimating the dissimilarity of natural sequences. *Journal of Molecular Evolution* **29:** 526–537.

Blanchette, M., T. Kunisawa, and D. Sankoff. 1996. Parametric genome rearrangement. *Gene* **172:** GC11-GC17.

Blanchette, M., G. Bourque and D. Sankoff. 1997. Breakpoint phylogenies. pp. 25–34 in *Genome Informatics 1997*, ed. S. Miyano and T. Takagi, Universal Academy Press, Tokyo.

Bodlaender, H., M. Fellows, and T. Warnow. 1992. Two strikes against perfect phylogeny. pp. 273–283 in *Automata, languages, and programming. 19th International Colloquium, Wien, Austria, July 13–17, 1992, Proceedings*, ed. W. Kuich. Lecture Notes in Computer Science, no. 623. Springer-Verlag, Berlin.

Bollback, J. 2002. Bayesian model adequacy and choice in phylogenetics. *Molecular Biology and Evolution* **19:** 1171–1180.

Bonet, M., M. A. Steel, T. Warnow, and S. Yooseph. 1998. Better methods for solving parsimony and compatibility. *Journal of Computational Biology* **5:** 391–407.

Bonet, M., C. Phillips, T. Warnow, and S. Yooseph. 1999. Constructing evolutionary trees in the presence of polymorphic characters. *SIAM Journal on Computing* **29:** 103–131.

Bourque, G., and P. A. Pevzner. 2002. Genome-scale evolution: Reconstructing gene orders in the ancestral species. *Genome Research* **12:** 26–36.

Bourque, M. 1978. *Arbres de Steiner et reseaux dont certains sommets sont à localisation variable.* Ph.D. Dissertation, Université de Montréal, Montréal, Quebec.

Brauer, M. J., M. T. Holder, L. A. Dries, D. J. Zwickl, P. O. Lewis, and D. M. Hillis. 2002. Genetic algorithms and parallel processing in maximum-likelihood phylogeny inference. *Molecular Biology and Evolution* **19:** 1717–1726.

Bremer, K. 1988. The limits of amino-acid sequence data in angiosperm phylogenetic reconstruction. *Evolution* **42:** 795–803.

Bremer, K. 1990. Combinable component consensus. *Cladistics* **6:** 369–372.

British Museum (Natural History). 1980. *Man's Place in Evolution.* Cambridge University Press, Cambridge.

Brodal, G. S., R. Fagerberg, and C. N. S. Pedersen. 2001. Computing the quartet distance between evolutionary trees in time $O(n \log^2 n)$. pp. 731–742 in *Algorithms and Computation, Proceedings of the 12th International Symposium, ISAAC 2001, Christchurch, New Zealand, December 19–21, 2001,* ed. P. Eades and T. Takaoka. Lecture Notes in Computer Sciences, No. 2223. Springer-Verlag, Berlin.

Bromham, L., D. Penny, A. Rambaut, and M. D. Hendy. 2000. The power of relative rates tests depends on the data. *Journal of Molecular Evolution* **50:** 296–301.

Bron, C. and J. Kerbosch. 1973. Algorithm 457: Finding all cliques of an undirected graph. *Communications of the Association for Computing Machinery* **16:** 575–577.

Brooks, D. R. 1981. Hennig's parasitological method: A proposed solution. *Systematic Zoology* **30:** 229–249.

Brooks, D. R. 1990. Parsimony analysis in historical biogeography and coevolution: Methodological and theoretical update. *Systematic Zoology* **39:** 14–30.

Brooks, D. R. and D. A. McLennan. 1991. *Phylogeny, Ecology, and Behavior: A Research Program in Comparative Biology.* University of Chicago Press, Chicago.

Brooks, D. R., M. G. P. van Veller, and D. A. McLennan. 2001. How to do BPA, really. *Journal of Biogeography* **28:** 345–358.

Brown, J. K. M. 1994a. Bootstrap hypothesis tests for evolutionary trees and other dendrograms. *Proceedings of the National Academy of Sciences USA* **91:** 12293–12297.

Brown, J. K. M. 1994b. Probabilities of evolutionary trees. *Systematic Biology* **43:** 78–91.

Brown, W. M., E. M. Prager, A. Wang, and A. C. Wilson. 1982. Mitochondrial DNA sequences of primates: Tempo and mode of evolution. *Journal of Molecular Evolution* **18:** 225–239.

Bruno, W. J. and A. L. Halpern. 1999. Topological bias and inconsistency of maximum likelihood using wrong models. *Molecular Biology and Evolution* **16:** 564–566.

Bruno, W. J., N. D. Socci, and A. L. Halpern. 2000. Weighted Neighbor Joining: A likelihood-based approach to distance-based phylogeny reconstruction. *Molecular Biology and Evolution* **17:** 189–197.

Bryant, D. and P. Waddell. 1998. Rapid evaluation of least-squares and minimum-evolution criteria on phylogenetic trees. *Molecular Biology and Evolution* **15:** 1346–1359.

Bryant, D. and V. Moulton. 1999. A polynomial time algorithm for constructing the refined Buneman tree. *Applied Mathematics Letters* **12 (2):** 51–56.

Bryant, D. and M. Steel. 2001. Constructing optimal trees from quartets. *Journal of Algorithms* **38:** 237–259.

Bryant, D. 2003. A classification of consensus methods for phylogenetics. pp. 163–183 in *Bioconsensus,* ed. M. F. Janowitz, F.-J. Lapointe, F. R. McMorris, B. Mirkin, and F. S. Roberts. DIMACS Series in Discrete Mathematics and Theoretical Computer Science, Volume 61. American Mathematical Society, Providence, Rhode Island.

Bryant, H. N. 1992. The role of permutation probability tests in phylogenetic systematics. *Systematic Zoology* **41:** 258–263.

Buckley, T. R. 2002. Model misspecification and probabilistic tests of topology: Evidence from empirical data sets. *Systematic Biology* **51:** 509–523.

Bull, J. J., J. P. Huelsenbeck, C. W. Cunningham, D. L. Swofford, and P. J.

Waddell. 1993. Partitioning and combining data in phylogenetic analysis. *Systematic Biology* **42:** 384–397.

Bulmer, M. 1991. Use of the method of generalized least squares in reconstructing phylogenies from sequence data. *Molecular Biology and Evolution* **8:** 868–883.

Buneman, P. 1971. The recovery of trees from measurements of dissimilarity. pp. 387–395 in *Mathematics in the Archeological and Historical Sciences,* ed. F. R. Hodson, D. G. Kendall, and P. Tautu. Edinburgh University Press, Edinburgh.

Buneman, P. 1974. A characterization of rigid circuit graphs. *Discrete Mathematics* **9:** 205–212.

Bustamante, C. D., R. Nielsen, and D. L. Hartl. 2002. A maximum likelihood method for analyzing pseudogene evolution: Implications for silent site evolution in humans and rodents. *Molecular Biology and Evolution* **19**: 110–117.

Calabrese, P. P., R. T. Durrett, and C. F. Aquadro. 2001. Dynamics of microsatellite divergence under stepwise mutation and proportional slippage/point mutation models. *Genetics* **159:** 839–852.

Camin, J. H. and R. R. Sokal. 1965. A method for deducing branching sequences in phylogeny. *Evolution* **19:** 311–326.

Cao, Y., J. Adachi, A. Janke, S. Pääbo, and M. Hasegawa. 1994. Phylogenetic relationships among eutherian orders estimated from inferred sequences of mitochondrial proteins: Instability of a tree based on a single gene. *Journal of Molecular Evolution* **39:** 519–527.

Caprara, A. 1999. Sorting permutations by reversals and Eulerian cycle decompositions. *SIAM Journal on Discrete Mathematics* **12:** 91–110.

Carpenter, J. 1992. Random cladistics. *Cladistics* **8:** 147–153.

Carpenter, J. 1996. Uninformative bootstrapping. *Cladistics* **12:** 177–181.

Carpenter, J. M., P. A. Goloboff, and J. S. Farris. 1998. PTP is meaningless, T-PTP is contradictory: A reply to Trueman. *Cladistics* **14:** 105–116.

Carter, M., M. Hendy, D. Penny, L. A. Székely, and N. C. Wormald. 1990. On the distribution of lengths of evolutionary trees. *SIAM Journal on Discrete Mathematics* **3:** 38–47.

Cavalli-Sforza, L. L. and A. W. F. Edwards. 1965. Analysis of human evolution. pp. 923–933 in *Genetics Today. Proceedings of the XI International Congress of Genetics, The Hague, The Netherlands, September 1963.,* Volume 3. ed. S. J. Geerts. Pergamon Press, Oxford.

Cavalli-Sforza, L. L. and A. W. F. Edwards. 1967. Phylogenetic analysis: Models and estimation procedures. *American Journal of Human Genetics* **19:** 233–257. *Evolution* **21:** 550–570.

Cavender, J. A. 1978. Taxonomy with confidence. *Mathematical Biosciences* **40:** 271–280.

Cavender, J. A. and J. Felsenstein. 1987. Invariants of phylogenies in a simple case with discrete states. *Journal of Classification* **4:** 57–71.

Cavender, J. A. 1989. Mechanized derivation of linear invariants. *Molecular Biology and Evolution* **6:** 301–316.

Cavender, J. A. 1991. Necessary conditions for the method of inferring phylogeny by linear invariants. *Mathematical Biosciences* **103:** 69–75.

Cayley, A. 1857. On the theory of analytic forms called trees. *Philosophical Magazine* **13:** 19–30. Also published in his *Collected Mathematical Papers,* Cambridge Vol. 3 (1891): pp. 242–246.

Cayley, A. 1889. A theorem on trees. *Quarterly Journal of Mathematics* **23:** 376–378. Also published in his *Collected Papers,* Cambridge **13 (1897):** 26–28.

Chakraborty, R. 1977. Estimation of time of divergence from phylogenetic studies.

Canadian Journal of Genetics and Cytology **19:** 217–223.

Chakraborty, R. and M. Nei. 1982. Genetic differentiation of quantitative characters between populations or species. I. Mutation and random drift. *Genetical Research* **39:** 303–314.

Chambers, G. K. and E. S. MacAvoy. 2000. Microsatellites: Consensus and controversy. *Comparative Biochemistry and Physiology Part B: Biochemistry and Molecular Biology* **126:** 455–476.

Chan, K. M. A. and B. R. Moore. 1999. Accounting for mode of speciation increases power and realism of tests of phylogenetic asymmetry. *American Naturalist* **153:** 332–346.

Chang, J. T. 1996a. Inconsistency of evolutionary tree topology reconstruction methods when substitution rates vary across characters. *Mathematical Biosciences* **134:** 189–215.

Chang, J. T. 1996b. Full reconstruction of Markov models on evolutionary trees: Identifiability and consistency. *Mathematical Biosciences* **137:** 51–73.

Charleston, M. A. 1998. Jungles: A new solution to the host/parasite phylogeny reconciliation problem. *Mathematical Biosciences* **149:** 191–223.

Charleston, M. A. 2001. Hitch-hiking: A parallel heuristic search strategy, applied to the phylogeny problem. *Journal of Computational Biology* **8:** 79–91.

Chen, F.-C. and W.-H. Li. 2001. Genomic divergences between humans and other hominoids and the effective population size of the common ancestor of humans and chimpanzees. *American Journal of Human Genetics* **68:** 444–456.

Chen, K., D. Durand, and M. Farach-Colton. 2000. NOTUNG: A program for dating gene duplications and optimizing gene family trees. *Journal of Computational Biology* **7:** 429–447.

Cheverud, J. M., M. M. Dow, and W. Leutenegger. 1985. The quantitative assessment of phylogenetic constraints in comparative analyses: Sexual dimorphism in body weight among primates. *Evolution* **39:** 1335–1351.

Chor, B., M. D. Hendy, B. R. Holland, and D. Penny. 2000. Multiple maxima of likelihood in phylogenetic trees: An analytic approach. *Molecular Biology and Evolution* **17:** 1529–1541.

Churchill, G. A. 1989. Stochastic models for heterogeneous DNA sequences. *Bulletin of Mathematical Biology* **51:** 79–94.

Clark, A. G. and C. M. Lanigan. 1993. Prospects for estimating nucleotide divergence with RAPDs. *Molecular Biology and Evolution* **10:** 1096–1111 .

Clayton, G. and A. Robertson. 1955. Mutation and quantitative variation. *American Naturalist* **89:** 151–158.

Clement, M., D. Posada, and K. A. Crandall. 2000. TCS: A computer program to estimate gene genealogies. *Molecular Ecology* **9:** 1657–1659.

Clutton-Brock, T. and P. H. Harvey. 1977. Primate ecology and social organisation. *Journal of Zoology* **183:** 1–33.

Clutton-Brock, T. and P. H. Harvey. 1979. Comparison and adaptation. *Proceedings of the Royal Society of London, Series B* **205:** 547–565.

Clyde, W. C. and D. C. Fisher. 1997. Comparing the fit of stratigraphic and morphologic data in phylogenetic analysis. *Paleobiology* **23:** 1–19.

Coddington, J. A. 1988. Cladistic tests of adaptational hypotheses. *Cladistics* **4:** 3–22.

Colless, D. H. 1980. Congruence between morphometric and allozyme data for *Menidia* species: A reappraisal. *Systematic Zoology* **29:** 288–299.

Colless, D. H. 1982. Phylogenetics, the theory and practice of phylogenetic systematics (book review). *Systematic Zoology* **31:** 100–104.

Congdon, C. B. 2001. Gaphyl: A genetic algorithms approach to cladistics. pp.

67–78 in *Principles of Data Mining and Knowledge Discovery*, ed. L. DeRaedt and A. Siebes. Lecture Notes in Computer Science, No. 2168. Springer-Verlag, Berlin.

Cosner, M. E., R. K. Jansen, B. M. E. Moret, L. A. Raubeson, L.-S. Wang, T. Warnow, and S. Wyman. 2000. An empirical comparison of phylogenetic methods on chloroplast gene order data in Campanulaceae. pp. 99–121 in *Comparative Genomics. Empirical and Analytical Approaches to Gene Order Dynamics, Map Alignment, and the Evolution of Gene Families.* ed. D. Sankoff and J. Nadeau. Kluwer Academic Publishers, Dordrecht, Netherlands.

Cox, D. R. 1961. Tests of separate families of hypotheses. *Proceedings of the 4th Berkeley Symposium on Mathematical Statistics and Probability,* Volume 1, pp. 105–123

Critchlow, D. E., D. K. Pearl, and C. L. Qian. 1996. The triples distance for rooted bifurcating phylogenetic trees. *Systematic Biology* **45:** 323–334.

Cryan, M., L. A. Goldberg, and P. W. Goldberg. 2001. Evolutionary trees can be learned in polynomial time in the two-state general Markov model. *SIAM Journal on Computing* **31:** 375–397.

Csűrös, M. and M. Y. Kao. 2001. Provably fast and accurate recovery of evolutionary trees through harmonic greedy triplets. *SIAM Journal on Computing* **31:** 306–322.

Csűrös, M. 2002. Fast recovery of evolutionary trees with thousands of nodes. *Journal of Computational Biology* **9:** 277–297.

Culberson, J. and P. Rudnicki. 1989. A fast algorithm for constructing trees from distance matrices. *Information Processing Letters* **30:** 215–220.

Cunningham, C. W., K. E. Omland, and T. H. Oakley. 1998. Reconstructing ancestral character states: A critical appraisal. *Trends in Evology and Evolution* **13:** 361–366.

Cutler, D. J. 2000a. Estimating divergence times in the presence of an overdispersed molecular clock. *Molecular Biology and Evolution* **17:** 1647–1660.

Cutler, D. J. 2000b. The index of dispersion of molecular evolution: Slow fluctuations. *Theoretical Population Biology* **57:** 177–186.

Darlu, P. and G. Lecointre. 2002. When does the incongruence length difference test fail? *Molecular Biology and Evolution* **19:** 432–437.

DasGupta, B., T. Jiang, S. Kannan, M. Li, and Z. Sweedyk. 1997. On the complexity and approximation of syntenic distance. pp. 99–108 in *RECOMB '97: Proceedings of the First Annual International Conference on Computational Molecular Biology, January 20–23, 1997, Eldorado Hotel, Santa Fe, New Mexico.* Association for Computing Machinery, New York.

Day, W. H. E. 1983. Computationally difficult parsimony problems in phylogenetic systematics. *Journal of Theoretical Biology* **103:** 429–438.

Day, W. H. E., D. S. Johnson, and D. Sankoff. 1986. The computational complexity of inferring rooted phylogenies by parsimony. *Mathematical Biosciences* **81:** 33–42.

Day, W. H. E. 1986. Computational complexity of inferring phylogenies from dissimilarity matrices. *Bulletin of Mathematical Biology* **49:** 461–467.

Day, W. H. E. and D. Sankoff. 1986. Computational complexity of inferring phylogenies by compatibility. *Systematic Zoology* **35:** 224–229.

Day, W. H. E. and D. Sankoff. 1987. Computational-complexity of inferring phylogenies from chromosome inversion data. *Journal of Theoretical Biology* **124:** 213–218.

Dayhoff, M. O. and R. V. Eck. 1968. *Atlas of Protein Sequence and Structure 1967–1968.* National Biomedical Research Foundation, Silver Spring, Maryland.

Dayhoff, M. O., R. M. Schwartz, and B. C. Orcutt. 1979. A model of evolutionary change in proteins. pp. 345–352 in *Atlas of Protein Sequence and Structure, Volume 5, Supplement 3, 1978,* ed. M. O. Dayhoff. National Biomedical Research Foundation, Silver Spring, Maryland.

DeBry, R. W. and N. A. Slade. 1985. Cladistic analysis of restriction endonuclease cleavage maps within a maximum-likelihood framework. *Systematic Zoology* **34:** 21–34.

DeBry, R. W. 1992. The consistency of several phylogeny-inference methods under varying evolutionary rates. *Molecular Biology and Evolution* **9:** 537–551.

De Laet, J. and E. Smets. 1998. On the three-taxon approach to parsimony analysis. *Cladistics* **14:** 363–381.

De Queiroz, A. 1993. For consensus (sometimes). *Systematic Biology* **42:** 368–372.

De Queiroz, A., M. J. Donoghue, and J. Kim. 1995. Separate versus combined analysis of phylogenetic evidence. *Annual Review of Ecology and Systematics* **26:** 657–681.

De Queiroz, K. and S. Poe. 2001. Philosophy and phylogenetic inference: A comparison of likelihood and parsimony methods in the context of Karl Popper's writings on corroboration. *Systematic Biology* **50:** 305–321.

De Soete, G. 1983. A least squares algorithm for fitting additive trees to proximity data. *Psychometrika* **48:** 621–626.

Desper, R. and O. Gascuel. 2002. Fast and accurate phylogeny reconstruction algorithms based on the minimum-evolution principle. *Journal of Computational Biology* **19:** 687–705.

Di Battista, G., P. Eades, R. Tamassia, and I. G. Tollis. 1994. Annotated bibliography on graph drawing algorithms. *Computational Geometry: Theory and Applications* **4:** 235–282.

Dimmic, M. W., D. P. Mindell, and R. A. Goldstein. 2000. Modeling evolution at the protein level using an adjustable amino acid fitness model. *Pacific Symposium on Biocomputing 2000,* pp. 18–29.

Di Rienzo, A., A. C. Peterson, J. C. Garza, A. M. Valdes, M. Slatkin, and N. B. Freimer. 1994. Mutational processes of simple-sequence repeat loci in human populations. *Proceedings of the National Academy of Sciences, USA* **91:** 3166–3170.

Dollo, L. 1893. Le lois de l'évolution. *Bulletin de la Societé Belge de Géologie de Paléontologie et d'Hydrologie* **7:** 164–167.

Dolphin, K., R. Belshaw, C. D. L. Orme, and D. L. J. Quicke. 2000. Noise and incongruence: Interpreting results of the incongruence length difference test. *Molecular Phylogenetics and Evolution* **17:** 401–406.

Donoghue, M. J. and J. W. Kadereit. 1992. Walter Zimmerman and the growth of phylogenetic theory. *Systematic Biology* **41:** 74–85.

Dopazo, J. 1994. Estimating errors and confidence intervals for branch lengths in phylogenetic trees by a bootstrap approach. *Journal of Molecular Evolution* **38:** 300–304.

Dowling, A. P. G. 2002. Testing the accuracy of TreeMap and Brooks parsimony analyses of coevolutionary patterns using artificial associations. *Cladistics* **18:** 416–435.

Dowton, M. and A. D. Austin. 2002. Increased congruence does not necessarily indicate increased phylogenetic accuracy—The behavior of the incongruence length difference test in mixed-model analyses. *Systematic Biology* **51:** 19–31.

Doyle, J. J. 1992. Gene trees and species trees: Molecular systematics as one-character taxonomy. *Systematic Botany* **17:** 144–163.

Dress, A. and M. Krüger. 1987. Parsimonious phylogenetic trees in metric spaces and simulated annealing. *Advances in Applied Mathematics* **8:** 8–37.

Dress, A., V. Moulton, and M. Steel. 1997. Trees, taxonomy, and strongly compatible multi-state characters. *Advances in Applied Mathematics* **19:** 1–30.

Dress, A., D. H. Huson, and V. Moulton. 1996. Analyzing and visualizing sequence and distance data using SPLITSTREE. *Discrete Applied Mathematics* **71:** 95–109.

Drolet, S. and D. Sankoff. 1990. Quadratic tree invariants for multivalued characters. *Journal of Theoretical Biology* **144:** 117–129.

Drouin, G., F. Prat, M. Ell, G. D. P. Clarke. 1999. Detecting and characterizing gene conversion between multigene family members. *Molecular Biology and Evolution* **16:** 1369–1390.

Eck, R. V. and M. O. Dayhoff. 1966. *Atlas of Protein Sequence and Structure 1966.* National Biomedical Research Foundation, Silver Spring, Maryland.

Edwards, A. W. F. and L. L. Cavalli-Sforza. 1963. The reconstruction of evolution. *Annals of Human Genetics* **27:** 105–106 (also published in *Heredity* **18:** 553).

Edwards, A. W. F. and L. L. Cavalli-Sforza. 1964. Reconstruction of evolutionary trees. pp. 67–76 in *Phenetic and Phylogenetic Classification*, ed. V. H. Heywood and J. McNeill. Systematics Association Publ. No. 6, London.

Edwards, A. W. F. 1970. Estimation of the branch points of a branching diffusion process. *Journal of the Royal Statistical Society B* **32:** 155–174.

Edwards, A. W. F. 1971. The likelihood for multinomial proportions under stereographic projection. *Biometrics* **28:** 618–620.

Edwards, A. W. F. 1972. *Likelihood.* Cambridge University Press, Cambridge.

Edwards, A. W. F. 1996. The origin and early development of the method of minimum evolution for the reconstruction of phylogenetic trees. *Systematic Biology* **45:** 79–91.

Edwards, S. V. and P. Beerli. 2000. Perspective: Gene divergence, population divergence, and the variance in coalescence time in phylogeographic studies. *Evolution* **54:** 1839–1854.

Eernisse, D. J. and A. G. Kluge. 1993. Taxonomic congruence versus total evidence, and amniote phylogeny inferred from fossils, molecules, and morphology. *Molecular Biology and Evolution* **10:** 1170–1195.

Efron, B. 1979. Bootstrap methods: Another look at the jackknife. *Annals of Statistics* **7:** 1–26.

Efron, B. 1985. Bootstrap confidence intervals for a class of parametric problems. *Biometrika* **72:** 45–58.

Efron, B. 1987. Better bootstrap confidence intervals. *Journal of the American Statistical Institute* **82:** 171–185.

Efron, B., E. Halloran, and S. Holmes. 1996. Bootstrap confidence levels for phylogenetic trees. *Proceedings of the National Academy of Sciences, USA* **93:** 7085–7090.

Eigen, M., R. Winkler-Oswatitsch, and A. W. M. Dress. 1988. Statistical geometry in sequence space: A method of comparative sequence analysis. *Proceedings of the National Academy of Sciences, USA* **85:** 5913–5917.

Eldredge, N. and J. Cracraft. 1980. *Phylogenetic Patterns and the Evolutionary Process.* Columbia University Press, New York.

Elemento, O., O. Gascuel, and M. P. Lefranc. 2002. Reconstructing the duplication history of tandemly repeated genes. *Molecular Biology and Evolution* **19:** 278–288.

Ellegren, H. 2000. Microsatellite mutations in the germline: implications for evolutionary inference. *Trends in Genetics* **16:** 551–558.

Elston, R. C. and J. Stewart. 1971. A general model for the genetic analysis of pedigree data. *Human Heredity* **21:** 523–542.

Emerson, B. C., K. M. Ibrahim, and G. M. Hewitt. 2001. Selection of evolutionary models for phylogenetic hypothesis testing using parametric methods. *Journal of Evolutionary Biology* **14:** 620–631.

Erdős, P. L., M. A. Steel, L. Székely, and T. J. Warnow. 1997a. Constructing big trees from short sequences. pp. 827–837 in *Automata, Languages and Programming. 24th International Colloquium, ICALP'97, Bologna, Italy, July 7–11, 1997, Proceedings*, ed. P. Degano, R. Gorrieri, and A. Marchetti-Spaccalmela. Lecture Notes in Computer Science No. 1256, Springer-Verlag, Berlin.

Erdős, P. L., M. A. Steel, L. A. Székely, and T. Warnow. 1997b. Local quartet splits of a binary tree infer all quartet splits via one dyadic inference rule. *Computers and Artificial Intelligence* **16:** 217–227.

Erdős, P. L., M. A. Steel, L. A. Székely, and T. J. Warnow. 1999. A few logs suffice to build (almost) all trees (I). *Random Structures and Algorithms* **14:** 153–184.

Estabrook, G. F. 1968. A general solution in partial orders for the Camin-Sokal model in phylogeny. *Journal of Theoretical Biology* **21:** 421–438.

Estabrook, G. F. and L. Landrum. 1976. A simple test for the possible simultaneous evolutionary divergence of two aminoacid positions. *Taxon* **24:** 609–613.

Estabrook, G. F., C. S. Johnson, Jr., and F. R. McMorris. 1976a. An algebraic analysis of cladistic characters. *Discrete Mathematics* **16:** 141–147.

Estabrook, G. F., C. S. Johnson, Jr., and F. R. McMorris. 1976b. A mathematical foundation for the analysis of cladistic character compatibility. *Mathematical Biosciences* **29:** 181–187.

Estabrook, G. F. and F. R. McMorris. 1980. When is one estimate of evolutionary relationships a refinement of another? *Journal of Mathematical Biology* **10:** 367–373.

Estabrook, G. F., F. R. McMorris, and C. A. Meacham. 1985. Comparison of undirected phylogenetic trees based on subtrees of 4 evolutionary units. *Systematic Zoology* **34:** 193–200.

Estabrook, G. F. 1992. Evaluating undirected positional congruence of individual taxa between two estimates of the phylogenetic tree for a group of taxa. *Systematic Biology* **41:** 172–177.

Eulenstein, O. and M. Vingron. 1998. On the equivalence of two tree mapping measures. *Discrete Applied Mathematics* **88:** 103–128.

Evans, S. N. and T. P. Speed. 1993. Invariants of some probability models used in phylogenetic inference. *Annals of Statistics* **21:** 355–377.

Evans, S. N. and X. Zhou. 1998. Constructing and counting phylogenetic invariants. *Journal of Computational Biology* **5:** 713–724.

Fack, V., S. Lievens, and J. Van der Jeugt. 1999. On rotation distance between binary coupling trees and applications for 3nj-coefficients. *Computer Physics Communications* **119:** 99–114.

Faith, D. P. 1985. Distance methods and the approximation of most-parsimonious trees. *Systematic Zoology* **43:** 312–325.

Faith, D. P. and P. S. Cranston. 1991. Could a cladogram this short have arisen by chance alone? On permutation tests for cladistic structure. *Cladistics* **7:** 1–28.

Faith, D. P. 1991. Cladistic permutation tests for monophyly and nonmonophyly. *Systematic Zoology* **40:** 366–375.

Faith, D. P. 1992. On corroboration: A reply to Carpenter. *Cladistics* **8:** 265–273.

Faith, D. P. and J. W. O. Ballard. 1994. Length differences and topology-dependent tests: A response to Källersjö et al. *Cladistics* **10:** 57–64.

Faith, D. P. and J. W. H. Trueman. 1996. When the topology-dependent permutation test (T-PTP) for monophyly returns significant support for monophyly, should that be equated with (a) rejecting a null hypothesis of nonmonophyly, (b) rejecting a null hypothesis of "no structure", (c) failing to falsify a hypothesis of monophyly, or (d) none of the above? *Systematic Biology* **45:** 580–586.

Faith, D. P. and J. W. H. Trueman. 2001. Towards an inclusive philosophy for phylogenetic inference. *Systematic Biology* **50:** 331–350.

Faivovich, J. 2002. On RASA. *Cladistics* **18:** 324–333.

Falconer, D. S. 1965. The inheritance of liability to certain diseases, estimated from the incidence among relatives. *Annals of Human Genetics* **29:** 51–76.

Falconer, D. S. and T. F. MacKay. 1996. *Introduction to Quantitative Genetics*, 4th edition. Longman Scientific and Technical, Harlow, U.K.

Falush, D. and Y. Iwasa. 1999. Size-dependent mutability and microsatellite constraints. *Molecular Biology and Evolution* **16:** 960–966.

Farach, M., S. Kannan, and T. Warnow. 1995. A robust model for finding optimal evolutionary trees. *Algorithmica* **13:** 155–179.

Farach, M. and S. Kannan. 1999. Efficient algorithms for inverting evolution. *Journal of the ACM* **46:** 437–449.

Farris, J. S. 1969a. On the cophenetic correlation coefficient. *Systematic Zoology* **18:** 279–285.

Farris, J. S. 1969b. A successive approximations approach to character weighting. *Systematic Zoology* **18:** 374–385.

Farris, J. S. 1970. Methods for computing Wagner trees. *Systematic Zoology* **19:** 83–92.

Farris, J. S., A. G. Kluge, and M. J. Eckardt. 1970. A numerical approach to phylogenetic systematics. *Systematic Zoology* **19:** 172–189.

Farris, J. S. 1972. Estimating phylogenetic trees from distance matrices. *American Naturalist* **106:** 645–668.

Farris, J. S. 1973a. On comparing the shapes of taxonomic trees. *Systematic Zoology* **22:** 50–54.

Farris, J. S. 1973b. A probability model for inferring evolutionary trees. *Systematic Zoology* **22:** 250–256.

Farris, J. S. 1976. Expected asymmetry of phylogenetic trees. *Systematic Zoology* **25:** 196–198.

Farris, J. S. 1977a. Phylogenetic analysis under Dollo's law. *Systematic Zoology* **26:** 77–88.

Farris, J. S. 1977b. On the phenetic approach to vertebrate classification. pp. 823–850 in *Major Patterns in Vertebrate Evolution*, ed. M. K. Hecht, P. C. Goody, and B. M. Hecht. Plenum, New York.

Farris, J. S. 1978. Inferring phylogenetic trees from chromosome inversion data. *Systematic Zoology* **27:** 275–284.

Farris, J. S. 1981. Distance data in phylogenetic analysis. pp. 3–23 in *Advances in Cladistics. Proceedings of the First Meeting of the Willi Hennig Society.*, ed. V. A. Funk and D. R. Brooks. New York Botanical Garden, Bronx.

Farris, J. S. 1983. The logical basis of phylogenetic analysis. pp. 7–36 in *Advances in Cladistics, Volume 2. Proceedings of the Second Meeting of the Willi Hennig Society*, ed. N. H. Platnick and V. A. Funk. Columbia University Press, New York.

Farris, J. S. 1985. Distance data revisited. *Cladistics* **1:** 67–85.

Farris, J. S. 1986. Distances and statistics. *Cladistics* **2:** 144–157.

Farris, J. S. and N. I. Platnick. 1989. Lord of the flies—the systematist as a study animal. *Cladistics* **5:** 295–310.

Farris, J. S., M. Källersjö, A. G. Kluge, and C. Bult. 1994a. Permutations. *Cladistics* **10:** 65–76.

Farris, J. S., M. Källersjö, A. G. Kluge, and C. Bult. 1994b. Testing significance of incongruence. *Cladistics* **10:** 315–319.

Farris, J. S., M. Källersjö, V. A. Albert, M. Allard, A. Anderberg, B. Bowditch, C. Bult, J. M. Carpenter, T. M. Crowe, J. De Laet, K. Fitzhugh, D. Frost, P. Goloboff, C.

J. Humphries, U. Jondelius, D. Judd, P. O. Karis, D. Lipscomb, M. Luckow, D. Mindell, J. Muona, K. Nixon, W. Presch, O. Seberg, M. E. Siddall, L. Struwe, A. Tehler, J. Wenzel, Q. Wheeler, and W. Wheeler. 1995a. Explanation. *Cladistics* **11:** 211–218.

Farris, J. S. 1995. Conjectures and refutations. *Cladistics* **11:** 105–118.

Farris, J. S., M. Källersjö, A. G. Kluge, and C. Bult. 1995b. Constructing a significance test for incongruence. *Systematic Biology* **44:** 570–572.

Farris, J. S., V. A. Albert, M. Källersjö, D. Lipscomb, and A. G. Kluge. 1996. Parsimony jackknifing outperforms neighbor-joining. *Cladistics* **12:** 99–124.

Farris, J. S. and A. G. Kluge. 1998. A/the brief history of three-taxon analysis. *Cladistics* **14:** 349–362.

Farris, J. S., M. Källersjö, T. M. Crowe, D. Lipscomb. 1999. Frigatebirds, tropicbirds, and Ciconiida: Excesses of confidence probability. *Cladistics* **15:** 1–7.

Farris, J. S. 1999. Likelihood and inconsistency. *Cladistics* **15:** 199–204.

Farris, J. S. 2000a. Corroboration versus "strongest evidence". *Cladistics* **16:** 385–393.

Farris, J. S. 2000b. Diagnostic efficiency of three-taxon analysis. *Cladistics* **16:** 403–410.

Farris, J. S., M. Källersjö, and J. E. De Laet. 2001. Branch lengths do not indicate support—even in maximum likelihood. *Cladistics* **17:** 298–299.

Farris, J. S. 2002. RASA attributes highly significant structure to randomized data. *Cladistics* **18:** 334–353.

Feldman, M. W., A. Bergman, D. D. Pollock, and D. B. Goldstein. 1997. Microsatellite genetic distances with range constraints: Analytic description and problems of estimation. *Genetics* **145:** 207–216.

Feller, W. 1971. *An Introduction to Probability Theory and Its Applications, Volume II.* 2nd edition. John Wiley and Sons, New York.

Felsenstein, J. 1968. *Statistical Inference and the Estimation of Phylogenies.* Ph.D. Thesis, Department of Zoology, University of Chicago.

Felsenstein, J. 1971. The rate of loss of multiple alleles in finite haploid populations. *Theoretical Population Biology* **2:** 391–403.

Felsenstein, J. 1973a. Maximum likelihood estimation of evolutionary trees from continuous characters. *American Journal of Human Genetics* **25:** 471–492.

Felsenstein, J. 1973b. Maximum likelihood and minimum-steps methods for estimating evolutionary trees from data on discrete characters. *Systematic Zoology* **22:** 240–249.

Felsenstein, J. 1978a. The number of evolutionary trees. *Systematic Zoology* **27:** 27–33. (Correction, Vol. 30, p. 122, 1981)

Felsenstein, J. 1978b. Cases in which parsimony or compatibility methods will be positively misleading. *Systematic Zoology* **27:** 401–410.

Felsenstein, J. 1979. Alternative methods of phylogenetic inference and their interrelationship. *Systematic Zoology* **28:** 49–62.

Felsenstein, J. 1981a. A likelihood approach to character weighting and what it tells us about parsimony and compatibility. *Biological Journal of the Linnean Society* **16:** 183–196.

Felsenstein, J. 1981b. Evolutionary trees from DNA sequences: A maximum likelihood approach. *Journal of Molecular Evolution* **17:** 368–376.

Felsenstein, J. 1981c. Maximum likelihood estimation of evolutionary trees from continuous characters. *American Journal of Human Genetics* **25:** 471–492.

Felsenstein, J. 1983a. Inferring evolutionary trees from DNA sequences. pp. 133–150 in *Statistical Analysis of DNA Sequence Data,* ed. B. S. Weir. M. Dekker, New York.

Felsenstein, J. 1983b. Parsimony in systematics: Biological and statistical issues. *Annual Review of Ecology and Systematics* **14:** 313–333.

Felsenstein, J. 1984. Distance methods for inferring phylogenies: A justification. *Evolution* **38:** 16–24.

Felsenstein, J. 1985a. Phylogenies and the comparative method. *American Naturalist* **125:** 1–15.

Felsenstein, J. 1985b. Confidence limits on phylogenies: An approach using the bootstrap. *Evolution* **39:** 783–791.

Felsenstein, J. 1985c. Confidence limits on phylogenies with a molecular clock. *Systematic Zoology* **34**: 152–161.

Felsenstein, J. 1985d. Phylogenies from gene frequencies: A statistical problem. *Systematic Zoology* **34:** 300–311.

Felsenstein, J. 1986. Distance methods: Reply to Farris. *Cladistics* **2:** 130–143.

Felsenstein, J. 1988a. Phylogenies and quantitative characters. *Annual Review of Ecology and Systematics* **19:** 445–471.

Felsenstein, J. 1988b. Phylogenies from molecular sequences: Inference and reliability. *Annual Review of Genetics* **22:** 521–565.

Felsenstein, J. 1991. Counting phylogenetic invariants in some simple cases. *Journal of Theoretical Biology* **152:** 357–376.

Felsenstein, J. 1992a. Estimating effective population size from samples of sequences: Inefficiency of pairwise and segregating sites as compared to phylogenetic estimates. *Genetical Research* **59:** 139–147.

Felsenstein, J. 1992b. Phylogenies from restriction sites: A maximum likelihood approach. *Evolution* **46:** 159–173.

Felsenstein, J. and H. Kishino. 1993. Is there something wrong with the bootstrap on phylogenies? A reply to Hillis and Bull. *Systematic Biology* **42:** 193–200.

Felsenstein, J. and G. A. Churchill. 1996. A hidden Markov model approach to variation among sites in rate of evolution. *Molecular Biology and Evolution* **13:** 93–104.

Felsenstein, J. 1996. Inferring phylogenies from protein sequences by parsimony, distance, and likelihood methods. *Methods in Enzymology* **266:** 418–427.

Felsenstein, J. 1997. An alternating least squares approach to inferring phylogenies from pairwise distances. *Systematic Biology* **46:** 101–111.

Felsenstein, J., M. K. Kuhner, J. Yamato, and P. Beerli. 1999. Likelihoods on coalescents: A Monte Carlo sampling approach to inferring parameters from population samples of molecular data. pp. 163–185 in *Statistics in Molecular Biology and Genetics*, ed. F. Seillier-Moiseiwitsch. IMS Lecture Notes–Monograph Series, Volume 33. Institute of Mathematical Statistics and American Mathematical Society, Hayward, California.

Felsenstein, J. 2001a. The troubled growth of statistical phylogenetics. *Systematic Biology* **50:** 465–467.

Felsenstein, J. 2001b. Taking variation of evolutionary rates between sites into account in inferring phylogenies. *Journal of Molecular Evolution* **53:** 447–455.

Felsenstein, J. 2002. Quantitative characters, phylogenies, and morphometrics. pp. 27–44 in *Morphology, Shape, and Phylogenetics*, ed. N. MacLeod. Taylor and Francis, London.

Feng, D.-F. and R. F. Doolittle. 1987. Progressive sequence alignment as a prerequisite to correct phylogenetic trees. *Journal of Molecular Evolution* **25:** 351–360.

Ferretti, V. and D. Sankoff. 1993. The empirical discovery of phylogenetic invariants. *Advances in Applied Probability* **25:** 290–302.

Ferretti, V., B. F. Lang, and D. Sankoff. 1994. Skewed base compositions, asymmetric transition matrices, and phylogenetic invariants. *Journal of Computational Biology* **1:** 77–92.

Ferretti, V. and D. Sankoff. 1995. Phylogenetic invariants for more general evolutionary models. *Journal of Theoretical Biology* **173:** 147–162.

Ferretti, V., J. H. Nadeau, and D. Sankoff. 1996. Original synteny. pp. 159–167 in *Combinatorial Pattern Matching, 7th Annual Symposium, CPM 96, Laguna Beach, California, USA, June 10–12, 1996, Proceedings.* ed. D. Hirschberg and G. Myers. Lecture Notes in Computer Science No. 1075, Springer-Verlag, New York.

Ferretti, V. and D. Sankoff. 1996. A remarkable nonlinear invariant for evolution with heterogeneous rates. *Mathematical Biosciences* **134:** 71–83.

Finarelli, J. A. and W. C. Clyde. 2002. Comparing the gap excess ratio and the retention index of the stratigraphic character. *Systematic Biology* **51:** 166–176.

Fisher, D. C. 1991. Phylogenetic analysis and its application in evolutionary paleobiology. pp. 103–122 in *Analytical Paleobiology,* ed. N. L. Gilinsky and P. W. Signor. Short courses in paleontology no. 4. Paleontological society, Lawrence, Kansas.

Fisher, D. C. 1992. Stratigraphic parsimony. pp. 124–129 in W. P. Maddison and D. R. Maddison, *MacClade: Analysis of Phylogeny and Character Evolution.* Sinauer Associates, Sunderland, Massachusetts.

Fisher, R. A. 1912. On an absolute criterion for fitting frequency curves. *Messenger of Mathematics* **41:** 155–160.

Fisher, R. A. 1918. The correlation between relatives on the supposition of Mendelian inheritance. *Transactions of the Royal Society of Edinburgh* **52:** 399–433.

Fisher, R. A. 1921. On the "probable error" of a coefficient of correlation deduced from a small sample. *Metron* **1:** 3–32.

Fisher, R. A. 1922. On the mathematical foundations of theoretical statistics. *Philosophical Transactions of the Royal Society of London, A* **222:** 309–368.

Fitch, W. M. and E. Margoliash. 1967. Construction of phylogenetic trees. *Science* **155:** 279–284.

Fitch, W. M. and E. Markowitz. 1970. An improved method for determining codon variability in a gene and its application to the rate of fixation of mutations in evolution. *Biochemical Genetics* **4:** 579–593 .

Fitch, W. M. 1971. Toward defining the course of evolution: Minimum change for a specified tree topology. *Systematic Zoology* **20:** 406–416.

Fitch, W. M. 1975. Toward finding the tree of maximum parsimony. pp. 189–230 in *Proceedings of the Eighth International Conference on Numerical Taxonomy,* ed. G. F. Estabrook. W. H. Freeman, San Francisco.

Fitch, W. M. 1977. Phylogenies constrained by the crossover process as illustrated by human hemoglobins and a thirteen-cycle, eleven-amino-acid repeat in human apolipoprotein A-I. *Genetics* **86:** 623–644.

Fitch, W. M. 1979. Cautionary remarks on using gene-expression events in parsimony procedures. *Systematic Zoology* **28:** 375–379.

Fitch, W. M. 1981. A non-sequential method for constructing trees and hierarchical classifications. *Journal of Molecular Evolution* **18:** 30–37.

Fitch, W. M. 1984. Cladistic and other methods: Problems, pitfalls, and potentials. pp. 221–252 in *Cladistics: Perspectives on the Reconstruction of Evolutionary History,* ed. T. Duncan and T. F. Stuessy. Columbia University Press, New York.

Fogel, D. B. (editor). 1998. *Evolutionary Computation: The Fossil Record.* IEEE Press, New York.

Foote, M. and D. M. Raup. 1996. Fossil preservation and the stratigraphic ranges of taxa. *Paleobiology* **22:** 121–140.

Ford, B. J. 1992. Brownian movement in *Clarkia* pollen: A reprise of the first observations. *The Microscope* **40:** 235–241.

Forster, M. 1986. Statistical covariance as a measure of phylogenetic relationship. *Cladistics* **2:** 297–319.

Foulds, L. R., M. D. Hendy, and D. Penny. 1979. A graph theoretic approach to the development of minimal phylogenetic trees, *Journal of Molecular Evolution* **13:** 127–149.

Foulds, L. R. and R. L. Graham. 1982. The Steiner problem in phylogeny is NP-complete. *Advances in Applied Mathematics* **3:** 43–49.

Freckleton, R. P., P. H. Harvey, and M. Pagel. 2002. Phylogenetic analysis and comparative data: A test and review of evidence. *American Naturalist* **160:** 712–726.

Friedman, N., M. Ninio, I. Pe'er, and T. Pupko. 2002. A structural EM algorithm for phylogenetic inference. *Journal of Computational Biology* **9:** 331–353.

Fu, Y.-X. and W.-H. Li. 1992. Necessary and sufficient conditions for the existence of linear invariants in phylogenetic inference. *Mathematical Biosciences* **108:** 203–218.

Fu, Y.-X. and W.-H. Li. 1993. Statistical tests of neutrality of mutations. *Genetics* **133:** 693–709.

Fu, Y.-X. 1994. A phylogenetic estimator of effective population size or mutation rate. *Genetics* **136:** 685–692.

Fu, Y.-X. 1995. Linear invariants under Jukes' and Cantor's one-parameter model. *Journal of Theoretical Biology* **173:** 339–352.

Furnas, G. W. 1984. The generation of random, binary, unordered trees. *Journal of Classification* **1:** 187–233.

Gaffney, E. S. 1979. An introduction to the logic of phylogeny reconstruction. pp. 79–111 in *Phylogenetic Analysis and Paleontology*, ed. J. Cracraft and N. Eldredge. Columbia University Press, New York.

Gallut, C., V. Barriel, and R. Vignes. 2000. Gene order and phylogenetic information. pp. 123–132 in *Comparative Genomics. Empirical and Analytical approaches to Gene Order Dynamics, Map Alignment, and the Evolution of Gene Families.* ed. D. Sankoff and J. Nadeau. Kluwer Academic Publishers, Dordrecht, Netherlands.

Gallut, C. and V. Barriel. 2002. Cladistic coding of genomic maps. *Cladistics* **18:** 526–536.

Galtier, N. and M. Gouy. 1995. Inferring phylogenies from DNA-sequences of unequal base compositions. *Proceedings of the National Academy of Sciences, USA* **92:** 11317–11321.

Galtier, N. and P. Boursot. 2000. A new method for locating changes in a tree reveals distinct nucleotide polymorphism vs divergence patterns in mouse mitochondrial control region. *Journal of Molecular Evolution* **50:** 224–231.

Galtier, N. 2001. Maximum-likelihood phylogenetic analysis under a covarion-like model. *Molecular Biology and Evolution* **18:** 866–873.

Garey, M. R. and D. S. Johnson. 1976. *Computers and Intractability. A Guide to the Theory of NP-completeness.* W. H. Freeman, San Francisco.

Garland, T. and R. Díaz-Uriarte. 1999. Polytomies and phylogenetically independent contrasts: Examination of the bounded degrees of freedom approach. *Systematic Biology* **48:** 547–558.

Garner, C. and M. Slatkin. 2002. Likelihood-based disequilibrium mapping for two-marker haplotype data. *Theoretical Population Biology* **61:** 153–161.

Garza, J. C., M. Slatkin, and N. B. Freimer. 1995. Microsatellite allele frequencies in humans and chimpanzees, with implications for constraints on allele size. *Molecular Biology and Evolution* **12:** 594–603.

Gascuel, O. 1994. A note on Sattath and Tversky's, Saitou and Nei's, and Studier and Keppler's algorithms for inferring phylogenies from evolutionary distances.

Molecular Biology and Evolution **11:** 961–963.

Gascuel, O. and D. Levy. 1996. A reduction algorithm for approximating a (nonmetric) dissimilarity by a tree distance. *Journal of Classification* **13:** 129–155.

Gascuel, O. 1997. BIONJ: An improved version of the NJ algorithm based on a simple model of sequence data. *Molecular Biology and Evolution* **14:** 685–695.

Gascuel, O. 1997. Concerning the NJ algorithm and its unweighted version, UNJ. pp. 149–170 in *Mathematical Hierarchies and Biology,* ed. B. Mirkin, F. R. McMorris, and A. Rzhetsky. American Mathematical Society, Providence, Rhode Island.

Gascuel, O. 2000. On the optimization principle in phylogenetic analysis and the minimum-evolution criterion. *Molecular Biology and Evolution* **17:** 401–405.

Gascuel, O., D. Bryant, and F. Denis. 2001. Strengths and limitations of the minimum evolution principle. *Systematic Biology* **50:** 621–627.

Gascuel, O., M. D. Hendy, A. Jean-Marie, and R. McLachlan. 2003. The combinatorics of tandem duplication trees. *Systematic Biology* **52:** 110–118.

Gatesy, J. 2000. Linked branch support and tree stability. *Systematic Biology* **49:** 800–807.

Gaut, B. S. and B. S. Weir. 1994. Detecting substitution-rate heterogeneity among regions of a nucleotide sequence. *Molecular Biology and Evolution* **11:** 620–629.

Gaut, B. S. and P. O. Lewis. 1995. Success of maximum likelihood in the four-taxon case. *Molecular Biology and Evolution* **12:** 152–162.

Gibbs, A. J., M. B. Dale, H. R. Kinns, and H. G. MacKenzie. 1971. The transition matrix method for comparing sequences; its use in describing and classifying proteins by their amino acid sequences. *Systematic Zoology* **20:** 417–425.

Gillespie, J. H. and C. H. Langley. 1979. Are evolutionary rates really variable? *Journal of Molecular Evolution* **13:** 27–34.

Gillespie, J. H. 1984. The molecular clock may be an episodic clock. *Proceedings of the National Academy of Sciences, USA* **81:** 8009–8013.

Gillespie, J. H. 1991. *The Causes of Molecular Evolution*. Oxford University Press, Oxford.

Gingerich, P. D. 1979a. The stratophenetic approach to phylogeny reconstruction in vertebrate paleontology. pp. 41–77 in *Phylogenetic Analysis and Paleontology*, ed. J. Cracraft and N. Eldredge. Columbia University Press, New York.

Gingerich, P. D. 1979b. Paleontology, phylogeny, and classification; an example from the mammalian fossil record. *Systematic Zoology* **28:** 451–464.

Gingerich, P. D. 1992. Analysis of taxonomy and phylogeny: Stratophenetics. pp. 437–442 in *Palaeobiology: A Synthesis,* ed. D. E. G. Briggs and P. R. Crowther. Blackwell Scientific Publications, Oxford.

Gittleman, J. L. and M. Kot. 1990. Adaptation: Statistics and a null model for estimating phylogenetic effects. *Systematic Zoology* **39:** 227–241.

Gladstein, D. S. 1997. Efficient incremental character optimization. *Cladistics* **13:** 21–26

Gogarten, J. P., H. Kibak, P. Dittrich, L. Taiz, E. J. Bowman, M. F. Manolson, R. J. Poole, T. Date, T. Oshima, J. Konishi, and M. Yoshida. 1989. Evolution of the vacuolar H^+-ATPase: Implications for the origin of eukaryotes. *Proceedings of the National Academy of Sciences, USA* **86:** 6661–6665.

Goldberg, L. A., P. W. Goldberg, C. A. Phillips, E. Sweedyk, and T. Warnow. 1996. Minimizing phylogenetic number to find good evolutionary trees. *Discrete Applied Mathematics* **71:** 111–136.

Goldman, N. 1990. Maximum likelihood inference of phylogenetic trees, with special reference to a Poisson process model of DNA substitution and to parsimony analysis. *Systematic Zoology* **39:** 345–361.

Goldman, N. 1993. Statistical tests of models of DNA substitution. *Journal of Molecular Evolution* **36:** 182–198.

Goldman, N. and Z. Yang. 1994. A codon-based model of nucleotide substitution for protein-coding DNA sequences. *Molecular Biology and Evolution* **11:** 725–736 .

Goldman, N., J. L. Thorne, and D. T. Jones. 1996. Using evolutionary trees in protein secondary structure prediction and other comparative sequence analyses. *Journal of Molecular Biology* **263:** 196–208.

Goldman, N., J. P. Anderson, and A. G. Rodrigo. 2000. Likelihood-based tests of topologies in phylogenetics. *Systematic Biology* **49:** 652–670.

Goldman, N. and S. Whelan. 2002. A novel use of equilibrium frequencies in models of sequence evolution. *Molecular Biology and Evolution* **19:** 1821–1831.

Goldstein, D. B., A. R. Linares, L. L. Cavalli-Sforza, and M. W. Feldman. 1995a. An evaluation of genetic distances for use with microsatellite loci. *Genetics* **139:** 463–471.

Goldstein, D. B., A. R. Linares, L. L. Cavalli-Sforza, and M. W. Feldman. 1995b. Genetic absolute dating based on microsatellites and the origin of modern humans. *Proceedings of the National Academy of Sciences, USA* **92:** 6723–6727.

Goloboff, P. A. 1991. Random data, homoplasy and information. *Cladistics* **7:** 395–406.

Goloboff, P. A. 1993a. Estimating character weights during tree search. *Cladistics* **9:** 83–91.

Goloboff, P. A. 1993b. Character optimization and calculation of tree lengths. *Cladistics* **9:** 433–436.

Goloboff, P. A. 1997. Self-weighted optimization: Tree searches and state reconstructions under implied transformation costs. *Cladistics* **13:** 225–245.

Goloboff, P. A. 1999. Analyzing large data sets in reasonable times: Solutions for composite optima. *Cladistics* **15:** 415–428.

Gomberg, D. 1968. "Bayesian" postdiction in an evolution process. Unpublished manuscript, Istituto di Genetica, University of Pavia, Italy.

Gonnet, G. H., M. A. Cohen, S. A. Benner. 1992. Exhaustive matching of the entire protein sequence database. *Science* **256:** 1443–1445.

Gonnet, G. H. and S. A. Benner. 1996. Probabilistic ancestral sequences and multiple alignments. pp. 380–391 in *Proceedings of the 5th Scandinavian Workshop on Algorithm Theory, Reykjavik, Iceland, July 1996.* ed. R. Karlsson and A. Lingas. Lecture Notes in Computer Science No. 1097, Springer-Verlag, Berlin.

Gonnet, G. H., C. Korostensky, and S. A. Benner. 2000. Evaluation measures of multiple sequence alignments. *Journal of Computational Biology* **7:** 261–276.

Goodman, M., J. Czelusniak, G. Moore, A. Romero-Herrera, and G. Matsuda. 1979. Fitting the gene lineage into its species lineage: A parsimony strategy illustrated by cladograms constructed from globin sequences. *Systematic Zoology* **28:** 132–168.

Gordon, A. D. 1980. On the assessment and comparison of classifications. pp. 149–160 in *Analyse de Donées et Informatique,* ed. R. Tomassone. INRIA, Les Chesnay.

Gordon, A. D. 1986. Consensus supertrees: The synthesis of rooted trees containing overlapping sets of labelled trees. *Journal of Classification* **3:** 335–348.

Gordon, A. D. 1987. A review of hierarchical classification. *Journal of the Royal Statistical Society A (General)* **150:** 119–137.

Gotoh, O., J.-I. Hayashi, H. Yonekawa, and Y. Tagashira. 1979. An improved method for estimating sequence divergence between related DNAs from changes in restriction endonuclease cleavage sites. *Journal of Molecular Evolution* **14:** 301–310.

Grafen, A. 1989. The phylogenetic regression. *Philosophical Transactions of the Royal Society of London, Series B* **326:** 119–157.

Grafen, A. 1992. The uniqueness of the phylogenetic regression. *Journal of Theoretical Biology* **156:** 405–423.

Graham, J. and E. A. Thompson. 1998. Disequilibrium likelihoods for fine-scale mapping of a rare allele. *American Journal of Human Genetics* **63:** 1517–1530.

Graham, R. L. and L. R. Foulds. 1982. Unlikelihood that minimal phylogenies for a realistic biological study can be constructed in reasonable computational time. *Mathematical Biosciences* **60:** 133–142.

Grassly, N. C. and E. C. Holmes. 1997. A likelihood method for the detection of selection and recombination using nucleotide sequences. *Molecular Biology and Evolution* **14:** 239–247.

Griffiths, R. C. 1989. Genealogical tree probabilities in the infinitely-many-site model. *Journal of Mathematical Biology* **27:** 667–680.

Griffiths, R. C. and S. Tavaré. 1994a. Sampling theory for neutral alleles in a varying environment. *Philosophical Transactions of the Royal Society of London, Series B* **344:** 403–410.

Griffiths, R. C. and S. Tavaré. 1994b. Ancestral inference in population genetics. *Statistical Science* **9:** 307–319.

Griffiths, R. C. and S. Tavaré. 1994c. Sampling probability distributions in the coalescent. *Theoretical Population Biology* **46:** 131–159.

Griffiths, R. C. and P. Marjoram. 1997. Ancestral inference from samples of DNA sequences with recombination. *Journal of Computational Biology* **3:** 479–502.

Griffiths, R. C. and S. Tavaré. 1999. The ages of mutations in gene trees. *Annals of Applied Probability* **9:** 567–590.

Grishin, N. V. 1999. A novel approach to phylogeny reconstruction from protein sequences. *Journal of Molecular Evolution* **48:** 264–273.

Gu, X., Y.-X. Fu, and W.-H. Li. 1995. Maximum likelihood estimation of the heterogeneity of substitution rate among nucleotide sites. *Molecular Biology and Evolution* **12:** 546–557.

Gu, X. and W.-H. Li. 1996. A general additive distance with time-reversibility and rate variation among nucleotide sites. *Proceedings of the National Academy of Sciences* **93:** 4671–4676.

Gu, X. 1999. Statistical models for testing functional divergence after gene duplication. *Molecular Biology and Evolution* **16:** 1664–1674.

Gu, X. 2001. Maximum-likelihood approach for gene family evolution under functional divergence. *Molecular Biology and Evolution* **18:** 453–464.

Guigó, R., I. Muchnik, and T. F. Smith. 1996. Reconstruction of ancient molecular phylogeny. *Molecular Phylogenetics and Evolution* **6:** 189–213.

Gusfield, D. 1991. Efficient algorithms for inferring evolutionary trees. *Networks* **21:** 19–28.

Hagedorn, T. R. 2000. Determining the number and structure of phylogenetic invariants. *Advances in Applied Mathematics* **24:** 1–12.

Hagedorn, T. R. and L. F. Landweber. 2000. Phylogenetic invariants and geometry. *Journal of Theoretical Biology* **205:** 365–376.

Hallett, M. T. and J. Lagergren. 2000. New algorithms for the duplication-loss model. pp. 138–146 in *RECOMB '00. Proceedings of the Fourth Annual International Conference on Computational Molecular Biology, April 8–11, 2000, Tokyo, Japan*, ed. R. Shamir, S. Miyano, S. Istrail, P. Pevzner, and M. Waterman. Association for Computing Machinery, New York.

Hannenhalli, S. 1995. Polynomial-time algorithm for computing translocation distance between genomes. pp. 162–176 in *Proceedings of the 6th Annual Symposium on Combinatorial Pattern Matching* ed. Z. Galil and E. Ukkonen. Springer-Verlag, Berlin.

Hannenhalli, S., C. Chappey, E. Koonin, and P. Pevzner. 1995. Genome sequence comparison and scenarios for gene rearrangements: A test case. *Genomics* **30:** 299–311.

Hannenhalli, S. and P. A. Pevzner. 1995. Transforming men into mice (polynomial algorithm for genomic distance problem). pp. 581–592 in *Proceedings of the 36th Annual Symposium on Foundations of Computer Science (FOCS'95), October 23 - 25, 1995, Milwaukee, Wisconsin.* IEEE Computer Society Press, Los Alamitos, California.

Hannenhalli, S. and P. A. Pevzner. 1999. Transforming cabbage into turnip: Polynomial algorithm for sorting signed permutations by reversals. *Journal of the ACM* **46:** 1–27.

Hansen, T. F. and E. P. Martins. 1996. Translating between microevolutionary process and macroevolutionary patterns: The correlation structure of interspecific data. *Evolution* **50:** 1404–1417.

Harding, E. F. 1971. The probabilities of rooted tree shapes generated by random bifurcation. *Advances in Applied Probability* **3:** 44–77.

Harper, C. W., Jr. 1979. A Bayesian probability view of phylogenetic systematics. *Systematic Zoology* **28:** 547–553.

Harshman, J. 1994. The effect of irrelevant characters on bootstrap values. *Systematic Zoology* **43:** 419–424.

Harvey, P. H. and G. M. Mace. 1982. Comparisons between taxa and adaptive trends: Problems of methodology. pp. 343–361 in *Current Problems in Sociobiology,* ed. King's College Sociobiology Group. Cambridge University Press, Cambridge.

Harvey, P. H. and M. D. Pagel. 1991. *The Comparative Method in Evolutionary Biology.* Oxford University Press, Oxford.

Harvey, P. H., R. M. May, and S. Nee. 1994. Phylogenies without fossils. *Evolution* **48:** 523–529.

Harvey, P. H., A. F. Read, and S. Nee. 1995a. Why ecologists need to be phylogenetically challenged. *Journal of Ecology* **83:** 535–536.

Harvey, P. H., A. F. Read and S. Nee. 1995b. Further remarks on the role of phylogeny in comparative ecology. *Journal of Ecology* **83:** 733–734.

Hasegawa, M., H. Kishino, and T. Yano. 1985. Dating of the human-ape splitting by a molecular clock of mitochondrial DNA. *Journal of Molecular Evolution* **22:** 160–174.

Hasegawa, M., H. Kishino, and T. Yano. 1987. Man's place in Hominoidea as inferred from molecular clocks of DNA. *Journal of Molecular Evolution* **26:** 132–147.

Hasegawa, M. 1990. Phylogeny and molecular evolution in primates. *Japanese Journal of Genetics* **65:** 243–265.

Hasegawa, M., H. Kishino, K. Hayasaka, and S. Horai. 1990. Mitochondrial DNA evolution in primates: Transition rate has been extremely low in the lemur. *Journal of Molecular Evolution* **31:** 113–121.

Hasegawa, M. and H. Kishino. 1994. Accuracies of the simple methods for estimating the bootstrap probability of a maximum-likelihood tree. *Molecular Biology and Evolution* **11:** 142–145.

Hastings, W. K. 1970. Monte Carlo sampling methods using Markov chains and their applications. *Biometrika* **57:** 97–109.

Heard, S. B. 1992. Patterns in tree balance among cladistic, phenetic, and randomly generated phylogenetic trees. *Evolution* **46:** 1818–1826.

Hein, J. 1989. A method that simultaneously aligns, finds the phylogeny and reconstructs ancestral sequences for any number of ancestral sequences. *Molecular Biology and Evolution* **6:** 649–668.

Hein, J. 1990. A unified approach to phylogenies and alignments. *Methods in Enzymology* **183:** 625–644.

Hein, J. 1993. A heuristic method to reconstruct the history of sequences subject to recombination. *Journal of Molecular Evolution* **36:** 396–406.

Hein, J., C. Wiuf, B. Knudsen, M. Moller, and G. Wibling. 2000. Statistical alignment: Computational properties, homology testing and goodness-of-fit. *Journal of Molecular Biology* **302:** 265–279.

Hein, J. 2001. An algorithm for statistical alignment of sequences related by a binary tree. *Pacific Symposium on Biocomputing 2001,* pp. 179–190.

Hendy, M. D., L. R. Foulds, and D. Penny. 1980. Proving phylogenetic trees minimal with *l*-clustering and set partitioning. *Mathematical Biosciences* **51:** 71–88.

Hendy, M. D. and D. Penny. 1982. Branch and bound algorithms to determine minimal evolutionary trees. *Mathematical Biosciences* **59:** 277–290.

Hendy, M. D. and D. Penny. 1989. A framework for the study of evolutionary trees. *Systematic Zoology* **38:** 297–309.

Hendy, M. D. 1989. The relationship between simple evolutionary tree models and observable sequence data. *Systematic Zoology* **38:** 310–321.

Hendy, M. D. 1991. A combinatorial description of the closest tree algorithm for finding evolutionary trees. *Discrete Mathematics* **91:** 51–58.

Hendy, M. D. and D. Penny. 1993. Spectral analysis of phylogenetic data. *Journal of Classification* **10:** 5–24.

Hendy, M. D. and M. A. Charleston. 1993. Hadamard conjugation: A versatile tool for modelling nucleotide sequence evolution. *New Zealand Journal of Botany* **31:** 231–237.

Hendy, M. D., D. Penny, and M. A. Steel. 1994. A discrete Fourier analysis for evolutionary trees. *Proceedings of the National Academy of Sciences, USA* **91:** 3339–3343.

Hendy, M. D. and D. Penny. 1996. Complete families of linear invariants for some stochastic models of sequence evolution, with and without the molecular clock assumption. *Journal of Computational Biology* **3:** 19–31.

Henikoff, S. and J. G. Henikoff. 1992. Amino acid substitution matrices from protein blocks. *Proceedings of the National Academy of Sciences, USA* **89:** 10915–10919.

Hennig, W. 1950. *Grundzüge einer Theorie der phylogenetischen Systematik.* Deutscher Zentralverlag, Berlin.

Hennig, W. 1953. Kritische Bermerkungen zum phylogenetischen System der Insekten. *Beitrage zur Entomologie* **3** (Special volume) 1–85.

Hennig, W. 1966. *Phylogenetic Systematics.* translated by D. D. Davis and R. Zangerl. University of Illinois Press, Urbana.

Henzinger, M. R., V. King, and T. Warnow. 1999. Constructing a tree from homeomorphic subtrees, with applications to computational molecular biology. *Algorithmica* **24:** 1–13.

Heuch, I. and F. H. F. Li. 1972. PEDIG: A computer program for calculation of genotype probabilities using phenotype information. *Clinical Genetics* **3:** 501–504.

Hey, J. 1992. Using phylogenetic trees to study speciation and extinction. *Evolution* **46:** 627–640.

Hey, J. and R. M. Kliman. 1993. Population genetics and phylogenetics of DNA sequence variation at multiple loci within the *Drosophila melanogaster* species complex. *Molecular Biology and Evolution* **10:** 804–822.

Higgins, D. G. and P. M. Sharp. 1988. CLUSTAL—a package for performing multiple sequence alignment on a microcomputer. *Gene* **73:** 237–244.

Higgins, D. G., J. D. Thompson, and T. J. Gibson. 1996. Using CLUSTAL for multiple sequence alignments. *Methods in Enzymology* **266:** 383–402.

Hilden, J. 1970. GENEX, an algebraic approach to pedigree probability calculus. *Clinical Genetics* **1:** 319–348.

Hillis, D. M. 1991. Discriminating between phylogenetic signal and random noise in DNA sequences. pp. 278–294 in *Phylogenetic Analysis of DNA Sequences,* ed. M. M. Miyamoto and J. Cracraft. Oxford University Press, Oxford.

Hillis, D. M. and J. J. Bull. 1993. An empirical test of bootstrapping as a method for assessing confidence in phylogenetic analysis. *Systematic Biology* **42:** 182–192.

Hillis, D. M., J. P. Huelsenbeck, and C. W. Cunningham. 1994. Application and accuracy of molecular phylogenies. *Science* **264:** 671–677.

Hochbaum, D. S. and A. Pathria. 1997. Path costs in evolutionary tree reconstruction. *Journal of Computational Biology* **4:** 163–175.

Hogeweg, P. and P. Hesper. 1984. The alignment of sets of sequences and the construction of phyletic trees: An integrated method. *Journal of Molecular Evolution* **20:** 175–186.

Holland, J. H. 1975. *Adaptation in Natural and Artificial Systems.* University of Michigan Press, Ann Arbor.

Holmes, I. and W. J. Bruno. 2001. Evolutionary HMMs: A Bayesian approach to multiple alignment. *Bioinformatics* **17:** 803–820.

Horne, S. L. 1967. Comparisons of primate catalase tryptic peptides and implications for the study of molecular evolution. *Evolution* **21:** 771–786.

Horner, W. G. 1819. A new method of solving numerical equations of all orders, by continuous approximation. *Philosophical Transactions of the Royal Society of London* **109:** 308–335.

Housworth, E. A. and E. P. Martins. 2001. Random sampling of constrained phylogenies: Conducting phylogenetic analyses when the phylogeny is partially known. *Systematic Biology* **50:** 628–639.

Howson, C. and P. Urbach. 1993. *Scientific Reasoning: The Bayesian Approach,* 2nd edition, Open Court, Chicago. Hudson, R. R. 1983. Testing the constant-rate neutral allele model with protein sequence data. *Evolution* **37:** 203–217.

Hudson, R. R. 1983. Properties of a neutral allele model with intragenic recombination. *Theoretical Population Biology* **23:** 183–201.

Hudson, R. R. and N. L. Kaplan. 1985. Statistical properties of the number of recombination events in the history of a sample of DNA sequences. *Genetics* **111:** 147–164.

Hudson, R. R. and N. L. Kaplan. 1988. The coalescent process in models with selection and recombination. *Genetics* **120:** 831–840.

Hudson, R. R. 1992. Gene trees, species trees and the segregation of ancestral alleles. *Genetics* **131:** 509–513.

Hudson, R. R. and J. A. Coyne. 2002. Mathematical consequences of the genealogical species concept. *Evolution* **56:** 1557–1565.

Huelsenbeck, J. P. 1991. Tree-length distribution skewness — an indicator of phylogenetic information. *Systematic Zoology* **40:** 257–270.

Huelsenbeck, J. P. and D. M. Hillis. 1993. Success of phylogenetic methods in the four-taxon case. *Systematic Biology* **42:** 247–264.

Huelsenbeck, J. P. 1994. Measuring and testing the fit of the stratigraphic record to phylogenetic trees. *Paleobiology* **20:** 470–483.

Huelsenbeck, J. P., D. M. Hillis, and R. Jones. 1996. Parametric bootstrapping in molecular phylogenetics: Applications and performance. pp. 19–45 in *Molecular Zoology: Advances, Strategies, and Protocols,* ed. J. D. Ferraris and S. R. Palumbi. Wiley-Liss, New York.

Huelsenbeck, J. P., J. J. Bull, and C. W. Cunningham. 1996. Combining data in phylogenetic analysis. *Trends in Ecology and Evolution* **11:** 152–158.

Huelsenbeck, J. P. and J. J. Bull. 1996. A likelihood ratio test to detect conflicting phylogenetic signal. *Systematic Biology* **45:** 92–98.

Huelsenbeck J. P. and K. A. Crandall. 1997. Phylogeny estimation and hypothesis testing using maximum likelihood. *Annual Review of Ecology and Systematics* **28:** 437–466.

Huelsenbeck, J. P., B. Rannala, and Z. H. Yang. 1997. Statistical tests of host-parasite cospeciation. *Evolution* **51:** 410–419.

Huelsenbeck, J. P. and B. Rannala. 1997. Maximum likelihood estimation of phylogeny using stratigraphic data. *Paleobiology* **23:** 174–180.

Huelsenbeck, J. P. 1998. Systematic bias in phylogenetic analysis: Is the Strepsiptera problem solved? *Systematic Biology* **47:** 519–537.

Huelsenbeck, J. P., B. Larget, and D. Swofford. 2000. A compound Poisson process for relaxing the molecular clock. *Genetics* **154:** 1879–1892.

Huelsenbeck, J. P., B. Rannala, and B. Larget. 2000. A Bayesian framework for the analysis of cospeciation. *Evolution* **54:** 352–364.

Huelsenbeck, J. P. and B. Rannala. 2000. Using stratigraphic information in phylogenetics. pp. 165–191 in *Phylogenetic Analysis of Morphological Data*, ed. J. J. Wiens. Smithsonian Institution Press, Washington.

Huelsenbeck, J. P., B. Rannala, and J. P. Masly. 2000. Accommodating phylogenetic uncertainty in evolutionary studies. *Science* **288:** 2349–2350.

Huelsenbeck, J. P. and F. Ronquist. 2001. MRBAYES: Bayesian inference of phylogenetic trees. *Bioinformatics* **17:** 754–755.

Huelsenbeck, J. P., F. Ronquist, R. Nielsen, and J. P. Bollback. 2001. Bayesian inference of phylogeny and its impact on evolutionary biology. *Science* **294:** 2310–2314.

Huelsenbeck, J. P. and J. P. Bollback. 2001. Empirical and hierarchical Bayesian estimation of ancestral states. *Systematic Biology* **50:** 351–366.

Huelsenbeck, J. P., J. P. Bollback, and A. M. Levine. 2002. Inferring the root of a phylogenetic tree. *Systematic Biology* **51:** 32–43.

Huey, R. B. and A. F. Bennett. 1987. Phylogenetic studies of coadaptation: Preferred temperatures versus optimal performance temperatures of lizards. *Evolution* **41:** 1098–1115.

Hull, D. L. 1988. *Science as Process: An Evolutionary Account of the Social and Conceptual Development of Science.* University of Chicago Press, Chicago.

Huson, D., S. Nettles, L. Parida, T. Warnow, and S. Yooseph. 1998. The disk-covering method for tree reconstruction. pp. 62–75 in *Proceedings of "Algorithms and Experiments" (ALEX98), Trento, Italy, Feb. 9–11, 1998*, ed. R. Battiti and A. A. Bertossi.

Huson, D., S. Nettles, and T. Warnow. 1999. Disk-covering, a fast converging method for phylogenetic tree reconstruction. *Journal of Computational Biology* **6:** 369–386.

Inger, R. F. 1967. The development of a phylogeny of frogs. *Evolution* **21:** 401–410.

Iwabe, N., K.-I. Kuma, M. Hasegawa, S. Osawa, and T. Miyata. 1989. Evolutionary relationship of archaebacteria, eubacteria, and eukaryotes inferred from phylogenetic trees of duplicated genes. *Proceedings of the National Academy of Sciences, USA* **86:** 9355–9359.

Jakobsen, I. B. and S. Easteal. 1996. A program for calculating and displaying compatibility matrices as an aid in determining reticulate evolution in molecular sequences. *Computer Applications in the Biosciences (CABIOS)* **12:** 291–295.

Jakobsen, I. B., Wilson, and S. Easteal. 1997. The partition matrix: Exploring variable phylogenetic signals along nucleotide sequence alignments. *Molecular Biology and Evolution* **14:** 474–484.

Jensen, J. L. and A.-M. K. Pedersen. 2000. Probabilistic models of DNA sequence evolution with context dependent rates of substitution. *Advances in Applied Probability* **32:** 499–517.

Jiang, T., E. L. Lawler, and L. Wang. 1994. Aligning sequences via an evolutionary tree: Complexity and approximation. pp. 760–769 in *Proceedings of the twenty-sixth annual ACM symposium on Theory of Computing (STOC), Montreal, Quebec, Canada.* ACM Press, New York.

Jin, L. and M. Nei. 1990. Limitations of the evolutionary parsimony method of phylogenetic analysis. *Molecular Biology and Evolution* **7:** 82–102.

Johnson, K. P., D. M. Drown, and D. H. Clayton. 2001. A data based parsimony method of cophylogenetic analysis. *Zoologica Scripta* **30:** 79–87.

Jones, D. T., W. R. Taylor, and J. M. Thornton. 1992. The rapid generation of mutation data matrices from protein sequences. *Computer Applications in the Biosciences (CABIOS)* **8:** 275–282.

Jones, D. T., W. R. Taylor, and J. M. Thornton. 1994a. A model recognition approach to the prediction of all-helical membrane protein structure and topology. *Biochemistry* **33:** 3038–3049.

Jones, D. T., W. R. Taylor, and J. M. Thornton. 1994b. A mutation data matrix for transmembrane proteins. *FEBS Letters* **339:** 269–275.

Jow H., C. Hudelot, M. Rattray, and P. G. Higgs. 2002. Bayesian phylogenetics using an RNA substitution model applied to early mammalian evolution. *Molecular Biology and Evolution* **19:** 1591–1601.

Jukes, T. H. and C. R. Cantor. 1969. Evolution of protein molecules. pp. 21–132 in *Mammalian Protein Metabolism,* Vol. III, ed. M. N. Munro. Academic Press, New York.

Källersjö, M., J. S. Farris, A. G. Kluge, and C. Bult. 1992. Skewness and permutation. *Cladistics* **8:** 275–287.

Kannan, S. and T. Warnow, 1994. Inferring evolutionary history from DNA sequences. *SIAM Journal on Computing* **23:** 713–737.

Kannan, S. and T. Warnow, 1997. A fast algorithm for the computation and enumeration of perfect phylogenies when the number of character states is fixed. *SIAM Journal on Computing* **26:** 1749–1763.

Karlin, S. and I. Ladunga. 1994. Comparisons of eukaryotic genome sequences. *Proceedings of the National Academy of Sciences* **91:** 12832–12836.

Karlin, S., I. Ladunga, and B. E. Blaisdell. 1994. Heterogeneity of genomes: Measures and values. *Proceedings of the National Academy of Sciences* **91:** 12837–12841.

Kaplan, N. and C. H. Langley. 1979. A new estimate of sequence divergence of DNA using restriction endonuclease mappings. *Journal of Molecular Evolution* **13:** 295–304.

Kaplan, N. and K. Risko. 1981. An improved method for estimating sequence divergence of DNA using restriction endonuclease mappings. *Journal of Molecular Evolution* **17:** 156–162.

Kaplan, N. L., T. Darden, and R. R. Hudson. 1988. The coalescent process in models with selection. *Genetics* **120:** 819–829.

Kaplan, N. L., R. R. Hudson, and C. H. Langley. 1989. The "hitchhiking effect" revisited. *Genetics* **123:** 887–899.

Kaplan, N., R. R. Hudson, and M. Iizuka. 1991. The coalescent process in models with selection, recombination and geographic subdivision. *Genetical Research* **57:** 83–91.

Kaplan, H., R. Shamir, and R. E. Tarjan. 1997. Faster and simpler algorithm for sorting signed permutations by reversals. pp. 581–592 in *Proceedings of the Eighth Annual ACM-SIAM Symposium on Discrete Algorithms, 5–7 January 1997, New Orleans, Louisiana, USA.* ACM Press, New York.

Kashyap, R. L. and S. Subas. 1974. Statistical estimation of parameters in a phylogenetic tree using a dynamic model of the substitutional process. *Journal of Theoretical Biology* **47:** 75–101.

Katoh, K., K. Kuma, and T. Miyata. 2001. Genetic algorithm-based maximum-likelihood analysis for molecular phylogeny. *Journal of Molecular Evolution* **53:** 477–484.

Kececioglu, J. and D. Sankoff. 1995. Exact and approximation algorithms for sorting by reversals, with application to genome rearrangement. *Algorithmica* **13:** 180–210.

Kelly, C. and J. Rice. 1996. Modeling nucleotide evolution: A heterogeneous rate analysis. *Mathematical Biosciences* **133:** 85–109.

Kendall, D. G. 1948. On the generalized "birth-and-death" process. *Annals of Mathematical Statistics* **19:** 1–15.

Kendall, M. G. and A. Stuart. 1973. *The Advanced Theory of Statistics,* Vol. 2, 3rd. edition. Hafner, New York.

Kidd, K. K. and L. A. Sgaramella-Zonta. 1971. Phylogenetic analysis: Concepts and methods. *American Journal of Human Genetics* **23:** 235–252.

Kim, J. 1996. General inconsistency conditions for maximum parsimony: Effects of branch length and increasing the number of taxa. *Systematic Biology* **45:** 363–374.

Kim, J. 2000. Slicing hyperdimensional oranges: The geometry of phylogenetic estimation. *Molecular Phylogenetics and Evolution* **17:** 58–75.

Kimura, M. and T. Ohta. 1972. On the stochastic model for estimation of mutational distance between homologous proteins. *Journal of Molecular Evolution* **2:** 87–90.

Kimura, M. 1980. A simple model for estimating evolutionary rates of base substitutions through comparative studies of nucleotide sequences. *Journal of Molecular Evolution* **16:** 111–120.

Kimura, M. 1981. Estimation of evolutionary distances between homologous nucleotide sequences. *Proceedings of the National Academy of Sciences, USA* **78:** 454–458.

Kingman, J. F. C. 1982a. The coalescent. *Stochastic Processes and Their Applications* **13:** 235–248.

Kingman, J. F. C. 1982b. On the genealogy of large populations. *Journal of Applied Probability* **19A:** 27–43.

Kingman, J. F. C. 1982c. Exchangeability and the evolution of large populations. pp. 97–112 in *Exchangeability in Probability and Statistics. Proceedings of the International Conference on Exchangeability in Probability and Statistics, Rome, 6th-9th April, 1981, in honour of Professor Bruno de Finetti.* ed. G. Koch and F. Spizzichino. North-Holland Elsevier, Amsterdam.

Kirkpatrick, M. and M. Slatkin. 1993. Searching for evolutionary patterns in the shape of a phylogenetic tree. *Evolution* **47:** 1171–1181.

Kishino, H. and M. Hasegawa. 1989. Evaluation of the maximum likelihood estimate of the evolutionary tree topologies from DNA sequence data, and the branching order in Hominoidea. *Journal of Molecular Evolution* **29:** 170–179.

Kishino, H., T. Miyata, and M. Hasegawa. 1990. Maximum likelihood inference of protein phylogeny and the origin of chloroplasts. *Journal of Molecular Evolution* **31:** 151–160.

Kishino, H. and M. Hasegawa. 1990. Converting distance to time: Application to human evolution. *Methods in Enzymology* **183:** 550–570.

Kishino, H., J. L. Thorne, and W. J. Bruno. 2001. Performance of a divergence time estimation method under a probabilistic model of rate evolution. *Molecular Biology and Evolution* **18:** 352–361.

Kjer, K. M., R. J. Blahnik, and R. W. Holzenthal. 2001. Phylogeny of the Trichoptera (caddisflies): Characterization of signal and noise within multiple datasets. *Systematic Biology* **50:** 781–816.

Klotz, L. C., N. Komar, R. L. Blanken, and R. M. Mitchell. 1979. Calculation of evolutionary trees from sequence data. *Proceedings of the National Academy of Sciences, USA* **76:** 4516–4520.

Kluge, A. G. and J. S. Farris. 1969. Quantitative phyletics and the evolution of anurans. *Systematic Zoology* **18:** 1–32.

Kluge, A.G. 1989. A concern for evidence and a phylogenetic hypothesis of relationships among *Epicrates* (Boideae, Serpentes). *Systematic Zoology* **38:** 7–25.

Kluge, A. G. and A. J. Wolf. 1993. Cladistics: What's in a word? *Cladistics* **9:** 183–199.

Kluge, A. G. 1994. Moving targets and shell games. *Cladistics* **10:** 403–413.

Kluge, A. G. 1997a. Testability and the refutation and corroboration of cladistic hypotheses. *Cladistics* **13:** 81–96.

Kluge, A. G. 1997b. Sophisticated falsification and research cycles: Consequences for differential character weighting in phylogenetic systematics. *Zoologica Scripta* **26:** 349–360.

Kluge, A. G. 1998. Total evidence or taxonomic congruence: Cladistics or consensus classification. *Cladistics* **14:** 151–158.

Kluge, A. G. 1999. The science of phylogenetic systematics: Explanation, prediction, and test. *Cladistics* **15:** 429–436.

Kluge, A. G. 2001. Philosophical conjectures and their refutation. *Systematic Biology* **50:** 322–330.

Kluge, A. G. 2002. Distinguishing "or" from "and" and the case for historical identification. *Cladistics* **18:** 585–593.

Knowles, L. L. and W. P. Maddison. 2002. Statistical phylogeography. *Molecular Ecology* **11:** 2623–2635.

Knuth, D. E. 1973. *The Art of Computer Programming. Vol. 1. Fundamental Algorithms,* 2nd edition. Addison-Wesley, Reading, Massachusetts.

Konings, D. A., P. Hogeweg, and B. Hesper. 1987. Evolution of the primary and secondary structures of the E1a mRNAs of the adenovirus. *Molecular Biology and Evolution* **4:** 300–314.

Koshi, J. M. and R. A. Goldstein. 1995. Context-dependent optimal substitution matrices. *Protein Engineering* **8:** 641–645.

Koshi, J. M., R. A. Goldstein. 1996. Probabilistic reconstruction of ancestral protein sequences. *Journal of Molecular Evolution* **42:** 313–320.

Koshi, J. M. and R. A. Goldstein. 1997. Mutation matrices and physical-chemical properties: Correlations and implications. *Proteins* **27:** 336–344.

Koshi, J. M. and R. A. Goldstein. 1998. Mathematical models of natural amino-acid site mutations. *Proteins* **32:** 289–295.

Krone, S. M. and C. Neuhauser. 1997. Ancestral processes with selection. *Theoretical Population Biology* **51:** 210–237.

Kruglyak, S., R. T. Durrett, M. D. Schug, and C. F. Aquadro. 1998. Equilibrium distributions of microsatellite repeat length resulting from a balance between slippage events and point mutations. *Proceedings of the National Academy of Sciences, USA* **95:** 10774–10778.

Kruglyak, S., R. Durrett, M. D. Schug, and C. F. Aquadro. 2000. Distribution and abundance of microsatellites in the yeast genome can be explained by a balance between slippage events and point mutations. *Molecular Biology and Evolution* **17:** 1210–1219.

Kubo, T. and Y. Iwasa. 1995. Inferring the rates of branching and extinction from molecular phylogenies. *Evolution* **49:** 694–704.

Kuhner, M. K. and J. Felsenstein. 1994. A simulation comparison of phylogeny algorithms under equal and unequal evolutionary rates. *Molecular Biology and Evolution* **11:** 459–468 (Erratum **12:** 525 1995).

Kuhner, M. K., J. Yamato, and J. Felsenstein. 1995. Estimating effective population size and mutation rate from sequence data using Metropolis-Hastings sampling. *Genetics* **140:** 1421–1430.

Kuhner, M. K., J. Yamato, and J. Felsenstein. 1998. Maximum likelihood estimation of population growth rates based on the coalescent. *Genetics* **149:** 429–434.

Kuhner, M. K. and J. Felsenstein. 2000. Sampling among haplotype resolutions in a coalescent-based genealogy sampler. *Genetic Epidemiology* **19**(Supplement 1): S15–21.

Kuhner, M. K., J. Yamato, and J. Felsenstein. 2000. Maximum likelihood estimation of recombination rates from population data. *Genetics* **156:** 1393–1401.

Kuhner, M. K., P. Beerli, J. Yamato, and J. Felsenstein. 2000. Usefulness of single nucleotide polymorphism data for estimating population parameters. *Genetics* **156:** 439–447.

Kumar, S. 1996. A stepwise algorithm for finding minimum evolutionary trees. *Molecular Biology and Evolution* **13:** 584–593.

Künsch, H. R. 1989. The jackknife and the bootstrap for general stationary observations. *Annals of Statistics* **17:** 1217–1241.

Lake, J. A. 1987. A rate-independent technique for analysis of nucleic acid sequence characters: Evolutionary parsimony. *Molecular Biology and Evolution* **4:** 167–191.

Lake, J. A. 1994. Reconstructing evolutionary trees from DNA and protein sequences: Paralinear distances. *Proceedings of the Natonal Academy of Sciences, USA* **91:** 1455–1459.

Lake, J. A. 1995. Calculating the probability of multitaxon evolutionary trees: Bootstrappers Gambit. *Proceedings of the National Academy of Sciences, USA* **92:** 9662–9666.

Lamping, J. and R. Rao. 1994. Laying out and visualizing large trees using a hyperbolic space. pp. 13–14 in *Proceedings of the 7th Annual ACM Symposium on User Interface Software and Technology, Marina del Ray, California.* ACM Press, New York.

Lande, R. 1976. Natural selection and random genetic drift in phenotypic evolution. *Evolution* **30:** 314–334.

Lanave, C., G. Preparata, C. Saccone, and G. Serio. 1984. A new method for calculating evolutionary substitution rates. *Journal of Molecular Evolution* **20:** 86–93.

Langley, C. H. and W. M. Fitch. 1973. The constancy of evolution: A statistical analysis of the α and β haemoglobins, cytochrome c, and fibrinopeptide A. pp. 246–262 in *Genetic Structure of Populations,* ed. N. E. Morton. University of Hawaii Press, Honolulu.

Langley, C. H. and W. M. Fitch. 1974. An examination of the constancy of the rate of molecular evolution. *Journal of Molecular Evolution* **3:** 161–177.

Lanyon, S. M. 1985. Detecting internal inconsistencies in distance data. *Systematic Zoology* **34:** 397–403.

Lapointe, F.-J. and G. Cucumel. 1997. The average consensus procedure: Combination of weighted trees containing identical or overlapping sets of taxa. *Systematic Biology* **46:** 306–312.

Larget, B. and D. L. Simon. 1999. Markov chain Monte Carlo algorithms for the Bayesian analysis of phylogenetic trees. *Molecular Biology and Evolution* **16:** 750–759.

Larget, B., D. L. Simon, and J. B. Kadane. 2002. Bayesian phylogenetic inference from animal mitochondrial genome arrangements. *Journal of the Royal Statistical Society B* **64:** 681–693.

Larribe, F., S. Lessard, and N. J. Schork. 2002. Gene mapping via the ancestral recombination graph. *Theoretical Population Biology* **62:** 215–229.

Legendre, P., Y. Desdevises, and E. Bazin. 2002. A statistical test for host-parasite coevolution. *Systematic Biology* **51:** 217–234.

Le Quesne, W. J. 1969. A method of selection of characters in numerical taxonomy. *Systematic Zoology* **18:** 201–205.

Le Quesne, W. J. 1974. The uniquely evolved character concept and its cladistic application. *Systematic Zoology* **23:** 513–517.

Lemmon, A. R. and M. C. Milinkovitch. 2002. The metapopulation genetic algorithm: An efficient solution for the problem of large phylogeny estimation. *Proceedings of the National Academy of Sciences* **99:** 10516–10521.

Lewis, P. O. 1998. A genetic algorithm for maximum-likelihood phylogeny using nucleotide sequence data. *Molecular Biology and Evolution* **15:** 277–283.

Li, M. and L. X. Zhang. 1999. Twist-rotation transformations of binary trees and arithmetic expressions. *Journal of Algorithms* **32:** 155–166.

Li, P. and J. Bousquet. 1992. Relative-rate test for nucleotide substitutions between two lineages. *Molecular Biology and Evolution* **9:** 1185–1189.

Li, S., D. K. Pearl, and H. Doss. 2000. Phylogenetic tree construction using Markov Chain Monte Carlo. *Journal of the American Statistical Association* **95:** 493–508.

Li, W.-H. 1981. Simple method for constructing phylogenetic trees from distance matrices. *Proceedings of the National Academy of Sciences, USA* **78:** 1085–1089.

Li, W.-H., C.-I Wu, and C.-C. Luo. 1985. A new method for estimating synonymous and nonsynonymous rates of nucleotide substitution considering the relative likelihood of nucleotide and codon changes. *Molecular Biology and Evolution* **2:** 150–174.

Li, W.-H. 1986. Evolutionary change of restriction cleavage sites and phylogenetic inference. *Genetics* **113:** 187–213.

Li, W.-H. 1989. A statistical test of phylogenies estimated from sequence data. *Molecular Biology and Evolution* **6:** 424–435.

Li, W.-H. and M. Gouy. 1991. Statistical methods for testing molecular phylogenies. pp. 249–277 in *Phylogenetic Analysis of DNA Sequences*, ed. M. M. Miyamoto and J. Cracraft. Oxford University Press, New York.

Li, W.-H. and A. Zharkikh. 1994. What is the bootstrap technique? *Systematic Biology* **43:** 424–430.

Li, Y. J., Y. Satta, and N. Takahata. 1999. Paleo-demography of the *Drosophila melanogaster* subgroup: Application of the maximum likelihood method. *Genes and Genetic Systems* **74:** 117–127.

Liò, P. and N. Goldman. 1999. Using protein structural information in evolutionary inference: Transmembrane proteins. *Molecular Biology and Evolution* **16:** 1696–1710.

Lockhart, P. J., M. A. Steel, M. D. Hendy, and D. Penny. 1994. Recovering evolutionary trees under a more realistic model of sequence evolution. *Molecular Biology and Evolution* **11:** 605–612.

Lorentzen, S. and J. Sieg. 1991. Phylip, Paup, and Hennig 86—How reliable are computer parsimony programs used in Systematics? *Zeitschrift für Zoologische Systematik und Evolutionsforschung* **29:** 466–472.

Losos, J. B. and F. R. Adler. 1995. Stumped by trees? A generalized null model for patterns of organismal diversity. *American Naturalist* **145:** 329–342.

Lundy, M. 1985. Applications of the annealing algorithm to combinatorial problems in statistics. *Biometrika* **72:** 191–198.

Lush, J. L. 1937. *Animal Breeding Plans.* Iowa State College Press, Ames.

Lyons-Weiler, J., G. A. Hoelzer, R. J. Tausch. 1996. Relative apparent synapomorphy analysis (RASA) I: The statistical measurement of phylogenetic signal. *Molecular Biology and Evolution* **13:** 749–757.

Lyons-Weiler, J. and M. C. Milinkovitch. 1997. A phylogenetic approach to the problem of differential lineage sorting. *Molecular Biology Evolution* **14:** 968–975.

Lyons-Weiler, J. and G. A. Hoelzer. 1997. Escaping from the Felsenstein zone by detecting long branches in phylogenetic data. *Molecular Phylogenetics and Evolution* **8:** 375–384.

Lyons-Weiler, J., G. A. Hoelzer, and R. J. Tausch. 1998. Optimal outgroup analysis. *Biological Journal of the Linnean Society* **64:** 493–511.

Lynch, M. and W. G. Hill. 1986. Phenotypic evolution by neutral mutation. *Evolution* **40:** 915–935.

Lynch, M. 1991. Methods for the analysis of comparative data in evolutionary biology. *Evolution* **45:** 1065–1080.

Lynch, M. and B. Walsh. 1998. *Genetics and Analysis of Quantitative Traits.* Sinauer Associates, Sunderland, Massachusetts.

Lynch, M. and J. S. Conery. 2000. The evolutionary fate and consequences of duplicate genes. *Science* **290:** 1151–1155.

Lynch, W. 1965. More combinatorial properties of certain trees. *Computer Journal* **71:** 299–302.

Ma, B., M. Li, and L. Zhang. 2000. From gene trees to species trees. *SIAM Journal on Computing* **30:** 729–752.

Maddison, D. R. 1990. *Phylogenetic Inference of Historical Pathways and Models of Evolutionary Change.* Ph. D. Thesis, Department of Organismic and Evolutionary Biology, Harvard University, Cambridge, Massachusetts.

Maddison, D. R. 1991. The discovery and importance of multiple islands of most-parsimonious trees. *Systematic Zoology* **40:** 315–328.

Maddison, W. P., M. J. Donoghue, and D. R. Maddison. 1984. Outgroup analysis and parsimony. *Systematic Zoology* **33:** 83–103.

Maddison, W. P. 1990. A method for testing the correlated evolution of two binary characters: Are gains and losses concentrated on certain branches of a phylogenetic tree? *Evolution* **44:** 539–557.

Maddison, W. P. and M. Slatkin. 1990. Parsimony reconstructions of ancestral states do not depend on the relative distances between linearly-ordered characters. *Systematic Zoology* **39:** 175–178.

Maddison, W. P. 1991. Squared-change parsimony reconstructions of ancestral states for continuous-valued characters on a phylogenetic tree. *Systematic Zoology* **40:** 304–314.

Maddison, W. P. 1995. Calculating the probability distributions of ancestral states reconstructed by parsimony on phylogenetic trees. *Systematic Biology* **44:** 474–481.

Maddison, W. P. 1997. Gene trees in species trees. *Systematic Biology* **46:** 523–536.

Maddison, W. P. 2000. Testing character correlation using pairwise comparisons on a phylogeny. *Journal of Theoretical Biology* **202:** 195–204.

Mahmoud, H. 1992. *Evolution of Random Search Trees.* Wiley, New York.

Makarenkov, V. and B. Leclerc. 1999. The fitting of a tree metric to a given dissimilarity with the weighted least squares criterion. *Journal of Classification* **16:** 3–26.

Margush, T. and F. R. McMorris. 1981. Consensus *n*-trees. *Bulletin of Mathematical Biology* **43:** 239–244.

Martins, E. P. 1994. Estimating rates of change from comparative data. *American Naturalist* **144:** 193–209.

Martins, E. P. 1996a. Conducting phylogenetic comparative studies when the phylogeny is not known. *Evolution* **50:** 12–22.

Martins, E. P. 1996b. Phylogenies, spatial autoregression, and the comparative method: A computer simulation test. *Evolution* **50:** 1750–1765.

Martins, E. P. and T. F. Hansen. 1997. Phylogenies and the comparative method: A general approach to incorporating phylogenetic information into the analysis of interspecific data. *Evolution* **45:** 534–557.

Martins, E. P. 1999. Estimation of ancestral states of continuous characters: A computer simulation study. *Systematic Biology* **48:** 642–650.

Martins, E. P., J. A. Diniz-Filho, and E. A. Housworth. 2002. Adaptive constraints and the phylogenetic comparative method: A computer simulation test. *Evolution* **56:** 1–13.

Martins, E. P. and E. A. Housworth. 2002. Phylogeny shape and the phylogenetic comparative method. *Systematic Biology* **51:** 873–880.

Matsuda, H. 1996. Protein phylogenetic inference using maximum likelihood with a genetic algorithm. *Pacific Symposium on Biocomputing 1996,* pp. 512–523.

Mau, B. and M. A. Newton. 1997. Phylogenetic inference for binary data on dendrograms using Markov chain Monte Carlo. *Journal of Computational and Graphical Statistics* **6:** 122–131.

Mau, B., M. A. Newton, and B. Larget. 1999. Bayesian phylogenetic inference via Markov chain Monte Carlo methods. *Biometrics* **5:** 1–12.

Maynard Smith, J. and N. H. Smith. 1998. Detecting recombination from gene trees. *Molecular Biology and Evolution* **15:** 590–599.

Mayo, D. G. 1996. *Error and the Growth of Experimental Knowledge.* University of Chicago Press, Chicago.

McArdle, B. and A. G. Rodrigo. 1994. Estimating the ancestral states of a continuous-valued character using squared-change parsimony: An analytical solution. *Systematic Biology* **43:** 573–578.

McGuire, G., F. Wright, and M. J. Prentice. 2000. A Bayesian model for detecting past recombination events in DNA multiple alignments. *Journal of Computational Biology* **7:** 159–170.

McGuire, G., M. C. Denham, and D. J. Balding. 2001. Models of sequence evolution for DNA sequences containing gaps. *Molecular Biology and Evolution* **18:** 481–490.

McPeek, M. A. 1995. Testing hypotheses about evolutionary change on single branches of a phylogeny using evolutionary contrasts. *American Naturalist* **145:** 686–703.

Meacham, C. 1981. A manual method for character compatibility analysis. *Taxon* **30:** 591–600.

Meacham, C. 1983. Theoretical and computational considerations of the compatibility of qualitative taxonomic characters. pp. 304–314 in *Numerical Taxonomy. Proceedings of the NATO Advanced Study Institute on Numerical Taxonomy held at Bad Windsheim, Germany, July 4–16, 1982.* ed. J. Felsenstein. NATO ASI Series G: Ecological Sciences, No. 1. Springer-Verlag, Berlin.

Metropolis, N., A. W. Rosenbluth, M. N. Rosenbluth, A. H. Teller, and E. Teller. 1953. Equation of state calculations by fast computing machines. *Journal of Chemical Physics* **21:** 1087–1092.

Metzler, D., R. Fleissner, A. Wakolbinger, and A. von Haeseler. 2001. Assessing variability by joint sampling of alignments and mutation rates. *Journal of Molecular Evolution* **53:** 660–669.

Metzler, D. 2003. Statistical alignment based on fragment insertion and deletion models. *Bioinformatics* **19:** 490–499.

Michener, C. D. and R. R. Sokal. 1957. A quantitative approach to a problem in classification. *Evolution* **11:** 130–162.

Mickevich, M. F. and M. S. Johnson. 1976. Congruence between morphological and allozyme data in evolutionary inference and character evolution. *Systematic Zoology* **25:** 260–270.

Mickevich, M. F. and J. S. Farris. 1981. The implications of congruence in *Menidia*. *Systematic Zoology* **30:** 351–370.

Mickevich, M. F. and C. Mitter. 1981. Treating polymorphic characters in systematics: A phylogenetic treatment of electrophoretic data. pp. 45–60 in *Advances in Cladistics. Proceedings of the First Meeting of the Willi Hennig Society.*, ed. V. A. Funk and D. R. Brooks. New York Botanical Garden, Bronx.

Mickevich, M. F. 1981. Quantitative phylogenetic biogeography. pp. 209–222 in *Advances in Cladistics. Proceedings of the first meeting of the Willi Hennig Society.*, ed. V. A. Funk and D. R. Brooks. New York Botanical Garden, Bronx.

Mickevich, M. F. 1982. Transformation series analysis. *Systematic Zoology* **31:** 461–478.

Mickevich, M. F. and C. Mitter. 1983. Evolutionary patterns in allozyme data: A systematic approach. pp. 169–176 in *Advances in Cladistics, Volume 2. Proceedings of the Second Meeting of the Willi Hennig Society*, ed. N. H. Platnick and V. A. Funk. Columbia University Press, New York.

Mickevich, M. F. and S. J. Weller. 1990. Evolutionary character analysis: Tracing character change on a cladogram. *Cladistics* **6:** 137–170.

Mickevich, M. F. and D. Lipscomb. 1991. Parsimony and the choice between different transformations for the same character set. *Cladistics* **7:** 111–139.

Miklós, I. 2002. An improved algorithm for statistical alignment of sequences related by a star tree. *Bulletin of Mathematical Biology* **64:** 771–779.

Miller, J. A. 2003. Assessing progress in systematics with continuous jackknife function analysis. *Systematic Biology* **52:** 55–65.

Mirkin, B., I. Muchnik, and T. F. Smith. 1995. A biologically consistent model for comparing molecular phylogenies. *Journal of Computational Biology* **2:** 493–507.

Mitchison, G. J. and R. M. Durbin. 1995. Tree-based maximal likelihood substitution matrices and hidden Markov models. *Journal of Molecular Evolution* **41:** 1139–1151.

Mitchison, G. J. 1999. A probabilistic treatment of phylogeny and sequence alignment. *Journal of Molecular Evolution* **49:** 11–22.

Miyamoto, M. M. and W. M. Fitch. 1995. Testing species phylogenies and phylogenetic methods with congruence. *Systematic Biology* **44:** 64–76.

Miyata, T. and T. Yasunaga. 1980. Molecular evolution of mRNA: A method for estimating evolutionary rates of synonymous and nonsynonymous amino acid substitution from homologous sequences and its application. *Journal of Molecular Evolution* **16:** 23–26.

Moilanen, A. 1999. Searching for most parsimonious trees with simulated evolutionary optimization. *Cladistics* **15:** 39–50.

Mooers, A. O. and S. B. Heard. 1997. Inferring evolutionary process from phylogenetic tree shape. *Quarterly Review of Biology* **72:** 31–54.

Moon, J. W. 1970. *Counting Labelled Trees.* Canadian Mathematical Monographs No. 1, Canadian Mathematical Congress.

Moret, B., S. Wyman, D. Bader, T. Warnow, and M. Yan. 2001. A new implementation and detailed study of breakpoint analysis. pp. 583–594 in *Pacific Symposium on*

Biocomputing 2001, ed. R. B. Altman, A. K. Dunker, L. Hunter, K. Lauderdale, and T. E. Klein. World Scientific, Singapore.

Morris, A. P., J. C. Whittaker, and D. J. Balding. 2002. Fine-scale mapping of disease loci via shattered coalescent modeling of genealogies. *American Journal of Human Genetics* **70:** 686–707.

Mueller, L. D. and F. J. Ayala. 1982. Estimation and interpretation of genetic distance in empirical studies. *Genetical Research* **40:** 127–137.

Müller, T. and M. Vingron. 2000. Modeling amino acid replacement. *Journal of Computational Biology* **7:** 761–776.

Müller, T., R. Spang, and M. Vingron. 2002. Estimating amino acid substitution models: A comparison of Dayhoff's estimator, the resolvent approach, and a maximum likelihood method. *Molecular Biology and Evolution* **19:** 8–13.

Muse, S. V. and B. S. Weir. 1992. Testing for equality of evolutionary rates. *Genetics* **132:** 269–276.

Muse, S. V. and B S. Gaut. 1994. A likelihood method for comparing synonymous and nonsynonymous nucleotide substitution rates, with application to the chloroplast genome. *Molecular Biology and Evolution* **11:** 715–724.

Muse, S. V. 1995. Evolutionary analysis of DNA sequences subject to constraints on secondary structure. *Genetics* **129:** 1429–1439.

Nadeau, J. and D. Sankoff. 1998. Counting on comparative maps. *Trends in Genetics* **14:** 495–501.

Nastansky, L., S. M. Selkow, and N. F. Stewart. 1973. The enumeration of minimal phylograms. *Bulletin of Mathematical Biology* **35:** 525–533.

Nastansky, L., S. M. Selkow, and N. F. Stewart. 1974. An improved solution to the generalized Camin-Sokal model for numerical cladistics. *Journal of Theoretical Biology* **48:** 413–424.

Navidi, W. C., G. A. Churchill, and A. von Haeseler. 1991. Methods for inferring phylogenies from nucleic acid sequence data by using maximum likelihood and linear invariants. *Molecular Biology and Evolution* **8:** 128–143.

Navidi, W. C., G. A. Churchill, and A. von Haeseler. 1992. Sample size for a phylogenetic inference. *Molecular Biology and Evolution* **9:** 753–769.

Nee, S., R. M. May, and P. H. Harvey. 1994. The reconstructed evolutionary process. *Philosophical Transactions of the Royal Society of London, Series B* **344:** 305–311.

Nee, S., E. C. Holmes, A. Rambaut, and P. H. Harvey. 1995. Inferring population history from molecular phylogenies. *Philosophical Transactions of the Royal Society of London* **349:** 25–31.

Nei, M. 1972. Genetic distance between populations. *American Naturalist* **106:** 283–292.

Nei, M., R. Chakraborty, and P. A. Fuerst. 1976. Infinite allele model with varying mutation rate. *Proceedings of the National Academy of Sciences, USA* **73:** 4164–4168.

Nei. M. and W.-H. Li. 1979. Mathematical model for studying genetic variation in terms of restriction endonucleases. *Proceedings of the National Academy of Sciences, USA* **76:** 5269–5273.

Nei, M. and F. Tajima. 1981. DNA polymorphism detectable by restriction endonucleases. *Genetics* **97:** 145–163.

Nei, M. and F. Tajima. 1983. Maximum likelihood estimation of the number of nucleotide substitutions from restriction sites data. *Genetics* **105:** 207–217.

Nei, M., J. C. Stephens, and N. Saitou. 1985. Methods for computing the standard errors of branching points in an evolutionary tree and their application to molecular date from humans and apes. *Molecular Biology and Evolution* **2:** 66–85.

Nei, M. and F. Tajima. 1985. Evolutionary change of restriction cleavage sites and phylogenetic inference for man and apes. *Molecular Biology and Evolution* **2:** 189–205.

Nei, M. and T. Gojobori. 1986. Simple methods for estimating the numbers of synonymous and nonsynonymous nucleotide substitutions. *Molecular Biology and Evolution* **3:** 418–426.

Nei, M. 1987. *Molecular Evolutionary Genetics.* Columbia University Press, New York.

Nei, M. and A. L. Hughes. 1992. Balanced polymorphism and evolution by the birth-and-death process in the MHC loci. pp. 27–38 in *11th Histocompatibility Workshop and Conference,* Vol. 2, ed. K. Tsuji, M. Aizawa, and T. Sasazuki. Oxford University Press, Oxford.

Nelson, G. 1979. Cladistic analysis and synthesis: Principles and definitions, with a historical note on Adanson's *Familles des Plantes* (1763–1764). *Systematic Zoology* **28:** 1–21.

Nelson, G. and N. Platnick. 1981. *Systematics and Biogeography. Cladistics and Vicariance.* Columbia University Press, New York.

Nelson, G. and N. I. Platnick. 1984. Systematics and evolution. pp. 143–158 in *Beyond Neo-Darwinism. An introduction to the new evolutionary paradigm,* ed. M. Ho and P. T. Saunders. Academic Press, London.

Nelson, G. and N. I. Platnick. 1991. 3-Taxon statements — a more precise use of parsimony. *Cladistics* **7:** 351–366.

Neuhauser, C. and S. M. Krone. 1997. The genealogy of samples in models with selection. *Genetics* **145:** 519–534.

Neuhauser, C. 1999. The ancestral graph and gene genealogy under frequency-dependent selection. *Theoretical Population Biology* **56:** 203–214.

Neumann, D. A. 1983. Faithful consensus methods for *n*-trees. *Mathematical Biosciences* **63:** 271–287.

Newton, M. A. 1996. Bootstrapping phylogenies: Large deviations and dispersion effects. *Biometrika* **83:** 315–328.

Neyman, J. 1971. Molecular studies of evolution: A source of novel statistical problems. pp. 1–27 in *Statistical Decision Theory and Related Topics,* ed. S. S. Gupta and J. Yackel. Academic Press, New York.

Nielsen, R. 1997. A likelihood approach to populations samples of microsatellite alleles. *Genetics* **146:** 711–716 (Erratum: *Genetics* 1997 **147:** 348).

Nielsen, R., J. L. Mountain, J. P. Huelsenbeck, and M. Slatkin. 1998. Maximum--likelihood estimation of population divergence times and population phylogeny in models without mutation. *Evolution* **52:** 669–677.

Nielsen, R. and Z. Yang. 1998. Likelihood models for detecting positively selected amino acid sites and applications to the HIV-1 envelope gene. *Genetics* **148:** 929–936.

Nielsen, R. 1998. Maximum likelihood estimation of population divergence times and population phylogenies under the infinite sites model. *Theoretical Population Biology* **53:** 143–151.

Nielsen, R., P. J. Palsboll. 1999. Single-locus tests of microsatellite evolution: Multi-step mutations and constraints on allele size. *Molecular Phylogenetics and Evolution* **11:** 477–484 .

Nielsen, R. 2000. Estimation of population parameters and recombination rates from single nucleotide polymorphisms. *Genetics* **154:** 931–942.

Nielsen, R. and J. Wakeley. 2001. Distinguishing migration from isolation: A Markov Chain Monte Carlo approach. *Genetics* **158:** 885–896.

Nielsen, R. 2002. Mapping mutations on phylogenies. *Systematic Biology* 51: 729–739.

Nieselt-Struwe, K. 1997. Graphs in sequence spaces: A review of statistical geometry. *Biophysical Chemistry* **66:** 111–131.

Nixon, K. C. 1999. The parsimony ratchet, a new method for rapid parsimony analysis. *Cladistics* **15:** 407–414.

Noor, M. A., R. M. Kliman, and C. A. Machado. 2001. Evolutionary history of microsatellites in the obscura group of *Drosophila. Molecular Biology and Evolution* **18:** 551–556.

Norell, M. A. and M. J. Novacek. 1992. The fossil record and evolution: Comparing cladistic and paleontologic evidence for vertebrate history. *Science* **255:** 1690–1693.

Olsen, G. J. 1987. Earliest phylogenetic branchings: Comparing rRNA-based evolutionary trees inferred with various techniques. *Cold Spring Harbor Symposia on Quantitative Biology* **52:** 825–837.

Olson, E. C. and R. L. Miller. 1958. *Morphological Integration.* University of Chicago Press, Chicago.

Ota, R., P. J. Waddell, M. Hasegawa, H. Shimodaira, and H. Kishino. 2000. Appropriate likelihood ratio tests and marginal distributions for evolutionary tree models with constraints on parameters. *Molecular Biology and Evolution* **17:** 798–803.

Ota, S. and W. H. Li. 2000. NJML: A hybrid algorithm for the neighbor-joining and maximum-likelihood methods. *Molecular Biology and Evolution* **17:** 1401–1409.

Ota, S. and W. H. Li. 2001. NJML+: An extension of the NJML method to handle protein sequence data and computer software implementation. *Molecular Biology and Evolution* **18:** 1983–1992.

Page, R. D. M. 1988. Quantitative cladistic biogeography: Constructing and comparing area cladograms. *Systematic Zoology* **37:** 254–270.

Page, R. D. M. 1989. Comments on component-compatibility in historical biogeography. *Cladistics* **5:** 167–182.

Page, R. D. M. 1990. Component analysis: A valiant failure? *Cladistics* **6:** 119–136.

Page, R. D. M. 1994a. Parallel phylogenies: Reconstructing the history of host-parasite assemblages. *Cladistics* **10:** 155–173.

Page, R. D. M. 1994b. Maps between trees and cladistic analysis of historical associations among genes, organisms, and areas. *Systematic Biology* **43:** 58–77.

Page, R. D. M. 1996. On consensus, confidence, and "total evidence." *Cladistics* **12:** 83–92.

Page, R. D. M. and M. A. Charleston. 1997. From gene to organismal phylogeny: Reconciled trees and the gene tree/species tree problem. *Molecular Phylogenetics and Evolution* **7:** 231–240.

Page, R. D. M. 2003. Modified mincut supertrees. pp. 537–552 in *Algorithms in Bioinformatics. Proceedings of the Second International Workshop, WABI 2002, Rome, Italy, September 17–21, 2002*, ed. R. Guigó and D. Gusfield. Lecture Notes in Computer Science No. 2452, Springer-Verlag, Berlin.

Pagel, M. D. 1992. A method for the analysis of comparative data. *Journal of Theoretical Biology* **156:** 431–442.

Pagel, M. 1994. Detecting correlated evolution on phylogenies: A general method for the comparative analysis of discrete characters. *Proceedings of the Royal Society of London, Series B* **255:** 37–45.

Pagel, M. 1999a. Inferring the historical patterns of biological evolution. *Nature* **401:** 877–884.

Pagel, M. 1999b. The maximum likelihood approach to reconstructing ancestral character states of discrete characters on phylogenies. *Systematic Biology* **48:** 612–622.

Pamilo, P. and M. Nei. 1988. Relationships between gene trees and species trees. *Molecular Biology and Evolution* **5:** 568–583.

Panchen, A. L. 1992. *Classification, Evolution, and the Nature of Biology.* Cambridge University Press, New York.

Paradis, E. 1997. Assessing temporal variations in diversification rates from phylogenies: Estimation and hypothesis testing. *Proceedings of the Royal Society of London, Series B* **264:** 1141–1147.

Paradis, E. 1998a. Detecting shifts in diversification rates without fossils. *American Naturalist* **152:** 176–187.

Paradis, E. 1998b. Testing for constant diversification rates using molecular phylogenies: A general approach based on statistical tests for goodness of fit. *Molecular Biology and Evolution* **15:** 476–479.

Paradis E. and J. Claude. 2002. Analysis of comparative data using generalized estimating equations. *Journal of Theoretical Biology* **218:** 175–185.

Patterson, H. D. and R. Thompson. 1971. Recovery of inter-block information when block sizes are unequal. *Biometrika* **58:** 545–554.

Pearson, W. R., G. Robins, and T. Zhang. 1999. Generalized neighbor-joining: More reliable phylogenetic tree reconstruction. *Molecular Biology and Evolution* **16:** 806–816.

Penny, D., L. R. Foulds, and M. D. Hendy. 1982. Testing the theory of evolution by comparing phylogenetic trees constructed from 5 different protein sequences. *Nature* **297:** 197–200.

Penny, D. and M. D. Hendy. 1985. Testing methods of evolutionary tree construction. *Cladistics* **1:** 266–278.

Penny, D. and M. Hendy. 1986. Estimating the reliability of evolutionary trees. *Molecular Biology and Evolution* **3:** 403–417.

Penny, D. and M . D. Hendy. 1987. TurboTree: A fast algorithm for minimal trees. *Computer Applications in the Biosciences (CABIOS)* **3:** 183–187.

Penny, D., M. D. Hendy, and M. A. Steel. 1991. Testing the theory of descent. pp. 155–183 in *Phylogenetic Analysis of DNA Sequences*, ed. M. M. Miyamoto and J. Cracraft. Oxford University Press, New York.

Penny, D., E. E. Watson, and M. A. Steel. 1993. Trees from languages and genes are very similar. *Systematic Biology* **42:** 382–384.

Penny, D., P. J. Lockhart, M. A. Steel, and M. D. Hendy. 1994. The role of models in reconstructing evolutionary trees. pp. 211–230 in *Models in Phylogeny Reconstruction*, ed. R. W. Scotland, D. J. Siebert, and D. M. Williams. Systematic Association Special Volume No. 52. Clarendon Press, Oxford.

Penny, D., M. D. Hendy, P. J. Lockhart, and M. A. Steel. 1996. Corrected parsimony, minimum evolution, and Hadamard conjugations. *Systematic Biology* **45:** 596–606.

Penny, D., B. J. McComish, M. A. Charleston, and M. D. Hendy. 2001. Mathematical elegance with biochemical realism: The covarion model of molecular evolution. *Journal of Molecular Evolution* **53:** 711–723.

Peres-Neto, P. R. and F. Marques. 2000. When are random data not random, or is the PTP test useful? *Cladistics* **16:** 420–424.

Perler, F., A. Efstratiadis, P. Lomedica, W. Gilbert, R. Kolodner, and J. Dodgson. 1980. The evolution of genes: The chicken proinsulin gene. *Cell* **20:** 565–566.

Phillips, C. and T. J.Warnow. 1996. The asymmetric median tree—a new model for building consensus trees. *Discrete Applied Mathematics* **71:** 311–335.

Phipps, J. B. 1971. Dendrogram topology. *Systematic Zoology* **20:** 306–308.

Platnick, N. I., C. J. Humphries, G. Nelson, and D. M. Williams. 1996. Is Farris optimization perfect?: Three-taxon statements and multiple branching. *Cladistics* **12:** 243–252.

Pol, D. and M. E. Siddall. 2001. Biases in maximum likelihood and parsimony: A simulation approach to a 10-taxon case. *Cladistics* **17:** 266–281.

Pol, D. and M. A. Norell. 2001. Comments on the Manhattan stratigraphic measure. *Cladistics* **17:** 285–289.

Pollock, D. D. and W. R. Taylor. 1997. Effectiveness of correlation analysis in identifying protein residues undergoing correlated evolution. *Protein Engineering* **6:** 647–657.

Pollock D. D., A. Bergman, M. W. Feldman, and D. B. Goldstein. 1998. Microsatellite behavior with range constraints: Parameter estimation and improved distances for use in phylogenetic reconstruction. *Theoretical Population Biology* **53:** 256–271.

Pollock, D. D., W. R. Taylor, and N. Goldman. 1999. Coevolving protein residues: Maximum likelihood identification and relationship structure. *Journal of Molecular Biology* **287:** 187–198.

Popper, K. R. 1959. *The Logic of Scientific Discovery.* Harper and Row, New York.

Popper, K. R. 1968a. *The Logic of Scientific Discovery,* 2nd ed. Harper and Row, New York.

Popper, K. R. 1968b. *Conjectures and Refutations: The Growth of Scientific Knowledge.* Harper and Row, New York.

Posada, D., K. A. Crandall, and A. R. Templeton. 2000. GeoDis: A program for the cladistic nested analysis of the geographical distribution of genetic haplotypes. *Molecular Ecology* **9:** 487–488.

Posada, D. and K. A. Crandall. 2001. Selecting the best-fit model of nucleotide substitution. *Systematic Biology* **50:** 580–601.

Prager, E. M. and A. C. Wilson. 1988. Ancient origin of lactalbumin from lysozyme: Analysis of DNA and amino acid sequences. *Journal of Molecular Evolution* **27:** 326–335.

Pupko, T., I. Pe'er, R. Shamir, and D. Graur. 2000. A fast algorithm for joint reconstruction of ancestral amino acid sequences. *Molecular Biology and Evolution* **17:** 890–896.

Purdom, P. W., Jr., P. G. Bradford, K. Tamura, and S. Kumar. 2000. Single column discrepancy and dynamic max-mini optimizations for quickly finding the most parsimonious evolutionary trees. *Bioinformatics* **16:** 140–151.

Purvis, A. and T. Garland. 1993. Polytomies in comparative analyses of continuous characters. *Systematic Zoology* **42:** 569–575.

Purvis, A. 1995. A modification to Baum and Ragan's method for combining phylogenetic trees. *Systematic Biology* **44:** 251–255.

Pybus, O. and P. H. Harvey. 2000. Testing macro-evolutionary models using incomplete molecular phylogenies. *Proceedings of the Royal Society of London, Series B*, **267:** 2267–2272.

Quicke, D. L. J., J. Taylor, and A. Purvis. 2001. Changing the landscape: A new strategy for estimating large phylogenies. *Systematic Biology* **50:** 60–66.

Ragan, M. A. 1992a. Matrix representation in reconstructing phylogenetic relationships among the eukaryotes. *Biosystems* **28:** 47–55.

Ragan, M. A. 1992b. Phylogenetic inference based on matrix representation of trees. *Molecular Phylogenetics and Evolution* **1:** 53–58.

Rao, C. R. 1973. *Linear Statistical Inference and Its Applications,* 2nd edition. John Wiley and Sons, New York.

Rannala, B. and Z. Yang. 1996. Probability distribution of molecular evolutionary trees: A new method of phylogenetic inference. *Journal of Molecular Evolution* **43:** 304–311.

Rannala, B. 1997. Gene genealogy in a population of variable size. *Heredity* **78:** 417–423.

Rannala, B. and J. P. Reeve. 2001. High-resolution multipoint linkage-disequilibrium mapping in the context of a human genome sequence. *American Journal of Human Genetics* **69:** 159–178.

Ranwez, V. and O. Gascuel. 2002. Improvement of distance-based phylogenetic methods by a local maximum likelihood approach using triplets. *Molecular Biology and Evolution* **19:** 1952–1963.

Ratner, V. A., A. A. Zharkikh, N. Kolchanov, S. Rodin, V. V. Solovyov, and A. S. Antonov. 1995. *Molecular Evolution*. Biomathematics Series Vol. 24. Springer-Verlag, New York.

Read, A.F. and S. Nee. 1995. Inference from binary comparative data. *Journal of Theoretical Biology* **173:** 99–108.

Reeves, J. H. 1992. Heterogeneity in the substitution process of amino acid sites of proteins coded for by mitochondrial DNA. *Journal of Molecular Evolution* **35:** 17–31.

Ren, F., H. Tanaka, and T. Gojobori. 1995. Construction of molecular evolutionary phylogenetic tree from DNA sequences based on minimum complexity principle. *Computer Methods and Programs in Biomedicine* **46:** 121–130.

Reynolds, J., B. S. Weir, C. C. Cockerham. 1983. Estimation of the co-ancestry coefficient—basis for a short-term genetic-distance. *Genetics* **105:** 767–779.

Ricklefs, R. E. and J. M. Starck. 1996. Applications of phylogenetically independent contrasts: A mixed progress report. *Oikos* **77:** 167–172.

Ridley, M. 1983. *The Explanation of Organic Diversity: The Comparative Method and Adaptations for Mating.* Oxford University Press, Oxford.

Rieppel, O. 1988. *Fundamentals of Comparative Biology.* Birkhauser Verlag, Boston.

Rinsma, I., M. Hendy, D. Penny. 1990. Minimally colored trees. *Mathematical Biosciences* **98:** 201–210.

Riska, B. 1991. Regression models in evolutionary allometry. *American Naturalist* **138:** 283–299.

Robertson, A. and W. G. Hill. 1983. Population and quantitative genetics of many linked loci in finite populations. *Proceedings of the Royal Society of London, series B* **219:** 253–264.

Robinson, D. F. and L. R. Foulds. 1979. Comparison of weighted labelled trees. pp. 119–126 in *Combinatorial Mathematics VI. Proceedings of the Sixth Australian Conference on Combinatorial Mathematics, Armidale, Australia, August, 1978,* ed. A. F. Horadam and W. D. Wallis. Lecture Notes in Mathematics, No. 748. Springer-Verlag, Berlin.

Robinson, D. F. and L. R. Foulds. 1981. Comparison of phylogenetic trees. *Mathematical Biosciences* **53:** 131–147.

Robinson, M., M. Gouy, C. Gautier, and D. Mouchiroud. 1998. Sensitivity of the relative-rate test to taxonomic sampling. *Molecular Biology and Evolution* **15:** 1091–1098.

Rodin, A. and W.-H. Li. 2000. A rapid heuristic algorithm for finding minimum evolution trees. *Molecular Phylogenetics and Evolution* **16:** 173–179.

Rodrigo, A. G. 1993. Calibrating the bootstrap test of monophyly. *International Journal for Parasitology* **23:** 507–514.

Rodrigo, A. G., M. Kelly-Borges, P. R. Bergquist, and P. L. Bergquist. 1993. A randomization test of the null hypothesis that 2 cladograms are sample estimates of a parametric phylogenetic tree. *New Zealand Journal of Botany* **31:** 257–268.

Rodrigo, A. G. 1996. On combining cladograms. *Taxon* **45:** 267–274.

Rodrigo, A. G. and J. Felsenstein. 1999. Coalescent approaches to HIV-1 population genetics. pp. 233–272 in *The Evolution of HIV*, ed. K. A. Crandall. Johns Hopkins University Press, Baltimore.

Rodríguez, F., J. L. Oliver, A. Marin, and J. R. Medina. 1990. The general stochastic model of nucleotide substitution. *Journal of Theoretical Biology* **142:** 485–501.

Rogers, J. S. 1994. Central moments and probability distribution of Colless's coefficient of tree imbalance. *Evolution* **48:** 2026–2036.

Rogers, J. S. 1996. Central moments and probability distributions of three measures of phylogenetic tree imbalance. *Systematic Biology* **45:** 99–110.

Rogers, J. S. 1997. On the consistency of maximum likelihood estimation of phylogenetic trees from nucleotide sequences. *Systematic Biology* **46:** 354–357.

Rogers, J. S. and D. L. Swofford, 1999. Multiple local maxima for likelihoods of phylogenetic trees: A simulation study. *Molecular Biology and Evolution* **16:** 1079–1085

Rogers, J. S. 2001. Maximum likelihood estimation of phylogenetic trees is consistent when substitution rates vary according to the invariable sites plus gamma distribution. *Systematic Biology* **50:** 713–722.

Rohlf, F. J. 1965. A randomization test of the nonspecificity hypothesis in numerical taxonomy. *Taxon* **14:** 262–267.

Rohlf, F. J. 1982. Consensus indices for comparing classifications. *Mathematical Biosciences* 59: 131–144.

Rohlf, F. J. 2001. Comparative methods for the analysis of continuous variables: Geometric interpretations. *Evolution* **55:** 2143–2160.

Ronquist, F. and S. Nylin. 1990. Process and pattern in the evolution of species associations. *Systematic Zoology* **39:** 323–344.

Ronquist, F. 1994. Ancestral areas and parsimony. *Systematic Biology* **43:** 267–274.

Ronquist, F. 1996a. Reconstructing the history of host-parasite associations using generalised parsimony. *Cladistics* **11:** 73–89.

Ronquist, F. 1996b. Matrix representation of trees, redundancy, and weighting. *Systematic Biology* **45:** 247–253.

Ronquist, F. 1997. Dispersal-vicariance analysis: A new approach to the quantification of historical biogeography. *Systematic Biology* **46:** 195–203.

Ronquist, F. 1998a. Fast Fitch-parsimony algorithms for large data sets. *Cladistics* **13:** 387–400.

Ronquist, F. 1998b. Phylogenetic approaches in coevolution and biogeography. *Zoologica Scripta* **26:** 313–322.

Ronquist, F. 2002. Parsimony analysis of coevolving species associations. pp. 22–64 in *Tangled trees: Phylogeny, Cospeciation and Coevolution*, ed. R. D. M. Page. University of Chicago Press, Chicago.

Rosenberg, N. 2002. The probability of topological concordance of gene trees and species trees. *Theoretical Population Biology* **61:** 225–247.

Rosenkrantz, R. D. 1977. *Inference, Method, and Decision: Towards a Bayesian Philosophy of Science.* D. Riedel, Boston.

Roux, M. 1988. Techniques of approximation for building two tree structures, pp. 151–170 in *Recent Developments in Clustering and Data Analysis*, ed. C. Hayashi, E. Diday, M. Jambu, and N. Ohsumi. Academic Press, New York.

Royall, R. 1997. *Statistical Evidence—a Likelihood Paradigm.* Chapman and Hall, Boca Raton.

Rzhetsky, A. and M. Nei. 1992. Statistical properties of the ordinary least-squares, generalized least-squares, and minimum-evolution methods of phylogenetic inference. *Journal of Molecular Evolution* **35:** 367–375.

Rzhetsky, A. and M. Nei. 1993. Theoretical foundations of the minimum-evolution method of phylogenetic inference. *Molecular Biology and Evolution* **10:** 1073–1095.

Rzhetsky, A. and M. Nei. 1994. METREE: A program package for inferring and testing minimum-evolution trees. *Computer Applications in the Biosciences (CABIOS)* **10:** 409–412.

Rzhetsky, A. 1995. Estimating substitution rates in ribosomal RNA genes. *Genetics* **141:** 771–783.

Rzhetsky, A. and T. Sitnikova. 1996. When is it safe to use an oversimplified substitution model in tree-making? *Molecular Biology and Evolution* **13:** 1255–1265.

Sackin, M. J. 1972. "Good" and "bad" phenograms. *Systematic Zoology* **21:** 225–226.

Saitou, N. and M. Nei. 1987. The neighbor-joining method: A new method for reconstructing phylogenetic trees. *Molecular Biology and Evolution* **4:** 406–425.

Saitou, N. 1988. Property and efficiency of the maximum likelihood method for molecular phylogeny. *Journal of Molecular Evolution* **27:** 261–273.

Salisbury, B. 1999. Strongest evidence: Maximum apparent phylogenetic signal as a new cladistic optimality criterion. *Cladistics* **15:** 137–149.

Salisbury, E. J. 1942. *The Reproductive Capacity of Plants; Studies in Quantitative Biology.* G. Bell, London.

Salter, L. and D. K. Pearl. 2001. Stochastic search strategy for estimation of maximum likelihood phylogenetic trees. *Systematic Biology* **50:** 7–17.

Sanderson, M. J. 1989. Confidence-limits on phylogenies—the bootstrap revisited. *Cladistics* **5:** 113–129.

Sanderson, M. J. and G. Bharathan. 1993. Does cladistic information affect inferences about branching rates? *Systematic Biology* **42:** 1–17.

Sanderson, M. J. and M. J. Donoghue. 1994. Shifts in diversification rate with the origin of angiosperms. *Science* **264:** 1590–1593.

Sanderson, M. J. 1995. Objections to bootstrapping phylogenies: A critique. *Systematic Biology* **44:** 299–320.

Sanderson, M. J. 1997. A nonparametric approach to estimating divergence times in the absence of rate constancy. *Molecular Biology and Evolution* **14:** 1218–1231.

Sanderson, M. J. 2002. Estimating absolute rates of molecular evolution and divergence times: A penalized likelihood approach. *Molecular Biology and Evolution* **19:** 101–109.

Sankoff, D., C. Morel, and R. J. Cedergren. 1973. Evolution of 5S RNA and the non-randomness of base replacement. *Nature New Biology* 245: 232–234.

Sankoff, D. 1975. Minimal mutation trees of sequences. *SIAM Journal of Applied Mathematics* **28:** 35–42.

Sankoff, D. D. and P. Rousseau. 1975. Locating the vertices of a Steiner tree in arbitrary space. *Mathematical Programming* **9:** 240–246.

Sankoff, D. and R. J. Cedergren. 1983. Simultaneous comparison of three or more sequences related by a tree. pp. 253–263 in *Time Warps, String Edits, and Macromolecules: The Theory and Practice of Sequence Comparison*, ed. D. Sankoff and J. B. Kruskal. Addison-Wesley, Reading, Massachusetts.

Sankoff, D. and M. Goldstein. 1989. Probabilistic models of genome shuffling. *Bulletin of Mathematical Biology* **51:** 117–124.

Sankoff, D. 1990. Designer invariants for large phylogenies. *Molecular Biology and Evolution* **7:** 255–269.

Sankoff, D. 1992. Edit distance for genome comparison based on non-local operations. pp. 121–135 in *Proceedings of the Third Symposium on Combinatorial Pattern Matching*, ed. A. Apostolico, M. Crochemore, Z. Galil, and U. Manber. Lecture Notes in Computer Science No. 644. Springer-Verlag, New York.

Sankoff, D., Y. Abel, and J. Hein. 1994. A tree · a window · a hill; generalization of nearest-neighbour interchange in phylogenetic optimization. *Journal of Classification* **11:** 209–232.

Sankoff, D., G. Sundaram, and J. Kececioglu. 1996. Steiner points in the space of genome rearrangements. *International Journal of Foundations of Computer Science* **7:** 1–9.

Sankoff, D. and M. Blanchette. 1998. Multiple genome rearrangement and breakpoint phylogeny. *Journal of Computational Biology* **5:** 555–570.

Sankoff, D. and M. Blanchette. 1999. Phylogenetic invariants for genome rearrangements. *Journal of Computational Biology* **6:** 431–445.

Sankoff, D. 1999. Genome rearrangement with gene families. *Bioinformatics* **15:** 909–917.

Sarich, V. M. 1969. Pinniped phylogeny. *Systematic Zoology* **18:** 416–422.

Sarich, V. M. and A. C. Wilson. 1967. Rates of albumin evolution in primates. *Proceedings of the National Academy of Sciences, USA* **58:** 142–148.

Sarich, V. M. and A. C. Wilson. 1973. Generation time and genomic evolution in primates. *Science* **179:** 1144–1147.

Sattath, S. and A. Tversky. 1977. Additive similarity trees. *Psychometrika* **42:** 319–345.

Sawyer, S. 1989. Statistical tests for detecting gene conversion. *Molecular Biology and Evolution* **6:** 526–538.

Schadt, E. E., J. S. Sinsheimer, and K. Lange. 1998. Computational advances in maximum likelihood methods for molecular phylogeny. *Genome Research* **8:** 222–233.

Schadt, E. E., J. S. Sinsheimer, and K. Lange. 2002. Applications of codon and rate variation models in molecular phylogeny. *Molecular Biology and Evolution* **19:** 1550–1562.

Schluter, D., T. Price, A. O. Mooers, and D. Ludwig. 1997. Likelihood of ancestor states in adaptive radiation. *Evolution* **51:** 1699–1711.

Schöniger, M. and A. von Haeseler. 1994. A stochastic model for the evolution of autocorrelated DNA sequences. *Molecular Phylogenetics and Evolution* **3:** 240–247.

Schröder, E. 1870. Vier combinatorische Probleme. *Zeitschrift für Mathematik und Physik* **15:** 361–376.

Schultz, T. R. and G. A. Churchill. 1999. The role of subjectivity in reconstructing ancestral character states: A Bayesian approach to unknown rates, states, and transformation asymmetries. *Systematic Biology* **48:** 651–664.

Schwikowski, B. and M. Vingron. 1997a. The deferred path heuristic for the generalized tree alignment problem. *Journal of Computational Biology* **4:** 415–431.

Schwikowski, B. and M. Vingron. 1997b. A clustering approach to Generalized Tree Alignment with application to Alu repeats. pp. 115–124 in *Bioinformatics: Proceedings of the German Conference on Bioinformatics (GCB '96), Leipzig, Germany, September 30-October 2, 1996,* ed. R. Hofestädt. Lecture Notes in Computer Science No. 1278, Springer-Verlag, Berlin.

Scotland, R. W. and M. A. Carine. 2000. Classification or phylogenetic estimates? *Cladistics* **16:** 411–419.

Semple, C. and M. Steel. 2000. A supertree method for rooted trees. *Discrete Applied Mathematics* **105:** 147–158.

Shao, K.-T. and R. R. Sokal. 1990. Tree balance. *Systematic Zoology* **39:** 266–276.

Shedlock, A. M., M. C. Milinkovitch, and N. Okada. 2000. SINE evolution, missing data, and the origin of whales. *Systematic Biology* **49:** 808–817.

Shimodaira, H. 1998. An application of multiple comparison techniques to model selection. *Annals of the Institute of Statistical Mathematics* **50:** 1–13.

Shimodaira, H. and M. Hasegawa. 1999. Multiple comparisons of log-likelihoods with applications to phylogenetic inference. *Molecular Biology and Evolution* **16:** 1114–1116.

Shimodaira, H. and M. Hasegawa. 2001. CONSEL: For assessing the confidence of phylogenetic tree selection. *Bioinformatics* **17:** 1246–1247.

Shimodaira, H. 2002. An approximately unbiased test of phylogenetic tree selection. *Systematic Biology* **51:** 492–508.

Sibly, R. M., J. C. Whittaker, and M. Talbot. 2001. A maximum-likelihood approach to fitting equilibrium models of microsatellite evolution. *Molecular Biology and Evolution* **18:** 413–417.

Siddall, M. E. and A. G. Kluge. 1997. Probabilism and phylogenetic inference. *Cladistics* **13:** 313–336.

Siddall, M. E. 1998a. Stratigraphic fit to phylogenies: A proposed solution. *Cladistics* **14:** 201–208.

Siddall, M. E. 1998b. Success of parsimony in the four-taxon case: long-branch repulsion by likelihood in the Farris zone. *Cladistics* **14:** 209–220.

Siddall, M. E. 2001. Philosophy and phylogenetic inference: A comparison of likelihood and parsimony methods in the context of Karl Popper's writings on corroboration. *Cladistics* **17:** 395–399.

Sillén-Tullberg, B. 1988. Evolution of gregariousness in aposematic butterfly larvae: A phylogenetic analysis. *Evolution* **42:** 293–305.

Sillén-Tullberg, B. 1993. The effect of biased inclusion of taxa on the correlation between discrete characters in phylogenetic trees. *Evolution* **47:** 1182–1191.

Simmons, M. P., C. D. Bailey, and K. Nixon. 2000. Phylogeny reconstruction using duplicate genes. *Molecular Biology and Evolution* **17:** 469–473.

Simmons, M. P., C. P. Randle, J. V. Freudenstein, and J. W. Wenzel. 2002. Limitations of relative apparent synapomorphy analysis (RASA) for measuring phylogenetic signal. *Molecular Biology and Evolution* **19:** 14–23.

Sinsheimer, J. S., J. A. Lake, and R. J. Little. 1996. Bayesian hypothesis testing of four-taxon topologies using molecular sequence data. *Biometrics* **52:** 193–210.

Sinsheimer, J. S., J. A. Lake, and R. J. Little. 1997. Inference for phylogenies under a hybrid parsimony method: evolutionary-symmetric transversion parsimony. *Biometrics* **53:** 23–38.

Sitnikova, T., A. Rzhetsky, and M. Nei. 1995. Interior-branch and bootstrap tests of phylogenetic trees. *Molecular Biology and Evolution* **12:** 319–333.

Sitnikova, T. 1996. Bootstrap method of interior-branch test for phylogenetic trees. *Molecular Biology and Evolution* **13:** 605–611.

Slade, P. F. 2001a. Simulation of selected genealogies. *Theoretical Population Biology* **57:** 35–49.

Slade, P. F. 2001b. Simulation of 'hitch-hiking' genealogies. *Journal of Mathematical Biology* **42:** 41–70.

Slatkin, M. and W. P. Maddison. 1989. Cladistic measure of gene flow inferred from the phylogenies of alleles. *Genetics* **123:** 603–613.

Slatkin, M. and R. R. Hudson. 1991. Pairwise comparisons of mitochondrial DNA sequences in stable and exponentially growing populations. *Genetics* **129:** 555–562.

Slatkin, M. 1995. A measure of population subdivision based on microsatellite allele frequencies. *Genetics* **139:** 457–462 (erratum, p. 1463).

Slatkin, M. and B. Rannala. 1997. Estimating the age of alleles by use of intraallelic variability. *American Journal of Human Genetics* **60:** 447–458.

Slatkin, M. and B. Rannala. 2000. Estimating allele age. *Annual Review of Genomics and Human Genetics* **1:** 225–249.

Slatkin, M. 2002. A vectorized method of importance sampling with applications to models of mutation and migration. *Theoretical Population Biology* **62:** 339–348.

Slowinski, J. B. and C. Guyer. 1989. Testing the stochasticity of patterns of organismal diversity: An improved null model. *American Naturalist* **134 :** 907–921.

Slowinski, J. B. and C. Guyer. 1993. Testing whether certain traits have caused amplified diversification: An improved method based on a model of random speciation and extinction. *American Naturalist* **142:** 1019–1024.

Slowinski, J. B., A. Knight, and A. P. Rooney. 1997. Inferring species trees from gene trees: A phylogenetic analysis of the Elapidae (Serpentes) based on the amino acid sequences of venom proteins. *Molecular Phylogenetics and Evolution* **8:** 349–362.

Slowinski, J. B. and B. I. Crother. 1998. Is the PTP test useful? *Cladistics* **14:** 297–302.

Slowinski, J. and R. D. M. Page. 1999. How should species phylogenies be inferred from sequence data? *Systematic Biology* **48:** 814–825.

Smouse, P. E. and W.-H. Li. 1987. Likelihood analysis of mitochondrial restriction-cleavage patterns for the human-chimpanzee-gorilla trichotomy. *Evolution* **41:** 1162–1176.

Sneath, P. H. A. 1957a. Some thoughts on bacterial classification. *Journal of General Microbiology* **17:** 184–200.

Sneath, P. H. A. 1957b. The application of computers to taxonomy. *Journal of General Microbiology* **17:** 201–226.

Sneath, P. H. A., M. J. Sackin and R. P. Ambler. 1975. Detecting evolutionary incompatibilities from protein sequences. *Systematic Zoology* **24:** 311–332.

Sneath, P. H. A. 1986. Estimating uncertainty in evolutionary trees from Manhattan-distance triads. *Systematic Zoology* **35:** 470–488.

Sober, E. 1985. A likelihood justification of parsimony. *Cladistics* **1:** 209–233.

Sober, E. 1988. *Reconstructing the Past. Parsimony, Evolution, and Inference.* MIT Press, Cambridge, Massachusetts.

Sokal, R. R. and C. D. Michener. 1958. A statistical method for evaluating systematic relationships. *University of Kansas Science Bulletin* **38:** 1409–1438.

Sokal, R. R. 1983. A phylogenetic analysis of the Caminalcules. I. The data base. *Systematic Zoology* **32:** 159–184.

Sokal, R. R. and P. H. A. Sneath. 1963. *Numerical Taxonomy.* W. H. Freeman, San Francisco.

Stebbins, G. L. 1950. *Variation and evolution in plants.* New York, Columbia University Press.

Steel, M. A., M. D. Hendy, L. A. Székely, and P. L. Erdős. 1992. Spectral analysis and a closest tree method for genetic sequences. *Applied Mathematics Letters* **5 (6):** 63–67.

Steel, M. 1992. The complexity of reconstructing trees from qualitative characters and subtrees. *Journal of Classification* **9:** 91–116.

Steel, M. and T. Warnow. 1993. Kaikoura tree theorems: The maximum agreement subtree problem. *Information Processing Letters* **48:** 77–82.

Steel, M. A. and D. Penny. 1993. Distributions of tree comparison metrics—some new results. *Systematic Zoology* **42:** 126–141.

Steel, M. A., L. A. Székely, P. L. Erdős, and P. J. Waddell. 1993. A complete family of phylogenetic invariants for any number of taxa under Kimura's 3ST model. *New Zealand Journal of Botany* **31:** 289–296.

Steel, M. A., M. D. Hendy, and D. Penny. 1993. Parsimony can be consistent. *Systematic Zoology* **42:** 581–587.

Steel, M. A. 1994a. Recovering a tree from the Markov leaf colourations it generates under a Markov model. *Applied Mathematics Letters* **7 (Issue 2):** 19–23.

Steel, M. A., L. A. Székely, and M. D. Hendy. 1994. Reconstructing trees when sequence sites evolve at variable rates. *Journal of Computational Biology* **1:** 153–163.

Steel, M. A. 1994b. The maximum likelihood point for a phylogenetic tree is not unique. *Systematic Biology* **43:** 560–564.

Steel, M. A. and Y.-X. Fu. 1995. Classifying and counting linear phylogenetic invariants for the Jukes-Cantor model. *Journal of Computational Biology* **2:** 39–47.

Steel, M. A., A. C. Cooper, and D. Penny. 1996. Confidence intervals for the divergence time of two clades. *Systematic Zoology* **45:** 127–134.

Steel, M., M. D. Hendy, and D. Penny. 1998. Reconstructing probabilities from nucleotide pattern probabilities: A survey and some new results. *Discrete Applied Mathematics* **88:** 367–396.

Steel, M. A. and L. A. Székely. 1999. Inverting random functions. *Annals of Combinatorics* **3:** 103–113.

Steel, M. and P. J. Waddell. 1999. Approximating likelihoods under low but variable rates across sites. *Applied Mathematics Letters* **12 (10):** 13–19.

Steel, M. and D. Penny. 2000. Parsimony, likelihood, and the role of models in molecular phylogenetics. *Molecular Biology and Evolution* **17:** 839–850.

Steel, M. A., A. W. M. Dress, and S. Böcker. 2000. Simple but fundamental limitations on supertree and consensus methods. *Systematic Zoology* **49:** 363–368.

Steel, M. and J. Hein. 2001. Applying the TKF model to sequence evolution on a star tree. *Applied Mathematics Letters* **14:** 679–684.

Steel, M. A. and A. McKenzie. 2001. Properties of phylogenetic trees generated by Yule-type speciation models. *Mathematical Biosciences* **170:** 91–112.

Steel, M. 2001. Sufficient conditions for two tree reconstruction techniques to succeed on sufficiently long sequences. *SIAM Journal on Discrete Mathematics* **14:** 36–48.

Steel, M. A. and L. A. Székely. 2002. Inverting random functions II: Explicit bounds for discrete maximum likelihood estimation, with applications. *SIAM Journal on Discrete Mathematics* **15:** 562–575.

Stephens, J. C. 1985. Statistical methods of DNA sequence analysis: Detection of intragenic recombination or gene conversion. *Molecular Biology and Evolution* **2:** 539–556.

Stephens, M. and P. Donnelly. 2000. Inference in molecular population genetics. *Journal of the Royal Statistical Society B* **62:** 605–635.

Stephens, M. 2000. Times on trees, and the age of an allele. *Theoretical Population Biology* **57:** 109–119.

Stinebrickner, R. 1984. *s*-consensus trees and indices. *Bulletin of Mathematical Biology* **46:** 923–945.

Stinebrickner, R. 1986. *s*-consensus method: An additional axiom. *Journal of Classification* **3:** 319–327.

Stone, J. and J. Repka. 1998. Using a nonrecursive formula to determine cladogram probabilities. *Systematic Biology* **47:** 617–624.

Strimmer, K. and A. von Haeseler. 1996. Quartet puzzling: A quartet maximum likelihood method for reconstructing tree topologies. *Molecular Biology and Evolution* **13:** 964–969.

Strimmer, K., N. Goldman, and A. von Haeseler. 1997. Bayesian probabilities and quartet puzzling. *Molecular Biology and Evolution* **14:** 210–211.

Studier, J. A. and K. J. Keppler. 1988. A note on the neighbor-joining algorithm of Saitou and Nei. *Molecular Biology and Evolution* **5:** 729–731.

Suchard, M. A., R. E. Weiss, and J. S. Sinsheimer. 2001. Bayesian selection of continuous-time Markov chain evolutionary models. *Molecular Biology and Evolution* **18:** 1001–1013.

Sullivan, J. and D. L. Swofford. 1997. Are guinea pigs rodents? The importance of adequate models in molecular phylogenetics. *Journal of Mammalian Evolution* **4:** 77–86.

Susko, E., Y. Inagaki, C. Field, M. E. Holder, and A. J. Roger. 2002. Testing for differences in rates-across-sites distributions in phylogenetic subtrees. *Molecular Biology and Evolution* **19:** 1514–1523.

Suzuki, Y., G. V. Glazko, and M. Nei. 2002. Overcredibility of molecular phylogenies obtained by Bayesian phylogenetics. *Proceedings of the Natonal Academy of Sciences, USA* **99:** 16138–16143.

Swofford, D. L. and D. R. Maddison. 1987. Reconstructing ancestral character states under Wagner parsimony. *Mathematical Biosciences* **87:** 199–229.

Swofford, D. L. and S. H. Berlocher. 1987. Inferring evolutionary trees from gene-frequency data under the principle of maximum parsimony. *Systematic Zoology* **36:** 293–325.

Swofford, D. L. and G. J. Olsen. 1990. Phylogeny reconstruction. Chapter 11, pp. 411–501 in *Molecular Systematics,* ed. D. M. Hillis and C. Moritz. Sinauer Associates, Sunderland, Massachusetts.

Swofford, D. L. 1991. When are phylogeny estimates from molecular and morphological data incongruent? pp. 295–333 in *Phylogenetic Analysis of DNA Sequences*, ed. M. M. Miyamoto and J. Cracraft. Oxford University Press, New York.

Swofford, D. L. 1995. PAUP*. *Phylogenetic Analysis Using Parsimony (* and Other Methods).* Sinauer Associates, Sunderland, Massachusetts.

Swofford, D. L., G. J. Olsen, P. J. Waddell, and D. M. Hillis. 1996. Phylogenetic inference. pp. 407–514 in *Molecular Systematics,* 2nd ed., ed. D. M. Hillis, C. Moritz, and B. K. Mable. Sinauer Associates, Sunderland, Massachusetts.

Swofford, D. L., J. L. Thorne, J. Felsenstein, and B. M. Wiegmann. 1996. The topology-dependent permutation test for monophyly does not test for monophyly. *Systematic Biology* **45:** 575–579.

Swofford, D. L. and M. E. Siddall. 1997. Uneconomical diagnosis of cladograms: Comments on Wheeler and Nixon's method for Sankoff optimization. *Cladistics* **13:** 153–159.

Swofford, D. L., P. J. Waddell, J. P. Huelsenbeck, P. G. Foster, P. O. Lewis, and J. S. Rogers. 2001. Bias in phylogenetic estimation and its relevance to the choice between parsimony and likelihood methods. *Systematic Biology* **50:** 525–539.

Symonds, M. R. E. 2002. The effects of topological inaccuracy in evolutionary trees on the phylogenetic comparative method of independent contrasts. *Systematic Biology* **51:** 541–553.

Tajima, F. and M. Nei. 1982. Biases of the estimates of DNA divergence obtained by the restriction enzyme technique. *Journal of Molecular Evolution* **18:** 115–120.

Tajima, F. 1983. Evolutionary relationship of DNA-sequences in finite populations. *Genetics* **105:** 437–460.

Tajima, F. 1992. Statistical method for estimating the standard errors of branch lengths in a phylogenetic tree reconstructed without assuming equal rates of nucleotide substitution among different lineages. *Molecular Biology and Evolution* **9:** 168–8

Tajima, F. 1993. Simple methods for testing the molecular evolutionary clock hypothesis. *Genetics* **135:** 599–607.

Takahata, N. and M. Nei. 1985. Gene genealogy and variance of interpopulational nucleotide differences. *Genetics* **110:** 325–344.

Takahata, N. 1986. An attempt to estimate the effective size of the ancestral species common to two extant species from which homologous genes are sequenced. *Genetical Research* **48:** 187–190.

Takahata, N. 1988. The coalescent in two partially isolated diffusion populations. *Genetical Research* **52:** 213–222.

Takahata, N. 1989. Gene genealogy in three related populations: Consistency probability between gene and population trees. *Genetics* **122:** 957–966.

Takahata, N., Y. Satta, and J. Klein. 1995. Divergence time and population size in the lineage leading to modern humans. *Theoretical Population Biology* **48:** 198–221.

Takezaki, N., A. Rzhetsky, and M. Nei. 1995. Phylogenetic test of the molecular clock and linearized trees. *Molecular Biology and Evolution* **12:** 823–833.

Takezaki, N. 1998. Tie trees generated by distance methods of phylogenetic reconstruction. *Molecular Biology and Evolution* **15:** 727–737.

Tamura, K. and M. Nei. 1993. Estimation of the number of nucleotide substitutions in the control region of mitochondrial DNA in humans and chimpanzees. *Molecular Biology and Evolution* **10:** 512–526.

Tanaka, H., F. Ren, T. Okayama, and T. Gojobori. 1997. Inference of molecular phylogenetic tree based on minimum model-based complexity method. pp. 319–328 in *Proceedings of the Fifth International Conference on Intelligent Systems for Molecular Biology*, ed. T. Gaasterland, P. Karp, K. Karplus, C. Ouzounis, C. Sander, and A. Valencia. AAAI Press, Menlo Park, California.

Tanaka, H., F. Ren, T. Okayama, and T. Gojobori. 1999. Topology selection in unrooted molecular phylogenetic tree by minimum model-based complexity method. *Pacific Symposium on Biocomputing 1999,* pp. 326–337.

Tang, M., M. Waterman, and S. Yooseph. 2002. Zinc finger gene clusters and tandem gene duplication. *Journal of Computational Biology* **9:** 429–446.

Tateno, Y., M. Nei, and F. Tajima. 1982. Accuracy of estimated phylogenetic trees from molecular data. I. Distantly related species. *Journal of Molecular Evolution* **18:** 387–404.

Tavaré, S., D. J. Balding, R. C. Griffiths, and P. Donnelly. 1997. Inferring coalescence times from DNA sequence data. *Genetics* **145:** 505–518.

Templeton, A. R. 1983a. Phylogenetic inference from restriction endonuclease cleavage site maps with particular reference to the evolution of humans and the apes. *Evolution* **37:** 221–244.

Templeton, A. R. 1983b. Convergent evolution and nonparametric inferences from restriction data and DNA sequences. pp. 151–179 in *Statistical Analysis of DNA Sequence Data*, ed. B. S. Weir. Marcel Dekker, Inc., New York.

Templeton, A. R., C. F. Sing, A. Kessling, and S. Humphries. 1988. A cladistic-analysis of phenotype associations with haplotypes inferred from restriction endonuclease mapping. 2. The analysis of natural-populations. *Genetics* **120:** 1145–1154.

Templeton, A. R. 1998. Nested clade analyses of phylogeographic data: Testing hypotheses about gene flow and population history. *Molecular Ecology* **7:** 381–397.

Thiele, K. 1993. The Holy Grail of the perfect character: The cladistic treatment of morphometric data. *Cladistics* **9:** 275–304.

Thompson, E. A. 1972. Distances between populations on the basis of gene frequencies. *Biometrics* **27:** 873–881.

Thompson, E. A. 1973. The method of minimum evolution. *Annals of Human Genetics* **36:** 333–340.

Thompson, E. A. 1975. *Human Evolutionary Trees.* Cambridge University Press, Cambridge.

Thompson, J. D., D. G. Higgins, and T. J. Gibson. 1994. Clustal-W—improving the sensitivity of progressive multiple sequence alignment through sequence weighting, position-specific gap penalties and weight matrix choice. *Nucleic Acids Research* **22:** 4673–4680.

Thorne, J. L., H. Kishino, and J. Felsenstein. 1991. An evolutionary model for maximum likelihood alignment of DNA sequences. *Journal of Molecular Evolution* **33:** 114–124.

Thorne, J. L., H. Kishino, and J. Felsenstein. 1992. Inching toward reality: An improved likelihood model of sequence evolution. *Journal of Molecular Evolution* **34:** 3–16.

Thorne, J. L. and H. Kishino. 1992. Freeing phylogenies from artifacts of alignment. *Molecular Biology and Evolution* **9:** 1148–1162.

Thorne, J. L. and G. A. Churchill. 1995. Estimation and reliability of molecular sequence alignments. *Biometrics* **51:** 100–113.

Thorne, J. L., N. Goldman, and D. T. Jones. 1996. Combining protein evolution and secondary structure. *Molecular Biology and Evolution* **13:** 666–673.

Thorne, J. L., H. Kishino, and I. S. Painter. 1998. Estimating the rate of evolution of the rate of molecular evolution. *Molecular Biology and Evolution* **15:** 1647–1657 .

Tillier, E. R. M. 1994. Maximum likelihood with multi-parameter models of substitution. *Journal of Molecular Evolution* **39:** 409–417.

Tillier, E. R. M. and R. A. Collins. 1995. Neighbor-joining and maximum likelihood with RNA sequences: Addressing the interdependence of sites. *Molecular Biology and Evolution* **12:** 7–15.

Tillier, E. R. M. and R. A. Collins. 1998. High apparent rate of simultaneous compensatory base-pair substitutions in ribosomal RNA. *Genetics* **148:** 1993–2002.

Trueman, J. W. H. 1993. Randomization confounded: A response to Carpenter. *Cladistics* **9:** 101–109.

Trueman, J. W. H. 1996. Permutation tests and outgroups. *Cladistics* **12:** 253–261.

Tufféry, P. and P. Darlu. 2000. Exploring a phylogenetic approach for the detection of correlated substitutions in proteins. *Molecular Biology and Evolution* **17:** 1753–1759.

Tuffley, C. and M. Steel. 1997. Links between maximum likelihood and maximum parsimony under a simple model of site substitution. *Bulletin of Mathematical Biology* **59:** 581–607.

Turelli, M., J. H. Gillespie, and R. Lande. 1988. Rate tests for selection on quantitative characters during macroevolution and microevolution. *Evolution* **42:** 1085–1089.

Upholt, W. B. 1977. Estimation of DNA sequence divergence from comparison of restriction endonuclease digests. *Nucleic Acids Research* **4:** 1257–1265.

Uzzell, T. and K. W. Corbin. 1971. Fitting discrete probability distribution to evolutionary events. *Science* **172:** 1089–1096.

Vach, W. 1989. Least squares approximation of additive trees. pp. 230–238 in *Conceptual and Numerical Analysis of Data,* ed. O. Opitz. Springer-Verlag, Berlin.

Vach, W. and P. O. Degens. 1991. Least squares approximation of additive trees to dissimilarities—characterisations and algorithms. *Computational Statistics Quarterly* **3:** 203–218.

Valdes, A. M., M. Slatkin, and N. B. Freimer. 1993. Allele frequencies at microsatellite loci: The stepwise mutation model revisited. *Genetics* **133:** 737–749.

Vingron, M. and A. von Haeseler. 1997. Towards integration of multiple alignment and phylogenetic tree construction. *Journal of Computational Biology* **4:** 23–34.

von Haeseler, A. and M. Schöniger. 1995. Ribosomal RNA phylogeny derived from a correlation model of sequence evolution. pp. 330–338 in *From Data to Knowledge,* ed. W. Gaul and D. Pfeffer. Springer-Verlag, Berlin.

Waddell, P. J., D. Penny, M. D. Hendy, and G. Arnold. 1994. The sampling distributions and covariance matrix of phylogenetic spectra. *Molecular Biology and Evolution* **11:** 630–642.

Waddell, P. J. 1995. Statistical methods of phylogenetic analysis, including Hadamard conjugations, LogDet transforms, and maximum likelihood. Ph.D. Thesis, Massey University.

Waddell, P. J. and M. A. Steel. 1997. General time-reversible distances with unequal rates across sites: Mixing Γ and inverse Gaussian distributions with invariant sites. *Molecular Phylogenies and Evolution* **8:** 398–414.

Waddell, P. J., D. Penny, and T. Moore. 1997. Extending Hadamard conjugations to model sequence evolution with variable rates across sites. *Molecular Phylogenetics and Evolution* **8:** 33–50 (erratum, p. 446).

Wagner, P. J. 1998. A likelihood approach for evaluating estimates of phylogenetic relationships among fossil taxa. *Paleobiology* **24:** 430–449.

Wagner, W. H., Jr. 1961. Problems in the classification of ferns. pp. 841–844 in *Recent Advances in Botany. From lectures and symposia presented to the IX International Botanical Congress, Montreal, 1959.* University of Toronto Press, Toronto.

Wakeley, J. and J. Hey. 1997. Estimating ancestral population sizes. *Genetics* **145:** 847–855.

Wakeley, J. 1998. Segregating sites in Wright's island model. *Theoretical Population Biology* **53:** 166–174.

Wald, A. 1949. Note on the consistency of the maximum likelihood estimate. *Annals of Mathematical Statistics* **20:** 595–601.

Wall, J. D. 2003. Estimating ancestral population sizes and divergence times. *Genetics* **163:** 395–404.

Wang, L. and T. Jiang. 1993. On the complexity of multiple sequence alignment. *Journal of Computational Biology* **1:** 337–348.

Wang, L., T. Jiang, and E. L. Lawler. 1996. Approximation algorithms for tree alignment with a given phylogeny. *Algorithmica* **16:** 302–315.

Wang, L. and D. Gusfield. 1997. Improved approximation algorithms for tree alignment. *Journal of Algorithms* **25:** 255–273.

Wareham, H. T. 1995. A simplified proof of the NP- and MAX SNP-hardness of multiple sequence tree alignment. *Journal of Computational Biology* **2:** 509–514.

Waterman, M. S., T. F. Smith, M. Singh, and W. A. Beyer. 1977. Additive evolutionary trees. *Journal of Theoretical Biology* **64:** 199–213.

Waterman, M. S. and T. F. Smith. 1978. On the similarity of dendrograms. *Journal of Theoretical Biology* **73:** 789–800.

Watterson, G. A. 1975. On the number of segregating sites in genetical models without recombination. *Theoretical Population Biology* **7:** 256–276.

Watterson, G. A., W. J. Ewens, T. E. Hall, and A. Morgan. 1982. The chromosome inversion problem. *Journal of Theoretical Biology* **99:** 1–7.

Webster, A. J. and A. Purvis. 2002. Testing the accuracy of methods for reconstructing ancestral states of continuous characters. *Proceedings of the Royal Society of London, Series B* **269:** 143–149.

Weir, B.S. 1990. *Genetic Data Analysis*. Sinauer Associates, Sunderland, Massachusetts.

Weiss, G. and A. von Haeseler. 1998. Inference of population history using a likelihood approach. *Genetics* **149:** 1539–1546.

Wedderburn, J. H. M. 1922. The functional equation $g(x^2) = 2\alpha x + [g(x)]^2$. *Annals of Mathematics (2nd series)* **24:** 121–140.

Westoby, M., M. R. Leishman, and J. M. Lord. 1995a. On misinterpreting the 'phylogenetic correction'. *Journal of Ecology* **83:** 531–534.

Westoby, M., M. Leishman, and J. Lord. 1995b. Further remarks on phylogenetic correction. *Journal of Ecology* **83:** 727–729.

Westoby, M., M. Leishman, and J. Lord. 1995c. Issues of interpretation after relating comparative datasets to phylogeny. *Journal of Ecology* **83:** 892–893.

Wheeler, W. C. and K. Nixon. 1994. A novel method for economical diagnosis of cladograms under Sankoff optimization. *Cladistics* **10:** 207–213.

Wheeler, W. C. 1991. Congruence among data sets: A Bayesian approach. pp. 334–346 in *Phylogenetic Analysis of DNA sequences*, ed. M. M. Miyamoto and J. Cracraft. Oxford University Press, New York.

Wheeler, W. C. and D. G. Gladstein. 1994. MALIGN: A multiple nucleic and sequence alignment program. *Journal of Heredity* **85:** 417–418.

Wheeler, W. C. 1996. Optimization alignment: The end of multiple sequence alignment in phylogenetics? *Cladistics* **12:** 1–9.

Whelan, S. and N. Goldman. 2001. A general empirical model of protein evolution derived from multiple protein families using a maximum-likelihood approach. *Molecular Biology and Evolution* **18:** 691–699.

Wiens, J. J. 1995. Polymorphic characters in phylogenetic systematics. *Systematic Biology* **44:** 482–500.

Wiens, J. J. 2000. Reconstructing phylogenies from allozyme data: Comparing method performance with congruence. *Biological Journal of the Linnean Society* **70:** 613–632.

Wiens, J. J. 2001. Character analysis in morphological phylogenetics: Problems and solutions. *Systematic Biology* **50:** 689–699.

Wilcox, T. P., D. J. Zwickl, T. A. Heath, and D. M. Hillis. 2002. Phylogenetic relationships of the dwarf boas and a comparison of Bayesian and bootstrap measures of phylogenetic support. *Molecular Phylogenetics and Evolution* **25:** 361–371.

Wilkinson, M. 1994. Common cladistic information and its consensus representation: Reduced Adams and reduced cladistic consensus trees and profiles. *Systematic Zoology* **43:** 343–368.

Wilkinson, M. 1996. Majority-rule reduced consensus trees and their use in bootstrapping. *Molecular Biology and Evolution* **13:** 437–444.

Wilkinson, M, P. R. Peres-Neto, P. G. Foster, and C. B. Moncrieff. 2002. Type 1 error rates of the parsimony permutation tail probability test. *Systematic Biology* **51:** 524–527.

Wiley, E. O. 1975. Karl R. Popper, systematics, and classification: A reply to Walter Bock and other evolutionary taxonomists. *Systematic Zoology* **24:** 233–243.

Wiley, E. O. 1981. *Phylogenetics. The Theory and Practice of Phylogenetic Systematics.* John Wiley and Sons, New York.

Williams, S. A. and M. Goodman. 1989. A statistical test that supports a human/chimpanzee clade based on noncoding DNA sequence data. *Molecular Biology and Evolution* **6:** 325–330.

Williams, W. T., and H. T. Clifford. 1971. On the comparison of two classifications of the same set of elements. *Taxon* **20:** 519–522.

Wills, M. A. 1999. Congruence between phylogeny and stratigraphy: Randomization tests and the gap excess ratio. *Systematic Biology* **48:** 559–580.

Willson, S. 1998. Measuring inconsistency in phylogenetic trees. *Journal of Theoretical Biology* **190:** 15–36.

Wilson, E. O. 1965. A consistency test for phylogenies based on contemporaneous species. *Systematic Zoology* **14:** 214–220.

Wilson, I. J. and D. J. Balding. 1998. Genealogical inference from microsatellite data. *Genetics* **150:** 499–510.

Wiuf, C. and J. Hein. 1999. Recombination as a point process along sequences. *Theoretical Population Biology* **55:** 248–259.

Woese, C. R. 1987. Macroevolution in the microscopic world. pp. 177–202 in *Molecules and Morphology in Evolution: Conflict or Compromise?* ed. C. Patterson. Cambridge University Press, Cambridge.

Wollenberg, K., J. Arnold, and J. C. Avise. 1996. Recognizing the forest for the trees: Testing temporal patterns of cladogenesis using a null model of speciation. *Molecular Biology and Evolution* **13:** 833–849.

Wollenberg, K. R. and W. R. Atchley. 2000. Separation of phylogenetic and functional associations in biological sequences by using the parametric bootstrap. *Proceedings of the National Academy of Sciences, USA* **97:** 3288–3291.

Wright, S. 1931. Evolution in Mendelian populations. *Genetics* **16:** 97–159.

Wright, S. 1934. An analysis of variability in number of digits in an inbred strain of guinea pigs. *Genetics* **19:** 506–536.

Wu, C. F. J. 1986. Jackknife, bootstrap and other resampling plans in regression analysis. *Annals of Statistics* **14:** 1261–1295.

Wu, C.-I and W.-H. Li. 1985. Evidence for higher rates of nucleotide substitutions in rodents than in man. *Proceedings of the National Academy of Sciences, USA* **82:** 1741–1745.

Wu, C.-I. 1991. Inferences of species phylogeny in relation to segregation of ancient polymorphisms. *Genetics* **127:** 429–435.

Wu, S. and X. Gu. 2003. Algorithms for multiple genome rearrangement by signed reversals. *Pacific Symposium on Biocomputing 2003*, pp. 363–374.

Yang, Z. 1993. Maximum-likelihood estimation of phylogeny from DNA sequences when substitution rates differ over sites. *Molecular Biology and Evolution* **10:** 1396–1401.

Yang, Z. 1994a. Maximum likelihood phylogenetic estimation from DNA sequences with variable rates over sites: Approximate methods. *Journal of Molecular Evolution* **39:** 306–314.

Yang, Z., N. Goldman, and A. Friday. 1994. Comparison of models for nucleotide substitution used in maximum-likelihood phylogenetic estimation. *Molecular Biology and Evolution* **11:** 316–324.

Yang, Z. 1994b. Statistical properties of the maximum likelihood method of phylogenetic estimation and comparison with distance matrix methods. *Systematic Biology* **43:** 329–342.

Yang, Z. 1995. A space-time process model for the evolution of DNA sequences. *Genetics* **139:** 993–1005.

Yang, Z., S. Kumar and M. Nei. 1995. A new method of inference of ancestral nucleotide and amino acid sequences. *Genetics* **141:** 1641–1650.

Yang, Z. 1996. Phylogenetic analysis using parsimony and likelihood methods. *Journal of Molecular Evolution* **42:** 294–307.

Yang Z. H. 1997a. On the estimation of ancestral population sizes of modern humans. *Genetical Research* **69:** 111–116.

Yang, Z. 1997b. How often do wrong models produce better phylogenies? *Molecular Biology and Evolution* **14:** 105–108.

Yang, Z. and B. Rannala. 1997. Bayesian phylogenetic inference using DNA sequences: A Markov Chain Monte Carlo method. *Molecular Biology and Evolution* **14:** 717–724.

Yang, Z. and R. Nielsen. 1998. Synonymous and nonsynonymous rate variation in nuclear genes of mammals. *Journal of Molecular Evolution* **46:** 409–418.

Yang, Z. 1998. Likelihood ratio tests for detecting positive selection and application to primate lysozyme evolution. *Molecular Biology and Evolution* **15:** 568–573.

Yang, Z., R. Nielsen, and M. Hasegawa. 1998. Models of amino acid substitution and applications to mitochondrial protein evolution. *Molecular Biology and Evolution* **15:** 1600–1611.

Yang, Z. and R. Nielsen. 2000. Estimating synonymous and nonsynonymous substitution rates under realistic evolutionary models. *Molecular Biology and Evolution* **17:** 32–43.

Yang, Z., R. Nielsen, N. Goldman, and A.-M. K. Pedersen. 2000. Codon-substitution models for heterogeneous selection pressure at amino acid sites. *Genetics* **155:** 431–449.

Yang, Z. 2002. Likelihood and Bayes estimation of ancestral population sizes in hominoids using data from multiple loci. *Genetics* **162:** 1811–1823.

Yang, Z. and R. Nielsen. 2002. Codon-substitution models for detecting molecular adaptation at individual sites along specific lineages. *Molecular Biology and Evolution* **19:** 908–917.

Yee, C. C. and L. Allison. 1993. Reconstruction of strings past. *Computer Applications in the Biosciences (CABIOS)* **9:** 1–7.

Yule, G. U. 1924. A mathematical theory of evolution, based on the conclusions of Dr. J. C. Willis, F.R.S. *Philosophical Transactions of the Royal Society of London, Series B* **213:** 21–87.

Zander, R. 2001. A conditional probability of reconstruction measure for internal cladogram branches. *Systematic Biology* **50:** 425–437.

Zaretskii, K. A. 1965. Postroenie dereva po naboru rasstoyanii mezhdu visyachimi vershinami (in Russian). *Uspekhi Matematicheskikh Nauk* **20 (6):** 90–92.

Zhang, L. 1997. On a Mirkin-Muchnik-Smith conjecture for comparing molecular phylogenies. *Journal of Computational Biology* **4:** 177–188.

Zharkikh, A. A. 1977. Algoritm postroeniya filogeneticheskikh drev po amino-kislotnym posledovatel'nostyam (in Russian). pp. 5–52 in *Matematicheskie Modeli Evolyutsii i Selektsii*, ed. V. A. Ratner. Institut Tsitologii i Genetiki, Novosibirsk.

Zharkikh, A. and W.-H. Li. 1992. Statistical properties of bootstrap estimation of phylogenetic variability from nucleotide sequences. I. Four taxa with a molecular clock. *Molecular Biology and Evolution* **9:** 1119–1147.

Zharkikh, A. and W.-H. Li. 1993. Inconsistency of the maximum-parsimony method: The case of five taxa with a molecular clock. *Systematic Biology* **42:** 113–125.

Zharkikh, A. 1994. Estimation of evolutionary distances between nucleotide sequences. *Journal of Molecular Evolution* **39:** 315–329.

Zharkikh, A. and W.-H Li. 1995. Estimation of confidence in phylogeny: The complete-and-partial bootstrap technique. *Molecular Phylogenetics and Evolution* **4:** 44–63.

Zhivotovsky, L. A. and M. W. Feldman. 1995. Microsatellite variability and genetic distance. *Proceedings of the National Academy of Sciences, USA* **92:** 11549–11552.

Zhivotovsky, L. A., M. W. Feldman, and S. A. Grishechkin. 1997. Biased mutation and microsatellite variation. *Molecular Biology and Evolution* **14:** 926–933.

Zhivotovsky, L. A., D. B. Goldstein, and M. W. Feldman. 2001. Genetic sampling error of distance $(\delta\mu)^2$ and variation in mutation rate among microsatellite loci. *Molecular Biology and Evolution* **18:** 2141–2145.

Zheng, Q. 2001. On the dispersion index of a Markovian molecular clock. *Mathematical Biosciences* **172:** 115–128.

Zimmerman, W. 1931. Arbeitsweise der botanischen Phylogenetik under anderer Gruppierungswissenschaften. pp. 941–1053 in *Handbuch der biologischen Arbeitsmethoden*, ed. E. Aberhalden. Abt. 3, 2, Teil 9. Urban and Schwarzenberg, Berlin.

Zmasek, C. M. and S. R. Eddy. 2001. A simple algorithm to infer gene duplication and speciation events on a gene tree. *Bioinformatics* **17:** 821–828.

Zuckerkandl, E. and L. Pauling. 1962. Molecular disease, evolution, and genetic heterogeneity. pp. 189–225 in *Horizons in Biochemistry*, ed. M. Kasha and B. Pullman. Academic Press, New York.

Zuckerkandl, E. and L. Pauling. 1965. Evolutionary divergence and convergence in proteins. pp. 97–116 in *Evolving Genes and Proteins*, ed. V. Bryson and H. J. Vogel. Academic Press, New York.

Index